AF517404

Crop Management

(2nd EDITION)

The Authors

Dr U S Walia, Sr. Agronomist-cum-Head (Retd.,), Department of Agronomy, Punjab Agricultural University, Ludhiana has teaching, research and extension experience for about thirty seven years. He handled nine adhoc research projects funded by national and international agencies especially in the field of weed science. His publications inclue six books, six book chapters, five research bulletins, four extension bulletins, two practical manuals, 325 publications (research and extension). Apart from these 93 recommendations (as co-worker) have been included in the package of practices for the benefit of Punjab farmers. He has guided 9 M.Sc. and 3 Ph.D. students. His contributions have been recognized through various awards *viz.*, FAI Silver Jubilee Award-1984 for outstanding Doctgoral Research in Fertilizer Usage, Fellow as well as Gold Medal of India Society of Weed Science during 2004 and 2007 respectively.

Dr S S Walia, is working as Senior Agronomist in the Department of Agronomy, Punjab Agricultural University, Ludhiana. He has published 244 (research and extension) articles, five book chapters and four manuals. He had 62 recommendtions for the benefit of Punjab farmers to his credit. Based on his significant contributions in the field of Agronomy, he received P S Deshmukh Young Agronomist award (2005). He is working of farming system and organic farming. He guided 11 M.Sc and 4 Ph.D students.

Related Books

Farming System and Sustainable Agriculture

S.S. Walia | U.S. Walia

The land holding of the farmer is decreasing day by day due to urbanisation and there is no chance for horizontal increase in agricultural land. To increase the income of the farmers, few steps for vertical increase in agricultural production have been discussed in this book. A detailed and comprehensive information regarding the historical background of farming system, farming system and its components, integrated farming system and allied enterprises, integrated farming system models in different agroclimatic zones, role of integrated farming system in agriculture and livelihood security, resource cycling and flow of energy in different farming systems, role of crop residues in agriculture, farming system of dryland agriculture, role of agroforestry in farming system, scope of organic farming in farming system have been given. A detailed information regarding the latest concepts of agronomy like conservation agriculture strategies in cropping system, sustainable agriculture, scope of hydroponics techniques in agriculture have also been discussed. In the end tools for determining production and efficiencies in cropping and farming system have also been given.

ISBN: 9789388812900 | 280 Pages

Crop Management

(2nd Edition)

Dr. U S Walia
Senior Agronomist cum Head (Retd.)
Department of Agronomy
Punjab Agricultural University, Ludhiana (Punjab)

Dr. S S Walia
Senior Agronomist
Department of Agronomy
Punjab Agricultural University, Ludhiana (Punjab)

Published by:

SCIENTIFIC PUBLISHERS (INDIA)
5A, New Pali Road, P.O. Box 91
Jodhpur 342 001, India
E-mail: info@scientificpub.com
Website: www.scientificpub.com

© 2021, Authors

All rights reserved. No part of this publication or the information contained herein may be reproduced, adapted, abridged, translated, stored in a retrieval system, computer system, photographic or other systems or transmitted in any form or by any means, electronic, mechanical, optical, digital, by photocopying, recording or otherwise, without written prior permission from the publisher. Any breach will attract legal action and prosecution without further notice.

Disclaimer: While every effort has been made to avoid errors and omissions, this publication is being sold and marketed on the understanding and presumption that neither the editors (or authors) nor the publishers nor the printers would be liable in any manner whatsoever, to any person either for an error or for an omission in this publication, or for any action to be taken on the basis of this work. Any inadvertent discrepancy noted may be brought to the attention of the publisher, for rectifying it in future editions, if published.

This book contains information obtained from authentic and highly regarded sources. Reasonable efforts have been made to publish reliable data and information, but the editors and publisher cannot assume responsibility for the validity of all materials or the consequences of their use. The editors and publisher have attempted to trace and acknowledge the copyright holders of all material reproduced in this publication and apologize to copyright holders if permission and acknowledgement to publish in this form have not been obtained. If any copyright material has not been acknowledged please write and let us know so that we may rectify it.

Trademark Notice: Publications or corporate names may be trademarks, and are used only for identification and explanation in bonafide intent without intent to infringe.

ISBN: 978-93-88449-00-7
eISBN: 978-93-88449-02-1

Printed in India

Dedicated to

My Late Parents

Sardar: Sarwan Singh

(November 1916 to January, 1971)

and

Sardarni : Parsin Kaur

(April 1921 to January, 2005)

With Respect and Love

ACKNOWLEDGEMENT

I am grateful to the almighty who blessed me with patience and fortitute for the completion of this noble job. Secondly, I feel privileged to express my heartful gratitude and indebtedness to late Dr. H.S. Gill, Sr. Agronomist (weed control) cum Head (Retd.) Department of Agronomy, PAU, Ludhiana and major advisor during my M.Sc. and Ph.D. programmes, who shaped my mind set and existence among weed science scientists to serve farmes, teachers, researches and students. Also, I express my sincere regards and thankfulness to Dr. L.S. Brar Sr. Agronomist (weed control) cum Head (Retd.), Department of Agronomy, PAU, Ludhiana, who inspired me to undertake an arduous job. It is beyond my access to acknowledge in words, the over encouraging moral support and co-operation which I enjoyed from Sardarni Parmjit Kaur Walia (wife) Er. Gurpreet Singh Walia and Er. Gaganpreet Singh Walia (sons) as well as Neena Walia and Rakhi Walia (daughter-in-law) throughout the completion of this revised manuscript. Above all, adorable and cuddly (Manvik Walia and Mehransh Walia), my grandsons, whose cheerful presence has provided a perfect milieu to my writing.

Dr. U.S. Walia

PREFACE

Agriculture is the largest enterprise in India which has been and will continue to be the lifeline of the Indian economy in the foreseeable future. However, due to urbanization, agricultural land is shrinking and human population is increasing year by year. So, there is a need for vertical increase in agricultural produce to feed the increasing population. Due to changing climatic conditions, there is a need for reorientation of presently practiced agricultural technologies. At the same time there is a need to save/conserve the natural resources. Crop yields can be improved with the adoption of improved production and protection technologies for raising field crops. In order to increase profit in agriculture, the farm inputs like tillage practices, fertilizers, irrigation water, pesticides etc. must be used judiciously and more stress should be laid on conservation agriculture.

The book covers basic but very comprehensive information on history of agriculture and relationship of Agronomy with other disciplines, tillage practices, nutrient elements for plant growth, weed and their management, irrigation management, crop physiology, crop ecology, integrated farming system and organic farming. A detailed information on history and origin, improved varieties, agronomic practices and plant protection techniques for important field crops viz. cereals, oilseeds, pulses, sugar crops and fiber crops has been given. Also information on cultivation practices for important medicinal, aromatic and spice crops as well as plantation crops along with their uses/medicinal values has been provided. Apart from this, information on dry land agriculture, crop production under special situations and hints for achieving higher yield of field crops are also given in details. This book will be very helpful for B.Sc. Agriculture as well as M.Sc. Agronomy students throughout the country as it covers nerly the entire syllabus for Agronomy courses framed by ICAR.

Authors

CONTENTS

1

HISTORY OF AGRICULTURE AND ROLE OF AGRONOMY

Agriculture is the science, art and industry of managing the growth of plants and animals to make use by human beings. In general, agriculture includes soil cultivation, growing and harvesting field crops, raising and breeding livestock, dairy and forest trees etc. The word agriculture is the English adaptation of Latin word i.e. agricultūra, from ager, "a field" and cultura, "cultivation" in the strict sense of "tillage of the soil". Thus, a literal reading of the word yields "tillage of a field/ fields".

Agriculture, is the largest enterprise in India, has been and will continue to be the lifeline of the Indian economy at least in the foreseeable future. It contributes nearly 13.6 percent during 2017 to the national GDP, sustain livelihood of about two – thirds (2/3) of the population, accounts for 52% of the national workforce and forms the backbone of the agro-based industry. India through modern agricultural technologies has moved from an era to the chronic food shortages and 'begging bowl' status during 1960's, when our annual food imports were around 8-10 millions tonnes, to a level of food self-sufficiency - buffer stocks and even food exports from 1990's. With reference to 1950, the productivity gains are nearly 3.5 times in foodgrains, 1.6 times in fruits, 2.1 times in vegetables, 5 times in fish, 1.8 times in milk and 4.8 times in egg production at present. The production of total food grains in India was 176.39 million tonnes during 1990-1991 and it increased to 275.11 million tonnes during 2016-17. However, the target for 2018-19 is 284.00 million tonnes. Total food grain production of Punjab during 2015-16 was 28.4 million tonnes. In India, geographical area is 329 mha, gross cropped area 195 mha and net sown area 141 mha. The available land per person (ha) is declining every year. During 1961, it was 0.34 ha which decreased to 0.234, 0.15 and 0.12 ha during 1980, 2001 and 2015 respectively. (World Bank Data).

Modern agriculture

Engineering and technology is heavily utilized in modern agriculture. Biological and physical sciences also help in raising agricultural production. Agriculture also involves specialized understanding of agricultural engineers especially for mechanization, drainage, irrigation etc. Precision farming is becoming very popular system of farming which includes laser land levelling, site specific nutrient management, need based (on soil or plant test basis) use of farm inputs etc.

One area of agriculture where plant production is advancing rapidly is Hydroponics. Hydroponics is a method for growing plants without soil. The plants are grown in an inert media such as Rockwool and supplied with nutrient rich water solution. Hydroponics are not dependent on soil, this advance plant cultivation method will allow many people in extremely dry regions like parts of Africa, for example, to cultivate plants for food and other resources. Hydroponics can also increase food production.

Agriculture used to be a laborious task and still is, however, from late 19th and 20th century technological advancement has simplified practices. Moreover, mechanization has greatly increased the efficiency and productivity of farming in the developed and developing nations. Although agriculture has evolved and advanced in many ways, still animals like oxen etc. are still utilized on farms to cultivate and harvest crops, and transport products to markets. However mechanical power is replacing animal power day by day. Now the computers have also become an important agricultural management tool. Hydroponics and greenhouse cultivation are two examples of agriculture where computers are involved for managing lighting, feeding, temperature setting etc.

In the U.S. and Australia, helicopters are generally used in agriculture for seeding purpose and spraying applications against insect and disease, fertilizer applications, transporting perishable commodities. Television and radio have played a vital role in agriculture by making available important weather reports and market reports both of which are concerns for most farmers. Television and radio also help in communicating research in agriculture i.e. latest recommendations to the farmers.

In 2007, one third of the world's workers were employed in agriculture. The services sector has overtaken agriculture as the economic sector employing the most people worldwide. Despite the size of its work-force, agricultural production accounts for less than five percent of the gross world product (an aggregate of all gross domestic products). With the increase in profit margin in other enterprises the profit in agriculture is approximately 5 times less than profit in industry per unit time.

Overview

Agriculture has played a key role in the development of human civilization as majority of the human population laboured in agriculture. Development of agricultural techniques has steadily increased agricultural productivity, and the widespread diffusion of these techniques during a time period is often called an agricultural revolution. A remarkable shift in agricultural practices has occurred over the past century in response to new technologies.

Synthetic nitrogen, mined rock phosphate, pesticides and mechanization, have greatly increased crop yields in the early 20th century. The global yield increases were experienced later in the 20th century when high-yielding varieties of common staple grains such as rice, wheat, and maize were introduced as a part of the green revolution. The green revolution exported the technologies (including pesticides and synthetic nitrogen) of the developed world to the developing world. Many governments have subsidized agriculture to ensure an adequate food supply. Proponents of organic farming such as Sir Albert Howard argued in the early 20th century that the overuse of pesticides and synthetic fertilizers damages the long-term fertility of the soil. While this feeling lay dormant for decades, as environmental awareness has increased in the 21st century there has been a movement towards sustainable agriculture by agricultural scientists, some farmers, consumers, and policymakers.

HISTORY OF AGRICULTURE

Agricultural practices such as irrigation, crop rotation, fertilizers, and pesticides were developed long ago, but have made great strides in the past century. The history of agriculture has played a major role in human development, as agricultural progress has been a crucial factor in worldwide socio-economic change. The ancient man started living near source of water and after exhausting eatable materials, he moved to other places in search of food. Slowly and slowly he identified economical plants and started raising those plants which later on called as crops. When farmers (ancient man) became capable of producing food beyond the needs of their own families, others in their society were freed to devote themselves to projects other than food acquisition. Historians and anthropologists have long argued that the development of agriculture made civilization possible.

Ancient origins

The Fertile Crescent of Western Asia, Egypt, and India were sites of the earliest planned sowing and harvesting of plants that had previously been gathered in the wild. Independent development of agriculture occurred in northern and southern China, Sahel in Africa, New Guinea and several regions of the Americas. The eight so-called Neolithic founder crops of agriculture that appeared were: first emmer wheat and einkorn wheat, then hulled barley, peas, lentils, bitter vetch, chick peas and flax.

By 7000 BC, small-scale agriculture reached Egypt. Also at that time, the Indian subcontinent saw farming of wheat and barley, as attested by archaeological excavation at Mehrgarh in Balochistan which is present day Pakistan. By 6000 BC, mid-scale farming was entrenched on the banks of the Nile. About this time, agriculture was developed independently in the Far East, with rice, rather than wheat, as the primary crop. Chinese and Indonesian farmers went on to domesticate taro and beans including mung, soya and azuki. To complement these new sources of carbohydrates, highly organized net fishing of rivers, lakes and ocean shores in these areas brought in great volumes of essential protein. Collectively, these new methods of farming and fishing inaugurated a human population boom that dwarfed all previous expansions and continues today.

By 5000 BC, the Sumerians had developed core agricultural techniques including large-scale intensive cultivation of land, monocropping, organized irrigation, and the use of a specialized labor force, particularly along the waterway now known as the Shatt al-Arab, from its Persian Gulf Delta to the confluence of the Tigris and Euphrates.

Domestication of wild aurochs and mouflon into cattle and sheep, respectively, ushered in the large-scale use of animals for food/fibre and as beasts of burden. The shepherd joined the farmer as an essential provider for sedentary and seminomadic societies. Maize, manioc, and arrowroot were first domesticated in the Americas as far back as 5200 BC.

The potato, tomato, pepper, squash, several varieties of bean, tobacco, and several other plants were also developed in America, as was extensive terracing of steep hillsides in much of Andean South America. The Greeks and Romans built on techniques pioneered by the Sumerians, but made few fundamentally new advances. Southern Greeks struggled with very poor soils, yet managed to become a dominant society for years. The Romans were noted for an emphasis on the cultivation of crops for trade.

In the same region, a parallel agricultural revolution occurred, resulting in some of the most important crops grown today. In Mesoamerica wild teosinte was transformed through human selection into the ancestor of modern maize, more than 6000 years ago. It gradually spread across North America and was the major crop of Native Americans at the time of European exploration. Other Mesoamerican crops include hundreds of varieties of squash and beans. Cocoa was also a major crop in domesticated Mexico and Central America. The turkey, one of the most important meat birds, was probably domesticated in Mexico or the U.S. Southwest. In the Andes region of South America the major domesticated crop was potatoes, domesticated perhaps 5000 years ago. Large varieties of beans were domesticated, in South America, as well as animals, including llamas, alpacas, and guinea pigs. Coca, still a major crop, was also domesticated in the Andes.

A minor centre of domestication, the indigenous people of the Eastern U.S. appear to have domesticated numerous crops. Sunflowers, tobacco varieties of squash and Chenopodium, as well as crops no longer grown, including marshelder and little barley were domesticated. Other wild foods may have undergone some selective cultivation, including wild rice and maple sugar. The most common varieties of strawberry were domesticated from Eastern North America.

By 3500 BC, the simplest form of the plough was developed, called the ard. Before this period, simple digging sticks or hoes were used. These tools would have also been easier to transport, which was a benefit as people only stayed until the soil's nutrients were depleted. However, trough excavations in Mexico it has been found that the continuous cultivating of smaller pieces of land would also have been a sustaining practice. Additional research in central Europe later revealed that agriculture was indeed practiced at this method. For this method, ards were thus much more efficient than digging sticks.

Middle ages

During the Middle Ages, farmers in North Africa, the Near East, and Europe began making use of agricultural technologies including irrigation systems based on hydraulic and hydrostatic principles, machines such as norias, water-raising machines, dams, and reservoirs. This combined with the invention of a three-field system of crop rotation and the moldboard plough greatly improved agricultural efficiency. In the European medieval period, agriculture was considered part of the set of seven mechanical arts.

Modern era

After 1492, a global exchange of previously local crops and livestock breeds occurred. Key crops involved in this exchange included the tomato, maize, potato, manioc, cocoa bean and tobacco going from the New World to the Old, and several varieties of wheat, spices, coffee, and sugar cane going from the Old World to the New. The most important animal exportation from the Old World to the New were those of the horse and dog (dogs were already present in the pre-Columbian Americas but not in the numbers and breeds suited to farm work). Although not usually food animals, the horse (including donkeys and ponies) and dog quickly filled essential production roles on western-hemisphere farms.

The potato became an important staple crop in northern Europe. Since being introduced by Portuguese in the 16th century, maize and manioc have replaced traditional African crops as the continent's most important staple food crops.

By the early 19th century, agricultural techniques, implements, seed stocks and cultivated plants selected were given a unique name because of its decorative or useful characteristics had improved the yield per land unit many times that was seen in the middle ages. Although there is a vast and interesting history of crop cultivation before the dawn of the 20th century, there is little question that the work of Charles Darwin and Gregor Mendel created the scientific foundation for plant breeding that led to its explosive impact over the past 150 years.

With the rapid rise of mechanization in the late 19th century and the 20th century, particularly in the form of the tractor, farming tasks could be done with a speed and on a large scale which was previously impossible. These advances have led to efficiencies enabling certain modern farms in the United States, Argentina, Israel, Germany, and a few other nations to output volumes of high-quality produce per land unit at highest rates.

The Haber-Bosch method for synthesizing ammonium nitrate repres-ented a major breakthrough and allowed crop yields to overcome previous constraints. In the past century agriculture has been characterized by enhanced productivity, the substitution of labor for synthetic fertilizers and pesticides, water pollution, and farm subsidies. In recent years there has been a backlash against the external environmental effects of conventional agriculture, resulting in the organic movement.

The cereals i.e. rice, corn, and wheat provide 60% of human food supply. Between 1700 and 1980, the total area of cultivated land worldwide increased 466% and yields increased dramatically, particularly because of selectively bred high-yielding varieties and due to judicious use of fertilizers, pesticides, irrigation, and machinery. For example, irrigation increased corn yields in eastern Colorado by 400 to 500% from 1940 to 1997.

However, concerns have been raised over the sustainability of intensive agriculture. Intensive agriculture has become associated with decreased soil quality in India and Asia, and there has been increased concern over the effects of fertilizers and pesticides on the environment, particularly as population increases and food demand expands. The monocultures typically used in intensive agriculture increase the number of pests, which are controlled through pesticides. Integrated pest manage-ment (IPM), which "has been promoted for decades and had some notable successes" has not significantly affected the use of pesticides because policies encourage the use of pesticides and IPM is knowledge-intensive.

Although the "green revolution" significantly increased rice yields in Asia, yield increases have not occurred in the past 15–20 years. The genetic "yield potential" has increased for wheat, but the yield potential for rice has not increased since 1966, and the yield potential for maize has "barely increased in 35 years". It takes a decade or two for herbicide-resistant weeds to emerge, and insects become resistant to insecticides within about a decade. Crop rotation helps to prevent resistances.

Agricultural exploration expeditions, since the late 19th century, have been mounted to find new species and new agricultural practices in different areas of the world. Two early examples of expeditions include Frank N. Meyer's fruit - and nut-collecting trip to China and Japan from 1916-1918 and the Dorsett-Morse Oriental Agricultural Exploration Exp-edition to China, Japan, and Korea from 1929-1931 to collect soybean ger-mplasm to support the rise in soybean agriculture in the United States.

In 2009, the agricultural output of China was the largest in the world, followed by the European Union, India and the United States, according to the International Monetary Fund. Economists measure the total factor productivity of agriculture and by this measure agriculture in the United States is roughly 2.6 times more productive than it was in 1948.Six countries - the US, Canada, France, Australia, Argentina and Thailand - supply 90% of grain exports. The United States controls almost half of world grain exports. Water deficits, which are already spurring heavy grain imports in numerous middle-sized countries, including Algeria, Iran, Egypt, and Mexico, may soon do the same in larger countries, such as China or India.

Role of breeders in agriculture

Crop alteration has been practiced by humankind for thousands of years, since the beginning of civilization. Altering crops through breeding practices changes the genetic make-up of a plant to develop crops with more beneficial characteristics for humans, for example, larger fruits or seeds, drought-tolerance, resistance to pests, high quality produce etc. Significant advances in plant breeding ensued after the work of geneticist Gregor Mendel. His work on dominant and recessive alleles gave plant breeders a better understanding of genetics and brought great insights to the techniques utilized by plant breeders. Crop breeding includes techniques such as plant selection with desirable traits, self-pollination and cross-pollination, and molecular techniques that genetically modify the organism. Domestication of plants has, over the centuries increased yield, improved disease resistance and drought tolerance, eased harvest and improved the taste and nutritional value of crop plants. Careful selection and breeding have had enormous effects on the characteristics of crop plants.

Genetic engineering

Genetically Modified Organisms (GMO) are organisms whose genetic material has been altered by genetic engineering techniques generally known as recombinant DNA technology. Genetic engineering has expanded the genes available to breeders to utilize in creating desired germlines for new crops. After the development of mechanical tomato-harvesters in the early 1960s, agricultural scientists genetically modified tomatoes to be more resistant to mechanical handling. More recently, genetic engineering is being employed in various parts of the world, to create crops with other beneficial traits.

Herbicide-tolerant GMO crops

Roundup ready seed has a herbicide resistant gene implanted into its genome that allows the plants to tolerate exposure to Round up (glyphosate). Roundup is a trade name for a glyphosate-based product, which is a systemic, non-selective herbicide used to kill weeds. Roundup ready seeds allow the farmer to grow a crop that can be sprayed with glyphosate to control weeds without harming the resistant crop. Herbicide-tolerant crops are used by farmers worldwide. Today, 92% of soybean acreage in the US is planted with genetically modified herbicide-tolerant plants. With the increasing use of herbicide-tolerant crops, the use of glyphosate increased many folds. In some areas glyphosate resistant weeds have developed, causing farmers to switch to other herbicides.

Role of climate on agriculture

Climate change has the potential to affect agriculture through changes in temperature, rainfall (timing and quantity), CO_2 concentr-ation, solar radiation and the interaction of these elements. Agriculture can both mitigate or worsen global warming. Some of the increase in CO_2 in the atmosphere comes from the decomposition of organic matter in the soil, and much of the methane emitted into the atmosphere is caused by the decomposition of organic matter in wet soils such as puddle field for rice cultivation.. Further, wet or anaerobic soils also lose nitrogen through denitrification, releasing the greenhouse gas nitric oxide. Changes in management can reduce the release of these greenhouse gases, and soil can further be used to sequester some of the CO_2 in the atmosphere.

Definition of Agronomy: Agronomy is a branch of agricultural science dealing with the principles and practices of soil, water and crop production. This term is derived from greek words '*agros*' meaning field and '*nomos*' meaning to manage. In recent times Agronomy has assumed newer dimensions and can be defined as a branch of agricultural science that deals with methods that provide favourable environment to the crop for higher productivity. Norman (1980) has defined Agronomy as a science of manipulating the crop environment complex with dual aims of improving agricultural productivity and gaining a degree of understanding of the processes involved.

Agronomy is the science and technology of using plants for food, fuel, feed, fibre, and reclamation. Agronomy encompasses work in the areas of plant genetics, plant physiology, meteorology, and soil science. Agronomy is the application of a combination of sciences like biology, chemistry, ecology, earth science, and genetics. Agronomists today are involved with many issues including producing food, creating healthier food, managing environmental impact of agriculture, and creating energy from plants. Agronomists often specialize in areas such as crop production, irrigation and drainage, plant physiology, soil fertility; weed control, crop ecology, cultivation of medicinal, aromatic and spice crops etc.

RELATIONSHIP OF AGRONOMY WITH OTHER DISCIPLINES

Agronomy is synthesis of several disciplines like soil science, agril- chemistry, crop physiology, crop ecology, biochemistry, economics, agricultural meteorology, plant breeding, entomology, plant pathology, animal science, extension education, food technology etc.

Plant Breeding: This involves selective breeding of plants to produce the high yielding and high quality varieties under various conditions and under different cropping systems. Plant breeding has increased crop yields and has improved the nutritional value of numerous crops, including corn, soybean, wheat, rice etc. It also has led to the development of new types of plants. For example, a hybrid grain called triticale was produced by cross breeding rye and wheat. Triticale contains more usable protein than does either rye or wheat. Agronomists and plant breeder have close associations as agronomists are to work out the agronomic requirements of newly evolved varieties under different environments.

Biotechnology: Biotechnology is often a lab activity requiring field testing of the new crop varieties that are developed under controlled conditions. In addition to increasing crop yields, biotechnology is increasingly being applied for novel uses other than food. For example, oilseed is at present used mainly for margarine and other food oils, but it can be modified to produce fatty acids for detergents, substitute fuels and petrochemicals. The success of plants developed by biotechnological approaches is studied under field conditions by the agronomists.

Soil Science: Agronomists study sustainable ways to make soils more productive and profitable. They classify soils and reproduce them to determine whether they contain substances vital to plant growth such as compounds of nitrogen, phosphorus, and potassium. If a certain soil is deficient in these substances, need based fertilizers may be provided. Soil science also involves investigation of the movement of nutrients through the soil, the amount of nutrients absorbed by the plant roots, and the development of roots and their relation to the soil. In addition, agronomists develop methods to preserve the soil and to decrease the effects of erosion by wind and water. For example, a technique called contour ploughing may be used to prevent soil erosion and conserve rainfall. Researchers in Agronomy also seek ways to use the soil more effectively in solving other problems. Such problems include the disposal of human and animal wastes; water pollution; and the build-up in the soil of pesticides.

Agroecology: Agroecology is the management of agricultural systems with an emphasis on ecological and environmental perspectives. This area is closely associated with work in the areas of sustainable agriculture, organic farming, alternative food systems and the development of alternative cropping systems.

Plant Protection: Plant protection includes Entomology and Plant Pathology. These disciplines prevent the crop plants from insect-pests and diseases so ultimately prevent yield losses.

Economics: The economist takes into account each and every input spent on raising crops. The economic analysis also determines the benefit cost ratio, net profit fetched.

Agrometeorology: Agrometeorology is a branch of meteorology which deals with the response of crops with the physical environment. The various climatic factors play important role in crop production. Its knowledge is very important in crop production. Agronomists can modify crop cultivation technologies as per changing environmental conditions suggested by the agrometerologists.

Animal Science: Animal Science deals with scientific rearing of livestock. These enterprises are combined with cropping systems to form farming system.

Agricultural Chemistry: Agricultural Chemistry is associated with some important farming concerns like fertilizer application, insecticides or pest control, herbicides, fungicides, soil composition, agricultural products analysis, and nutritional requirements of farm animals.

Biochemistry: It is the branch of science which deals to study the composition of plant residues, effect of different herbicides, their residues in grain, straw, soil and drinking water etc. The nutrient content of plant parts viz., roots, shoots, grain and straw and soil is determined using different analytical methods.

Extension Education: Extension Education plays an important role in transfer of the advanced technology to the farmers. The extension wrokers act as a bridge between the farmers problems and scientists as the farmers problems are brought to the scientists and new advanced technologies are given to the farmers by extension workers.

Horticulture: It deals with raising of fruit plants. It is also called as pomolgy. During initial stages, intercropping of field crops in the fruit plants can be done for higher profits and their suitability is worked out by the agronomists.

Vegetable Crops: It deals with raising of vegetable plants. It is also called as olericulture. For higher economics, intercropping of vegetables in the field crops can be done.

Floriculture: It deals with raising of ornamental flower and plants. Agronomist can study the selective herbicide for weed control in ornamental plants and their suitability as intercrop can be worked out.

Forestry: Which deals with the raising of sole tree plantations and along with crop plants. Agronomist play very important role for evaluation of field crops which are best suited in forestry plantations of variable ages.

Plant Physiology: It is very closely related with Agronomy as growth regulators or growth retardants can be used to modify plant growth for higher yields.

Food Science Technology: Post harvest technologies for value additions for enhancing income of the farmers with the adoption of any value addition techniques. Food Science has also revolutionized the way we pack, process and market agricultural products. For example, dehydration and quick-freezing methods have increased farm products markets (food processing, preservation, meat packing etc). The refrige-rated carriers are used to transport the perishable agricultural products to airport and then these high value crops and vegetables are transported to the other countries.

LIST OF ICAR INSTITUTIONS

Deemed Universities - 4

1. ICAR-Indian Agricultural Research Institute, New Delhi
2. ICAR-National Dairy Research Institute, Karnal
3. ICAR-Indian Veterinary Research Institute, Izatnagar
4. ICAR-Central Institute on Fisheries Education, Mumbai

Institutions - 64

1. ICAR-Central Island Agricultural Research Institute , Port Blair
2. ICAR-Central Arid Zone Research Institute, Jodhpur
3. ICAR-Central Avian Research Institute, Izatnagar
4. ICAR-Central Inland Fisheries Research Institute, Barrackpore
5. ICAR-Central Institute Brackishwater Aquaculture, Chennai
6. ICAR-Central Institute for Research on Buffaloes, Hissar
7. ICAR-Central Institute for Research on Goats, Makhdoom
8. ICAR-Central Institute of Agricultural Engineering, Bhopal
9. ICAR-Central Institute for Arid Horticulture, Bikaner
10. ICAR-Central Institute of Cotton Research, Nagpur
11. ICAR-Central Institute of Fisheries Technology, Cochin
12. ICAR-Central Institute of Freshwater Aquaculture, Bhubneshwar
13. ICAR-Central Institute of Research on Cotton Technology, Mumbai
14. ICAR-Central Institute of Sub Tropical Horticulture, Lucknow

15. ICAR-Central Institute of Temperate Horticulture, Srinagar
16. ICAR-Central Institute on Post harvest Engineering and Technology, Ludhiana
17. ICAR-Central Marine Fisheries Research Institute, Kochi
18. ICAR-Central Plantation Crops Research Institute, Kasargod
19. ICAR-Central Potato Research Institute, Shimla
20. ICAR-Central Research Institute for Jute and Allied Fibres, Barrackpore
21. ICAR-Central Research Institute of Dryland Agriculture, Hyderabad
22. ICAR-National Rice Research Institute, Cuttack
23. ICAR-Central Sheep and Wool Research Institute, Avikanagar, Rajasthan
24. ICAR- Indian Institute of Soil and Water Conservation, Dehradun
25. ICAR-Central Soil Salinity Research Institute, Karnal
26. ICAR-Central Tobacco Research Institute, Rajahmundry
27. ICAR-Central Tuber Crops Research Institute, Trivandrum
28. ICAR-ICAR Research Complex for Eastern Region, Patna
29. ICAR-ICAR Research Complex for NEH Region, Barapani
30. ICAR-Central Coastal Agricultural Research Institute, Ela, Old Goa, Goa
31. ICAR-Indian Agricultural Statistics Research Institute, New Delhi
32. ICAR-Indian Grassland and Fodder Research Institute, Jhansi
33. ICAR-Indian Institute of Agricultural Biotechnology, Ranchi
34. ICAR-Indian Institute of Horticultural Research, Bengaluru
35. ICAR-Indian Institute of Natural Resins and Gums, Ranchi
36. ICAR-Indian Institute of Pulses Research, Kanpur
37. ICAR-Indian Institute of Soil Sciences, Bhopal
38. ICAR-Indian Institute of Spices Research, Calicut
39. ICAR-Indian Institute of Sugarcane Research, Lucknow
40. ICAR-Indian Institute of Vegetable Research, Varanasi
41. ICAR-National Academy of Agricultural Research & Management, Hyderabad
42. ICAR-National Institute of Biotic Stresses Management, Raipur
43. ICAR-National Institue of Abiotic Stress Management, Malegaon, Maharashtra
44. ICAR-National Institute of Animal Nutrition and Physiology, Bengaluru
45. ICAR-National Institute of Research on Jute & Allied Fibre Technology, Kolkata
46. ICAR-National Institute of Veterinary Epidemiology and Disease Informatics, Hebbal, Bengaluru
47. ICAR-Sugarcane Breeding Institute, Coimbatore
48. ICAR-Vivekananda Parvatiya Krishi Anusandhan Sansthan, Almora

49. ICAR-Central Institute for Research on Cattle, Meerut, Uttar Pradesh
50. ICAR-National Institute of High Security Animal Diseases, Bhopal
51. ICAR-Indian Institute of Maize Research,New Delhi
52. ICAR- Central Agroforestry Research Institute , Jhansi
53. ICAR-National Institute of Agricultural Economics and Policy Research, New Delhi
54. ICAR- Indian Institute of Wheat and Barley Research, Karnal
55. ICAR- Indian Institute of Farming Systems Research, Modipuram
56. ICAR- Indian Institute of Millets Research, Hyderabad
57. ICAR- Indian Institute of Oilseeds Research, Hyderabad
58. ICAR- Indian Institute of Oil Palm Research, Pedavegi, West Godawari
59. ICAR- Indian Institute of Water Management, Bhubaneshwar
60. ICAR-Indian Institute of Rice Research, Hyderabad
61. ICAR- Central Institute for Women in Agriculture, Bhubaneshwar
62. ICAR-Central Citrus Research Institute, Nagpur
63. ICAR-Indian Institute of Seed Research, Mau
64. ICAR-Indian Agricultural Research Institute, Post Box No. 48, Hazaribag 825 301, Jharkhand

National Research Centres - 15

1. ICAR-National Research Centre for Banana, Trichi
2. ICAR-National Research Centre for Grapes, Pune
3. ICAR-National Research Centre for Litchi, Muzaffarpur
4. ICAR-National Research Centre for Pomegranate, Solapur
5. ICAR-National Research Centre on Camel, Bikaner
6. ICAR-National Research Centre on Equines, Hisar
7. ICAR-National Research Centre on Meat, Hyderabad
8. ICAR-National Research Centre on Mithun, Medziphema, Nagaland
9. ICAR-National Research Centre on Orchids, Pakyong, Sikkim
10. ICAR-National Research Centre on Pig, Guwahati
11. ICAR-National Research Centre on Plant Biotechnology, New Delhi
12. ICAR-National Research Centre on Seed Spices, Ajmer
13. ICAR-National Research Centre on Yak, West Kemang
14. ICAR-National Centre for Integrated Pest Management, New Delhi
15. National Research Centre on Integrated Farming (ICAR-NRCIF),Motihari

National Bureaux - 6

1. ICAR-National Bureau of Plant Genetics Resources, New Delhi

2. ICAR-National Bureau of Agriculturally Important Micro-organisms, Mau, Uttar Pradesh
3. ICAR-National Bureau of Agricultural Insect Resources, Bengaluru
4. ICAR-National Bureau of Soil Survey and Land Use Planning, Nagpur
5. ICAR-National Bureau of Animal Genetic Resources, Karnal
6. ICAR-National Bureau of Fish Genetic Resources, Lucknow

Directorates/Project Directorates - 13

1. ICAR-Directorate of Groundnut Research, Junagarh
2. ICAR-Directorate of Soybean Research, Indore
3. ICAR-Directorate of Rapeseed & Mustard Research, Bharatpur
4. ICAR-Directorate of Mushroom Research, Solan
5. ICAR-Directorate on Onion and Garlic Research, Pune
6. ICAR-Directorate of Cashew Research, Puttur
7. ICAR-Directorate of Medicinal and Aromatic Plants Research, Anand
8. ICAR-Directorate of Floricultural Research, Pune, Maharashtra
9. ICAR-Directorate of Weed Research, Jabalpur
10. ICAR-Project Directorate on Foot & Mouth Disease, Mukteshwar
11. ICAR-Directorate of Poultry Research, Hyderabad
12. ICAR-Directorate of Knowledge Management in Agriculture (DKMA), New Delhi
13. ICAR-Directorate of Cold Water Fisheries Research, Bhimtal, Nainital

2

TILLAGE PRACTICES

Tillage is as old as agriculture. Primitive man used to disturb the soil for placing seeds. "Jethro Tull" who is considered as father of tillage, proposed a theory that plants absorb minute particles of soils. Therefore, he suggested that thorough ploughing and other operations were necessary so as to make soil into fine particles, to destroy weeds competing with crops and to keep the soil in between the rows in a good crumbly and friable condition for water infiltration. However, during the present time his theory i.e. tillage operations are carried out to prepare a fine seedbed for sowing crops is not correct due to the introduction of zero till technology for reducing cost of cultivation.

Over the years, growers have become interested in developing tillage practices to give greater protection to the soil against various losses of soil and water by degradation and erosion. The amount of surface residues and surface roughness both have an impact on agriculture. Crop residue management has been developed either to incorporate the residues in soil or to leave more of the harvest residues on or near the surface.

Tillage is the physical manipulation of soil with tools and implements to result in good tilth for better germination and subsequent growth of crops. While the tilth refers to the physical condition of the soil in relation to plant growth resulting from tillage. Tilth depends not only on aggregate formation and stability, but also on such factors as bulk density, soil moisture content, degree of aeration, rate of water infiltration, drainage and capillary water holding capacity. A soil with good tilth is quite porous and has free drainage upto water table. The capillary and non-capillary pores should be in equal proportion so that sufficient amount of water as well as free air is retained in the soil.

Objectives of tillage

The objectives of tillage are as follows:

- To enhance the metabolic processes of crop plants and microorganisms.
- To make the soil loose and porous.
- To have repeated exchange of air between atmosphere and soil.
- To increase the soil temperature.
- To control weeds.
- To remove stubbles and making a clean seed bed.

- To destroy insect pests.
- To break hard pan.
- To incorporate organic or other bulky manures.
- To invert soil to improve its fertility status.

EFFECT OF TILLAGE ON SOIL PROPERTIES

Tillage has considerable influence on soil properties. However, the effects vary greatly depending on soil, crop and weather conditions.

I. Soil structure: It is well recognized that tillage can cause deterioration of soil structure. However, the conservation tillage is useful to control soil degradation but it may lead to excessive soil compaction, thus, negatively affecting crop growth. Soil compaction is also a serious yield limiting factor for the mechanized no tilled systems especially in the soil predominantly low in clay content.

However, soils with crumbly and granular clods are considered to be with good structure. When the soil is subjected to tillage at optimum moisture, crumb structure is developed so that loss of soil by erosion is greatly reduced. Soil structure is destroyed when tillage is carried out at inappropriate soil moisture.

II. Pore space: Soils are made up of particles of different sizes. Air filled spaces between these particles constitute pore space. When a field is ploughed, the soil particles are loosely stacked in a random manner and pore space is increased.

When the soil is in good tilth, the capillary and non-capillary pores would be roughly equal. This facilitates free movement of air and moisture in the soil and increases infiltration.

III. Bulk density: When the soil is loosened, the soil volume increases without any effect on weight. Therefore, bulk density of tilled soil is less than the untilled soil.

IV. Soil colour: Tillage increases oxidation and decomposition of organic matter which is responsible for the dark brown to dark grey colour of the soil, thus, resulting in slight change of colour.

V. Soil moisture: Tillage improves soil moisture in different ways. The amount of available water depends on soil porosity, soil depth and random roughness. All these characters are increased by the tillage. Tillage also increases rate of infiltration, water holding capacity and hydraulic conductivity.

VI. Soil temperature: Tillage helps soil temperature to be optimum for seed germination and seedling establishment. Tillage looses the soil surface resulting in decrease of thermal conductivity and heat capacity. However, under no tillage, slight moderation of soil temperature takes place.

VII. Soil acidity: The soil acidity increases under no tilled conditions as the soil is not disturbed and fertilizer remains upto upper layer. In temperate humid zone but apparently not in tropics, there is marked tendency towards acidification of upper layer of soil under no tillage. It is caused by :

(a) Increased leaching experience under no tillage system.

(b) The soil is not turned and mixed.

(c) Accelerated acidification with no tillage has been attributed partly to decomposition of the concentrated layer of organic residues at the surface and with subsequent leaching of resultant organic acids into mineral soils and further to mineralization and subsequent nitrification of organic nitrogen and losses of basic cations as Ca^{2+} and Mg^{2+} by leaching. The decrease in soil pH is associated with no tillage described here, however, does not occur in all ecological regions.

VIII. Soil fauna: Living things in the soil consist of microorganisms like bacteria, fungi, algae, nematode, protozoa and large organisms like insects, earthworms and arthropods. These organisms break down organic matter making nutrients available to plants. Tillage can change abundance and composition of earthworms population. However, the actual impact is dependent on soil factors, climatic conditions and tillage operations. The decline in earthworms population is often reported in conventional tillage system which is associated with undesirable changes in soil conditions resulting from excessive tillage.

TYPE OF TILLAGE

Tillage operations are grouped into two types based on the time at which they are carried out. They are preparatory cultivation, which is carried out before sowing the crop, and cultivation that is practiced after sowing the crop for controlling weeds is called interculture or post sowing tillage.

Preparatory cultivation: It is the operations that are carried out from the time of harvest of a crop to the sowing of the next crop. In other words, preparatory cultivation is defined as the tillage operations that are carried out in any cultivable land to prepare seedbed for sowing of crops. This term preparatory cultivation is synonymous with seedbed preparation. It consists of three distinct operations.

Primary tillage or ploughing

Secondary tillage

Layout of seed bed

1. Primary tillage

Primary tillage of soil is mainly for the cutting and loosening of soil to a depth of 0 to 30 cm or more depending upon type of implement. It is the opening of the compacted soil with the help of different ploughs. It is done mainly to open the hard soil. Moreover, primary tillage also ensures inversion of soil (whenever necessary), uprooting of weeds and stubbles. The correct time for primary tillage or ploughing depends on soil moisture. The optimum range of soil moisture for effective primary tillage is 25 to 50 per cent depletion of available soil moisture. Light soils can be ploughed at a wide range of soil moisture conditions while the range is narrow for heavy soils. The number of tillage operations required to obtain good tilth depends upon many factors like soil type, weed problem and crop residues on the soil surface.

Light soils require 1 to 3 ploughings while the heavy soils require 3 to 5 ploughings to obtain proper tilth of the soil.

Implements used for primary tillage: Depending upon the purpose, soil condition and nature of weed problem, different types of ploughs are used. The mouldboard plough is the most common primary tillage tool in the world, and has the capacity to break up many types of soil. This type of plough leaves no unploughed land as the furrow slices are cut clean and inverted to one side resulting in better pulverization. It has the further ability to turn over and cover roots, crop residues and weeds. Mouldboard ploughs are used where soil inversion is necessary. Victory plough is an animal drawn mouldboard plough with a short shaft. For special applications there are many types of ploughs used under different situations such as fields with stubbles, clay soil, stiff-rod, black land, chisel general purpose and slatted ploughs. They may be used singly or in groups of two to a large number of bottoms (shares), with the width of each plough share being between 25 and 45 cm or more.

Another common primary tillage tool is the disc plough, comprising a hardened steel round concave disk of 50 to 95 cm in diameter. The discs have sharpened and sometimes serrated edges, and are often fitted with self-cleaning scrapers. The disc plough bears little resemblance to the common mouldboard plough, however, a large revolving concave steel disc replaces the share and the mouldboard. The disc plough is more suitable for land in which there is much fibrous growth of weeds as the disc cuts and incorporates the weeds. The disc plough works well in soils free from stones. No harrowing is necessary to break the clods of the upturned soil as in a mouldboard plough. The disc plough performs better in sticky, non-scouring (poor plough cleaning) soils, in hard dry ground, in many organic peat soils and where it is necessary to accomplish deep ploughing.

Chisel ploughs also find use in many areas. They are tools with long shanks and double-ended chisel points, usually about 6.4 cm in width. Chisels are usually mounted on a frame in gangs of 5 to 10 or more at spacings of 30 cm or so. Chisels can cut, loosen and stir the soil, but do very little turning over. They are well adapted to loosening hard dry soils, shattering hard pans and soles and conserving the mulch of crop residues on the field surface, which is useful for soil and water conservation in some areas. It is mainly used for deep ploughing (60-70 cm) with less disturbance to the top layers. Its body is thin with replaceable cutting edge so as to have minimum disturbance to the top layers. It contains a replaceable share to shatter the lower layers.

The sub-soiler is similar in principle to the chisel, but it is more heavily built and rigid for operation at depths of 40 to 90 cm to loosen deep soil layers for the promotion of water movement and root growth. The subsoil plough is designed to break up hard layers or pans without bringing them to the surface. The body of the subsoil plough is wedge shaped and narrow while the share is wide so as to shatter the hard pan and making only a slot on the top layers. Sometimes, a torpedo or bullet-shaped "mole" drain former is attached to the rear of the sub-soiler, and wings can be attached to the sides of the sub-soiler plough leg to increase the effective working width upto 90 cm on very heavy models. A tractor of 40 to 60 KW power is needed to pull one sub-soiler shank at a depth of 45 cm in heavy soil, while a larger track-laying tractor is needed for three winged sub-soiler operating at 90 cm depth.

The rotary plough is another primary tillage instrument, requiring a mechanical power source, usually provided by an auxiliary drive on the tractor. It consists of a set of knives, tynes or rods which are rotated on a horizontal shaft and covered by a sheet metal hood. Soil is chopped up by the knives and thrown against the inside of the hood, resulting in as fine, loose soil structure, depending on the forward speed of the implement. It cuts the soil and pulverizes it. The depth of cut is upto 12 to 15 cm. It is suitable for light soils. A good level seedbed of very loose soil is created, but at the expense of increased capital, maintenance and energy costs compared to other primary tillage implements. The rotary plough or rotary hoe is used extensively in intensive vegetable production, but is not recommended in areas where severe soil erosion due to wind or water flow are prevalent, especially in light or organic soils.

Basin lister is a heavy implement with one or two mouldboards or shovels. These shovels are mounted on a special type of frame on which they act alternatively. This implement is used to form listed furrows (broken furrows with small dams and basins) to prevent free run-off of rainfall and blowing off the soil in low rainfall areas.

Ridge plough has two mouldboards, one for turning the soil to the right and another to the left. The share is common for both the mouldboards i.e. double winged. These mouldboards are mounted on a common body. The ridge plough is used to split the field into ridges and furrows and for earthing up of crops. Ridge ploughs are used to make broadbed and furrows by attaching two ridge ploughs on a frame at 150 cm spacing between them.

Wooden plough or indigenous plough is an implement which is made of wood with an iron share point. It consists of body, shaft pole, share and handle. It is drawn with bullocks. It cuts a V-shaped furrow and opens the soil but there is no inversion. The ploughing operation with wooden plough is also not perfect because some unploughed strip is always left between furrows. However, this unploughed strip is reduced by cross ploughing, but even then small squares remain unploughed.

2. Secondary tillage

Secondary tillage operations are those which are performed after a primary treatment on a field for one or more purposes. After ploughing, the fields are left with large clods with some weeds and stubbles partially uprooted. The secondary tillage operations aim at improved seedbed in levelness and structure, increased soil pulverization, conservation of moisture, to crush the clods, destruction of weeds, chopping of crop residues etc. Planking is done to crush the hard clods, to smoothen the soil surface and to compact the soil tightly. The harrow is a common term for many of the secondary tillage implements. Thus, the field is made ready for sowing after ploughing by harrowing and planking. Generally sowing operations are also included in secondary tillage.

Implements used for secondary tillage: The different types of implements that are used for secondary tillage are cultivators, harrows, plankers and rollers etc. The harrows are used for shallow cultivation in operations such as seedbed preparation covering seeds and destroying weed seedlings. The frame of a harrow is fitted with anyone of a number of types of soil moving tools. The names of some of these are disc, spike-tooth, spring-tooth or rotary cross harrow etc. Sometimes harrows are combined with rollers comprised of V-shaped wheels or hollow cage wheels for improved fineness and smoothness of the soil surface. The harrows are of two types i.e. disc harrow and blade harrow. The disc harrow consists of a number of concave discs of 45 to 55 cm in diameter. These discs are smaller in size than disc plough, but more number of discs are arranged on a frame. These discs are fitted 15 cm apart on axles. Two sets of discs are mounted on two axles. All the discs revolve together with axles. The discs cut through the soil and effectively pulverize the clods. The blade harrows are used for different purposes like removal of weeds and stubbles, crushing of clods, working of soil to shallow depth, covering the seeds, inter-cultivation and harvesting of groundnut etc. Other popular secondary tillage tools are cultivators with V-shaped sweeps or blades for the removal of weeds at shallow depths between crop rows, or rod weeders which have vertical rods that turn and gather up weeds. Cultivators are used for finer operations like breaking clods and working the soil to a fine tilth in the preparation of seedbed. Cultivator is also known as tiller or tooth harrow. It has two rows of tynes attached to its frame in staggered form. The main objective of providing two rows and staggering the position of tynes is to provide clearance between tynes so that clods and plant residues can freely pass through without blocking. Provision is also made in the frame by drilling holes so that tynes can be set close or apart as desired. The number of tynes ranges from 7 to 13. The shares of the tynes can be replaced when they are worn out. The sweep cultivator is another kind of cultivator which is used in stubble mulch farming where it is difficult to prepare the land with ordinary implements due to clogging. The sweep cultivator is used to cut upto 12 to 15 cm depth of soil during first operation after harvest but shallower during subsequent operations. It is used frequently to control weeds and to prepare seedbed.

3. Layout of seedbed and sowing

After the preparation of seedbed, its layout is very important for the proper sowing or establishment of seedlings, control of weeds, application of irrigation or other cultural operations and management practices which are crop specific. For some crops like wheat, groundnut, bajra, cotton, pulses etc. flat field is necessary for their sowing, so these crops are sown without any land treatments in the form of ridges or furrows after the secondary tillage while other crops like maize, vegetables etc. are sown on ridges or furrows. Sugarcane is planted in furrows or trenches.

Some crops like tobacco, chillies, tomato etc. are planted with equal intra and inter-row spacing so as to facilitate two way inter-cultivation. For the plantation of these crops after field preparation, a marker is run in both the directions and seedlings are transplanted at the intercepts.

In Gujarat, for sowing of cotton and groundnut, the setline planting technique is adopted, where the seed rows are in the same place every year, thus, the seed lines are set permanently at wider spacings and the inter-row space is not cultivated. This method of planting has several advantages as it reduces tillage, conserves moisture in seed rows and also increases fertilizer use efficiency.

Some of the requirements of ideal seedbed are:

I. It must permit rapid infiltration of water.

II. There is sufficient retention of useable rain water.

III. It must afford an adequate air capacity and ready exchange of soil air with that of atmosphere.

IV. It must offer only little resistance to root penetration.

V. It must resist erosion of the soil.

VI. It must facilitate the placement of surface residue.

After the preparation of field, sowing is done with drills. The seeds are covered by running blade harrow to a shallow depth followed by planking so as to level and impart necessary compaction. Sowing is also done by dropping seeds behind the country plough, by *pora* or *kera* method, by broadcasting the seed over field, dibbling method or by transplanting the seedlings in the field.

Implements used for layout of seedbed: For the layout of seedbed, the bunds are made for irrigation by manual labour using spades or with tractor using bund former. The bund maker is designed to replace manual labour. Bunds are also formed across the contous in the low rainfall regions to conserve soil moisture. Country plough and ridge plough are used for laying out the field into ridges and furrows or to lay out irrigation channels. Ridge ploughs, when attached to a frame can be used for making broadbed furrows. Marker is used to mark intercepts for transplanting seedlings by square/cross planting method. The seedlings are transplanted at the intercepts.

Implements used for sowing: The implements used for sowing are country plough, seed drill and seed-cum-fertilizer drill. In the country plough, the seeds are dropped by hand in the furrows by *kera* or *pora* method. However, the seeds fall at uneven depths due to falling at random in furrows slice. In seed drill, 3 to 6 tynes are fixed on a wooden, beam, these tynes open the furrows into which the seeds are dropped. Through seed-cum-fertilizer drill, the fertilizers are placed at a depth of 5 cm and 5 cm away from seed rows for effective utilization of fertilizers. Both operations viz. drilling seeds and fertilizers are done simultaneously by seed-cum-fertilizer drill. It is similar to seed drill, but with extra tynes and hopper for drilling fertilizers. Sowing under zero tillage conditions can be done with zero till drill or with happy seeder and if weeds are present these may be killed with the spray of contact herbicide like paraquat or glyphosate.

After cultivation: The tillage operations that are carried out when the crop is standing are known as after tillage or post sowing tillage. The frequency of post-sowing tillage should be minimum but at the same time weeds must be controlled. Objectives of post-sowing tillage or after cultivation are :

I. To control weeds.
II. To increase the infiltration rate for the irrigation or rain water in the soil.
III. To enhance aeration in soil.
IV. To reduce the rate of evaporation through the creation of soil mulch on surface.
V. For thinning the crop when the plant population is higher than the required.
VI. For drilling or side dressing of fertilizers, earthing up and inter-cultivation.

Implements used for post-sowing tillage: The country plough or ridge plough are used for the operation of earthing up so as to form ridges at the base of the crop. It is done either to provide extra support against lodging as is done in sugarcane or to provide more soil volume for better growth of tubers as in potato or to facilitate irrigation as is done in vegetables. Inter-cultivation is done in between the crop rows to control weeds. It also conserves moisture by closing deep cracks in black soils. It is done with *danthi,* blade harrows, rotary hoes, ploughs and weeders. Country plough is run to a shallow depth to control weeds in widely spaced crops and fruit trees. Small sized blade harrows are widely used for inter-cultivation. They are simple in design, cheap and serve the purpose excellently. Inter-cultivation in close growing crops is also done with *danthi* which is a small implement with a blade length of 10 to 15 cm and each *danthi* covers one row only. A number of *danthis* are attached to a yoke for covering more area. Star weeder is a small implement pushed by manual labour and is used to control small weeds in close growing crops like groundnut, foxtail, millet etc. while Japanese rotary weeder is used for weeding and incorporation of fertilizers in line sown low land rice.

MODERN CONCEPTS OF TILLAGE

In order to overcome the constraints involved in conventional tillage like wastage of energy and sometimes destruction of soil structure, several new concepts have been introduced, namely minimum tillage, zero tillage, stubble mulch tillage, conservation tillage etc.

1. Conventional tillage

In this system, the frequency of ploughing is very high and depending upon the soil type, the field is ploughed 5 to 7 times. The soil becomes very loose and friable. After harvest of crop, one or two disc ploughing is done followed by tilling and planking. In all, 7 to 10 operations are required for seed led preparation.

Advantages

I. Perennial weeds are controlled very effectively.
II. The seed zone becomes very loose and friable.
III. The germination percentage of crop is very high.

IV. Root development is more.

V. Infiltration rate of water becomes high.

VI. Crop residues can be incorporated in the soil.

VII. Helps in conserving moisture especially in dry land agriculture.

Disadvantages

I. Cost on seed bed preparation is high

II. There is more wear and tear of farm machinery.

III. More soil erosion with wind or water

IV. A hard pan may be formed beneath the plough layer.

V. Decomposition of organic matter during hot weather.

2. Minimum tillage

The aim of minimum tillage is to disturb the soil to a minimum extent so as to ensure good seedbed, rapid germination of seeds, a satisfactory stand and favourable growing conditions. It can be defined as the short cut method involving the use of modified equipments to prepare seedbed with less energy than the conventional methods in less time. The suitable methods are based on the combination of factors like soil type, climate, crop rotation, labour supply and availability of machinery. The tillage operations can be reduced in two ways.

- By omitting operations which do not give much benefit as compared to the cost.
- By combining agricultural operations like seedling and fertilizer application with the use of seed-cum-fertilizer drill.

The minimum tillage involves a single pass with a cultivator/disc or a power harrow or a rotary cultivator prior to drilling. The important aim of minimum tillage is to reduce cost of cultivation.

Advantages of minimum tillage

I. It improved the soil conditions due to decomposition of plant residues *in situ.*

II. There is no burning of organic matter and less loss of nitrogen by volatilization.

III. It improves infiltration due to vegetation present on the soil and channels formed by the decomposition of dead roots.

IV. It imparts less resistance to root growth due to improved soil structure.

V. It causes less soil compaction due to reduced movement of heavy tillage vehicles.

VI. Soil erosion caused by wind is also reduced as compared to conventional tillage.

VII. It is cheaper as compared to conventional tillage due to minimum number of operations.

VIII. It improves the nutrient and water use efficiency.

IX. It also helps in reducing water pollution.

Disadvantages of minimum tillage

I. The seed germination is lower with minimum tillage as compared to conventional tillage.

II. More nitrogenous fertilizers have to be added in minimum tillage as the rate of decomposition of organic matter is slow.

III. Nodulation is affected in some leguminous crops.

IV. The continuous use of herbicides causes pollution problems as well as dominance of perennial problematic weeds.

V. Sowing is done with difficulty.

Different methods involved in minimum tillage are

- **Strip ploughing:** In this method, a narrow strip of about 8-9 inches is prepared by tilling upto 5-6 inches depth ahead of the drill opener for sowing of the crop like cotton but the soil between the sown rows remains undisturbed.
- **Plough planting:** After the soil is ploughed, a special planter is used and is run over the field, the row zone is pulverized and seeds are sown.
- **Bed planting:** It is done with the help of bed-maker with planter. The machine makes beds and planter is mounted on bed maker to sow the seeds of crop like wheat on beds.
- **Wheel track planting:** Ploughing is done as usual. Tractor is used for sowing and the wheels of the tractor pulverize the row zone.
- **Row zone tillage:** After primary tillage with mouldboard plough, secondary tillage operations like discing and harrowing are reduced. The secondary tillage is done in the row zone only.
- **Lister planter:** Listing is done with lister planter of 2-4 rows and the crop is sown behind the lister.

3. Zero tillage

Zero tillage or no tillage is an extreme form of minimum tillage. In this, primary tillage is completely avoided and secondary tillage is restricted to seed bed preparation in the row zone only. It is mostly practiced on soils which are subjected to wind and water erosion, timing of tillage operations is too less and requirements of energy and labour for tillage are too high. In zero tillage, the vegetation if present has to be destroyed before sowing of crop with the broad spectrum, non-selective herbicides like paraquat; glyphosate etc. which has no or short residual effect. The herbicides applied should not cause injury to the succeeding crop. During subsequent stages, selective and persistent herbicides are used.

Till planting is one method of practicing zero tillage. In this planting, four tasks are accomplished in one operation i.e. clean a narrow strip over the crop row, open the soil for seed insertion, place the seed and cover the seed properly. A wide sweep and task bars clear a strip over the previous crop row and planter-shoe opens a narrow strip into which seeds are planted and covered.

Advantages of zero tillage

I. Zero tillage soils are homogenous in structure with more number of earthworms.
II. The organic matter content of soil increases due to less mineralization.
III. The surface runoff of water tends to reduce due to the presence of surface mulch.
IV. It helps to sustain natural resources and productivity.
V. There is no erosion of soil either by wind or water.
VI. It is practiced in areas where the requirements of energy and labour for tillage are too high.
VII. It reduces the cost of production.
VIII. Sowing gets advanced by 5-7 days.
IX. Intensity of annual weeds gets reduced in zero tillage.
X. Modernization of soil temperature takes place.

Disadvantages of zero tillage

I. The seedling establishment in zero tillage is comparatively less than in conventional tillage.
II. The higher dose of nitrogenous fertilizers has to be applied as mineralization of organic matter is slow in zero tillage.
III. Large population of perennial weeds particularly *Cynodon dactylon, Cyperus rotundus* and *Sorghum halepense* appears in zero tilled plots.
IV. Higher number of volunteer plants and build up of pests is there.
V. There is clogging in zero till drill by rice straw, so there is difficulty in sowing the next crop especially wheat.
VI. There may be more infestation of rodents.

4. Stubble mulch tillage

Stubble mulch tillage is a new approach to protect the soil at all times either by growing a crop or by leaving the crop residues on the surface during fallow periods. In other words, it is a year round management of stubbles in such a way that a protective amount of stubble is left on soil surface even after sowing the crops. Soil is tilled as often as necessary to control weeds during the interval between two crops. For successful stubble mulching 8 to 10 quintals of stubbles/ha is sufficient. Good management of stubble mulch farming system begins with the harvest of the crop. Sweeps or blades are generally used to cut the soil upto 12 to 15 cm depth in the first operation after harvest and the depth of cut reduced during subsequent operations. When unusually large amount of residues are present, a disc type implement is used for the first operation to incorporate some of the residues into the soil. This hastens decomposition, but still keeps enough residue on the soil.

Methods involved for sowing crops in stubble mulch tillage are:

- A wide sweep and trash bars are used to clear a strip and a narrow planter-shoe opens a narrow furrow where seeds are placed.

- A narrow chisel of 5 to 10 cm width is worked through the soil at a depth of 15 to 30 cm leaving all plant residues on the surface. Planting is done through residues with special planters.

Advantages of stubble mulch tillage

I. Unnecessary repeated harrowings can be avoided after harvesting of crops in dryland areas.
II. It reduces the number of ploughings to the minimum level.
III. It covers the risk due to burning of stubles.
IV. It controls wind erosion.
V. It helps in reducing water erosion.
VI. Water loss through evaporation is reduced due to presence of stubbles on soil surface.
VII. It helps in decreasing the soil crust.
VIII. It also helps in maintaining the soil tilth.
IX. It improves infiltration of water.
X. It is economical as less number of cultivations are required.
XI. It helps in controlling weeds as the weed germination is reduced due to mulch presence on surface.

Disadvantages of stubble mulch tillage

I. There is difficulty in management of stubbles.
II. Poor germination of crop seed is generally experienced as compared to conventional tillage.
III. More diseases, insects and pests harbour on soil surface under the mulches.

5. Conservation tillage

In recent years, land-management systems have been developed in such a way so as to minimize the need for soil tillage. Since these systems leave considerable plant residues on or near the soil surface they protect the soil from erosion. Conservation tillage is the collective term commonly given to no tillage or minimum tillage or ridge tillage to denote the specific practice as this practice has conservation goal of some nature, means usually the retention of 30 per cent of surface residue cover which characterizes the lower limit of classification for conservation tillage. The conservation tillage is defined as any tillage system that reduces soil and/or water loss compared to clean tillage, where all residues are incorporated into the soil. The other conservation objectives for the practice include conservation of time, fuel, soil water, maintenance of soil structure and nutrients. Thus, residue levels alone do not adequately describe all conservation tillage practices. In simple words, it has objective to incorporate 30 per cent of residue with other benefits obtained from no tillage practices. Conservation tillage is an ecological approach in relation to soil surface management and seed bed preparation.

Advantage of conservation tillage

I. Higher crop yields are attained.
II. Less soil erosion by wind and water.
III. Improved infiltration and more efficient use of water.
IV. Increased acreage of sloping land that can safely be used for row crops.
V. Conservation tillage sustains the crop productivity.
VI. It improves the timing of planting and harvesting.
VII. It involves lower labour, machinery and fuel costs.
VIII. It decreases the fluctuation in soil temperature of surface layer.
IX. It protects the ground water from contamination.
X. It improves soil organic matter and carbon content in soil.
XI. It improves the structure of soil.

Disadvantages of conservation tillage

I. More potential for rodents, insects and diseases.
II. Cooler soil temperatures in spring, result in slower germination and more stand problems.
III. Greater management ability is required.

3

NUTRIENT ELEMENTS FOR PLANT GROWTH

Essential elements for plant growth

A large number of mineral elements are present in plants, but they do not require all these for growth. Out of the 107 elements in the periodic table, 90 elements are naturally occuring in plants and out of which only 17 elements are essential for plant growth.

Although common sense goes a long way in defining the concept of an essential element, Arnon and Stout in 1939, stated that an essential element should fulfill following criteria:

1. A deficiency in absence of the elements should make it impossible for the plant to complete its lifecycle.
2. Must not be replaceable by another element.
3. Must be directly involved in plant metabolism, that is, it must be required for a specific physiological function.

To Arnon and Stout's three requirements for essential elements should be added a fourth:

4. The element must be required by a substantial number of plant species, not just a single species or two.

Primary and secondary nutrients

All essential elements are required for plant growth and for completion of the plant life cycle from seed to seed. Some essential elements are needed in large quantities and others in much smaller quantities. However, from a practical standpoint, three of the six essential macro-nutrients are most often "managed" by the addition of fertilizers to soils, while the others are most often found in sufficient quantities in most soils and no soil amendments are required to supply adequate supplies.

From a management perspective only, the **primary nutrients** are N, P, and K, because they are most often limiting from a crop production standpoint. Apart from these calcium (Ca), magnesium (Mg) and sulphur (S) are also required by the plants in large quantity. All of the other essential macronutrient elements are **secondary nutrients** because they are rarely limiting, and more rarely added to soils as fertilizers.

The ability of soils to supply secondary nutrients to plants indefinitely is subject to the law of conservation of matter and is therefore dependent upon nutrient cycling. Continued crop removal of Ca, Mg, and S requires replenishment just as surely as primary nutrients, but most likely less frequently. Calcium and magnesium are often supplied by mineral weathering, either of natural soil materials or ground limestone added to correct soil acidity. Sulphur is often added to soil as either atmospheric deposition (associated with air pollution) or as impurities in fertilizers, particularly common Phosphatic fertilizers.

To demonstrate that this classification is more responsive to soil ability to supply nutrients than plant requirements, it should be noted that plant requirements for Ca, a secondary nutrient element, is greater than for P. Calcium is found as a principle exchangeable cation in most soils and an important soluble cation in the soil solution. Phosphorus, on the other hand, is only slightly soluble in most soils, and many soils (particularly acidic soils and alkaline soils) have the potential for causing phosphorus deficiencies.

Whether a macronutrient or micronutrient, or whether a primary or secondary nutrient, the "Law of the Minimum" holds: the most growth-limiting nutrient will limit growth, no matter how favourable the nutrient supply of other elements. For example, a deficiency of Fe or Mn (most common in soils containing calcium carbonate) can severely limit plant growth in spite of adequate N, P, and K.

Nutrients so far discovered

Elements	Essentiality established by	Year
A. Essential elements for all groups of plants		
Carbon, Hydrogen, Oxygen		Since time immemorial
Nitrogen	Theodore de Saussure	1804
Phosphorus	C. Sprengel	1839
Potassium	C. Sprengel	1839
Magnesium	C. Sprengel	1839
Sulphur	Salon Horstmas	1911/1857
Iron	E. Gris	1844
Manganese	J.S. McHargue	1922
Zinc	A.L. Sommer and C.B. Lipman	1926
Copper	A.L. Sommer, C.P. Lipman and C. McKinney	1931
Molybdenum	D.L. Arnon and P.R. Stout	1939
Sodium	P.P. Brownell and J.G. Wood	1957
Cobalt	S. Ahmad and H.J. Evans	1959
B. Essential elements for higher plants only		
Calcium	C. Sprengel	1839
Boron	K. Warington	1923
Chlorine	T.C. Broyer, Carlton, Johnson and Stout	1954

C. Beneficial elements required by a group of plants

Elements	Groups of plants required for	Scientist	Year
Vanadium	*Scenedesmus obliquus*	Arnon and Wessel	1953
Silicon	*Diatoms sp.*	J.C. Lewin	1962
Iodine	*Polysiphonia sp.*	L. Fries	1966
Selenium	*Astragalus sp.*	S.P. Trelease and H.M. Trlease	1938
Gallium	*Aspergillus niger*	R.A. Stunberg	1938
Aluminium	Ferns	K. Tanbock	1942

Besides above mentioned elements there are few more elements like rubidium, strontitium, nickel, chromium, arsenic etc. which are found to stimulate the growth and to have beneficial effects at very low concentrations. These nutrients are called as "beneficial elements"; potential micro-nutrients or micro nutrients.

ESSENTIAL ELEMENTS FOR PLANT GROWTH

Ionic charge

All but one of the essential elements is absorbed by plants from either soil solutions or nutrient solutions as either a **cation**, a positively-charged ion, or an **anion**, a negatively-charged ion. Cations and anions may differ not only in the sign of the ionic charge (positive/negative), but also its magnitude (1, 2, or 3).

In the Table below, the essential elements are listed in their approximate order of critical concentrations, from greatest to least, from top to bottom. Interestingly, several elements are absorbed as oxyanions, ions composed of the element surrounded by three or four oxygen atoms, giving a net negative charge. Also of interest is the fact that several elements are taken up in different forms.

Cation(3+)	Cation(2+)	Cation(1+)	Neutral(0)	Anion(-1)	Anion(-2)
Macronutrients:					
	Ca^{2+}	K^+		NO_3^-	HPO_4^{2-}
	Mg^{2+}	NH_4^+		$H_2PO_4^-$	SO_4^{2-}
Micronutrients:	Fe^{3+}				
	Fe^{2+}		$B(OH)_3$	Cl^-	MoO_4^{2-}
	Mn^{2+}				
	Zn^{2+}				
	Cu^{2+}				
	Ni^{2+}				

From this Table, a number of points can be made:

- Almost all elements are taken up as charged ions, with charges of +2, +1, -1, and -2.
- Absent are all organic compounds of the nutrient elements, indicating that all organic fertilizers must undergo *mineralization* of nutrients to become available to plants.

- Nitrogen may be taken up by plants as either an anion NO_3^- (nitrate) or a cation NH_4^+ (ammonium). Plant preferences for nitrate or ammonium are species-specific and are usually related to the natural environment of the species, with plants adapted to bogs, marshes, cold, or acid conditions often preferring ammonium-N, and most other dryland species preferring nitrate-N.
- The ionic charge on the orthophosphate ion, whether $H_2PO_4^-$ or HPO_4^{2-}, is dependent upon pH conditions prevailing in solution.
- Iron can be found as either Fe^{3+} or Fe^{2+}, with a difference in redox state as well as ionic charge. Evidence exists that mandatory reduction of Fe^{3+} to Fe^{2+} is an essential part of the uptake of Fe in many plants.

List of essential elements

The essential mineral elements are:

Nitrogen, Phosphorus, Potassium, Calcium, Magnesium, Sulphur, Boron, Chlorine, Iron, Manganese, Zinc, Copper, Molybdenum, and Nickel.

In addition to the essential mineral elements, there are some other **beneficial elements**, elements which promote plant growth in many plant species but are not absolutely necessary for completion of the plant life cycle, or fail to meet Arnon and Stout's criteria on other grounds. Recognized beneficial elements are:

Silicon, Sodium, Cobalt and Selenium

Other elements that have been proposed as candidates for essential or beneficial elements include Chromium, Vanadium and Titanium, although strong evidence is lacking at this time.

Another group is the **essential nonmineral elements**, elements taken up as gas or water, which are:

Hydrogen, Oxygen and Carbon

Out of all of the many natural elements, essential mineral elements, essential nonmineral elements, and beneficial elements are not randomly scattered, but instead cluster in several groups on the periodic chart.

Essential and beneficial elements in higher plants

Essential Mineral Element

Potassium (K), Magnesium (Mg), Calcium (Ca), Manganese (Mn), Molybdenum (Mo), Iron (Fe), Nickel (Ni), Copper (Cu), Zinc (Zn), Boron (B), Nitrogen (N), Phosphorus (P), Sulphur (S) and Chlorine (Cl)

Beneficial Mineral Element

Sodium (Na), Cobalt (Co), Silicon (Si) and Selenium (Se)

Essential Non-Mineral Element

Hydrogen (H), Carbon (C) and Oxygen (O)

Various classification schemes for essential elements include:

Macronutrients and micronutrients based on the relative concentrations in plant tissue.

Cations and anions: based on the positive or negative charge on the ionic form of the element as it is absorbed by plant roots.

What is plant nutrition

Plants use inorganic minerals for nutrition, whether grown in the field or in a container. Complex interactions involving weathering of rock minerals, decaying organic matter, animals, and microbes take place to form inorganic minerals in soil. Roots absorb mineral nutrients as ions from soil water. Many factors influence nutrient uptake for plants. Ions can be readily available to roots or could be "tied up" by other elements or the soil itself. Soil too high in pH (alkaline) or too low (acid) makes minerals unavailable to plants.

Fertility or Nutrition

The term "fertility" refers to the inherent capacity of a soil to supply nutrients to plants in adequate amounts and in suitable proportions. The term "nutrition" refers to the interrelated steps by which a living organism assimilates food and uses it for growth and replacement of tissue. Previously, plant growth was thought of in terms of soil fertility or how much fertilizer should be added to increase soil levels of mineral elements. Most fertilizers were formulated to account for deficiencies of mineral elements in the soil. The use of soilless mixes and increased research in nutrient cultures and hydroponics as well as advances in plant tissue analysis have led to a broader understanding of plant nutrition. Plant nutrition is a term that takes into account the interrelationships of mineral elements in the soil or soilless solution as well as their role in plant growth. This interrelationship involves a complex balance of mineral elements essential and beneficial for optimum plant growth.

Essential vs Beneficial

The term essential mineral element (or mineral nutrient) was proposed by Arnon and Stout (1939). They concluded that three criteria must be met for an element to be considered essential.

These criteria are important guidelines for plant nutrition but exclude beneficial mineral elements. Beneficial elements are those that can compensate for toxic effects of other elements or may replace mineral nutrients in some other less specific function such as the maintenance of osmotic pressure. The omission of beneficial nutrients in commercial production could mean that plants are not being grown to their optimum genetic potential but are merely produced at a subsistence level. This discussion of plant nutrition includes both the essential and beneficial mineral elements.

MINERAL ELEMENTS

There are actually 20 mineral elements necessary or beneficial for plant growth. Carbon (C), hydrogen (H), and oxygen (O) are supplied by air and water. The six macronutrients, nitrogen (N), phosphorus (P), potassium (K), calcium (Ca), magnesium (Mg), and sulphur (S) are required by plants in large amounts. The rest of the elements are required in trace amounts (micronutrients). Essential trace elements include boron (B), chlorine (Cl), copper (Cu), iron (Fe), manganese (Mn), sodium (Na), zinc (Zn), molybdenum (Mo), and nickel (Ni). Beneficial mineral elements include silicon (Si) and cobalt (Co). The beneficial elements have not been deemed essential for all plants but may be essential for some. The distinction between beneficial and essential is often difficult in the case of some trace elements. Cobalt for instance is essential for nitrogen fixation in legumes. It may also inhibit ethylene formation and extend the life of cut roses. Silicon, deposited in cell walls, has been found to improve heat and drought tolerance and increase resistance to insects and fungal infections. Silicon, acting as a beneficial element, can help compensate for toxic levels of manganese, iron, phosphorus and aluminum as well as zinc deficiency. A more holistic approach to plant nutrition would not be limited to nutrients essential to survival but would include mineral elements at levels beneficial for optimum growth. With developments in analytical chemistry and the ability to eliminate contaminants in nutrient cultures, the list of essential elements may well increase in the future.

Mineral elements in plant production

The use of soil for greenhouse production before the 1960s was common. Today a few growers still use soil in their mixes. The bulk of production is in soilless mixes. Soilless mixes must provide support, aeration, nutrient and moisture retention just as soils do, but the addition of fertilizers or nutrients are different. Many soilless mixes have calcium, magnesium, phosphorus, sulphur, nitrogen, potassium and some micronutrients incorporated as a pre-plant fertilizer. Nitrogen and potassium still must be applied to the crop during production. Difficulty in blending a homogenous mix using pre-plant fertilizers may often result in uneven crops and possible toxic or deficient levels of nutrients. Soilless mixes that require addition of micro and macronutrients applied as liquid throughout the growth of the crop, may actually give the grower more control of his crop. To achieve optimum production, the grower can adjust nutrient levels to compensate for other environmental factors during the growing season. The absorption of mineral ions is dependent on a number of factors in addition to weather conditions. These include the cation exchange capacity or CEC and the pH or relative amount of hydrogen (H+) or hydroxyl ions (OH-) of the growing medium, and the total alkalinity of the irrigation water.

Cation exchange capacity (CEC)

The cation exchange capacity refers to the ability of the growing medium to hold exchangeable mineral elements within its structure. These cations include ammonium, nitrogen, potassium, calcium, magnesium, iron, manganese, zinc and copper. Peat moss and mixes containing bark, sawdust and other organic materials all have some level of cation exchange capacity.

pH

The term pH refers to the alkalinity or acidity of a growing media water solution. This solution consists of mineral elements dissolved in ionic form in water. The reaction of this solution whether it is acid, neutral or alkaline will have a marked effect on the availability of mineral elements to plant roots. When there is a greater amount of hydrogen (H^+) ions, the solution will be acid (<7.0). If there is more hydroxyl (OH^-) ions, the solution will be alkaline (>7.0). A balance of hydrogen to hydroxyl ions yields a pH neutral soil (= 7.0). The range for most crops is 5.5 to 6.2 or slightly acidic. This creates the greatest average level for availability for all essential plant nutrients. Extreme fluctuations of higher or lower pH can cause deficiency or toxicity of nutrients.

THE ELEMENTS OF COMPLETE PLANT NUTRITION

The following is a brief guideline of the role of essential and beneficial mineral nutrients that are crucial for growth. Eliminate any one of these elements, and plants will display abnormalities of growth, deficiency symptoms, or may not reproduce normally.

1. Macronutrients

Nitrogen: It is a major component of proteins, hormones, chlorophyll, vitamins and enzymes essential for plant life. Nitrogen metabolism is a major factor in stem and leaf growth (vegetative growth). Too much application can delay flowering and fruiting. Deficiencies can reduce yields, cause yellowing of the leaves and stunt growth.

Phosphorus: It is necessary for seed germination, photosynthesis, protein formation and almost all aspects of growth and metabolism in plants. It is essential for flower and fruit formation. Low pH (<4) results in phosphate being chemically locked up in organic soils. Deficiency symptoms are purple stems and leaves; maturity and growth are retarded. Yields of fruit and flowers are poor. Premature drop of fruits and flowers may often occur. Phosphorus must be applied close to the plant's roots in order for the plant to utilize it. Large applications of phosphorus without adequate levels of zinc can cause a zinc deficiency.

Potassium: It is necessary for formation of sugars, starches, carbohydrates, protein synthesis and cell division in roots and other parts of the plant. It helps to adjust water balance, improves stem rigidity and cold hardiness, enhances flavor and color on fruit and vegetable crops, increases the oil content of fruits and is important for leafy crops. Deficiencies result in low yields, mottled, spotted or curled leaves, scorched or burned look to leaves.

Sulphur: It is a structural component of amino acids, proteins, vitamins and enzymes and is essential to produce chlorophyll. It imparts flavor to many vegetables.

Deficiencies show as light green leaves. Sulphur is readily lost by leaching from soils and should be applied with a nutrient formula. Some water supplies may contain sulphur.

Magnesium: It is a critical structural component of the chlorophyll molecule and is necessary for functioning of plant enzymes to produce carbohydrates, sugars and fats. It is used for fruit and nut formation and essential for germination of seeds. Deficient plants appear chlorotic, show yellowing between veins of older leaves; leaves may droop. Magnesium is leached by watering and must be supplied. It can be applied as a foliar spray to correct deficiencies.

Calcium: It activates enzymes, is a structural component of cell walls, influences water movement in cells and is necessary for cell growth and division. Some plants must have calcium to take up nitrogen and other minerals. Calcium is easily leached. Calcium, once deposited in plant tissue, is immobile (non-translocatable) so there must be a constant supply for growth. Deficiency causes stunting of new growth in stems, flowers and roots. Symptoms range from distorted new growth to black spots on leaves and fruit. Yellow leaf margins may also appear.

2. Micronutrients

Iron: Iron is necessary for many enzyme functions and as a catalyst for the synthesis of chlorophyll. It is essential for the young growing parts of plants. Deficiencies are pale leaf color of young leaves followed by yellowing of leaves and large veins. Iron is lost by leaching and is held in the lower portions of the soil structure. Under conditions of high pH (alkaline), iron is rendered unavailable to plants. When soils are alkaline, iron may be abundant but unavailable. Applications of an acid nutrient formula containing iron chelates, held in soluble form, should correct the problem.

Manganese: Is involved in enzyme activity for photosynthesis, respiration, and nitrogen metabolism. Deficiency in young leaves may show a network of green veins on a light green background similar to an iron deficiency. In the advanced stages, the light green parts become white, and leaves are shed. Brownish, black, or greyish spots may appear next to the veins. In neutral or alkaline soils plants often show deficiency symptoms. In highly acid soils, manganese may be available to the extent that it results in toxicity.

Boron: It is necessary for cell wall formation, membrane integrity, calcium uptake and may aid in the translocation of sugars. Boron affects at least 16 functions in plants. These functions include flowering, pollen germination, fruiting, cell division, water relationships and the movement of hormones. Boron must be available throughout the life of the plant. It is not translocated and is easily leached from soils. Deficiencies kill terminal buds leaving a rosette effect on the plant. Leaves are thick, curled and brittle. Fruits, tubers and roots are discolored, cracked and flecked with brown spots.

Zinc: It is a component of enzymes or a functional cofactor of a large number of enzymes including auxins (plant growth hormones). It is essential to carbohydrate metabolism, protein synthesis and internodal elongation (stem growth). Deficient plants have mottled leaves with irregular chlorotic areas. Zinc deficiency leads to iron deficiency causing similar symptoms. Deficiency occurs on eroded soils and is least available at a pH range of 5.5 - 7.0. Lowering the pH can render zinc more available to the point of toxicity.

Copper: Copper is concentrated in roots of plants and plays a part in nitrogen metabolism. It is a component of several enzymes and may be part of the enzyme systems that use carbohydrates and proteins. Deficiencies cause die back of the shoot tips, and terminal leaves develop brown spots. Copper is bound tightly in organic matter and may be deficient in highly organic soils. It is not readily lost from soil but may often be unavailable. Too much copper can cause toxicity.

Molybdenum: It is a structural component of the enzyme that reduces nitrates to ammonia. Without it, the synthesis of proteins is blocked and plant growth ceases. Root nodule (nitrogen fixing) bacteria also require it. Seeds may not form completely, and nitrogen deficiency may occur if plants are lacking molybdenum. Deficiency signs are pale green leaves with rolled or cupped margins.

Chlorine: It is involved in osmosis (movement of water or solutes in cells), the ionic balance necessary for plants to take up mineral elements and in photosynthesis. Deficiency symptoms include wilting, stubby roots, chlorosis (yellowing) and bronzing. Odours in some plants may be decreased. Chloride, the ionic form of chlorine used by plants, is usually found in soluble forms and is lost by leaching. Some plants may show signs of toxicity if levels are too high.

Nickel: It has just recently won the status as an essential trace element for plants. It is required for the enzyme urease to break down urea to liberate the nitrogen into a useable form for plants. Nickel is required for iron absorption. Seeds need nickel in order to germinate. Plants grown without additional nickel will gradually reach a deficient level at about the time they mature and begin reproductive growth. If nickel is deficient, plants may fail to produce viable seeds.

Sodium: It is involved in osmotic (water movement) and ionic balance in plants.

Cobalt: Cobalt is required for nitrogen fixation in legumes and in root nodules of non legumes. The demand for cobalt is much higher for nitrogen fixation than for ammonium nutrition. Deficient levels could result in nitrogen deficiency symptoms.

Silicon: It is found as a component of cell wall. Plants receiving soluble silicon produce stronger, tougher cell wall making them a mechanical barrier to piercing and sucking insects. This significantly enhances plant heat and drought tolerance. Foliar sprays of silicon have also shown benefits reducing populations of aphids on field crops. Tests have also found that silicon can be deposited by the plants at the site of infection by fungus to combat the penetration of the cell walls by the attacking fungus. Improved leaf erectness, stem strength and prevention or depression of iron and manganese toxicity have all been noted as effects from silicon. Silicon has not been determined essential for all plants but may be beneficial for many.

BIOLOGICAL FUNCTIONS, NUTRIENT DEFICIENCIES AND SOURCES

Deficiencies of nutrients in plants have various visual symptoms that are usually similar regardless of the species. The most common deficiency symptom is reduced growth, which is difficult to detect and diagnose at a glance. Other visual symptoms usually involve changes in colouration following a specific pattern, such as from the leaf tip down the midrib towards the base of the leaf or from the leaf margin toward the midrib, or between the veins of the leaf. Such symptoms may appear in new leaves or old leaves, indicating the phloem-mobility of the deficient nutrient and the ability of the plant to translocate existing stocks of the deficient nutrient. In many cases, internodal distances will shorten as well.

It may be remembered that plants are autotrophs, gathering solar energy for fixation of atmospheric carbon into energy–rich compounds consumed by heterotrophs, which gather the energy–rich plants for consumption and completion of their (and our) life cycles. In addition to energy, plants also harvest the mineral nutrients required by animal life. For the most part, the essential elements required by higher animals are similar to those of plants. However, animals require sodium (only beneficial in plants), selenium (beneficial to only a small group of Se-hyperacc umulating plants, iodine (essential only to certain marine alga), silicon (beneficial to a number of plant species), and cobalt (essential to N_2-fixing Rhizobium symbiotically associated with leguminous plants). Yet other trace elements essential to animal and human nutrition, but apparently neither essential nor beneficial for plants, are chromium, lithium, fluorine and vanadium (although some propose that chromium and vanadium).

1. Nitrogen

Biological function of N

- Protein synthesis (one or more N per amino acid)
- Base pairs for RNA/DNA
- Prosthetic groups for protein (ex.: heme group of chlorophyll)
- Hormones (ABA, cytokinins)
- Metal uptake (phytosiderophores) and transport in xylem & phloem (ex: Cu with amines)
- Osmoregulation
- Chemical defenses, alkaloids, miscellaneous biochemicals (ex: mescaline, cocaine, morphine, nicotine, caffeine, quinine)

Deficiency

Mild N deficiency will restrict plant growth, but often in a subtle manner that can only be assessed by comparison to plants grown with an adequate N supply. Moderate N deficiency will cause leaves to be light green or yellowish. Severe symptoms include necrosis (tissue death) starting at the tips of older leaves, with the tissue death developing a V-pattern down the midrib toward the base of the leaf.

Sources

In natural systems, N sources are:

- **Mineralization of organic-N:** Microbial decomposition of plant and animal residues in soil releases mineral nitrogen, nitrate and ammonium, forms of N that plants can absorb from the soil solution. Soil organic matter is also constantly slowly decomposing (even as new soil organic matter is being formed), releasing a slow trickle of nitrate and ammonium into the soil solution.
- **Biological fixation of atmospheric Nitrogen:** Even though the atmosphere is mostly Nitrogen, this N source is unavailable to plants without catalytic conversion to mineral N (NH_4^+) by highly specified micro-organisms. Some of these microorganisms live in association with plants, principally the legumes, in symbiotic relationships while other nonsymbiotic microorganisms also fix Nitrogen, which is released in mineral form only upon the decomposition of the microbial biomass upon its death.
- **Rainfall:** Precipitation contains both nitrate and ammonia. The total input from precipitation is small, A certain amount of the dissolved N is due to reactions of atmospheric Nitrogen during lightning events, although volatilization of ammonia and atmospheric pollution (as nitric acid) is a significant part of this input.

All systems, including natural systems, "leak" nitrogen because nutrient cycling is never absolutely perfect. Agricultural systems, because of their expectations of large biomass production and crop removal of N in the harvest, require N inputs greater than those provided by natural systems in order to be sustainable. These human inputs include:

- Addition of manures and organic matter. This is basically a redirection of the natural process, but in human hands includes the disposal of many of the wastes generated by agriculture.
- Synthetic fertilizers, among them ammonia, ammonium nitrate, and urea. These fertilizers have been proposed since the days of Justus von Liebig but have only become available since the development of the Haber-Bosch process of synthetic nitrogen fixation.

N Testing

Soil

- **Nitrate:** a simple water extract will extract nitrate from the soil since most soils have only a small anion exchange capacity and nitrate is not preferred on that anion exchanger.
- **Nitrate+Ammonium** (="mineral nitrogen"): strong salt solutions to displace exchangeable ammonium (NH_4^+). Most commonly used, 2 *M* KCl.
- **Organic N:** either determined by acid digestion and Kjeldahl distillation or determined as fixed percentage of organic matter content
- **Mineralizable N:** usually an incubation study that measures increase of mineral N over time.

Plant

- **Tissue test**: usually a quick test for nitrate, as in laboratory
- **Plant analysis**: either whole plant or specific plant portions at specified level of maturity. Digestion and total N analysis, as by Kjeldahl distillation in laboratory

Nitrogen fertilizers

The Haber-Bosch process produces NH_3 from atmospheric N_2 and H_2 produced from fossil fuel source, usually natural gas. It is a high pressure/high temperature reaction (130-680 atm, 340-420°C) using an iron catalyst. Ammonia from the industrial fixation of N_2 is the starting material for almost all synthetic N fertilizers.

I. Anhydrous ammonia (NH_3): First among N fertilizers in terms of synthesis path is anhydrous ammonia, since it is the product of the Haber-Bosch process. It contains 82% N and is a volatile gas at atmospheric pressure and ambient temperature but can be easily liquified with pressure and refrigeration. Agronomic use of anhydrous ammonia therefore, requires proper storage infrastructure. Application must be subsurface (to avoid immediate volatilization), preferably into a soil that is not too dry so that the ammonia gas can dissolve into the soil water.

II. Aqua ammonia, ammonium hydroxide (NH_4OH): Ammonia gas is dissolved in water to give a 33% solution (saturated). Ammonia is still volatile if not contained. Bulk of fertilizer weight is water, not N.

III. Ammonium nitrate (NH_4NO_3): Solid, white powder usually pelletized and is conditioned to improve handling, 33 or 34-0-0. Potentially explosive because of presence of both reducing and oxidizing agents (NH_4 and NO_3) in same compound.

IV. Urea [$CO(NH_2)_2$]: White powder, usually pelletized, 45 or 46-0-0. First organic compound ever synthesized inorganically. Soil micro-organisms cause urea to hydrolyze to ammonium carbonate, which may then spontaneously reform as (volatile) ammonia gas and ammonium bicarbonate if pH remains high, by the following reaction:

$$CO(NH_2)_2 \longrightarrow 2\ NH_4^+ + CO_3^{2-} \leftrightarrow NH_3 + NH_4^+ + HCO_3^-$$

Incorporation of urea into the soil is done so that the pH buffer capacity reduces the potential volatilization loss of ammonia which is essential for efficient use of urea. It is the most common fertilizer in many regions with infrastructure inadequate to transport and store liquified ammonia. Urea is produced industrially by high temp/high pressure reaction of ammonia and CO_2:

$$2\ NH_3 + CO_2 \longrightarrow CO(NH_2)_2$$

V. Ammonium sulphate, [$(NH_4)_2SO_4$]: 21-0-0, It is produced either by addition of ammonia to sulphuric acid using virgin materials or as by product of coke industry in which the flue gases containing ammonia from the organic N in coal are reacted with sulphuric acid solutions.

VI. Calcium nitrate, [$Ca(NO_3)_2$]: 15-0-0. Not particularly popular in U.S. as fertilizer except for nutrient solutions/greenhouses, but more popular in Europe, perhaps because of its suitability for soils in danger of acidification.

VII. Potassium nitrate, (KNO_3): 14-0-44. Commonly used in greenhouses and sometimes for high value crops. Produced by reaction of KCl with nitric acid, followed by volatilization of HCl:

$$KCl + HNO_3 \leftrightarrow KNO_3 + HCl(g)$$

VIII. Ammonium bicarbonate, (NH_4HCO_3): 17-0-0. Relatively unstable N fertilizer source that spontaneously reforms as ammonia, carbon dioxide, and water, causing major volatilization losses:

$$NH_4HCO_3 \leftrightarrow NH_3 + CO_2 + H_2O$$

Overall N fertilizer recovery rates for this source are low ~30%, but this is the major synthetic N source used in China.

IX. Ammonium phosphates: MAP, monoammonium phosphate, $NH_4H_2PO_4$, typically 12-52-0, and DAP, diammonium phosphate, $(NH_4)_2HPO_4$, typically 18-46-0. Formulated by reacting various quantities of ammonia with phosphoric acid. More often regarded as P fertilizer than N fertilizer since fertilizer application of DAP to achieve desired N rates exceed usual P rates, in most cases.

X. Sodium nitrate, ($NaNO_3$): Chilean saltpeter". Salt deposits of natural origin, accumulated in high desert valley of Chile, deposits discovered in 1809. Mined by stripping off overburden, blasting the ore, and refining to 96% pure product. Principal N fertilizer of commerce until replaced by Haber-Bosch products. Sodium content generally considered to have negative influence on soil structure.

The "Big Three" N fertilizers in the U.S. are anhydrous ammonia, ammonium nitrate, and urea. Anhydrous ammonia is a gas that is liquified for ease of handling by low temp and/or high pressure. Ammonium nitrate and urea are both solids, but soluble in water, thereby making a liquid N fertilizer. Liquid N fertilizers may be subdivided as:

- **High pressure:** Anhydrous ammonia. Pressure required to keep liquified.
- **Low pressure:** Aqua ammonia. Pressure required to retain ammonia gas dissolved in water.
- **No pressure:** Concentrated mixtures of urea and ammonium nitrate in water, UAN or URAN. Analyses are typically 28-0-0 (1°F), 30-0-0 (15°F), and 32-0-0 (32°F), where temperature is below when the fertilizer salts fall out of solution to form a hard-to-redissolve mass at the bottom and sides of tank.

Slow-release N fertilizers

These fertilizers are high value-added products, that is a relatively cheap material (fertilizer N), is remanufactured into a more expensive product for a specific purpose i.e., fertilizer availability to the gradual and continual plant nutrient needs, also minimizing sudden leaching of excess fertilizer. These products are typically used for high-value agricultural needs such as turf and greenhouses, rather than row crops. The basic strategy is to either make the N insoluble or trap it with some sort of barrier.

Urea form fertilizers are manufactured by the reaction of formaldehyde, CH_2O, with urea, $CO(NH_2)_2$, to produce compounds such as methylene diurea:

$NH_2C(=O)NH-CH_2-NHC(=O)NH_2$ or dimethylene triurea:

$NH_2C(=O)NH-CH_2-NHC(=O)NH-CH_2-NHC(=O)NH_2$

Using urea with isobutyraldehyde, $(CH_2)_2CHCHO$, instead of with formaldehyde, yields isobutylidene diurea (IBDU):

$NH_2C(=O)NH-CH(CH\ (CH_2)_2)-NHC(=O)NH_2$

About 90% of the N in IBDU is not water-soluble.

Another technique in the manufacture of controlled-release fertilizers is coating the fertilizer. Sulphur-coated urea (SCU) typically consists of 30-40% N (as urea), 10-30% S, 2-3% conditioner, and 2-3% sealant. Soil microorganisms oxidize the sulphur to sulphate, breaching the coating and allowing the urea to be released. Other coatings are formulated with micropores and cracks to permit slow diffusion of the fertilizer out of the fertilizer granule. Yet another group of techniques "trap" the N fertilizer in a matrix, either a gel or fritted solid or a cation exchanger such as zeolites or vermiculite, to slow the release of N.

2. Phosphorus

Notable in the ubiquitious ATP/ADP energy transport and storage compounds. Additionally, sugar phosphates form the "rails" of the nucleic acids DNA and RNA (which N-containing bases forming the "rungs"). Phospholipids are an important constituent of membrane chemistry and phosphoproteins are essential for life functions.

Results of P deficiency are spread more or less evenly around the plant, usually with no glaring visual deficiency symptom except for stunted growth and late maturity. Grassy species, including corn, will show reddening of leaves if P is severely deficient.

The forms of P in soil are: soluble, mineral, and organic.

Soluble P in soils is often in the micromolar concentration range and is an almost negligible percent of total soil P except for the fact that it is the only P that plants can use. The forms of soluble P are $H_2PO_4^-$ and HPO_4^{2-}, which the proportion between them dependent upon pH. At pH 7.2, they are present in equal concentrations. At pH 6.2, the ratio $H_2PO_4^-$: HPO_4^{2-} is 10:1, and at pH 8.2, 1:10.

Mineral P constitutes 20 to 85% of total soil P, depending upon the soil itself, particularly the soil organic matter content. Generally, formation of Al and Fe phosphates, either by precipitation or sorption, limits P solubility in acid soils and Ca phosphates limit P solubility in alkaline soils. This picture creates a "sweet spot" for maximum P solubility in equilibrium with P minerals at pH 6.5. By the same token, the minimum amount of P retention, or fixation, (within the range of common soil pH values) is at pH ~6.5.

Organic P constitute 15 to 80% of total P (and exactly complements the mineral P since soluble P is numerically negligible.) Soil organic matter typically contains

C/N/P/S in a ratio of 140:10:1.3:1.3, so that knowledge of the amount of organic matter will provide an estimate of the amount of organic P. Organo-P is cleaved by enzymes called phosphatases, which are present in many soil organisms and produced by some plants as well. In some soils, P is the most limiting plant nutrient, not N. In such soils, usually in the tropics, the pH is in the acid range, soils have a great P-fixing capacity, soil organic matter levels are very low, and no organic residues are returned to soils.

Another way of looking at P, particularly from a management point of view, is:

soluble-P <==> labile-P <==> nonlabile-P

Soluble P is fairly well defined, of course, but the chemistry of the labile P is more complicated, being a combination of mineral and organic P, and both are absorbed and precipitated. In addition, the mineralization of organic-P is a particularly valuable source of bioavailable P, particularly in strongly-P-fixing soils. The soluble P may be thought of as an intensity and labile P as a quantity, so that the P buffer capacity is quantity/ intensity.

P Inputs to soil

- Plant and animal residues
- Fertilizers
- Aeolian dust

P Outputs from soil

- Crop removal
- Runoff/erosion of the particulates which contain organic-P (mainly loss of top soil)
- Leaching to subsoil or groundwater is of little or no consequence

P Testing

The most common methods of testing for P availability are the Bray-1, Mehlich III, and Olsen methods. Bray-1 uses a mixture of HCl and NH_4F to dissolve adsorbed and poorly crystallized phosphate minerals into an acid extractant that increases the solubility of phosphate for purposes of extractant; fluoride is a good complexing agent for Al^{3+} to improve phosphate solubility. The Mehlich method is similar and uses a mixture of NH_4F, acetic acid, and NH_4Cl/HCl. Both are applicable for measuring labile phosphate in acid and neutral soils. For calcareous soils (containing free $CaCO_3$, with pH>7.5) such acid extractants would simply dissolve the carbonates and consume the acid in the extractant. Instead, the Olsen method uses 0.5 M $NaHCO_3$ at pH 8.5, which suppresses Ca^{2+} by both the high HCO_3^- concentration and high pH, allowing phosphates to dissolve out of calcium phosphate minerals (by the common ion principle).

Yet another P-test that is becoming increasingly popular is iron-oxide impregnated filter paper, which uses to advantage the strong sorption of phosphate on iron oxides. A solution of ferric chloride is neutralized to form iron oxides on filter paper. The iron-oxide impregnated filter paper can either be shaken with soil and water or buried in soil for a period of time to absorb soil phosphate and then recovered. The phosphate is then desorbed from the filter paper and analyzed. Results generally show a strong correlation between labile P by this method and by the more conventional methods.

Results of these analyses are used with tables of low, medium, high and very high values, based on probability and degree of plant response to added P. Such tables are collected soil-by-soil and crop-by-crop using field trials and may be distributed also as computer-computed recommendations.

Phosphorus fertilizer analysis

Fertilizer analyses of phosphorus in the U.S. is in the form of P_2O_5. [To convert % P to % P_2O_5, multiply by (2x31 + 5x16)/2x31 [= 2.29]. Analyses are divided into water-soluble P_2O_5 and citrate-soluble P_2O_5. Monocalcium phosphate, the major portion of superphosphate, is water-soluble. Dicalcium phosphate is largely citrate-soluble. Rock phosphate has low citrate-soluble P and very low water-soluble P.

Phosphatic fertilizers

Rock Phosphate: Source of all mineral P for fertilizers. Ore grade contains 6 to 15% P. Mined from deposits in Florida, Tennessee, North Carolina, Morocco, and elsewhere. Very insoluble. Effectively used as fertilizer material only on very acid soils or if acidulated or partially acidulated.

Single Superphosphate: Basically the product of the reaction of rock phosphate, usually a fluorapatite, with sulphuric acid according to the reaction:

$$2\ Ca_5(PO_4)_3F + 7\ H_2SO_4 \rightarrow 3\ Ca(H_2PO_4)_2 + 7\ CaSO_4 + 2\ HF$$

producing a mixture of monocalcium phosphate, a water-soluble phosphate, and gypsum, along with the volatile, and very nasty, hydrofluoric acid gas as a by-product. The commercial product contains between 7 to 9.5% P, or 16 to 22% P_2O_5.

Phosphoric Acid, (H_3PO_4): Produced by the same process as single superphosphate, but using more sulphuric acid.

$$2\ Ca_5(PO_4)_3F + 10\ H_2SO_4 \rightarrow 6\ H_3PO_4 + 10\ CaSO_4 + 2\ HF$$

Phosphoric acid is then separated mechanically from the gypsum for use. Phosphoric acid is rarely used directly as a fertilizer except for purposes of fertigation and commercial hydroponics, particularly when the source water contains (calcium) bicarbonate alkalinity (=high pH) that would otherwise cause phosphate precipitation. This material is, however, the acid used to manufacture the ammonium phosphates and the next fertilizer. Pure phosphoric acid is a solid that melts at 42°C. Phosphoric acid of commerce is typically ~23%P, or 53-54% 16 to 22% P_2O_5.

Triple Superphosphate: The active ingredient is calcium monophosphate, as in the single superphosphate, but the product carries less gypsum content because it was made with phosphoric acid instead of sulphuric acid:

$$2\ Ca_5(PO_4)_3F + 12\ H_3PO_4 + 9\ H_2O \rightarrow 9\ Ca(H_2PO_4)_2 + 2\ HF$$

This material contains 17 to 23% P or 44 to 52% P_2O_5. It was the most popular P fertilizer until surpassed in usage by ammonium phosphates. Much of the gypsum generated as a byproduct is diverted into construction.

Ammonium Phosphates: MAP, monoammonium phosphate, $NH_4H_2PO_4$, typically 12-52-0, and DAP, diammonium phosphate, $(NH_4)_2HPO_4$, typically 20-50-0. Formulated by reacting various quantities of ammonia with phosphoric acid. More often regarded as P fertilizer than N fertilizer since fertilizer application of DAP to achieve desired N rates exceeds usual P rates, in most cases. Completely water-soluble P.

Superphosphoric Acid: A phosphoric acid mixture containing a substantial proportion of one or more polyphosphoric acids (pyrophosphoric acid, $H_4P_2O_7$; tripyrophosphoric acid, $H_5P_3O_{10}$; tetrapyropho-sphoric acid, $H_6P_4O_{13}$. Manufactured by "roasting" phosphoric acid to remove some water, thereby polymerizing two or more phosphates. Typically 30 to 36% P, or 68 to 83% P_2O_5

Ammonium Polyphosphates, APP: Product of reacting polyphosphoric acid, with ammonia to form the final product, with a typical fertilizer composition of 10-34-0, 10% N and 15% P. The product is very water soluble, and has the capability to chelate some micronutrients, such as Fe and Zn.

Organic sources of P: The very same sources of organic N mentioned previously, for example, manure, whey, sludge, etc., also contain organic P that can serve as a P source as the organic matter mineralizes. Typical values are 4 lbs P/1000 gal in (solid) manure and 6.3 lbs P/1000 gal in sewage sludge. The use of organic sources as both N and P sources is sometimes problematic because the amount required to meet N needs may also add P far in excess of crop needs; the extra P in the top soil may increase the hazard of P runoff to surface waters.

Other so called "organic" P sources are bonemeal, in which the active P ingredient is closely related to a carbonatic rock phosphate, and rock phosphate (which is not really organic at all). Both are generally acceptable to organic farmers and gardeners. The availability of P in both is increased by acidulation, either partial or complete, to convert the apatite minerals to more soluble phosphates, Guano, bird and bat dung accumulations, are also popular as organic fertilizer but wreak a heavy cost to the environments from which they are extracted.

3. Potassium

Biological systems. Only one oxidation state, soluble with all biologically significant anions (except one) and at all pH values. Nontoxic, even at soil and plant levels far in excess of minimum requirements for optimum growth.

K Inputs to soil

- Plant and animal residues
- Fertilizers
- Aeolian dust

K Outputs from soil

- Crop removal
- Leaching to subsoil or groundwater is of little or no consequence
- Runoff/erosion of topsoil

Soluble K^+, ~0.1% of total soil K, does not form any chelates, complexes, or ion pairs in the soil solution. Typical concentrations soil solutions in alfisols and mollisols is 150 µM, which is equivalent to 6 mg/L. In humid soils, the typical values are even lower.

Exchangeable K, 0.1 to 2% of total soil K, is the major bioavailable form of K in the soil. Cation exchange capacities of soils typically range from 5 to 60 C mol(+) kg^{-1} (SI, =5 to 60 milliequivalents/100grams), and K is typically 1 to 3% of the total CEC, rarely more.

Fixed K, 1-10% of total K, is somewhat less well defined, although fairly easy to imagine. High levels of soluble and exchangeable K, particularly with wetting and drying cycles, will cause K^+ to dehydrate at the internal surfaces of 2:1 layer silicates (illite, vermiculite, and smectite) and become part of that mineral, or fixed. This K, although it is not displaced quickly by strong salts such as ammonium acetate, is released when soluble K levels are low for a protected period of time and can meet part of plant K requirements as slowly available K.

Mineral K, 90 to 98% of total K, is locked in soil minerals such as micas, feldspars, and illite, and only is released to plants with the actual weathering and destruction of the minerals themselves.

K testing

Soluble K (Intensity) could be measured in a water extract of soil but more often labile K (*Quantity*), mainly exchangeable K, is measured using either ammonium acetate or Bray-I (NH_4F) as extractants capable of displacing exchangeable K.

Potassium fertilizers

Potassium Chloride, (KCl): This single K source amounts to ~90% of all K fertilizers consumed in the U.S. For the most part, it is mined as sylvite ore (KCl, which is the potassium analog of halite, or rock salt, NaCl) mixed with NaCl, beneficiated to remove some of the contaminants, and sold. These products are often red because of residual contaminants which, although not water-soluble, are not harmful for soil application. Some KCl is the result of crystallization from brine, either from solution mining of KCl ore or precipitation from hypersaline lakes (e.g., Searles Lake and Dead Sea), often with carnallite $KClMgCl_2$ as an intermediate precipitate. The final form of KCl from crystallization processes is white crystals. Typical fertilizer analysis of KCl is 0-0-60 (=52% K), where ideal composition is 63%).

Potassium Sulphate, (K_2SO_4): About 5% of the U.S. market, this fertilizer is used particularly for horticultural crops in which chloride uptake is a problem, as in tobacco. 0-0-50.

Potassium Magnesium Sulphate, (K_2SO_4+2 $MgSO_4$): This is the primary K fertilizer produced by German and French mines from langbeinite ore and historically was called "double manure salts." The analysis is 0-0-18, but it also contains the secondary nutrients S (23%) and Mg (11%).

Potassium Nitrate, (KNO_3): 13-0-44 (less the 14-0-47 of pure salt because of conditioners added against hygroscopic moisture). Formulated from KCl and nitric acid. Expensive for agronomic use, used for horticulture and greenhouses. Identical to "saltpeter" known to the ancients and recognized by Glauber in the 17th Century as the "principle of vegetation".

Other potassium sources include wood ash, which contains K mainly as a carbonate or hydroxide, and powdered rocks (e.g., granite) and minerals (alunite, orthoclase, microcline, etc.). The release of K from the rocks and minerals requires weathering over long periods of time, although calcining the rocks can "crack" the K out of the structure (but makes the K more expensive than that from the evaporite ores).

4. Calcium

Biological functions

Calcium is present in the cytoplasm at levels that would indicate that it is a micronutrient, ~0.1 μM, in order to prevent interference with the high levels of P. Nonetheless, small fluctuations in Ca levels in the cytoplasm are part of signalling mechanisms for environmental stress. Ca pumps are directed out of the cytoplasm, either to vacuoles, where it may be precipitated as calcium oxalate, or across the plasmalemma. The function of the majority of plant Ca is structural, in the cell walls of shoots and roots.

Soil calcium

- Soluble Ca^{2+}
- Exchangeable Ca^{2+}
- Mineral Ca^{2+}: gypsum, carbonates, feldspars, etc.

Calcium inputs to soil

- Aeolian dust and precipitation
- Fertilizers, liming materials, manures, and residues

Calcium outputs from soil

- Crop removal
- Leaching

Soil tests for calcium

The most common soil test for plant-available Ca^{2+} is 1 M ammonium acetate (pH 7), which is capable of displacing exchangeable Ca^{2+} into the extracting solution, where it can be measured by either AAS, ICP, or even titrated with EDTA using a colorimetric endpoint. This method is not appropriate for calcareous soils (containing free $CaCO_3$)

or gypsiferous soils (containing $CaSO_4{\cdot}2H_2O$) because Ca^{2+} will also dissolve from these soil minerals into the extracting solution. Calcium can also be measured in water extracts of soil, particularly for calculation of SAR (sodium adsorption ratio) and PAR (potassium adsorption ratio).

Calcium fertilizers

- Calcium carbonate, $CaCO_3$, or calcium magnesium carbonate, $CaMg(CO_3)_2$, as agricultural limestone are the primary sources of added Ca to soils, although they are used primarily or additionally to neutralize soil acidity.
- Gypsum, $CaSO_4{\cdot}2H_2O$. No remediation of soil acidity should be expected from gypsum, but Ca will become available.
- Calcium nitrate, $Ca(NO_3)_2$, contains 19% Ca.
- Chelated calcium, usually as CaEDTA or Ca forms of hydrolzyed protein. If applied, often foliar application.
- Single superphosphate (18-21% Ca) and triple superphosphate (12-14% Ca)

5. Magnesium

Biological function

- The most familiar function of Mg, as the central ion in chlorophyll molecule, accounts for <25% of total plant Mg
- Mg in the cytoplasm is related to enzyme activation: for example, the substrate for ATPases is MgATP.
- Additional Mg is present in the cell wall

Magnesium deficiencies in soils are relatively rare as far as plants are concerned, but animals fed on Mg-deficient forage may develop hypomagesameia (grass tetany). "Excess" Mg is known almost only in soils derived from serpentine or olivene, both Mg-rich rocks, in which the exchangeable Ca/Mg ratio is ~1/5 instead of the more common 4/1 to 1/1; such soils often also contain toxic, high levels of bioavailable metals, among them Ni, which together with high Mg limit the distribution of "normal" vegetation, leading to "serpentine barrens".

Fertilizers

- Dolomitic agricultural limestone, $CaMg(CO_3)_2$, contains up to 13% Mg. Even calcitic limestone will contain some Mg as Mg substitution for Ca (magnesian calcite).
- Potassium magnesium sulphate, $K_2SO_4{\cdot}2MgSO_4$
- Magnesium sulphate, $Mg(SO_4){\cdot}7H_2O$, Epsum salt
- Magnesium nitrate, $Mg(NO_3)_2$, often applied as foliar spray.

6. Sulphur

Biological functions

- Amino acids: cysteine, methionine. organic-N/organic-S in plants typically has a 30-40:1 molar ratio
- coenzymes and prosthetic groups
- glutathionine: antioxidant and precursor to phytochelatins
- alliins in allium (with allinase produce pungency in onions) and glycosinolates in brassica (produces isothiocynates) for chemical defenses and S storage.

No toxicity to sulphate beyond salinity, S uptake beyond needs stored in part as sulphate. Deficiency symptoms are yellowing of whole plant, moderately phloem mobile. Response to S fertilizer in many crops and regions. With the reduction of sulphur at the smokestacks, agriculture may be losing the only benefit of acid rain.

Sulphur inputs to soil

- Precipitation and dry deposition (source: volcanoes, marine aerosols, coal burning)
- Fertilizers (both intentionally used to fertilize for S and incidentally contained in N-P-K fertilizers)

Sulphur outputs

- Crop removal
- Leaching
- Volatilization

Soil testing

Water extract for sulphate; occasionally use $CaCl_2$ or $Ca(H_2PO_4)_2$ solutions in order to displace sulphate from anion exchange sites (if present). Captures existing sulphate levels, but does not account for sulphate losses due to leaching or accumulation from mineralization of organic-S.

Sulphur fertilizers

- Potassium magnesium sulphate, $K_2SO_4 \cdot 2MgSO_4$: 22%S
- Sulphur-coated urea: 10-20% S
- Single Superphosphate (13.9% S as gypsum; only 1% S in triple superphosphate)
- Ammonium sulphate, $(NH_4)_2SO_4$, 24% S
- Elemental Sulphur, S (100%S): oxidizes to sulphuric acid, used also as acidifying agent
- Ammonium thiosulphate (ATS), $(NH_4)_2S_2O_3$, 26% S, 12% N
- Also $Al_2(SO_4)_3$, and $FeSO_4$, also used as acidifying agents

7. Boron

Functions of Boron

(i) Boron plays an important role in the development and differentiation of tissue, carbohydrate metabolism and translocation of sugar in plants.

(ii) It makes up the calcium deficiency to some extent.

(iii) It helps in the normal growth of plant and in absorption of nitrogen in soil.

(iv) It helps in root development and flower and pollen grain formation. Warington has established boron as essential element for plants in 1923.

Deficiency symptoms

(i) Plant growth is retarded and the leaves turn yellow or red.

(ii) Boron deficiency is often found to be associated with sterility and malformation of reproductive organs.

(iii) Each crop produces its characteristic growth abnormalities associated with boron deficiency.

(iv) Boron deficiency decreases the rate of water absorption and of translocation of sugar in plants.

(v) Seeds also fail to mature without boron.

Visual deficiency and toxicity symptoms

Although knowing the symptoms associated with nutrient deficiencies or toxicities is essential for every grower, it is important to remember that once visual symptoms are present reductions in crop yield or quality have often already occurred. Rigorous soil testing programs should be followed to help decrease the incidence of nutritional disorders.

In most cases, symptoms of nutritional disorders occur in defined patterns and are specific for each nutrient. Elements that are mobile in plants generally induce deficiencies on the older (lower) leaves first, while immobile elements induce deficiencies on the younger (upper) leaves first. In some cases, pesticide toxicity or disease symptoms may resemble nutrient deficiencies or toxicities. In addition, symptoms of nutritional disorders are often species or variety dependent. Use of soil and plant tissue analysis should be used to help confirm whether the symptoms truly are nutritional.

ORGANIC MANURES

Farm Yard Manure(FYM), compost, green manuring and crop residues are the important sources of organic manures. Few advantages of organic manures are:

1. Improve physical and chemical properties of soil.
2. Provide macro and micro nutrients for plant growth.
3. Helps in moderization of soil temperature

4. Improve water retention capacity of soil.
5. Improve microbial growth.
6. Make soil porous for better plant growth.
7. Improve crop yield.

Farm Yard Manure(FYM) is a balanced diet for crop plants. When good quality FYM is applied to soil, it brings considerable improvement in the soil environment which leads to enhanced productivity. On the other hand, use un-decomposed(raw) FYM may lead to more weed problem and more attack of termites. Generally mature weeds plants are cut along with fodders and weed seeds come in the intact form along with cattle dung. In the villages, cattle dung is stored in open due to which weed seeds remain viables and nutritional value of FYM also declines due to the discharge of water soluble nutrients with rainfall.

Method of preparing FYM

Cattle dung and cattle urine are very important component of FYM. It should be made a practice to spread some good absorbing bedding material under the animals to absorb sufficient urine which is very rich in nutrients. It has been seen that one kg of paddy straw can soak about 1.5 kg urine Dig a trench of 6-7 metre long, 1 to 1.5 metre wide and 3 feet deep. The depth should be 3 feet at one end and 3.5 feet at the other end. The trenches should be dug at such a place where rain water does not seep to avoid loss of humus from FYM. The trench should be filled with mixture of dung and urine from lower depth and should be filled approximately one and a half feet higher from the soil surface. A soil layer should be spread on the top of it. Then start filling the second pit and so on. With this techniques all weed seeds will be destroyed due to high temperatures of pit or due to fermentation and hence high quality FYM will be obtained within 3 to 4 months. A well rotted farm yard manures contains

Nitrogen - 0.5 - 1.5%
Phosphorus - 1.2 - 1.8%
Potassium - 1.2 - 2.0%
Calcium - 0.2 - 3.8%
Magnesium - 0.2 - 0.7%
Sulphur - 0.1 - 0.3%

Green manuring is another of organic manures because FYM is not available in sufficient quantity. Few crops like Sunhemp/Saan (*Crotolaria juncea*), Dhaincha (*Sesbania aculeata*) and cowpea (*Vigna unguiculata*) are best suited for green manuring due to quick growth and leguminous in nature. Any one of these crops can be sown when a period of 40-50 days is available between harvest of one crop and sowing of next crop. These crops are less input responsive crops and acquires considerable biomass within a short period of time. Bury these crops just before puddling rice and 7 to 10 days before sowing other crops like maize, Wheat, Potato and vegetable crops.

The residues of crops which are less economical can be incorporated in situ or can be used as surface mulching or compost can be made after piling them at one place. The management of rice straw in Punjab is big problem and can be used for improving soil fertility. Incorporate rice straw in soil if heavy machinery is available otherwise sow wheat in standing rice stubbles with happy seeders. The third options available for the farmer is to collect rice straw and compost can be prepared. Other waste material(bi-products or crop materials) can also be used for compost making.

Photographs of Important Field Crops

RICE *(Oryza sativa)*

Rice crop

Rice grains

WHEAT *(Triticum aestivum)*

Wheat crop

Wheat grains

MAIZE (*Zea mays*)

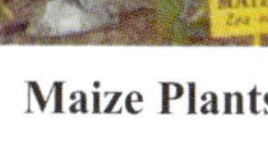

Maize Plants

Maize Cob

BARLEY (*Hordeum vulgare*)

Barley Crop

Barley Grains

SORGHUM (*Sorghum bicolor*)

Sorghum crop

Sorghum ear head

PEARL MILLET (*Pennisetum glaucum*)

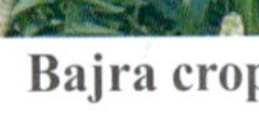

Bajra crop

Bajra seed

GROUNDNUT (*Arachis hypogaea*)

Groundnut crop

Plant with pods

SOYBEAN (*Glycine max*)

Soybean crop

Soybean pod

RAPESEED AND MUSTARD

Raya (*Brassica juncea*)

Gobhi sarson (*Brassica napus*)

LINSEED (*Linum usitatissimum*)

Crop **Seed**

SUNFLOWER (*Helianthus annuus*)

Crop Plant Flower

CHICKPEA (*Cicer arietinum*)

Chickpea crop **Chickpea seed**

LENTIL (*Lens esculenta Moench*)

Lentil Crop

PEA (*Pisum sativum*)

Field pea plant

Field pea crop

PIGEON PEA (Cajanus cajan L. Millsp.)

Pigeon pea crop

Pigeon pea pods

GREEN GRAM (*Vigna radiata L. Wilczek*)

Green gram pod

Green gram seeds

BLACKGRAM (*Vigna mungo L. Hepper*)

Blackgram plant

Blackgram seeds

COWPEA (*Vigna unguiculata L. Walp*)

Cowpea fruiting and flowring plants

Cowpea seeds

SUGARCANE (*Saccharum officinarum*)

Plant

Crop

SUGARBEET (Beta vulgaris)

Sugarbeet Crop

Mature plant

COTTON (*Gossypium hirsutum*)

Cotton plant

Cotton crop

JUTE (Corchorus capsularis and C. olitorius)

Jute plant

Jute crop

SUNHEMP (*Crotalaria juncea*)

Pods of sunhemp

Nodule of sunhemp

MESTA (*Hibiscus sabdariffa*)

Crop

Flower

4

WEEDS AND THEIR MANAGEMENT

The association of weeds with crops is as old as agriculture. These natural growth of weed plants are to be kept under check with the adoption of any method of weed control in order to attain economical crop yield.

HARMFUL AND BENEFICIAL EFFECTS OF WEEDS

Weeds are plants that are unwanted in a given situation and may be harmful, dangerous or economically detrimental. Weeds are a serious threat to crop production and biodiversity. They reduce farm and forest productivity, displace native species and contribute significantly to land and water degradation.

Weeds are those plants which grow out of their proper place, interfers with the utilization of natural resources, prolific, persistent, resistant, competitive, harmful and even poisonous in nature and can grow under adverse climatic conditions.

Weed management is an important component of plant protection improving the production potential of crops. It includes management of the weeds in a way that the crop sustains its production potential without being harmed by the weeds. Weed management is done through the mechanical, cultural and chemical means. Use of biological control methods in field crops is being considered, but still not much in use. Use of herbicides is an important method in the modern concept of weed management technology. New hand tools and implements have also been designed to assist in weed management programme.

Characteristics of weeds

Weeds are also like other plants but have special characteristics that tend to put them in the category of unwanted plants (Walia, 2014).

- Most of the weeds especially annuals produce enormous quantity of seeds, e.g. wild oats (*Avena ludoviciana*), produces 250 seeds per plant, whereas wild amaranthus (*Amaranthus viridis*) produces nearly 11 million seeds. It has been observed that among 61 perennial weeds, the average seed production capacity was 26,500 per plant.
- Weeds have the capacity to withstand adverse conditions in the field, because they can modify their seed production and growth according to the availability

of moisture and temperature. They can germinate under adverse soil moisture conditions, have short period of plant growth, generally grow faster and produce seed earlier than most of the crops growing in association.

- Weed seeds remain viable for longer period without losing their viability, e.g. annual meadow grass (*Poa annua*) and scarlet pimpernel (*Anagallis arvensis*) remain viable for about 8 years; creeping thistle (*Cirsium arvense*) for 20 years and field bind weed (*Convolvulus arvensis*) for about 50 years.
- Weed seeds have a tremendous capacity to disperse from one place to another through wind, water and animals including man. Many of times, weed seeds mimic with the crop seeds due to their size and get transported form one place to another along with them.
- Weed plants can tolerate more to adverse climatic conditions than the crop.
- They germinate early than the crop and take lead in competition.
- Most of the weeds resemble with the crop and are difficult to control.

Harmful effects

- Weeds have serious impacts on agricultural production. It is estimated that in general weeds cause 5% loss in agricultural production in most of developed countries, 10% loss in less developed countries and 25% loss in least developed countries.
- In India, yield losses due to weeds are more than those from pest and diseases. Yield losses due to weeds vary with the crops.Every crop is exposed to severe competition from weeds. Most of these weeds are self sown and they provide competition caused by their faster rate of growth in the initial stages of crop growth. In some crops, the yields are reduced by more than 50% depending upon the intensity of weeds.
- Weeds compete with crops for soil moisture, nutrients, light, and space, and thus reduce the crop yields.
- Weeds also act as alternate hosts to harbor insects, pests and diseases and other micro-organisms.
- Some weeds release into the soil inhibitors or poisonous substances that may be harmful to the crop plants, human brings and livestock.
- Weeds seeds like *Avena ludoviciana, Medicago denticulate* etc. reduce the quality of marketable agricultural produce. Contamination of weed seeds of *Datura stramonium,* Argemone *maxicana, Brassica spps* etc., are harmful to human health.
- Weeds not only reduce yield but also interfere with agricultural operations. Weeds make mechanical sowing a difficult process and render harvesting difficult, leading to increased expenditure on labour, equipment etc.
- In aquatic environment, weeds block the flow of water in canals, water transport system and drainage system, rendering navigation difficult. The dense growth of aquatic weeds pollutes water by deoxygenating it and killing the fish.

- Weeds are also a nuisance and a fire hazard along railway lines, roads, airports, forest and industrial sites.

Beneficial effects

In spite of all the difficulties caused by weeds, they can offer some beneficial properties, particularly when occurring at low densities. Some of the potential benefits of weeds are listed below:

- Helping to conserve soil moisture and prevent erosion. A ground cover of weeds will reduce the amount of bare soil exposed helping to prevent erosion e.g., *Cynodon dactylon.*
- Food and shelter can be provided for natural enemies of pests and even alternative food sources for useful crop pests. The actual presence of weed cover may be a factor in increasing effectiveness of biological control of pests and reducing pest damage.
- Weeds can also be valuable indicators of growing conditions in a field, for example of water levels, compaction and pH.
- Weeds can be an important source of food for wildlife, especially birds. Bird populations have been declining on farmland over the last few decades and leaving weeds as a resource has been shown to help revive bird populations.
- Few weeds have medicinal values e.g. *Fumaria parviflora, Cleome viscosa* etc.
- Weeds were used in scientific studies e.g. *Saccharum spontaneum.*
- Weeds are good source of fodder e.g. *Cynodon dactylon, Phalaris minor.*

CLASSIFICATION OF WEEDS

1. Based on life span

Based on life span (Ontogeny), weeds are classified as annual weeds, biennial weeds and perennial weeds.

I. Annual weeds

Weeds that live only for a season or a year and complete their life cycle in that season or year are called as annual weeds.

These are small herbs with shallow roots and weak stem. Produces seeds in profusion and the mode of propagation is commonly through seeds. After seeding the annuals die away and the seeds germinate and start the next generation in the next season or year following.

Most common field weeds are annuals. The examples are

a. Monsoon annual - *Commelina benghalensis, Boerhavia erecta, Eleusine spp.*

b. Winter annual - *Chenopodium album, Phalaris minor*

II. Biennial weeds

It completes the vegetative growth in the first season, flower and set seeds in the succeeding season and then dies. These are found mainly in non-cropped areas. e.g. *Alternanthera echinata, Daucus carota.*

III. Perennial weeds

Perennials live for more than two years and may live almost indefinitely. They adapted to withstand adverse conditions. They propagate not only through seeds but also by underground stem, root, rhizomes, tubers etc. and hence they are further classified into

a. *Simple perennials:* Plants propagated only by seeds. e.g. *Sonchus arvensis.*
b. *Bulbous perennials:* Plants which possess a modified stem with scales and reproduce mainly from bulbs and seeds. e.g. *Allium* sp.
c. *Corm perennials:* Plants that possess a modified shoot and fleshy stem and reproduce through corm and seeds. e.g. Timothy (*Phleum pratense*)
d. *Creeping perennials:* Reproduced through seeds as well as with one of the following.
    - Rhizome: Plants having underground stem – *Sorghum halapense*
    - Stolon: Plants having horizontal creeping stem above the ground – *Cynodon dactylon*
    - Roots: Plants having enlarged root system with numerous buds – *Convolvulus arvensis*
    - Tubers: Plants having modified rhizomes adapted for storage of food – *Cyperus rotundus*

2. Based on ecological affinities

I. Wetland weeds

They are tender annuals with semi-aquatic habit. They can thrive as well under waterlogged and in partially dry condition. Propagation is chiefly by seed. e.g. *Ammania baccifera, Eclipta alba*

II. Garden land weeds (Irrigated lands)

These weeds neither require large quantities of water like wetland weeds nor can they successfully withstand extreme drought as dryland weeds. e.g. *Trianthema portulacastrum, Digera arvensis.*

III. Dry lands weeds

These are usually hardy plants with deep root system. They are adapted to withstand drought on account of mucilaginous nature of the stem and hairiness. e.g. *Tribulus terrestris, Argemone mexicana.*

3. Based on soil type (Edaphic)

(I) **Weeds of black cotton soil:** These are often closely allied to those that grow in dry condition. e.g. *Aristolochia bracteata*

(II) **Weeds of red soils:** They are like the weeds of garden lands consisting of various classes of plants. e.g. *Commelina benghalensis*

(III) **Weeds of light, sandy or loamy soils:** Weeds that occur in soils having good drainage. e.g. *Leucas aspera*

(IV) **Weeds of laterite soils:** e.g. *Lantana camara, Spergula arvensis*

4. Based on place of occurrence

(I) **Weeds of crop lands:** The majority of weeds infests the cultivated lands and cause hindrance to the farmers for successful crop production. e.g. *Phalaris minor* in wheat

(II) **Weeds of pasture lands:** Weeds found in pasture / grazing grounds. e.g. *Indigofera enneaphylla*

(III) **Weeds of waste places:** Corners of fields, margins of channels etc., where weeds grow in profusion. e.g. *Parthenium hysterophorus*, *Calotropis gigantea*

(IV) **Weeds of playgrounds, road sides:** They are usually hardy, prostrate perennials, capable of withstanding any amount of trampling. e.g. *Alternanthera echinata, Tribulus terrestris*

5. Based on origin

(I) **Indigenous weeds:** All the native weeds of the country are coming under this group and most of the weeds are indigenous. e.g. *Acalypha indica, Abutilon indicum.*

(II) **Introduced or Exotic weeds:** These are the weeds introduced from other countries. These weeds are normally troublesome and control becomes difficult. e.g. *Parthenium hysterophorus, Phalaris minor, Acanthospermum hispidum.*

6. Based on cotyledon number

Based on number of cotyledons these can be classified as dicot and monocot.

(I) *Monocot - Panicum flavidum, Echinochloa colona*

(II) *Dicot - Crotalaria verucosa, Indigofera viscosa*

7. Based on soil pH

Based on pH of the soil, the weeds can be classified into three categories.

(I) Acidophile – Acid soil weeds eg. *Rumex acetosella*

(II) Basophile – Saline & alkaline soil weeds eg. *Taraxacum* sp.

(III) Neutrophile – Weeds of neutral soils eg *Acalypha indica*

8. Based on morphology

Based on the morphology of the plant, the weeds are also classified into three categories. This is the most widely used classification by the weed scientists.

(I) **Grasses:** All the weeds come under the family Poaceae are called as grasses which are characteristically having long narrow leaves. The examples are *Echinocloa colonum, Cynodon dactylon.*

(II) **Sedges:** The weeds belonging to the family Cyperaceae come under this group. The leaves are mostly from the base having modified stem with or without tubers. The examples are *Cyperus rotundus, Fimbrystylis miliaceae.*

(III) **Broad leaved weeds:** This is the major group of weeds as all other family weeds come under this except the weeds belonging to Poaceae and Cyperaceae families. All dicotyledon weeds are broad leaved weeds. The examples are *Digera arvensis, Amaranthus viridis, Chenopodium album.*

9. Based on nature of stem

Based on development of bark tissues on their stems and branches, weeds are classified as woody, semi-woody and herbaceous species.

(I) **Woody weeds:** Weeds include shrubs and undershrubs and are collectively called brush weeds. e.g. *Lantana camera, Prosopis juliflora*

(II) **Semi-woody weeds:** e.g. *Croton sparsiflorus*

(III) **Herbaceous weeds:** Weeds have green, succulent stems are of most common occurrence around us. e.g. *Amaranthus viridis, Chenopodium album.*

10. Based on specificity

Besides the various classes of weeds, a few others deserve special attention due to their specificity. They are: a. Poisonous weeds, b. Parasitic weeds and c. Aquatic weeds.

I. Poisonous weeds

The poisonous weeds cause ailment on human beings and livestock resulting in death and cause great loss. These weeds are harvested along with fodder or grass collected for fodder or while grazing the cattle consume these poisonous plants. e.g. *Datura fastuosa, D. stramonium* and *D. metel* proves poisonous to animals and human beings. The berries of *Withania somnifera* and seeds of *Abrus precatorius* are also poisonous.

II. Parasitic weeds

Some weeds are either total or partial parasitic which means, the weeds that depend completely on the host plant are termed as total parasites while the weeds that partially depend on host plant for minerals and capable of preparing its food from the green leaves are called as partial parasites. Those parasites which attack roots are termed as root parasites and those which attack shoot of other plants are called as stem parasites. The typical examples are:

a. Total root parasite – *Orabanche cernua* on Tobacco

b. Partial root parasite - *Striga lutea* on sugarcane and sorghum

c. Total stem parasite - *Cuscuta chinensis* on leucerne and onion

d. Partial stem parasite - *Cassytha filiformis* on orange trees and *Loranthus longiflorus* on mango and other trees.

III. Aquatic weeds

Unwanted plants, which grow in water and complete at least a part of their life cycle in water are called as aquatic weeds. They are further grouped into four categories as submersed, emersed, marginal and floating weeds.

a. *Submersed weeds:* These weeds are mostly vascular plants that produce all or most of their vegetative growth beneath the water surface, having true roots, stems and leaves. e.g. *Utricularia stellaris, Ceratophyllum demersum.*
b. *Emersed weeds:* These plants are rooted in the bottom mud, with aerial stems and leaves at or above the water surface. The leaves are broad in many plants and sometimes like grasses. These leaves do not rise and fall with water level as in the case of floating weeds. e.g. *Nelumbium speciosum, Jussieua repens.*
c. *Marginal weeds:* Most of these plants are emersed weeds that can grow in moist shoreline areas with a depth of 60 to 90 cm water. These weeds vary in size, shape and habitat. The important genera that comes under this group are: *Typha, Polygonum, Cephalanthus, Scirpus, etc.*
d. *Floating weeds:* These weeds have leaves that float on the water surface either singly or in cluster. Some weeds are free floating and some rooted at the mud bottom and the leaves rise and fall as the water level increases or decreases. e.g. *Eichhornia crassipes, Pistia stratiotes, Salvinia, Nymphaea pubescens.*

PROPAGATION AND DISSEMINATION OF WEEDS

Propagation

Propagation is the process of multiplying or increasing the number of plants of the same species and at the same time perpetuating their desirable characteristics. Plants may be propagated under two general categories: sexual and asexual propagation.

Reproduction by seed

Reproduction by seed is called sexual reproduction. It requires pollination and fertilization of an egg which results in seed that is capable of producing a new plant. Seed production varies greatly among and within weed species in part due to environmental variability between years, competition from neighboring plants, and genetic variability. Seed production capacity of annual weeds is very high.

Vegetative reproduction

In vegetative (asexual) reproduction, a new plant develops from a vegetative organ such as a stem, root, or leaf. Several modifications of these organs are common in perennial weeds, such as underground stems (rhizomes), above-ground stems (stolons), bulbs, corms, and tubers. Although vegetative structures generally do not survive as long in the soil as do seeds, very small structures can result in a new plant.

Vegetative reproduction can be as prolific as seed production. Yellow nut-sedge (*Cyperus esculentus*) has been reported to produce more than 1,900 new plants and more than 6,800 tubers in 1 year.

Dissemination / Dispersal

The chances of a seed developing are generally enhanced if there is a mechanism for dispersing to an appropriate habitat some distance from the parent plant. The reason for dispersal is that closely related organisms have similar ecological requirements. Seeds have no way to move on their own, but they are excellent travelers. Plants have evolved various mechanisms that disperse their seeds effectively. Many species of plants have seeds with anatomical structures that make them very buoyant, so they can be dispersed over great distances by the winds. In the absence of proper means of their dispersal, weeds could not have moved from one country to another.

1. Wind

Many seeds are well adapted to wind travel. Cottony coverings and parachute-like structures allow seeds to float with the wind. Examples of wind-dispersed seeds include common milkweed (*Asclepias syriaca),* common dandelion, Canada thistle, and perennial sowthistle (*Sonchus arvensis).* Weed seeds and fruits that disseminate through wind possess special organs to keep them afloat. Such organs are

I. Pappus – It is a parachute like modification of persistent calyx into hairs. e.g. Asteraceae family weeds - *Tridax procumbens*

II. Comose - Some weed seeds are covered with hairs, partially or fully e.g. *Calotropis* sp.

III. Feathery, persistent styles - Styles are persistent and feathery e.g. *Anemone* sp.

IV. Baloon - Modified papery calyx that encloses the fruits loosely along with entrapped air. e.g. *Physalis minima*

V. Wings - One or more appendages that act as wings. e.g. *Acer macrophyllum*

Factors that influence wind dispersal:

a. seed weight
b. seed shape
c. structures (wings or pappus)
d. height of release
e. wind speed and turbulence

2. Water

Aquatic weeds disperse largely through water. They may drift either as whole plants, plant fragments or as seeds with the water currents. Terrestrial weed seeds also disperse through irrigation and drainage water.Weed seed often moves with surface water runoff into irrigation water and ponds, where it is carried to other fields. Weeds growing in ditch banks along irrigation canals and ponds are the major source of weed seed contamination of irrigation water.

Weed seed often remains viable in water for several years, creating a "floating seedbank" and allowing weeds to disperse over large areas in moving water. Field bindweed seed, for example, remains over 50 percent viable after being submerged in water for more than 4 years. Some seeds have special adaptations that aid in water travel.

3. Animals

Several weed species produce seeds with barbs, hooks, spines, and rasps that cling to the fur of animals or to clothing and then can be dispersed long distances. Farm animals carry weed seeds and fruits on their skin, hair and hooves. This is aided by special appendages such as Hooks (*Xanthium strumarium*), Stiff hairs (*Cenchrus* spp), Sharp spines (*Tribulus terrestris*) and Scarious bracts (*Achyranthus aspera*). Even ants carry a huge number of weed seeds.

Weed seed often is ingested and passed through the digestive tracts of animals. Animal droppings provide an ideal nutrient and moisture environment for weed germination. While only a small percentage of the seed remains viable after exposure to an animal's digestive enzymes. The ingested weed seeds are passed in viable form with animal which is dropped wherever the animal moves. This mechanism of weed dispersal in called endozoochory. e.g. *Lantana* seeds by birds. *Loranthus* seeds stick on beaks of birds. Viable weed seeds are present in the dung of farm animals, which forms part of the FYM. Besides, addition of mature weeds to compost pit as farm waste also act as source of weed seeds.

4. Dispersal by man

Man disperses numerous weed seeds and fruits with raw agricultural produce. Weeds mature at the same time with crop, and are harvested together. Such weeds are called "Satellite weeds" e.g. *Avena fatua, Phalaris minor.*

5. Dispersal by machinery

Weed seeds often are dispersed by tillage and harvesting equipment. Seeds move from field to field on the soil that sticks to tractor tires, and vegetative structures often travel on tillage and cultivation equipment and latter dropping them in other fields to start new infestation. Disc type cultivation equipment is less likely to drag vegetative plant parts than are shovels or sweeps.

6. Intercontinental movement of weeds

Introduction of seeds of few noxious weeds from one continent to another is through crop seed, feed stock, packing material and nursery stock. Eg. *Phalaris minor, Parthenium hysterophorus* seeds.

7. As admixtures with crop seed, animal feed, hay, and straw

Weeds probably are spread more commonly during the seeding of infested crop seed with weed seeds or animal feed containing weed seeds or annual bedding containing weed plants.

WEED BIOLOGY AND ECOLOGY

Weed biology

Weed biology is the study of the establishment, growth, reproduction, and life cycles of weed species and weed societies/vegetation. Weed biology is an integrated science with the aim of minimizing the negative effects, as well as using and developing the positive effects, of weeds.

Life Cycle: Based on life cycle weeds are classified as annuals, biennials or perennials.

Annuals

Annuals complete their life cycle from seed to seed in less than one year. There are two types: summer and winter annuals. Summer annuals germinate in the spring, mature, produce flowers and seeds and die before fall. Winter annuals germinate in the fall, overwinter in a seedling or rosette stage, mature, produce flowers and seeds, and die in the spring or early summer. Because of the seedling stage, annual weeds are generally easy to control.

Biennials

Biennials generally complete their life cycle over two years. During the first year the seeds germinate and form a basal cluster of leaves and a tap root. The plant overwinters in this stage. During the second year the weed produces a flower stalk, sets seed and dies. Examples of biennial weed are *Dacus carota, Launea nudicaulis* etc.

Perennials

Perennial weeds live for more than two years. These weeds are the most common in blueberry fields and generally the most difficult to control. Perennial weeds may reproduce primarily by seed (daisy); by both seed and roots (sheep sorrel); or primarily be vegetative means (bunch-berry). Perennial weeds include both woody and herbaceous species.

Growth characters: Based on growth characters weeds are classified as grasses, broadleaf weeds, ferns and herbaceous or woody weeds.

Broadleaf

Broadleaf weeds are annual, biennial or perennial plants which generally have two leaves (cotyledons) emerging upon germination. The leaves normally have a branching network of veins and the flowers have distinct petals.

Grasses, sedges and rushes

Grasses can be annual or perennial plants. They generally have narrow, upright, parallel-veined leaves. Grasses have jointed stems, usually hollow at the internodes and are circular in cross section.

Sedges are a large group of perennial (rarely annual) grass like plants which are common in wet, poorly drained soils. Sedge stems are triangular in cross section, solid, and not jointed.

Rushes are annual or perennial plants similar in appearance to sedges with grass-like tufted leaves common at the plant base. Rush stems are hollow, circular in cross section, and not jointed. Like the sedge, this plant is also common in wet areas or poorly drained soil.

Ferns

Ferns are primitive perennial plants that do not produce flowers and seeds. Ferns consist of a leaf or frond, a stalk and an expanded blade which may then be further subdivided several times. Ferns spread by long creeping rhizomes and/or by spores.

Reproductive strategy: Based on reproductive strategy weeds are classified as seed, vegetative reproduction.

Reproduction by seed

Reproduction by seed is called sexual reproduction. It requires the fertilization of an egg by sperm, usually in the form of pollen. Pollination of the egg in a flower results in seed that is capable of producing a new plant. Seed production varies greatly among and within weed species in part due to environmental variability between years, competition from neighboring plants, and genetic variability.

Through sexual reproduction abundant and small seeds are produced. Annual and biennial weeds survival depend on seed production, as the sole means of propagation and survival of perennial weeds are less dependent on this mechanism. For example, white Canada thistle has been observed to produce as few as 680 seeds per plant, curly dock often produces more than 30,000 seeds per plant.

Vegetative reproduction

In vegetative (asexual) reproduction, a new plant develops from a vegetative organ such as a stem, root, or leaf. Several modifications of these organs are common in perennial weeds, such as underground stems (rhizomes), above-ground stems (stolons), bulbs, corms, and tubers. Although vegetative structures generally do not survive as long in the soil as do seeds, very small structures can result in a new plant. Vegetative reproduction can be as prolific as seed production.

Weed ecology

Weed ecology is the study of the interaction or relationship between a weed and its environment (other living organisms as well as abiotic factors). Ecology is concerned with growth characteristics and adaptations that enable weeds to survive the change in the environment. Man plays an important role in changing the environment by altering the crop husbandry practices and by maintaining weed free monocrop or multicrop culture. For effective weed control, the study on both biology and ecology of a weed species are important.

The weed seed bank and seed dormancy

Not only can weed seed and vegetative tissue travel great distances to infest new fields, but once in the soil, weed seed can remain viable for many years. In any given location, the weed seedbank contains a vast library of weed species and ecotypes that are adapted to a great range of environmental conditions and are ready to germinate when given the proper signal. A study reported that a square foot of soil, 6 inches deep, contained from 98 to 3,068 viable weed seeds. This represents between 4.3 million and 133 million viable seeds per acre.

The amount of time that a seed is capable of producing a seedling, or its viability, varies with weed species. In the extreme, lotus (*Nelumbo nucifera)* seeds found in a Manchurian lakebed were viable after 1,000 years. More commonly, the annual plant jimsonweed (*Datura stramonium)* has over a 90 percent germination rate after 40 years in the soil.

Additionally, many weed seeds remain dormant in the soil until the conditions for germination and survival are appropriate for that particular seed. Dormancy is the seed's resting stage and is the primary method of weed seed dispersal in time. Some weed seeds have seed coats that are impermeable to water and/or oxygen or are mechanically resistant. Others contain immature embryos or have a waiting period (called after-ripening) that must be completed before the seed will germinate. Seed dormancy is affected by environmental conditions, including temperature, light, oxygen, and the presence of chemical inhibitors.

Seed dormancy as survival mechanism

Weed seeds possess a variety of special germination mechanisms adapted to changes in temperature, moisture, aeration, exposure to light, depth of burial of seeds etc., when conditions are unfavourable for germination, they can remain dormant or delay germination.

Conditions favourable for weeds seed germination are:

- Seeds of many weeds require an exposure to light for germination. This is regulated by bluish-green protein pigment called phytochrome.
- Many weed seeds germinate under aerobic conditions while some require anaerobic condition. Soil turnover during ploughing and other operations exposes the seeds to light and induces germination.
- Periodicity of germination is another specialised germination mechanism. *Amaranths* spp have a definite pattern of peaks of germination at regular intervals.
- Summer annuals favour higher temperature and winter annuals germinate at lower temperatures whereas some weeds germinate freely throughout the year.

Seed dormancy: Dormancy is a state of seeds and buds of weeds in which they are alive but do not germinate. If all weed seeds germinate at one time, their seedlings could be destroyed. Dormancy allows storage of millions of weed seeds in soil and enables them to grow in flushes over years. In this context, the old gardeners saying

"*One year seeding seven years weeding*" is very appropriate. In fact, weed seeds have been found viable even after 20-40 years of burial in soil. Weed seeds exhibit three types of dormancy.

(I) Enforced dormancy: It is due to deep placement of weed seeds in soil during ploughing of the field. Weed seeds germinate readily when they are restored to top 3 to 5 cm. Enforced dormancy is a non-specific character of seed. Cultivation encounters enforced dormancy by bringing the weeds to surface where they are exposed to light besides better aeration. High soil temperature and NO_3 content of surface soil may further help in breaking seed dormancy.

(II) Innate dormancy: It is a genetically controlled character and it is a feature of specific weed seeds which fail to germinate even if they are present in the top 3-5 cm soil and adequate soil moisture and temperature provided to them. The possible reasons are the presence of

(i) Hard seed coats e.g., *Setaria spp., Ipomoea spp, Xanthium* spp.

(ii) Immature embryos e.g., *Polygonum spp.*

In certain weed seeds particularly of xerophyttic origin, presence of inhibitors is responsible for innate dormancy. It can be overcome with passage of time, or under the influence of some climatic pressure.

(III) Induced dormancy: Induced dormancy results from some sudden physiological change in normally non-dormant weed seeds under the impact of marked rise in temperature and or CO_2 content of soil, low O_2 pressure, water logging etc. Wild oat *(Avena fatua)* seeds exhibit all three kinds of dormancy.

Persistence of weeds (adaptation)

Persistence is an adaptive potential of a weed that enables it to grow in any environment. In an agricultural situation, the cropping system with its (associated habitat) management practices, determines the persistence of weed species. It is largely influenced by climatic, edaphic (soil) and biotic factors, which affect its occurrence, abundance, range and distribution.

Factors affecting persistence

1. Climatic factors

The important climatic factors are light, temperature, rainfall, wind and humidity.

Light

Light intensity, quality and duration are important in influencing the germination, growth, reproduction and distribution of weeds. Photoperiod governs flowering time, seed setting and maturation and on the evolution of various ecotypes within a weed species. Tolerance to shading is a major adaptation that enables weeds to persist. Depending up on the light requirements, weeds can be classified as sciophytes and heliophytes.

Temperature

Temperature of atmosphere and soil affects the latitudinal and longitudinal distribution of weeds. Soil temperature affects seed germination and dormancy, which is a major survival mechanism of weeds. Depending upon temperature requirements, weeds can be grouped into two catagories.

I. ***Rabi seasson***: Germinates at temperature of 10 to 20° C e.g., *Phalaris minor*.

II. ***Kharif season***: Germinates at temperature above 20° C e.g., *Trianthem portulacastrumn*.

Rainfall

Rainfall has a significant effect on weed persistence and distribution. More rainfall or less rainfall determines reproduction and survival. Depending upon water requirements, weeds can be classified as Hydrophytes, Mesophytes and Xerophytes.

Wind

Wind is a principal factor in the dissemination of weeds. Climate can effect variations in cuticle development, pubescence, vegetative growth, vigour, competitiveness *etc.* Climate thus has a profound effect on the persistence of weeds which can adapt to a wide variety of climates.

2. Soil factors

Soil factors are soil water, aeration, temperature, pH and fertility level and cropping system. Some weed species are characteristically alkali plants, known as basophilic (pH 8.5) which can grow well in alkali soils and those grow in acidic soil is known as Acidophiles.

Basophiles – *Puccinalia* spp., *Agropyron repens*

Acidophiles – *Cynodon dactylon, Digitaria sanguinalis*

Neutophiles — common weed

Several weed species of compositae family grow well in saline soils. A shift in soil pH, towards acid side due to continuous use of Ammonium sulphate as a 'N' source could cause a shift in the weed spectrum.

Many weeds can grow well in soils of low fertility level however, can adapt well to soils of high fertility also. Weeds also has adaptation to moist soil, drought condition etc.

3. Biotic factors

In a cropping situation, the major effects on weeds are those exerted by the crop as it competes for available resources. Once, a particular weed species is introduced, its persistence is determined by the degree of competition offered by the crop and also the agricultural practices associated with the growing of a crop may encourage or discourage specific weeds.

e.g. Ponding of water – *Cynodon dactylon* weed dies
Repeated cultivation – discourage nut sedge.

Crops that serve as hosts to parasitic weeds (Sorghum host for Striga) and crop-induced stimulants are examples of other biotic factors.

Crop weed association

Weeds posseses many growth characteristics and adaptations which enable them to exploit successfully the numerous ecological niches left unocccupied by crop cultures. Weeds compete with themselves and with crop plant. Among the more important adoptations relevent to competitive advantage are properly synchronized germination, rapid establishment and growth of seedlings, tolerance to shading effects by the crop or by other weeds at the time of establishment, quick response to available soil moisture and nutrients, adaptation to the most severe climatic situations of the habitat, adaptations to the edaphic regime, relative immunity to post seeding soil disturbance, practices and resistance to herbicides that are used. In the initial stages of invasion by weeds of exposed ecological niches, only a very limited competition for resources by the crop and weed may occur, but as establishment of the crop-weed association is completed, competition for the available reources is more obvious.

Plant competition is a natural force whereby crop and weed plants tend to attain a maximum combined growth and yield, with the development of each species being to some extent at the expense of the other. It occurs when the demands of the plants for moisture, nutrients, light, and possibly carbon dioxide exceed the available supply. Competition may develop between crop and weed plants and also between individual plants of each. The ultimate outcome of competition usually results in the development of a characteristic crop-weed association.

A principle of plant competition is that the first plants to occupy an area have an advantage over late comers. This principle is of foremost consideration in practical weed control, where cropping practices are always directed to the establishment of the crop ahead of the weeds.

Competition and allelopathy are the main interactions, which are of importance between crop and weed. Allelopathy is distinguished from competition because it depends on a chemical compound being added to the environment while competition involves removal or reduction of an essential factor or factors from the environment, which would have been otherwise utilized.

CROP WEED COMPETITION

Weeds appear much more adapted to agro-ecosystems than our crop plants. Without interference by man, weeds would easily wipe out the crop plants. This is because of their competition for nutrients, moisture, light and space which are the principle factors of production of crop.

1. Competition for nutrients

Weeds usually absorb mineral nutrients faster than many crop plants and accumulate them in their tissues in relatively larger amounts.

- The associated weeds are more responsive to nitrogen and they utilizes more of the applied 'N' than the crop. e.g. The 'N' uptake by *Echinochloa crusgalli* is more than rice.
- Nutrient removal by weeds leads to huge loss of nutrients in each crop season, which is often twice that of crop plants. For instance at early stages of maize cultivation, the weeds found to remove 9 times more of N, 10 times more of P and 7 times more of K, than crop if remains uncontrolled.

2. Competition for moisture

- In general, for producing equal amounts of dry matter, weeds transpire more water than do most of our crop plants. It becomes increasingly critical with increasing soil moisture stress, as found in arid and semi-arid areas.
- As a rule, C_4 plants utilize water more efficiently resulting in more biomass per unit of water. *Cynodon dactylon* had almost twice as high transpiration rate as pearl millet.
- In weedy fields soil moisture may be exhausted by the time the crop reaches the fruiting stage, i.e. the peak consumptive use period of the crop, causing significant loss in crop yields.

3. Competition for light

- It may commence very early in the crop season if a dense weed growth smothers the crop seedlings.
- Presence of weeds reduces photosynthetic activity of crop plants.

4. Competition for space

Crop-weed competition for space is the requirement for CO_2 and the competition may occur under extremely crowded plant community condition. A more efficient utilization of CO_2 by C_4 type weeds may contribute to their rapid growth over C_3 type of crops. Crop plants growing under competitive conditions will have weak and lanky stems.

Allelopathy

Allelopathy is the detrimental effects of chemicals or exudates produced by one (living) plant species on the germination, growth or development of another plant species (or even microorganisms) sharing the same habitat.

Allelopathy does not form any aspect of crop-weed competition, rather, it causes crop-weed interference, it includes competition as well as possible allelopathy.

Allelo chemicals are produced by plants as end products, by-products and metabolites liberalised from the plants; they belong to phenolic acids, flavanoides, and other aromatic compounds viz., terpenoids, steroids, alkaloids and organic cyanides. Leaves and inflorescence of *Parthenium* sp. affect the germination and seedling growth of maize.

Mechanism of action of allelochemicals

- Interfere with cell elongation
- Interfere with photosynthesis
- Interfere with respiration
- Interfere with mineral ion uptake
- Interfere with protein and nucleic acid metabolism

Effect of weed competition on crop growth and yield

- Crop growth and yield is affected
- Crop suffers from nutritional deficiency
- Leaf area development is reduced
- Yield attributes will be lowered
- Reduce the water use by the crop
- Affect the dry matter production
- Lowers the input response
- Pest and disease incidence will be more

Factors affecting the competitive ability of crops against weeds

I. Density of weeds: Increase in density of weed decrease in yield is a normal phenomena. However, it is not linear as few weeds do not affect the yields so much as other weed does and hence, it is a sigmoidal relationship.

II. Crop density: Increase in plant population decreases weed growth and reduce competition until they are self competitive. Crop density and rectangularity are very important in determining the quantum and quality of crop environment available for the growth of weeds. Wide row spacing with simultaneous high, intra-row crop plant population may induce dense weed growth. In this respect, square planting of crops in which there are equal row and plant spacing should be ideal in reducing intra-crop plant competition.

III. Type of weeds species: The type of weeds that occur in a particular crop influences the competition. Occurrence of a particular species of weed greatly influences the competition between the crop and weed. For example. *E. crusgalli* in rice, *Trianthema portulacastrum* in corn.

IV. Type of crop species and their varieties: Crops and their varieties differ in their competing ability with weeds e.g., the decreasing order of weed competing ability is as: barley, wheat and oat. High tolerance of barley to competition from weeds is assigned to its ability to develop more roots and more leaf area that are extensive during initial three weeks growth period than the others.

Fast canopy forming and tall crops suffer less from weed competition than the slow growing and short stature and crops. Dwarf and semi-dwarf varieties of crops are usually more susceptible to competition from weeds than the tall varieties became they grow slowly and initial stage. In addition, their short stature covers the weeds less effectively. Rapeseed and mustard crops are more competitive than wheat and among wheat cultivars, aestivum are more competitive than durum cultivars.

V. Soil factors: Soil type, soil fertility, soil moisture and soil reaction influences the crop weed competition. Elevated soil fertility usually stimulates weeds more than the crop, reducing thus crop yields. Fertilizer application of weedy crop could increase crop yields to a much lower level than the yield increase obtained when a weed free crop is applied with fertilizer.

Weeds are adapted to grow well and compete with crops, in both moisture stress and ample moisture conditions. Removal of an intense moisture stress may thus benefit crops more than the weeds leading to increased yields. If the weeds were already present at the time of irrigation, they would grow so luxuriantly as to completely over power the crops. If the crop is irrigated after it has grown 15 cm or more in a weed free environment irrigation could hasten closing in of crop rows, thus suppressing weeds.

Abnormal soil reactions often aggravate weed competition. It is therefore specific weed species suited to different soil reactions exist with us, our crops grow best only in a specified range of soil pH. Weeds would offer more intense competition to crops on both low or high pH soils than on normal pH soils.

VI. Climate: Adverse weather condition, e.g. drought, excessive rains, extremes of temperature, will favour weeds since most of our crop plants are susceptible to climatic stresses. All such stresses weaken crops inherent capacity to fight weeds.

VII. Time of germination: In general, when the time of germination of crop coincides with the emergence of first flush of weeds, it leads to intense crop-weed competition. Sugarcane takes about one month to complete its germination phase while weeds require very less time to complete its germination.

Weed seeds germinate most readily from 1 to 2 cm depth of soil and few weeds can germinate even from 15 cm depth. Therefore, planting method that dries the top 3 to 5 cm of soil rapidly enough to deny weed seeds opportunity to absorb moisture for their germination usually postpones weed emergence until the first irrigation. By this time the crop plants are well established to compete with late germinating weeds.

VIII. Cropping practices: Cropping practices, such as method of planting crops, crop density and geometry and crop species and varieties have pronounced effects on crop-weed competition.

IX. Crop maturity: Maturity of the crop is yet another factor which affects competition between weeds and crop. As the age of the crop increases, the competition for weeds decreases due to its good establishment. Timely weeding in the early growth stages of the crop enhances the yield significantly.

Critical period of weed competition

Critical period of weed competition is defined as the shortest time span during the crop growth when weeding results in highest economic returns.

The critical period of crop-weed competition is the period from the time of sowing up to, which the crop is to be maintained in a weed free environment to get the highest economical yield. The crop weed competition is invariably severe in early stages of crop growth than at later stages. Generally in a crop of 100 to 120 days duration, the first 35 days after sowing should be maintained in a weed free condition. There is no need to attempt for a weed free condition throughout the life period of the crop, as it will entail unnecessary additional expenditure without proportionate increase in yield.

Depending upon the type of the crop, weed free condition for 3-8 weeks in general are required and there is a need for timely weed control without which the crop yield gets drastically reduced.

METHODS OF WEED CONTROL

For designing any weed control programme in a given area, one must know the nature and habitat of the weeds in that area, how they react to environmental changes and how they respond to herbicides. Before selecting a method of weed control one much have information on the reproductive potential, nature of dispersal of seeds, dormancy of seeds, longevity of buried seeds and ability to survive under adverse conditions, life span of the weed, soil textures, soil moisture etc.

Principles of weed control are:

I. Prevention

II. Eradication

III. Control

IV. Management

I. Preventive weed control

It encompasses all measures taken to prevent the introduction and/or establishment and spread of weeds in a new area. No weed control programme is successful if adequate preventive measures are not taken to reduce weed infestation. It is a long term planning so that the weeds could be controlled or managed more effectively and economically. Following preventive control measures are suggested for adoption wherever possible and practicable.

- Avoid using seed of crops which are infested with weed seeds for sowing
- Avoid feeding material containing weed seeds to the farm animals.
- Avoid adding weeds to the manure pits or feeding to animals.
- Always use well decomposed farm yard manure (FYM) and never use raw FYM.
- Clean the farm machinery thoroughly before moving it from one field to another. This is particularly important for combines and farm implements.

- Avoid the use of gravel sand and soil from weed infested fields.
- Inspect nursery stock for the presence of weed seedlings, tubers, rhizomes, etc.
- Keep field boundaries, irrigation channels, fence-lines, and uncropped areas clean.
- Use vigilance. Inspect your farm frequently for any strange looking weed seedlings. Destroy such patches of a new weed by digging deep and burning the weed along with its roots. Sterilize the spot with suitable chemical.
- Quarantine regulations are available in almost all countries to deny the entry of weed seeds and other propagules into a country through airports and shipyards.

Weed free crop seeds

It may be produced by adopting following precautionary measures.

- Separating crop seeds from admixture of crop and weed seeds using physical differences like size, shape, colour, weight / texture and electrical properties.
- Using air-screen cleaners and specific gravity separators, which differentiate seeds based on seed size, shape, surface area & specific gravity.
- Always use seeds certified by seed certification authority as these are free from weed seed contamination.
- Weed laws are helpful in reducing the spread of weed species.
- Quarantine laws enforce isolation of an area in which a severe weed has become established and prevent the movement of the weed into an uninfected area.
- Use of pre-emergence or post-emergence herbicides are also helpful in prevention of weeds because herbicides will not allow the germination of weeds.

II. Eradication

Eradication means complete removal of weed species, its seed and vegetative part and that weed will not reappear unless reintroduced to the area. Because of its difficulty and high cost, eradication is usually attempted only in smaller areas. Eradication is often used in high value areas such as green houses, ornamental plant beds, nursery sites, vegetable fields on small scale etc. This may be desirable and economical when the weed species is extremely noxious and persistent as to make cropping difficult and economical.

III. Control

It encompasses those processes where by weed infestations are reduced but not necessarily eliminated. It is a matter of degree ranging from poor to excellent control of weeds. Weed control does not mean cent percent control of weeds but due to this method weed population is lowered to such a level that economic yields are ensured. Weeds can be controlled with the adoption of any weed control method.

IV. Weed management

Weed control aims at only putting down the weeds present by some kind of physical or chemical means while weed management is a system approach whereby whole land use planning is done in advance to minimize the very invasion of weeds in aggressive forms so that crop plants can compete strongly with weeds.

Weed control methods are grouped into cultural, physical, chemical and biological. Every method of weed control has its own advantages and disadvantages. No single method is successful under all weed situations. Many a time, a combination of these methods gives effective and economic control than a single method.

Few important weed management methods are as follows.

1. Mechanical weed control

Mechanical or physical methods of weed control are being employed ever since man began to grow crops. The mechanical methods include tillage, hoeing, hand weeding, digging, interculture, mowing, burning, flooding, mulching etc.

I. Tillage: Tillage uproots weeds from the soil resulting in their death. It may weaken plants through injury of root and stem pruning, reducing their competitiveness or regenerative capacity. Tillage also buries weeds. Tillage operation includes ploughing, discing, harrowing and leveling which is used to promote the germination of weeds through soil turnover and exposure of seeds to sunlight, which can be destroyed effectively later on in case of perennials, both top and underground growth is injured and destroyed by tillage.

II. Hoeing: Hoe has been the most appropriate and widely used weeding tool for centuries. It is however, still a very useful implement to obtain results effectively and cheaply. It supplements the cultivator in row crops. Hoeing is particularly more effective on annuals and biennials weeds as their growth can be completely destroyed. In case of perennials, it destroyed the top growth with little effect on underground plant parts resulting in re-growth.

III. Hand weeding: It is done by physical removal or pulling out of weeds by hand or removal by implements called khurpi, which resembles sickle. It is probably the oldest method of controlling weeds and it is still a practical and efficient method of eliminating weeds in cropped and non-cropped lands. It is very effective against annuals, biennials and controls only upper portions of perennials.

IV. Digging: Digging is very useful in the case of perennial weeds to remove the underground propagating parts of weeds from the deeper layer of the soil.

V. Sickling and mowing: Sickling is also done by hand with the help of sickle to remove the top growth of weeds to prevent seed production and to starve the underground parts. It is popular in sloppy areas where only the tall weed growth is sickled leaving the root system to hold the soil in place to prevent soil erosion. Mowing is a machine operated practice mostly done on roadsides and in lawns.

VI. Burning: Burning or fire is often an economical and practical means of controlling weeds in cultivated areas or in orchards. It is used to (a) dispose of vegetation (b) destroy dry tops of weeds that have matured (c) kill green weed growth in situations where cultivations and other common methods are impracticable.

VII. Flooding: Flooding is successful against weed species sensitive to longer periods of submergence in water. Flooding kills plants by reducing oxygen availability for plant growth. The success of flooding depends upon complete submergence of weeds for longer periods. It is important method of weed control in transplanted paddy.

Merits of mechanical method

- Oldest, effective and economical method
- Large area can be covered in shorter time
- Safe method for environment
- Does not involve any skill
- Weeding is possible in between crop rows.
- Deep rooted weeds can be controlled effectively

Demerits of mechanical method

- Labour consuming
- Possibility of damaging crop
- Requires ideal and optimum condition which are fit for hoeing.
- Inadequate avaibility of required labour.

2. Cultural weed control

Several cultural practices like tillage, planting pattern, fertiliser application, irrigation etc., are employed for creating favourable condition for the crop. These practices if used properly, help in controlling weeds. Cultural methods, alone cannot control weeds, but help in reducing weed population. They should, therefore, be used in combination with other methods. In cultural methods, tillage, fertiliser application. and irrigation are important. In addition, aspects like selection of variety, time of sowing, cropping system, cleanliness of the farm etc., are also useful in controlling weeds.

I. Field preparation: The field has to be kept weed free. Flowering of weeds should not be allowed. This helps in prevention of build up of weed seed population.

II. Summer tillage: The practice of summer tillage or off-season tillage is one of the effective cultural methods to check the growth of perennial weed population in crop cultivation. Initial tillage before cropping should encourage clod formation. These clods, which have the weed propagules, upon drying desiccate the same. Subsequent tillage operations should break the clods into small units to further expose the shriveled weeds to the hot sun.

III. Maintenance of optimum plant population: Lack of adequate plant population is prone to heavy weed infestation, which becomes, difficult to control later. Therefore, practices like selection of proper seed, right method of sowing, adequate seed rate, protection of seed from soil borne pests and diseases etc. are very important to obtain proper and uniform crop stand capable of offering competition to the weeds.

IV. Crop rotation: The possibility of occurrence of certain weed species or group of species is greater if the same crop is grown year after year. In many instances, crop rotation can eliminate or reduce difficult weed problems. The obnoxious weeds like *Cyperus rotundus* can be controlled effectively by including low land rice in crop rotation.

V. Growing of intercrops: Inter cropping suppresses weeds better than sole cropping and thus provides an opportunity to utilize crops themselves as tools of weed management. Many short duration pulses viz., green gram and soybean effectively smother weeds without causing reduction in the yield of main crop. Also cowpea as fodder is very good intercrop for conrolling weeds.

VI. Mulching: Mulch is a protective covering of material maintained on soil surface. Mulching has smothering effect on weed control by excluding light from the photosynthetic portions of a plant and thus inhibiting the top growth. It is very effective against annual weeds and some perennial weeds like *Cynodon dactylon.* Mulching is done by dry or green crop residues, plastic sheet or polythene film. To be effective the mulch should be thick encugh to prevent light transmission and eliminate photosynthesis.

VII. Solarisation: This is another method of utilisation of solar energy for the desiccation of weeds. In this method, the soil temperature is further raised by 2 to 5°C by covering à pre-soaked fallow field with thin transparent plastic sheet. The plastic sheet checks the long wave back radiation from the soil and prevents loss of energy by hindering moisture evaporation.

VIII. Stale seedbed: A stale seedbed is one where initial one or two flushes of weeds are destroyed before planting of a crop. This is achieved by soaking a well prepared field with either irrigation or rain and allowing the weeds to germinate. At this stage a shallow tillage or non- residual herbicide like paraquat may be used to destroy the dense flush of young weed seedlings. This may be followed immediately by sowing. This technique allows the crop to germinate in almost weed-free environment.

IX. Blind tillage: The tillage of the soil before the emergence of crop (those take more time for emergence) is known as blind tillage. It is also extensively employed to minimise weed intensity in drill sown crops where emergence of crop seedling is hindered by soil crust formed on receipt of rain or irrigation immediately after sowing.

X. Crop management practices: Good crop management practices that play an important role in weed control are

(a) Vigorous and fast growing crop varieties are better competitors with weeds.

(b) Proper placement of fertilizers ensures greater availability of nutrients to crop plants, thus keeping the weeds at a disadvantage.

(c) Better irrigation practices like drip irrigation helps in reducing weed pressure.

(d) Proper crop rotation programme

(e) Higher plant population per unit area results in smothering effect on weed growth

Merits of cultural method

- Low cost for weed control
- Easy to adopt
- No residual problem
- Technical skill is not involved
- No damage to crops
- Effective weed control
- Crop-weed ecosystem is maintained

Demerits of cultural method

- Immediate and quick weed control is not possible
- Weeds are kept under suppressed condition
- Perennial and problematic weeds can not be controlled
- Practical difficulty in adoption

3. Chemical control

Using chemicals, generally referred as herbicides, for the control of weeds is called chemical weed control. In 1944 - discovery of 2,4-D Na salt as a land mark in herbicide usage.

Merits

- Herbicide can be recommended for adverse soil and climatic conditions, as manual weeding is highly impossible during monsoon season.
- Herbicide can control weeds even before they emerge from the soil so that crops can grow in completely weed-free environment at early stages. It is usually not possible with physical weed control methods.
- Weeds, which resemble with crop in vegetative phase, may escape in manual weeding. However, these weeds are controlled by herbicides.
- Herbicides are highly suitable for broadcasted and closely spaced crops.
- Controls the weeds without any injury to the root system of the associated standing crop especially in plantation crops like Tea and Coffee.
- Reduces the need for pre planting tillage.
- Controls many perennial weed species.
- Herbicides control the weed in the field *in situ* where as mechanical method may lead to dispersal of weed species especially in heavy textured soils.
- It is profitable where labour is scarce and expensive.
- Suited for minimum tillage concept.
- Highly economical.

Demerits

- Pollutes the environment.
- Affects the soil microbes if the dose exceeds.
- Herbicide causes drift effect to the adjoining field.
- Continuous use of same group may lead to development of resistance in weeds or shift in weed flora.
- Leaves residual effects in soil as well as in crop produce.
- Some herbicide is highly costlier.
- Suitable herbicides are not available for mixed and inter-cropping system.

4. Biological control

Use of living organism's viz., insects, disease organisms, herbivorous fish, snails or even competitive plants for the control of weeds is called biological control. In biological control method, it is not possible to eradicate weeds but weed population can be reduced. This method is not useful to control all types of weeds. Introduced weeds are best targets for biological control.

Qualities of bioagents

- The bioagents must feed or affect only one host and not other useful plants.
- It must be free of predators or parasites.
- It must readily adapt to environment conditions.
- The bioagents must be capable of seeking out itself to the host.
- It must be able to kill the weed or at least prevent its reproduction in some direct or indirect way.
- It must possess reproductive capacity sufficient to overtake the increase of its host species, without too much delay.

Merits

- Least harm to the environment.
- No residual effect.
- Relatively cheaper and comparatively long lasting effect.
- Will not affect non-targeted plants and safer in usage.

Demerits

- Multiplication is costlier.
- Control is very slow.
- Success of control is very limited.
- Very few host specific bioagents are available at present.

5. Integrated weed management (IWM)

An integrated weed management may be defined as the combination of two or more weed control methods at low input levels to reduce weed competition in a given cropping system below the economical threshold level. Integrated Weed Management (IWM) approach aims at minimizing the residue problem in plant, soil, air and water. An IWM involves the utilization of a combination of mechanical, chemical and cultural practices of weed management in a planned sequence, so designed as not to affect the ecosystem. The nature and intensity of the species to be controlled, the sequence of crops that are raised in the rotation, the standard of crop husbandry, and the ready and timely availability of any method and the economics of different weed management techniques are some of the potent considerations that determine the success for the exploitation of the IWM approach.

Why IWM

- One method of weed control may be effective and economical in one situation and it may not be so in other situation.
- No single herbicide is effective in controlling wide range of weed flora.
- Continuous use of same herbicide creates resistance in weed flora or causes shift in the weed flora.
- Continuous use of only one practice may result in some undesirable effects. e.g. in rice–wheat cropping system, with the continuous use of isoproturon *Phalaris minor* may develop resistance to this herbicide.
- Only one method of weed control may lead to increase in population of particular weed.
- Indiscriminate herbicide use may result in adverse effects on the environment and human health.

FAO definition

It is a method whereby all economically, ecologically and toxicologically justifiable methods are employed to keep the harmful organisms below the threshold level of economic damage, keeping in the foreground the conscious employment of natural limiting factors.

IWM is the rational use of direct and indirect control methods to provide cost-effective weed control. Such an approach is the most attractive alternative from agronomic, economic and ecological point of view.

Advantages of IWM

- It shifts the crop-weed competition in favour of crop
- Prevents weed shift towards perennial nature
- Prevents resistance in weeds to herbicides
- No danger of herbicide residue in soil or plant
- No environmental pollution

- Gives higher net return
- Suitable for high cropping intensity

HERBICIDE CLASSIFICATION, FORMULATIONS AND METHODS OF APPLICATION

Herbicide: It is a chemical used to kill some targeted plants (weeds) whereas it is very safe to crop plants

Principles of chemical weed control

The selectivity exhibited by certain chemicals to cultivated crops in controlling its associated weeds without affecting the crops, forms basis for the chemical weed control. Such selectivity may be due to differences in the morphology, differential absorption, differential translocation, differential deactivation processes etc.

Classification of herbicides

1. Based on method of application

(*i*) *Soil applied herbicides:* Herbicide act through root and other underground parts of weeds. e.g. fluchloralin, pendimethalin etc.

(*ii*) *Foliage applied herbicides*: Herbicide primarily active on the plant foliage e.g. glyphosate, paraquat, 2, 4-D etc.

2. Based on mode of action

(*i*) *Selective herbicide:* A herbicide is considered as selective when applied on mixed growth of plant species, it kills some species without injuring the others. e.g. atrazine.

(*ii*) *Non-selective herbicide:* It destroys majority of treated vegetation including crop and weed plants. e.g. paraquat.

3. Based on mobility

(*i*) *Contact herbicide*: A contact herbicide kills those plant parts with which it comes in direct contact e.g. paraquat.

(*ii*) *Translocated herbicide*: Herbicide which tends to move from treated part to untreated areas through xylem / phloem depending on the nature of its molecule. e.g. 2,4-D, clodinafop.

4. Based on time of application

(*i*) *Pre - plant incorporation (PPI)*

Application of herbicides before the crop is planted or sown. For example, fluchloralin can be applied to soil and incorporated before sowing rainfed groundnut.

(*ii*) *Pre – emergence*

Application of herbicides before a crop or weed has emerged. In case of annual crops application is done after the sowing of the crop but before the emergence of weeds and this is referred as pre-emergence to the crop and weeds while in the case perennial crops it can be said as pre-emergence to weeds. For example, pre-emergence application of atrazine to plant crop of sugarcane is termed are pre-emergence to crop and weeds while its application to ratoon crop of sugarcane is called pre-emergence to weeds and post-emergence to crop.

(*iii*) *Post – emergence*

Herbicide application after the emergence of crop and weed is referred as post-emergence application. When the weeds grow before the crop plants have emerged from the soil and are killed with a herbicide then it is called as early post-emergence. For example spraying 2,4-D Na salt to control parasitic weed striga in sugarcane is called as post-emergence while spraying of paraquat to control emerged weeds after 10-15 days after planting potato can be called as early post-emergence.

5. Based on molecular structure

a. Inorganic compounds
b. Organic compounds

Formulations

Herbicides in their natural state may be solid, liquid, volatile, non-volatile, soluble or insoluble. Hence these have to be made in forms suitable and safe for their field use. An herbicide formulation is prepared by the manufacturer by blending the active ingredient with substances like solvents, inert carriers, surfactants, stickers, stabilizers etc

Objectives in herbicide formulations are

- Ease of handling.
- High controlled activity on the target plants.

Need for preparing herbicide formulation

- To have a product with physical properties suitable for use in a variety of types of application equipment and conditions.
- To prepare a product which is effective and economically feasible to use.
- To prepare a product which is suitable for storage under local conditions.

Types of formulation

(*i*) *Emulsifiable concentrates (EC):* A concentrated herbicide formulation containing organic solvent and adjuvants to facilitate emulsification with water eg. butachlor.

(*ii*) *Wettable powders (WP):* A herbicide is absorbed by an inert carrier together with an added surface acting agent. The material is finely ground so that it may form a suspension when agitated with a required volume of water eg., atrazine.

(*iii*) *Granules (G):* The inert material (carrier) is given a granular shape and the herbicide (active ingredient) is mixed with sand, clay, vermiculite, finely ground plant parts (ground corn cobs) as carrier material. eg. alachlor granules.

(*iv*) *Water soluble concentrates (WSC):* eg. paraquat.

Methods of application

I. Spraying

II. Broadcasting

MODE OF ACTION OF HERBICIDES

The term mode of action refers to the sequence of events from absorption of herbicide into plants to plant death. The mode of action of the herbicide influences how the herbicide is applied. For example, contact herbicides that disrupt cell membranes, such as paraquat (Gramoxone Extra), need to be applied post-emergence to leaf tissue in order to be effective. Seedling growth inhibitors, such as trifluralin (Treflan) and alachlor (Lasso), need to be applied to the soil to effectively control newly germinated seedlings.

To be effective, herbicides must (1) adequately come in contact with foliage; (2) be absorbed by plants; (3) move within the plants to the site of action, without being deactivated; and (4) reach toxic levels at the site of action. The application method used, whether pre-plant incorporated, pre-emergence, or post-emergence, determines whether the herbicide will contact germinating seedlings, roots, shoots, or leaves of plants.

The herbicide families listed below are grouped on the basis of how they affect weed plants:

I. The growth regulator herbicides: Examples are: 2,4-D, MCPA, dicamba, and triclopyr. These are mostly foliar applied herbicides which are systemic and translocate in both the xylem and phloem of the plant. They mimic natural plant auxins, causing abnormal growth and disruption of the conductive tissues of the plant. The injury from this family of herbicides consists of twisted, malformed leaves and stems.

II. The inhibitors of amino acid synthesis: Examples are: glyphosate, halosulfuron, imazethapyr, and sulfometuron. Both foliar and soil applied herbicides are in this family. Glyphosate translocates in the phloem with photosynthate produced in the leaves. Others in this family move readily after root or foliar absorption. These herbicides inhibit certain enzymes critical to the production of amino acids. Amino acids are the building blocks of proteins. Once protein production stops, growth stops.

III. Cell membrane disrupters: Examples : oxyfluorfen, lactofen, and acifluorfen. Soil and foliar applied with limited movement in soil. These herbicides enter the plant through leaves, stems, and roots, but are limited in their movement once they enter the plant. Membrane damage is due to lipid peroxidation. Symptoms are necrosis of leaves and stem.

IV. Lipid biosynthesis inhibitors: These are diclofop, fluazifop, sethoxydim, and clethodim. Foliar applied diclofop has both soil and foliar activity. Herbicides in this family move in both the xylem and phloem of the plant and inhibit enzymes critical in the production of lipids. Lipids are necessary to form plant membranes which are essential to growth and metabolic processes. Symptoms include stunting and death of tissue within the growing points of plants.

V. Pigment inhibitors: These are norflurazon, fluridone, and amitrol. Soil applied and move in the xylem except amitrol, which moves in both phloem and xylem. These herbicides inhibit carotinoid biosyntehsis, leaving chlorophyll unprotected from photooxidation. This results in foliage which lacks color. Symptoms include albino or bleached appearance of foliage.

VI. Growth inhibitors of shoots: Thiocarbamate herbicides including: EPTC, cycloate, pebulate, and molinate. Soil applied and somewhat volatile, requiring incorporation. Enter the plant through the roots and translocated through the xylem with the transpiration stream to the growing points in the shoot. Mode of action is unclear, but affects developing leaves in growing points of susceptible plants. Symptoms include stunting and distortion of seedling leaves.

VII. Herbicides which disrupt cell division: Examples are: trifluralin, DCPA, fluchloralin, oryzalin, pronamide, pendimethalin, and napropamide. All are soil applied, with limited movement in the soil. Absorbed through roots or emerging shoot tips. Once absorption takes place, movement is limited (site of action is near the site of absorption). These herbicides inhibit cell division or mitosis, except pronamide and napropamide which stop cell division before mitosis. Symptoms include stunting and swollen root tips.

VIII. Cell membrane disrupters: These have no soil activity and examples are paraquat, diquat, glufosinate, acids, oils, soaps. These herbicides are foliar applied with no soil activity. They enter the plant through the leaves and stems and do not move significantly within the plant once absorbed. These herbicides either act directly on cell membranes (acids, soaps. oils) or react with a plant process to form destructive compounds which result in membrane damage. Symptoms include rapid necrosis of the leaves and stem.

IX. Inhibitors of photosynthesis: Atrazine, simazine, metribuzin, cyanazine, prometryn, diuron, linuron, tebuthiuron, and bromacil are the common herbicides of this group. These are soil applied herbicides, however, all except simazine also have foliar activity. They move readily in the plant in the xylem with the transpiration stream where they concentrate in the leaves at the site of photosynthesis. Once there they block the electron transport system of photosynthesis, causing a build up of destructive high energy products which destroy chlorophyll and ultimately the leaf tissues. Symptoms include chlorotic (yellowed) leaves which become necrotic.

AQUATIC AND PROBLAMATIC WEEDS AND THEIR MANAGEMENT

Aquatic weeds

Almost all the water bodies have plants growing in them. Presence of plants in water bodies is essential for the conversion of solar energy into chemical energy for the development of aquatic fauna like fish, prawns etc. and for the continuous addition of oxygen to water during photosynthesis. If the water plants due to overgrowth make such water bodies unfit and take the shape of noxious aquatic vegetation, these may be referred as aquatic weeds. Aquatic weeds are the greatest problem in fishing, irrigation and efficient water supply. Because of scarce water supply and high population it has almost become imperative in every country to save water from the ravages of aquatic weeds. Since the beginning of this century, greater efforts are being made using the variety of implements, chemicals and bio-agents.

Problems of aquatic weeds: All these weeds develop dense mats on the surface of the water and becoming a major weed problem. The main problems arising from the growth of these weeds are:

- An enormous water loss through evapotranspiration, that alters the water balance of entire regions
- Clogging rivers and canals and related problems
- The impediment to water flow, that increases sedimentation, causing flooding and soil erosion
- Hampering fishing and dramatically reducing the catch and the source of food and income for local populations
- A drastic change in the physical and chemical properties of water and in the environment in the water bodies invaded, with detrimental effects on plants and animals;
- Interfering with the activity of hydroelectric power stations
- A serious threaten to agricultural production, following the blockage of irrigation canals and drainage systems.
- Hampering fishing
- Hampering navigation
- Indirectly affects human health, since it provides a suitable breeding habitat for pests and vectors of diseases.

Management

An integrated and to the extent possible an environment friendly approach is to be employed for management of aquatic weeds. Aquatic weed control measures can broadly be grouped into the following categories.

Preventive

The success of preventive weed management programmes varies with the weed species, its means of dissemination and the amount of efforts applied. Preventive weed management programmes usually require community action through the enactment and enforcement of appropriate laws and regulations.

Mechanical

This has both, advantages and limitations. Advantages include utilization of available manpower; is environment friendly, yields immediate results, is non-selective with fewer chances of permitting ecological shifts in aquatic flora; lessens mass nutrient load of eutrophic water bodies, helping indirectly in diminishing the future weed populations; reduces dependence on import of herbicides; harvested weeds may have various utilities as feed, manure, energy source etc; and most importantly can be exercised in any localized areas of water bodies. The limitations include limited effectiveness as in some cases the weeds regrow from their rootstocks, rhizomes and the spreading weeds infest new areas; labour-intensive and expensive and sometime removal of weeds may deplete water bodies of their nutrients limiting growth of planktons. The methods include, netting, erecting barriers, chaining, dredging, draining, use of water-weed cutters, submergence, shading, cleaning of irrigation waters etc.

Biological

Biological methods of management require the use of organisms that have been used for biological control, are diverse and include various types of animals and plants like insects, fishes, pathogens, nematodes etc. Use of several species of herbivorous fishes which feed on submerged aquatic weeds include *Tilapa* sp., *Ctenopharyngodon idella*, and other species. Observations are also available for rodents, snails etc. The use of insects like *Neochetina bruchi* and *N. eichhorniae* for control of water-hyacinth, and *Cyrtobagous salviniae* for control of *Salvinia molesta* has been found effective in India.

Problamatic weeds

Weed control practices often have an effect on the weeds, on a year by year basis. Before the development of herbicides, growers relied heavily on tillage as a tool for controlling and suppressing weeds. Once herbicides became a valuable tool, some of the problem weeds found in predominantly tillage based management practices began to fade and new problematic weeds began to fill the gap. As our habits change, specific weeds will exploit the new niches we create and become the more dominant species.

Some of the weeds like *Cyperus rotundus*, *Cynodon dactylon*, *Eleusine indica* etc., are listed as world's worst weeds. *Cyperus rotundus* is the most problamatic weed present in 92 countries, followed by *Cynodon dactylon* in 80 countries. Both weeds are perennial, mainly propagated by vegetative means and also by seeds. *Cyperus rotundus* is a problem weed in 52 crops while *Eleusine indica* in 46 crops.

Perennial weed management

Prevention

The most basic and effective of all methods to control perennial weeds is prevention. As discussed earlier, there are several means of weed seed dispersal, most of which can be prevented. Ensuring clean crop seed, animal feed, and hay is the most important measure in preventing seed dispersal. Other methods of prevention include cleaning field machinery and harvest equipment when moving between fields, proper long-term manure storage to reduce seed viability after passing through animals' digestive tracts, and maintenance of weed-free irrigation water.

Crop rotation can be another effective method to prevent the establishment of perennial weeds. The most effective crop rotations for this purpose include not only crops that compete well with perennial weeds, but also those that allow the use of herbicides to control perennial seedlings.

Mechanical weed control

Cultivation, when combined with other management tactics, can be used to control seedlings before energy-storing vegetative tissue has accumulated. Mechanical control no longer is effective after energy has been stored in underground vegetative tissue. In fact, cultivation of established perennials can spread weeds by cutting roots and moving them to new areas.

Perennial weeds are more common in reduced tillage fields, where there is little soil disturbance to disrupt the development of below-ground storage organs. Once perennial weeds are established in reduced tillage fields, cultivation is ineffective and might increase the spread of vegetative roots.

In pasture and forage crops, frequent mowing or cutting can prevent weed seed production and reduce the amount of energy stored in below-ground structures. Most important, maintenance of a vigorous crop stand through proper fertility and water management, seeding density, and variety selection will allow the competitive ability of the crop to suppress perennial weed growth. This simple "hands-off" approach requires little additional input or management, but can greatly reduce weed seed production and root growth.

Chemical weed control

Perennial weed control with herbicides must be repeated for 2 to 3 years and combined with other management tactics such as mowing. The key is to get the herbicide into the roots. Herbicide activity relies on foliar absorption and transport from the leaves to the root system. Young leaves move nutrients from the root in an upward, above ground direction, while more mature leaves transport photosynthetic products to the root system for storage. Thus, the most effective herbicide activity occurs as the product is transported to the roots with the products of photosynthesis.

Herbicides are most effective on perennial weeds in the early fall, when weeds are transporting energy to the roots before winter dormancy. Treatment just before flower bud initiation is effective. Repeated sprays of 2,4-D or glyphosate (depending upon the type of weed) at lower doses is very effective.

Biological weed control

Biological control is a slow process, and results are not guaranteed. Therefore, it is used most appropriately as a component of an integrated weed management system that relies on multiple tactics for perennial weed control. For instance *Plusia verticillata* (an insect) can be used effectively for the control of *Lantana camara* - a perennial weed.

Integrated weed management

Management of perennial weeds is most successful when multiple tactics are employed, such as the combination of chemical, mechanical, and cultural control. Integrated weed management, when combined with prevention and control of weeds outside of crop production areas, provides the best long term management of perennial weeds. The adoption of chemical method is which weeds have developed resistance to applied herbicide (s) should be integrated with mechanical method i.e., hand pulling. Also proper crop stand along with good agronomic practices acts as weed smothering approach and perform as integrated weed management method along with other methods of weed control i.e., chemical or mechanical.

5

IRRIGATION MANAGEMENT

IRRIGATION

Irrigation is generally defined as the artificial application of water to soil for the purpose of supplying the moisture essential for plant growth. (Israelson & Hensen, 1940).

However, a broader and more inclusive definition of the term irrigation is the application of water to the soil for any number of the following six purposes —

I. To add water to soil to supply the moisture essential for plant growth.

II. To provide crop insurance against short duration drought.

III. To cool the soil and atmosphere, thereby making more favourable environment for crop growth.

IV. To wash out or dilute salts in the soil.

V. To reduce the hazard of the soil piping.

VI. To soften tillage pans (big clods are soften for sowing).

In short irrigation is the science of harnessing and controlling various natural sources of water for the benefit of agriculture.

Objectives: Water is indispensable for human, animal and plant life. It is a major component of all organisms some of which contain more than 90% of water. Water is a constituent of protoplasm. It maintains turgidity of cells which is essential for stomatal opening, expansion of cells and for different physiological processes. About 400 to 500 litres of water is necessary for the production of a kilogram of plant dry matter. Water is also required for translocation of nutrients and dissipation of heat.

As population increases we need more production, so it is necessary to manage water resource for the purpose of crop production and other activities like industrialisation, power generation etc. which in turn provide employment opportunity and ultimately good living conditions of people. We also need irrigation management for :

I. To store and regulate water resources for further use and non season use.

II. To allocate water with proper proportion based on area and crop under cultivation. It is called balanced equity in distribution.

III. To convey water without much loss through percolation, seepage. It is called efficiency in use.

IV. To apply sufficient quantity of water to field crops i.e. optimization of use of water. If we apply excess there will be leaching of plant nutrients and wastage of water. With the availability of less or insufficient water, there will be reduction in quantity of food grains, loss in crop yields.

V. To utilize the water considering cost benefit ratio, we have to adopt economically viable management.

VI. To distribute available water without any social problem i.e. for judicious distribution of water.

VII. To meet the further requirement for other purposes like domestic use and to protect against famines i.e. for resource management.

VIII. To protect environment from over use or misuse of water (i.e. environment friendly use)

Importance: Classification given by Krishnamurthy (1965) about the world area:

I. 1/3rd of land surface receives less than 25 cm annual rainfall

II. 1/3rd of land surface receives between 25-50 cm annual rainfall

III. And remaining 1/3rd receives more than 50 cm annual rainfall

Total precipitation/rainfall is not important but its distribution throughout crop season is more important. A well distributed rainfall of about 60-65 cm (25") in a year is much more useful than 150-200 cm rainfall (60-70") receives during 1 or 2 months.

A well distributed rainfall is important not the intensity.

Classification of area according to rainfall*

Area	Annual Rainfall (Cm)	% of total area of land surface
Arid	<25	25
Semi arid	25–50	30
Sub humid	50–100	20
Humid	100–150	11
V. Humid	150–200	9
Wet	>200	5

*Krishnamurthy (1965)

According to this Table about 55% of land mass belongs to arid and semi arid zones and rainfall is about 0-50 cm and these areas lies between 10-35°L of the Northern and Southern hemisphere. In Northern hemisphere, like South Western USA, Mexico, S. France, Italy, Greece, Asia Minor, West Indies and China and in Southern hemisphere-Western regions of the Andes, Southern portion of South America, South Africa and Australia fall into these zones.

Sources of water supply in agriculture

I. Main source - The basic source of surface water supply is precipitation. This is in form of rainfall, snowfall/hails/sleet.

II. Minor sources -

(a) *Dew* - The water is absorbed by plant leaves.

- Reduce the loss of water due to transpiration.

(b) Fog: Water is absorbed through leaves of plants

(c) High humidity- Reduce loss of water and lower the demand of evapotranspiration.

III. Ground water - It is the water below the soil surface. It is available to plant by the phenomenon of capillarity action. The water table should not be low and also should not be too high so that supply of moisture continue and doesn't affect aeration.

IV. Flood water:-Only a small quantity of flood water can be utilized by the crop but it higher the underground water table.

Why irrigation is necessary?

I. It controls soil moisture and overcome the drought situation.
II. It makes multiple cropping system possible.
III. It increases the crop yields because if irrigation facilities are available, we can grow long duration high yielding crop varieties instead of ordinary ones
IV. Degree of famines are reduced
V. Addition of manures and fertilizers are possible if water is available which ultimately increase efficient use of inputs.
VI. Labour will be engaged throughout the year.
VII. It is possible to grow crops like sugarcane, cotton, potato and vegetables.
VIII. It ensures the returns all around the year.
IX. It softens the soil clods and dissolve the plant nutrients.
X. It makes tillage operations possible (pre-sowing – for soil preparation, post-sowing- for intercultural practices).
XI. It helps in leaching down the excess salts.
XII. It improves the efficiency of soil applied pesticides and herbicides. Movement of herbicide is fast and reaches the lower layers of soil and is protected from sun.

HISTORY

Irrigation is our age old art. Historically, civilization took place at locations where water was available. When supply of water failed due to failure of rains, or any other resons people migrated to other places near the source of water. About 5000 years back Mohanjo Daro civilization (Indus valley), Harappa at River Ravi had well designed water supply and drainage systems, public swimming pools, water works, tanks, canals etc. Egypt claims to have the world's oldest dam 355 ft. long and 40 ft. high built about 500 years ago to store water for drinking and irrigation purposes. Irrigation in India has been practiced since time immomorial, the earliest evidence is available in Atharva Veda in which norms for digging of canals are mentioned. The Sudarshan Lake was constructed about 300 B.C. near Kathiawar (Gujarat) during the region of Chandragupta and Ashoka.

Irrigation in Northern India: The oldest Yamuna canal or Ferozeshah Tughalak canal was build by Ferozeshah Tughlak at the end of 14th century. It was 140 miles long and was dug to irrigate gardens at Hissar. Gradually this canal suffered from lack of maintenance. Then Akbar got it repaired and it got the name of Nahr-i-shah. Later on Shahjahan branch canal was constructed from Yamuna to irrigate the gardens at Delhi. With collapse of Mughals, these canals deteriorated and eventually went out of use, It was in 1817 for the first time the attention of the British government was drawn to the restoration of these canals. The Shahjahan canal was reopened and water reached Delhi again after a lapse of 67 years. In 1823 the main canal to Hisar was received, branches remodeled and the whole system which is know as Western Yamuna Canal system came into full operation. With this is associated the name of Sir Colvin, who is known as the father of irrigation in Northern India. Sir Contloy (of the late Bengal Artillary) figured most prominently in the digging and construction of Gangas Canal system. The first investigation for the site of the head works was made in the year 1836 near Hardwar. Solani aquaduct head-works was completed in 1848 and Gangas canal was opened during the famine in 1865-66. The Kon (Botawa) and Sadar Canal system in U.P., the upper Bari Doab Canal, Sirhind Canal and lower Chonab canal in the Punjab needs special mention. The inhabitants of Sind used to tap the water of river Indus by small canals which silted up every year and thus required frequent re-excavation. The construction of Sakhar Barrago in Sind nearly a mile in length was started in 1923 and opened by Lord Wallington in 1932, irrigates almost 7.5 million acres of land.

Southern India: Conditions in southern India differ fundamentally from those in Northern India. In North India the canal system is found by a net work of Perenial Rivers which have their source in the snow clad Himalayas Mountains. During the whole winter season, the rivers carry enormous quantity of water got for feeding the canals. But in the south, the rivers are rainfed and in off season particularly, no water is left in rivers to do irrigation in winter. So agriculture here entirely depends on monsoon or the water stored during the rainy season in the form of tanks. In 1804, it was thought to tame the river Kavery to protect the city of Tanjore. At this stage to control floods, Sir Arther Collin built a big barrago in 1834 known as Upper Anicut. This was for diverting the required quantity of water into Kavery. To control the silt problem in Kavery, an other dam called Kavery dam and remodelling of Upper Anicut was done in 1843-45. Similar works were under taken on Godavari and Krishna.

Central India: The story of irrigation works in the central India is closely linked with the history of Telongana people. Telongana is known as the land of thousand tanks. In the Muslim period, architectural beauty to the irrigation works was introduced by the Perssians. Hussian Sagar lying between the cities of Hyderbad and Secondrabad and Mir Alam Tanks are the famous examples. Many other tanks were constructed viz. Usman Sgar, Himayat Sagar, Nizan Sagar etc. Nizan Sagar is one of the biggest irrigation tanks in the country.

Major canal systems in Punjab: After partition, Bhakra Nangal complex was taken up and was completed in 1962. Two large dams on the Beas river one at Pandoh near Mandi in H.P. and other at the foot hills at Pong (Talwara) in Punjab have been completed (1976-77). Thein Dam, 147 metres high on Ravi has been constructed at village Thein 20 km above Madhopur head-works. This is a storage dam to stablise the supply in Uppar Bari Doab Canal and Madhopur Beas Link.

Important reservoirs in India

(More than 2 thousand million cubic meter capacity)

Name	Location	Name	Location
Nagar Juna Sagar	Andhra Pradesh	Gandhi Sagar	Madhya Pradesh
Sri Ram Sagar	Ahdhra Pradesh	Jaya Kawadi	Maharashtra
Sri Sailam	Andhra Pradesh	Koyna	Maharashtra
Ukai	Gujarat	Hirakund	Orissa
Gobind Sagar	Himachal Pradesh	Balimela	Orissa
Pong Dam	Himachal Pradesh	Mettar	Tamilnadu
Linganamakki	Karnataka	Ramganga	Uttar Pradesh
Tungabhadra	Karnatka	Rihand	Uttar Pradesh

Major canal systems of Punjab

Name of Canal	Year of Opening	Source River	Head–Works	Disch-arge (cusec)	Distt. Irrigated
Sirhind	1954	Sutlej	Ropar	12,400	Ludhiana, Patiala, Sangrur, Bhatinda and Ferozepur.
Upper Bari Doab	1954	Ravi	Madhopur	6,00	Gurdaspur & Amritsar
Eastern Canal	1938	Sutlej	Hussainiwala	2,756	Ferozepur,
Bhakra Nangal canal	1954	Sutlej	Nangal	12,450	Ludhiana, Patiala, Sangrur
Bist Doab Canal	1954	Sutlej	Ropar	800	Hoshiarpur, Jalandhar, Kapurthala
Gang Canal (Bikaner Canal)	1927	Sutlej	Hussainiwala	2,720	Gangasagar (Rajasthan)
Rajasthan Canal	1954	Sutlej	Hari Ka Pattan	18,000	Rajasthan

The Rajasthan Canal 692 km long with full supply is the largest irrigation canal in the world.

ROLE OF WATER IN CROP PRODUCTION

Role of water in crop production can be discussed under two main heads.

1. Ecological
2. Physiological

1. Ecological importance

The distribution of vegetation over the surface of earth is controlled more by the availability of water than by any other single factor. Regions where rainfall is abundant and fairly distributed over the growing season, have dense vegetation e.g. forest in hills. Where summer droughts are frequent and severe, forests are replaced by grassland. Further decrease in the rainfall results is semi-desert with scattered shrubs and finally deserts. Ecological significance of water can be discussed as:

I. Water as a factor affecting plant characters: Inspite of the limitations imposed by heredity, the morphological and physiological characters of plant are markedly affected by water balance in the plant system. The plant have developed varying characters to fit in varying water conditions. On one hand, plants have their habitat in wet conditions as Hydrophytes while on other hand, have their habitat in regions of extreme dryness, are called xerophytes and those having their habitate in moderately wet conditions are called mesophytes. While heliophytes are those plants which are capable of growing or surviving in water or water logged habitats.

Trophophytic plants are such deciduous trees which are alternately mesophytic and xerophytic. But none of the plants are as versatile as many species of alage, Lichens and Mosses which can survive very prolonged state of intense desication and again resume activity when water becomes available.

Characteristics of xerophytic plants

a. Thick cuticle. (the outer corky or waxy covering of the plant)
b. Rolling or folding of leaves of narrow or prickly leaves.
c. Pubescent (hairy) leaves and stems.
d. Extensive root system.
e. Reduced number of stomata or sunken or covered stomata.
f. Resistance to protoplasmic dehydration under high moisture stress conditions.

Hydrophytes have the following characters

a. Poorly developed and less extensive root system.
b. Large number of air cavities and more of the mesophyll cell
c. Cuticle (soft cellular tissue) outside the epidermis is absent
d. More number of stomata (an opening in the epidermis for the passage of gasses and water vapours).

II. Water as a factor affecting the soil properties: Water plays a major role in the geological processes leading to soil formation, its nature, properties and processes that govern their agricultural usefulness. Soils of dry regions are generally of light type as compared with those of humid and sub-humid regions.

Soil reaction is also controlled by water. Soils of hilly areas are generally acidic because of excessive leaching of salts by high rainfall. Soils of arid areas are generally alkaline. The availability of nutrients in a soil is also affected by water.

III. Ecological distribution of plants - Each species varies in its essential requirements for successful reproduction and growth. In respect of every habitat, however, there are optimum conditions and limits within which a plants species can exist and these are referred to as their **'range of tolerance'**.

2. Physiological importance of water

Almost every plant process is affected directly or indirectly by the water supply. The importance of water in the physiological process of plants can be discussed as under:

I. **Constituent of protoplasm:** Water constitutes 80-89% of the fresh weight of the herbaceous plants and 50% of the woody plants. It is an important part of the protoplasm. Reduction of water content below some level changes structure and ultimately causes the death of protoplasm. The death is due to the fact that most of the organic substances in protoplasm including proteins and carbohydrates are hydrated in the natural state and if the water is removed, their physical and chemical properties are affected. Protoplasm of very few cells can survive if their water content drops below 10% of their saturation level. Most plants are killed if it is 30-50% below the saturation level.

II. **Water act as solvent:** Water has high dissolving power in which gases, minerals and other solutes are dissolved enter plant cells and move from cell to cell and organ to organ, on account of permeability of most cell walls and membrance in plant system; Water acts as a medium for movement of dissolved substance in plant system.

III. **Water act as reagent:** Water participates in many chemical reactions occurring in protoplasm. Hydrolysis and condensation reactions in which water is added, and removed from molecules are important in metabolic processes. Water plays important role in the interconversion of carbohydrates in the process of photosynthesis and respiration.

IV. **Water act as a buffer against high and low temperature:** By transpiration water cools down the vegetation. This is because water has high value of latent heat of vaporization.

At 100°C – 540 calories

By transpiration water evaporate from plant surface, much of the solar energy is utilized which otherwise would have given much heat to the plant and may not be tolerated sometimes. Because of the high latent heat of fusion, 1gm water at 0°C releases 80 calories of heat during freezing. Therefore, it releases some heat during freezing because of this it protects the vegetation from low temperature injury.

V. **Water act as maintenance of turgidity:** Turgidity is essential for cell enlargement and growth and maintaining the form of herbrebaceous plants. The shape of cell, is dependent upon water. In turgid cells, the water accumulates in the vacuoles of the cell and builds up a pressure, which forced the protoplasm against the cell walls.

VI. **Water helps in germination:** Seeds absorb water for swelling and initiation of growth activities. Generally the quantity absorbed is double their weight. Most of the seeds remain dormant till adequate water is not absorbed. Seed germination is rapid when soil is at field capacity. Most seeds, however, germinate even when the water content of the soil is much lower, provided the other conditions are satisfactory. Seeds can absorb water from humid air and will germinate in saturated air.

Seed absorbs water by imbibition process, gets swollon and seed coat becomes week, permeable to gases and enzymes get activated, dissolved in water and diffused to stored food material (cotyledons). By enzymes, starch, fats and proteins are converted to sugar, amino acids and fatty acids translocated and dissolved cells gain maximum size development.

Composition of soil on volume basis

In soil moisture relations – we are more concerned with the soil solution

Soil solution percentage limit vary with the soil type e.g. sandy soil have more soil air and less soil solution. However, clay soil have more soil solution and less soil air.

Soil moisture retention

The retention of soil moisture, its movement in the soil, storage capacity of soil and its availability to the crop plants is directly or indirectly related to the –

I. Size and distribution of soil pores

II. Attraction of soil solids for moisture

Simple case of rainfall or irrigation can explain the retention of soil–moisture.

Composition of soil on volume basis in relation to weight of components

Diagram showing the volume of solids, water and air in a well granulated, silt loam soil at different moisture constants.

When we irrigate a field or there is a rainfall

I. Saturated Point (SP)

100 g solids	40 g water	0 g air

All the pore space is filled with water (macro & micro there is no air present)

II. Field capacity (FC)

100 g solids	20g water	20 g air

After saturation water will soon move out of the macropores (large pores) and soil reaches at field capacity.

III. Permanent Wilting Point or Coefficient (PWP)

100 g solids	10g H_2O	30 g air

Plants remove water rapidly from soil and soil reaches at a point called Permanent Wilting Point. At this point even though there is still considerable moisture in the soil but availability to plants reduces.

IV. Hygroscopic Coefficient (HC)

100 g solids	8g H_2O	32 g air

With further decrease in moisture, soil reaches at a point called Hygroscopic Coefficient. At this point water is held very tightly on the soil colloids.

The term colloid was derived from two Greek words

I. Kolla – glue

II. Eidos – like

Forces of retention of soil water

Mechanisms by cohesion and adhesion

The main forces which account for the retention of water by soil solids are —

I. **Cohesion:** The forces by which molecules of the same substance are held together e.g.: water – water.

II. **Adhesion:** The force by which molecules of different substances held together e.g. –

glass – water, glass – mercury, soil solid – H_2O

Adhesion depends upon the property of liquid and solid.

So these two forces adhesion and cohesion largely account for the retention of moisture by soil solids and both these forces make it possible for the soil solids to retain water to control its movements and utilization.

Retention can also be explained by angle of contact. More the angle of contact less will be retention and vice versa. In case of sandy soil angle of contact is more because of macropores. If it is >90° than soils are not wettable soils. When soil organic matter is more, angle is less and are called wettable soils. Different forces influencing soil moisture retention and movement of water in soil are as follow.

I. **Capillary water:** It is that water which is retained within soil pores or held by the surface tension forces as a continuous film around the soil particles against the force of gravity.

II. **Surface tension (Gravitational potential):** The phenomen occurs at the interfaces may be solid – liquid interface, liquid – liquid interface, liquid – gas interface. Surface tension occurs due to the both cohesion and adhesion forces.

III. **Matric potential forces:** It is due to difference in electric potential of soil solids and water.

Matric forces influence the soil moisture retention and movement of water in soil. Matric potential force is the result of adsorption and capillary phenomena occurring in the soil. Due to the matric potential forces, water is adsorbed by the soil particles and their exchangeable ions. Matric potential is always negative.

IV. **Osmotic potential:** This results due to the presence of solutes / salts in the soil system. Whenever there is more concentration of salts in an area, the osmotic potential will be increased so to neutralize this osmotic pressure water rashes to this area.

Important terms

I. **Field capacity or capillary capacity:** It is the percentage of soil moisture expressed on dry weight basis after the soil has been completely saturated with water and downward movement of water has practically ceased. The tension (force with which water is held in the soil) at this stage corresponds to 1/10 to 1/3 atmosphere. At this stage, only micropores hold water.

II. **Saturation point of soil:** At this point interspaces in the soil are completely filled with water (both types of pores i.e. macro and micro). At this point run off occurs. Here the tension is zero atmosphere.

III. **Run off :** Water leaving the land surface and entering the stream channel system.

IV. **Permanent wilting point:** It is the moisture content in percentage of a soil at which nearly all plants wilt and do not recover in a humid dark chamber unless water is added from an outside source. It corresponds to 15 atmosphere tension.

V. **Hygroscopic coefficient:** Amount of moisture present in dry soil when it is equilibrated with some standard relative humidity near a saturated atmosphere (about 98%). It is expressed in terms of percentage on the basis of oven dry soil.

Units of expressing energy levels of soil water

I. **Height of water column:** The height in cm. of a unit water column, whose weight just equals the tension under consideration, is a common unit of expressing tension or potential of soil water. Greater the height then greater the tension.

II. **Bar:** It is the pressure exhorted by the column of water 1023 cm. in height. This approximates the standard atmosphere, which is the air pressure at sea level. The standard atmosphere being 14.7 lb / sq. inch or 760 mm Hg.

III. **Millibar:** The term millibar (m bar) 1/100 bar. The suction of water 10,100 and 1000 cm height in about 10,100 and 1000 (1 bar) m bars respectively.

IV. **Atm:** One atm suction force is the pressure equivalent to the pressure exerted by 1032 cm water per square centimeter.

V. **pF:** It is the logarithm of height in cm of a column of water.

VI. **Mega Pascal:** To get soil water tension in the Mega Pascal unit (MPa) unit, multiply the number of bars by 0.1

Soil tension and water content

I. Silty loam soils contain more available soil moisture than sandy loam soils.

II. At the same tension silty loam soil contains more moisture than sandy loam.

Range of available water holding capacity of different soils

Soil Type	% age Moisture		Depth of available water per unit of soil (cm / metre soil depth)
	FC	PWP	
Fine Sand	3 – 5	1 – 3	2 – 4
Sandy Loam	5 – 15	3 – 8	4 – 11
Silty Loam	12 – 18	6 – 10	6 – 13
Clay Loam	15 – 30	7 – 16	10 – 18
Clay	25 – 40	12 – 20	16 – 20

Classification of soil water

Soil water can be classified into two categories:

1. Physical – Based on relative degree of retention
2. Biological

1. Physical classification

I. Free or gravitational water

a. Water in the soil above that held at F.C.

b. It is held loosely between <0.1 – 0.5 atm tension.

c. Removed from the soil in drainage.

d. Undesirable for plants.

e. It moves in response to film tension and gravitational forces.

f. Nutrients leached with this water.

II. Capillary water

a. Not all water available to plants.

b. Held between F.C. and H.C. in microscopes.

c. Held between tension 0.1 – 31 atm.

d. It acts as the solution.

e. It moves by film adjustment from thick to thin.
(thick film where tension is low).

When H_2O comes in contact with the surface of soil particles it will be attracted by surface of the soil by adhesive force and gravitational force. At the same time there is repulsion for this attract due to cohesive force of water molecule. Hence there is no equilibrium at this stage and there is tangently force along the liquid surface. Due to this surface tension the liquid tries to move tangently along the water surface. This movement is known as capillary water movement and water held by surface tension in soil capillaries against pull of gravity, so available water to plant is declared by capillary water which will be function of pore space which depends upon soil texture, structure and organic matter.

III. Hygroscopic water

a. Held at H.C. (Hygroscopic coefficient)
b. Tension varies from 31 – 10,000 atm.
c. Held mainly by soil colloids.
d. It is present mostly in the vapour forms.

Capillary potential: It is a measure of attraction forces with which water is held by a soil. It is expressed in terms of work that must be done to remove water against the capillary force of the soil.

Capillary fringe: It is a vertical distance above a water table along which moisture content varies from full saturation to field capacity.

2. Biological classification

There is certain relationship between soil moisture retention and its utilization by plants. On this basis biological classification is given.

I. Superfluous water: The moisture excess of that held in the soil at the field capacity is of no use for higher plants also called as free water Harmful effects of superfluous water on plants growth are very common.

- Due to poor aeration of the soil. Roots and aerobic organisms deprived oxygen.
- Adverse biochemical changes may be encouraged.
- Nutrients may leach down especially from course textured soil.

II. Available water: The available moisture is generally considered as water held between field capacity and permanent wilting point.

When 50-85% of this water is used by the plants then apply irrigation.

i.e. When water content reaches near PWP availability is less .

III. Unavailable water: Held at PWP and that part of capillary water which is utilized by plants too slowly to prevent wilting.

Factors affecting the availability of soil water

A number of plant, soil and climatic factors effect the amount of water which plant can absorb efficiently from a given soil.

1. Soil factors

I. **Moisture tension relation:** It influences the amount of available moisture in a soil by influencing the FC and PWP. These two factors are mainly influenced by soil texture, structure and organic matter.

(a) **Soil texture:** Fineness of soil increases the FC and PWP and ultimately available water.

(b) **Soil structure:** Fine particle soils contains more water than the loose soil i.e. with coarse particles.

(c) **Organic matter:** The entire effect is not through increasing water holding capacity of soil but also due to its favourable influence on soil structure, texture and soil porosity.

II. **Osmotic tension (salt conc.):** The presence of salts in soil solution reduces the range of available moisture by increasing the PWP percentage. The effect of salts on PWP percentage are significant in salvia souls of arid and semi arid areas.

III. **Soil depth:** The deep soils have higher water holding capacity as compared to shallow soils. For deep rooted crops, deep soils are found suitable for their cultivation particularly in semi arid areas where supplements are not available. Hard pans especially in upper layers not only slow down drastically the rate of movement of H_2O but also influence unfavorably the penetration of plant roots. They sometime restrict root growth and effectively reduce soil depth from which moisture is drawn.

IV. **Soil layering:** Soil layering influences the available moisture range and water movement in the soil. Impervious layers present in the soil slow down the rate of movement of water into deep layers and penetration of plant roots. Movement of water is highest in soils with no hard pan

V. **Soil temp:** It affects the viscosity of water and water become less immobile if temp lower down. The water absorption at 25°C is about 5 times greater than at 10°C.

VI. **Soil aeration:** Proper aeration enhances the water absorption. Aeration effects root growth and root activity. It adversely effects water absorption.

2. Plant factors

I. Type of plant / crop grown – Sorghum VS Maize: Sorghum can perform better at lower moisture content because root system is better as it has more rootlets. Volume of soil in contact with rootlets of sorghum is more as compare to maize and hence sorghum plants can draw more water as compared to maize plants.

II. Root characters: Rooting habits of plant also effects water absorption.

a. Root depth – More deep – more available water

b. Spread – More – more absorption

c. Rate of root growth – More – more absorption
d. Life of root hairs – More – more absorption

3. Climatic factors

I. Sun shine – If more sun shine hours are available then absorption will be more to meet requirement of ET losses.
II. Cloudy days – less requirement for water and hence less absorption.
III. Humidity – More – less requirement – less absorption.
IV. Wind – More speed – more loss of water – more absorption.
V. Rainfall – More rainfall – more absorption.

SCHEDULING IRRIGATION

Irrigation scheduling is defined as frequency with which water is applied based on needs of the crop and nature of soil. It is total number of irrigations and their frequency required to meet the crop water requirement.

It can also be defined as scientific management technique of allocating irrigation water based on the indication of crop water requirement with an aim to achieve maximum crop production per unit of water applied over a unit area in a unit time. In irrigation scheduling, there are three major question:

- How much irrigation to apply?
- When to apply irrigation?
- How to apply irrigation?

Three major considerations which influence these questions are:

- Water needs of the crop
- Availability of water with which to irrigate
- Capacity of the root zone soil to store the water

Time of irrigation is most important concept of irrigation

The broad basic principle

The broad basic principle governing irrigation is that water should be applied at longest possible interval but before the crop receives a set back in terms of economic yield. The quality of water should be such that there is no wastage of water as well as entire root zone is sufficiently supplied with soil water. Irrigation scheduling is a effective management of supplying water in accordance with crop needs. Criteria most suitable for scheduling irrigation vary from one situation to another. For example where water availability is less and expensive, irrigation should be applied at most critical crop stage to maximum crop production per unit of applied water. The criteria selected for irrigation should permit favourable crop yields and optimum use of water and proper attention to other factors involved. The scheduling of irrigation can be done on following basis:

1. Plant basis

2. Soil factors
3. Climatological approval

1. Plant basis

It is now widely recognised that plant growth is directly related to the water balance in plant tissues.

General appearance of plant

(a) **Colour of leaves:** A dark green/bluish green colour in gram, berseem and cotton show the deficiency of moisture while light green colour is generally indication of the adequate moisture and satisfactory growth. In wheat yellow colour show need for watering. The change in colour, however, may also be due to deficiency of certain nutrient especially Nitrogen or due to diseases etc.

(b) **Angle or posture of leaves or degree of wilting or rolling of leaves:** Grain crops indicate need for water by temporary wilting particularly during the warmest part of the day. Sugarbeet and clovers readily indicate need for water by temporary wilting.

(c) **Degree of opening and closing of stomata:** The loss of water from the intercellular spaces to the atmosphere through the stomata reduces the turgor of guard cells decrease the stomata opening. So the degree of stomata may be taken as basis of scheduling irrigation.

(d) **Exudation:** Exudation from topped plants depends directly on root pressure. Exudation from crop plants stops at soil moisture content above wilting point. eg: It is suggested by some workers that irrigation to cotton crop before the exudation from topped plants ceases permanently.

(e) **Use of indicator plant:** Growing indicator plant in association with main crop that exhibit water stress symptoms earlier than the main crop plants (eg. Sunflower plant)

(f) **Plant growth–Leaf growth:** Thickness of sugar beet leaves, stem or trunk growth – shortening of internodes of sugarcane, cotton plants etc.

(g) **Leaf reflection and temperature:** Turgid cells reflect more of infra-red light and have low temperature as compared to fluccid cells.

Limitations

I. In fruit trees, it is difficult to detect need for irrigation on appearance basis.

II. When there is a lag of water absorption to transpiration, plant develop water stress. If this lag further increases the leaves loose turgor and droop. Wilting is an expression of internal water deficit. Drooping is an indicator of deficit. Drooping occurs in some plants having parenchyma cells where as plants having schernchyma cells do not exhibit drooping at all although they are wilted.

III. At noon evaporation is more than absorption and wilting appears even if there is sufficient moisture in the soil and this basis thus frequency of irrigation will be high.

IV. If we schedule irrigation on basis of appearance of wilting, then the crop might have already been adversely affected.

Internal plant water deficit

Plant growth is controlled directly by plant water stress and only indirectly by soil-water stress. The important methods of measurement of plant water stress are –

Direct methods

I. Water content: The water content or moisture percentage in fact varies with species, organs, age, season and time of the day. So fluctuations in the fresh weight are measured.

$$\text{Water content} = \frac{\text{Wf} - \text{Wd} \times 100}{\text{Wd}}$$

Wf – Fresh weight; Wd – Dry weight.

II. Water deficit (Water saturation deficit): An improvement over use of water content as an indicator of water stress is the use of water saturation deficit and relative turgidity. Water saturation deficit was proposed by Stocker (1929). It is the amount of water taken up by leaves or twigs when they are allowed to become saturated, expressed as percentage of water content at saturation.

$$\text{Saturation deficit} = \frac{\text{Wt} - \text{Wf} \times 100}{\text{Wt} - \text{Wd}}$$

Wt – Turgid weight of leaves

Wf – Fresh weight of leaves

Wd – Dry weight of leaves

III. Relative turgidity/Relative leaf water content: Leaf discs about one centimeter diameter cut from leaves, weighed then floated on water until they gain water and then weighed. Dry in an oven and take dry weight.

$$\text{Relative Turgidity} = \frac{\text{Wf} - \text{Wt} \times 100}{\text{Wt} - \text{Wd}}$$

Wt – Turgid weight of discs

Wf – Fresh weight of discs

Wd – Dry weight of discs

IV. Critical stage concept: We can schedule the irrigation by classifying crops in two groups

- Crops which have clear cut stages
- Crops which do not have clear cut stages

Limitations

- This concept does not holds good for crops which have no marked growth stages
- Concept does not give information on quantity of water to be applied

Important stages of different crops

Rice –	Tillering	10-20 days after transplanting (DAT)
	Jointing	(35-45 DAT)
	Boot	(52-58 DAT)
	Flowering	(63-68 DAT)
	Grain Development	(75-85 DAT)
Maize –	Seeding	15 + 2 days after sowing (DAS)
	Tasseling –	(55 ± 3 DAS)
	Silking –	(74 ± 3 DAS)
Wheat –	CRI	(20-30 DAS)
	Tillering	(40 -45 DAS)
	Jointing	(55-60 DAS)
	Boot or Flag Leaf	(70-75 DAS)
	Flowering	(85-90 DAS)
	Milking	(100-105 DAS)
	Dough	(115-120 DAS)
Gram –	Pre-flowering	(45 ± 3 DAS)
	Flowering	(70 ± 2 DAS)
Pigeon Pea		
	Flower initiation	(75 ± 5 DAS)
	Peak flowering	(80 ± 5 DAS)

Indirect method

Cell sap concentration (CSC)

The CSC increases with the decrease in plant water content which depend upon the amount of water content present in soil. CSC is weighed with the help of refractometer.

Diffusion pressure deficit/Suction force : Fully turgid leaves cells has zero diffusion pressure deficit (DPD). As decreasing amount of water in tissues its DPD increases.

Measurement of stomatal apperature i.e. degree of opening and closing of stomata: The loss of water from intercellular spaces to atmosphere through stomata reduces turgor of guard cells and decreases the stomatal opening may be taken as basis of scheduling irrigation. A pure silicon rubber in applied to lower side of leaf and allowed to hardened and stripped off then cellulose acetate film is made from the rubber impression and studied under microscope to determine extent of stomatal opening.

Electrical resistance method – by Box and Lemon (1953) has reported this method for measuring moisture stress in cotton. This is based on assumption that electrical conductivity of living plant tissue is related to internal moisture stress experienced by plant.

Beta Ray Gauging – By Mederski (1961) to measure plant water deficit and it is non destructive, continuous and automatic way of measuring plant water status.

Leaf water Potential/Xylem pressure – (Stegman 1983 in USA in Maize).

Leaf water potential/Xylen pressure in non stress treatments typically declined. It is always negative (-1.20 MPa).

Canopy Temperature Clawson and Blad (1982) in Maize at USA. Canopy temperature and air temperatures are measured daily with the help of Infra-red Thermometer. The difference (Tc-Ta) between these when added yield a factor over a time called "Stress degree day (SDD) and based on SDD value we can schedule irrigation.

2. Soil factors

I. **Soil appearance and feel method:** Determination of soil moisture in the absence of more sophisticated method. Feel the soil with fingers. Examine wheather bolls forms or not etc. Though long experience is required in this case.

II. **Soil water content:** A common criteria. Allows depletion of various fractions of available water in the crop rooting depth. i.e. 25%, 50% and 75% prior to irrigating a crop. Water content can be determined by direct and indirect methods.

Direct methods: Thermo gravimetric method, Volumetric method and Neutron Scattering Probe.

Indirect methods: Tensiometer, Gypsum Block method.

In selecting the water deficit to be permitted prior to irrigation, consideration must be given to the effects of other soil factors, plant factors, climatic conditions and crop management practices. We can calculate absolute deficit.

If we compare heavy soil with 12% soil moisture and light soil having 10% moisture, it does not mean that heavy soil will be good as it will be judged from percentage available water i.e. FC & PWP.

Heavy Soil	Light soil
Soil moisture 12%	10%
FC -20%	15%
PWP-8%	5%

$$\text{Available water \% (Heavy soil)} = \frac{12-8}{20-8} \times 100 = 33\%$$

15 – 5

$$\text{Available water \% (Light soil)} = \frac{10-5}{15-5} \times 100 = 50\%$$

Though moisture is less in light soil but more fraction of water i.e. 5% of H_2O is available out of 10% in light soil as compared to 4% of available water out of 12% in heavy soil.

***Ease of availability*:** It is measured by Tensiometer. We can fix the treatments on tension basis i.e. 0.3, 0.6, 0.8 atm etc. At FC tension is 1/3 atm and at PWP tension is 15 atmosphere. We can form a curve of moisture and tension and then schedule irrigation on moisture basis.

***Consumptive use*:** Scheduling of irrigation may be done by knowing the consumptive use of a crop for a particular period. Suppose a storage capacity of a soil is 3 inch and consumptive use is 0.3 inch per day, it will take 10 days to deplete entire amount. We can irrigate when 50-60% deficit has taken place. Irrigation efficiency should also be taken into account. Say we have deficit of 3 inch in root zone and irrigation efficiency is 60% than water to apply

$$\text{Water to apply} = \frac{3 \times 100}{60} = 5 \text{ inch}$$

Irrigation efficiency

Soil type	Irrigation Efficiency %age
Sandy	60
Sandy Loam	65
Loam	70
Clay Loam	75
Clay	80

3. Climatological approach

The moisture in the soil is regarded as being a balance between what enters in it as a result of precipitation or irrigation and what amount is lost through evaporation and transpiration. Precipitation can be measured by means of rain gauge. It is not easy to measure the ET but the climatological approach provides a way to determine it. Consequently we can determine the daily moisture loss from the soil through ET and can compare it with daily rainfall. An irrigation schedule thus can be set up in a book keeping procedure. The moisture is the soil as a bank account ET will drain from it. We surely need to keep pace of the ET and restore by irrigation whatever, is not promptly returned by precipitation.

I. **Day degree concept (Dass, 1936):** In Hawii, he used this concept in sugarcane. He accumulated day degrees above 70°F daily and when the total day degrees exceeds 350°F give irrigation. In this case maximum temperature was to be taken. Infact it is not temperature only but its duration, radiation effects etc also counts.

II. **Use of evapormetry:** ET depends primarily on climatic factors such as temperature, radiation, wind velocity, Relative humidity and so on. Each of these elements can be correlated to ET by crop plants. An evaporimeter is an instrument which integrates the effect of all the different climatic factors. It is relatively simple, cheaper and thus more useful in irrigation practices but require a good knowledge of soil moisture relationship.

a. US Weather Bureau Class I Pan – for measuring Evaporation

b. Sunken screened Pan Evaporimeter – for estimating ET

These evaporimeters give close co-relation of the values of consumptive use of water by field crop and evaporation from the evaporimeter. An even simpler evaporimeter can be constructed for field use which is called Pan Evaporimeter. This could be 1 kg capacity or more with a pointer inside at 1.5 cm below the brim to facilitate recording of water levels.

Penman formula

The penman formula is a semi-empirical equation combining mass transfer (E_a) and energy budget (H) methods. The formula was developed by Penman in 1948 and is still widely used for calculating the potential evaporation using synoptic meteorological data.

According to Penman the potential evaporation E_o (in mm/day) can be calculated as:

$$E_o = \frac{\left(\frac{\Delta}{\gamma} H + E_a\right)}{\frac{\Delta}{\gamma} + 1}$$

where $\frac{\Delta}{\gamma}$ is an empirical parameter depending on temperature (Appendix 11.2.9).

H is calculated as $H = (1 - r)R_{in} - R_o$ where R_{in} (incoming radiation) is given by

$$(1 - r)R_{in} = 0.95\, R_a(0.18 + 0.55\text{n/N})$$

where R_a is the solar radiation (Appendix 11.1.1), Ro is the outgoing radiation, r is the albedo (0.05 for water), and *n/N* is the ration between actual sunshine hours and possible sunshine hours. The term *n/N* can also be estimated using the cloudiness, e.g., a cloudiness of 60 % gives an n/N of 40 % (= 100 - 60). R_o is calculated by

$$\frac{\text{Wf} - \text{Wt} \times 100}{\text{Wt} - \text{Wd}}$$

where e_d is the actual vapor pressure, and σT_a^4 is the theoretical black body radiation (Appendix 11.2.4).

E_a is calculated by

$$E_a = 0.35(0.5 + u_2/100)\,(e_a - e_d)$$

where u_2 is the wind speed in miles/day (1 mile = 1609 m) and ea is the saturation vapor pressure (Appendix). Remember that the relative humidity $RH = e_d/e_a$.

Appendix. Ra Expressed in Equivalent Evaporation (mm day^{-1})

(Reproduced from J.S.G. McCulloch (1995) E. African Agric. Forest. J., xxx (3) 286-295, by permission.)

	Lat.	Jan.	Feb.	Mar.	Apr.	May	June	July	Aug.	Sept.	Oct.	Nov.	Dec.
	NORTHERN HEMISPHERE												
	60	1.4	3.6	7.0	11.1	14.6	16.4	15.6	12.6	8.5	4.7	2.0	0.9
	50	3.7	6.0	9.2	12.7	15.5	16.6	16.1	13.7	10.4	7.1	4.4	3.1
	40	6.2	8.4	11.1	13.8	15.9	16.7	16.3	14.7	12.1	9.3	6.8	5.6
	30	8.1	10.5	12.8	14.7	16.1	16.5	16.2	15.2	13.5	11.2	9.1	7.9
	20	10.8	12.4	14.0	15.2	15.7	15.8	15.8	15.4	14.4	12.9	11.3	10.4
	10	12.8	13.9	14.8	15.2	15.0	14.8	14.9	15.0	14.8	14.2	13.1	12.5
Equator	0	14.6	15.0	15.2	14.7	13.9	13.4	13.6	14.3	14.9	15.0	14.6	14.3
	10	15.9	15.7	15.1	13.9	12.5	11.7	12.0	13.1	14.4	15.4	15.7	15.8
	20	16.8	16.0	14.5	12.5	10.7	9.7	10.1	11.6	13.6	15.3	16.4	16.9
	30	17.2	15.8	13.5	10.9	8.6	7.5	7.9	9.7	12.3	14.8	16.7	17.5
	40	17.3	15.1	12.2	8.9	6.4	5.2	5.6	7.6	10.7	13.8	16.5	17.8
	50	16.9	14.1	10.4	6.7	4.1	2.9	3.4	5.4	8.7	12.5	16.0	17.6
	60	16.5	12.6	8.3	4.3	1.8	0.9	1.3	3.1	6.5	10.8	15.1	17.5
	SOUTHERN HEMISPHERE												

Appendix. Values of s T^4 (mm of water)

(Ministry of Agriculture, Fisheries and Food (1967) *Potential Transpiration,* Reproduced by permission of the Controller, Her Majesty's Stationery Office, © Crown copyright.)

°F	0	1	2	3	4	5	6	7	8	9
30	11.0	11.1	11.2	11.3	11.4	11.5	11.6	11.6	11.7	11.8
40	11.9	12.0	12.1	12.2	12.3	12.4	12.5	12.6	12.7	12.8
50	12.9	13.0	13.1	13.2	13.3	13.4	13.5	13.6	13.7	13.9
60	14.0	14.1	14.2	14.3	14.4	14.5	14.6	14.7	14.8	14.9
°C										
–0	11.2	11.0								
0	11.2	11.4	11.5	11.7	11.9	12.0	12.2	12.3	12.5	12.7
10	12.9	13.1	13.3	13.5	13.7	13.9	14.0	14.2	14.4	14.6
20	14.8	15.0								

Appendix. Air Temperature and Saturation Vapour Pressure (mm)

(Ministry of Agriculture, Fisheries andFood (1967) *Potential Transpiration.* Reproduced by permission of the Controller, Her Majesty's Stationery Office, © Crown copyright)

Aur temp. (°F)	30+	40+	50+	60+	Air temp. (°C)	0+	10+	20+
0.1	4.20	6.29	9.21	13.26	-0.5	4.40		
0.5	4.30	6.42	9.38	13.49	0.0	4.58	9.21	17.53
1.0	4.39	6.54	9.56	13.73	0.5	4.75	9.52	
1.5	4.48	6.67	9.74	13.98	1.0	4.93	9.84	18.68
2.0	4.58	6.80	9.92	14.23	1.5	5.11	10.18	
2.5	4.67	6.94	10.10	14.48	2.0	5.30	10.52	19.8
3.0	4.77	7.07	10.29	14.73	2.5	5.49	10.87	
3.5	4.87	7.21	10.48	15.00	3.0	5.69	11.23	21.1
4.0	4.97	7.35	1067	15.27	3.5	5.89	11.61	
4.5	5.07	7.50	10.87	15.54	4.0	6.10	11.99	
5.0	5.17	7.64	11.07	15.81	4.5	6.32	12.38	
5.5	5.27	7.79	11.28	16.08	5.0	6.54	12.79	
6.0	5.38	7.93	11.48	16.36	5.5	6.77	13.13	
6.5	5.49	8.09	11.69	16.65	6.0	7.02	13.63	
7.0	5.60	8.24	11.90	16.95	6.5	7.26	14.08	
7.5	5.71	8.40	12.12		7.0	7.52	14.53	
8.0	5.82	8.55	12.34	17.53	7.5	7.79	15.00	
8.5	5.94	8.71	12.56		8.0	8.05	15.49	
9.0	6.05	8.88	12.79	18.16	8.5	8.33	15.97	
9.5	6.17	9.05	13.02		9.2	8.62	16.47	
					9.5	8.91		

Appendix. Weight Factors Δ/γ and Temperature

(Ministry of Agriculture, Fisheries and Food (1967) *Potential Transpiration,* Reproduced by permission of the Controller, Her Majesty's Stationery Office, © Corwn copyright)

°F	30	40	50	60	°C	0	10	20
0	0.66	0.91	1.26	1.75	0	0.69	1.26	2.23
0.5	0.66	0.92	1.28	1.77	0.5	0.71	1.30	2.28
1.0	0.67	0.94	1.31	1.80	1.0	0.73	1.34	2.35
1.5	0.68	0.96	1.33	1.83	1.5	0.76	1.38	2.41
2.0	0.69	0.97	1.35	1.86	2.0	0.78	1.342	2.48
2.5	0.70	0.99	1.37	1.88	2.5	0.80	1.47	
3.0	0.71	1.01	1.40	1.91	3.0	0.83	1.51	
3.5	0.72	1.02	1.42	1.94	3.5	0.85	1.56	
4.0	0.74	1.04	1.44	1.97	4.0	0.88	1.60	
4.5	0.75	1.06	1.47	2.00	4.5	0.91	1.65	
5.0	0.76	1.07	1.49	2.03	5.0	0.94	1.69	
5.5	0.77	1.09	1.52	2.06	5.5	0.97	1.74	
6.0	0.79	1.11	1.54	2.10	6.0	1.00	1.79	
6.5	0.80	1.13	1.57	2.13	6.5	1.03	1.84	
7.0	0.82	1.15	1.59	2.16	7.0	1.06	1.89	
7.5	0.83	1.17	1.62	2.20	7.5	1.09	1.94	
8.0	0.84	1.19	1.64	2.23	8.0	1.13	1.99	
8.5	0.86	1.20	1.67	2.27	8.5	1.16	2.04	
9.0	0.87	1.22	1.69	2.32	9.0	1.19	2.11	
9.5	0.89	1.24	1.72		9.5	1.23	2.17	

Practical consideration in scheduling of irrigation

Wise use of the criteria discussed above for scheduling irrigation in accordance with actual needs can substantially increase crop yields, save water and labour minimize deterioration of soil structure and loss of soluble plant nutrients. Lesser crop production problems and facilitate other cultural operations. These consideration, so important in successful irrigation again indicate the need to give careful intention for achieving optimum irrigation schedule for the conditions prevailing.

METHODS OF IRRIGATION

Irrigation method refers to the manner in which irrigation water is applied to the field. There are many methods of irrigation which are discussed as below:

1. Surface irrigation

Water is applied from the field ditches to the field without bunds to guide the flow, very inefficient method as only small per cent of water is used by the plants. The rest goes as waste in the form of run-off, seepage and evaporation.

Advantages

- Installation or operation cost is low.
- No interference with the use of farm implements.
- No risk of damage by live stock.

Disadvantages

- Excessive loss of water.
- Uneven distribution of water.
- Soil erosion losses.
- Fertilizer and farm yard manure gets eroded.
- Loss of N by deep percolation.
- Harmful to crops sensitive to water logging.

2. Border strip method

It consists of dividing the field into a number of strips separated by low ridges. Water turned into the upper end of each strip flows slowly towards the lower end in a thin sheet guided by the ridges. Depending upon the situation, strips may be straight or contour. The strip has a uniform gentle slopes (0.05–5%) in direction of irrigation.

Width: It depends upon size of stream and varies from 3–15 meters.

Length: It depends upon

- Size of stream.
- Slope.

Advantages

- The ridges can be constructed cheaply with simple farm implements.
- Large area can be covered as compared to check method.
- Uniform distribution of water.
- High water application efficiency.
- Large irrigation streams can be used.
- Operation of the system easy and simple.
- Surface drainage can be provided.

Disadvantages

- Requires, more precise leveling than other methods.
- Large irrigation streams are required.

3. Check basin method

It consists of running water into relatively level plots surrounded by small ridges. It is the most commonly used method of irrigation in India. The check may be rectangular or ring or contour. In the rectangular check basin method, the field is divided into series of small rectangular plots surrounded by ridges to hold water within the compartments. The method is suited to nearly level areas.

In this method there is usually one field distribution ditch for every two lines of check basins. It supplies water to the plots on either side.

The size of check basins depends upon

- Flow of water source
- Infiltration rate of the soil.
- The slope of field.

As the infiltration rate increases, as in light sandy soil, the stream size must be increased or the size of basin reduced in order to cover the area in a short period. If the size of stream is small, the basin size may also be reduced.

The size of check basins may vary from 1 meter square for vegetable growing to as large as one acre for growing rice under wet land farming.

Advantages

- It is familiar to an Indian farmer.
- It can be used for small discharges.
- Uniform application of water and no run-off losses.
- Efficient use of water.

Disadvantages

- High initial cost on labour in leveling.
- Much area is wasted in bunds and channels.
- Forbids the use of bullock drawn implements.
- Difficult to practice on the soil with slopes.
- Not suitable for crops sensitive to wet conditions.
- Impedes surface drainage.

4. Ring basin irrigation

Ring basin irrigation is suitable for orchards and other high value crop where size of the plot to be irrigated is very small. Variable ring basins are commonly used for irrigating fruit trees depending upon the tree spacing and land leveling.

Advantages

- No water is lost by run-off
- Rapid irrigation possible
- Almost no loss of applied fertilizer and farmyard manure through run-off.

Disadvantages

- Initial cost is high.

5. Furrow irrigation

The water is applied by running small streams in furrows between the crop rows. The water soaks into the soil and spreads out into the crop root zone laterally to irrigate areas in between the furrows. Duration will depend upon the deficit and infiltration rates. The method is suitable for crops like maize, potato, sorghum, sugarcane, cotton, tobacco, potato, pulses, oilseeds and vegetables. The method is good where crops are sensitive to water ponding like maize, potato and beans. The furrows can be

I. Graded furrows
II. Level furrows
III. Contour furrows
IV. Corrugations
V. Alternate furrows which require limited water and check salt accumulation.

Length: Depends upon

I. Permeability
II. Slope of furrow
III. Size of stream

On a porous sandy soil the length is shorter than on clay soil. The length can be as short as 45 m on soils with infiltration rate of more than 5 cm / hour or as much as 300 m or more on heavy soils having infiltration rates less than 3 cm / hour.

6. Corrugation

It is commonly used for close growing crops such as small grains, vegetables as carrots and for pasture growing on steep slopes. Corrugations are very shallow furrows and are generally combined with border irrigations. This may be straight or contour.

7. Sub–surface irrigation

In sub-surface irrigation, water is applied beneath the ground surface usually by creating and maintaining an artificial water table at some pre-determined depth (10–100 cm) depending upon root characteristics of the crop, texture and layering of soil. Moisture reaches plant roots through capillary movement.

I. Natural: Impervious layer, water is led into series of ditches dug upto the impervious layer which then moves laterally and wets the root zone.

II. Artificial: Raising water table by laying porous tubes where artificial water table could be maintained. It is suitable for irrigation of vegetable, crops and most of field crops, small grains and forage crops.

Advantages

- Maintenance is cheap.

- Efficient use of water.
- No evaporation losses.
- Soil ever fit for cultivation and weed removal.
- No wastage of area in bunds and channels.
- No Interference to farm machinery.
- Very suitable for ornamental gardens and high value crops.
- Weed seeds are not carried with water.
- Channels serve as drainage channels also.

Limitations

- Costly.
- Risk of soil getting saline.
- Neighbouring land damaged due to heavy seepage.
- Deep rooted crops difficult to grow.

In India, this method is used in areas around Dal Lake in Kashmir for vegetable production and in soils of Kuttanad (Kerala) for irrigation of coconut palms but on a limited scale.

8. Sprinkler irrigation

Water sprayed into air and allowed to fall on land surface in a uniform pattern with rates lesser than the infiltration rate of the soil.

Advantages

- Applicable under odd conditions where other methods cannot be used.
- Very efficient in areas with excessive slopes.
- Suitable for irregular topography.
- Very effective method for erosive soil.
- Suitable method for excessively permeable soil.
- No harm to crop in excessively impermeable soil.
- Land preparation cost less.
- No wastage of land in channels.
- Quantity of water can be controlled.
- Rate of application according to infiltration rate.
- No deep percolation and run off losses.
- Efficient use of water.

Disadvantages

- High initial cost.
- High degree of technical skill.

- More evaporation losses on hot windy days.
- Water is required under pressure (0.5 to 10 kg/cm^2) and that increases the cost of pumping.
- Little scope under ordinary farming conditions, scope only under high value crops.
- Interference with use of inter–cultivation operations
- Not useful on ripening stage of plants having soft fruits
- Dirty water can not be used.

Types of sprinklers

(i) Rotating Head System.

(ii) Perforated Pipe System.

9. Drip irrigation

This is also called trickle irrigation. The irrigation is given by using small diameter PVC tubing with devices called emitters or drippers at selected places near the base of plants. The losses by deep percolation and evaporation are minimum. The amount of water applied daily is equal to consumptive use and it keeps the soil at low tension. It was originally developed by Simca Blass a water engineer in Israel in 1959. The method is good for economizing on water use, where poor quality water is used and is gaining popularity in arid areas.

It has been found successful in musk melon, cucumber, vines and other orchard plants, sugarcane, maize, tomato etc. Salt distribution is better than sprinkler when poor quality waters are used.

Advantages

- No deep percolation losses.
- We can maintain good moisture.
- Economizing water use saving.
- We can maintain good quality waters.
- Fertilizer can also be applied along with water.

Disadvantages

- High cost.
- Rats and other animals may damage the system.
- Dirty water cannot be used.
- Does not work under very low pressure.

WATER AND IRRIGATION REQUIREMENTS OF CROPS

Wheat (*Triticum aestivum*): Water requirement of wheat crop is 50-55 cm and irrigation requirement is 35-40 cm. Sow wheat after a heavy pre-sowing irrigation

(10 cm) except when follows rice. In case wheat sowing is delayed due to late harvest of rice, the pre-sowing irrigation of wheat can be given to the standing rice 5-10 day (depending upon soil type) before its harvest except where the crop is to be harvested with combine. This practice advances the sowing of wheat by about a weak.

The first irrigation should be relatively light and be given about three weeks after sowing to the normal sown crop and after four weeks of sowing to late sown crop. The subsequent irrigations are determined by the prevailing weather and soil conditions. In general 4 to 5 irrigations at CRI, jointing, ear head initiation, milk and dough stages are required. However, irrigation at CRI (crown root initiation) stage is very impotant for getting higher yields.

Gram (*Cicer arietinum*): Water requirement of gram is 25-30 cm and irrigation requirement is 17.5-25 cm. Where irrigation facilities are available give a heavy pre-sowing irrigation (*rauni*). It will ensure deep rooting for proper utilization of soil moisture. Afterwards, give one more irrigation between mid December and end January depending upon the date of sowing and the rainfall. This irrigation reduces the incidence of wilt disease. In no case should this irrigation be given earlier than four weeks after sowing. If early rains are received delay the irrigation. Excess of irrigation results excessive vegetative growth and depresses grain yield.

Raya (*Brassica juncea*): Water requirement = 25-32 cm. Irrigation requirement = 17.5-25 cm. Crop sown after a heavy pre-sowing irrigation (10-12 cm) should be irrigated 3-4 weeks after sowing. Irrigation at this stage promotes deeper rooting which helps in the use of water stored in the lower soil layers. If necessary, another irrigation at flowering stage may be given. If the crop is threatened by frost damage, the second irrigation may be given earlier. Taramira (*Eruca sativa*) crop can be raised without any irrigation.

***Gobhi Sarson* (*Brassica napus*):** First irrigation like raya. Second irrigation may be given at the end of December or beginning of January. The third and last irrigation be given during second fortnight of February. The crop should not be applied any irrigation, thereafter, as it lodges.

Sunflower (*Helianthus anus*): Water requirement = 50-55 cm, irrigation requirement 45-50 cm. This crop generally requires 6-9 irrigations depending upon the soil type, rain and weather. Apply the first irrigation about a month after sowing and reduce the intervals during the subsequent months to 3 to 2 weeks during March and 8-10 days during hot summer months of April-May. Stop irrigations about 12-14 days before harvest. Ensure that crop does not suffer moisture stress at flower initiation and seed development stage.

Potato (*Solanum tuberosum*) : The first irrigation should be given immediately after sowing, as it ensures better germination. The potato crop, responds better to light and repeated irrigations. While applying irrigation, avoid the over-flooding of the ridges to avoid consequent hardening of the soil surface which interfere with germination and the development of tubers. Water requirement 70-80 cm, irrigation requirement 60-70 cm.

Berseem (*Trifolium alexandrinum*): Water requirement = 120-125 cm, irrigation requirement =100-115 cm. Berseem is considered to be rice of winter. The first irrigation is important and should be applied early before the soil surface gets dried so as to obtain a good stand. On light soils, the first irrigation may be given within 3-5 days after sowing. Whereas on heavy soils, it may be delayed for 8-10 days. Afterwards, irrigate the crop at 15-20 days interval depending upon the weather consideration. For seed crop, apply frequent irrigation during seed formation and ripening.

Sugarcane (*Saccharum officinarum*): Water requirement = 150-160 cm. Irrigation requirement = 120-130 cm. Because of dry hot season, April to June is the most critical period for the growth of sugarcane. During this period, irrigate the crop at 7-12 days intervals. During the rainy season, adjust the frequency of irrigation according to rainfall. Drain away excess water from the sugarcane fields if they got flooded during the rainy season.

During winter (November to January), irrigate the crop at monthly intervals. To save the crop from frost, apply one irrigation in mid-December and another in the first week of January.

Autumn sugarcane: First irrigation one month after planting, followed by three irrigations upto February and subsequent irrigations as per the recommendations for the spring crop. Other recommendations are the same as for the spring crop.

Cotton (*Gossypium arboreum*): Water requirement = 75 cm. Irrigation requirement = 40-55 cm. Cotton requires four to six irrigations, depending upon the seasonal rainfall. The first irrigation should be given about one month after sowing and the subsequent ones at intervals of two to three weeks. The crop must not be allowed to suffer for want of water during the flowering and fruiting stages. Otherwise a lot of shedding of flowers and bolls will take place, resulting in low yields. Cotton during its early growth is very sensitive to water stagnation. To hasten boll opening, give the last irrigation to *Desi* cotton and American cotton by the end of September.

Rice (*Oryza sativa*): Water requirement = 150-180 cm. Irrigation requirement 120-130 cm. Keep the water standing continuously in the crop for two weeks after transplanting so that the seedling establish properly. Afterwards apply irrigation two days after the pounded water has infiltrated into the soil. In this way, irrigation water can be saved without causing any reduction in yield. The depth of standing water should not exceed 10 cm. Drain away excess water before applying fertilizers, interculturing or weeding and irrigate the field after these operations. Stop irrigation about a fortnight before maturity to facilitate easy harvesting and the timely sowing of the succeeding *rabi* crops.

Maize (*Zea mays*): Water requirement = 75 cm, irrigation requirement = 40-55 cm. Generally 4-6 irrigations are required, depending on the rainfall. Adequate water supply is essential throughout the crop season, particularly during the pre-tasseling and silking stages. Maize can tolerate heavy rains, provided water does not stand in the field for long period. Therefore, drain away excess water by making a drain of adequate capacity at the lower end of the field. Do not grow maize in fields from which it is not possible to drain away excess water.

DEFINITIONS OF IMPORTANT TERMS

Transpiration ratio: It is the amount of water transpired to produce a unit amount of dry matter. This ratio was taken as water requirement of crops in the beginning of this century. This concept has now been rejected for field use.

Potential evaporation: It is the evaporation from a large body of free water surface which is not normally affected by adjective energy and depends on evaporative demands of climate.

Transpiration: It is the amount of water lost by plant mainly through leaves.

Evapo-transpiration: It is the total water lost through evaporation from the soil and transpiration from the crop for a particular area during a specified time.

Consumptive use: It is the water lost through evaporation, transpiration and that used by a plant in metabolic activities. It is generally expressed as depth of water in inches or millimeters. It is often called as evapo-transpiration as the water used for growth is only 1-2 percent of the total water lost.

Irrigation efficiency: It is the ratio expressed in percentage of water stored in desired profile to water applied.

Leaching requirements: It is the fraction of water of the total water requirement which must percolate through a given soil profile to move down the salts to a sufficient depth for safer crop growth.

Water requirement: It is defined as the water needed for raising a crop in a given period. It includes consumptive use and other economically unavoidable losses and that applied for special operations such as land preparation, transplanting etc. It is usually expressed as depth of water for a given time. Some workers use the term water needs synonymous with water requirement.

Water expense: Gross water needs (consumptive use + economical unavoidable losses + special purpose + fraction of inefficient rainfall even if it is very less).

Irrigation requirement: It refers to the quantity of water exclusive of precipitation required for successful growth of crops. This in other words can be stated as net irrigation requirement plus other economically unavoidable losses. It is expressed as depth of water for a given time.

Effective rainfall: It is a total precipitation or a part thereof which forms a part of the water requirement. Moisture deficit in the rooting depth of a crop before precipitation mostly decides the value of effective rainfall. According to the climatological concept of water requirement of crops, even light showers also come under effective rainfall. Only the rainfall which is lost by run-off and deep percolation is ineffective.

Critical period: It is the period during which yield attribute of a crop is affected to the maximum extant due to lack of adequate moisture. This term is often confused with the period of peak rate of water use.

Duty: It is an area of a crop commanded by one cusec flow in a season.

Delta: Delta is the total volume of water delivered to a crop or it is total depth of water required by the crop during the entire life cycle on the field.

Open pan evaporation: It is the value of evaporation from open pan evaporimeter.

Irrigation frequency: Interval between two irrigations i.e. the number of days between two consecutive irrigations.

Moisture use efficiency (MUE): It is the ratio of dry matter produced per unit of water used. It is expressed as under:

$$\text{MUE} = \frac{\text{Dry matter produced (DW)}}{\text{Consumptive use (cm)}} = \text{kg/cm/ha}$$

Permanent wilting point: It is the moisture content in percentage of a soil at which nearly all plants wilt and do not recover even if they are kept in a dark humid chamber, unless water is added from a outside source. It corresponds to 15 atm tension.

Field capacity or capillary capacity: It is the percentage of soil moisture expressed on dry weight basis after the soil has been completely saturated with water and downward movement of water has practically ceased. The tension (force with which water is held in the soil) at this stage corresponds to 1/10 to 1/3 atmosphere. At this stage, only micro-pores hold water.

Head: The height of water above any point or plane of reference.

Broad irrigation: It is the irrigation with sewage water where the primary purpose is to dispose off the sewage.

Gross commanded area: It is the total area which can be irrigation by an irrigation channel/canal project/tubewell.

Drainage: Removal of excess gravitational water from the soil profile by artificial means to enhance crop production.

Drainage requirement: Volume of water which must be drained in a given time for satisfactory crop growth.

Hygroscopic coefficient: It is the amount of moisture in dry soil when it is in equilibrium with some standard relative humidity near a saturated atmosphere (About 98%) expressed in terms of percentage on the basis of oven dry soil.

Advective energy: The energy developed from horizontal heterogeneity in climatic parameters. It is, thus, the energy brought by winds which is of considerable importance especially in tropical hot summers.

Aridity: It refers to the climate characterized by excess of evaporation over precipitation

Capacity factor of a channel: It is the ratio of actual supply to the capacity of a channel.

Capillary fringe: It is the vertical distance above a water table along which moisture content varies from full saturation to field capacity.

Capillary potential: It is usually expressed in terms of work that must be done to remove water against the capillary forces of the soil.

Clothes line effect: It is the horizontal heat transfer from outside area.

Dead storage: Capacity of a reservoir below the outlet level.

Infiltrability: It is the maximum rate at which a soil under a given condition and at a given time can absorb water when there is no divergent flow at borders.

Live storage: Capacity stored of the reservoir above the outlet level

Microclimate: It is the local climatic condition near the ground.

Mulch: It is a natural or artificially applied layer of plant residue or other material on the surface of soil with the objective of moisture conservation, temperature control, prevention of surface compaction or crusting, reduction of run-off and erosion, improvement of soil structure or weed control.

Oasis effect: It is the exchange of heat where by air over the crop is cooled to supply heat for evaporation.

Osmotic pressure: It is the pressure exerted on account of unequal concentrations of salt in relations on both sides of a cell wall or semi permeable membrane.

Real density: It is the density of oven-dry soil particles when air is excluded. It is also called particle density.

Potential evapro-transpiration: It is the amount of water evaporated in a unit-time from a short uniform green crop, growing actively and covering an extended surface and never short of water.

Soil moisture stress: It is the sum of the effect of soil moisture tension and osmatic pressure of the soil solution.

Submerged flow: It is the flow in which there is a back pressure as a result of which the discharge is lower than the free flow.

Weir: It is the general term for a continuous solid barrier (wall of stone or masonary) built across a channel over which water is allowed to flow. It raises the water level up steam in order to supply a canal take-off above it. It is a device which is also used by measuring the water discharges of small channel flows.

Anicut: Is a Tamil name for weir.

Tail water: The water just down stream the structure.

Parabolic weir: A measuring device (weir) whose notch is bounded on the sides by parabolas such that the flow is proportional to the head.

Barrage: Is a structure (weir) provided with a series of gates erected across the stream to regulate the water flow. A barrage is sometimes called a regulator or Diversion dam also.

Nappe:

(i) A sheet or curtain of water over flowing the nappe has an upper and lower surface.

(ii) It also means a stream discharging over a crest.

Cusec/(22500 gallon/hr): It is the measure of the rate of flow of water as cubic feet per second. It can approximately cover one acre to a depth of tone inch in one hour. A day-cusec is the rate of water flow @ 1 cusec for 24 hours and it equals 1.98 acre ft. or 86249 cu. ft. of water volume.

Bulk density: It is the ratio of the weight of oven dry soil to its maximum bulk column. It is expressed in grams per cube cm (g/cc).

Acre-inch: 1 inch of water over an area of 1 acre OR

3630 c.ft. of water OR

102.8 cubic metres of water

1 cubic ft. 28.31 litres or 6.23 imperial gallons or 7.84 US gallons.

(a) 1 m^3 of water = 1000 litre of water or 35.314 c.ft.

(b) 1 m^3/sec a day = 3.64 ha-meter

g (acceleration due to gravity) = 9.81 meter/sec^2 or 32.2 ft./sec.

1 USA gallon = 0.833 imperial gallon or 3.7854 litres

1 imperial gallon = 1.201 USA gallon or 4.5436 litres

1 standard atmosphere = 14.71 sq. inch or 1036 cm of water column or 1.013 x 10^6 dynes/sq.cm.

1 bar = 1023 cm of water column or 10^6 dynes/sq.cm.

QUALITY OF IRRIGATION WATER

The quality of irrigation water is determined by composition and concentration of salts in water. The different types of hazards are :

I. Salinity hazard

II. Sodium hazard

III. Boron hazard

The rating limits fixed for irrigation water in respect of these hazards for Punjab are given as under:

Texture of the Soil to be irrigated	Conductivity (micromphos/cm)		Residual (m.e./1)	
	Tolerant crops	Semi-tolerant	Sodium carbonate	Boron
Sandy	Up to 8,000	Up to 6,000	< 5.0	< 2.0
Sandy Loam	Up to 6,000	Up to 4,000	< 5.0	< 2.0
Loam	Up to 4,000	Up to 3,000	< 5.0	< 2.0
Clay Loam	Up to 3,000	Up to 2,000	< 5.0	< 2.0
Clay	Up to 2,000	Up to 1,500	< 5.0	< 2.0

For judging the quality of water required for water logged conditions, half of the above values in each category are taken.

A marginal type of water which is suitable for light textured soils should not be used on heavy textured soils.

1. Salinity hazard

The rating limits of salinity for different types of soils are given in the above Table. The relative tolerance of crops is as under:

Semi-salt tolerant crops : Rice, wheat, oats, jowar, maize, castor, turnip, senji, metha, berseem, cowpeas, tomato, cabbage, cauliflower, potato, onion, carrot, peas, cucumber, bitter gourd.

Salt-tolerant crops: Dhaincha, sugarcane, barley, tobacco, cotton, taramira, dubgrass, sugarbeet, spinach.

2. Sodium hazard

Sodium is harmful when present in soil in exchangeable form, even at low concentrations. It causes adverse chemical and physical conditions. Sodium makes the moist soil impermeable to air and water and on drying it makes it hard and difficult to cultivate. The crust interferes with germination and emergence. Excessive concentration of sodium is directly toxic to plants. Sodium hazard is measured in different terms:

I. Sodium adsorption ratio (SAR)

$$SAR = \frac{Na^+}{\sqrt{\dfrac{Ca^{++} + Mg^{++}}{2}}} \quad \text{all expressed in ml/1.}$$

II. Sodium percentage

$$Na\% = \frac{Na^+}{Ca^{++} + Mg^{++} + Na^+} \times 100$$

Generally it is considered that if water contains more than 60% of sodium, it is not safe to use for irrigation purpose. Harmful effect is greatly influenced by the amount of other associated ions.

III. Residual sodium carbonate (RSC)

Eaton (1950) has used RSC

$$RSC = (Co_3^{=} + HCO_3^{-}) - (Ca^{++} + Mg^{++})$$

The limits are given in the above table.

3. Boron hazard

All natural waters contain boron in minor concentrations. Though boron is an essential nutrient for plant growth but is toxic if present in soil solution more than a few ppm. In Punjab, 2 ppm limit is given for all soils. General classification of crops in respect of boron tolerance is as under:

Sensitive up to 1 ppm : Plum, grape, orange.

Semi-tolerant up to 2 ppm : Potato, sunflower, cotton, barley, wheat, maize, oats

Tolerant up to 4 ppm : Sugarbeat, alfalfa, onion, carrot, cabbage.

Improving poor quality water

I. Dilution with good quality water.

II. Alternate irrigation with good and poor quality water in different ratios.

III. Treatment with amendments: Gypsum is either mixed with water or gypsum powder is sprinkled on soil surface. It is a source of calcium. The added Ca would tend to correct the SAR or reduce the HCO_3 by precipitation as $CaCO_3$. It reduces pH. All waters testing more than 2.5 RSC would bring forth the need of applying gypsum to the soil or in water.

IV. Growing salt tolerant crops and crop varieties.

PREVENTING WASTAGE OF IRRIGATION WATER

Losses during conveyance

Canals	Distributaries	Field channels
17%	8%	30%

Different methods for preventing irrigation water losses are:

I. **Lining of water courses:** Helps in reducing/eliminating water losses.

II. **Discharge:** It should be maximum, but easy to control.

III. **Maintenance of water courses:** Water courses should be properly maintained. Properly cleaned from weeds and silt.

IV. **Size and shape of plot:** Not that big that there is wastage of water. No that small that wastage of land in bunds and channels. It should be rectangular rather than square.

V. **Land preparation for irrigation:** Land leveling is very important. There is need for use of scrapper and float. Float does micro-levelling. Planker does not perform that operation so efficiently as it just follows contours.

VI. **Use of water control equipment:** For regulation of water at the farm. The following equipment needs to be used.

(a) Head gate: For opening and closing of water supply

(b) Diversion boxes: For diverting water in different directions.

(c) Check dams: Canvas check dams, iron check demands are useful

(d) Turn out boxes: To divert water from supply channel to field.

(e) Syphones: For conveying water from channels to beds or furrows.

(f) Spill tubes: Spill tubes made of bamboo, iron, can be used to take out water from secondary ditch to field.

(g) Drops A drop is used to discharge water in a channel from one land to a lower one.

VII. **Scheduling irrigation properly:** Adequate water leads to maximum yield and limited water leads to maximum water use efficiency.

VIII. **Optimum quantity of water to be applied:** Different soils hold different quantities of water. Water applied in excess of that a soil can hold will go as deep percolation loss.

IX. **Tapping shallow water table:** Irrigations can be reduced under these conditions.

X. **Irrigation in relation to manuring:** The response to fertilizer is of higher order under irrigated conditions than unirrigated. Water use efficiency increases with fertilizer.

Crop growth under excess water conditions

Causes of excess water

(i) Over irrigation (ii) Seepage from canals, water courses (iii) Excessive rains. The movement of excess water being restricted by canals, railway lines (iv) Geological formations as rising of springs and (v) Water coming from higher areas.

Effects on crop raised in excess water conditions

I. **Soil aeration:** Less oxygen as small amount of oxygen dissolved in water and extremely slow rate of gaseous diffusion. Gaseous exchange is confined only to top fraction of an inch of soil.

II. Increase in CO_2 concentration.

III. **Decrease infiltration rate:** With continuous standing water due to blocking of channels with decomposition of products from anerobic microbiological activity.

IV. Formation of incompletely oxidized and reduced organic compounds like methane, marsh gas, methyl compounds, complex aldehydes.

Oxidized form	Reduced form
Nitrate ions	Ammonia, nitrite, nitrous oxide and nitrogen gas.
Sulphate ions	Sulphide ions or Hydrogen sulphate
Ferric ions	Ferrous ions
Manganic	Manganus
Phosphate	Phosphine

V. Accumulation of O.C. - N remain locked up.

VI. Lack of nitrogen - Fixation

VII. pH increases and decreases with submergence and drainage respectively due to sulphide and sulphate ions respectively.

VIII. Toxic concentration of certain nutrients.

Mn, manganous- toxic

Ferric – Ferrous-toxic

IX. Leaching of nutrients

X. Degradation of soils to leaching of salts under prolonged submergence.

XI. Poor soil structure as conditions alternate wetting and drying do not exists.

XII. Accumulation of salts under water logged conditions.

XIII. Transpiration, absorption, photosynthesis, respiration and nutrient uptake are ultimately reduced.

XIV. Low temperature conditions.

Plant characters adopted to water logged conditions

I. Develop superficial rooting system by specialized roots functional in water. They are massed just below soil surface and supply sufficient O_2 because of their large absorbing surface.

II. Deep penetrating roots on stems containing large intercellular spaces. Oxygen of shoot released in photo synthesis is used by the roots. Rice roots maintain oxidizing conditions around them due to O_2 excretion.

III. Development of adventious roots – observed in beans, tomato.

IV. Many crops have roots with large intercellular spaces like barley, oat, corn, tomato. These spaces increase by collapse of cells due to lack of O_2.

V. Inherent characteristics of the plant.

AGRICULTURAL DRAINAGE

Agricultural drainage is the removal of the excess water, known as free water or gravitational water, from the surface or below the surface of farm land so as to create favourable soil conditions for plant growth.

Types of drainage

I. Surface drainage.

II. Sub surface drainage

The process of removing the excess water from the land surface is known as surface drainage. If the excess water saturates the pore spaces of the soil, the process of its removal by downward flow through the soil, is known as sub-surface drainage or internal drainage. The excess water that is drained away is the gravitational water which is not available for plant growth. Drainage does not remove the capillary water and there is no danger of over-draining the soil by placing the drains too close together.

Soil may need artificial drainage for either or both of the following reasons.

(a) Where there is a high water table and should be lowered.

(b) When excess surface water cannot more downward through the soil or over the surface of the soil.

A permeable soil seldom needs artificial drainage, except when there is high water table. Slowly permeable soils often need artificial drainage especially when land surface is level and rainfall is high.

Particle size affects the pore size distribution and drainage property of the soil. The finer the texture, smaller the pore size or greater the need for drainage. Conversely the need for removal of excess water by artificial drainage is reduced as particle size increase from clay to silt to loam to sand.

Platy structure of the soil indicates poor drainage whereas the granular structural offers favourable conditions for drainage.

Benefits of drainage

I. Drainage provides a better environment for plant growth. Excess water fills the pore spaces and air is excluded. The air is essen-tial for root respiration, growth of beneficial soil bacterial etc.

II. Drainage improves structure, infiltration capacity of soil. Higher infiltration capacity reduces soil erosion.

III. Drainage promotes leaching of salts and prevents their accumulation in the soil.

Surface drainage is required in (a) heavy soils (b) shallow soils (c) No outlet available (d) Larger quantity rain water to be drained in short time.

The surface drainage is done through open ditches drainage system. They are like open channels, however, the capacity of drainage channel should increase downstream as against irrigation channel. The size of the drain is fixed in relation to the quantity of 'runoff' which varies with rainfall, soil type and of 'drainage coefficient'. The drainage coefficient is defined as the rate of drainage expressed as the depth in cm of water drained off from a given area in 24 hours.

Methods of sub surface drainage: (I) Pole type; (II) Stone drain; (III) Tile drain. Tile drain may be of following types:

(a) Random (b) Herring bone

(c) Gridiron and (d) Interceptor or cut off system.

Water management in saline soils

Points to be considered for water management in saline soils

(1) Method of irrigation

(2) Amount and frequency of irrigation

(3) Planting practices in relation to irrigation system.

1. Methods of irrigation

The method of irrigation depends upon many factors like:-

I. Salinity status

II. Types of salts

III. Topography

IV. Soil characteristics

V. Availability of water

Points to remember

- Flood irrigation method may be used where salinity is a serious problem for leaching, soluble salts like Ca, Mg are present.
- When sodium is in excess flooding to be avoided unless gypsum is added.
- Sub surface, method of irrigation is not suitable as it leads to further rise of salts.
- Levelling for uniform application of water so that leaching from whole field is also uniform.
- Higher salt content water should not be used. This may aggravate the problem.

2. Amount and frequency of irrigation

Under irrigation is harmful. Frequency should be more to keep the salts under check i.e. to reduce osmotic pressure. Total quantity of water should be more as some quantity is needed for leaching salts and some to meet the water needs of the crops.

Increasing soil permeability

When evaporation rate is more than soil permeability salt conc. increases. If low soil permeability is due to hard pan it may be broken with deep ploughing/chiselling. If the permeability is low due to high exchangeable sodium, it can be improved by replacing excessive exchangeable sodium content with calcium by the application of gypsum.

3. Planting practices in relation to irrigation

I. Selection of salt tolerant crops and varieties.

II. Keeping soil moist at emergence stage. At germination stage special care is needed. Pre-emergence irrigation for better imbibitions of water. This will make surface soil moist, reduce soil strength and emergence with improved sprinkling irrigation at this stage will also be useful. This will keep down the salt concentration. If moisture content is less the salt concentration increases.

III. Planting on the sides of ridges is better than on the top. As the salts move inside with movement of water.

IV. Planting on northern side. Utilization of solar energy for desalinization. Row direction evaporation will take place from southern side due to sun's rays and salts will move to that side. The northern side will be free of salts.

V. Alternate row irrigation and planting in those rows. The salts with irrigation water will move to the other side.

VI. Puddling for rice - In alkali soil puddling is not needed.

VII. Increasing soil permeability.

FROST

When the temperature of air falls below 0°C the water vapours change into solid form i.e. frost.

Conditions favourable for frost

I. A prevailing stable air mass with cool surface temperature.

II. Clear sky to allow loss of heat by radiation.

III. Little or no surface wind to mix the cold air near the ground with the warmer air above.

IV. Lengthy nights.

V. High pressure area-air comes down from above.

Measures

I. Irrigation – temperature of air and soil is increased.

0°C → 80 Cal released

II. Wind machines

Big fans- air is put into circulation which does not allow the temperature to fall below 0°C.

III. Smookers - Paddy straw

↓

air become hot

IV. Sprinklers - directly increase the temperature of air.

V. Polythene or Straw covers - in such a way they cover NW direction and exposed to SE direction

(NW Cold wind)

covers from NW e.g. tomato

Irrigation for frost protection

Frost can cause serious damage to plants. Frost occurs if the temp. falls to 32°F (0°C) or below after dew has formed. Dew formation depends on the R.H., which is the %age of moisture in the air at any temperature compared with the maximum moisture that the air could hold at that temperature. Remember that, as air temperature drops, the absolute amount of moisture the air can hold decreases when R.H. reaches 100%, dew is formed, if this dew freezes white frost forms.

Sometimes the air is so dry that dew is not formed at 32°F (0°C), however, plants still suffer frost damage (usually burning of leaves or flower damage), this type of dry frost damage is referred to as 'black frost'.

Winter burn is another form of cold damage. It occurs if the above ground parts of plants start active growth (break dormancy) while the ground is still frozen. The new growth requires water, but because the plants cannot remove water from frozen soil, a drought like condition is established that causes leaf burn, bud death and death of the entire plant if the conditions persist for a long enough period. Water absorption by plants reduces sharply below 10°C of soil temperature. Because of : (i) reduced root growth/activity, increased viscosity of water (twice viscous at 0°C than at 25°C), decreased permeability of cell membrance of root cells resulting into increased resistance to the entry of water and decreased metabolic activities of root cells.

Effect of frost on plant cells

If the temperature drops rapidly to below freezing, cytoplasm freezes and cellular organization is disrupted; enzymes may be precipitated and cell die.

Measures for frost protection

There are a few practices a crop producer can rely on to control temperature within a limited range in order to reduce damage from frost.

I. The use of large fans installed permanently on poles (wind mills) above orchards and vineyards. By stirring the air, they force higher, warmer air down into the crops atmosphere. The higher air can be as much as 5°F (2.8°C) warmer than the air surrounding crop. Such fans can increase temperature around plants from 2-5°F (1-3°C).

II. Smudging (outdoor fire with dense smoke) can be used especially to warm air in orchards. Smudging requires some type of petroleum burner to serve as a heater. The heaters are lighted as the temperature approaches freezing (32°F or 0°C). Smudging can raise temperature in an orchard more than 7°F (4°C). In some areas, the burning of oil-soaked rags and old tires has been used in place of commercial heaters. Such smudge pots create some heat at tree level and form a dense pall of smoke which spreads over the surrounding area and reduces the loss of heat by radiation in much the same way that a cloud cover does. But during day a smoke cover retards warming at the surface of earth and also creates air pollution. More efficient heaters which give off less smoke are in common use. Crop residue (Paddy straw burning) such as crop straw can also be used to heat the air.

III. Use of polythene or straw mulch can reduce the loss of heat from soil (advection) during night but it will also check gaining of heat by the soil during day. Sensitive vegetable and plantation crops can be protected from frost by covering them from N-W direction and keeping them exposed to S-W direction.

IV. Use of irrigations (light and frequent protective irrigation).

Frost protection by irrigation

Overhead irrigation can be used in two ways to guard crops against damage by frosts: by wetting the surface soil in advance, and thus relea-sing warmth from the ground to combat the frost and by continuously forming a protective ice coating on tender leaves and buds, so, that their temperature cannot falls to damaging level.

Advance anti-frost irrigation

This method makes use of the fact that a wet, compact mineral soil is a better conductor of heat than a dry, loose one. Hence, irrigated soil absorbs more heat on a warm day and releases more of the ground's heat on a cold night than dry soil. Soil temperature at a depth of 5 cm is observed to be higher by 2.5-4.0°C than air temperature, when the soil is at field capacity. These effects make the air immediately above wet soil warmer by about 1–2°F on a frosty night than it would be if left in a drystate. Hence, irrigation affords limited protection against night frosts for young crops. This pre-wetting technique is less applicable to organic soils, because their thermal conductivity is very low and wetting may not make an effective difference to temperature above the soil. Using conventional irrigation, an hour's application of 1/3 inch of water per acre every 3 or 4 days, while the risk of frost persists, should give useful protection in marginal frosts, but success depends upon the day time weather being sunny or warm, so as to add warmth to the soil. When frost occurs after a cool, overcast day, the pre-wetting technique is less likely to afford effective protection.

Continuous anti-frost sprinkling

This method of frost protection depends on keeping the plants warm by means of the latent heat released when water changes into ice (In the reverse process, when ice melts the same amount of heat has to be added to bring about the change). Fine misty sprays are desirable and the water has to be applied continuously and fast enough to make sure that it is continuously freezing on the trees and bushes. So long this is going on, their temperature cannot fall below freezing point (32°F). The sap in blossoms, buds and young tender leaves does not freeze until the temperature is several degrees below this, so they remain undamaged as long as fresh ice continuously to form outside them.

Sprinkler systems have been used for frost protection for several years on barriers, almonds, citrus, vegetables, fruits and flowers. Crops have been saved in temperature as low as -8°C by sprinkler irrigation. Sprays should be started at or above 30°F for frost protection. Water should be applied until the ice is melted.

6

CROP PHYSIOLOGY

It is very important for an agronomist to get knowledge about the important physiological processes of crop plants so that yields could be improved with the adoption of suitable agronomic practices.

PHYSIOLOGICAL PROCESS IN CROP GROWTH AND DEVELOPMENT

Growth

Growth is irreversible increase in size or weight. Growth in plants has various phases which start with activation of embryo and ends with imoluration of seed. This process is very complex and its complexity also increases because it is accompanied by changes in the nature.

Germination

Seeds when sown in moist soil, germinates under favourable temperature and sufficient aeration. Seeds of some plant species do not germinate even under favourable conditions and these seeds are said to be dormant. The period of dormancy may be a few days, weeks, or months. Dormancy may be due to several reasons such as improper development of embryo, high concentration of growth inhibiting substances, hard and impermeable seed coat. Based on the light requirement for germination, seeds are grouped as (i) Positive photoblastic, those that require light for inducing germination, (ii) Negative photoblastic, those that do not germinate when exposed to light and (iii) non-photoblastic, those that germinate either in light or dark. Seeds of many cultivated crops belong to non-photoblastic group.

Biochemical process

Germination can be grouped into biochemical and morphological processes. When the seed is sown into the moist soil, it absorbs water twice its weight. As a result, the seed bulges and breaks the seed coat. Subsequently, embryo secretes gibberllin which in turn induces the secretion of hydrolytic enzymes which break down complex food materials like starches and proteins into simple sugars and amino acids respectively. The simple products thus formed are soluble and can be translocated to growing plants until it becomes autotropic.

Morphological process

The radicle is the first organ to emerge from the seed in nearly all cases. The radicle is soon followed by the plumule of young shoots. In many dicotyledonous plants, such as groundnut, beans etc. the cotyledons emerge from the soil and function as first leaves. The cotyledons are brought out of the seed by curved elongation of hypocotyl. This type of germination where cotyledons emerge out of the soil is called epigeal germination. In monocotyledonous plants, the cotyledon remains in the soil. The plumule grows or is pushed upward by the elongation of epicotyl. This type of germination is called hypogeal germination since the cotyledon is below or inside the soil. A slit develops at vent on the tip of the coleoptile and leaf emerges through it. The radicle develops seminal roots.

Emergence

The emergence of crops is critical for good stand establishment. Generally it is hindered by soil crust, high salt concentration, water logging etc. Plants with bigger seeds and with epigeal germination can overcome the soil crust problem and emerge out. After emergence, the cotyledons become green and support the seeding in dicotyledonous plants. In the case of monocotyledonous plant like wheat, the emerged seeding become independent when the second leaf has fully expanded and the third just appears. Seeding is able to take up nitrogen, phosphorus, zinc and possibly most minerals from the soil. The reserves present in the seed usually have more of carbohydrates or lipids which are sufficient till plant become autotrophic, while minerals present in the seed are not sufficient up to that stage.

Root growth

In monocotyledonous plants, radicle elongates and constitute primary root. Primary roots grow downwards while secondary roots they appear during the seeding stage grow horizontally for a few cms before turning downwards. These roots are called seminal roots. At the end of seeding stage, fresh roots development from lower nodes of stem, so called nodel roots (adventitious) - a major portion of total root mass and become important for the rest of plants, while the seminal roots system is taproot type compared to fibrous root system of most of the monocotyledonous plants. The rate of elongation is related to temperature, soil moisture, photosynthates and nutrient supply. The appearance and death of roots hairs occurs continuously.

Inflorescence

The growth of inflorescence in length usually shows two phases. An initial period of slow elongation during the time of formation of new spikelets, followed by a period of rapid elongation of the internodes of the rachis. Moisture stress at early stage of growth of inflorescence delays the formation of spikelets. In panicles of cereals moisture stress at panicle emergence causes death of spikelets and at anthesis germination and growth of pollen tube is effected.

Seed and fruit growth

Growth of seed/fruit continue at lower rates up to anthesis and picks up after fertilization. If fertilization does not occur the ovary gradually shrinks, and flower is dropped down. In certain cases, seed growth continues for long time even after the stoppage of fruit growth e.g seed of groundnut.

Development

Development processes relate to varying degrees of expansion or growth in weight of an organ. A characteristic of development process is that they are discrete.

Development stages

The crop growth is expressed as crop enters into different growth stages. The life cycle of the crop is divided into different stages based on the morphological characters, developmental stages etc. These stages are helpful in identifying the time for scheduling irrigation, application of fertilizers, herbicides etc. Certain stages of crop growth are very sensitive to moisture stress. For example moisture stress at CRI stage reduces the grain yield of wheat. The crop developmental stages are; germination, emergence, seedling stage, maximum vegetative growth stage, primordial differentiation, flowering, fruit growth, physiological maturity and harvest maturity.

Germination and emergence

When seed absorb moisture and put fourth radical and plumule it is called germination, when it come out of soil and this is known as emergence.

Seedling growth

During this stage, the plants establish by developing sufficient number of seminal roots and leaves. This stage ends when the plant put forth branches.

Maximum vegetative growth stage

During this stage crop grow at a faster rate. Plants start producing branches or tillers and the crop covers the ground, to intercept more radiation.

Primordial differentiation

This stage is called as panicle initiation stage in cereal and millets, squaring in cotton, flower bud initiation in sunflower etc.

Flowering stage

This stage is also called as 50 per cent flowering, anthesis, blooming etc. The crop is said to be at 50 per cent flowering when 50 per cent of plants put forth flowers. Opening of flowers and shedding of pollen is called anthesis. Blooming also indicates opening of flowers. This stage is very sensitive to moisture stress. Fruit set is development of flowers into fruits and this is a very important stage.

Fruit growth

During this stage, photosynthates are accumulated in the fruits. Fruit growth is also called as grain growth in cereals, pod growth in groundnut and pulses, boll growth in cotton etc.

Physiological maturity

When fruit growth is complete and photosynthares are no longer translocated to fruits, it is known as physiological maturity. An abscission layer forms between fruit and peduncle. Crop, if necessary, can be harvested at this stage.

Harvest maturity

At physiological maturity, the moisture percentage of grains is about 20 percent. In addition, all the grains might not be at physiological maturity. Therefore, it is a general practice, to allow the crop for 10 or more days after physiological maturity to reduce the moisture percentage and also to provide time for the development of late formed grains.

Photoperodism

Photoperodism is the phenomenon of physiological change occurring in plants in response to relative length of day and night. Crop developmental processes like rate of leaf production, flowering etc. are influenced by duration of sunlight. Most plants are influenced by relative length of day and night, especially for floral initiation. The duration of the night or of complete darkness is more important than the daylight. This effect of light on plants is known as photoperodism.

Root development under variable environments

Roots are effected in activity, external form and structure by change in environment.

Heredity

The primary form of the root is governed by the hereditary growth characters of the species or variety. Some species have tap roots that develop rapidly and penetrate deeply, others branch near the soil surface and tap root soon looses its dominance.

Environment

Roots are influenced by water content, aeration, soil structure and nutrients. The character of the root system is an indicator of soil conditions. Some plants become adapted to soil conditions due to modifications in root system and other plants do not. Similarly, grafting on other root stocks is done because roots of such plants can adopt according to situations. Such grafting is done to make the plant tolerant to alkali conditions, excessive moisture or dry conditions or disease resistance.

Low water content

A relatively low water content, stimulates the roots for greater development, resulting in a greatly increased absorbing surface. Corn grown for 5 weeks in a moist, rich soil with available soil moisture content of 19 per cent (moist soil) had a total root area which was 1.2 times more than that of its transpiring surface of stems & leaves. Corn with similar hereditary characters, grown in same type of soils with available soil moisture content of only 9 per cent (dry soil) had a root area 2.1 times greater than the tops. The total length and diameter of the main roots in the two cases was the same. The main roots made up only 11% of the total absorbing area in both moist and drier soil. Thus, a low water content, within a certain limit, stimulated and increased root development.

High water content

Due to lack of aeration, roots remains very shallow and form a mat above the water level. Roots are superficially placed but they would die if extend vertically. Therefore, they are horizontal and above the water level. Most cultivated plants require well aerated soil.

Irrigation and drainage

To make plants deep feeder, water is avoided in early stages of the crop. The proportion of roots to tops may be definitely increased by decreasing the soil moisture. The amount of shedding of leaves and bolls are directly proportional to the extent of the root system that was submerged and died as a result of a rising water table in cotton. Sugarbeets, alfalfa, and other plants with taproots 5 to 20 feet deep require a deeper and better drained soil than melons or other cucurbits where the main portion of the roots occupies the surface on all sides of the plant to a distance of 15 to 20 feet.

Nutrients

Crops grown in rich soil have roots that are shorter, more branched and more compact than in poorer soil. Roots are also shorter in more concentrated solutions. Plants grown in soils with alternate layers having more nutrients, branched more in such layers. In general enriched soils promotes better shoot development, which in turn, helps a better food supply for higher root growth.

Temperature

For most of the temperate crops, soil temperature between 60° to 80° F is quiet good. However, shallow rooted plants like cucurbits grow better at higher temperature than the deep rooted plants. The shallow root habit of certain desert plants has been thought to be result of sub-soil temperature which are too low to promote root growth. The decrease in depth has a great influence on root activities. Alternate freezing and thawing of a soil tears the roots and sometimes causes losses of 25 to 50% of the roots.

Aerial environment

The environmental factors that affect the shoot also affect the root through top portions of the plant. The plant height, degree of humidity, temperature etc., affect root development as they affect the shoot through water loss, photosynthates etc. Diffuse light causes shorter plants with longer roots and full light produces short stocky plants with long, branching roots as the carbohydrates supply to roots is more due to more photosynthates which leads to more development of roots.

Cultural practices

The distribution and development of roots is greatly affected by various cultural practices. In transplantation, there is possibility of injury or breakage of roots. Therefore, nursery men transplant trees, shrubs and crops with two or three times more seedlings than the seed hills to force root development near the stem and thus to ensure the preservation of more young roots when the plants are lifted for shipment. Transplanting in general is not beneficial to a plant but it is done to give special care in cultivation. Grass develop better root system if undisturbed or moved only once or twice a year than frequent grazing. Oxygen is necessary for proper development of roots. Proper supply of nutrients is also essential. Phosphorus promotes root growth in length, so beneficial in drought conditions. Potassium helps in shorter and thicker sweet potatoes. Nitrogen tends to produce long sweet potatoes. Excessive acidity affects the roots of plants by partially or wholly retarding growth in length. The roots are often thicken and soon become dull white or yellowish brown in colour. Mulching the soil and lack of tillage result in a marked growth of fibrous roots in the surface layers. Roots of fruit trees and other plants are often very superficial, under a straw mulch.

ROLE OF GROWTH REGULATORS

Plant growth substances are the chemicals produced in plants (endogenous) or synthetic substances applied to plants externally (exogenous) which cause modifications in plant growth and development. Plant growth substances produced by the plant are referred to as phytohormones. Growth substances initiate bio chemical processes which help in increased production of roots, shoots, flowers etc.

Plant growth substances are mainly grouped into growth promoters (which enhance growth) and growth retardants (which reduce growth). There are five major groups of endogenous growth substances present in plants viz. auxins, cytokinins, gibberellins, abscissic acid and ethylene. Auxius, cytokinins and gibberllins are growth promoters while abscissic acid and ethylene are growth retardants. Growth promoters exist in multiple forms while abscissic acid and ethylene are in single form. The mechanism of growth substances is not well understood because of multiplicity of active molecules.

Auxins

Auxins stimulate elongation of cells of stems and coleoptiles. The endogenous auxins are Indoleacetic acid and phenol acetic acid. Some of the synthetic are NAA. Indole-3 acetic acid, Indole-3 -proponic acid and 2, 4-D, Auxins promote cell enlargement resulting in elongation of coleoptiles, stem etc.

Cell division and root formation are some of the functions of auxins. Auxins are used as herbicide when applied at higher doses. 2, 4-D and 2, 4, 5-T are used as herbicides e.g. NAA induces flowering in Pineapple. IBA is used for inducing rooting in cuttings.

Cytokinins

Cytokinins stimulate cell division. Cytokinins promote orderly development of embryo's of seed, break dormancy of seed and buds. They delay senescence.

Gibberellins

Gibberellins is first isolated from the fungus *Gibberella fujikurai.* Different Gibberellins are $GA_{1,}$ $GA_{2,}$ $GA_{3,}$ GA_4 etc. They generally increase cell division and cell elongation. Dormancy is broken and flowering is induced by gibberellins.

Abscissic acid

Abscissic Acid (ABA) induces leaf and fruit abscission. Dormancy is induced due to accumulation of abscissic acid.

Ethylene

Ethylene stimulates the swelling or isodiametric growth of stems and roots. Regulation of ethylene can trigger ripening of fruits.

PHYSIOLOGICAL MATURITY

When fruit growth is complete and photosynthates are no longer translocated to fruits, it is known as physiological maturity. An abscission layer forms between fruit and peduncle. Crops, if necessary, can be harvested at this stage. Removal of entire plants or economic parts after maturity from the field is called harvesting. Portion of the stem that is left on the field is known as stubble. The economic product may be grain, seed, leaf, root or entire plant.

Time of harvesting

If the crop is harvested early, the produce contains high moisture and more immature grains. The yields will be low due to unfilled grains. It is very difficult to store the produce as shriveled grains with high moisture are prone to primary infestation of pests. The quality of grain as well as germination percentage are reduced. Late harvesting results in shattering of grains, germination even before harvesting during rainy season and breakage during processing. Hence, harvesting at correct time is essential to get good quality grains and higher yield.

Crops can be harvested at physiological maturity or at harvest maturity. Crop is considered to be at physiological maturity when the translocation of photosynthates are stopped to economic part. In other words, physiological maturity refers to a developmental stage after which no further increase in dry matter occur in the economic part. In cereals, moisture content of grains is very high during milking stage and it gradually decreases due to accumulation of photosynthates. The moisture content falls steeply from 40% to 20% which is an indication of attaining physiological maturity.

At this stage, translocation of carbohydrates is stopped due to formation of abscission layer between rachis and grain.

Harvest maturity generally occurs seven days after physiological maturity. The important processes during this period is loss of moisture from the plants. The general symptoms of harvest maturity are yellowing of leaves, drying of grains or pods. Crop is harvested at physiological maturity when there is need to vacate the field for sowing another crop. Time for harvesting can be known approximately by the duration of crop. As maturity depends on climate, maturity symptoms are good indicators for deciding the time for harvesting.

For deciding harvesting date of fodder crops, additional aspects have to be considered. They are: toxins present in the crop, nutritive value, purpose of harvest and single or multicut. When toxins are present, they are generally high in early stages and in case of sorghum their level is high upto 30 days after sowing. The nutritive value of fodder crop especially protein content decreases and fibre content increases with the advancement of the crop stage. The time of harvesting also depends on the purpose for harvesting. For stall feeding, crop is harvested when protein content is high and also when the fodder is succulent and leafy. Harvesting is delayed by a few more days to get more dry matter if the purpose is hay making. Good quality silage with long preservative characters is obtained when fodder contains more carbohydrates and less proteins at the time of harvest. Crops with ratooning ability are harvested at periodical intervals while others are harvested as a single cut.

Method of harvesting

Harvesting is done either manually or by mechanical means. In manual harvesting, sickle is the most important tool. The other tool used for harvesting is knife especially for harvesting of plants with woody stems.

When there is labour shortage or in periods of peak labour demand when planting and harvesting coincides or for quicker harvesting, mechanical methods are adopted. Combine harvester or simply combines are used for harvesting. Combines perform several operations: cut the crop, separates the grain from straw, clean it from chaff and transports grains to the storage tank.

Harvesting of crops is done by several ways depending on the crop. Cutting of the crop with sickles leaving small stubbles near the ground is the common method of harvesting. This method is followed in crops whose maturity is almost uniform and occur at a time. In indeterminate plants, harvesting is done at intervals as the economic product comes to maturity at different periods.

Storage

Harvesting of crops is seasonal, but consumption of the food grains is continuous. The market value of the produce is generally low at harvesting time. There is, therefore, necessity to store the produce for different periods. The different categories of agricultural produce needing storage are food grains, oilseeds and fodder.

Storage of foodgrains and oilseeds

During storage, foodgrains are subjected to several losses. The losses due to different pests during storage and estimated to be about 6.5%. Respiratory losses occur which depend on moisture content of grain. At 13% grain moisture the respiratory losses (mg/kg/year) are: wheat and rice 0.98, sorghum 1.76 and maize 2.92.

In addition to loss in quantity, quality of food grains is also affected in storage. The quality losses depend on duration of storage and place of storage. Quality losses are more in the produce stored in coastal areas compared to inland areas due to higher humidity in coastal areas.

Factors affecting storage: Several factors that influence the storage of foodgrains are moisture content, quality of produce, climate and storage conditions. The most important factor deciding the storability of the produce is moisture content of grains. Higher moisture content of grains results in severe attack of insects and micro organisms in addition to heating and germination. Grains with high moisture respire at higher rate than dry seeds. When the moist seeds are stored, the moisture from upper layers moves downwards and deterioration of grains in lower layers take place. Grains with high moisture are prone to higher insect attack than dry seeds. Sometimes moist grains may even germinate and become unfit for consumption. Moisture content for safe storage of grains of most crops is above 14%. Early harvested crop contains ill filed and shriveled grains. Mechanical harvesting through combines often results in higher proportion of broken grains. Shriveled and broken grains are the main causes for insect attack. Among the climatic factors, temperature, light and relative humidity are important factors influencing storage of food grains. Respiration of grains increases with increase in temperature. In addition temperature influences metabolism growth development, reproduction, behaviour and distribution of insects. Insect development in generally limited below 10°C and above 45°C.

Storage facilities

Foodgrains are stored either in bags or in bulk. The storage of produce in bags is called bag storage and without bagging is called bulk storage. The storage facility meant for bulk storage has to satisfy certain conditions:

I. Protection of grain from excessive moisture, insects and rodents.
II. Provision of safety and convenience while moving grains in or out of storage.
III. Facilities for inspecting the grain without removal from storage.
IV. Provision for controlled aeration.
V. Provision for making sufficiently airtight for fumigation.

Principles of storage

Dunnage, stacking and pest control are more important aspects of storage. Dunnage is any material like crates, mats, wooden beams, stones which are placed over the ground and below the bags so as to avoid direct contact of grains with the floor and for providing aeration. Wooden crates provide sufficient space between floor and produce and allow free air circulation. If the bags are placed on the floor itself, moisture from the top layers migrates to lower layers and accumulates at the bottom as there is no escape. This process causes caking up, chassing and development of heat in lower layers of the stock. Mats should be spread on crates before placing the bags. The second important aspect of storage is stacking. In case of bag storage, stacking is done up to 13 bags high. The stack should be brought to pyramidal shape.

Several pests attack the produce during storage. They can be controlled by adopting different methods of pest control like prevention, spraying and fumigation. Moisture contact of grains and pest intensity are directly related. It can be reduced by thorough drying of the produce to a safe moisture limit. Godowns or stores have to be kept clean by sweeping periodically so as to avoid pest build up. When the pest population increased, spraying with insecticides is done on the walls of godowns, alley ways and bags. The pest is killed by the poisonous gas released from the fumigant. Generally fumigation is done with aluminium phosphate or methylbromide. Aluminium phosphide tablets are placed at different heights of the stack and it is covered with gas proof cover. The tablets absorb moisture from the stmosphere and release phosphine gas which kills the insects.

BIOLOGICAL NITROGEN FIXATION

The conversion of atmospheric nitrogen into inorganic or organic useable forms through agency of living organism is called biological nitrogen fixation. The process is carried by two main types of micro organisms: those which are "free living" or asymbiotic and these which live in close symbiotic association with other plants. However, a third category which fix nitrogen in association with roots of cereals and grasses has been recognized. It is called associative symbiotic nitrogen fixation.

1. Asymbiotic biological nitrogen fixation

Soil contains a number of free living nitrogen fixing organisms. They include a number of aerobic and anaerobic bacteria and blue green algae (cyanobacteria). Recent studies, based on the use of radio active isotopes of nitrogen (N^{15}) have shown that several other organisms also fix atmospheric nitrogen. The asymbiotic nitrogen fixers can be classified as follows:

I. Free living aerobic nitrogen fixing bacteria eg. Azotobacter, Derxia.

II. Free living anaerobic nitrogen fixing bacteria eg. Clostridium, Bacillus.

III. Free living photosynthetic bacteria eg. Chlordium, Rhodop

IV. Free living chemosynthetic bacteria eg. Desulfovib

V. Free living fungi eg. Phullularia and Yeasts

VI. Hetero bearing blue green algae (Cyanch) eg. Nosta, Anabaena, calothrix etc.

The asymbiotic free living nitrogen fixers are quite primitive. The fixation is a reduction process independent of respiration. These organisms fix nitrogen more actively under poor aeration, provided no hydrogen gas is being produced. Some of blue green algae are important nitrogen fixers of Indian rice fields. Nearly 40 spices of blue green algae are known to fix nitrogen. These plants generally bear thick walled, hyaline heterocysts, which is the site of nitrogen fixation.

2. Associative symbiotic nitrogen fixation

Certain bacteria, living in close contact with the roots of cereals and grasses fix nitrogen. This association is a loose mutualism called associative symbiosis. The bacteria reside in the transition zone between soil and root (rhizosphere) and sometimes enter the roots. Some of the fixed nitrogen is absorbed by the roots and in return the bacteria get nourishment from the carbohydrates released by the roots. Some of the examples are given below:

(*i*) *Azospirillum brasilense* in association with cereal roots.

(*ii*) *Azotobacter paspali* in association with roots of tropical grass *Paspalum notatum.*

3. Symbiotic biological nitrogen fixation

Symbiotic biological nitrogen fixation includes all those associations where two organisms live together symbiotically and the microbial partner fixes atmospheric nitrogen. Recently a more specific term mutualism has been given for such symbiotic associations where the two organisms live together and benefit each other. However, the various examples of symbiotic biological nitrogen fixation can be grouped under the following three categories:

I. Nitrogen fixation through nodule formation in leguminous plants: A large number of species (more than 2500) belonging to family leguminosae are known to produce root nodules in association with a well known soil bacteria – *Rhizobium* spp. The bacteria fix nitrogen only when they are present inside the nodules. The association is regarded as symbiotic because the host plant supplies the nodule to bacteria and the required organic carbon (carbohydrates). In return the microorganisms supply fixed nitrogen to the host plant.

II. Nitrogen fixation through nodule formation in non-leguminous plants: Many plants belonging to families other than leguminosae are known to produce root nodules. The important among them are primarily trees and shrubs. The important examples of non –leguminous plants that produce root nodules and fix nitrogen are:

(*i*) *Alnus* (alder) in association with *Frankia* (actinomycete)

(*ii*) *Casuarina equisetifolia* in association with *Frankia*

(*iii*) *Myrica gale* in association with *Frankia*

(*iv*) *Parasponia* in association with *Rhizobium*

(*v*) Root nodules of *Podocarpus* (a gymnosperm)

(*vi*) Leaf nodules of *Pavetta zinumermanniana*

III. Nitrogen fixation through non-nodulation: There are a large number of examples where root nodules are not formed but symbiotic nitrogen fixation occurs. Some of the examples are given below:

(*i*) Lichens – An association of fungi and algae.

(*ii*) Anthoceros (Bryophyte) associated with nostoc.

(*iii*) Azolla (fern) in association with *Anabaena azollae*.

(*iv*) Roots of *Digitaria,* sorghum and maize associated with *Spirillum notatum*.

(*v*) Some nitrogen fixers grown on the surface of leaves of shrubs and trees. This is termed as Phyllosphere association.

PHOTOSYNTHESIS

Life is not possible without sun except for a few bacteria, they derive their energy from inorganic compounds. Mostly life at any level depends on the photosynthesis, by which plant use light energy to synthesize fact molecular from CO_2 and H_2O. The sugar produced is then further converted to structural materials and other cell constituents.

Over 90% of dry matter of crop plants depends on photosynthetic activity and the crop yield depends on the rate and duration of photosynthesis.

In the ecosystem, plants are called primary producers because they are the ultimate source of food for all life.

- Starch crops : Potato, sugarcane
- Protein crops : Soybean, all pulses
- Oilseed crops : Brassica, rapeseed and mustards
- Cellulose : Potato

Leaf is principle plant organ of Eukaryotes in which photosynthesis occur. The assimilation process occur in chloroplast which are found mainly in leaf. Chloroplast are the plastids that contains green pigment chlorophyll which is an enzyme and a protein making it the most abundant protein on earth. Mainly 2 types of chlorophyll are there :

Chlorophyll 'a' and Chloropyll 'b'.

Chlorophyll 'a': It occur in all photosynthesizing Eukaryotes, however, vascular plants and Bryophytes (non-vascular plants) also require chlorophyll 'b' for photosynthesis.

Apart from chlorophyll, photosynthesis involves two other pigments i.e. carotenoids and phycobilins.

Phycobilins is water soluble accessory pigment not found in higher plants.

Principle reaction for photosynthesis :

$$6\,CO_2 + 12\,H_2O \longrightarrow C_6\,H_{12}\,O_6 + 6O_2 + 6H_2O$$

The CO_2 is obtained from air while H_2O is obtained from soil. The ultimate result of photosynthesis is the reduction of CO_2 to carbohydrates.

Photosynthesis can be discussed with reference to the fate of CO_2 in 3 steps:

I. Entry of CO_2 into plant organ by the process of diffusion activity.

II. Harvesting of light by photochemical processes.

III. Assimilation and reduction of CO_2 by biochemical process in C_3, C_4 and CAM plants.

C_3 Plants : Wheat, rice

C_4 Plants: Maize, sorghum

CAM Plants: Pineapple.

In C_3 plants, photorespiration occurs. C_4 plants are more efficient as they have more efficiency for CO_2 reduction.

Different physiological processes envolved in photosynthesis are as follows:

1. Co_2 entry in plant tissue: Diffusion is movement of molecules or substances from a place of higher concentration to a place of lower concentration e.g. across the conc. gradient. The main principle part in plant organ is stomata. Diffusion occur because of CO_2 gradient created by low conc. of CO_2 in the leaf than the air on the surface. The diffusion occurs across a series of barriers formed by tissues, cell wall and finally through the protoplasm of the mesophyll cells to the chloroplast and CO_2 availability for fixation is affected by such barriers.

Opening and closing of stomata control the amount of CO_2 entering the leaf through the stomata because opening of the aperture of the stomata pores is done by H_2O present in the guard cells and through that will cause stomata to close. Plant also differ in structure and organization of leaf cells.

2. Light Reactions: Harvesting of light is a photochemical process and the ultimate source of energy in ecosystem is sunlight. Visible light constitutes only a small proportion of the electromagnetic spectrum and within this range the plant pigments are associated with photosynthesis absorbed light energy. Each of the pigment has a specific action spectrum where they are more effective for different light dependent processes such as photosynthesis and these pigments are absorbed within the range of wavelength of light is called absorption spectrum.

Chlorophyll a is absorbed maximally at 430 and 662 nm and is the primary photosynthetic pigment. Together chl 'a' and 'b' absorb maximally at 400-500 nm i.e. violet blue part and 600-700 nm i.e. orange red part. Because all the green light coming toward them is reflected back so plants looks green. While some plants have only carotenoid pigments so they look red or orange. Carotenoids are accessory pigments for photosynthesis and they absorb maximaly wavelength between 460-550 nm. The light absorbing pigments are packet in thyllakoids of chloroplasts in discrete units called photosystems. Photosystems are energy collecting and energy transferring system that operate in chloroplast. There are two photosystems i.e. Photosystem I (PS-I) & Photosystem II (PS-II): PS-II occurs at 680 and PS-I at 700 nm wave length.

Only a pair of chlorophyll molecule called a reaction centre. Chlorophyll out of all pigments present is capable of utilizing light energy. The other pigments form the antennae pigment. The reaction centre chlorophyll in PS-I is a special molecule of chlorophyll and is called P_{700} because it has a peak absorption at 700 nm. In PS-II similarly P_{680} is more efficient in harvesting light energy than PS-I. The light energy absorbed is transferred in a relay fashion from one photosynthetic pigment to another until it reaches the reaction. The energy is used in forming the chemical bonds for formation of ATP i.e. Adenosine triphosphate from ADP and Inorganic Phosphorus and the process is called photophosphorylation. In the light reaction co-enzyme $NADP^+$ is also reduced to NADPH. Overall the light reaction can be represented by the equation given as :

$$\text{Light quanta} + 2H_2O + NADP^+ + 3ADP + 3iP \longrightarrow 2NADPH + 2H + 3ATP + 3\,H_2O + O_2$$

Photophosphorylation: Formation of ATP (usable energy)

I. In light reaction, ATP i.e. reducible energy is formed along with their reducing potential i.e. NADPH and these two products they enter the next phase of photosynthetic cycle i.e. light independent or dark reaction.

II. This process is CO_2 fixation process and occurs by 2 different processes, depending upon the plant species and based on first stage product of pathway, these are called C_3 and C_4 pathway.

C_3 Pathway

- In C_3 plants, C_3 pathway or calvin cycle or calvin-benson cycle occurs.
- The first stable product of C_3 Pathway is a 3-C compound called Phosphoglycerate.
- C_3 Pathway is also called calvin cycle and plants using this pathway for CO_2 assimilation are called C_3 plants.
- C_3 plants include both monocots line barley, wheat, rice, oat and dicots like pea, soybean, sunflower etc.
- This assimilation process is catalyzed by Rubisco or RUBP or Ribulose 1,5-bisphosphate carboxylase and the overall reaction is represented by the following equation i.e.
- $6\,Co_2 + 12\,NADPH + 12H\text{ ions} + 18\,ATP \longrightarrow \text{Glucose} + 12\,NADP + 18\,ADP + 18\,iP + 6\,H_2O$

C_4 Pathway

The first stable product resulting from this pathway is a 4-carbon molecule compound called oxaloacetate. C_4 plant include maize, sorghum, sugarcane etc.

These plants produce two type of chloroplast, one in the bundle sheath and another in mesophyll tissue which are smaller in size.

PEP carboxylase catalyses the assimilation of CO_2 into PEP which is then transported to the chloroplast in bundle sheath where the malate is converted into pyruvate releasing CO_2 in the process and that CO_2 is fixed via C_3 cycle.

Photorespiration: C_3 plants experience a loss of energy through a process called photorespiration which is the oxidation of carbohydrates produced during synthesis and release of CO_2 unlike the respiration produces no ATP so this reduces overall efficiency of C_3 pathway making it less efficient than C_4 pathway and this energy loss mainly occur in hot and dry conditions. C_3 pathway is a sort of wasteful energy process.

The hot and dry conditions mostly occur in tropics and subtropics and under these conditions stomata of plant closes to conserve moisture and therefore, entry of CO_2 into leaf is restricted and these conditions promote the phenomena of photorespiration.

C_4 plants have a mechanism that enables CO_2 to pump into sheath cells to keep the internal conc. of CO_2 about 100 times higher than normal level. This allows the enzyme to continue to fix and it is virtually impossible to light saturate C_4 plants and consequently C_4 plants continue to photosynthesize even in full sunlight.

Factors affecting photosynthesis

The rate of photosynthesis in higher plants is effected by number of environmental conditions like radiations, CO_2 concentration temperature, H_2O, length of day, plant age and genetic make up.

1. H_2O: Photosynthesis in higher plants is highly limited by H_2O. So, plants in desert are highly unproductive while tropical rainforest are most productive. When water potential inside the plant become too -ve, cellular expansion is retarded and growth is reduced and with still higher H_2O stress stomata begin to close and CO_2 uptake is restricted. So, photosynthesis is less under H_2O stress conditions because of reduced leaf expansion and restricted absorption of CO_2. Enzymetic activities are also reduced under H_2O stress.

2. Radiations: Of the total radiance only 30% of energy that reaches the plant or in the wavelength 400-700 nm which is capable of photosynthesis, is called photosynthetic active radiation. The actual amount of radiant energy in the par range varies with atmospheric conditions depending upon absorption by atmospheric vapours, cloud cover. In any environment radiance varies with season, time of the day and specially with elevation or a particular point within plant canopy. About 80% of par is absorbed by a leaf depending on cloud cover, location, leaf structure and leaf age. The remaining radiation is transmitted to lower leaves or the ground below or is reflected to the surrounding (i.e. short wave and green wavelength). The radiance at which photosynthesis balance or equals respire i.e. net carbon gain is zero, is called light compensation point only when the radiance is above the carbon compensation point of plants dry weight of plant increases. Increasing the light intensity, increase the rate of photosynthesis upto a certain level and after the system is saturated, increasing the irradiance does not result increasing photosynthesis.

At high light intensity CO_2 and other factors may become limiting factors. However, at low light intensity the rate of CO_2 fixation is balanced by rate of CO_2 released through respiration resulting in no net fixation.

C_4 plants have a lower rate of CO_2 assimilation at low light intensity than C_3 plants but as the radiations increases reverse occurs and C_4 plants show higher rate of CO_2 assimilation than C_3 plants. These differences are due to photorespiration that occurs in C_3 pathway.

3. CO_2 : As light intensity increases, increasing CO_2 conc. results in increase in CO_2 compensation plants because they are more efficient in trapping CO_2 than C_3 plants. Increasing the conc. of CO_2 is also called as CO_2 fertilization, is sometime used a strategy in green houses for increasing the crop productivity. In these CO_2 fertilization atmosphere is enriched from 0.035 to 0.1% CO_2 conc., productivity increases because of reduced photorespiration. However, the benefits of increased concentration of CO_2 are limited because stomata closes and photosynthesis stops, if CO_2 conc is increased beyond 0.15%.

C_3 and C_4 plants differ significantly in the way they respond to increasing CO_2 concentration. The increase in net CO_2 assimilation is more graduate in C_3 plants than in C_4 plants because they respond more to increases CO_2 concentration. This difference is due to the fact that Rubisco, the primary CO_2 acceptor in C_3 plants has a lower affinity for CO_2 than PEP carboxylase in C_4 plants. So this lower affinity of Rubisco means that it requires comparatively high CO_2 conc. to become saturated.

4. Temperature: CO_2 assimilation is a biochemical process involving enzymes which are temperature sensitive and increase in temp cause denaturation of protein, so activity of enzymes decreases. The photosynthetic rate decreases as temperature increases. Plants of tropics have generally higher optima than plants of temperate region. In general optimum temperature for photosynthesis are similar to the day time temperature at which the plant normally grow except that in cold environment. The optima are usually higher than the prevailing air temperature.

C_4 plants generally have higher temperature optima than C_3 plants because of higher rate of photorespiration in C_3 plants. Increase in temp. usually increases in photosynthetic rate until enzymes denaturation and photosystem destruction begins. However, respiratory Co_2 loss also increases with increase in temp and simultaneously photorespiration also increases.

5. Day length: When all factors are equal and within optional limits plants are benefited from a longer photo period or day length due to increase in photosynthesis so this strategy is also used sometimes to provide supplementary light intensity from artificial source.

6. Leaf age: Leaf's ability to photosynthesize increases with age for certain times and then it begins to decreases even before maturity. Old senescene leaves become yellow and are unable to photosynthesize because of chlorophyll breakdown and loss of functional chloroplast.

Carbohydrate translocation

An internal control of photosynthesis is done by the rate of translocation from source leaves to various sinks. Removal of strong sinks inhibit photosynthesis after a few days especially in adjescent leaves which normally translocate CHO to their organs. Severe infection of leaves by pathogens also inhibit photosynthesis. Build up of starch grain in chloroplast press thylakoids together thereby preventing the light reaching from the thylakoids.

CAM or Crassulaean Acid Metabolisms

CO_2 is generally assimilated by this process in succulents such as cacti and pineapple etc. In this process during CO_2 react with PEP to form oxaloacetate which is reduced to malate by malate dehydragenase. Malate is stored in vacuoles as maleic acid until break. During day light, maleic acid diffuses autovacoules and is decarboxylated. CO_2 released by decarboxylation is refixed by Rubisco into phosphoglyceraldehyde by calvin cycle for the formation of CHO.

Respiration

In this process, CHO are oxidized to release energy. The sequence of enzyme catalyzed reactions that results in oxidation of organic substances usually CHO to CO_2 and H_2O and production of energy in the form of ATP is termed as cellular respiration. Photosynthesis occur in chloroplast while the respiration occur in mitochondria. It is reverse of photosynthesis and usually requires O_2 to occur and therefore, it is called aerobic respiration. However, under certain conditions respiration may occur in O_2 deficient conditions is called anaerobic respiration. In crop production we usually ensure to avoid conditions that allow the plants to respire anaerobically. However, certain process like silage making the anaerobic respiration occur.

Respiration is generally of two types:

1. Maintenance respiration
2. Growth respiration

1. Maintenance respiration: Some of the energy from respiration is used for maintaining the plant and to support its functioning. This house keeping activity is called maintenance respiration and it does not directly impact plant biomass. The maintenance include replacement or repair of plant chemical constituents. So maintenance energy required by plants is equal to utilized to replace or repair plant constitutents and energy required to maintain concentration gradient because energy is required for the uptake of nutrients and maintenance of osmotic potential in cells. The amount of maintenance energy required depends on the amount of biomass and the composition of biomass. Since maintenance activities drain the energy and do not reduce biomass so it is important in agricultural product that maintenance energy is minimized. The amount of energy and materials needed for maintenance is affected by several factors like plant size, age, nitrogen content of existing substances and climate of the region. Molecular activity increases as temperature increases. So there is increase in respiration with increase in temperature also as plant grow older they increase in size and accumulate large amount of tissue and materials associated with urging or maturity such as lignins, cellulose and starch. These materials generally require less maintenance but older plants have a large proportion of non-photosynthetic tissues that require maintenance. So this results in the larger proportion of new assimilates that are diverted for maintenance. This generally happens when crop attain full canopy cover. Plants at this stage photosynthesize at constant rate while biomass and maintenance activities continue to increases. The rate of maintenance respiration increases linearly with N content of the plant. Proteins and lipid hydrolyze and breakdown and similarly their resynthesize require energy in the forms of ATP. Such activities occur more in warmer climate than in the cooler climates.

2. Growth respiration: For crop productivity the respiration that directly effects biomass synthesize is called growth respiration. So catabolism provides the energy required for biosynthesis and for the formation of products required for growth. The amount of assimilates required for biosynthesis and growth activities depend on the desired economic product. Overall respiration reaction is summarized by the equation:

$$C_6H_{12}O_6 + 6O_2 \longrightarrow 6\,CO_2 + 6H_2O \;\; + \text{energy}$$

There are 3 stages in respiration.

I. Glycolysis
II. Kreb Cycle
III. Electron Transport

I. Glycolysis

Mean sugar splitting and involves breaking down of glucose molecules into a mole of pyruvic acid. It is a 10 step process that occurs in cytosol and in this process chemical bond energy from substrate is used to bond phosphate to ADP to form ATP. In glycolysis, no O_2 is used and no O_2 is evolved.

Functions of glycolysis

(a) Glycolysis involves oxidative breakdown of glucose into two molecules of three carbon compound pyruvic acid ($CH_3COCOOH$).

(b) The end product of this phase is pyruvic acid (2 molecules), NADH + H^+ (2 molecules) and ATP (2 molecules).

(c) Formation of molecules that can be used from this pathway to synthesis several other constituents for plant composition.

(d) Pyruvate produced is oxidized in mitochondria to yield large amount of ATP. Overall reaction of glycolysis is

$$C_6H_{12}O_6 + 2NAD^+ + 2ADP + 2P \longrightarrow 2\text{ Pyruvic acid} + 2NADH^+ + 2ATP + 2H^+$$

II. Kreb cycle or Tricarboxylic acid cycle

Pyruvic acid formed from the glycolysis enters in Kreb cycle in mitochondria. So Kreb cycle is also involved in substrate phosphorylation. So 7 out of 8 step occur in mitochondrial matrix while one step occur in mitochondrial membrane i.e. oxidation of succinic acid to fumaric acid.

III. Electron transport chain (ETC)

In respiration total 36 ATP are produced from one molecule of glucose. The ETC is the final step in respiration. It involves the transfer of electron generated in glycolysis and Kreb cycle downfall the electron gradient and energy released is used to form ATP from ADP in the form of oxidative phosphorylation because in this a series of oxidation- reduction reaction of the ETC uses energy from NADH or NADPH to phosphorylate ADP to ATP is required only at end of ETC as an electron acceptor.

Factors effecting respiration

Many environmental factors effect the rate of respiration and its importance in plant maintenance and growth.

(*i*) Respiration depends on the available substrate i.e glucose and that will be utilized for maintenance and growth. Plant with low sugar respire at low rate. Respiration rate of leaves are often higher just after sun set when sugar level are high then just before sunrise when sugar level are low. Shaded leaves usually respire slow than the leaves exposed to high light level. The differences in sugar content of the shaded leaves is responsible for low respire rates in those leaves

(*ii*) O_2 supply also influences respiration but magnitude of its influence differ greatly with plant spp. and even in different organs of plant, diffusion of oxygen through specialized/tissue intercellular spaces from leaves to root is important in providing O_2 needed for root respiration in water logged soils.

Flooding for long periods is toxic to all mesophytes because it creates anaerobic conditions and no O_2 is present near the roots. Among crop plants only rice can tolerate these O_2 deficient condition for long time. Insufficient conditions or reduced O_2 result in less absorption of mineral nutrient especially N, reduced permeability to H_2O, slower CHO translocation and accumulation of toxins around the roots.

(*iii*) **Temperature:** Respiration occurs in narrow temperature range (between 5° and 25°C). the optimum temperature for most of the plants is 35°C. However this limit varies with species (temperate, tropical). Rise in temperature upto optimum rises respiration rate upto 2–3 times (Q_{10} = 2,3). But beyond this limit respiration rate declines, as enzymes catalysing respiration get decreased.

(*iv*) **Type and age of plant:** Bacteria and fungi respire higher than plants because of small food reserve and similarly root tip and other meristematic region of plant with large protoplasmic content have high respiration rate because of high use of ATP & NADP for synthesis of cell wall material, membrane contents and nucleic acid. Similarly in seeds the changes in protoplasm due to dessication almost shuts down the metabolism as proper conditions are not present. Respiration remain high during rapid vegetative growth but it declines prior to flowering in mature plants much of respiration is by young leaves roots and by flowers.

(*v*) Changes in respiration also occur during the development of fruit and are high when fruit are young because the cells are actively dividing and growing.

Phasic development in crop production

Photoperodism

Photoperodism is the phenomenon of physiological change occuring in plants in response of relative length of day and night.

Crop developmental processes like rate of leaf production, flowering etc. are influenced by duration of sunlight. Most plants are influenced by relative length of day and night, especially for floral initiation. The duration of the night or of complete darkness is more important than the daylight. This effect of light on plants is known as Photoperodism. Depending on the length of photoperiod required for floral initiation, plants are classified as long day, short day and day neutral plants. Long day plants require comparatively long days (usually more than 14 hours) for floral initiation. They put forth more vegetative growth when days are short most of the temperate crops like wheat, barley, and oats fall under this category. In short day plants, flower initiation takes place when the days are short (less than 10 hours) or when dark period is long. Most of the tropical crops like rice, sorghum and maize are short day plants. The short day plants and long day plants are further divided into absolute and facultative types. Those that have an absolute requirement for a given day length (either short day or long day) will remain in vegetative phase for an indefinite period (months to year) as long as they are not exposed to floral inductive day length. Cocklebur is a classic qualitative short day plants initiate flowering even in long day but initiation of flowering is delayed by a few days to months. Most of the tropical crops are quantitative short plants. Day neutral plants do not require either long or short dark period. The crops that belong to this group are cotton, sunflower, buck wheat etc.

The rate of initiation of flowering depends on day length. Shorter the days, more rapid is initiation of flowering in short-day plants. Longer the days, more rapid is the initiation of flowering in long day plants. Relative length of day and night are important for photoperiodic response and light intensity has less influence. However, alterative light and low light intensity influence photoperiodism. Low light intensity following high intensity, is more effective on photoperiodism than high light intensity following low light intensity.

Significance of photoperiodism on crop production

Varietal difference: In the same crop, photoperiodic response differs with varieties. For example, long duration hybrids of maize give 15 per cent more yield with short length compared to short duration hybrids.

Optimum date of sowing: Source and sink should be in proper proportion to obtain high yield. A short- day plant, when exposed to short days earlier, produces less number of leaves and thus source is less. If it is subjected to short days at very late stage, more foliage is developed which may not be necessary and increases duration of the crop without increasing the yield. Thus, optimum date of sowing assumes importance.

Vernalization

Vernalization has been defined as a process which promotes flower initiation by a previous cold treatment. It has been successfully applied in a number of plants, mostly belonging to the families of Gramineae and Leguminoseae such as wheat, oats, rye, pea etc. Simply, vernalization is the acquisition or acceleration of ability to flower by a chilling treatment. The perception of stimulus of cold treatment is normally received by the shoot apex in contrast to leaves in photoperiodism. In seeds, the stimulus is perceived by the embryo.

Requirement of vernalization: To achieve vernalization several condition must be met.

I. *Low temperature* – Usually between 0 to 10°C and minimum up to -6°C.

II. *Active embryo or shoot* – vernalization does not occur in dry seeds. Seeds should have an active embryo.

III. *Aerobic respiration* – vernalization is an aerobic process. Therefore, oxygen is absolutely required. It also requires an active metabolism process has, however, a high temperature co-efficient.

IV. *Water* – for activation of embryo, the seeds are soaked in water for 22-24 hrs to bring the moisture content to 30-40 percent.

Devernalization: The effect of vernalization can be prevented by high temperature of 35°-40°, reduced water supply and unfavourable photoperiods. In general, devernalization can be carried out only immediately after vernalization treatment.

Application of vernalization: It has been observed that short days could substitute to some extent, for the chilling effect in promoting flowering. A practical importance of vernalization may be found in the sugar beet industry where only a slight cold exposure was require to induce flowering in some varieties. They were found to bolt and produce seed stacks in high percentage, particularly in winter planting with sufficiently low temperature. Also, it was found that seed production could still be successful in warm winter climates, provided certain precautions were taken. Early planting and dense stands favored profuse bolting and good seed production.

In wheat and barley, a short day can be substituted by low temperature to vernalize winter varieties. An early flowering type of seeds can be artificially produced directly from seeds of late flowering varieties by vernalization during ripening period. In spinach also, vernalization has helped to produced good seed through smooth and complete development of flower stalk. In *Brassica* crops, it was found that seed vernalization types showed dominance to the green plant vernalization types.

Thermoperiodism: The response of plants to rhythmic diurnal fluctuations in temperature is called thermoperiodism. The term thermoperiodicity or thermoperiodism designates the effects of alteration of temperature between the day and night periods upon reaction of plants. The temperatures of any place changes constantly from hour to hour in a diurnal cycle. Many plants have, in course of time, got adjusted to rhythmic diurnal cycle in temperature changes for several phonological events. Thus, regulation of phenology to periodic thermal changes is also the result of thermoperiodism.

Although, the seeds of *Daucus carota* and *Lolium perenne* germinate well under constant temperatures as under fluctuating temperatures, and those of most species such *Apium graveolens* germinate much better under certain alternating temperatures. It is interesting to note that for seed germination, the particular levels for alteration, are not so important as for other process, and still other conditions can eliminate the need for rhythmic temperature fluctuations.

Normal function in actively growing plants may be benefited by rhythmic changes in temperature that have rather distinctive optima at different stages of development. Stem elongation and fruiting of tomatoes are distinctly better when the day temperature is 26.5°C and night temperature of 15-20°C than when the temperature is uniform at 26.5°C. Similar response has been observed in grasses. Also the opening of flowers and the development of hardiness proceed most normally under alternating temperatures. Germination, establishment and development to maturity may be restricted to favourable temperatures, thus, diurnal range of temperatures may be a factor of considerable importance in controlling plant distribution and helping to divide areas into zones of main group of plants depending upon their climatic requirements. Thus, the most favourable areas for the production of various field and horticultural crops could be determined. Different plants vary in their tempera-ture requirements, accordingly they have been classified as eurythermic and stenothermic plants.

I. **Eurythermic plants:** Most of the land plants thrive in quite a wide range of temperature from almost freezing to about 55°C and these are called eurythermic plants. But their optimum range is not so broad and show best growth between 20°-30°C.

II. **Stenothermic plants:** Aquatic environments usually show less fluctuations in temperature and, hence, many aquatic plants thrive in a narrow range of temperatures and these are called stenothermic plants.

7

CROP ECOLOGY

Agricultural ecology has a strong theoretical basis, yet is eminently practical. It deals with processes in nature and relates these processes to man in intimate fashion. It recognizes the ties that exist to economics, politics and other fields of human activities. Man's agricultural activities have become the dominant ecological force over nearly 1/3rd of the land area of the earth. Both the intentional and inadvertent consequences of actions of man, benefit and damage the food production activities upon which man depends absolutely. Imbalance between population and resources also create many problems.

Man has ability to exploit the environment i.e. a variety of food supply and production systems were developed. Careful examination of the ecological basis for long term success of some systems and failure of others should help us to develop ecologically sound approaches to agriculture. Attempts to achieve a revolutionary increase in food production through the expansion of mechanized agriculture, without a full awareness of ecological and social repercussions are premature decisions. Modern agriculture technology is not time tested, even in areas where it has led to greatly improved yields with few apparent ill effects. It can be stated with the reasonable certainty that the intensive mechanized agriculture is causing some degree of long term ecological damage. So, one should not look blindly to current agricultural technology for a quick solution to the problems of over population and under production. Our objective must be to achieve an ecologically sound approach with the potential of meeting long range needs of a stabilized human population. Intensive cultivation of soil, disruption of decomposer food chains, depletion of organic matter in the soil, exhaustion of native nutrients pool, exposure of surface to the sun and precipitation have led to the severe deterioration of the soil structure and fertility. Synchronized cultivation of single, genetically homogeneous crop species over large areas has increased the vulnerability of crops to outbreak in insect, pest and diseases. All these in turn have led to the massive use of pesticides on some crops which had encouraged rapid resurgence of pests and outbreak of secondary pests and even resistance in them. These have increased instability within the cropping systems in addition to the problem of resource use and supply. So, total relationship between man and systems of his food and fibre is needed to be known carefully. Thus, we should recognize agriculture system as very important ecosystem and must apply ecologically sound agricultural techniques to increase the production without destroying in long run the agricultural lands and damaging global economy.

CROP ECOLOGY AND PRINCIPLES

Ecological principles

I. Enhance recycling of biomass and optimizing nutrient availability and balancing nutrient flow.

II. Securing favourable soil conditions for the plant growth particularly by managing organic matter and enhancing soil biotic activity.

III. Minimizing losses of carbon and energy due to flows of solar radiation, air and water by way of microclimate management, water harvesting and soil management through increased soil cover.

IV. Species and genetic diversification of the agroecosystem in time and space.

V. Enhance beneficial biological interactions and synergisms among agrobiodiversity components. Thus, resulting in the promotion of the key ecological processes and mechanisms.

Ecological processes

I. Strengthen the immune system i.e. proper functioning of natural pest control.

II. Decreased toxicity through elimination of agrochemicals.

III. Optimize metabolic function i.e. organic matter decomposition and nutrient cycling.

IV. Balance regulatory systems i.e. nutrient cycles, water balance, energy flow, population regulation etc.

V. Enhance conservation and regeneration of soil-water resources and biodiversity.

VI. Increase and sustain long term productivity.

Mechanisms

I. Increase of plant species and genetic diversity in time and space.

II. Enhancement of functional biodiversity i.e. natural enemies, antagonists etc.

III. Enhancement of soil organic matter and biological activity.

IV. Increase of soil cover and crop competition ability.

V. Elimination of toxic inputs and residues.

Basic concepts of ecology

I. The organisms and the environment reacting against each other form a complex unit which should be studied as a whole.

II. Organism are elastic and get adapted in time and space to a range of environmental variations.

III. Organism living together develop an economy of their own and organize into communities.

IV. The environment is a complex of interrelated factors with temporal and spatial variations.

V. Plant activities such as dispersal, growth, competition, reproduction, death and decay lead to development of new vegetation and change in environmental characteristics. It leads to succession of plant communities.

VI. Plant succession leads to development of a climax community which is relatively stable and in equilibrium with the environment.

Crop ecology: The concept of crop ecology date back to when agriculture first began more than 10000 years ago. Since, early humans first started to settle in one area and actively identified plants as economical crops. They have paid close attention to organisms within their systems, how they interact and how they relate to their environment. In short, they have become familiar with ecology of system. There are as many definitions of ecology as per different ecologist.

Odum (1971) defined ecology as the study of structure and function of nature. Krebs (1985) defined ecology as the scientific study of the interactions that determine the distribution and abundance of organisms. Ehrlich and Roughgarden (1987) defined ecology as the study of relationship between the organisms and their physical and biological environment.

Agricultural ecology or agroecology is the ecology of agricultural production systems and natural resources required to sustain them. We can define it as the study of interactions among organisms in an agricultural setting and the interactions between these organisms and their environment. Agroecology includes all of the ecological components commonly studied within the natural systems while it may be convenient to examine these components within the well defined boundaries of agriculture field, the setting is not limited to single field or a portion of field. Thus, crop ecology deals with the crops as community and agroecology is the similar and same science while the plant ecology deals with only single individual plant. Depending upon the questions involved, the context may range from only small, even microscopic portion of a field to a much larger scale such as whole farm, region or world.

The agricultural systems as well as other kind of ecological systems are made up of both biotic and abiotic components. The biotic factors are biological components of system or the living organisms while abiotic factors are non-living or physical components and include such factors as energy, water, climate and nutrients. Both biotic and abiotic factors have significant effects on plant growth and productivity within agriculture systems. We can study agriculture systems from many different levels of organization. We can start at individual level i.e. single plant of maize or single animal may be cow and looking at how an individual maize plant is adapted to a region. We may look at the physiology of plant to determine that it grows best in the full sun. We can examine the molecular structure of plant to determine that it will have resistance to a particular pathogen. While groups of individuals of the same species make up a population i.e. maize field or herd of cows. In an agriculture system, we generally speak of population of maize plant. If we study system at this level, we might look at the questions concerning competition among maize plants. All the populations in a given area are referred to as a community which includes many different type of organisms. Community in a maize field not only include maize plants but also all weeds in the field, insects that feed on plants, the fungi below ground, soil nematodes, earthworms and all other living organisms.

Most crop ecological research is conducted to some degree at the ecosystem level and agricultural ecosystems are referred to as agroecosystems. Agroecosystems can be studied from many different view points and within the variety of disciplines. The abiotic factors can be studied within the field of Chemistry, Aghrometeorology, Soil science and Hydrology while biotic components can be studied within the disciplines of Entomology, Agronomy especially weed science, Plant pathology, Plant physiology and Horticulture. Of course, the system as a whole must be examined from Economics, Sociology and Political science point of view also which are important in managed system since choices that growers make are always driven by feasibility and economics profitability. Management is one of the main characteristics of an agroecosystem that distinguishes it from natural system. Natural systems, while almost always affected directly or indirectly by human activity are not often managed by humans where as agriculture systems are directed by human activity and choice. Humans decide what plants to grow in a field, whether or not to spray insecticide, how much water to be applied through irrigation, whether or not to add nutrients etc. Though many choices depend upon nature are mostly influenced by economical and political factors.

Agroecosystems and natural ecosystems also differ from each other in regard to their boundaries. The boundaries of natural systems tend to blend together without well defined edges. For example forests blend into shrub lands which further blend into grasslands. In this way, the borders of each system tend to blend with borders of adjacent systems but agricultural boundaries are often economically defined and have very distinct borders. The end of the managed area is usually considered to be the end of agroecosystem, even though the natural systems around it may be source of beneficial insects, weed seeds and vertebrate pests. Due to these reasons, hedge rows and planned borders are often included in agroecosystem. However, the regulation system within them is the most important in the natural systems.

Difference between natural ecosystem and agroecosystem

S.N.	Natural Ecosystem	Agroecosystem
1.	There is more biotic diversity and less synchronization in growth.	Diversity is less, more synchronization in growth and maturity owing to monocultures.
2.	Trophic structure is complex	Trophic structure is simple.
3.	Less vulnerable to catastrophic changes	More vulnerable to catastrophic changes
4.	Less influenced by weather changes	More influenced by weather
5.	Pest and disease outbreak are uncommon, all populations being in balance.	Pest and disease outbreaks are of common occurrence.

6.	They are self perpetuating and more stable.	They are controlled by man through artificial practices and are less stable.
7.	They are continuous in space and time and vegetation is naturally selected.	They are discontinuous in space and time and vegetation is selected by man.
8.	They are of long duration, usually 50 years and are of permanent nature.	They are of short duration, many times only of 3-4 months or a year or so and are of temporary nature.
9.	Natural balance exists	Natural balance does not exist
10.	Moderately productive	Highly productive
11.	Herbivore pressure – more	Herbivore pressure – less
12.	They are more mature from succession point of view	They are not allowed to mature. Succession is hated by man.

HISTORICAL EVOLUTION

Early humans lived in hunter-gatherer societies that moved across a landscape in search of food and water. Their demands were low. They knew their land well and were well adapted to environment. They depleted food resources at one place and moved on to other, so they were not good conservationists. However, their nomadic existence prevented the complete destruction of ecology in a local area. As civilization grew, division of labour began and cultivation of plants started. The population also increased and its impact on the natural environment also increased which was felt on forests, grasslands and pastures. As a result degradation and desertification was caused. Similarly over cultivation caused soil erosion and land scarring. Improper use and mismanagement of irrigation caused salinization in arid region and depletion of regional ground water. As agricultural societies developed, environmental destruction became common due to more land area and less population and no body cared for conservation. With increase in population, pollution and effects of climatic change, the environmental destruction related to agricultural systems has become very serious. With low population and abundant resources, sustainability of ecosystem is not affected much by agriculture. But as population increased, productivity is forced to increase beyond the sustainable limits. To increase productivity, monoculture pest control, fertilizers and irrigation have played role and also had their disruptive influences on the agroecosystem.

Diversification can lead to stability and sustainability not only on small family owned farms but also on larger farms that are production oriented and directed by economics. Diversification can also increase population of natural enemies of pests, allow nutrients to be used more efficiently, allow recycling of nutrients and keep the soil covered to prevent erosion. Multiple crop production also leads to stability in some agricultural systems. Integrating livestock and crop production on the same farm closes the nutrient loop, decreasing the need for fertilizers and eliminating problems with pollution. Unfortunately we are subsidizing undiversified cropping systems, highly productive as monocultures e.g. rice, wheat which is away from highly diversified feed the family systems. Thus, the study of crop ecology is very important. Conventional agricultural systems do not mimic natural systems and must be managed carefully with expensive inputs to maintain ecosystem health and sustainability. Natural systems are more stable as nutrients are recycled, natural enemies prey on pests, nutrients are harvested efficiently by a variety of plants with slightly different requirements. The soil organisms enable decomposition to occur, providing a wealth of nutrients and organic matter to the system. If an agroecosystem is designed to mimic a natural ecosystem, we can minimize expensive inputs of fossil fuels and other limited resources. The challenge with such a system is to maintain a level of agricultural production efficient for an expanding human popu-lation with still conserving resources. However, agroecosystems are most sustainable if they function as natural systems.

Crop ecology as a field of study looks at how the agroecosystem functions naturally and examines how adjustments can be made to the agroecosystem without destroying other facts of the system.

The evolution of agricultural ecosystem passed through the different ages like :

1. Palaeolithic age or old stone age

In this about 5 lacs years ago, man learnt the use of fire and natural stone tools like hand axe, chopper, scrapper, knife and some bone made harpoons, knives and flake implements. He was a food gatherer, hunter and fisherman. There was no evidence of agriculture and domestication of animals.

2. Mesolithic age or middle stone age

In the beginning of this, dog was domesticated to help in hunting and after dog it was goat to be tamed. There was wide use of small tools like spear heads, arrow heads and sledge. There is evidence that he harvested food grains from wild plants or he practiced limited shifting cultivation. So, from food gatherer, he became very gradually food producer. In this age, agriculture was first started in Thailand and well watered valleys on the borders of Arabian and Iranian deserts which is known as fertile crescent where sheep, goat, pig and cattle were reared while wheat and barley were first crops to be grown. After this, all progress moved to Palestine to North Syria, Iraq and Iran. Thus, domestication of animals and cereals spread from Palestine to Egypt during 5000 B.C. The change over from hunting and food gathering to farming and stock breeding took place in India at a very later stage.

3. Neolithic age or the new stone age

In this age, man kept large herds of cattle for food and he tended plants and began to live settled life in the permanent pit dwellings, mud houses with wooden posts and thatched roofs near the cultivated fields, mostly near the bank of rivers and thus developed villages and small towns. In this way, organized social life started in this Neolithic culture which started in Asia and spread much later to Europe. But first of all in Europe, the lake dwellings were found in Switzerland.

Domestication of most of the animals and plants started in 1500 B.C. The practice, which combines agriculture with the raising of animals is known as mixed farming, thus, came into being. It was between 10000 and 20000 years back that cultivation of plants started at the following 9 regions.

(a) The plateau of Central America – Bananas.

(b) North east Africa – Barley, millets, wheat and coffee.

(c) North east Asia – Beans, rice, peas and tea.

(d) Southwest Asia – Barley, wheat and millets.

(e) The Andean region of South America – Corn, potatoes, sweet potatoes, tobacco and pineapple.

(f) South-Central parts of North America – Sunflower.

(g) South-Central Europe – Oats and rye.

(h) South-East Asia – Rice, sugarcane and millets.

(i) The Mediterranean region of Africa and Asia – Fig and olive.

Time being an environmental dimension acted throughout through historical evolution in various phases to modern agriculture. From stone age to new stone age was brought about by time. Environment played its role through various tolerance theories or otherwise from introductions, domestications and cultivation of plants to modern high yielding crops. The various tolerance theories helped plants in their distribution. Thus, plant distribution and further survival from original centre to various regions of the world occurred subject to tolerance range of the species as well as various physiological and ecological limits besides the means of transportation and harmony with the environment. The brief account of such theories, factors and limits has been attempted.

Tolerance range theory: It is given by Livington and Shrev (1831) It deals with physiological limits and tells about compensation or substitution of factors. For example, the aquatic plants which cannot survive in the absence of water are seen growing successfully on the boundaries of ponds or lakes due to more moist and humid conditions near the water bodies.

Biotic potential or biological potential or total capacity of an organism can be judged only if an organism is provided with adequate set of environment. Vegetation is a good indicator of environment because plant is a product of environment and its combination with the gene i.e. interaction of environment x gene has given rise to present day plants and man helps to distribute and introduce these plants. An organism may be growing successfully in a particular month of a year. The physiological limits suggested that for every vital function, there is a maximum and minimum zero point with respect to any conditioning factor beyond which the function ceases. Further for every distinct climatic area, there appears to be a corresponding type of vegetation and this principle is probably of primary importance in study of plant distribution. In this way, physiological limits are important to describe the limiting factors.

Odum (1959) combined the several theories with the impression that under natural conditions, plants appear to be controlled mainly by several conditions and out of them, 3 kind of forces are very important :

I. Quantity and variability of materials for which there are minimal requirements. It is synonymous to limiting factors.

II. Physical factors which are critical. It is also related to limiting factors but not exactly.

III. Limits of tolerance of plants themselves to these above 1 and 2 conditions and other factors of the environment which are related.

Trigger factors: Trigger is a thing which explores on a little push and sets into motion a chain of reactions. When a limiting factor is removed, then it acts as a trigger factor. For example, drought or limiting moisture is a limiting factor. Water is provided through any means like canal, tubewell, rainfall etc. Vegetation will emerge at a very fast speed, even those types of vegetation which were not available earlier will be growing. In shortage of water, limited but selective vegetation and wildlife was available. Grasses would emerge first of all for animals and roots of grasses will improve the soil structure. The trees would come up to provide shade and then population will be shifted. So, water has changed so many things. For example, in Haryana maize, bajra, sorghum, gram and barley were grown in water deficient areas and eaten by the people. When Bhakhra canal provided the water, the wheat and rice replaced all other *rabi* and *Kharif* crops respectively.

Animal and human population is always thin under water deficiency period. Ecosystem is always in a balanced state but its delicate balance is disturbed due to drought, flood, pollutants etc. With the multiple cropping, first of all major nutrients, then micronutrients become deficient with more crops sown on soil with long cycle of many years. For example, Punjab soils are alluvial soils rich in nutrients with all the facilities of modern agriculture, so it became thickly populated state and soils are eroded of nutrient status due to multiple cropping and now the water table is going down. In this way, ecosystem is being disturbed as biomass of crops we are taking from the soil is not being returned into the soils.

Levels of distribution of plants: The existence of any organism any where is subject to the approval of local environment i.e. climate determines the crop adaptation where the plant will grow. Anyone species has two levels of distribution.

I. *Macro or geographic distribution*: It includes the wide tolerance range species that can grow upto generalized distribution boundary lines.

   The wide ranging species have 6 distribution types as Arctic-alpine, Temperate, Pan tropical, Broad endemic, Narrow endemic and Discontinuous

II. *Micro or ecological distribution*: It includes the narrow tolerance range species. For example, North facing slopes or edges have different species than south facing slopes. Also the, species especially vascular plants having no tolerance for cold weather are growing in tropics or sub-tropics.

Tolerance range: Tolerance range is that range of environmental conditions in which plant of a given species will grow if its seeds, spores or other propagules can get there. Tolerance range affects the plant distribution. If a plant is growing and reproducing in a certain location, it indicates that environment lies within the tolerance range of that plant. Tolerance ranges are based on physiological requirements.

Ecologic range: Ecologic range affects plant distribution as it is that range of environmental conditions in which plants of a species actually do grow.

Geographic range: Geographic range indicates the actual boundary limits of plant distribution because it is the geographic limit of ecologic range.

Potential geographic range: It is the geographic limit of possible environments open to species as determined by its tolerance range and assuming that it can get there by some means or the other.

Free transportation of many species is geographically denied, such species are known as restricted species which either had broad tolerance range or were geographically isolated from great areas of their potential geographic range. It caused great explosive migration of weedy vascular plants, fungi, bacteria, insects, birds and even mammals. These migrations and invasions by broadly tolerant species represent a serious threat to the existence of native but narrowly tolerant species. Man must control these uncontrolled migrations because they are not beneficial.

As a rule, crops are not profitable unless adapted to region grown but crops are grown even if they are not adapted or profitable in that region due to social or artificial environments.

Structure of species: The structure of species also affects the plant distribution. Individuals have some degree of ability to grow in different environments, even though in the adaptation process they may change their appearance and also the physiology within limits. These changes are phenotypic and not genotypic. The environment selects the successful individual from the local population adapted to that range of particular environment which is called environment fitness. A group of local population from the same type of environment and showing similar inherited morphological characters is known as ecotype where as ecological race is group of local population adapted to a given type of environment. Local populations and ecological races can be detected most easily by growing plants of given species from the various sources in a homogenous environment. This can be like a garden or a controlled environment room called phytotrone.

Paleoecological events and their relation with plant distribution : Distribution of most species is dynamic and constantly changing. A paleoecological event is any environmental change occurring in the prehistoric times which had an effect on living organisms. Thus, very marked effects are preserved in the fossil records, continental glaciation, rise of mountains, severe climatic changes, rise of land from sea and vice versa.

The presence of a species in a place indicates that an environment is suitable and species has been able to get and grow there.

The absence of a species from a place may be for one or more reasons:

I. The environment can be beyond the tolerance range.

II. Barriers may have prevented the seeds from getting there.

III. It may once have been there but was eliminated by paleoeco-logical event which also indicates how the climate had changed.

Holocoenotic principle of environment: As one limiting factor is removed, another takes place. If one factor is changed, it may cause shift in another environment component. For example, temperature of green house is increased by 10°C, then air holds more water vapour. The vapour pressure of liquid surface will increase, there will be more evaporation and transpiration rate, more absorption of soil moisture, reduced free water in the soil and air enters into the soil which increases dryness of soil. This chain of shifts shows the complexity of environment which is a complex of interacting factors. If one factor is changed, all others are changed eventually.

A German ecologist Karl Friederich (1927), described that the community – environment relationships are holocoenotic. There are no barriers between the factors of an environment and organism. Ecosystem reacts as a whole. One single factor cannot be controlled without affecting other factors in an ecosystem. So, any change is reflected throughout the ecosystem. This principle concerns the interactions in an environment, so, it is an environment centred.

Holocoenotic principal of environment (Billings 1952)

Tolerance theory: Every individual plant species exists and reproduces only under a definite range of climate and other conditions when plant migration also occurs. While climate pattern is changing due to tolerance limit, some species are suppressed while others will come up which are favoured by environment.

Tolerance theory ranges show that a species flourish under optimum ranges and the species may vanish/die as the range increases/decreases to the extremes.

Tolerance range		
Minimum	Optimum	Maximum
Absent	Flourish	Absent

Shelford (1913) had following view points:

- Organism with wide ranges of tolerance for all factors of environment is likely to be widely distributed. The stenothermal or oligothermal – those plants like gram that tolerate, survive and distribute in narrow limit and Eurythermal or polythermal – that would tolerate wide range like wheat is distributed widely.
- Organism may have wide range of tolerance for one factor and narrow range for another.
- When conditions are not optimum for one factor, the limits of tolerance may be reduced with respect to other.
- The period of reproduction is more critical stage when all the environmental factors are not within the favourable range.

Good (1931) had the opinion that morphologically similar species may show wider differences in tolerance limits while the species with similar tolerance may show little morphological similarity and relative distribution of a species with similar tolerance is finally determined by result of competition between them.

As per Mason (1936) extremes of climatic conditions are more significant than their means. Both dispersal and establishment are essential to plant migration and tolerance theory should emphasize factor – function relationship.

Physical factors: As a general rule, the crops are not profitable unless they are adapted to the region in which they are grown. If the average farmer is to raise crops successfully, crop must be productive and easy to grow under existing conditions. Normally, three groups of environmental factors affect the crop growth and production i.e., social or cultural, physical and biotic environments. Physical environment is created by nature such as physiography, climate and soil. Climate largely determines the type of vegetation that grows naturally in any part of the world and kinds of agriculture production that are possible. However, the interaction between the plant or crop and the physical environment which affect the plant growth may be studied through temperature, light, atmospheric and root environment.

All the factors of environment are interrelated. The knowledge of the effects of climatic factors including geography and elevation upon plant responses is called bioclimatics.

The bioclimatic law deals with the changes in the flowering time of same species of plant as affected by latitude, longitude and altitude and states that

- For each degree of latitude, north and south of equator, flowering is retarded by 4 calendar days.
- For each 5 degree of longitude, from east to west on land areas, flowering is advanced by 4 calendar days.
- For each 400 feet increase in altitude, flowering is retarded by 4 calendar days.

PHYSICAL ENVIRONMENTS FOR PLANT GROWTH

Besides the above, climate, soils, vegetation, diversity, adaptations, food sources, human effects etc. are influencing various bioclimatic zones. The physical environment has been discussed here (Dhaliwal and Kler, 1993).

1. Temperature

There are 3 cardinal growth temperatures for each species i.e. minimum, optimum and maximum. Crops are distributed as per climatic/ temperature zones. Temperature provides energy in the form of heat for various plant functions and processes like respiration. Temperature is the index of heat energy available to plants. It is the final temperature of plant with sub-normal temperature, the maturity of the plant is delayed. In some regions like polar and tundra lands, low temperature is a major limiting factor in crop production as a result of snow and ice. The response of plant to rhythmic diurnal fluctuation in temperature is called thermoperiodism. The growth of crop plant is restricted to a temperature between 10 to 40°C. High temperature combined with high humidity favours plant diseases. High temperature also affects plant metabolism and ripening of fruits. At high temperature, dessication of plants occurs and thus maturity of crops is hastened. High temperature and longer frost free season near the equator cause greater diversity of crops. For every increase of temperature by 10°C, respiration rate increases 2 to 4 times. Thus, an important part of organic matter produced in day is consumed even upto 50 per cent in night when nights are long and warm. The chief cause of low yields in the tropics is high night temperature. When temperature becomes extremely high, sun scald takes place due to death of cambium called high temperature injury. The optimum range of temperature for most crops is 16° to 27°C and a few crops' seeds germinate at below 5°C. Abnormal temperature during growth of a crop adversely affects the rate of photosynthesis, cell wall permeability, protein synthesis, enzyme activity, absorption of nutrients, respiration and transpiration.

2. Light

Light affects development of crop plants by affecting their structure and time required for certain species to produce seed. The climatic requirements for crops are more important than the soil. As per availability of climatic factors, various crops are grown and are distributed. Light affects crop growth and production through its use in photosynthesis and photoperiodic reactions. The amount of incoming light which is suitable and can be utilized by the crop for photosynthesis is important. The amount of light determines theoretical upper limit to crop productivity but crop plant is less perfect in utilizing sunlight. Light also affects the absorption of CO_2, vegetative shapes and production of chlorophyll, leaves and flowers. In continued darkness, etiolation condition is found in plants, xylem is poorly developed and the mechanical tissue becomes almost absent. However, the plants grown in strong sunlight remain short. The effect of length of the day light period (photoperiod) on development and growth of plants, flowers and fruits is called photoperiodm. In many plants flowering will not occur, if the light conditions (duration of day length) are not optimum. Short day, long day and intermediate or photoneutral plants will bear flowers only if their light requirements are according to their needs. In general, not the light energy but its distribution in crop canopy becomes a limiting factor in photosynthetic process. When successive layers absorb and transmit just the right amount of light that would be the most efficient plant canopy.

3. Atmospheric environment

Photosynthesis involves the photolytic splitting of water by the reduction of CO_2 from the atmosphere. The availability of CO_2 affects the rate of photosynthesis and the utilization of incoming energy. The average CO_2 in the atmosphere is about 0.03 per cent and constitute a major limiting factor to photosynthesis of crops. Besides these, wind increases evaporation of moisture from the soil and amount of water used by plant and is often responsible for distributing organisms that cause diseases. Wind may be beneficial in carrying rain and spreading pollen. Air movement and wind affect the distribution of CO_2 in crop canopy and thus influence photosynthesis. Low velocities of air movement would be sufficient to ensure adequate air renewal at the leaf surface in all but the very densest crops but damage caused to plants by strong winds is a considerable factor in increasing crop risk as leaf damage, crown distribution, lodging etc. Saturation deficit of air also imposes moisture stress on the plants and if high may cause death of plants.

4. Root environment

The root environment consists of soil of one kind or another which may be treated in various ways to improve or decrease its suitability for the establishment, growth and harvesting of crops. The suitability of a soil for crop production is very closely linked with the climatic factors of rainfall, nature of soil, water holding capacity and production. The soil must be deep and capable of holding sufficient water to grow crops successfully. The varieties/crops must be suitable for a particular soil. The tilth (structure, porocity, heaviness, etc.), nutrient content, nutrient holding ability, soil reaction and the nutrient availability, organic matter and cultural practices of the soil are important for penetration of plant roots. The nutrient uptake in soil by plant is very closely related to climate and rainfall. Under low moisture stress conditions, higher leaf area can be maintained in the plant canopy through nitrogen nutrition but potassium nutrition is more important in crop growth under moisture stress situation. Thus, the interaction with climatic and soil factors affecting moisture availability may cause the differences in varietal adaptation. Certain crops are adapted to certain type of soil and give best yield. For example small grains and grass crops grow well on heavier soils as clay or clay loam. Rice grows well on loamy soils. Corn prefers loam and silt loam. Potato and truck crops grow well on loam and sandy loam but truck crops prefer light soils. Thus, crops are grown on best soils for higher production.

Precipitation: We divide plants on the basis of this as :

(a) **Hydrophytes:** Ones having high water requirements. For example lowland type of rice.

(b) **Mesophytes:** Ones having medium water requirements. For example most of the crop plants like wheat, maize, cotton etc.

(c) **Xerophytes:** Ones having lowest water requirements. For example dwarf grain sorghum.

Water requirements: It is defined as the grams of water used per gram of dry matter produced. Crops grow well if their water requirement are met, otherwise yield will be less or plant will die.

Rainfall: According to rainfall, crop regions are :

Region	Annual rainfall (inches)
Arid	<24
Semiarid	25 to 49
Sub-humid	50 to 99
Humid	100 to 149
Wet	150 and above

As production, consumption and decomposition are important features of an ecosystem. Oxygen is taken in by oxidation process in respiration for the energy of metabolism while oxygen is given out by production process in photosynthesis, which results in build-up of oxygen in the atmosphere for the evolution and continuation of life for higher forms. Excess production is buried without respiration or decomposition resulting in the fossil fuels and more O_2 and less CO_2 in the atmosphere. These changes keep an ecosystem living and possible to survive.

In a disturbed ecosystem, burning of more fossil fuels have raised CO_2 in the atmosphere If this is not maintained, increased CO_2 will lead to higher temperature causing a chain affect of changes, resulting in an imbalance in the atmosphere and ecosystem. In an agroecosystem, there is a seasonal use of CO_2 through photosynthesis only when crops are growing in the field. Crop yield primarily depends on the photosynthetic production.

Requirements of photosynthetic production: There are some essentials for the basic process of photosynthesis as:

- Chlorophyll – For trapping light photons primarily.
- Visible light – Especially 400-490 nm (violet, blue), 580-700 nm (orange, red) which are the peaks of spectrum for photosynthesis. However, the visible light spectrum is 390-760 nm.
- CO_2 – For assimilation of energy.
- H_2O – As reductant.
- Oxidant – Fe^{2+}, Mg^{2+} in the form of P as electron acceptors.
- Protein synthesis – C, O, H, N, S from soil or water.
- Environment – Particular temperature, RH, moisture and other gases.
- Simple inorganic nutrients – About 17 essential elements due to autotrophic nature of green plant.
- Respiration – Utilizing food through metabolism

CROP YIELD – ITS PHYSIOLOGICAL LIMITS

Productivity in an ecosystem: Energy enters in an ecosystem at the rate at which primary producers can convert solar radiation into chemical energy. Basic or primary productivity is defined as rate at which energy is stored by photosynthetic activity of green plants in the form of organic substances can be used as food materials where as biological productivity means rate of production per day or per year i.e. amount of materials at the end of production season.

Productivity of an ecosystem also refers to its richness i.e. organisms present there. So, yield or productivity of crop is different from biological productivity. In biological productivity, the word rate is very important along with time i.e. gram or kcal/unit time/unit area.

There are four successive steps in the production process or these may be called as types of productivity :

I. Gross primary productivity (GPP) – Total rate of production or photosynthesis including respiration. NPP + CF = GPP; NPP is the net primary productivity and CF is the amount used in respiration.

II. Net primary productivity (NPP) – It includes only assimilation. Also called apparent photosynthesis or net assimilation as per hour per day, per week or per month, K Cal/m²/day. It is the rate of storage of organic matter in plant tissue, in excess of respiratory utilization.

GPP – C.F. = NPP

Where, CF is amount used in respiration

III. Net community productivity (NCP) – It is given as NCP = NPP – HC.

Where, HC is heterotrophic consumption.

IV. Secondary productivity (SP) – It is the rate of energy storage at the consumer level which may be herbivores feeding on vegetation. It can not be divided into gross and net because it involved only consumption and transformation.

Basic process of production: Photosynthesis is the basic process in production. The green plants plays the major and important role to do this process. Light energy is converted to chemical energy or bound energy of food or potential energy. This process of photosynthesis involves oxidation reduction reactions while the H_2O acts as a reductant.

$$2H_2O \longrightarrow 4H^+ + O_2$$

$$4H^+ + CO_2 \longrightarrow (CH_2O) + H_2O$$

$C_2O + H_2O$ in the presence of light energy and of enzyme system associated with chlorophyll produce glucose + O_2.

$$6\ CO_2 + 12\ H_2O \xrightarrow[\text{Chlorophyll enzyme}]{\text{Light}} C_6H_{12}O_6 + 6\ O_2 + 6\ H_2O$$

$$\text{or} \quad \underset{\text{1920 K Cal}}{6\ CO_2 + 6\ H_2O} \xrightarrow[\text{Chlorophyll}]{\text{Light}} \underset{\text{673 K Cal}}{C_6H_{12}O_6 + 6\ O_2}$$

The process of photosynthesis operates theoretically at an efficiency of 35 per cent because 673 k cal are evolved out of 1920 k cal for 6 CO_2 and one mole of CO_2 it should be 320 k cal. One quanta considered equal to 40 k cal and 8 quanta equal to 1 mole CO_2.

Standing biomass or standing crop biomass at a given time is not productivity because in biomass production, so many phases come and result in harvest phase leading to grains and straw or the yield or production at the end. Organisms are not accounted in this, so, it is only production or yield of a crop and not productivity of that system.

Productivity is the rate of production i.e. K Cal per unit time per unit area while production is yield i.e. grain and straw at the end at harvest. In ecology, the interest is focused on the rate of production i.e. productivity i.e. how much dry weight has been produced by a crop/tree/forest in a given period in an ecosystem in a unit time.

Methods to measure productivity

I. **Harvest method :** In this normally NCP is measured. Rate of production starts from zero at planting and reaches the maximum at harvest and if not harvested, it decreases.

II. **O_2 measurement** : In this, gross production is measured through the relation between the O_2 production and the food produced. Light and dark bottle method for both NCP and GPP, respectively is used.

III. **CO_2 method** : By IRGA (Infra-red gas analysis) photosynthesis is measured in terms of CO_2 uptake. Measure absorption of CO_2 on KOH column and estimate uptake of CO_2 if not done by IRGA. This method of CO_2 measurement is more practical than O_2 measurement.

IV. **The pH method** : In this, how much CO_2 is dissolved in the H_2O is known by measurement of pH. So, calibration curve is prepared between CO_2 and pH because pH is a function of CO_2 which is dissolved in water. Generally, CO_2 is decreased by photosynthesis and is increased by respiration.

V. **Disappearance of raw materials**: N and P accumulate in H_2O during winter season and rate of their use can be measured during spring season. This method also measures NCP.

VI. **Productivity determination with radioactive materials or biomass method**: C^{14} or P^{32} is used in this method. This method is used in the standing crop only where the organisms do not disappear shortly.

VII. **Chlorophyll method** : In this, rate of production per gram of chlorophyll i.e. g O_2 produced/hour/g of chlorophyll or g CO_2 assimilated/hour/g of chlorophyll is measured. Shade adapted plants or plant parts tend to have a higher concentration of chlorophyll than light adapted plants or plant parts. This property enables them to trap higher light photons. So, efficiency of light utilization is high in shaded system but the photosynthetic yield and assimilation rate are low. If assimilation rate and the available light are known, gross production (GPP) can be estimated.

Efficiency of light = Dry matter produced/light absorbed = g MJ^{-1}

Productivity in a standing crop or biomass method: Solar energy coming from the sun is received by the earth which is only 47% and rest 53% is absorbed within the atmosphere and only 1-5% is converted to gross photosynthesis under the most favourable conditions resulting in NPP and GPP. There are much losses at each trophic level, thus, net to gross productivity ratio varies from 50-80%. It means the food available to heterotrophs or man is very low. The relationship of total energy with GPP or NPP varies greatly with variant situations.

Situations	Percentage transfers of solar energy during a growing season				
	Total solar radiation	Absorbed by autotrophs	GPP	NPP	NPP less than GPP (%)
Maximum favourable conditions	100	50	5	4	20
Available favourable conditions	100	50	1	0.5	50
Average for biosphere	100	<50	0.2	0.1	50

It is clear from the Table that NPP is less by 20-50% as compared to GPP. However, the NPP is least for the biosphere.

Situation	Kcal/m²/day			Harvest index		
	Solar radia-tion	GPP	NPP	GPP/ Solar rad. (%)	NPP/ Solar Rad. (%)	NPP/ GPP (%)
Sugarcane (Hawaii islands)	4000	306	190	7.6	4.8	62
Irrigated maize (Israel)	6000	405	190	6.8	3.2	47
Sugarbeet (England)	2650	202	144	7.7	5.4	72

So, more respiration, high temperature and higher water stress led to low NPP/GPP percent i.e. 47% in Israel. Similarly, rice production is more in north India than south India because more energy is spent in respiration in south. Thus, it costs more to maintain plant structure in hot climates. Day length may be another reason.

Trophic level, food chain and food web: The chemical energy contained in the tissues of green plants is the primary source of energy used by all other organisms in the ecosystem. So, biotic components of any ecosystem are related to each other through feeding or trophic existence.

Green plants form the base of the trophic structure called *primary producers* (P). The plant tissue is fed upon by various kinds of herbivores called as *primary consumers* (C_1). Some of the compounds manufactured by plants are converted to animal tissue and eaten by carnivores (C_2) and many carnivores are eaten by higher level carnivores (C_3). Thus, each species of producers, herbivores, carnivores etc. can be assigned to a trophic level but is related to others in the system like a link in a chain called a *Food Chain* concept. Trophic levels are abstract and energy does not merely flow along a simple species chain. Instead the pathways for energy flow become intricately interwoven in a net like arrangemet referred to as a *Food Web.* This type of trophic relationship is called graging food chain or grazing food web. About half the total energy passes through the entire system. Some dead plant and animal matter (detritus) accumulates and is gradually broken down by detritus-feeding animals and decomposers. The organisms that consume and break down this dead non-living (detritus) material make a detritus food chain or detritus food web. Decomposers do not belong to specific trophic level.

How to increase NPP

- Make sure that the rate of production is high.
- Solar energy be trapped more and utilized fully.
- Less energy is spent on respiration.
- Phytic losses may be minimum through insects, diseases, rats, birds, animals, etc.
- Physical conditions should be more favourable for high yield.

If rainfall is less, give irrigation as energy subsidy but to run a tubewell, we need fossil fuels as energy subsidy. In dryland farming, ploughing and planking for moisture conservation are energy subsidies. Insecticides, herbicides, fungicides etc. are all energy subsidies. It means in agroecosystem, various energy subsidies are used to increase the crop yield. So cultivation, irrigation, fertilization and better diet in case of animals, genetic selection (i.e. better variety and seed), control measures for weed, insect, disease etc. and fossil fuels added from outside the system are all energy subsidies to increase the production.

Any energy source that reduces the cost of internal or self maintenance of the system and thereby increases the amount of other energy that can be converted to production or yield is called auxillary energy flow or energy subsidy. It is provided from outside energy source and goes with the system enhancing the NCP of an ecosystem, though with 10 times consumption of energy subsidies, 3 times more production is realized in USA as compared to Asia and in India also as per ICAR estimates, we can raise yields i.e. food grain production can be doubled even if 60% of the technologies developed by agriculture scientist can be transferred to farmers field. However, be careful because too much use of energy subsidies can create the problem of environment pollution, imbalance in an ecosystem and rapid depletion of resources. So, use them wisely as and when required as per the recommendations depending upon the prevailing conditions so that energy subsidy may not act as energy drain or energy stress.

Sometimes, energy subsidy can act as energy drain or stress. For e.g.

I. **Evapotranspiration:** It reduces energy in dry climates but subsidies it in humid climates.

II. **Ploughing of soil:** As leaching of nutrients occur depending upon the soil type, crop etc.

III. **Use of sewage water:** If sewage water is treated, it will act as energy subsidy and if untreated, it is harmful due to more concentration of some nutrient elements.

Limitations to productivity or limits of productivity: Sometimes, we think that productivity can be increased even upto 100% or more but generally it may not be possible because nature has set certain limits to it i.e. there are various limits to NPP and very little of the available solar energy is converted to chemical energy by the vegetation due to many reasons.

1. Efficiency of photosynthesis

The process of photosynthesis operates theoretically at 35% but practically the efficiency of photosynthesis is not more than 20%, so quantum efficiency is low. This limited photosynthetic efficiency is responsible for low actual yields as compared to those theoretically predicted.

$$6CO_2 + 6H_2O \xrightarrow{\text{Light, Chlorophyll}} C_6H_{12}O_6 + 6O_2$$

Practically, one mole of CO_2 need 520 kcal for its reduction and stores only 105 kcal.

2. Rate of photosynthesis

Light saturation and photosynthesis are related i.e. photosynthesis rate is affected by light intensity. Normally photosynthesis increases with increase in light intensity in many land plants but after sometime it becomes independent and then decreases. So, photosynthesis cannot be increased indefinitely.

3. LAI and amount of chlorophyll

There is a limit to amount of chlorophyll or leaves that can be distributed over a given surface called LAI. In terrestrial ecosystem, we take leaf area over ground area equal to LAI. Many crops reach their maximum photosynthesis at a LAI of 4 or 5 where the production is high. If LAI goes above, competition for light is such that overall production is reduced. Generally LAI of other vegetation may be greater than crop plants. However, in some tropical crops like castor or cassava it may go upto 20, which is a upper limit called upper natural limit for LAI.

4. Solar spectrum utilization

Utilization of light spectrum depends upon sensitivity of chlorophyll i.e. in some plants violet and mostly red and blue spectrum is used. Photosynthesis utilizes only those wavelengths of light spectrum which are sharply absorbed by chlorophyll and these are at 430 nm and 670 nm. Infrared radiations that account for major portion of solar heat are not absorbed by terrestrial plants, thus they save themselves against lethal temperature. So, chlorophyll is sensitive to only about half of the visible solar spectrum and we cannot change the solar radiation and solar spectrum. Though light spectrum is quite long but its utilization and contribution is very low than received on the earth, so is the low efficiency. Thus the net primary productivity is the result of all such limits.

5. Plant architecture

In the agroecosystem, crop plant is the main that influences the primary production with its characteristics like LAI, leaf density, leaf inclination i.e. horizontal or vertical leaves, light absorption, leaves resistance to gaseous exchange, rate of photosynthesis of individual leaves at light saturation, reflection, transmission, absorption of light in the canopy. For higher yield, ideal crop canopy should have straight, short, numerous leaves which should be oriented with an acute angle to the sun's rays during noon and the angle should increase with an increase in LAI from top of the crop down until bottom leaves are at the right angle. These ideal conductions are not available, so they put limit to productivity.

6. Environment

Intensity of total radiation is diffused by wind, water vapours, dust particles, IFR radiations from sky, air temperature, humidity, soil temperature, etc. and all these are needed at optimum limits. Generally, the optimum conditions of an ideal environment are also lacking. Under high solar intensities, maximum production would occur in canopies having steeply inclined leaves (para heliotropism) with larger LAI. Similarly, maximum production would occur under low solar energy and horizontal leaves with small LAI (dia heliotropism).

Generally, the ecosystems in humid climates are more productive than the dry regions and those in warm areas are more productive than in cool areas. Generally, the ecosystems with a substantial inflow of water, nutrients and chemical energy are more productive than the ecosystems with low inflow of these materials.

Flow of energy in an ecosystem is governed by I and II law of thermodynamics.

First law states that energy can be converted from one form to another e.g.

solar energy → stored chemical energy in green plant → tissues heat energy by herbivore metabolism.

Second law states that no conversion of energy from one form to another is complete and a part of energy is always lost in the conversion process. There is always an upper limit on the number of possible energy transfers from one trophic level to next. Much energy is used to build/repair the tissues and dissipated as heat.

The efficiency of energy transfer varies from about 5% to 30%. The efficiency of energy flow depends upon biological composition and the physical environment. Flow of energy transfer works on Lindermann (1942) Law of Ten Percent. At each step in food chain when food energy is transferred about 90% is lost as heat and only 10% is passed on to next trophic level. e.g. 3000 k cal/m^2/day solar energy (incident energy) strikes on green plants and 1500 k cal/m^2/day (about 50% of incident light) is absorbed by green plants. 15 k cal (about 1% of absorbed light energy) is fixed in photosynthetic process as NPP at producers level and 1.5 k cal is passed on to herbivores and a similar degradation of 10% occurs and 0.15 kcal enters in carnivores and so on.

So, due to all these factors crop plants are less efficient in using incoming light energy in photosynthesis than supposed purely from the measurement of light absorbed by crop but sometimes high figures of dry matter production present an interesting point which could be attainable goal for agriculture scientist who is interested in improving crop productivity.

All ranks of an ecosystem are open and not closed. Energy must be taken in and energy must go out for the existence of an ecosystem. For example:

I. Sand culture – If we do sand culture, we have to apply nutrients from outside only then plant will accumulate photosynthates.

II. It takes oxygen from outside and gives CO_2 for its survival.

III. Energy and material exchange within the population, so a agroecosystem is interconnected.

Energy subsidies normally create modified environments as irrigation, shade, mulches, reflectors shelters, techniques to grow crops on suitable slopes etc.

AGRO-ECOLOGICAL MANIPULATIONS FOR EFFICIENT SOLAR ENERGY USE

Basically, agriculture is an energy conversion process. It is both a producer and consumer of energy. The photosynthetic reduction of carbon dioxide results in chemical storage of intrinsic energy in plants, which is greater than the energy supplied. This regenerated energy forms the basis of modern agriculture. It becomes essential to understand that light reaction in photosynthesis is very quick in plants and light cannot be stored for this process. So, it should be used immediately by the plants. In general, the upper layers of the crop canopy are light saturated or sometimes they get more light than required, while the lower layers of leaves may be saturated of light. For the proper benefit from incoming radiations, so that both the upper and lower leaves are not at loss, some agronomic manipulations are needed. Solar radiation interception depends upon.

1. Leaf area index

The horizontal type and vertical type of leaves behave differently in different crops. Similarly the optimum leaf area achievement is very difficult in relation to light interception. The optimum, LA, when the arrangement of leaves is in such a way that the parasitic efficiency of the crop as a whole is a minimum. It has been seen in wheat that the yield increases linearly upto LAI of 5 and after that decreases. But, as per the regression equation, 7.5 is an optimum LAI for wheat. However, 98% of the solar radiation interception indicates the possibilities of supporting further LAI with more erect leaves for increased productivity. So, the adjustment between LAI and the light interception must be there and besides these, extinction coefficient is also important. In case of winter maize with the highest LAI of 2.8 the PAR interception was 76% and extinction coefficient (k) was about 0.1, which indicates the possibility of supporting still higher LAI in this crop, for increased production. In Pigeonpea linear relationship between LAI and PAR interception was obtained indicates the solar radiations as a function of LAI. LAI of 3, produced the highest yield in Pigeonpea. In toria crop retained maximum number of leaves between 45 and 60 days of age. Significant positive correlation was observed between LAI and seed yield.

2. Orientation or Direction of rows

Crop sown in N-S direction may permit greater penetration of radiation than the E-W direction. Wheat, maize, sorghum, pearl millet, barley, maize + moong, raya

+ gram recorded higher grain yield when sown in N-S row orientation compared to E-W orientation. N-S rows enjoyed the better lightening than the E-W rows. PAR interception was also higher in N-S sown crops which resulted higher grain yield in different crops.

3. Bidirectional sowing

Earlier in the morning and late in the evening E-W rows admit more light, whereas during rest of the day, more light penetrates N-S rows. But, we can not rotate field according to such interception of light to get more light interception throughout the day. So, bidirectional sowing will help, being the rows on both sides, i.e. E-W and N-S directions. Thus, bidirection could intercept the solar radiation whole day and has given higher yields in various crops as compared to unidirectional sowing.

4. Population dynamics

It has been observed that changing plant population is an important tool to attain desired LAI for higher photosynthetic efficiency. So, obtaining uniform and optimum stand of crop is essential as it is prerequisite for getting higher yields. However, a population which is efficient in solar energy interception during early stage may become inefficient due to low crop growth with canopy development and increased mutual shading. Thus it becomes imperative that population dynamics for achieving higher efficiency through the crop growth. In some cases crop plasticity also is able to adjust a wide range variations in population. However, in general, more the LA, higher will be the photosynthesis i.e. more will be the photosynthetic efficiency. It has been observed that changing plant population is an important tool to attain desired LAI for higher parasitic efficiency. Hence, increasing seed rate under late sown conditions is the way to achieve more LA with more plant population. In chickpea, it has been seen that under late sown conditions 65 kg seed/ha resulted 7% more yield than normal seed rate of 45 kg/ha. Similarly 100 kg seed/ha in wheat gave maximum yield due to more LAI and light interception in closer row spacing of 15 cm as compared to wider row spacing of 22.5 cm.

5. Prismatic canopy

For the efficient use of solar energy and to avoid mutual shading a suitable canopy structure is needed. Efficiency of solar energy utilization can be increased by increasing the surface area by planting crop species in different canopy structural patterns. For example, prismatic canopy structure is obtained by sowing crops/crop varieties of different heights in a set sequence. Such heterogenomes mixtures may also be useful for creating diversity against certain diseases and may also help to reduce light reflection and increase transmittance with crop canopies. Moreover, flag leaf and earhead contribute a major portion of photosynthates that are translocated to grains in later stages of crop growth and any type of competition with ear heads leads to reduced yields. In wheat, prismatic canopy obtained by sowing one row each of four varieties gave significantly higher yield than the pure stand yield in flat canopies. U-shaped canopy formed through sowing five rows each of four varieties of different heights of wheat produced significantly higher grain yield than the pure stand yield due to efficient utilization of solar energy. Columnar canopy recorded significantly more grain yield than uneven and flat canopies through at par with U-shaped and columnar canopies. All the systematic canopies recorded higher PAR interception, relative crowding value and higher grain yield as compared to flat canopies. A cropping pattern of corn, rice, cassava and sorghum if arranged in rows for better mechanism approaches the light use of a tropical rain forest.

6. Planting on sunny slopes

Flow of solar energy could be properly directed in winter season to raise temperature. Sowing of crops on sunny slopes/ridges is aimed at manipulation of soil temperature and interception of solar radiation for better crop production, especially where low temperature is the limiting factor. Winter maize planted in southern slope of E-W ridges gave the highest grain yield. Planting on northern slope of the same ridge produced the lower yield. Wide differences of yield was due to difference in light penetration and soil temperature on two sides of the ridge. LAI and PAR interception were also considerably more on southern slope than on eastern slope with a linear relation with yield.

So, it may be concluded that agronomic manipulations help in consistent improvement in yield, with improvement in interception of solar radiation through modifying crop geometry. Wheat and moong sown with bidirectional sowing method gave higher yield than unidirectional sowing. Similarly, increased population of chickpea under late sown conditions significantly increased the yield. The direction of ridges/slopes for planting was found to be the most important factor for achieving high yield in winter maize. Some agrochemicals help to regulate the crop growth. TIBA, an anti auxin is known to produce triangular shaped canopy in a plant. Cycocel suppresses plant growth. Thus both these help penetration of PAR within crop canopy. Mutual shading may be reduced with detopping/defoliation which may help penetration of light as well as balance the source-sink relationship. The crop associations also help to use solar energy, water and nutrition in a better way to increase productivity. Farm-tree-crop associations also help to use natural resources like water and light for maximizing crop productivity.

AGRO-ECOLOGICAL REGIONS OF INDIA

An ecological region is an area of the earth's surface characterized by distinct ecological response to macro-climate as expressed by soils, vegetation, fauna and aquatic systems. Therefore, the agroecological region is the land unit carved out of agroclimatic zone when superimposed on land form and the kinds of soils and soil conditions that act as modifiers of climate and length of growing period. Agro-ecological region depends upon the whole environment/the ecological factors while the agroclimatic zone depends upon only climatic/the climatic factors.

An agroclimatic zone is a land unit in terms of major climates, suitable for a certain range of crops and cultivars. India has been divided into 15 agroclimatic zones based on physiography and climate as Eastern Himalayan Zones; Lower, Middle, Upper and Trans-Gangetic zones; Eastern, Central, Western and Southern Plateau and hills zones; East Gujarat and West Coastal Plains and Hills/Ghats zones; Western dry zone and Island zone. Then under NARP, these zones were divided into 120 Sub-zones based on rainfall, cropping pattern and administrative units. However, in agro-ecological region, due importance was given to edaphic factors by superimposing the bioclimatic (rainfall, temperature, vegetation and PET) cum length of growing period (rainfall, PET and soil storage) over the bare soil-scape.

Different types of ecosystems are:

1. Arid ecosystem

I. Western Himalayas, cold arid ecoregion with shallow skeletal soils and length of growing period (GP) < 90 days (A13E1).

II. Western Plain, Kutch Peninsula and part of Kathiawar Peninsula, hot arid eco-region, with desert and saline soils and GP <90 days (M9E1).

III. Deccan Plateau, hot arid ecoregion, with mixed red and black soils & GP <90 days (K6E1).

2. Semi-arid ecosystem

I. Northern Plain and central highlands including Aravallis, hot semi-arid eco-region, with alluvium derived soils and GP 90 to 150 days (N8D2)

II. Central (Malwa) Highlands, Gujarat Plains and Kathiawar Peninsula, hot semi-arid ecoregion, with medium and deep black soils and GP 90 to 150 days (I5D2)

III. Decca plateau, hot semi-arid ecoregion with shallow and medium (inclusion of deep) black soils and GP 90 to 150 days (K4D2)

IV. Deccan (Telangana) Plateau and Eastern Ghats, hot semi-arid ecoregion, with red and black soils and GP 90 to 150 days (K6D2)

V. Eastern Ghats (TN uplands) and Deccan (Karnataka) Plateau, hot semi-arid ecoregion with red loamy soils and GP 90 to 150 days (H1D2).

3. Subhumid ecosystem

I. Northern Plain, hot subhumid (dry) ecoregion, with alluvium-derived soils and GP 150 to 180 days (N8C3).

II. Central Highlands (Malwa & Bundelkhand) hot sub-humid (dry) ecoregion with medium and deep black soils and GP 150 to 180 days (to 210) days (I5C3)

III. Deccan Plateau and Centeral Highlands (Bundhelkhand), hot sub-humid ecoregion with mixed red and black soils and of 150 to 180 days (K6C3).

IV. Eastern Plateau (Chhatisgarh), hot subhumid ecoregion, with red and yellow soils and GP 150 to 180 (to 210) days (J3C3)

V. Eastern (Chhotanagpur) Plateau and Eastern Ghats, hot subhumid ecoregion, with red loamy and lateritic soils and GP 150 to 180 (to 210) days (J1C3).

VI. Eastern Plain, hot subhumid (moist) ecoregion, with alluvium derived soils and GP 180 to 210 days (O8C4).

VII. Western Himalayas, warm subhumid (to humid inclusion of perhumid) ecoregion with brown forest and podzolic soils and GP 180 to 210 (+) days (A15C4)

4. Humid perhumid ecosystem

I. Assam and Bengal Plains, hot subhumid (moist and inclusion Sub-humid) ecoregion, with alluvium-derived soils and GP > 210 days (Q8B5).

II. Eastern Himalayas, warm per humid ecoregion with brown and red hill soils and GP > 210 days (C11A5).

III. North-eastern Hills (Purvanchal), warm per humid ecoregion with red and lateritic soils and GP >210 days (D2A5).

5. Coastal ecosystem

I. Eastern Coastal Plains, hot subhumid to semi-arid ecoregion, with coastal alluvium – derived soils and GP 180 to 210 days (S7C4).

II. Western Ghats and Coastal Plains, hot humid-perhumid ecoregion with red, lateric and alluvium-derived soils and GP >210 days (E2A5).

6. Island ecosystem

Islands of Andaman-Nicobar and Lakshadweep, hot humid to perhumid island ecoregion, with red loamy and sandy soils and GP > 210 days (T1A5).

Photoperiodism: The difference in length of days and nights as related to latitude is a factor in the natural distribution of plants. Response of plants to relative length of day and night (light and dark period) is known as photoperiodism. The length of the day and night periods in the 24 hours cycle also regulates the leaf fall, flowering and other phonological processes in plants. Photoreceptor for the light wave-length which is effective in producing response in plants is phytochrome which functions above 550 nm (between 580 nm and 780 nm) and is found in all photosensitive organs. Phytochrome can be extracted from leaves.

Flowering hormone is named as florigen and it is believed that this flower inducing stimulus could be in the form of a chemical. Particular length of day required under normal conditions to initiate flowering is called photo-critical period or critical day length (CDL) which is about 15.6 hours in long day plants and 9 hours in short day plants. Generally, CDL can not exceed in short day plants but always exceed in long day plants but reverse is for critical dark period. Day length is a climatic factor and depends upon the latitude change from year to year at a particular location that does not change from year to year at a particular location. Day length is important for a farmer while introducing new varieties or changing the date of planting. Agronomists and plant breeders think that many morphological and developmental changes in plants are associated with day length. It has fixed value due to regular moment of earth. Minor variations only occur in day length due to clouds early in the morning or late in the evening.

Thermoperiodism: The rising and setting of the sun in each day set in motion complex rhythmic variations in several important environmental factors. As the sun rises, relative humidity decreases while light and temperature increase, and the direction of these changes reverses, sometimes in the afternoon. Many plants have become so adjusted to this regular diurnal sequence of events that they require lowered temperature at night and will not exhibit normal behaviour when grown in artificially uniform environments. The response of plants to rhythmic diurnal fluctuations in temperature is called thermoperiodism. The term thermoperiodicity or thermoperiodism designates the effects of alteration of temperature between the day and night periods upon the reaction of plants. Two forms of time/temperature combinations are known – daily thermoperiodism and seasonal thermoperiodism. The first corresponds to the day and night changes of temperature as they occur under natural conditions, and the latter to seasonal changes of temperature.

Applications of thermoperiodism : The physiological basis of thermoperiodism seems to rest upon the fact that separate and complementary processes like growth and photosynthesis go on under day and night conditions and these processes have different cardinal tempera-tures.

The temperature of any place changes constantly from hour to hour in a diurnal cycle. Many plants have, in course of time, got adjusted to rhythmic diurnal cycle in temperature changes for several phenological events. Thus, regulation of phenology to periodic thermal changes is also the result of thermoperiodism.

Although, the seeds of *Daucus carota, Phleum pretense* and *Lolium perenne* germinate well under constant temperature as under fluctuating temperatures, and those of most species such as a *Apium graveolens, Poa pratensis*, and *Dactylis glomerata* germinate much better under certain alternating temperatures. It is interesting to note that for seed germination, the particular levels for alteration, are not so important as for other processes, and still other conditions can eliminate the need for rhythmic temperature fluctuations.

Normal function in actively growing plants may be benefitted by rhythmic changes in temperature that have rather distinctive optima at small different stages of development. Cardinal temperatures differ for the same functions in different plants e.g. sorghum, wheat and opuntia growth. Similarly different functions of same plant may have different cardinal temperatures as respiration and photosynthesis. Also, various organs of the same plant may have different cardinal temperatures for the same functions e.g. root and shoot. Stem elongation and fruiting of tomatoes are distinctly better when the day temperature is 26.5°C and night temperature of 15-20°C than when the temperature is uniform at 26.5°C. Similar response has been observed in grasses. Also, the opening of flowers and the development of hardiness proceed most normally under alternating temperatures. Germination; establishment and development to maturity may be restricted to favourable temperatures, and thus, diurnal range of temperatures may be a factor of considerable importance in controlling plant distribution and helping to divide areas into zones of main groups of plant climates to predict further crop maturity dates. Thus, the most favourable areas for the production of various field and horticultural crops could be determined.

Different plants vary in their temperature requirements, accordingly they have been classified as eurythermic and stenothermic plants.

I. **Eurythermic plants:** Most of the land plants thrive in quite a wide range of temperature from almost freezing to about 55°C and these are called eurythermic plants. But their optimum range is not so broad and show best growth between 20-30°C.

II. **Stenothermic plants:** Aquatic environments usually show less fluctuations in temperature and hence, many aquatic plants thrive in a narrow range of temperature and these are called stenothermic plants.

Applications of photoperiodism: Photoperiod controls or affects several morpho-physiological responses which in turn affect crop yields. A number of important applications of the phenomenon of photoperiodic response are possible :

I. **Vegetative growth:** Light inhibits stem growth and promotes leaf expansion. Lettuce and radish gave a greater top to root ratio in short days, which is desirable in lettuce since the top is harvested, but undesirable in radish since the tap root is the economic organ. Growing season of a crop at different latitudes depends upon day length, and date of planting is decided accordingly. In the production of onions, long days promote bulb formation whereas short days promote flowering and seed production. This situation poses a problem, to the plant breeder who may wish to use germplasm from varieties not adapted to a particular latitude. Importance of knowledge of photoperiodic response is important at the time of introduction of new species or new varieties.

II. **Tuber, rhizome and bulb formation:** The vegetative reproduction of the potato is accomplished by the accumulation of carbohydrates, in the underground runners which thicken to form tubers. In some varieties, tubers form under short days and in others, tubers form only under long days and low temperature. Generally, potato is a qualitative short day plant for tuberization and qualitative long day plant for plant growth.

III. **Sex expression:** In cucurbits and other monoecious plants, abnormalities in sex ratios due to day length have been observed. In *Cucumis sativus*, long days increase the staminate flowers which decrease the fruit bearing flowers.

IV. **Fruit set:** Photoperiod sensitive varieties of rice need additional photoinductive cycles for the development of the flower primordia. Three varieties of soybeans flowered early under short days. Although they flowered under long days with prolonged exposure, short days were essential for fruit set since the flowers that developed under long days did not produce anthers with viable pollen. Photoperiod also influences seed germination in some plants, the formation of pigments like anthocyanin and the formation and size of root nodules in soybeans.

V. **Flower initiation and development:** Flower induction is most important response of the photoperiod. The flower, fruit or seed is most important structure in many agricultural crops because fruit or seed is harvested as an economic yield. This must develop fully to give high yields. In sugarcane, tobacco and forage plants, the reproductive development is delayed or prevented to get high yields. However, the control of reproduction is important and day length is usually the major environmental factor that controls flowering. The effect of day length on the flowering of plants was first shown in 1920 by Garner and Allard in tobacco.

Types of plants

1. Long day plants

Plants which develop and reproduce normally only when the photoperiod is greater than a critical minimum day length, are called long day plants. In such plants, flowering is promoted by long days and short nights. A long day plant will flower only when it has been exposed to a day length above the critical one, e.g., *Hyoscyamus niger* (Henbane). Critical day length for it is 11 hours. It will flower only when day length is more than 11 hours. Long day plants are of two types.

(i) **Qualitative or obligate long day plants:** These plants will flower only under long day conditions. They are absolute long day plants such as *Hyoscyamus niger,* cabbage, radish.

(ii) **Quantitative or facultative long day plants:** These plants will flower, preferably under long day conditions but they flower under short day conditions also, e.g., spring varieties of barley, wheat, oat.

2. Short day plants

Plants which flower only when they are exposed to a day length below the critical length of light, are called short day plants. For example critical day length for *Xanthium stramarium* is 15.5 hours and it will flower when day length is less than 15.5 hours. Short day plants are of two types.

(i) **Qualitative or obligate short day plants:** These will flower only under short day conditions, such as *Nicotiana tabacum, Glycine max, Xanthium stramarium.*

(ii) **Quantitative or facultative short day plants:** They will preferably flower under short day conditions, but they can also flower under long day conditions such as *Zea mays*, Millet, *Oryza sativa.*

In general, critical day length for long day plants is 13-16 hours and for short day plants is 11-12 hours. Long day plants are also called short night plants and if we interrupt dark period with a flash of light, they will not flower. Short day plants are also called long night plants and if we interrupt dark period with light, they will not flower.

3. Qualitative long day – short day plants

These plants will flower only when they are exposed to long day conditions followed by short day conditions. For example, *Cestrum nocturnum* will flower, when it is exposed to 15 long day cycles (critical day length of 11 hours), followed by 25 short day cycles (critical day length of 15 hours).

4. Qualitative short day - long day plants

Initially, these plants require short day conditions for floral initiation, after that they require long day cycles, for blossoming. Such plants are *Trifolium repens, Secale cereale*.

5. Day neutral plants

Those species in which phonological events are indifferent to length of light or dark period. They flower at a wide range of both temperature and photoperiodic conditions. They can flower comparatively under all photoperiodic situations. Tomato, pea, cotton, sweet potato are such plants.

6. Intermediate plants

These plants can flower only within a definite range of light hours, above and below which no flowering occurs as in case of some varieties of sugarcane (*Saccharum spontaneum*). These are near today neutral plants.

7. Amphiphotoperiodic plants

They behave just opposite to the intermediate plants. They can flower only at both, above and below an intermediate day length. These plants remain vegetative at intermediate day length. *Media elegans* is the example of this category.

Principles of plant distribution

Plant geographer Good, 1953 has given 6 principles in addition to evolution i.e. historical aspect

I. Primarily is the climatic (water, light, temperature etc.) controlled plant distribution.

II. Secondarily is the edaphic factor (soil) which controls plant distribution.

III. A great movement of flora has taken place in the past and is still continuing from one country to other.

IV. Plant migration by transport at dispersal phases is always taking place e.g. Akk (*Calotropis procera*), Pohli (*Carthymus oxycantha*) etc.

V. Variation in climate at higher latitudes in the history of angiosperms which cause distribution.

VI. Variation in relative distribution of area under sea and land where weathering play role since million/billion of years.

Plant distribution factors: It includes limiting factors, trigger factors, holocoenotic principle of environment and growth factors, social environment, physical environment and biogeochemical cycles.

Biogeochemical cycles: Energy does not circulate in the nature i.e. energy flow is not cyclic. It is used only once by an organism and is then converted into heat which is lost to the ecosystem. Thus, there is nothing like an energy cycle. In contrast to this, the materials i.e. chemical elements including all the essential elements of protoplasm like water, C, N etc. tend to circulate in the biosphere in its characteristic paths from environment to organisms and also back to environment. These more or less circular paths are known as biogeochemical cycles. The word 'Bio' means life or living and 'Geo' means soil.

Soil → Plant → Environment and back to soil and organism.

Life continues on the earth by inflow of sun's energy and only about 1% of solar radiation is converted into energy by photosynthesis. It is this energy which drives the mineral and hydrological cycles and also maintains the environmental temperature within the narrow range of tolerance by protoplasm. The movement of these elements and inorganic compounds that are essential to life can be conveniently designated as nutrient or chemical cycling which is a narrow term, for example, N or P cycling, whereas whole global cycle covering all the cycles like N, CO_2, O_2, H_2O, P, etc. is known as biogeochemical cycling.

Types of cyclic pools

Each cycle consists of two compartments or pools:

(a) **Reservoir pool:** This is a large, slow moving, generally non-biological component and slowly available pool. Nutrients are in large quantities but are not recycled and are released slowly into exchangeable pool. So, are not available immediately.

(b) **Exchangeable/cycling/available pool:** It is small but more active component. It is moving to and fro rapidly between the organisms and their environment. There is a slow flux between available and unavailable components.

BIOGEOCHEMICAL CYCLES

From the stand point of the biosphere as a whole, biogeochemical cycles fall into two basic groups:

Gaseous type: Reservoir is in the atmosphere or in the hydrosphere. For example CO_2 as gas available from carbonate and bicarbonate goes to air and ocean. The nutrients in air are quickly dissipated by air movement and the increased output is compensated by increased plant uptake as in case of CO_2 while the carbonate formation in the sea due to sea absorption is known as negative feedback. Thus, balance is maintained within the definite limits e.g. N 79% and CO_2 0.033%. So, these are more perfect cycles due to negative feedback.

Sedimentary type: Reservoir is in the earth's crust. For example P cycle. Phosphate from rocks is supplied to the soil. So, rocks are the source, hence called sedimentary type cycle. These are less perfect cycles e.g. P and Fe or imperfect cycles. These are more easily disrupted by local changes due to inactive and immobile form, though they are used for long time by crops, they by rains or floods go to streams to riverain and finally to sea. So, they cannot come back easily. Therefore it is less perfect cycle because its return or recycling mechanism is mainly biotic. In case of abiotic mechanism, it is out of cycle. Thus, the aim of conservation of natural resources is to make acyclic processes to more cyclic or imperfect to perfect ones.

Soluble and carboxylation cycles: Main reservoir of mineral elements is water because the mineral nutrients (e.g. P.) are water soluble but are not volatile gases. So, water cycling is very important. However, major abiotic reservoirs of C, H, O and N occur in atmosphere and are readily accessible to living things and the living organisms has limited access to sedimentary deposits. So, Deevey (1970) named them as soluble element cycles (e.g. P) and carboxylation (C, O, H and N) cycles.

The living organisms have limited access. However, major abiotic reservoirs of C, O, H and N occur in atmosphere and are readily accessible to living things. In short, pattern of movements of basic nutrients in biogeochemical cycles consists of organic phase (living organisms and their food chains) and a geologic phase (abiotic phase) depending upon the nature and basic properties of each element. The amount that is transferred from one nutrient pool to another pool per unit time is known as nutrient flux rate.

Radioactive carbon is used to see the unidirectional flow of energy in an ecosystem and relationship between available energy and decomposition which is essential to create balance in an ecosystem for its perfect development. In the biosphere (a large ecosystem) in its mature stage of development (climax) has all major biogeochemical cycles in a steady-state condition where out flow and inflow of energy is evenly balanced while a disturbed ecosystem (land slide) is out of balance.

Nitrogen cycle: The pattern of movement of N through an ecosystem differs in several respects from that of carbon. Main reservoir of N is like C in the atmosphere. Concentration of N in the atmosphere is higher than carbon but not available to most of the living organisms. There is distinct biotic subcycle for N but not in C that involves the biological break down of organic nitrogenous compounds to inorganic forms. Some gaseous N enters atmosphere through volcanic action which is balanced by loss through its incorporation into deep oceanic sediments. Atmospheric pool of N (79%) is virtually inexhaustible because molecular N is utilized by few specialized microorganisms. Most organisms obtained N in the form of soluble nitrate or ammonium ions. Some atmospheric N is converted by lightening-induced oxidation to nitrogen oxides that react with water to form nitrates. The carbon and nitrogen are gaseous type perfect cycle, and contain most of the basic features of nutrient cycling.

Hydrologic or water cycle : Water is the most abundant chemical compound in the biosphere. Water differs from basic nutrients as it retains its basic chemical state throughout the biotic and abiotic phases of its cycle. Water is a hydrogen source for living organisms. It is a prime mover of soluble nutrients from one place to another and is a medium in which these nutrients enter plants. The cyclical movement of water is by solar energy and gravity.

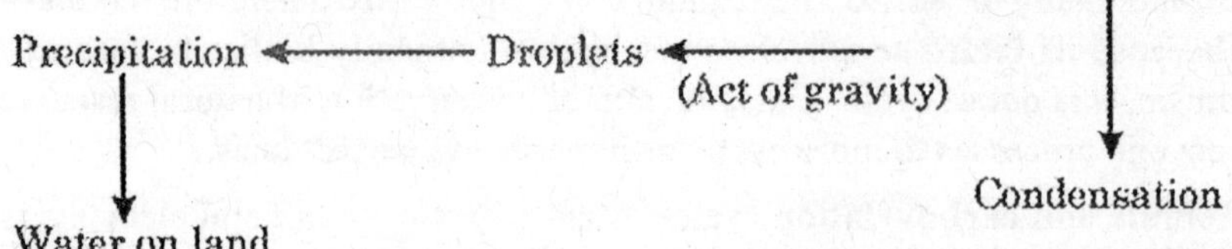

95% of water is bound chemically into a crystalline structure of rocks and it does not cycle. Most of the remainder is in the oceanic reservoirs and only small portion occurs as water in the cells, water vapours and free fresh water.

The world wide distribution of evaporation and precipitation is very uneven. There is more evaporation over the ocean and more precipitation over the land. Over oceans, the water cycle is very simple. Water leaves by evaporation and returns by precipitation. Water when falls on land is soon evaporated to atmosphere or retained by land ecosystems temporarily through surface runoff, percolation, living organisms by transpi-ration, perspiration, etc.

Transpiration is an important part of water cycle and to the functioning of biogenochemical cycling because all the mineral nutrients that enter the roots of plants through an aqueous solution. Gross movement of nutrients within and among ecosystems takes place through runoff and flow of water through the soil. Sometimes vast amount of materials are depleted but also occasionally carried by water from elsewhere.

The movement of water is an important weathering force that helps to uncover nutrient bearing rocks and to recycle chemicals that are otherwise trapped in sediments. The gravitational flow of water continually moves these nutrients from high to low lying ecosystems and then into deep ocean basins where they are again trapped in sedimentary deposits.

Phosphorus cycle: The cycling of phosphorus characterizes sedimentary type. Phosphorus in the environment is available as inorganic phosphate or soluble inorganic phosphate ions or as part of other organic or inorganic molecules of varying solubility. It is first made available to living organisms through the weathering of rocks. Plants obtain soluble phosphate ions through their root system for their living tissues, then it goes to herbivores or decomposers. Thus, phosphorus is passed along the food chain. Excess phosphate is excreted and forms phosphorus rich deposits. For example Guana beds in Chile and Peru by sea birds excreta.

Carbon cycle: The carbon cycle is unique in that its organic phase does not exist as a complete sub-cycle and both the biotic and abiotic phases are closely intervened. C enters the biotic portion of an ecosystem through the uptake of atmospheric CO_2 by tissues of green plants as organic molecules by photosynthesis. Energy is thus stored by living organisms in the form of fixed C. When any organic-C-containing compounds are used in the cell respiration, energy is released and C is returned to the atmosphere in the form of CO_2. Some of these C-containing molecules pass through grazing and detritus food chain and some are deposited as C-rich deposits as coal, oil and gas and are permanently stored. Large amounts of CO_2 react with water to form carbonates and bicarbonates which make clams with the combination of organisms. The dissolved Ca and $CaCO_3$ is formed in their shells, when animals die. This $CaCO_3$ enters as solution or sediment which is oxidized later on through weathering. Thus, C compounds are degraded to yield CO_2. The turnover rate of C is so rapid that although it is present in the atmosphere in a very low concentration but there is an adequate supply for living organisms.

Significance of biogeochemical cycle in crop production: Basic nutrients are used over and over again in a cyclical passage, though each chemical element follows a different route, all are restored to their inorganic form suitable for reuse by other organisms. Basic chemical properties of each element affect the pattern of its movement in a cycle. Their nutrient pool, nutrient flux rate, stage of development, in flow and out flow of nutrients, their balance and disturbance in nature is studied and if the cycles are acyclic, they are made to be cyclic for the conservation of natural resources. Relationship between available energy and decomposition is taken into account. Movement of energy and nutrients through biological communities is seen through these cycles. Radioactivity is used to study the pattern of energy flow through ecosys-tems. The gases, water and various minerals are absorbed by the green plants from the environment. They pass through several trophic levels of nature via food chain and are ultimately returned to environment in several ways and reutilized by green plants. Every nutrient enters the biotic community from abiotic environment and again returns to environment. In this way, atoms are used over and over again and are recycled.

Primary producers are green crop plants, trees, weeds, hedges etc. Consumers of Ist order (C_1) are herbivores, insects like thrip, ants, aphids etc. then rats, rabbits and man. Carnivores are insectivorous insects, frogs and birds. These are second order consumers (C_2). Parasitic fungi and bacteria cause diseases of crop plants. Snake and predatory birds like kite and hawk are top carnivores. These are the third order consumers (C_3). Then saprotrophs or decomposers are saprophytic soil fungi and bacteria which are aided by protozoa and other detrivorous animals on and in the soil. However, sun is the main stay of agroecosystem. Nutrients' materials and energy flow studies are very important in an agroecosystem from crop production point of view. How nutrients are depleted and added in fields and what are their effects on crop production, are very important points to predict and sustain crop yield. Recycling of water, N, C other nutrients and other organic and inorganic materials are of utmost importance to watch pollution in soil, water and atmosphere and their remedies for improving crop productivity from biochemical cycle point of view.

Quantitative aspect of biogeochemical cycles i.e. how much material passes along the routes in the cycles and how fast it moves. For comparing exchange rates between different components of an agroecosystem turn over rate or turn over time is useful. For example, if 1000 units are present in a component and 10 go out or enter each hour, the turn over rate will be 10/1000 or 0.01 or 1% per hour. The turn over time is 1000/100 or 100 hours. In aquatic ecosystems, the turn over is faster than terrestrial ecosystems.

Residence time: Residence time is that time, a given amount of substrate remains in a particular component of a system.

Studies of nutrient cycles have shown that over fertilization is as bad as under fertilization. If more material is added than the actual need of organisms active at that time, the excess is often quickly tied up in the soil or lost through leaching and becomes unavailable at the time when it is required for increased growth. It means that blind dumping of fertilizers or continued excessive fertilization is a wasteful proposition and also creates changes in the ecosystem as well as downstream pollution.

It is due to biochemical cycles that we know that root crops like radish, turnip, beets etc. take plenty of N, K, Ca from the soil, so they should be grown in rotation with the legume crops.

Geochemistry and biogeochemistry have their origin in biogeochemical cycles. To fulfill the requirements of elements, synthetic materials are used and they make the process acyclic because they speed up the movement of other materials and create a situation of too much or too little. For example: man mines and processes phosphate rocks so carelessly that it leads to pollution near mines and phosphate mills.

When about 2 billion years ago, life began on earth, there was more carbon dioxide than oxygen and only anaerobic life was possible. With the passage of time build up of oxygen and decline of carbon dioxide have resulted from more production (P) and less respiration (R). However, more O_2 and less CO_2 are limiting to photosynthesis.

Carbon: Carbon cycle is being affected by man to his disadvantage. Tremendous amount of CO_2 is being released by fossil fuel burning and green belt is decreasing, leading to decreased capacity for the removal of CO_2; affecting its content in atmosphere. The CO_2 content holds the radiant load/heat while the increasing dust and particulate matter, reflects incoming radiant energy. This is sure to affect the climate.

Fossil fuels: Fossil fuel is formed as a part of organic C. About 6 billion tonnes of CO_2 each year is returned in the atmosphere by only burning fossil fuels in homes and factories.

Limestone rocks: Limestone rocks are formed as a part of organic C which gets locked up in the earth's crust as limestone. $CaCO_3$ is sometimes released as a byproduct of photosynthetic assimilation by a number of green plants living in the alkaline water but in the long run $CaCO_3$ gets mixed up with clay and turns into limestone. About 100 million tonnes of CO_2 is annually released into the atmosphere by weathering of limestone rocks.

Biological fixation of nitrogen: About 140-700 mg/m^2/year nitrogen is released from air. Biological N is much greater in the fertile areas and it may be upto 20 g/m^2/day.

Non-biological fixation of nitrogen: It is the physico-chemical fixation which is not more than 35 mg/m^2/year in the temperate regions. In oceans rate of nitrogen fixation per square metre is less but total fixation is more than land areas.

Ammonia released by biological communities is dissolved in rain water and supports the epiphytic plants. This rainfall is useful in recycling available nitrogen. Some lichens are able to fix nitrogen because of the symbiotic blue green algae. The nitrogen cycle is affected by industrial air pollution as nitric oxide and nitrogen peroxide are transitory steps in the nitrogen cycle. Combustion of fossil fuels have increased concentration of these in air especially in the urban areas affecting the respiratory processes of organisms.

Phosphorus: Due to mining of phosphate rocks, 1-2 million tonnes of P is washed away to sea and in the form of fish about 60,000 tonnes is returned per year to land.

Calcium: Through rivers Ca goes to sea and raised in mountain ranges of limestone. About 7% of total sedimentary material in sea is Ca. For every 1000 atoms of Ca, 2.4 atoms of radioactive strontium moves to sea which is fission of uranium entering biosphere due to nuclear weapon tests and wastes of atomic energy uses. Radioactive strontium has now followed Ca from soil and water into the vegetation, animals and human food and bones.

Sulphur: Sulphur cycle is also affected by the air pollution. Fossil fuel, electrical generation plants and coal releases SO_2 into the atmosphere. Increased SO_2 in the atmosphere damages the photosynthetic process.

Water cycle: Water cycle is also being affected by pollution and other explained facts. So an International hydrological decade comparable to International Biological Programme (IBP) has been planned. More water evaporates from sea than returns via rainfall and vice versa for land. A part of rainfall that supports land ecosystem and food production comes from water evaporated over the sea. The difference between annual rainfall (1 geogram) and runoff (0.2 geogram) i.e. 0.08 geogram is the annual recharge rate of ground water. Thus, increasing rate of runoff is decreasing the ground water.

Limiting factors: A factor of growth which is of most immediate need to the plant or an organism and growth comes to stand still due to limitation of that factor which may be nutrient, water, light etc. Liebig (1840) a chemist, explained the law of limiting factors because plant growth is dependent upon factor which is limiting or present in minimum quantity. Tayler (1934) an ecologist, expanded the limiting law of Liebig and added that it is not only nutrient but all environmental factors which affected the plant growth, rather it is a condition of organism itself at the time of application of factor. Condition may be optimum temperature for germination, growth, reproductive phase, harvesting and it will be different for different plants and their phases. Distribution of plants is also limited by this condition. For example, wheat has wider limits of tolerance for growth factors and can grow widely in plains and mountains as compared to gram which has narrow limit of tolerance. Thus, Tayler emphasized the critical season, critical year and critical stage of development when an organism may have a narrower tolerance for certain factors intensity as water is more important at germination and anthesis as compared to vegetative phase in most of the plants.

Blackman (1905) developed the theory of optimum and limiting factors in which rate of reaction was emphasized. Theory says when a process was conditioned as to its rapidity by a number of factors, rate of process was limited by pace of slowest factor. For example, chain is as strong as its weakest linkage. So, entire process is dependent upon weakest link.

Cain (1944), elaborated that not only one factor but all the factors should be taken into account because with the limitation or excess of one factor, other factors may be accumulated by other limiting factors and as such they also become limited.

Both Blackman and Liebig laws are rather very rigid laws and there is no scope of any degree of interaction of factors and replace ability of factors. Sometimes one factor is replaced by other. For example : Sodium can play important role when potassium is limiting but apply it cautiously because Na can knock out the K^+, may be due to the similar size but if we will do so normally, the sodium application will lead to our soils saline and one day sodium will become toxic, so use it wisely. So, here potassium was limiting factor and replacing factor was sodium.

Sometimes, moisture is a limiting factor in dry land areas and if we apply water, limiting factor is removed through canal water, rainfall etc. and at same time water will act as a trigger factor as it brings vegetation very fast.

In coastal areas near the sea, humidity is much higher and fog in these areas replaces the rain. Sometimes, at night, negative transpiration takes place, in this way, xerophitic plants in arid conditions manage to grow where water is coming through leaves to roots for dilution of nutrients. This is called factor of replace ability i.e. some factor replaces the quantity factor of the other factor as fog can replace rainfall to some extent and sodium when potassium is limiting. Similarly altitude or elevation can be substituted by latitude and vice-versa. Slope and direction may be substituted for latitude. This is just a temperature adjustment. So, similar crops will be found where temperature is similar.

RESPONSE OF CROP PLANTS TO ENVIRONMENT

Crop growth and production depends upon the interaction within the plants and the physical environment i.e. a biological system. Generally, we take the environmental factors for granted and do not try to learn how to increase the production with modification of physical environment or by cultural treatments because there are interacting effects of cultural methods on the yield. Suppose different varieties of the same crop are sown in a particular region, some varieties will be resistant to pest and diseases and some varieties with vertical oriented leaves will be much superior at closer spacing as compared to wider spacing. The response to spacing may be modified by climate, soil factors of moisture available. Thus, physical environment should be studied properly to see such affects for increasing crop yield. A convenient division of environmental factors could be :

I. Light affecting photosynthesis and photoperiodic reaction.

II. Atmosphere i.e. the availability of CO_2 and O_2 and saturation deficit affecting ET. They all affect plant growth. Wind is also a constituent of atmosphere affecting plant growth by changing CO_2 availability in leaf zone and also by changing the saturation deficit of air at the leaf surface and through mechanical affects on the aerial plant parts.

III. **Root environment:** Root environment through soil physical structure, nutrients and moisture availability, aeration, salinity and so on affecting the plant growth.

IV. **Temperature:** Temperature is a major factor to define the tropics and it limits the extension of tropical crops to higher latitude and altitudes.

V. **Light:** Light affects the crop growth and production both through its use in photosynthesis and photoperiodic reactions. Photoperiodic effects generally occur in tropical rice varieties and in number of species. However, photosynthetic light affects crop production through:

   a. Total amount of incoming light which is suitable for photosynthesis.

   b. Amount of light which is available to crop and can be utilized by the crop for photosynthesis.

First is useful for determining the theoretical upper limit for crop growth. Since, about 90-95% of dry matter of a plant is derived from photosynthesis. To calculate the theoretical upper limit, we require various data as number of incoming quanta of natural sunlight in photosynthetic region of spectrum, number of quanta required to reduce each mole of CO_2 used in photosynthesis and respiratory losses of crop. Generally wavelengths of light which are effective in photosynthesis is in range of 400-700 nm. Incoming light is influenced by climatic conditions and if mean daily quanta of incoming solar radiation is 500 cal/cm²/day, then an estimate within the range of PAR 400-700 nm, of the energy quanta available for photosynthesis, the photosynthetic radiation will be 222 cal/cm²/day, then net daily production possible is 71 g/m²/day which comes to more than 100 tons of dry matter per year per acre.

The low efficiency of photosynthesis is due to :

I. The process of photosynthesis operates theoretically at 35% but practically the efficiency of photosynthesis is not more than 20% so quantum efficiency is low.

$$6\,CO_2 + 6\,H_2O \xrightarrow{\text{light + Chl}} C_6H_{12}O_6 + 6O_2$$

One mole of CO_2 need 520 kcal for its reduction practically and stores only 105 kcal.

II. Light saturation and photosynthesis i.e. photosynthesis and light intensity relations are very important. Normally photosynthesis increases with increase in light intensity in many land plants but after sometimes it becomes static and then decreases. So, photosynthesis cannot be increased indefinitely.

III. LAI and amount of chlorophyll: There is a limit to amount of chlorophyll or the leaf area that can be distributed over a given surface called LAI.

LAI = Leaf area/ground area

In terrestrial ecosystem, we take leaf area over ground area equal to LAI. Many crops reach their maximum photosynthesis at a LAI of 4 or 5 when the production is high. If LAI goes above, competition for light is such that overall production is reduced. Generally LAI of other vegetation may be greater than crop plants. However, in some tropical crops like castor or cassava it may got upto 20, which is a upper limit called upper natural limit for LAI.

IV. Utilization of a part of light spectrum i.e. in some plants violet spectrum is also helpful but mostly red and blue spectrum is used. Photosynthesis utilizes only those wavelengths of light spectrum which are sharply absorbed by chlorophyll and these are at 430 nm and 670 nm (4300 A° & 6700A°). So, chlorophyll is sensitive to about half of the visible solar spectrum and we cannot change the solar radiation and solar spectrum. As light spectrum is quite long but its utilization and contribution is very low than received on the earth, so, is the low efficiency and low net primary productivity is also the result of all such limits.

V. Plant architecture: In the agroecosystem, crop plant is the main that influences the primary production with its characteristics like LAI, leaf density, leaf inclination i.e. horizontal or vertical leaves, light absorption, leaves resistance to gaseous exchange, rate of photosynthesis of individual leaves at light saturation, reflection, transmission, absorption of light in the canopy. These ideal conditions are not available, so they put limit to productivity. Thus, due to all these factors crop plants are less efficient in using incoming light energy in photosynthesis than supposed purely from the measurement of light absorbed by crop but these high figures of dry matter production (100t/yr/acre) present an interesting point which could be attainable goal for agricultural scientist who is interested in improving crop productivity.

Atmospheric environment: Photosynthesis involves the photolytic spitting of water followed by reduction of CO_2 from atmosphere. Consequently, the availability of CO_2 affects rate of photosynthesis and utilization of incoming energy, if this is in short supply. In fact the average concentration of CO_2 in the atmosphere is about 0.03% and in soil atmosphere is 0.33% and this does constitute a major limiting factor to the photosynthesis of crops. The energy quanta in natural sunlight are generally in excess of those required to reduce all available atmospheric CO_2 at the surface of fully exposed leaves. So, normally CO_2 becomes limiting factor in photosynthetic process in fully exposed leaf surface. However, when the leaves are placed deep within crop canopy, the amount of photosynthetic radiation reaching these is reduced as light quantity may become major limiting factor in photosynthetic process. Increasing the level of atmospheric CO_2 will partially release the inhibition to photo-synthesis in natural sunlight and this fact is used in glasshouse conditions for the production of high value crops. Intensity of total radiation is diffused by wind, dust and water vapours and they affect photosynthesis. The air temperature, humidity, soil temperature and even the infrared radiation from the sky also affect photosynthesis, but these all are needed in optimum limits. Optimum conditions of an ideal environment are also lacking, so they put limit to productivity.

Root environment: It consists of soil of different types which affect the establishment, growth and harvesting of crops. So, suitability of soil for crop production is closely linked with climatic factors of rainfall and rain distribution such as saturation deficit of air. For example : a course sand or poor soil may yield excellent crops with adequate irrigation, fertilization and other cultural practices. In non-irrigated conditions, the climate with evenly distributed rainfall may give good yields even under poor soils. However, agriculture subsidy given in terms of agricultural inputs, irrigation, fertilization and good management can create good root environment for higher yields. Genetic factors also play role under moisture stress or other nutrient content, soil tilth conditions. So, soil structure, porosity, heaviness i.e. soil tilth, nutrient content should be according to the needs of the crops depending on climate and rainfall as per their growth phases. The rate of addition of organic residues and breakdown rate of organic matter in soil is influenced by cultural practices which affect the colloidal matter to affect nutrients.

I. **Root extension:** In case of sweet corn, root extend laterally more than half as much as stalk. The roots of field corn extend 4-feet laterally with depth of 8 feet.

II. **Rate of root growth:** The rate of root growth is closely correlated with the environment. In common grasses, it is more than 0.5 inches/day. The main vertical root of corn sometimes penetrate downward under very favourable condition at a rate of 2.0 to 2.25 inches/day during a period of 3-4 weeks.

III. **Activities of roots in subsoil:** The roots are active in the absorption of water and nutrients to the maximum depth of penetration in barley. During heading and ripening of the grains, the bulk of absorption is by younger portion of the roots in the deeper soil. As per Kramer's view relatively larger amounts of materials are taken from surface soil layer than deeper soil layer because roots occupy upper soil layer for longer time especially in the annuals. Little or no branching occurs in surface layers and structural changes like production of an abundance of cork, precludes absorption by the main root. So, favourable conditions for roots upto substratum are necessary especially in range grass management and improvement of natural vegetation. This portion is much within the control of farmer or plant grower to bring soil and plant into favourable relationship with root system.

IV. **Hereditary factors:** Hereditary growth characters of species or varieties affect the roots of a plant. For example : tap roots develop rapidly and penetrate deeply in some crops and in others branch near the soil and tap root loses its dominance very soon while in some other crops fibrous roots spread widely or penet-rate nearly downward. So, great variations occur within the genus, species or variety. The root characteristics are also fixed as in seed and other above ground parts for a genus, species or a variety.

V. **Environment:** Water content, aeration, soil structure, nutrients etc. affect root system. The character of root system is usually an indicator of soil conditions. Some plants due to modified root characteristics are adapted to particular environments but other plants cannot. Grafting on other stocks is done because roots of these can adapt accordingly. For example: to make alkali tolerant, excessive moisture, dry conditions or disease resistant plants.

VI. **Response to low water content:** A relatively low water content but enough to ensure good growth stimulates roots to greater development resulting in a greatly increased absorbing surface. For example: in dry soil, corn grown with 9% ASM (available soil moisture) had 2.1 times the total root area (both weight and length) than the tops i.e. transpiring surface or leaf area and when it was grown in moist soil at 19% ASM, it had 1.2 times transpiring surface than absorbing surface or roots. Although total length and diameter of main root was same in both the conditions of dry and moist soil.

	Moist Soil	Dry Soil
Main root (%)	11	11
Primary laterals (%)	38	75
Secondary laterals (%)	51	14
Total	100	100

Thus, a low water content within certain limits stimulated and increased root development. As soil became dry, developed less secondary laterals. Primary lateral roots will develop more in dry soil while secondary lateral roots will develop more in moist soil. However, the ideal root system is not one with the most extensive branching but one that fully occupied the soil to an adequate depth throughout a radius sufficient to secure at all times enough water and nutrients. Where the soil is very dry, root development is greatly retarded or even ceases and above ground parts are consequently dwarfed.

VII. **Response to high water content:** Due to lack of aeration, roots are very shallow as in bogs (marshy land). Roots and rhizomes form mat above the water level, superficially placed but if vertical, they will die at the water surface.

VIII. **Root habits and irrigation or drainage**: To make the plants deep feeder, water is avoided in the early growth stages. The proportion of root to tops may be increased with decreasing soil moisture. The amount of shedding of leaves and bolls was directly proportional to extent of root system that was submerged and died as a result of rising water table in cotton. So, proper drainage is must in cotton. The drainage of swamps and bog lands for cultivated crops should be determined with reference to root relations. If soil is course and capillary action is low, too great lowering of water table will lead to more dry soil and reduction in yields. Crops like sugarbeet and alfa alfa having tap roots require 15-20 feet deep water table i.e. better drained soils than melons or cucurbits where main portion of roots occupies surface soil on all sides of plant to a distance of 15-20 feet. So, far as cucurbits are concerned, sandy loams with a clay subsoil are ideal, since they warm quickly and promote rapid root growth but still retain moisture sufficiently for shallow root system.

IX. **Response to temperature:** For most of the temperate crops, soil temperature 60-80°F is quite good. However, cucurbits with shallow rooted system grow better at higher temperature than the deep rooted plants. In some plants, if atmospheric conditions are not favourable but the soil conditions are favourable, considerable vegetative growth is possible. Decrease in soil temperature or frozen soil may harm roots. Alternate freezing and thaw ing can have the roots out of the soil. A direct correlation between resistance to winter chilling and extensibility of roots is found is some plants.

X. **Aeration and Soil temperature:** As the oxygen supply in soil layer is decreased rate of root growth diminishes in a soil with high temperature Aerial environment and growth of top and roots also depends upon the amount of available light.

a. *Darkness*: Tall seedlings and poorly developed roots.

b. *Diffused light*: Shorter plants and longer roots.

c. *Full light*: Short stocky plants with long branching roots. Carbohydrates supply to roots is more due to more photosynthesis, thus root development will be more.

8

CULTIVATION PRACTICES FOR CEREALS

CEREALS

Cereals such as rice, wheat and maize are members of the grass family and they are particularly important to humans because of their role as staple food in many areas of the world. Cereals are also used to produce animal feed, oils, starch, flour, sugar, syrup, processed foods, malt, alcoholic beverages, gluten and renewable energy. Approximately 50% of the world's calories are provided by rice, wheat and maize, but in many parts of Africa and Asia, people rely mainly on grains such as sorghum or millet. Maize, sorghum and barley are important sources of livestock feed whereas barley and rice are used in the brewing industry. Grains of all kinds, oats in particular, have been shown to contain chemical compounds that help to reduce the chance of certain types of cancer and coronary heart disease.

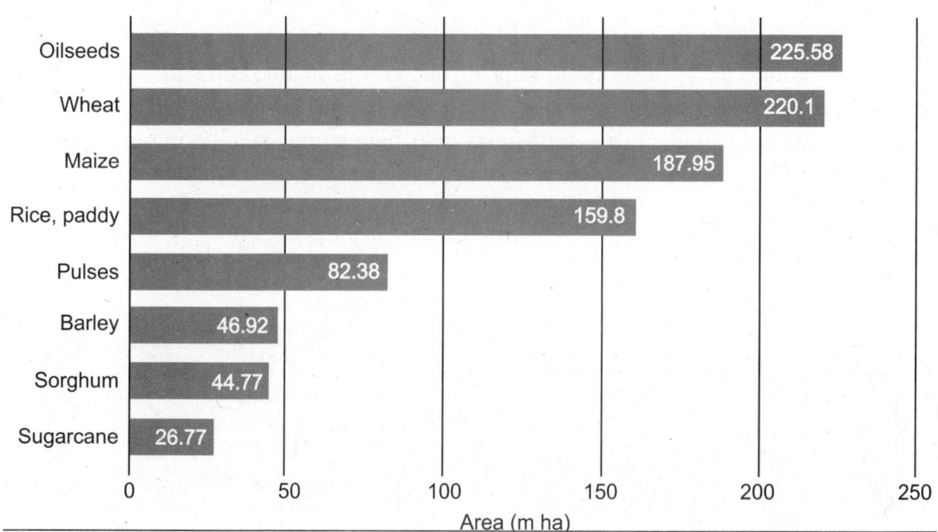

Fig. 8.1. Area under different field crops in the world (FAO STAT 2016)

Grasses are believed to have been evolved over 60 to 70 million years ago and the diverse members of this family vary widely in appearance, zones of adaptation, physiology, number of chromosomes and genome size. Despite these apparent differences, most grasses have a similar set of genes and show relative conservation of gene order along the chromosomes. Among cereal crops, wheat occupies highest area (220.1 m ha) followed by maize (187.95 m ha) and rice (159.8 m ha) (Fig. 8.1). Among the cereal crops, the production of maize is highest (1060.1 mt) which is followed by wheat (749.46 mt) and rice (740.96 mt) globally (Fig. 8.2). Globally area of sugarcane is less but its total production is highest due to more average yield per unit area.

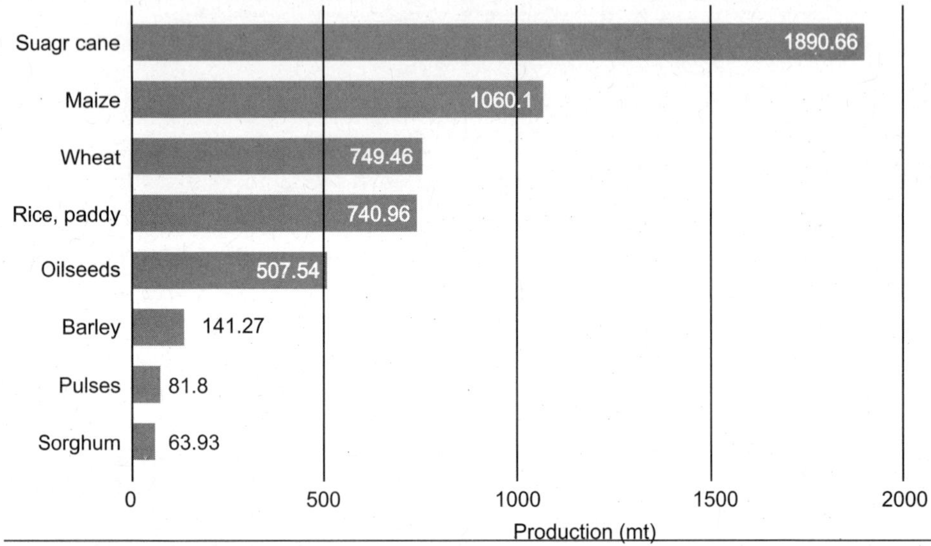

Fig. 8.2. Production of different field crops in the world (FAO STAT 2016)

The detailed description pertaining to origin and history, climate requirements, suitable soil type, method of sowing, seed rates, fertilizer requirements, water and weed management, plant protection measures, harvesting etc. for important cereal crops are discussed under.

RICE

Rice crop   Rice grains

Introduction

Rice (*Oryza sativa*) is the staple food for over 2.5 billion people and it covers about 9 per cent of the earth's arable land. Rice provides 21 per cent of global human per capita energy and 15 per cent of per capita protein. Calories from rice are particularly important in Asia, especially among the poor, where it accounts for 50-80 per cent of daily caloric intake. Asia accounts for over 90 per cent of the world's production of rice, with China, India and Indonesia are leading countries (Fig. 3). China produces 211.09 mt of rice followed by India (158.76 mt) which comes out to be 28.4 and 21.4 per cent share globally. Only 6-7 per cent of the world's rice crop is traded in the world market. Thailand, Vietnam, China and the United States are the world's largest exporters. About 85 per cent of the rice that is produced in the world is used for direct human consumption. However 90 per cent of all rice grown in the world is produced and consumed in the Asian regions. India is one of the major rice producing and consuming countries in the world. Rice is also the single most important crop in Indian agriculture. It is produced in all the states of the country and is a part of the staple diet of a very large section of the Indian population. Rice in India occupies about 42.96 million hectares of land with production of 158.76 mt. [FAO STAT 2016].

Rice is primarily a high energy or high calorie food. It contains less protein than wheat. The protein content of milled rice is usually 6 to 7 per cent. Rice however, compares favourably with other cereals in amino acid content. The biological value of its protein is high. The fat content of rice is low (2-2.5 per cent) and much of the fat is lost during milling. Rice contains a low percentage of calcium. Rice grain contains as much B group vitamin as wheat. Milled rice loses valuable protein, vitamins and minerals in the milling process. Much of the loss of nutrients can be avoided through parboiling process. The by-products of rice milling are used for a variety of purposes. Rice bran is used as cattle and poultry feed. Rice straw is good cattle feed as well as it is also used in making hats, mats and ropes.

Origin and history

Rice is regarded as a first cultivated crop of Asia. Preserved rice grains were found in China around 3000 BC. Paddy grains were found during excavation at Hastinapur (India) around 1000-750 BC considered as a oldest sample in the world. Vavilov (1926) concluded that south-west Himalayas has various types and varieties of rice that indicated probable centre of its origin. De Condolle (1886) mentioned south India as its centre of origin. Thus based on all these archaeological evidences India is considered to be the centre of origin of rice.

Area and distribution

Rice cultivation in the world extends from 39° S latitude (Australia) to 45° N latitude (Japan) and 50°N latitude (China). In India, it stretches from 8° N latitude to 34°N latitude and is grown on an area of 42.96 million hectares with production level of 158.76 million tones, (Fig. 8.3) during 2016. [FAO STAT 2016-17]. India leads the world in terms of area under rice but is ranked at second position in terms of production succeeding China. Rice is also grown even in areas below the sea level, as in the Kuttanad region of Kerala. It also grows well at altitudes above 1,979 m, as in parts of Jammu and Kashmir. It is cultivated as a purely rainfed upland crop in West Bengal, Bihar, Uttar Pradesh, etc., where the monsoon is precarious and its distribution is often erratic. In some parts of West Bengal and Bihar, under shallow and deep water (4.75-6 m) conditions, rice is also grown. The area under rice in Punjab was 3.04 m ha with production of 18.86 mt during 2017–18.

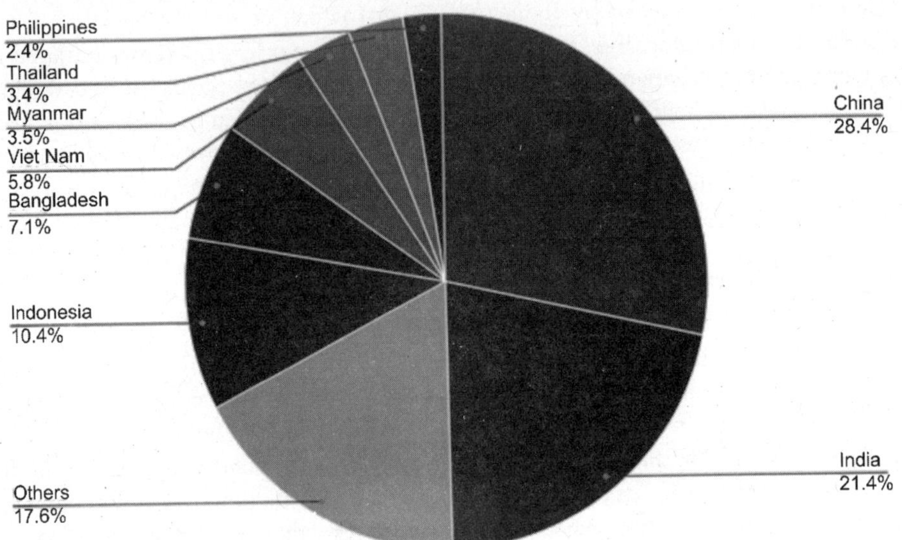

Fig. 8.3. Annual global production of rice during 2016 (Percentage) (FAO STAT 2016)

Improved Varieties of Rice released by ICAR

2017-18

Variety	Area of adoption	Salient features
HRI-183	Haryana, Uttarakhand, Uttar Pradesh and Gujarat	Early duration irrigated, bacterial leaf blight (BLB) resistant hybrid developed by deploying MarkerAssisted Selection (MAS), medium slender grain, yield 6,100 kg/ha
HRI-180	Punjab, Haryana and Uttarakhand	Early maturing hybrid, irrigated, resistant to bacterial leaf blight (BLB) and tolerant to blast, brown spot (BS) and false smut, yield 6,100 kg/ha

PAC 129	Madhya Pradesh	Short duration hybrid (115-120 days), less water requirement, fits well in rice - vegetable/potato/ mentha - wheat crop rotations, suitable for Direct Seeded Rice (DSR) system
BS 129G	Chhattisgarh	Early duration hybrid (115-120 days), BLB resistant, less water requirement and fits well in vegetable / potato/mentha and wheat crop rotations
28P09	Odisha, West Bengal, UP, Assam, Chhattisgarh, Maharashtra, Gujarat, TN and AP	Rice hybrid, irrigated, medium slender grain, moderately resistant to BLB, tolerant to neck blast (NB), BS, LF, yield 5,834 kg/ha
CO 51	Haryana, Uttarakhand, Odisha, Bihar, WB, UP, MP, Maharashtra, Gujarat, AP, Kerala, Karnataka and TN	Early maturing variety, moderately resistant to green leaf hopper (GLH), brown plant hopper (BPH), yield 6,620 kg/ha
27P36 Hybrid	Bihar, Madhya Pradesh, Jharkhand, Chhattisgarh and Odisha	Tolerant to BLB, grain - long slender, maturity 128- 132 days, average yield in rainfed/upland environment 7,200 -7,500 kg/ha
VL Dhan 158	Himachal Pradesh and Uttarakhand	Developed for hill ecology, resistant to leaf and NB under field and controlled condition, yield 2,750 kg/ha
Jalkunwari	Assam	Irrigated lowland (IRL), moderately resistant to blast, yield 4,000 kg/ha
Jalashree	Assam	Irrigated lowland (IRL), moderately resistant to blast and stem borer, yield 4,300 kg/ha
Kanaklata	Assam	Non-lodging, non-shattering, responsive to fertilizer, suitable for irrigated transplanted in boro ecology, yield 6,000 kg/ha
Gitesh	Assam	Irrigated lowland (IRL), resistant to sheath rot, moderately resistant to blast, brown spot and stem borer, yield 5,300 kg/ha
Malviya Sugandha Dhan HUR-917	Uttar Pradesh	Scented rice, tolerant to NB, moderately tolerant to bacterial leaf blight and stem borer, yield 3,355 kg/ha

Sabour Surbhit	Bihar	Irrigated medium maturity for upland, grain type slender, moderately resistant to major pests and diseases, yield 4,000-4,500 kg/ha
SHIATS Dhan-2	Uttar Pradesh	Ecosystem irrigated mid-early (IRME), long slender grains, suitable for rice-wheat cropping system, moderately resistant to sheath blight, brown spot (BS), yield 5,065 kg/ha
SHIATS Dhan-3	Uttar Pradesh	Ecology irrigated mid (IRM), long slender grains, suited for late sown wheat cropping system, moderately resistant to BS, sheath blight, yield 4,834 kg/ha
Dhruba	West Bengal	Irrigated lowland, short bold grain, moderately resistant to leaf and NB, BS, leaf folder (LF), yield 5,000-5,500 kg/ha
JKRH-3333 Hybrid	MP, WB, Bihar, Gujarat, Chhattisgarh and AP	Hybrid, suitable for DSR and machine sowing, medium slender grain, medium duration with 135-140 days maturity
CR Dhan 508	Assam, West Bengal and Odisha	Ecology - deep water, grain long bold, moderately resistant to NB, brown spot and sheath rot, yield 4,048 kg/ha
27P22	Punjab and Haryana	Hybrid, irrigated mid-early maturity, medium-slender grain, moderately resistant to BLB, yield 6,477 kg/ ha
NPH 8899	Uttar Pradesh, Bihar and Assam	Hybrid, suitable for boro ecology, short bold, moderately resistant to leaf blast, NB, BPH, yield 5,667 kg/ha
GK 5022	Bihar and Chhattisgarh	Hybrid suitable for aerobic ecology, long- slender grain, moderately resistant to leaf blast, NB, BPH, yield 4,232 kg/ha
PAN-2423	West Bengal	Hybrid, yield 5,500 kg/ha
Gujarat Anand Rice-3	Gujarat	Irrigated, resistant to BLB, NB and LB, moderately resistant to stem borer (SB), LF, long bold grain

Pant Dhan-23	Uttarakhand	Moderately resistant to leaf blight, sheath blight and stem borer, yield 4,453 kg/ha
Pant Dhan-26	Uttarakhand	Moderately tolerant to stem borer, BLB, sheath blight, Rice Tungro Disease (RTD), NB and sheath rot, yield 4,530 kg/ha
Chhattisgarh Sugandhit Bhog	Chhattisgarh	Resistant to gall midge biotype 1 and 5, moderately resistant to LB, sheath rot (SR), BS and RTD
Tarangini	Andhra Pradesh	Irrigated ecology, resistant to leaf blast (LB), BPH, long slender grains, yield 7,500 kg/ha
Bheema (Dheera)	Andhra Pradesh	Tolerates flash floods for 10 days at vegetative stage, stagnant flooding and submergence during germination for 2 weeks, resistant to BPH, moderately resistant to BS, yield 6,000 kg/ha
Rajdeep CN 1039-9	West Bengal	Tolerant to sheath blight and sheath rot
Sabour Deep	Bihar	Irrigated medium upland ecology, tolerant to BLB, SB, LB, BS, yellow stem borer, green leaf hoppers (GLH) and gundhi bug, grain-elongated slender, yield 3,500-4,000 kg/ha
Sabour Ardhjal	Bihar	Irrigated, medium upland and aerobic, moderately resistant to BLB, grain long slender, yield 4,500- 5,500 kg/ha
CARI Dhan 1	Andaman and Nicobar	Selection from Quinglivan 1, rainfed shallow lowland, grain medium slender, yield 4,800 kg/ha
CARI Dhan 5	Andaman and Nicobar	Somoclonal selection from Pokkali, suitable for rainfed lowland and coastal saline conditions, short-bold, yield 4,500-5,500 kg/ha
Hybrid Indam 200-022	Madhya Pradesh	Suitable for upland cultivation, early duration (110 days from seeding), grain-long bold, paddy-vegetable crop rotation, moderately resistant to LB, NB, BS and white backed plant hopper (WBPH)

Dehangi	Assam	Rainfed, upland rice, maturity 120 days, average yield 2,800 kg/ha
Inglongkiri	Assam	Rainfed, upland rice, maturity 120 days, average yield 2,800 kg/ha
Rongkhang	Assam	Rainfed, upland rice, maturity 120 days, average yield 2,800 kg/ha
Bauna Kalanamak 102	Uttar Pradesh	Aromatic rice
CR Dhan 506	Assam, AP and Karnataka under semi deep water	High yeilding variety for the rainfed ecology, tolerant to waterlogging
CR Sugandh Dhan 908	Odisha, West Bengal and Uttar Pradesh	Scented rice, shallow low line area, non-lodging, semi-dwarf plant type
Pusa Basmati 1718	Punjab, Haryana and Delhi	Marker assisted selection (MAS) derived near isogenic line of Pusa Basmati 1121, maturity 136-138 days, resistant to BLB, long slender grains with strong aroma
Gangavati Ageti	Karnataka	Suitable for normal, early and late conditions,resistant to shattering, good for idli and dosa preparation, medium slender grain type, yield 6,500-7,000 kg/ha
Rajendra Nilam	Bihar	Resistant to lodging and non-shattering, tolerant to LF, SB and gundhi bug, adapted to aerobic condition
Purna	Gujarat	Rainfed area of south Gujarat
PAN 802 Hybrid	West Bengal	Resistant to grain discoloration (GD) and sheath rot, tolerant to SB and LF, resistant to lodging and Shattering
Karjat 8	Maharashtra	Rainfed and irrigated areas in transplanted conditions during kharif season, non-lodging, non-shattering
Karjat 9	Maharashtra	Non-lodging, non-shattering, moderately resistant to NB, BLB and resistant to leaf blast, BPH, WBPH and moderately resistant to SB
RCM-30, IET-24200	Manipur	Resistant to gall midge (GM), tolerant to NB, duration 125 to 135 days

CN 1272-55-105	WB, Bihar, Odisha, AP, Maharashtra and Karnataka	Late duration, grain - short bold, resistant to NB,BS and sheath rot (SR)
CR Dhan 408	Odisha	Suitable for shallow lowland, grain - long bold, medium tolerance to LB, NB, BLB, SR, SB, LF and WBPH along with moderate submergence tolerance for one week, yield of 4,800 kg/ha
Surabhi	Maharashtra and Gujarat,	Suitable for irrigated kharif and rabi, medium slender super fine grain, moderately resistant to BLB, SBL, BS, maturity (kharif) 125-135 days
KPH-459	Gujarat and Maharashtra	Mid duration hybrid, grain type medium slender, moderately resistant to BLB, suitable for late sown condition
TKM 13	Gujarat and Maharashtra	Irrigated mid-early, moderately resistant to Bl, RTD, BS, SR, SB, LF, GLH, yield 5,940 kg/ha
KPH 473	Chhattishgarh, Madya Pradesh and Maharashtra	Hybrid with mid early duration, long bold grains, moderately resistant to LB
BIO 799 Hybrid	Odisha, Bihar, West Bengal, Uttar Pradesh, Jharkhand	Hybrid, long bold grain with multiple disease resistance to BLB, LBL, RTD and field tolerance to BPH
MRP 5408	Andhra Pradesh, Karnataka and Puducherry	Hybrid, long slender grains with moderately resistance to LB, NB, BS, RTD, and false smut
CR Dhan 909	Assam, Bihar, Uttar Pradesh and Maharashtra	Irrigated, aromatic, short grain, tolerant to LB, NB, SR, RTD, SB
Nandyala Sona	Andhra Pradesh	Suitable for late sown conditions, replacement to BPT 5204 (Kurnool Sona), recommended for scarce rainfall zone, tolerant to water stress, , maturity 135-140 days, non-lodging, fertilizer responsive, non-shattering, short slender grain, highly tolerant to BPH, LF and moderately tolerant to blast, yield: 6,500 - 7,000 kg/ha
SRIMATI CR Dhan 207	Odisha	Early maturing (110-115 days), non-lodging, intermediate grain shattering, moderately resistant to LBl, NBl, BS, SR, SB, LF, GLH, GM 1 and 4, fertlizer responsive, av. yield of 3,700 kg/ha

PRIYA	Odisha	Early maturing (110-115 days), non-lodging, intermediate grain shattering, fertilizer responsive, grain long slender. Moderately tolerant to LB, NB, BS, RTD, SB, LF, WBPH, GLH, average yield 3,570 kg/ha
CR Dhan 507	Odisha	One week submergence tolerance, tall plant type, intermediate grain shattering, fertlizer responsive,late maturity (160 days), moderately resistant to NB, BS, SB, SR, SB, LF, average yield 4,750 kg/ha.
CR Dhan 409	Odisha	One week, submergence tolerance, moderately resistant to NBl, LB, SR, SR, SB, LF
CR Dhan 800	Odisha	Tolerant to BLB
CR Sugandh Dhan 910	Odisha	Moderately resistant to blast, BL, NB, SR, RTD, SB, LF, WBPH

2016-2017

Rice Variety	Area of adoption	Salient features
Pusa Basmati 1609	Delhi, Uttarakhand, Punjab and western Uttar Pradesh	Moderately resistant to leaf blast, neck blast
Birsa Vikas Sugandha1	Jharkhand	Moderately resistant to blast, brown spot, stem borer
Pant Dhan 24	Odisha, Bihar	Long slender grains; moderately resistant to gall midge biotype-1, stem borer, brown spot
Kalachampa	Odisha	
KPH460	Andhra Pradesh, Tamil Nadu, Karnataka, Maharashtra and Gujarat	Medium slender grains; resistant to leaf blast
Nua Acharamati	Odisha	Short bold grains; resistant to brown plant hopper, leaf folder, moderately resistant to gall midge, sheath blight, whorl maggot, leaf blast, sheath rot, rice tungro virus, bacterial leaf blight, brown spot
Hiranmayee	Odisha	Medium slender grains; resistant to leaf folder, moderately resistant to gall midge, stem borer, whorl maggot, leaf blast, sheath rot, rice tungro virus, bacterial leaf blight, brown spot

Tanmayee	Odisha	Medium bold grains; resistant to leaf blast, moderately resistant to gall midge, stem borer, leaf folder, whorl maggot, sheath blight, rice tungro virus, bacterial leaf blight
DRR Dhan 44	Uttarakhand, Haryana, Bihar	Long slender grains; resistant to leaf blast, moderately resistant to bacterial blight, brown plant hopper, white- backed plant hopper
DRR Dhan 43	Tamil Nadu, Puducherry, Kerala, Karnataka	Long bold grains; resistant to leaf blast, moderately resistant to sheath rot, brown spot, neck blast, brown plant hopper
ADV 8301	Gujarat, Maharashtra, Andhra Pradesh, Tamil Nadu	Medium slender grains
Pant Basmati 2	Punjab, Haryana, Uttarakhand and Uttar Pradesh	Moderately resistant to bacterial leaf blight, brown spot, sheath blight, leaf blast
Pant Basmati 1	Delhi, Uttarakhand and Uttar Pradesh	Moderately resistant to leaf blast
DRR Dhan 45	Tamil Nadu, Karnataka and Andhra Pradesh	High zinc; long slender grains; moderately resistant to leaf blast
CR Dhan 310	Odisha, Madhya Pradesh and Uttar Pradesh	High protein variety with medium slender grains
Ajit	Uttarakhand, Maharashtra and Tamil Nadu	Medium slender grains; moderately resistant to neck blast, leaf blast, rice tungro disease, sheath rot, brown spot
KRH 4	Karnataka, Gujarat and Maharashtra	Good milling, cooking and puffing
PR 121	Punjab	Long slender and translucent grains; resistant to all 10 pathotypes of bacterial leaf blight prevalent in Punjab
PR 122	Punjab	Long slender and translucent grains; resistant to all 10 pathotypes of bacterial blight prevalent in Punjab
PR 123	Punjab	Long slender and translucent grains; resistant to all 10 pathotypes of bacterial blight prevalent in Punjab
GNR 3	Gujarat	Resistant to bacterial leaf blight, leaf blast, false smut, moderately resistant to stem borer, leaf folder

CR Dhan 101	Odisha	Moderately resistant to leaf blast, neck blast, sheath blight, brown spot, stem borer, leaf folder, green leaf hopper
CR Dhan 203	Odisha	Moderately resistant to leaf blast, sheath rot, brown spot, stem borer, leaf folder
CR Dhan 206	Odisha	Moderately resistant to leaf blast, sheath rot, brown spot, stem borer, leaf blight
CR Dhan 307	Odisha	Moderately resistant to leaf blast, sheath rot, brown spot, stem borer, leaf folder
Pusa 1592	Punjab, Haryana, Delhi and Jammu and Kashmir	Resistant to bacterial leaf blight
Chhattisgarh Zinc rice 1	Chhattisgarh	High zinc variety with long grains; moderately resistant to leaf blast, neck blast, stem borer
SHIATS Dhan 4	Uttar Pradesh	Medium slender grains; moderately resistant to blast, stem borer, brown spot, rice tungro disease
SHIATS Dhan 5	Uttar Pradesh	Moderately resistant to leaf blast, brown spot, bacterial leaf blight
Chandra	Punjab, Bihar, Madhya Pradesh, Chhattisgarh, Maharashtra, Tamil Nadu, Karnataka and Kerala	Long bold grains ; resistant to false smut, moderately resistant to leaf blast, neck blast, brown spot, sheath rot, rice tungro disease, white- backed plant hopper, green leaf hopper
DRR Dhan 46	Bihar, Madhya Pradesh and Maharashtra	Long slender grains; moderately resistant to leaf blast, neck blast, brown spot, bacterial leaf blight, sheath rot, brown plant hopper, white- backed plant hopper
BNKR 3	West Bengal	Moderately resistant to leaf blast, neck blast, brown spot, bacterial leaf blight, stem borer, leaf folder
Badshabhog Selection 1	Chhattisgarh	Moderately resistant to brown spot, sheath rot
Tarun bhog Selection 1	Chhattisgarh	Moderately resistant to leaf blast
Dubraj Selection 1	Chhattisgarh	Moderately resistant to leaf blast, sheath rot, bacterial leaf blight
PBNR 03-2	Maharashtra	Drought tolerant
JRH 19	Madhya Pradesh	Early maturity; medium slender grains; drought tolerant

Vishnu bhog Selection1	Chhattisgarh	Short grains, aromatic rice
Kunram Sannalu	Telangana	Long slender grains, less prone to shattering and lodging;tolerant to leaf blast, neck blast
Somnath	Telangana	Resistant to gall midge biotype-1, 5, leaf blast, neck blast
Telangana Sona	Telangana	Medium maturity; medium slender grains
JGL 18047	Telangana	Resistant to blast, bacterial leaf blight, brown plant hopper, green leaf hopper
Ratnagiri 5	Maharashtra	Moderately resistant to leaf blast, bacterial leaf blight
Sahayadri 5	Maharashtra	Long bold grains; moderately resistant to leaf blast, bacterial leaf blight
HKR 48	Haryana	Moderately resistant to bacterial leaf blight, leaf folder
Gujarat Anand Rice 3	Gujarat	Moderately resistant to leaf blast, neck blast, bacterial leaf blight, stem borer, leaf folder, whitebacked plant hopper
Chinsurah Nona 1	West Bengal	Resistant to bacterial leaf blight
Punjab Basmati 3	Punjab	Long slender scented rice; tolerant to blast
GNR4	Gujarat	Moderately resistant to leaf blast, neck blast, brown spot, sheath rot, bacterial leaf blight
NK 16520 Hybrid	Uttar Pradesh, Chhattisgarh, Bihar, Jharkhand, Odisha and Telangana	Moderately resistant to leaf blast, neck blast, brown spot, sheath rot, bacterial leaf blight
JKRH 3333 Hybrid	West Bengal, Bihar, Chhattisgarh, Gujarat and Andhra Pradesh	Direct seeded aerobic hybrid; medium slender grains
Swarna Shreya	Chhattisgarh, Madhya Pradesh and Bihar	Tolerant to water scarcity
KPH 272 Hybrid	Telangana, Karnataka and Tamil Nadu	Resistant to leaf blast.
TM 07278	Bihar and Odisha	Moderately resistant to leaf blast, neck blast, rice tungro disease, stem borer, leaf folder
KPH 467 Hybrid	Chhattisgarh, Madhya Pradesh and Maharashtra	Moderately resistant to leaf blast

Pusa Basmati 1637	Punjab, Haryana, Jammu, Uttarakhand and Uttar Pradesh	Long slender grains, strong aroma; resistant to blight
Pusa Basmati 1728	Punjab, Haryana, Delhi, Uttarakhand, Uttar Pradesh	Extra long slender grains, very strong aroma; resistant to bacterial blight
PAC 8744	Chhattisgarh	Moderately resistant to leaf blast, brown spot, sheath rot
BS 158	Madhya Pradesh	Moderately resistant to leaf folder
Chhattisgarh Madhuraj Dhan 55	Chhattisgarh	High zinc and low glycemic index
BNKR 3	West Bengal	Moderately resistant to leaf blast, neck blast, brown spot, bacterial leaf blight, sheath blight, leaf folder
MTU 1121	Andhra Pradesh	Resistant to blast, brown plant hopper
Bauna Kakanamak 101	Uttar Pradesh	Aromatic, semi dwarf; short slender grains
JR 767	Madhya Pradesh	Suitable for rainfed areas
CR 1009 Sub 1	Tamil Nadu	Moderately resistant to brown spot, leaf blast, brown plant hopper, white- backed plant hopper
TKM 13	Tamil Nadu	Moderately resistant to leaf folder, stem borer, leaf blast, green leaf hopper, rice tungro disease, brown spot, sheath rot
VL 7620	Uttarakhand	Cold tolerant, tolerant to leaf blast, brown spot
KPR1	Hilly Zone of Karnataka	Moderately resistant to leaf blast

Improved Varieties for Punjab

- **PR 127:** It is a medium maturing rice variety with an average height of 104 cm. It matures in about 137 days after seeding. It possesses long slender, clear translucent grains with high total and head rice recoveries. It is resistant to all the ten presently prevalent pathotypes of bacterial blight pathogen in the Punjab state. Its' average paddy yield is 75 quintals per ha. Do not grow this variety in alkali soils and under brakish water.
- **PR 126:** It is an early maturing rice variety. It's average plant height is 102 cm and matures in about 123 days after seeding. It possesses long slender, clear translucent grains. It is resistant to seven of the ten presently prevalent pathotypes of bacterial blight pathogen in the Punjab state. Its' average paddy yield is 75 quintals per ha.

- **PR 124:** Its' average plant height is 107 cm and matures in about 135 days after seeding. It possesses long slender, clear translucent grains.It is resistant to all the ten presently prevalent pathotypes of bacterial blight pathogen in the Punjab state. This variety is susceptible to brown leaf spot. For its management use recommended fungicides. Its' average paddy yield is 76.25 quintals per ha.
- **PR 123:** Its' average plant height is 105 cm and matures in about 143 days after seeding. It possesses extra long slender, clear translucent grains with excellent cooking and eating quality characteristics. It is resistant to all the ten presently prevalent pathotypes of bacterial blight pathogen in the Punjab state. Its' average paddy yield is 72.5 quintals per ha.
- **PR 122:** Its' average plant height is 108 cm and matures in 147 days after seeding. It possesses long slender translucent grains. It is resistant to all the ten presently prevalent pathotypes of bacterial blight pathogen in the Punjab state. Its' average paddy yield is 78.75 quintals per ha.
- **PR 121:** It is a short statured, stiff strawed lodging tolerant variety with dark green and erect leaves. It attains height of about 98 cm and matures in 140 days after seeding. Its' grains are long slender, translucent with good cooking quality. It is resistant to all the ten presently prevalent pathotypes of bacterial blight pathogen in the Punjab state. Its' average paddy yield is 76.25 quintals per ha.
- **PR 114:** It has narrow, dark green erect leaves. It attains an average height of about 102 cm and matures in about 145 days after seeding. It possesses extra long, clear translucent grains. It is resistant to five of the ten pathotypes of bacterial blight pathogen presently prevalent in Punjab state. Its' average yield is 68.75 quintals of paddy per ha.
- **PR 113:** Its' average plant height is about 105 cm. Its' grains are bold and heavy. It matures in about 142 days after seeding. It is resistant to eight of the ten pathotypes of bacterial leaf blight pathogen presently prevalent in Punjab state. Its' average paddy yield is 70 quintals per ha..

Climate

There are two main seasons for growing rice in India, though three crops are taken in Tamil Nadu and Kerala. The *kharif* season is characterized by a gradual fall in temperature, more numerous cloudy days, low light intensity, a gradual shortening of the photoperiod, high relative humidity and cyclonic weather. During *rabi*, there is a gradual rise in temperature, bright sunshine, near absence of cloudy days, a gradual lengthening of the photoperiod and a lower relative humidity. Rice is essentially a short day plant. However, a combination of temperature, photoperiod and light intensity determines the growth period, crop performance and productivity. In general, yields are higher in *rabi* than in *kharif.* Rice cultivation is conditioned by temperature parameters at the different phases of growth. The critical mean temperature for flowering and fertilization ranges from 16° to 20°C, whereas during ripening, the range is from 18° to 32°C. The temperatures beyond 35°C affect not only pollen shedding but also gain filling. Thus for vigorous vegetative growth, moderately high temperatures are required, whereas for panicle initiation, slightly lower temperatures (20-22°C) are ideal. High temperatures and high light intensity adversely affect grain filling.

Area of cultivation

India has been broadly classified into the following six agroclimatic zones:

(*i*) The humid western Himalaya region, which comprises of Jammu and Kashmir, Himachal Pradesh and Kumaon and the Garhwal division of Uttrakhand. The major soil groups are sub-montane, hills and terai soils.

(*ii*) The humid Bengal – Assam basin, which includes West Bengal and Assam. The major soil groups are riverine alluvium, terai soils, lateritic soils, red-yellow loams and red sandy soils.

(*iii*) The humid eastern Himalayan region and the Bay Islands, which include Arunachal Pradesh, Nagaland, Manipur, Mizoram, Tripura, Meghalaya and the Andaman and Nicobar Islands. The major soil groups are red loamy soils, lateritic soils, red yellow soils and alluvial soil.

(*iv*) The sub-humid Sutlej-Ganga plains which include Punjab, Uttar Pradesh, Bihar and Delhi. The major soil groups are calcareous alluvial soils, riverine alluviums, saline and alkaline soils, red-yellow loams, mixed red and black soils and red sandy soils.

(*v*) The sub humid to humid eastern and south eastern uplands comprising of Orissa, Andhra Pradesh and eastern Madhya Pradesh. The major soil groups are lateritic soils, red-yellow loams, mixed red and black soils, deltaic alluvium, deep and medium deep black soils, red loamy soils and coastal alluvium.

(*vi*) The arid western plains, which include Haryana, Rajasthan and Dadra and Nagar Haveli. The major soil groups are alluvial soils, red-yellow soils and medium to deep black soils.

Soil

Experts point out that in India rice is grown in such varied soil conditions that it is difficult to point out the soil on which it cannot be grown. However, soils having good water retention capacity with good amount of clay and organic matter are considered ideal for rice cultivation. Hence, clay and clay loams are most suited for rice cultivation. Rice being a semi aquatic plant grows best under submerged conditions. A major portion of land under rice cultivation in India is under 'low land' conditions. Rice plant is able to tolerate a wide range of soil reaction, but it does have a preference for acidic soils. It grows nicely in soils having a pH range between 5.5 and 6.5. It can also be grown on alkali soils after treating them with gypsum or pyrite.

Rotations

- Paddy - Wheat
- Paddy – Potato – Urd (black gram)
- Paddy – Toria –Wheat

- Paddy – Wheat – Moong (green gram)
- Paddy – Potato – Moong (green gram)
- Paddy – Pea (for pods) - Moong (green gram)
- Paddy – Gram
- Paddy – Lentil
- Paddy - Pea (for seed)
- Paddy - Raya

In case of intensive crop rotations where there are more than two crops in a year, a paddy variety of early duration should be selected.

Cultivation

The systems of rice cultivation in various rice growing areas of the country are largely dependent upon the rice growing conditions prevalent in the respective regions. The principal systems followed in India are 'dry', 'semi-dry' and the 'wet'. The dry and semi-dry systems of cultivation are mainly confined to tracts which depend on rains and do not have supplementary irrigation facilities. The wet system is practiced in areas with assured and adequate supply of water, either by way of rainfall or by irrigation.

Dry and Semi-dry systems: Basically, the steps involved are essentially the same under both these systems of rice culture. The fields are ploughed and harrowed in summer for achieving the required tilth. Farmyard manure is uniformly distributed 2-3 weeks before sowing. The seed is sown directly with the onset of the monsoon showers, either by broadcasting, dibbling behind the country plough or by drilling in lines. Line sowing is preferable, as it ensures an adequate stand establishment and facilitates easy weeding and interculture. The reduced seed rate requirement is another advantage. The row spacing may be suitably adjusted from 20 to 25 cm. Under the semi-dry system, the rain-water is impounded till the crop is about 1½-2 months old and thereafter it is converted into a wetland crop. By that time, major operations, such as weeding, interculturing and fertilizer application might have been completed.

***Wet system*:** Under this system, the land is ploughed thoroughly and puddled with 3-5 cm of standing water in the field with an optimum depth of puddling of about 10 cm in clay and clay-loam soils. The primary objective of puddling is to obtain a soft seedbed for the seedlings to establish themselves faster, to minimize the leaching losses of nutrients and thereby increase the availability of plant nutrients by achieving the reduced soil conditions which facilitate a better availability of nutrient elements, to incorporate the weeds and stubble into the soil and to minimize the weed problem. Puddling can be done with ploughs, tillers or any other implement, depending upon their availability and soil conditions. The land is levelled after puddling to facilitate a uniform distribution of water and fertilizers.

Sowing/Planting

Seeds may be sown after sprouting or the seedlings may be transplanted under this system. The weed problem is serious under direct-seeded conditions and is difficult to control completely in a broadcast crop. Water management also poses a major problem in unlevelled lands and, therefore, transplanting has been generally practiced. Transplanting in a puddled field has the following advantages:

- A good levelling of the land is ensured.
- The weeds are buried at the time of puddling and the weed problem is reduced.
- The plant population becomes more uniform.
- The availability of most plant nutrients, such as phosphorus, iron and potassium, is increased and nitrogen is conserved better.
- The seedlings transplanted in a soft puddle are able to establish themselves faster and start early tillering and growth.
- Nurseries occupy only on a small area of the main field and the cost of maintaining them (irrigation and plant protection) is reduced considerably as compared with a crop sown with broadcast method.
- Plant protection measures can be effectively used in the nursery.
- The community nurseries facilitate timely transplanting.
- The treatment of seedlings for nutrient deficiencies and for protection against pests and diseases are facilitated before transplanting.

Seed rate

The seed rate for direct sowing by broadcasting is 15 to 20 kg/ha and by drill sowing 10-15 kg/ha is sufficient. Only well filled viable seeds should be used for sowing. Lighter seeds that float on a solution of common salt (1.06 specific gravity) should be rejected. The choice of a suitable variety based on ecological conditions, soil type and duration is a prerequisite for obtaining high yield. Soak 20 kg seed of rice in 25 lit. of water containing 20 g Bavistin 50 WP (carbendazim) and 2.5 g of streptocycline for 8 to 10 hrs before sowing. The seeds should then be thoroughly dried in shade and used for sowing.

Nursery raising

Raise healthy as well as disease and insect free nursery in order to obtain good crop of transplanted rice. Mix 30 to 40 tons of well rotted farmyard manure or compost per hectare of land on which nursery is to be raised (calculate dose of FYM as per nursery area). Irrigate the field to permit the germination of weeds. Then plough the field to prepare proper seed bed. Apply 65 kg urea and 150 kg single super phosphate and 100 kg zinc sulphate heptahydrate (21% zinc) per hectare at sowing. Apply second dose of 65 kg urea 15 days after sowing nursery so as to get seedlings ready for transplanting in 25-30 days. If the seedlings in the nursery show yellowing of new leaves, spray three times with 0.5 to 1 % ferrous sulphate solution (.5 to 1 kg ferrous sulphate dissolved in 100 liter of water) at weekly intervals.

As per requirements, rice nursery can be raised with the following methods.

1. **Wet method:** It is very common method of raising rice nursery as its very easy method. In this method pre sprouted seeds are used. Dip the seeds of rice in water

for 12 hours and spread the seeds in 7-8 cm thick layer over wet gunny bags and cover them with other wet gunny bags. Keep the seed moist by sprinkling water on it. The seeds sprout in about 24 to 36 hrs. Puddle 400 sq.m in standing water and then broadcast 20 kg pre sprouted and treated seeds uniformly for transplanting 1 hectare of land. The nursery area should be kept wet by applying frequent and light irrigations. The nursery is ready for transplanting within 25 to 30 days of sowing.

2. **Dry method:** Prepare raised beds of about one meter width and of convenient length under dry conditions. Sow seeds of rice on top of the raised beds in lines and cover them with soil. Irrigate lightly at frequent intervals. The area of top of the beds (excluding furrows) and seed rate is same as discussed in wet method. The germination is more sure in this method but chances of iron deficiency are more. Flood the nursery area with water after germination or spray ferrous sulphate two to three times. Weeding is also very easy in this method.

3. **Mat type:** Select a location having fertile soil, assured irrigation and minimum transportation distance of location to the field. There should be no stones or other hard material in the soil. The field should be preferably laser leveled and 20 metres away from tubewell and trees. Prepare the field at proper moisture. Spread polythene sheet 50-60 gauge, 90 cm wide having 1-2 mm dia perforations over it. Polythene sheets weighing 270 g spread to a length of about 15 meters (for about 150 mats) is sufficient for preparing seedlings for one acre.

   Place one or more iron frames over the polythene sheet having compartments of size 45×21×2 cm for engine operated transplanter and 58×28×2 cm for self propelled transplanter. Number and size of compartments vary according to machine specifications. Fill the soil from both sides of the frames uniformly upto the top surface. Spread 50-60 g of pre-germinated seed evenly in each compartment to achieve uniform density of 2 or 3 seeds/sq cm in the mat. For uniform seed distribution, use drum type nursery sowing seeder. The length of nursery sowing seeder is to be equal to width of frame and has openings of 1 cm diameter on full length of the roller. About 25-30 kg seed is sufficient to sow about 375 mats required for transplanting in one hectare. Cover the seeds by a thin layer of soil and sprinkle water by hand sprayer for proper setting of the soil. Lift frames and put these at the next place and repeat the above procedure for sowing the required number of seedling mats.

   After sowing, irrigate the field on same day, but the flow of water for first 2-3 irrigations should be very mild and level should be uniform so that there is no damage to newly formed mats. Care must be taken that the seedling mats should be always wet. Spray the fertilizer after an interval of about 10 days with 200 g urea dissolve in 15 litre water to 150 mats. The seedling mats will be ready after 25-30 days of sowing (DAS). Drain water from the nursery field a few hours before uprooting of nursery. Give a cut with a sharp blade/ knife along the nursery boundaries of the mat. The uprooted nursery mats are ready for transport to the field.

4. **'Dapog' method:** Dapog nursery growing method is to sow rice seeds on a concrete floor or on a raised bed of soil covered with polythene sheets. This method is used especially in places where there is assured water-supply and when early transplanting is needed. A small area is required for raising this type of nursery, 30-40 m^2 being enough to raise seedlings for transplanting one hectare. The seedlings are ready after 14 days by using this method. Raised seed beds are prepared after final levelling and are packed, levelled and covered with polythene sheets. Banana leaves, with their midribs removed, can also be used instead of polythene sheets. Pre-germinated seeds should be sown on top of these sheets at the rate of 1 kg of seed per square metre of the nursery. The germinating seeds are sprinkled with water and pressed down gently with hand or with a wooden flat board twice a day for the first 3-6 days. This helps the roots of the seedlings to remain in contact with water retained on the surface and prevents drying. After six days, the seedbed could be irrigated up to a depth of 1-2 cm of water. The seedlings raised by using the 'dapog' method are then divided into convenient sizes and rolled like a mat with roots outwards. It is necessary to control the water level in the main field transplanted with 'dapog' seedlings, as they are too small and are liable to be damaged very easily by letting in too much of standing water. The levelling of the fields is very essential to avoid stagnation of water and mortality of seedlings. Six to eight seedlings are placed in a hill. This method of nursery is also useful for raising a post-flood rice crop in Assam and West Bengal.

Rice transplanting

The seedlings are ready for transplanting within 25 to 30 days (4-5-leaf stage). Two to three seedlings are planted at 20 × 10 cm or 20 × 15 cm spacing depending upon varieties and time of transplanting in levelled field. In regions of low fertility, and for late planting, closer planting is advocated. Plant 33 hills/sq. m under normal transplanting and 44 hills per sq in under late transplantings.

Weed management

In line planted or drilled rice, weeding can be done with a hand-hoe or with rotary weeders. In a broadcast-sown crop, weeds pose a major problem and hand pulling is still practiced. The best time to weed the crop is three weeks after sowing or planting. Herbicides, such as butachlor 1.5 kg/ha, anilophos 0.375 kg/ha, thiobencarb 1.5 kg/ha, pretilachlor 0.75 kg/ha, pendimethalin 0.75 kg/ha, oxadiargyl 90 g/ha, pyrazosulfuron ethyl 15 g/ha etc. provides a satisfactory control of grass weeds when used at the recommended doses and at the right time (within 3 days of transplanting). Irrespective of the method of weeding, it is ideal to maintain a weed free condition up to 40-45 days after sowing. Bispyribac at 25 g/ha can be sprayed as post-emergence (25 to 30 DAT) for controlling nearly all weeds of transplanted paddy (grasses and sedges).

In case of direct seeded rice, both paddy and non-paddy weeds appear. For the control of non-paddy weeds, pre-emergence application of pendimethalin 0.75 kg/ha can be done and for the control of typical paddy weeds, post-emergence (25-30 days after sowing) application of bispyribac 25 g/ha can be made.

For the control of sedges and broad leaf weeds from rice (both direct and transplanted), apply metsulfuron 20 WP at 15 g/ha or ethoxysulfuron 15 WDG at 20 g/ha or bensulfuron methyl 60 DF 60 g/ha or azimsulfuron 50 DF 20 g/ha or metsulfuron methyl + chlorimuron ethyl @ 4 g ha, 20-25 days after transplanting or sowing through spray applications. Drain out the ponded water before spray in transplanted crop to get good results.

Note:

I. Drain out excess water before spraying.

II. Do not delay herbicide application.

III. Always use herbicide in rotation.

Water management

The water requirement of rice is higher than that of other cereal crops of similar duration. Figures ranging from 37 to 75 acre inches of water have been reported from various locations in India as the water requirement of rice, and this variation is primarily due to different soil and environmental conditions available in different parts of the country. Losses due to percolation are more in submerged rice lands. In lighter soils, such losses amount to about 60 per cent of the total water requirement. Soil compaction and puddling help to reduce percolation losses. Losses due to transpiration account for about 40 per cent. Losses due to evaporation depend upon the climatic factors and range from 20 to 40 per cent. Considerable saving (30-40 per cent) in water is possible when shallow submergence (2-5 cm) is maintained from tillering to flowering and near saturation during other phases of growth.

Fertilizer and nutrient management

The new short statured varieties have a high fertilizer response and give high yields of grains. The major nutrients required by rice are nitrogen, phosphorus and potassium. Among the micro nutrients, zinc is becoming important. Among all the nutrient elements, nitrogen is most important for the rice crop in almost all the soils, particularly when nitrogen-responsive varieties are grown. The economic optimum dose of N for the new varieties is 100-120 kg/ha. About one third of the total fertilizer, now available in India, is used for the rice crop alone. However, the rice-growing conditions are equally conducive to rapid losses of applied nutrients. This is particularly true to nitrogen, and this element is usually deficient in most of the rice areas. Rice is a poor utilizer of applied nutrients, there is a need to improve the fertilizer-use efficiency. The added nitrogen is lost from the soil system through mineralization-nitrification-denitrification reactions. Therefore, the time and the method of nitrogen application based on the crop growing conditions and the physiological growth stages of the crop are more important. An adequate supply of N in the early stages of growth is, however, essential for growth, tillering and maintenance of root activity and dry matter production. In soils which are fertile, the basal dose of N may not be required, but in areas where the soil fertility is low and the average yields are less then 3 tonnes per hectare, a part of the N (about one-third) needs to be applied as a basal dose. An early application of moderate levels of N at planting and tillering are effective in increasing the formation of panicles whereas late applications just before the initiation of panicles increase the number and the size of the spikelets.

Nitrogen can be applied either as ammonium sulphate or as urea. The loss of nitrogen from urea, the principal source of nitrogen, now being manufactured in the country, may be 60 to 80 per cent. Apply basal dose of N as urea followed by top-dressing at tillering and about a week before the initiation of panicles is effective over a wide range of situations. The future development in fertilizer use would consist of a judicious combination of organic and inorganic components as well as biological N-fixation in an integrated nutrient-supply system. Rice also requires phosphorus for its normal growth and yield. Phosphorus and potash are added on soil test basis. However, the phosphorus requirement of lowland rice is not as pronounced as its nitrogen requirement. In the acid laterites, black clay soils and in coastal sandy soils, fertilization with phosphorus is a problem because of the soil fixation of large amounts of phosphates. The availability of P is increased under flooded conditions owing to the reduction of ferric phosphates to soluble ferrous phosphates on the one hand and the hydrolysis of iron and aluminum phosphates on the other hand.

The optimum rate of phosphorus for rice varies with the soil type, conditions of rice culture, the source of P used and the variety. However 40 to 60 kg of P_2O_5 per hectare can be used in soils which are deficient in phosphorus. Superphosphate is very efficient source of phosphorus in rice on especially highly acidic soils. Phosphatic fertilizers are best applied as a basal dose before planting. In acidic soils, rock phosphate can also be used. The application of rock phosphate, at least 15 days before planting, is essential for better utilization. Rice soils in India are fairly rich in potassium and its requirement by rice to a great extent is met from the irrigation water and by the incorporation into the soil of stubble and crop residues which contain a lot of potassium absorbed by the plant. However, with the high intensity cropping and the cultivation of high yielding varieties, potash fertilization can become necessary in light textured and coastal sandy soils.

Zinc deficiency

Zinc deficiency is commonly associated with soils of high pH above neutrality. It is believed that one-unit increase in pH makes zinc one hundred times less soluble. However, submerged acid soils may, at times, develop Zn deficiency because of the possible rise in pH of these soils after flooding. In alkaline soils, a high content of the organic matter of the soil tends to aggravate Zn deficiency. The continuous submergence of neutral or calcareous soils leads to zinc deficiency, as the concentration of zinc in soil solution gets depleted owing to continuous submergence. An excessive use of N and P fertilizers also promotes microbial activity and the microbial fixation of zinc. The symptoms of zinc deficiency appear 2-3 weeks after transplanting. The lower leaves become rusty brown near the base and ultimately dry up. The seedling with zinc deficiency remains stunted and tillerless. Apply 50 to 60 kg per hectare of zinc sulphate before or after transplanting or direct sowing of rice crop to check zinc deficiency.

Iron deficiency

Under scarcity of water, chlorosis among seedlings appears in the youngest leaf about three weeks after transplanting. Plants die and often the crop fails completely. Start giving copious irrigations as soon as chlorosis appears and give 2 to 3 sprays of 1.0% ferrous sulphate solution at weekly intervals (2.5 kg of ferrous sulphate in 250 litres of water per hectare).

Management of paddy straw

In Punjab, about 220 lakh tonnes paddy straw is produced annually and a large portion of the straw is being burnt by the farmers. The straw contains a significant amount of essential plant nutrients. One tonne of paddy burning causes loss of 400 kg organic carbon, 5.5 kg of N, 2.3 kg of P, 25 kg of K and 1.2 kg of S. The burning of residue causes complete loss of nitrogen and sulphur. Due to burning of paddy straw poisonous gases like carbon dioxide, carbon monoxide, methane and nitric oxide are produced which are harmful for human and animal health. To avoid burning of straw, the following techniques are recommended :

1. In-situ management of paddy straw

(a) **Wheat sowing with Happy Seeder:** Happy Seeder is recommended to sow wheat in combine harvested paddy field without any straw burning. It is a PTO driven machine operated by 45 HP tractor and it covers about 0.75-0.80 acre per hour. The loose straw needs to be spread uniformly in the field for the proper working of Happy Seeder. It can be done manually or harvest the paddy by combine harvester fitted with PAU super straw management system (SMS). After the harvesting of paddy with combine fitted with Super SMS, wheat can be sown with Happy Seeder or Spatial Zero Till Drill, without straw burning. The paddy straw acts as mulch which adversely affects the weed population.

In case the paddy is harvested with a combine without SMS system, use PAU Straw Cutter cum Spreader for chopping and spreading of paddy straw. After chopping of straw, sow wheat with PAU Happy Seeder (fitted with press wheels).

(b) **Incorporation:** Chop the straw and stubbles with Paddy Straw Chopper cum Spreader after combine harvesting of paddy. The chopped straw can be mixed in to the soil with rotary tillers after applying a light irrigation. The field comes in wattar condition in 2- 3 weeks depending on type of soil. Sow wheat seed with zero till drill or with conventional drill. Grow short duration rice varieties for increasing window period for straw management.

2. Management after removal of paddy straw

Use Straw Baler for baling the paddy straw after combine harvesting. It collects the loose straw or complete straw after chopping the standing stubbles with stubble shaver. These bales can be used for different purposes as follows:

(a) **Power generation:** Seven bio mass power plants have been established in Punjab for the generation of electricity from paddy straw bales. Heat produced by burning paddy straw is used to run steam turbine which further generates electricity.

(b) **Paddy straw based bio-gas plant:** Paddy straw can be used in a specially designed biogas plant for bio gas production. This biogas plant is filled one time with 16 quintal chopped paddy straw along with 4-5 quintal cattle dung and it provides 6-7 m^3 biogas daily for 3-4 months.

(c) **Paddy straw geyser:** A geyser has been developed by the University for heating water by using paddy straw bales. In this geyser about 100 litres of water can be heated to 45-500C in 3-4 hours. The water once filled in the geyser remains hot for up to 24 hours or even more.

(d) **Phospo-compost:** Collect rice-straw from fields and bring it to the composting site near the tubewell on the farm to have easy water availability. It can be made into bundles of convenient size (about 10-15 kilograms).

Prepare large quantity of a "soaking solution" by thoroughly mixing one kg cow dung for every 1000 litre of water in a big tank. (The volume of the tank can be calculated by measuring Length x Breadth x Height of the tank in metres. One cubic metre is equal to 1000 Litres of water). Dip the bundles one by one into the "soaking solution" for 2-3 minutes.

Drain the excess solution by placing the bundles on a slope lined with a plastic sheet. The drip should be collected and recycled into the tank again. Make 15 cm raised beds 5 metre long and 1.5 metre wide on the ground. This will help in assessing the exact watering of the heap later. Draining of water out of bed is a visual indication of excess watering.

Take the wet rice-straw to the location of the compost heap. Line the bed with 2-6 centimetre diameter tree branches/sticks. This helps in aeration in the heaped rice-straw. The wet rice straw will generally have 70 per cent moisture. Place the wet rice straw on the beds uniformly until 500 kilograms has been stacked. Powdered low-grade rock phosphate (low grade rock phosphate can be had from Rajasthan State Mines and Minerals Ltd 4, Meera Marg, Udaipur 313004) should be mixed @ 6 per cent on dry weight basis of the rice straw approximately. For 500 kilogram of the rice straw, 30 kg of the rock phosphate should be sprinkled uniformly while making the heap after wetting. This will give approximately 1% phosphorus in the final decomposed product. The height of 500 kg rice-straw stack is 1.5 metre approximately. Any quantity of rice-straw can be composted in multiple heaps of 500 kg at one time leaving a passage of 1 metre between the beds.

Cover the heap with a 20-30 centimetre thick layer of unsoaked rice-straw. This will minimize water loss while providing the necessary aeration. The major key to success is the ability to maintain about 70 per cent moisture in the heap. Any major error in this step will delay composting. Water the heaps using watering lance with the help of Tullu Pump. (Note : watering heaps with sprinklers does not work because water generally runs down the sides, instead of going inside the heap. Ensure that the water penetrates the heap by using a lance with a sharp point to pierce the heap of rice straw. Pierce the lance deepest possible with an aim to water uniformly). Composting can be terminated after 80-90 days by which time it is ready for processing or for field application. By this time its carbon and nitrogen ratio changes to 15:1. At this stage, strands of the rice straw are weak and twisting can readily break a hand full of it.

(e) **Prali Char:** Prali Char is a carbon rich porous product obtained after the partial combustion of rice straw at low temperatures in the presence of little or no oxygen. It can be prepared in a pyramid or dome type kiln, made up of bricks and clay. To prepare 8 quintals of Prali Char from 12 quintals of rice straw, the height of this dome should be 14 ft with 10 ft diameter. Two windows each at the top and at the bottom of the kiln are provided for loading of paddy straw. Six vents of 2 inch diameter in the upper portion and eight vents are provided at three heights on the remaining portion of the structure. Fill the kiln with rice straw up to top. Seal the lower loading window and ignite the straw from the top loading window and immediately seal it with clay. The partial combustion of rice straw will start from the top and proceeds towards the bottom. The emission of thin blue smoke from the vents indicates that the process of Prali Char formation is complete in this zone. To facilitate the combustion in the next zone, seal the vents located in the upper portion of the dome and likewise proceed to seal the vents of the middle and lower zone, respectively. The whole process usually takes about 10-12 hours. To seal the cracks developed during combustion and to cool the kiln, pour a diluted mixture of clay and water on the outer surface of the kiln. Normally it takes about two days to cool, however it can be cooled by sprinkling water for removing the Prali Char on the same day. On an average, it contains 30-36 % carbon, 0.5-0.6 % nitrogen, 0.16-0.22 % phosphorus and 1.6-2.2 % potassium. Its application to rice and wheat @ 5.0 t/ha saves 40 kg N (35 kg urea), increases crop productivity and improves soil health.

(f) **Straw Pellets:** The paddy straw can be used for making enriched fodder pellets and fuel pellets. The concept of turning straw into food pellets is not a new process and considered a successful animal feed. Countries like singapore and Thailand are already using this concept. Japan is importing paddy straw from China for feeding to cattle also there is a tremendous scope of export of enriched fodder pellets to Qatar and the middle east countries from there it can be exported to other countries.

Besides, paddy straw can also be used as mulch material in different crops, mushroom cultivation and as animals bedding.

Plant Protection

Insect pests

- **Rice stemborers:** The larvae of these insects bore into the stem and cause damage from July to October. The affected young plants show dead-hearts (yellowing and drying of central shoot) whereas the old ones produce empty earheads which turn white and stand erect.

The fields showing more than 5% dead hearts (Economic Threshold Level, ETL) should be sprayed with 50 ml Fame 480 SC (flubendiamide)* or 425 g of Mortar 75 SG (cartap hydrochloride) or 2.5 litre of Coroban/Dursban/Lethal/ Chlorguard/Durmet/Classic/ Force 20 EC (chlorpyriphos) in 250 litres of water per ha. Further application of any of these insecticides may be repeated as and when damage reaches economic threshold level.

- **Leaf folder:** The larvae fold the leaves, eat out the green tissue and produce whitestreaks. The damage is highest during August-October. When the leaf damage reaches 10% (ETL), adopt the following control measures:
- **Mechanical control:** The mechanical control of leaf folder can be done only before flowering by passing the 20-30 m long coir/jute rope, forwards and then backwards, both ways while touching the crop canopy. While passing the rope, ensure that water must be standing in the crop.
- **Chemical control :** Spray the crop with 50 ml Fame 480 SC (flubendiamide)* or 425 g Mortar 75 SG (cartap hydrochloride) or 2.5 litre of Coroban/Durmet/ Force 20 EC (chlorpyriphos) in 250 litres of water per ha.
- **Plant hoppers:** These hoppers include, whitebacked planthopper and brown plant hopper. Both nymphs and adults of these pests suck the cell sap particularly from the leaf-sheath from July to October. The crop dries up in patches. As the plants dry up, the hoppers migrate to the adjoining plants and kill them. In a few days, the area of the dry patches enlarge.

About one month after transplanting, a few plants in the field should be slightly tilted and tapped 2 or 3 times at the base at weekly interval. If minimum 5 plant hoppers per hill (ETL) are seen floating in the water, only then the crop should be treated with insecticide. The plant hoppers can be controlled by spraying 300 g Chess 50 WG (pymetrozine) or 100 ml Confidor 200 SL/Crocodile 17.8 SL (imidacloprid) or 2.0 litre Ekalux/Quinguard/ Quinalmass 25 EC (quinalphos) or 2.5 litre Coroban/Dursban 20 EC (chlorpyriphos) in 250 litres of water per ha. Repeat the spray if necessary. For better effectiveness, use knapsack sprayer while directing its spray towards the base of the plants. If the damage is noticed at hopper burn stage, treat the affected spots alongwith their 3-4 metre periphery immediately as these spots harbour high population of the insect.

- **Grasshoppers:** The adults and nymphs of the grasshoppers eat the leaves especially in nursery. Insecticides recommended for the control of plant hoppers are also effective for grasshoppers.
- **Rice hispa:** The grubs of this pest tunnel into the leaves, whereas the adults are exposed feeders. The grubs cause damage by producing bold, white streaks on the leaves. If the attack starts in nursery, clip-off and destroy the leaf tips of the affected seedlings before transplanting. On the transplanted crop, spray 2.0 litre Ekalux 25 EC (quinalphos) or 2.5 litre Dursban 20 EC (chlorpyriphos) in 250

litres of water per ha with a manually operated sprayer. Repeat the spraying if the attack persists.

- **Rice root weevil:** Its' white, legless grubs feed on roots in the soil from July to September. The attacked plants turn yellow, stunted and produce only a few tillers. Apply Furadan 3G @ 33 kg/ha or Thimet 10G @ 10 kg/ha for its control.
- **Rice-ear-cutting-caterpillar:** The larvae of this insect are gregarious in habit and are commonly known as 'armyworm'. The young larvae feed on leaves, leaving only the mid- ribs and stems. The old larvae cut off the panicles mostly at the base and hence the name "rice ear-cutting caterpillar". This stage of the insect causes serious loss to the paddy crop. The larvae are shy of sunlight and generally feed at night. The damage to paddy crop is caused mostly during September to November.

The insect can be controlled by spraying the crop with 1.0 litre of Ekalux/ Quinguard 25 EC (quinalphos) in 250 litres of water per ha. As the pest is nocturnal in behaviour, the spray should be done in evening hours for getting better results.

Swarming caterpillar: This pest cause severe damage in nursery beds. Newly hatched larvae cut leaves and cause plants to look sickly with withered tips. Older larvae feed voraciously and cause almost complete defoliation. In case of moderate to severe damage spray Ekalux 25 EC at the rate of 2 mililitre in 1 litre of water.

Mole cricket: This is a problem on raised nursery beds and in upland fields. The adults feed on young roots and basal portions of the plant below the ground. They may kill the plant by cutting at the base. Apply Furadan 3 per cent G at the rate of 33 kg per hectare or Thimet 10 percent G at the rate of 10 kg per hectare for the control of this pest.

Diseases

- **Sheath blight (*Rhizoctonia solani*):** Greyish green lesions with purple margin develop on the leaf-sheath above the water level. Later, the lesions enlarge and coalesce with other lesions. Its severe attack results in the poor filling of the grains. Destroy the rice straw and stubbles after harvesting the affected crop. Avoid the excessive use of nitrogenous fertilizers. Keep the bunds clean by removing the grass.

  Give two sprays of Amistar Top 325 SC or Tilt/Bumper 25 EC (propiconazole) or Folicur/Orius (tebuconazole) 25 EC @ 500 ml or Nativo 75 WG @ 200 g or Bavistin 50 WP @500 g in 500 litres of water/ha. First spray at boot stage of crop and second spray after 15 days.
- **False smut (*Ustilaginoidea virens*):** It is a fungal disease in which the individual grains transforms into large yellowish/greenish velvety spore-balls. High relative humidity, rainy and cloudy days during the flowering period increase the incidence of the disease. The application of organic manures and high dose of nitrogenous fertilizers also increase the intensity of attack.

  To control this disease, give first spray of Kocide 46 DF (copper hydroxide) @

1.25 kg in 500 litres of water per ha at boot stage of the crop in disease prone areas followed by second spray with Tilt 25 EC @ 500 ml in 500 liters of water after 10 days interval.

- **Brown leaf spot (*Drechslera oryzae*):** It produces oval, eye-shaped spots with a conspicuous dark-brown dot in the centre and light brown margin. Spots are also produced on the grains. This disease occurs in poor soils, therefore, give adequate and balanced nutrition to the crop. To control the disease, give two sprays of Nativo 75 WG (trifloxystrobin + tebuconazole) @ 200 g or Indofil Z-78 75 WP @ 1.25 kg in 500 litres of water/ ha. First spray at boot stage of crop and second spray after 15 days.
- **Blast (*Pyricularia grisea*):** The fungus causes spindle shaped spots with greyish centre and brown margin on the leaves at maximum tillering. It also causes brown lesions on the neck of the panicle, showing neck rot symptoms and the panicles fall over. The disease is more severe on Basmati cultivars particularly in the submontaneous areas and under application of heavy nitrogenous fertilizers. Spray the affected crop with Amistar Top 325 SC (azoxystrobin + difenconazole) @ 500 ml or Indofil Z-78, 75 WP (zineb) @ 1.25 kg per ha in 500 litres of water, at the boot and ear-emergence stages.
- **Bunt/Kernel Smut (*Neovossia horrida*):** Only a few grains in the panicle are infected. Frequently, only a part of the grain is replaced by a black powder. Sometimes, entire grain is also attacked and the black powder scatters on to other grains or leaves, and this is often the easiest way to detect the disease in the field. Also avoid heavy doses of nitrogenous fertilizers. To control the disease, give first spray of Tilt 25 EC @ 500 ml in 500 litres of water/ha at boot stage and second 10 days thereafter.
- **Sheath rot (*Fusarium moniliforme*):** The rot occurs on the uppermost leaf-sheaths where oblong to irregular and grey-brown to light-brown lesions develop. The lesions often coalesce to cover the entire sheath. In severe cases, young panicles either do not emerge or emerge partially. A white-powdery growth of the fungus appears on the panicle inside the sheath. The glumes of infected florets are discoloured, dark red or purple brown to black and often do not fill. The fungus over-winters in rice straw and grains. Destroy the rice straw after harvesting the infected crop. Use disease free seed for sowing. Give two sprays of Bavistin 50 WP @ 500 gm in 500 litres of water per ha. The first spray should be given at boot stage and second after 15 days interval.
- **Stem rot (*Sclerotium oryzae*):** The fungus affects the stem at earing and black lesions are produced on the sheath at water level. Later on, the stem gets infected and rots leading to withering and lodging of the plant. The incidence of this disease has declined on high yielding varieties due to improved cultural practices. Destroy the diseased debris of infected crop. Avoid excessive irrigation and use recommended dose of nitrogenous fertilizer.
- **Bacterial blight (*Xanthomonas oryzae pv. oryzae*):** Greenish-yellow stripes

appear along the leaf margins and extend both lengthwise and breadthwise. The leaf starts drying from the tip, becomes white in severe cases and dries up completely. The disease sometimes attacks the freshly transplanted seedlings which start wilting and in a few days the whole clump dries up. The bacterium perpetuates through seed, rice straw, and roots of non-host plants during the off-season. In order to mitigate the losses, adopt the following integrated measures:

- For the management of bacterial blight, grow resistant varieties.
- Do not apply excessive dose of nitrogen. Nitrogen should not be applied beyond six weeks after transplanting (except when LCC is used).
- Do not pond water in the field continuously.
- Treat seed before sowing to kill primary inoculum, see "Seed Treatment" under Agronomic Practices.
- Do not grow nursery under shade.
- Rice crop itself should not be grown in the shady area.
- The growing of rice in the area near wheat straw stack (Kup) should be avoided.

- **Bacterial leaf streak (*Xanthomonas oryzae pv. oryzicola*):** Small translucent streaks appear in the interveinal areas of the leaf. The streaks gradually enlarge and turn reddish, when the plants near maturity. Seed treatment can help to control this disease.

Tungro virus: Older leaves turn yellow orange starting from the tip and margin. Plants become stunted and bear poor panicles with empty glumes showing dark brown colour on young leaves. Interveinal chlorosis is observed. The green stubbles should be uprooted and burnt after harvest. Isolated plants showing disease symptoms in the beginning should be uprooted and destroyed so that the insect does not get incoculum to spread on healthy plants. The vector can be controlled effectively by spraying Diazinon at the rate of 1.5 kg (a.i.)/hectare. In all, 5 sprays should be given including one in nursery after 10 days of sowing followed by 15, 30, 45 and 60 days after transplanting in the main field. Dimecron 100 also controls the vector population. Furadan and Carbofurrn 75 wettable powder as seed dresser or pre-transplanting dip have also been found to give good control of the insect. Grow IR 20, Ratna which are tolerant to this disease.

Khaira disease: The disease is caused by zinc deficiency. Usually appears in nursery but may appear, in patches, after 10-15 days of transplanting. Growth of the diseased plant is stunted. High yielding varieties show chlorosis between the veins of new leaves, where brown spots are formed. On lower leaves, a large number of small, brown to bronze spots appear, which coalesce to form bigger spots and ultimately the entire leaf turns bronze coloured and dries. Root growth is also restricted and usually the main roots turn brown. The finer roots are destroyed. Spray a mixture of 5 kg zinc sulphate and 2.5 kg lime in 1000 litres of water if symptoms appear in the field, 10 days of sowing in nursery. Give second spray as above on 20th day in nursery. Give third spray as above after 15-20 days of transplanting.

Harvesting

Timely harvesting prevents losses in yield due to shedding of grains. The later stage of grain ripening is a dehydration process and maturity is hastened when water is withdrawn from the field at the hardening stage of the grains. The timely draining away of water makes the field hard to facilitate mechanical harvesting. The harvesting of early and medium varieties, 25-30 days after flowering and 35-40 days in the case of late varieties is desirable. Paddy should be harvested when the moisture content of the grain is about 20-22 per cent which is to be reduced to 13 to 14 per cent before milling. The direct drying under the sun leads to an increased breakage of grains during milling. Gradual drying in shade is essential for better recovery in mills. The produce after thrashing is known as rough rice, paddy or grain. The rough rice is milled before it is consumed.

WHEAT (Triticum aestivum)

Wheat crop

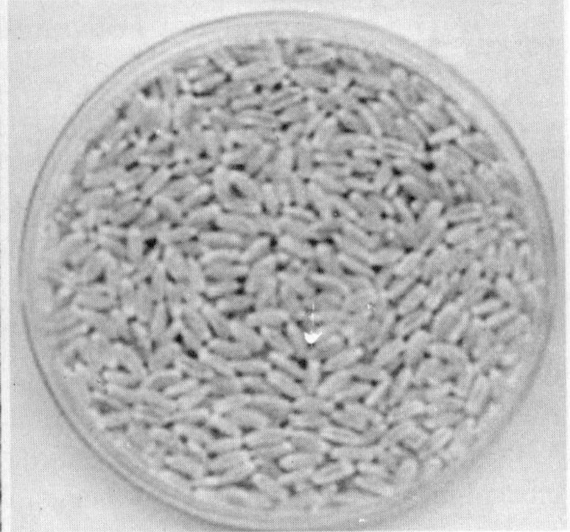

Wheat grains

Introduction

Among food crops in the world, wheat (*Triticum* spp.) occupies number one position. India is one of the principal wheat producing and consuming countries in the world. Its importance in Indian agriculture is second to only rice. The Green Revolution, initiated in the country in the late 1960s, played significant role in increasing the yield of wheat. Wheat is generally grown for intended food for humans, but lesser quality wheat and the nutrient dense by-products of flour refining are used for animal feed. Wheat is also used for whiskey and beer production, while the husk can be separated and ground into bran. Most breads, even rye and oat breads, are made with at least a portion of wheat flour because of two main quality characteristics of wheat that improve the breads - its gluten, and its alpha amylase activity. High gluten flours offer elasticity in the dough, allowing for it to rise without developing large air pockets. Tender pastries and biscuits, are best with low gluten flours. All wheat flours are best with low alpha amylase activity, because alpha amylase turns starch to sugar and prevents development of proper dough characteristics.

Wheat flour based products, such as the *chapati*, are part of the staple diet in most parts of India particularly in northern India. The by products derived from the milling of wheat into white flour include wheat bran, standard middlings, flour middlings, brown shots and grey shots. All of these by-products are higher in protein content than wheat itself and serve as a valuable protein supplement in many livestock rations. Wheat straw is also used as excellent feed for livestock in undeveloped countries whereas in developed countries it is used for bedding on farms and some is manufactured into paper. Most of the straw in the leading wheat producing areas is wasted, usually being burnt following threshing.

Origin and history

The growing of wheat began very early in pre-historic time, so long ago that the origin of wheat is still a matter of speculation. From all possible records, it seems that its centre of origin is South-Western Asia. It is believed that Aryans brought wheat grains to India, since then it has been cultivated in India. After independence research on development of high yielding varieties of wheat and improved management techniques has been carried out in India for the last several decades and many valuable varieties were bred and released for commercial cultivation. These varieties were tall and mainly suited to low management with low yield potential. However, a turning point came in the history of wheat in 1963 with the introduction of dwarf, photo insensitive high yielding Mexican wheat breeding material developed by Noble Laureate Dr. Norman E. Borlaug. In *rabi*, 1964 these varieties were tested at 155 locations under All India Coordinated Wheat Improvement Project, where Lerma Rojo and Sonora 64 out yielded the local tall wheat varieties. As a result in 1965 both these varieties were released for general cultivation in wheat growing regions of India. After sometime these varieties were replaced by amber seeded high yielding wheat varieties like Kalyan Sona and Sonalika. In 1970 triple dwarf varieties were released and since then so many dwarf varieties have been evolved for cultivation in different agro-climatic zones of India.

Area and distribution

Wheat is the most widely cultivated of all cereals. It is grown under various climatic conditions between 47°S and 57°N latitudes. Hence wheat crop is almost sown or harvested, all round the year in one country or the other. A systematic zonation of wheat culture on the globe includes the North America continent, Latin America, Europe, West Asia and North Africa, South and East Africa, South Asia and far East Australia. During 2016, India contributed 12.5 per cent (Fig. 8.4) whereas China contributed 17.6 per cent in the global production of wheat, whereas contribution of Russian Federation is 9.8 percent.

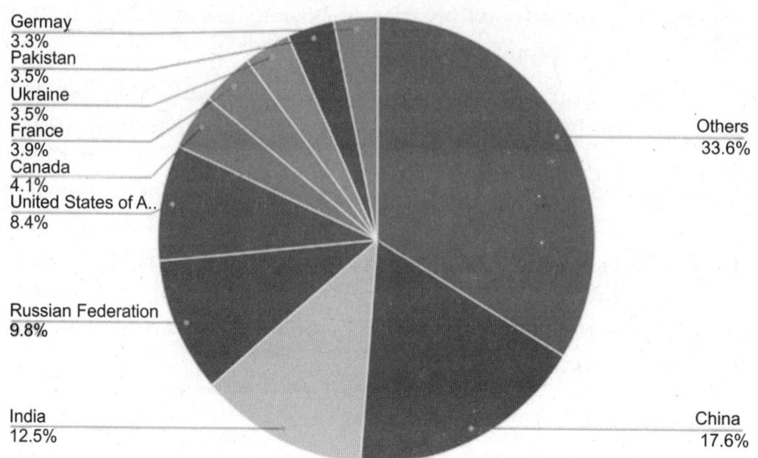

Fig. 8.4. Annual global production of wheat (Percentage) during 2016 (FAO STAT 2016)

Wheat cultivation in India spans between 10ºN in the peninsular India and 35ºN in the Northwestern Himalayan foothills. In India, it is cultivated on an area of 30.23 million hextares with production of 93.50 m tonne during 2015–16 with average yield of 3093 kg/ha/FAO STAT 2016). In Punjab its area is 3.51 m ha with production of 6.07 m tonnes during 2015–16 (Anouynous 2015–16) About 80% of the area under what is irrigated. Efforts were initiated in 1976 to extend the cultivation of wheat in the States where it had not been traditionally cultivated, such as West Bengal, Kerala, Assam, Orissa and Andhra Pradesh.

India is broadly divided into 5 wheat zones based on agroclimatic conditions:

(i) **The North-Western Plain Zone:** This is the most important zone consisting of the plains of Punjab, Haryana, Jammu, Rajasthan (except the south-eastern portion) and Western Uttrakhand. The wheat here is planted in late October – November and the harvesting usually begins by middle of April.

(ii) **The North Eastern Plain Zone:** This zone consists of Eastern Uttar Pradesh, Bihar, West Bengal, Assam, Orissa, Manipur, Tripura, Meghalaya, Nagaland, Mizoram, Arunachal Pradesh and Sikkim. As there is late harvesting of paddy in these area, wheat is mainly sown in late November or early December. Harvesting is done by March – April.

(iii) **The Central Zone:** This zone consists of Madhya Pradesh, Gujarat, South Eastern Rajasthan (Kota and Udaipur area) and the Bundelkhand area of Uttar Pradesh. Almost 75% of wheat cultivated in this area is rainfed. The best quality *durum* wheat are produced in this zone.

(iv) **The Peninsular Zone:** The peninsular zone consists of the Southern states of Maharashtra, Andhra Pradesh, Karnataka, and Tamil Nadu. Sowing in this zone is mostly complete by early November and the harvesting starts in second half of February. The duration of the wheat crop is the earliest in this region.

(v) The Northern Hill Zone: This zone consists mainly of the hilly areas of Kashmir, Himachal Pradesh, Uttar Pradesh, West Bengal, Assam and Sikkim. The sowing in this zone is done in October and the harvesting is done in May-June. The crop remains almost dormant in the cold months from November to March and as the temperature rises in April the crop starts growing.

Classification

There are several bases that may be used to classify wheat, but the most important are the groupings based on botanical characteristics. Wheat belongs to the grass family Gramineae (Poaceae) and the genus *Triticum*. The species of *Triticum* can be classified into three distinct groups; diploids, tetraploids and hexaploids, with 7, 14 and 21 pairs, respectively of chromosomes in their cells. Although as many as 18 species of wheat have been described and recognized, the 12 cultivated species of *Triticum* are as follows:

Group	Chromosome No. (2n)	Species	Genome formula	Common name
Diploid	7	*T. monococcum*	AA	Einkarn
Tetraploid	14	*T. dicoccum*	AABB	Emmer
		T. durum	AABB	Durum
		T. persicum	AABB	Persian
		T. turgidum	AABB	Poulard or Rivet
		T. polonicum	AABB	Polish
Hexploid	21	*T. aestivum*	AABBDD	Common
		T. compactum	AABBDD	Club
		T. spletra	AABBDD	Splet
		T. sphaerococcum	AABBDD	Shot
		T. macha	AABBDD	Macha
		T. vavilovi	AABBDD	Vavilovi

In India, only three species of wheat namely *Triticum aestivum* (common bread wheat), *T. durum* (macroni or durum wheat) and *T. dicoccum* (emmer wheat) are economically important. The bread wheat is the most important species accounting to a little over 87 per cent of the total wheat production followed by the durum wheat (about 12 per cent) and dicoccum wheat (about 1 per cent). The common wheat, which is good for chapati making and bakery products, is grown in whole of North India and also in central and south India. The durum wheat which is good for suji, semya, sphagetti, pasta and macroni is grown only in central and south India and that too under rainfed conditions. The dicoccum wheat is grown only on limited acreage in Tamil Nadu, Andhra Pradesh, Maharashtra and Gujarat. It is good for the south Indian dish 'uppumav' or 'upma'. The common varieties of wheat species are grown in different climatic zones:

Improved Varieties of Wheat released by ICAR

2017-18

Variety	Area of adoption	Salient features
WB-02	Punjab, Haryana, Delhi, Rajasthan, Uttar Pradesh, Jammu and Kashmir, Himachal Pradesh, Uttarakhand	Bread wheat, high zinc (42 PPM) variety, timely sown, irrigated condition, good level of yellow (stripe) rust resistance
Pusa Tejas	Madhya Pradesh, Chhattisgarh, Gujarat, Rajasthan and Uttar Pradesh	Suitable for timely sown irrigated conditions, Durum wheat, rust resistance, dual purpose variety suitable for making chapati, pasta and other traditional food products. high protein content (12%), and essential micronutrients like iron (42.1 ppm) and zinc (42.8 ppm), average yield 5,700 kg/ha
Pusa Ujala	Maharashtra, Karnataka and Tamil Nadu	Suitable for timely sown, restricted irrigation conditions, high levels of resistance to black and brown rust diseases, excellent chapati making quality, high protein (around 13%) and rich in micronutrients like iron (43 ppm) and zinc (35 ppm), average yield 3,000 kg/ha
HD 3171	Uttar Pradesh, Bihar, West Bengal and Assam	Rainfed, timely sown condition, resistance to leaf rust (LR) and leaf blight (LB)
Unnat PBW 343	Punjab, Haryana, Delhi, Rajasthan, UP, J and K, HP and Uttarakhand	Suitable for timely sown, irrigated condition. Bread wheat, developed through MAS, carrying five rust resistance genes, possesses resistance to yellow rust (YR) and brown rust (BR) diseases, maturity 155 days, average yield 5,480 kg/ha
PBW 1Zn	Punjab, Haryana, Delhi, Rajasthan, Uttar Pradesh, Jammu and Kashmir, Himachal Pradesh, Uttarakhand	Suitable for timely sown, irrigated condition, high zinc, resistant to BR and moderately resistant to YR, maturity 151 days, average yield 5,600 kg/ha
MACS 3949	Maharashtra and Karnataka	Suitable for timely sown, irrigated condition. Durum wheat, resistance to stem rust and LR and having good pasta quality (7.25/9), maturity 110-115 days, average yield 4,398 kg/ha.
Chhattisgarh Genhu 4	Chhattisgarh	Suitable for late sown, irrigated condition, heat tolerant,

Sabour Samriddhi	Bihar	Suitable for timely sown, irrigated condition, semi dwarf, erect type, strong waxiness on leaf sheath, maturity 120-125 days, yield 4,000-5,000 kg/ha.
Sabour Shreshtha	Bihar	Suitable for late sown, irrigated condition, resistant to important diseases and insect pests, good chapati making quality, high protein and zinc content, highly responsive to nitrogen up to 120 kg N/ha, lodging resistant, having easy threshability and non- shattering in nature, average grain yield 4,300 kg/ ha.
Sabour Nirjal	Bihar	Rainfed, lodging and drought tolerant, timely sown, strong waxiness on leaf sheath and ear, semi erect plant, maturity of 125- 130 days, yield 2,500-3,000 kg/ha.
Gujarat Junagadh Wheat 463	Gujarat	Suitable for timely sown irrigated conditions and early sown irrigated conditions, protein content 12.7% and micronutrients Fe, Zn, Mn and Cu content of 31.9, 23.8, 22.4 and 5.9 ppm, respectively, good chapati making quality, grain yield 5,575 kg/ha (Saurashtra) and 5,091 kg/ha (Gujarat)
CoW 3	Tamil Nadu	Suitable for limited irrigation, timely sown, short winter conditions, maturity 95-100 days, resistant to LR, SR, yield 4,076 kg/ha
K-1317	Eastern UP, Bihar, Jharkhand, Odisha, WB (excluding hills), Assam and plains of NE States	Recommended for rainfed, timely sown conditions, resistant to BR and leaf blight, good chapati quality (Score:8.05)

2016-17

Wheat Variety	Area of adoption	Salient features
UAS 347	Maharashtra and Karnataka	Suitable for timely sown, rainfed conditions; resistant to black and brown rusts
PBW 658	Punjab	Suitable for irrigated, late-sown conditions; resistant to stripe and leaf rust, leaf blight

Raj 4238	Madhya Pradesh, Chhattisgarh, Gujarat, Rajasthan and Uttar Pradesh (Jhansi and Chitrakoot division)	Suitable for irrigated, late-sown conditions; good for chapati-making
Central Wheat HS 562	Western Himalayan regions of Sikkim and hills of West Bengal and North–Eastern States	Suitable for both rainfed and irrigated, timely sown conditions; field resistance to yellow rust
Durum		
Pusa Malwi	Madhya Pradesh, Chhattisgarh, Gujarat, Rajasthan and Uttar Pradesh	Suitable for irrigated, timely sown conditions; resistant to brown and stem rusts; with large grains
HD 3117	National Capital Region	Suitable for conservation agriculture under latesown conditions; tolerant to Karnal bunt, moderately resistant to yellow rust
GW 451	Gujarat	Suitable for irrigated, timely sown conditions; resistant to brown and stem rusts
HDSCW 18	National Capital Region	Suitable for conservation agriculture under early sown conditions; resistant to brown rust, moderately resistant to yellow rust
PhuleSamadhan	Maharashtra	Suitable for both timely and late sown irrigated conditions; resistant to brown and black rusts
MP 3382	Madhya Pradesh	Timely sown, irrigated conditions; tolerant to brown and black rust;with good chapati- making
MPO (JW 1255)	Madhya Pradesh	Suitable for rainfed/ restricted irrigation, timely sown conditions; suitable for pasta- making; resistant to black and brown rust
DH 114	Himachal Pradesh (High altitude areas)	Developed through double haploid technique
PBW 660	Punjab, Haryana, Delhi, Rajasthan, western Uttar Pradesh, Jammu and Kashmir, Himachal Pradesh and Uttarakhand tarai region	Suitable for timely sown, rainfed conditions; resistant to yellow and brown rusts; with good chapati-making quality
PBW 677	Punjab	High degree of resistance to yellow and brown rusts
PBW 725	Punjab	Resistant to yellow and brown rusts

PDKV Sardar	Maharashtra	Suitable for late-sown; and is early maturing
VL 953	Uttarakhand (Hills and Plains)	Suitable for irrigated, organic farming conditions
UP 2784	Uttarakhand (Plains)	Resistance to leaf and stripe rust; moderately resistant to leaf blight in fields

Improved Varieties for Punjab

Timely Sown Irrigated Conditions

- **Unnat PBW 343:** It is improved version of PBW 343, has average plant height of 100 cm. It is resistant to brown rust, moderately resistant to yellow rust but susceptible to loose smut disease. It matures in about 155 days. Its average grain yield is 58.0 quintals per hectare.
- **Unnat PBW 550:** It is improved version of PBW 550, has average plant height of 86 cm. It is resistant to yellow and brown rust but susceptible to loose smut disease. It matures in about 145 days. It has bold grains and its average grain yield is 57.5 quintals per hectare. Use 45 kg seed per acre and sow this variety during 2nd to 4th week of November.
- **PBW 1 Zn:** It possesses higher grain zinc concentration which is good for human health. Its average plant height is 103 cm and is resistant to brown rust and moderately resistant to yellow rust but susceptible to loose smut disease. It matures in about 151 days. Its average grain yield is 56.25 quintals per hectare.
- **PBW 725:** Its average plant height is 105 cm and matures in about 154 days. It is resistant to yellow and brown rust, but susceptible to loose smut disease. Its average grain yield is 57.25 quintals per hectare. This variety is suitable for growing under poplar plantation.
- **PBW 677:** Average plant height of this variety is 107 cm and matures in about 157 days. It is resistant to yellow and brown rust, but susceptible to loose smut disease. Its average grain yield is 56 quintals per hectare. This variety is suitable for growing under poplar plantation.
- **HD 3086:** The average plant height of this variety is 96 cm and matures in about 156 days. It is resistant to yellow rust, moderately susceptible to brown rust but susceptible to loose smut disease. Its average grain yield is 57.5 quintals per hectare.
- **WH 1105:** Its plant height is 97 cm and matures in about 157 days. It is susceptible to new races of yellow rust and moderately resistant to brown rust but susceptible to loose smut disease. Its average grain yield is 55.25 quintals per hectare.
- **HD 2967:** Average plant height is 101 cm and matures in about 157 days. It is resistant to brown rust while susceptible to new races of yellow rust. It is also

susceptible to Karnal bunt and loose smut. Its average yield is 53.5 quintals per hectare.

- **PBW 621:** The average plant height of this variety is 100 cm and matures in about 158 days. It is moderately resistant to yellow rust and resistant to brown rust and susceptible to loose smut disease. Its average yield is 52.75 quintals per hectare.
- **WHD943:** It is durum wheat (Wadanak)with an average plant height of 93 cm and matures in about 154 days. It has low incidence of yellow berry. It possesses desirable quality characteristics and suitability for pasta making. It is resistant to yellow and brown rusts and less susceptible to leaf blight. It possesses field resistance to Karnal bunt disease. Its average yield is 49.5 quintals per hectare.
- **PDW 291:** It is also durum wheat (Wadanak) with an average plant height of 83 cm and matures in about 155 days. It has low incidence of yellow berry and possesses suitable for pasta making. It is resistant to yellow and brown rusts, loose smut and flag smut and less susceptible to powdery mildew, leaf blight and head scab diseases. It possesses field resistance to Karnal bunt disease. Its average yield is 48.5 quintals per hectare.
- **TL 2908:** It is variety of triticale with an average plant height of 113 cm and matures in about 153 days. It is highly resistant to yellow and brown rusts,Karnal bunt, loose smut and powdery mildew diseases. Triticale is suitable for poultry feed. Good quality biscuits, satisfactory breads and chapatis can be prepared from the flour of this variety particularly if blended with wheat to the extent of 50 per cent. Its average yield is 41 quintals per hectare.

Late Sown Irrigated Conditions

- **PBW 658:** The average plant height of this variety is 85 cm and matures in about 133 days. It is resistant to yellow rust, brown rust and leaf blight. Its average grain yield is 44 quintals per hectare.
- **PBW 590:** Its average plant height is 80 cm and matures in about 128 days. It is susceptible to new race of yellow rust, moderately resistant to brown rust and susceptible to Karnal bunt and loose smut diseases. Its average grain yield is 41 quintals per hectare.

Timely sown rainfed conditions

- **PBW 660:** Its plant height is 100 cm and matures in 162 days. It has an excellent chapati quality. It is resistant to yellow and brown rust and less susceptible to Karnal bunt and loose smut diseases. Its average grain yield is 42.75 quintals per hectare.
- **PBW 644:** The average plant height of this variety is 102 cm and matures in 159 days. It is less susceptible to yellow rust, moderately resistant to brown rust but susceptible to Karnal bunt and loose smut diseases. It average grain yield is 41 quintals per hectare.

Climate

Wheat is grown over a wide range of latitudes ranging between 60ºN and 60ºS and altitudes ranging from the sea level to an elevation up to 3500 metres in the tropics and subtropics. Normally, the most ideal conditions for wheat growth are cool and moist weather during the vegetative growth and warm and dry weather during grain formation. Because of this reason, major wheat regions of the world are found in the temperature zones between 30-60ºN and 25-40ºS. Growth and reproduction of the wheat plant is influenced by temperature, rainfall and light.

The optimum temperature for the germination of wheat is between 20-22ºC, though wheat can also germinate at 4ºC. The optimum temperature for vegetative growth ranges from 16 to 22ºC. During the grain development period wheat requires a mean maximum temperature of about 25ºC for at least four to five weeks. The temperatures above 25ºC reduce the test weight of grain due to hastened maturity.

Wheat crop thrives well in those areas where annual rainfall ranges between 12.5 to 100 cm, but most of it is received in summer during the monsoon period. However, heavy rains are detrimental to wheat growth as it causes delay in sowing, water logging, leaching of nutrients, lodging, infestation of insects-pests and diseases. Crop can not withstand extended periods of soil moisture stress. Rainfed wheat requires a minimum evenly distributed winter rainfall of 15 to 20 cm. Wheat normally flowers in the long days of spring or early summer. Short days increase the vegetative growth of the wheat, whereas longer days hasten the formation of inflorescence.

Rotation

The wheat is followed by *kharif* crops such as maize, jowar, bajra, cotton and arhar. Sometimes, some of the green-manure crops, such as sunhemp/dhaincha, moong, guar, or cowpea are sown immediately after the harvest of wheat crop which enrich the soil fertility. Three crops per year such as rice–potato–wheat, rice-wheat-moong beans can be taken successfully. In Punjab, Haryana and Western Uttar Pradesh rice has become an important crop in the *kharif* which is followed by wheat. In eastern India, wheat has become an important crop and is grown extensively after rice on land which is left either fallow or put under 'boro' rice. In some states like West Bengal, the rice-wheat-jute rotation has become common. The sugarcane-wheat rotation is also common in northern India. Where irrigation is available a legume crop is grown in between two cereal crops to enrich the soil as well as to get the needed pulses. The rainfed wheat is rotated with jowar, bajra or cotton in *kharif.*

Soil

Well drained loams and clayey loams are considered to be ideal for wheat. Good crop of wheat have also been raised on sandy loams and the black soils. Soils with nearly neutral pH are most suitable for wheat cultivation. However, it can be grown on soils with pH ranging from 6.0 to 8.5.

Cultivation

For good and uniform germination, the wheat crop requires a well pulverized but compact seedbed. Traditional system of land preparation for wheat includes one deep ploughing followed by two to three harrowings. After the harvest of *kharif* crops, the field is irrigated and at optimum workable moisture content, the field should be ploughed with disc or mould board plough followed by two to three harrowing and planking. Under late sown conditions, the pre-sowing irrigation for wheat may be applied in previous standing crop so that field may be prepared immediately after the harvest of previous crop. Alternately, a no tillage system may be followed and wheat is sown directly using zero-tillage seed drill.

In rainfed areas, collection and conservation of soil moisture and timely cultivation are key practices for the success of wheat crop. One deep chiseling once in three years should be done to break the hard pan created by the continuous use of the country plough. Two to three summer ploughings have been found useful for soaking the rain water in deeper layer. After cessation of monsoon all the possible efforts should be made to conserve moisture. The ploughing (3-4) should be done in the evening time, furrows should be kept open whole night and planking should be done after each ploughing early in the morning.

Sowing

Selection of seed

The seed used for sowing should be healthy with good germination capacity and free from seed borne diseases and seeds of weeds. The seed should be certified and obtained from a reliable seed agency. Seed should be treated with 4 ml Dursban/Durment 20 EC (Chlorphyrphos) or 6 ml Regent 5 SC (Fipronil) per kg seed dry in shade to control termites. For the control of loose smut and flag smut, treat 40 kg seed with 13 ml Raxil Easy 6 FS (tebuconazole) or 120 g Vitavax 75 WP (carboxin) with seed treating drum.

Time of sowing

It depends on the climatic conditions, soil conditions and other available facilities like irrigation. Under irrigated conditions, the first fortnight of November is considered to be the optimum time for sowing of medium long duration varieties. For short duration varieties, the second fortnight of November is the optimum time. In eastern India, the sowing of wheat is often delayed owing to the late harvesting of paddy. The sowing in such case may go up to even the third week of December. In north-western India also, wheat sowings get delayed due to late harvesting of paddy or sugarcane or potato. However, wheat sowings beyond the third week of December are not very profitable. Under rainfed conditions, second half of October is the optimum time. The sowing of wheat, when the temperatures are high, results in a poor stand, poor tillering, attack by insects-pests and the early onset of flowering. All these factors depress yield. The success of rainfed wheat depends upon the amount of conserved moisture in the soil.

Method of sowing

Sowing of wheat is generally done by following methods:

Broadcasting: It is the most prevailing and simple method of wheat sowing in India. Seeds are broadcasted on soil surface, then worked in by harrowing or ploughing with cultivator and covered by planking. However, all the broadcasted seeds do not get proper contact with moist soil. Secondly some seeds left uncovered and are eaten by birds or some are not sown at proper depth. This results in poor and uneven germination. To overcome these impediments, 25% higher seed rate is recommended. Boradcasting is crude method of sowing and should only be practiced under special situations such as under very clayey if soils or soil containing more stubles which hinder drill sowing.

Behind the plough: In this method, the seeds are dropped by hand into the furrows that have been opened with local plough. When seed is dropped in furrows by hand, it is called *kera* method and when seeds are dropped through a pore like attachment with local plough it is called *pora* method. In this method, seeds are dropped at a desired depth and germination is satisfactory.

Drilling: Drilling of seed with the help of seed drill or seed-cum-fertilizer drill is the best method of sowing. Seed drills are calibrated at a optimum seed rate. These seed drills drop the seed at an optimum depth (4 to 6 cm) and spacing while running over the field with the help of a tractor or bullocks.

Dibbling: It is used when quantity of seed is limited. Sowing is done with the help of small implement known as dibbler. In this method dibblers are pressed into the soil to make the holes and one or two seeds are dropped by hand in each hole and then seeds are covered.

Seed rate

Seed rate of wheat varies with varietal characters like seed size, tillering ability and germination percentage, time of sowing, moisture content in the soil at sowing time and method of sowing. Under normal conditions a seed rate of 100-125 kg/ha is sufficient. Under late sown conditions seed rate should be increased by 25 per cent. Similarly when soil moisture is less at the sowing time seed rate should be increased by 25 per cent. Method of sowing also affects the seed rate of wheat. Broadcasting requires 150 kg/ha while sowing with local plough and seed drill requires a seed rate of 125 and 100 kg/ha, respectively under normal conditions. Where seed is sown with dibbler a seed rate of 25-30 kg/ha is sufficient.

Spacing and depth

For timely sown and irrigated wheat, a row spacing of 15 to 22.5 cm is followed. For bi-direction or cross sowing 22.5 cm is considered to be the best spacing. For late but irrigated wheat, row spacing of 15-18 cm is considered to be the maximum. For dwarf varieties, planting depth should be between 4 to 6 cm. Usually deeper sowing is done in dry, rough and light soils, whereas shallow sowing is done in moist soils.

Weed management

For controlling non graminaceous broad leaved weeds, spray of 2,4-D at 0.4 kg a.i./ha or metsulfuron 5 g/ha (35 to 45 DAS) or carfentrazone 20 g/ha by dissolving in 500 litres of water between 30-35 days after sowing is effective. Spraying of especially 2, 4-D before or after this specific period will be harmful to the wheat crop. Prefer carfentrazone where *Malva neglecta* (Button booti) and metsulfuron where *Rumex spinosus* (Kandyali palak) are problem weeds, these herbicides also controls other broad leaf weeds of wheat. For the control of all hardy broad leaf weeds of wheat, apply metsulfuron + carfentrazone (50 DF) @ 25 g/ha] (30-35 DAS). Pre-emergence application of trifluralin 48 EC (Treflan) at 1.2 kg ha^{-1} or pendimethalin 30 EC (Stomp) at 0.75 kg ha^{-1} can be made for the control of *Phalaris minor* (canary grass/ gullidanda/ sitti/ mandusi) from wheat crop. Alternatively, post-emergence herbicides such as isoproturon 75 WP (0.94 kg ha^{-1}), sulfosulfuron 15 WP (25 g ha^{-1}), clodinafop 75 WP (60 g ha^{-1}), fenoxaprop-p-ethyl 10 EC (100 g ha^{-1}), pinoxaden 5 EC (50 g ha^{-1}). Also use of mesosulfuron + iodosulfuron 3.6 WDG (12 g ha^{-1}) or sulfosuluron + metsulfuron 75 WG (30 g ha^{-1}) or fenoxaprop 8% + metriluzin 14% at 1.25 lit./ha of commercial product is very effective for the control of all weeds of wheat especially *Phalaris minor* when applied 30-35 days after sowing.

Water management

Under irrigated conditions 4-6 irrigations are needed for wheat. The first irrigation which is very critical, should be given at the crown root initiation stage, 20-25 days after sowing. Other rounds of irrigation should be done at the time of late tillering, late jointing, flowering, milk and dough stages. Extra two or three irrigations are given if the soil is very light or sandy. Irrigation in case the water supply is limited should be given in the following way:

(i) If the water is available only for one irrigation, it should be given at the crown root initiation stage to the tillering stage.

(ii) If the water is available for two irrigations, the first irrigation should be applied between the crown root initiation and the late tillering stage and second irrigation 7 to 8 weeks after the first irrigation (flowering stage).

(iii) If the water is available for, three irrigations, the first should be given at the crown root initiation stage and the next two at intervals of six to seven weeks at the flowering and the milk stages, respectively.

It was considered necessary to optimize the irrigation schedule to obtain increased yield. The investigations revealed that first irrigation was most essential for high productivity, which must be applied 21 days after sowing. About 17 days after sowing, the crown node starts swelling and crown root develops from this node approximately 21 days after sowing. Tillers develop from the crown root and, therefore, this phase was directly relevant to the number of tillers and final heads and is supposed to be the most critical period of irrigation. Depending on the availability of water with the farmer a schedule for irrigation at different growth and development stages is very important. Keeping in view, the critical phenological phases, four irrigations, first at crown-root initiation, followed by at flowering, late tillering and milk stages, should be applied to normal sown crop particularly under high fertility conditions.

Fertilizer management

The rainfed wheat in the majority of cases is not fertilized by the farmers owing to economic reasons. It is desirable that 8 to 10 tonnes of farmyard manure per hectare or some other organic matter is applied 5 or 6 weeks before sowing. Nitrogen half dose and P_2O_5 full dose gives a considerable boost to the wheat yield, if applied 10 cm deep (3-4 cm below the seed) at or before sowing. The fertilizer requirements of the irrigated wheat are as follows:

a. With assured fertilizer supply

Nitrogen (N) 80-120 kg/ha

Phosphorus (P_2O_5) 40-60 kg/ha

Potash (K_2O) 40 kg/ha (This quantity may be adjusted according to the soil test results).

b. Under fertilizer constraints

Nitrogen (N) 60-80 kg/ha

Phosphorus (P_2O_5) 30-40 kg/ha

Potash (K_2O) 20-25 kg/ha (This quantity may be adjusted according to the soil test results).

The total quantities of phosphorus and potash and half the quantity of nitrogen should be placed 3 to 4 cm below the seed at the time of sowing. The remaining quantity of nitrogen should be applied at the time of first irrigation to be applied at crown root initiation. Above recommendation can vary as per soil type.

Zinc deficiency

Zinc deficiency generally appears in light soils under intensive cropping system. Zinc deficiency is particularly reported from Punjab, the *tarai* areas of Uttar Pradesh, some parts of Haryana, western Uttar Pradesh and Delhi. When recommended dose of zinc sulphate has been applied to the *kharif* crop, its application may be omitted to the following wheat crop. Paddy responds more to zinc application, therefore zinc should be applied to paddy and its residue is sufficient to meet the requirement of the following wheat crop. However, if zinc has not been applied to paddy and zinc deficiency is expected or appears showing symptoms of a stunted and bushy crop with leaves chlorotic in the middle, which later break and keep hanging, apply 62.5 kg of zinc sulphate per acre which will be enough for 2-3 years. Zinc deficiency can also be corrected by foliar spray of 0.5% zinc sulphate (21% zinc). Prepare the solution for spray by dissolving 2.5 kg zinc sulphate and 1.25 kg unslaked lime in 500 litres of water. This solution is sufficient for spraying an hectare of wheat once. Two or three sprays at 15 days interval are needed.

Manganese deficiency

Manganese deficiency generally appears in light soils under intensive cropping especially in rice-wheat rotation. The symptoms appear on the middle leaves as interveinal chlorosis with light greyish yellow to pinkish brown or buff coloured specks of variable size confined largely to 2/3 lower portion of the leaf. Later, the specks coalesce forming a streak or band in between the veins which remain green. In acute deficiency, whole of the plant may become dry. At earing stage, the symptoms become prominent on flag leaf. In manganese deficient soils, give one spray of 0.5% manganese sulphate solution (2.5 kg manganese sulphate in 500 litres of water for one hectare) 2-4 days before first irrigation and three sprays afterwards at weekly intervals on sunny days. Do not grow durum wheat in sandy soils as it is more prone to manganese deficiency. Manganese sulphate should be sprayed only as its soil application is not profitable.

Sulphur deficiency

Wheat crop suffers from sulphur deficiency when sown in sandy soils. The deficiency is more severe when the winter rains continue for a long time in the early growth period. The symptoms first appear on the younger leaves with fading of the normal green colour. This is followed in the veins. The topmost leaves become light yellow except for the tip, while the lower leaves retain green colour for a long time. This is distinctly different from the nitrogen deficiency where the yellowing starts with the lower leaves. The plants are short in size with fewer tillers. Where phosphorus was not applied as single superphosphate, apply 250 kg of gypsum per hectare before sowing to meet the sulphur requirement of the crop. Gypsum can also be applied in standing crop if deficiency of sulphur is observed.

Plant Protection

Insect Pests

Termite damage the crop soon after sowing and near maturity. The damaged plants dry up completely and are easily pulled out. The plants damaged at later stages give rise to white ears.

Treat the seed at the rate of 4 ml Dursban/Ruban/Durmet 20 EC (chlorpyriphos) or 6 ml Regent 5 SC (fipronil) per kg seed. Dilute 160 ml of Dursban/Ruban/Durmet or 240 ml of Regent in one litre of water and spray the same on 40 kg seed spread as a thin layer on the pucca ground or tarpaulin or polythene.

Note: Insecticides applied against termites would also control root aphid and the Gujhia weevil. Seed treated with chlorpyriphos is also less attacked by birds.

Aphids damage the crop resulting in discolouration of leaves. Spray 100 ml of Confidor 17.8 SL (imidacloprid) or 50 g of Actara/Taiyo 25 WG (thiamethoxam) or 30 g of Dantop 50 WDG (clothianidin) or 375 ml of Rogor 30 EC (dimethoate) in 200-250 litres of water per acre using knap sack sprayer or in 30 litres of water per hectare with power sprayer.

Note:

- Control aphids at earhead stage at economic threshold level of 5 aphids per earheads recorded from 10 randomly taken ear heads in each of the 4 quarters of one acre field.
- Since the aphids appear first on borders of the crop, spray only the infected strip to check their further spread.

Army worm attacks wheat and barley during March-April. It damages leaves and earheads. Spray 500 ml Nuvan 85 SL (dichlorvos) or 1 litre Ekalux 25 EC (quinalphos) in 200-250 litres of water per hectare with hand-operated knapsack sprayer or in 75 litres of water with a motorized knapsack sprayer.

Note: These insecticides will also control aphid and jassid.

Gram pod borer (*Helicoverpa armigera*) attacks wheat at maturity. It feeds on the grains in the earheads. Damage is more where wheat follows cotton. Spray 2 litre Ekalux 25 EC (quinalphos) in 250 litres of water per hectare with hand operated knap sack sprayer.

Pink stem borer (*Sesamia inferens*) generally attacks the wheat crop at seedling stage. The larva bore into the stem of young plant and kills the central shoot causing 'dead heart'. Spray the crop with Ekalux 25 EC (quinalphos) @ 2 litre/hectare in 250 litres of water.

Brown mite is a minor insect of wheat causing discolouration of leaves. To control it use same insecticides as recommended for the control of aphids.

Rodents

Field rats: They cause heavy loss to wheat crop and do considerable damage to the harvested crop lying in stacks in the field. For the control of field rates close all burrows in the evening and fumigate the reopened burrows of rats in the morning time with the following poison baits:

I. Zinc phosphide bait: Smear 1 kg of bajra or sorghum or cracked wheat or their mixture with 20 g of oil and mix it thoroughly with 25 g of zinc phosphide. Never add water in zinc phosphide bait and always use freshly prepared bait.

II. 0.005% Bromadiolone bait: Mix 20 g of 0.25% Bromadiolone powder, 20 g of oil and 20 g of powdered sugar in 1 kg of any cereal flour.

III. 0.0375% racumin bait: Mix 50 g of 0.75% racumin powder, 20 ml of groundnut or sunflower oil and 20 g of powdered sugar in 1 kg of cracked wheat or any other cereal.

Diseases

Yellowing in the seedling stage can be caused by nutrient deficiency, bad weather conditions, poor drainage, attack of Alternaria and Drechslera sp. or soil infestation with cereal cyst nematode.

A general yellowing of the leaves and necrosis occurs especially starting from the tips and along the edges. Later on, necrotic area develops within the chlorotic tissue, there is stunting and general decline. In case of zinc deficiency, the plants remain stunted and bushy; leaves become chlorotic, stocky and break in the middle. Yellowing of young leaves may be due to sulphur deficiency, whercas yellowing of older leaves could be due to nitrogen deficiency.

- In light soil, add 62.5 kg zinc sulphate per hectare. One such application will be enough for 2-3 years.
- Add muriate of potash along with N and P fertilizers to high yielding varieties, as recommended under fertilization.
- Give proper irrigation to counteract the effect of frost. Avoid excessive irrigation.
- If the cereal cyst nematode is present in the soil, apply Furadan 3 G @ 32.5 kg per hectare at sowing time.
- Apply 250 kg gypsum per ha for correcting sulphur deficiency.

Wheat rusts:

(i) Yellow or stripe rust (*Puccinia striiformis* f.sp.tritici): Yellow powdery pustules appear on leaves, forming stripes. The disease appears first in the sub-mountaineous areas of the state. Minimum temperature in the range of 7-13°C coupled with 85-100% relative humidity during night and maximum temperature in the range of 15-24°C during day are congenial for infection, development and spread of disease. Frequent rains along with wind during Feb-March may further enhance the disease severity and spread.

Integrated management of yellow rust:

- Grow rust resistant varieties.
- Don't grow unrecommended varieties especially in sub-mountainous zone.
- Do not sow the crop in the month of October particularly in the disease prone sub- mountainous areas.
- Monitor the crop from 2nd week of December onward particularly after irrigation or rain to locate the initial foci of infection.
- Spray the crop with Nativo 75 WG (trifloxystrobin+tebuconazole) @ 300 g or Tilt 25 EC/Shine 25EC/Bumper 25 EC/ Stilt 25 EC/Compass 25 EC/Markzole 25 EC (propiconazole) @ 500 ml in 500 litres of water/hectare as soon as the disease is noticed.
- Initially only spot application surrounding infected area should be done.
- Regularly monitor the crop and repeat the spray as per need.

(ii) Brown or leaf rust (*P. triticina*): Round, orange powdery pustules, irregularly arranged or in clusters on leaves, less common on the leaf sheath and stalk.

- Grow rust resistant varieties.
- Spray the crop with any of the fungicides mentioned under yellow rust as soon as the disease is noticed.

Karnal bunt (*Neovossia indica*): In the ear only a few grains are infected. The infected grains on pressing give out black powder of spore mass. This powder gives peculiar stinking smell. Grow resistant varieties. A single spray of Tilt (propiconazole)/ Folicur 25 EC (tebuconazole) @ 500 ml per hectare using 500 litres of water at ear emergence stage for the control of Karnal bunt is recommended in wheat meant for seed production only.

Powdery mildew: The fungus develops numerous superficial white floury spots on all the above ground parts of the plant. The white colour of the floury spots changes to grey or reddish brown when cleistothecia develop. Infected plants become stunted due to reduction in the size and number of leaves. Spray with Nativo 75 WG (trifloxystrobin + tebuconazole) @300 g or Karathane 40 EC (dinocap) @ 250 ml or Tilt 25 EC (propiconazole) @ 500 ml in 500 litres of water per hectare.

Loose smut (*Ustilago segetum tritici*): The fungus destroys the ears completely, turning them into a black loose powdery mass consisting of spores and leaving behind the rachis only.

- Grow resistant varieties.
- Soak the wheat seed in ordinary water from 8 a.m. to 12 noon on any calm and sunny day during May/June. After 4 hours soaking, spread out the moist seed in the sun in a thin layer on cemented floor (pucca), on tarpaulin or sheets of cloth. Dry the seed completely and store in a dry place till sowing.
- Alternatively the seed may be treated as per recommendations.

Flag smut or leaf-smut (*Urocystis agropyri*): Long narrow lead grey or black streaks or stripes running parallel to veins are formed on the leaves. The stripes eventually rupture and expose black sooty mass of spores. Practise shallow sowing. Rougue out the affected stools and destroy them by burning. Treat the seed with recommended fungicides as given under seed treatment.

Head blight or Scab (*Fusarium spp.*): Individual spikelets or portion of earheads show premature bleaching or brown discolouration. Characteristic pinkish mycelium of the fungus may be visible on infected spikelets during humid conditions. The grain formation is either completely inhibited or the infected spikelets produce shrivelled chalky grains showing pinkish discolouration. Seed lots affected by scab should be cleaned thoroughly to eliminate shrivelled seeds. The incidence of this disease is more on durum varieties, their cultivation in highly humid areas, particularly near rivers should be avoided.

Leaf blight, glume blight and black tip of grains: Elongated brown spots and blotches appear on leaves and glumes. Ears are poorly filled, grains shrivelled, discoloured and black tipped. Treat the seed with 3 g Thiram (tetramethyl thiurum disulphide) or Captaf (captan) per kg of seed.

Ear cockle (Mamni) and Yellow ear rot (tundu) (*Anguina tritici and Rathyi bacter*): Diseased plants have spreading tendency and swollen base, leaves become crinkled and twisted. Earheads contain dark-brown, hard and roundish galls (Mamni) instead of grains. Severely diseased plants are stunted and may die at seedling stage. Occurrence of yellow, slimy mass indicates ear rot phase which prevents grain and gall formation. Put wheat seed in ordinary water and agitate vigorously for few minutes.

Ear-cockle galls will float to the surface. These may be skimmed off with an ordinary sieve and burnt.

Molya (*Heterodera avenae*): Plants become stunted with yellowing of leaves, reduced tillering, absence of ears on some tillers or small ears with poorly filled grains. The root system of infected plants gets reduced in size and becomes bunchy with profuse development of thin rootlets. Cysts (shining white bodies of female nematode) are seen attached to the roots at the later stage of the crop. Expose the soil to the hot sun by cultivation during May and June. Practise rotation with non cereals in badly infested soils. Apply Furadan 3G (carbofuron) @32.5 kg per hectare at sowing time.

Harvesting

The wheat crop is harvested when the grains harden and the straw becomes dry and brittle. The harvesting time varies from zone to zone and also depends whether the wheat is under irrigated or rainfed conditions. The rainfed crop reaches harvesting stage much earlier than the irrigated crop. However, the yields under the rainfed crops are much lower compared to the irrigated crop.

Yield

The potential yield of semi-dwarf wheat varieties varies between 60-75 quintals per hectare. At farmer's field the grain yield of high yielding varieties generally ranges between 40-50 quintals per hectare. Generally, the yield of straw from rainfed crop is about the same as that of grain whereas from the irrigated crop it is about one and a half to two times the grain yield.

MAIZE (*Zea mays*)

Maize Plants

Maize Cob

Introduction

Maize (*Zea mays*) is one of the most important crop in world's agricultural economy as it is used both as food for man and feed for animals. It has yield potential far higher than any other cereal and that's why it is sometimes referred to as the miracle crop or the 'Queen of Cereals.'

Maize is grown in almost all the States of India. It occupies an area of about 10.2 million hectares in the country with production of 26.26 mt during 2016 (FAO STAT 2016). It is next to rice and wheat with regard to area and production in India.

Maize crop is also utilized in many ways like other grain crops. Over 85 per cent of maize produced in the country is consumed as human food. Several food dishes including *chapaties* are prepared out of maize flour and grains. Green cobs are roasted and eaten by people with great interest. The grains of a special variety called the 'Pop corn' are characterized by a hard corneous interior structure, are converted into the popped form to be a favourite food for children in cities. It is also a good feed for poultry, piggery and other animals. It ranks below wheat and sorghum but considerably above rice in nutrition. Maize grain contains about 10 per cent protein, 4 per cent oil, 70 per cent carbohydrate, 2.3 per cent crude fibre, 10.4 per cent albuminoides, 1.4 per cent ash. Maize protein called 'Zein' is deficient in tryptophane and lysine, the two essential amino acids. Maize grain has significant quantities of vitamin A, nicotinic acid, riboflavin and vitamin E. Maize is low in calcium, fairly high in phosphorus.

Maize crop furnishes huge quantities of green fodder for cattle. Both wet and dry milling industries separate the fraction of the grain into germ, hull and endosperm to produce various products and by products. In India with the growing demand of poultry feed, the demand for maize is also going up.

Origin and history

The primary centre of origin of maize is considered by most authorities to be the Central America and Mexico, where many diverse types of maize are found. The discovery of fossil maize pollen with other archaeological evidence indicates Mexico to be the native home of maize. The studies indicated that maize was a significant crop in Mexico 5000 years ago and perhaps earlier. American Indians grew and selectively improved maize from 3400 BC to 1500 AD.

The maize plant was unknown in the old world before 1492, by the time Columbus arrived in America. In Europe maize was first introduced in Spain sometimes after Columbus returned from his second voyage. The plant was grown in Europe as a garden curiosity. Later on, it spread from Spain to southern France and Italy. Its introduction into India probably occurred about the beginning of the seventeenth century, during the early days of the East India Company.

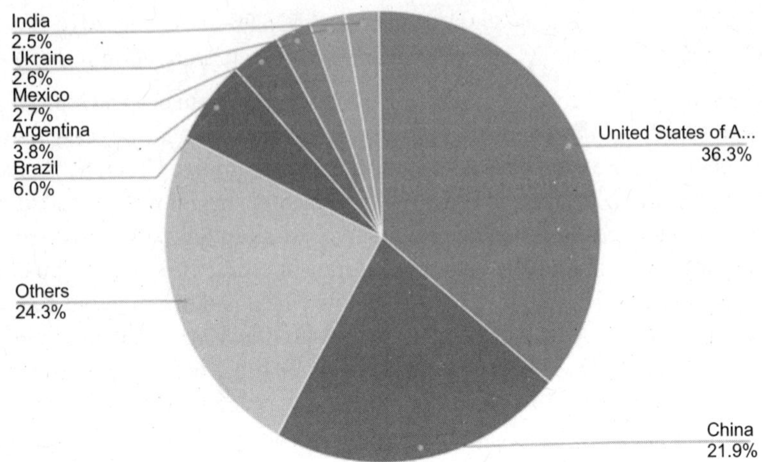

Fig. 8.5. Annual global production of maize during 2016 (Percentage) (FAO STAT 2016)

Area and distribution

Maize is one of the top three grains grown in the world. Fig. 8.5 shows maize production in metric tonnes. There are 163 countries producing maize, but only two countries together - the US and China - produced over half of the total amount accounting to 36.3 and 21.9 percent respectively. About 68 per cent of the land devoted to maize is located in the developing world, however only 46 per cent of maize production occurs there, indicating the need for improving yields in developing countries where it is a major source of direct human consumption for many of the poor.

Maize occupies an area of 10.20 m ha in India during 2016 with production of 26.26 m tonne (FAO STAT. 2016). The states of Uttar Pradesh, Bihar, Madhya Pradesh, Rajasthan and Punjab account for over 75 per cent of the area and production of this cereal in the country.

Classification

Maize (*Zea mays*) is an annual plant which belongs to family Poaceae and genus *Zea. Zea mays* L. has a normal chromosome complement of 10 pairs. It is divided into seven groups. The classification is based largely on the character of the kernels (Kipps, 1959).

1. ***Zea mays indurata* or Flint corn**: The endosperm in this type of maize kernel is soft and starchy in the centre and completely enclosed by a very hard outer layer. The kernels are usually rounded but are sometimes short and flat. Colour may be white or yellow. This is the most commonly cultivated type in India.
2. ***Zea mays indentata* or Dent corn**: In this type of maize, kernels have both hard and soft starches. The hard starch extends on the sides and the soft starch is in the center and extends to the top of the kernels. In the drying and shrinking of

the soft starch, various forms and degrees of indentation result. This is the most common type of maize grown in USA.

3. ***Zea mays everta* or Pop corn**: It possesses exceptional popping qualities. Size of the kernels is small and the endosperm is hard. When they are heated, the pressure built up within the kernel suddenly results in an explosion and the grain is turned inside out.
4. ***Zea mays saccharata* or Sweet corn**: Kernels possess a considerable amount of sugar which absorbs water, making the cells turgid, on drying these cells collapse, making the grains shrivelled or wrinkled. It has sweeter taste than other corns.
5. ***Zea mays amylacea* or Soft corn**: It possesses a soft endosperm. Kernels are soft and of all colours, but white and blue are the most common. They are like flint kernels in shape.
6. ***Zea mays tunicata* or Pod corn**: The pod corns are characterized by having each kernel enclosed within a pod or husk. It is a primitive type of corn and hence of no importance.
7. ***Zea mays ceratina* or Waxy corn**: The endosperm of the kernel when cut or broken gives a waxy appearance. It produces the starch similar to tapioca starch for making adhesive for articles.

Improved Varieties of Maize released by ICAR

2017-18

Variety	Area of adoption	Salient features
Vivek Hybrid 27 (Central	J and K, Uttarakhand, HP, Delhi,	Suitable for kharif, early duration high yielding
Maize VL Baby Cron	Punjab, Haryana, UP, Karnataka, TN, AP, Telangana, Maharashtra, Rajasthan, Gujarat, MP and Chhattisgarh	Babycorn hybrid, maturity (babycorn) in 52-54 days in hills and in 48-52 days in plains, moderately resistant to H. turcicum and H. maydis leaf blight, common rust and post-flowering stalk rot.
P3401	Andhra Pradesh, Telangana, Maharashtra, Karnataka and Tamil Nadu	Hybrid suitable for kharif season, moderately tolerance to turcicum leaf blight (TLB), Rajasthan downy mildew (RDM), brown stripe downy mildew (BSDM) and post flowering stalk rot (PFSR), Chilo partellus and Sesamia inferens
P3533 Hybrid	AP, Telangana, TN Maharashtra, Karnataka	Suitable for rabi season, resistance to TLB and C. rot
P3544 Hybrid	Punjab, Haryana, Delhi, UP, Bihar, Jharkhand, Odisha, WB, AP, Telangana, Maharashtra, Karnataka and TN	Suitable for kharif season, orange yellow, semi flint grain

Central Maize VL 55 Hybrid	J and K, Uttarakhand, HP and Karnataka, TN, AP, Telangana and Maharashtra	Suitable for kharif season, moderately resistant against H. turcicum and H. maydis, common rust and PFSR
VMH 4106 Sweet Corn Hybrid	Bihar, UP, Jharkhand, West Bengal and Odisha	Tolerant to MLB, TLB, banded leaf and sheath blight (BLSB), SDM, DM and rusts, tolerance to Chilo paretellus
HTMH 5402	AP, Telangana, Maharashtra, Karnataka and Tamil Nadu	Suitable for kharif season, resistant to RDM, sorghum downy mildew (SDM), bacterial stalk rot (BSR), Curvularia leaf spot (CLS), tolerance to Chilo paretellus, tolerant to mild moisture stress
GK 3150	Punjab, Haryana, Delhi and Uttar Pradesh	Suitable for rabi season, resistant to TLB, SDM, PFSR and moderately resistant to C. rot resistant to Sesamia inferens and Chilo partellus, tolerant to drought, heat and cold.
DRONA	Punjab, Haryana, Delhi, Uttar Pradesh, Andhra Pradesh, Telangana, Maharashtra, Karnataka and Tamil Nadu	Suitable for rabi season, resistant to common rust, PFSR and C.rot, tolerant to drought
Hy-brix 53 Hybrid	J and K, HP, Uttarakhand, North East Hills, Punjab, Haryana, Delhi, UP, Bihar, Odisha, Jharkhand, WB, TN, AP, Telangana, Maharashtra and Karnataka	Suitable for kharif, resistant to Fusarium stalk rot (FSR), RDM and moderately resistant to TLB and maydis leaf blight (MLB)
Hi-brix-39 (Sweet Corn)	J and K, HP, Uttarakhand, Punjab, Haryana, Delhi, UP, Bihar, Odisha, Jharkhand, WB, TN, AP, Telangana, Maharashtra, Karnataka, Assam, Manipur, Meghalaya, Mizoram, Arunachal Pradesh, Tripura and Nagaland	Suitable for kharif, resistant to common rust, BSR, Fusarium stalk rot, RDM and moderately resistance to TLB and MLB
Shalimar Pop Corn-1	J&K, HP, Uttarakhand, Punjab, Haryana, Delhi, UP, Bihar, Odisha, Jharkhand, WB, Rajasthan, Gujarat, MP, Chhattisgarh, Assam, Manipur, Meghalaya, Mizoram, Arunachal Pradesh, Tripura and Nagaland	Suitable for kharif, moderately resistant to banded leaf and sheath blight and FSR, resistant to stem borer
Pant Sankar Makka 4	Uttarakhand	Suitable for kharif, resistant to TLB, BLSB, BSDM and ESR

Bisco Champion 61	J and K, HP, Hills of Uttarakhand, and NE region and Maharashtra, Andhra Pradesh, Karnataka and TN	Suitable for kharif season, moderately resistant to prevailing diseases (TLB, MLB, DM, BSDM, PFSR, C. Rust and tolerant to pests)
Bisco X 5129	Maharashtra, Andhra Pradesh, Karnataka, Tamil Nadu (Zone 4)	Suitable for rabi, moderately resistant to prevailing diseases (TLB, MLB, DM, BSDM, PFSR, C. rust and tolerant to pests)
Karimnagar Makka	Telangana, Andhra Pradesh, Maharashtra, Karnataka and Tamil Nadu	Suitable for kharif, rainfed situation, grows in both high and low fertility conditions
Pusa HM-8 Improved	Maharashtra, Karnataka, Telangana, Andhra Pradesh and Tamil Nadu	QPM hybrid developed through MAS, nutria-rich with enhanced tryptophan (80 %), lysine (76%) and high protein (4.18 %)
Pusa HM-9 Improved	Bihar, Jharkhand, Odisha, Uttar Pradesh and West Bengal	QPM hybrid developed through MAS, nutria-rich with enhanced tryptophan, lysine and high protein
Pusa HM-4 Improved	Punjab, Haryana, Delhi, Uttarakhand and Uttar Pradesh	QPM hybrid developed through MAS, nutri-rich with enhanced tryptophan, lysine, high pro-vitamin A, high protein
Pusa Vivek QPM-9 Improved	J and K, HP, Uttarakhand, Meghalaya, Sikkim, Assam, Tripura, Nagaland, Manipur, Arunachal Pradesh, Maharashtra, TN, AP and Telangana	QPM hybrid developed through MAS, nutria-rich with enhanced tryptophan (74%), lysine (60%) and high protein (3.62 %)
JKMH-4848	Maharastra, Karnataka, Andhra Pradesh, Telangana and Tamilnadu	Resistant to Chilo partellus, FSR, charcoal rot and common rust, moderately resistant to MLB, TLB, charcoal rot, RDM and CLS
DMRH-1301 Hybrid	Bihar Jharkhand, Odisha, West Bengal and Uttar Pradesh (Eastern Region), Rajasthan, Madhya Pradesh, Chhattishgarh and Gujarat	Tolerant to Sesamia inferens, Chilo partellus and moderately resistant to TLB
DMRH-1402 Hybrid	Punjab, Haryana, Delhi, Uttarakhand (Plain), UP (Western Region), Rajasthan, Gujarat, MP and Chhattishgarh	Popcorn hybrid, moderately tolerant to Chilo partellus (Zone II) and (Zone V), and moderately resistant to FSR (Zone V)
DMRH-1308 Hybrid	Rajasthan, Madhya Pradesh, Chhattishgarh and Gujarat	Tolerant to Sesamia inferens and Chilo partellus

2016-17

Maize Variety	Area of adoption	Salient features
BPCH 6	All zones across the country	Early maturity; yellow, flint kernels
Palam Sankar Makka 1	Rajasthan, Gujarat, Chhattisgarh and Madhya Pradesh	Late maturity; yellow, flint kernels
KMH7 148	Punjab, Haryana, Delhi, western Uttar Pradesh for rabi	Late maturity; orange yellow, dent to semi dent with large kernels
Candy	All maize-growing areas in kharif	Early maturity; yellow, juicy, flint, medium-bold kernels; stable sweetness
D 2244	Jammu and Kashmir, Himachal Pradesh, Uttarakhand, North East Hills, Andhra Pradesh, Karnataka, Maharashtra, Tamil Nadu, Madhya Pradesh, Rajasthan, Gujarat and Chhattisgarh	Early maturity, yellow orange, semi-dent kernels
Shalimar Maize Composite 5	Jammu and Kashmir/2013	Early maturity; white, semi-dent kernels
Shalimar Maize Comp 6	Jammu and Kashmir/2013	Early maturity; yellow, dent kernels
Shalimar Maize Comp 7	Jammu and Kashmir/2013	Medium maturity; yellow, semi-flint kernels
LAXMI 3636	Andhra Pradesh, Telangana, Karnataka, Maharashtra and Tamil Nadu	Late maturity; orange yellow, semi-dent kernels
BIO 9782	Uttar Pradesh, Bihar, Jharkhand, Odisha, West Bengal, Rajasthan, Gujarat, Chhattisgarh and Madhya Pradesh for rabi	Late maturity; orange yellow, semi-dent kernels
Dragon	Punjab, Haryana, Delhi, western Uttar Pradesh for rabi	Late maturity; bright yellow, semi-dent, bold kernels with orange tinge
PMH	Punjab in spring	Medium maturity; yellow orange, flint kernels
HEMA	Karnataka	Late maturity, yellow orange, semi-dent kernels; resistant to Turcicum leaf blight (TLB), Polysora rust

Pratap Hybrid Maize 3	Rajasthan, Gujarat, Chhattisgarh and Madhya Pradesh	Medium maturity; yellow, semi-dent kernels; resistant to post flowering stalk rot (FSR), moderately resistant to Curvularia leaf spot (CLS) and Rajasthan downy mildew (RDM)
DKC 9126	All maize-growing areas	Late maturity, yellow orange, semi-flint kernels
Palam Sankar Makka 2	Himachal Pradesh, Jammu and Kashmir, Uttarakhand and North Eastern Hill states	Medium maturity; yellow orange, flint kernels; moderately resistance to Maydis leaf blight (MLB) and TLB
Uday	Maharashtra	Medium maturity; yellow orange, semi-dent kernels
Pratap Makka 9	Rajasthan	Medium maturity, yellow, bold, semi-dent kernels
Karimnagar Makka1 10131	Telangana	Medium maturity, orange, semi-flint kernels; resistant to TLB, CLS, RDM
Gujarat Anand Yellow Maize Hy 1	Gujarat	Early maturity, orange yellow, flint kernels; resistant to Maydis leaf blight
HTMH 5106 Hybrid	Telangana	Late maturity
RJ 2020	Rajasthan, Gujarat, Chhattisgarh and Madhya Pradesh for rabi	Late maturity; yellow orange, flint kernels
RMH 4620	Andhra Pradesh, Tamil Nadu, Karnataka and Maharashtra	Late maturity; orange yellow, semi-flint kernels
Central Maize VL Sweet Corn 1	Jammu and Kashmir, Uttarakhand, Himachal Pradesh, North Eastern hills, Delhi, Punjab, Haryana, western Uttar Pradesh, Karnataka, Tamil Nadu, Andhra Pradesh, Telangana, Rajasthan, Gujarat, Chhattisgarh and Madhya Pradesh	Early maturity; plant sturdy, medium in height with good and green husk cover; TSS 16%
PMH 7	Punjab in spring	Early maturity; orange, flint kernels
SHIATS Makka 3	Uttar Pradesh	Late maturity; white, dent kernels
PMH 10 Hybrid	Punjab in spring	Medium maturity; orange, flint kernels

Improved Varieties / Hybrid's of Maize for Punjab

Long Duration Varieties/Hybrids

PMH 1: This single cross hybrid has tall plants with well developed root system. The stem is zig-zag, sturdy and purple colored. The leaves are medium broad. Tassel is open and medium in size. Ears are medium long with yellow orange flint grains. The plants remains stay green at maturity. It matures in 95 days and average yield is 52.50 quintals per ha.

Parbhat: This composite has medium thick stem which resists lodging. Plant and ear height is medium tall. Husk covers are well developed. Ears are medium long and thick. The grains are medium bold, yellow-orange and semi-flint to flint. It matures in about 95 days. It yields about 43.75 quintals per ha.

Medium Duration Variety

Kesri: It is a medium maturing, high yielding composite with attractive orange-flint grains. Plant height and ear placement are medium. The ears are thick at the base, taper towards the tip and have good tip filling and husk cover. It matures in about 85 days and its average yield is 40 quintals per ha.

Short Duration Hybrid

PMH 2: It is a short duration single cross hybrid. It has medium plant height with medium ear placement. Leaves are medium sized and dark green in colour. Tassel is of medium size and semi-open. Silk is of green colour. It has medium long ears. Grains are yellow orange flint with yellow caps. The hybrid resists lodging and is tolerant to bacterial stalk rot. It matures in about 83 days and its average yield is 45 quintals per ha.

Special Purpose Varieties

Punjab Sweet Corn 1: This composite has tall plants with medium thick stem and medium ear placement. The leaves are broad and tassel is open with creamish anthers. Ears are medium long and the cob colour is white. Silk colour at the time of emergence is usually creamish. Husk cover is well developed and grains are orange in colour at maturity. This variety is highly suitable to use as sweet corn on commercial basis because its developing and immature grains in green ears possess high sugar content. It matures in about 95-100 days. Its average green ear and grain yield is 50 and 32.5 quintals per ha, respectively.

Pearl Popcorn: This is a composite variety of popcorn. Its ears are long and thin and grains are small and round. The commercial value of these grains is very high. It matures in about 88 days. Its average yield is 30 quintals per ha and popping quality is very good.

Varieties of Spring Maize

PMH 10: Plants of this single cross hybrid are medium tall with medium ear placement. Leaves are broad. Tassels are medium in size, semi-open with green anthers. Silks are pink in colour. Ears are medium long, conico-cylindrical with attractive orange flint grains. It matures in about 120 days. Its average yield is 78.75 quintals per hectare.

DKC 9108: Plants are medium tall with low ear placement. Leaves are broad and erect. Tassels are semi-open. Anthers and silks are green. Ears are medium long and cylindrical. Grain are dull yellow, semi-dent with caps. It matures in about 122 days having average grain yield of 80 quintals per hectáre.

PMH 8: This hybrid has medium tall plants with medium ear placement. Leaves are medium broad and curved. The tassel is open with strongly curved branches. The glumes, anthers and silks are green. Pollen shedding and silk emergence are synchronous. It has medium sized, conico-cylinderical ears with attractive yellow orange flint grains with slight yellow capping. It matures in about 117 days. Its average yield is 77.5 quintals per hectare.

PMH 7: This single cross hybrid has medium tall plants with medium ear placement. Leaves are medium broad and curved. The tassel is open with strongly curved branches. Glumes and anthers are green. Silks are green at emergence. Pollen shedding and silk emergence are synchronous. Ears are medium sized and conico-cylindrical with orange flint grains. It matures in 115 days and its average yield is 75 quintals per hectare.

PMH 1: It has tall plants with well developed root system. The stem has purple coloration and is zig-zag and sturdy. The leaves are medium broad. Tassel is open and medium in size. Ears are medium long with yellow orange flint grains. Its average yield is 69 quintals per hectare and matures in 118 days.

Rainfed Maize

PMH 2: It is a short duration drought tolerant hybrid. It has medium plant height with medium ear placement. Its leaves are medium sized and dark green. Tassel is of medium size and semi open. Silk is of green colour. Ears are medium long with orange flint grains having yellow caps. The cob colour is white. It matures in about 82 days. Its average yield is 41.25 quintals per ha under rainfed conditions with well distributed rains. The hybrid resists lodging and is tolerant to bacterial stalk rot.

Parkash: It is an early maturing single cross hybrid. It has medium tall plants with medium ear placement. Its leaves are dark green, medium sized and semi-erect. Tassel is of medium size and open. Short anthesis-silking interval confers drought tolerance to the hybrid. Ears are uniform and long with slightly blank tip. Grains are attractive orange flint. Cob is thin and white. Plants have stay green characteristic. It matures in about 82 days. Its average yield is 37.5 quintals per ha.

Megha: This composite combines early maturity with drought tolerance and is specifically suited for the rainfed area of the state. It matures in 82 days. It has medium tall plants with medium ear placement. The grains are attractive, medium sized, yellow-orange flint. It yields about 30 quintals per ha.

Climate

Maize is essentially a warm weather or *kharif* crop and as such is largely dependent upon the rains. It is grown under extremely divergent climatic conditions in different parts of the world, ranging from tropical to temperate regions. It is cultivated from sea level to 3000 metre altitudes under diverse conditions. It can be successfully grown where the night temperature does not go below 15.6ºC (60ºF). It cannot withstand frost at any stage of its growth. It requires considerable moisture and warmth from germination to flowering. The ideal temperature for germination is 21°C and for growth 32°C. Well distributed rainfall of 50-75 cm is conducive to growth. In India, its cultivation extends from the hot arid plains of Rajasthan and Gujarat to the wet hills of Assam and Bengal (receiving over 400 cm of rainfall).

There are three distinct seasons for the cultivation of maize: the main season is *kharif*; whereas its cultivation in *rabi* is done in Peninsular India and Bihar, and during spring in northern India. Higher yields have been recorded in the *rabi* and spring crops. The higher yields are primarily due to better water management and a lower incidence of disease and pests. In most parts of India, maize during *kharif* is sown with the break of monsoon, the actual dates varying from region to region. It is sown in early March in north eastern hills, in April to early May in north western hills, in May-June in Peninsular India and in the end of June to mid-July in the Indo-Gangetic Plains. The late sowing of maize may extend up to late August in certain irrigated tracts of Punjab. Spring maize is sown in late January to the end of February *Rabi* maize is generally sown in Bihar, Andhra Pradesh, Tamil Nadu and Karnataka in the end of October to mid-November. Both the spring and the *rabi* crops are raised, more or less, under irrigated conditions.

Soil

Maize requires fertile, deep and well drained soils. Although, it can be grown on any type of soil, ranging from deep heavy clays to light sandy ones, it is best adapted to well drained sandy loam to silty loam soils. It is, however, necessary that the pH of the soil does not deviate from the range 7.5 to 8.5. Maize plants, particularly in the seedling stage, are highly susceptible to salinity and water logging. Accordingly, the provi-sion of proper drainage is essential for the successful cultivation of this crop. The light sandy soils greatly facilitate drainage, but have a relatively poor water-holding capacity; on the contrary, very heavy soils, with excellent water holding capacity, have relatively poor drainage. Hence, soils ideally suited for maize cultivation should have adequate water holding capacity and should also provide for good drainage. Over 85% of the maize acreage is sown under rainfed conditions during the monsoon when over 80% of the annual rainfall is received. The alluvial soils of Uttar Pradesh, Bihar and Punjab are very suitable for growing maize crop.

Rotation

Since maize is a short duration crop, it conveniently fits into a wide range of crop rotations. In northern India, particularly in the states of Uttar Pradesh, Punjab, and

Bihar, it is rotated with wheat, potato, barley, etc. in a one year rotation under irrigated conditions. It is usually grown as a pure crop, but occasionally legumes, e.g. mung, arhar or beans, and quick growing vegetables, e.g. pumpkins and various types of gourds are taken as subsidiary crops. In central India, under rainfed conditions, however, cotton and arhar are grown mixed with maize. A usual kharif crop can also follow maize grown for green ears or fodder. In spring, very early maturing varieties of Moong and Mash (Sathi) are intercropped in maize. Given below are some of the most important crop rotations:

- Maize – Wheat
- Maize – Toria – Wheat
- Maize – Potato – Wheat
- Maize – Berseem
- Maize – Toria – Sugarcane
- Maize – Potato – Onion
- Maize – Senji – Sugarcane
- Maize – Raya
- Maize – Gram
- Maize – Lentil
- Maize – Pea

Crops like soybean, urd, moong, cowpea etc., are also grown mixed with maize. These legume crops are grown in the space between two rows of maize. In some parts of the country maize is also grown with pigeonpea

Cultivation

A seedbed for maize should be fine but compact. Maize kernels need a seedbed which is friable, well aerated, moist and free from weeds. It is desirable that the previous crop refuse is buried under with a mould-board plough. In due course, two or three ploughings with the wooden plough are given. In case the tractor is used, one ploughing followed by a couple of diskings is generally adequate. There is no need of preparing extremely fine seedbed for maize. The first ploughing for maize should be done with soil inverting plough so that 20-25 cm deep soil may become loose. It should be followed by two or three harrowings or three or four intercrossing ploughings with local plough. Planking should be done with each ploughing. It is important to note that while preparing the field for maize crop, levelling is not overlooked. A properly levelled and uniformly graded field is necessary for good water management.

Time of sowing

Time of sowing is indirectly determined by the soil and atmospheric temperature as well as by the supply of water. Thus, there is a general preference for sowing of the crop at the time of onset of monsoon in most parts of the country. The actual dates vary from region to region as indicated below

North Eastern Hills	Early March
North Western Hills	April to early May
Peninsular zone	Mid May to mid June

Under irrigated conditions in northern India, time of sowing varies from last week of May to end of June, depending on the prevailing conditions of the region. In fields which are prone to water stagnation, the crop is sown in the end of May/ early June so that the crop gets firmly established before the onset of monsoon. Maize cultivation has been extended to *rabi* season also i.e. from September end to beginning of November particularly in the Indo-Gangetic plains region. Trials conducted under All India Coordinated Maize Improvement Project revealed that yield potential of maize in *rabi* season was about twice that of the *kharif* season (Dayanand and Jain, 1994). Delay in sowing of *rabi* maize beyond first half of November in north western India results in a decline in its yield because of very cold conditions encountered during December and January. In eastern region and south India delayed sowing leads to reduction in yield due to shortening of the growing season.

Selection of seed

It is necessary to use certified seed of improved varieties. Before planting, the seed lot must be tested for its germination percentage and the seed quantity must be adjusted accordingly. For achieving maximum yield, every year a new hybrid seed must be used. The yield may be reduced drastically by using the seed from one's own hybrid crop, by 30 per cent or even more. However, the composite maize varieties do not have this characteristic and therefore, a farmer can use the seed from his own composite crop for next year also, provided no mixture has been allowed in the field and at the threshing floor. It is advisable to change the seed of composite every three to four years.

Seed rate and spacing

A population of 60,000-65,000 plants at harvest per hectare would be needed to attain maximum yield. It is very important that optimum plant density is maintained in maize, because unlike tillering plants such as rice and wheat, maize cannot compensate for lost space. Pal (1993) reported that under irrigated conditions of Delhi, the recommended plant density is 65,000-70,000 plants per hectare. In *rabi* season, the recommended plant density is 85,000-90,000 plants per hectare and this could give an yield advantage of 0.3-0.5 tonnes per hectare (Dayanand and Jain, 1994). Maize is sown in rows, 60-75 cm apart, whereas the plants in the row are spaced at 20 to 25 cm. Sawhney *et al.* (1989) reported that under Punjab conditions for high yield realization in winter maize for 25 October to 5 November sown crop, a spacing of 50 cm x 20 cm was ideal for sowing on flat beds. Under late sown conditions, best results were obtained by seeding the crop on southern slope of east-west ridges, spaced 60 cm apart with plant to plant spacing of 20 cm. Based on seed weight and required plant population, the seed rates are as follows:

Hybrids:	20-25 kg per hectare
Composites:	18-20 kg per hectare

Method of planting

Maize should be planted across the slope and the seed be sown 3-5 cm deep. The planting depth to a large extent will depend on the moisture status of the field and the type of soil. For example, if soil is dry and sandy, it is advisable to plant deeper. Normally planting is done in one of the following ways:

(a) Planting on the side of the ridge: A method adopted in high rainfall situation.

(b) Planting in narrow furrows: A method adopted in low rainfall areas.

(c) Planting in flat bed with no earthing up: In normal conditions.

(d) Planting on flat bed and earthing up after 40-50 days of planting: In areas where there is heavy storm during rainy season.

Weed management

During the initial stage, the growth of the maize plants is suppressed by weeds and weeds create an acute problem in rainy season crop. Lal and Saini (1985) reported that a reduction of 40% in yield can occur due to weed infestation. The main weeds found in maize are annual grasses, sedges and broad leaf weeds. Weeding may be done between the rows with bullock or tractor drawn implements, while the weeding within the rows is done by hand. Two to three hoeings are required at 15 and 30 days after sowing, following which the crop is earthed up to provide for better stand ability. No intercultivation after flowering should be done, as it is likely to damage the lateral roots.

Weeds can also be controlled by growing one or two rows of fodder cowpea in between maize rows and harvest it at 35-45 days after sowing, thereafter no weed control operation is required. Effective weed control can be obtained by spraying Atrazine 50 WP (@ 1.0 kg a.i. ha^{-1} before the crop germinates i.e., within two days of sowing in 500 litres of water. The dose of herbicides should be reduced by 40% in light soils. Where *Trianthema portulacastrum* is not a problem, Lasso can also be applied at 2.5 kg a.i. / ha in 500 litres of water within two days of sowing as pre-emergence spray. When hardy weeds like *Commelina benghalensis* (Kaun makki), *Acrachne racemose* (ghas), *Bracharia reptans* (Bans patta) are a problematic weeds, spray tank mix application of atrazine 0.75 kg/ha with Lasso 50 EC 1.25 kg a.i./ha or Stomp 30 EC 0.75 kg a.i./ha or Treflan 48 EC 1.2 kg a.i./ha as pre-emergence.

For the control of Dila/Motha (*Cyperus rotundus*) and broad leaf weeds from maize, apply 2,4-D amine salt 58 SL at 0.58 kg a.i./ha, 20-25 days after sowing.

Water management

Maize is extremely susceptible both to excess water and moisture stress. Maize can withstand heavy rain, however the water should not be allowed to stand in the field any time during maize growth. Water stagnation even for as small period as six hours can destroy the crop. Hence for *kharif* cultivation, it is essential that adequate drainage is provided. It may be provided in the form of very shallow surface drains at 40-50 m apart (depending on the slope and the texture of the soil) across the slope and connected to a main outlet. Shallow drains do not obstruct the movement of cattle or tractors during cultivation. Surface drains should be provided before sowing. In regions with about 60 cm of well distributed rainfall during the growing season, any additional irrigation is not necessary. A good crop of maize does require about 460-600 mm of water during its life cycle. It must be ensured that the maize plants never wilt because of water shortage any time during their life cycle. Tasseling and silking stage is very crucial. At this stage water shortage for even two days can reduce maize yield by 20 per cent. Also inadequate soil moisture during flowering and post-flowering particularly during the grain-filling period markedly reduce the yield. During the grain filling period, the most susceptible stage, additional irrigation, if needed, should be applied. The total number of irrigations will depend on the rainfall distribution pattern. The spring and *rabi* crops are entirely raised under irrigation; the number of irrigations may, however, vary from 5-10, depending upon the type of the soil and the prevailing temperature.

Fertilizer management

Fertilizers and manure both play a critical role in maize cultivation. For obtaining high yields, the maize crop should be heavily manured. Twenty five to thirty cartloads of farmyard manure or compost should be ploughed into the soil before sowing. For hybrid and composite varieties of maize, 100-120 kg nitrogen, along with 60 kg P_2O_5 and 40 kg K_2O per hectare, is recommended. The precise level of application of phosphorus and potash should be modified in the background of soil analysis. One-third of nitrogen and total quantity of potash and phosphorus should be applied before sowing, while the remaining nitrogen should be applied as side dressing at the knee high stage and at tasseling in two equal doses. Farmers planting local varieties may apply 40-60 kg per hectare nitrogen to the soil as per soil test. However, winter or *rabi* maize requires 50 % higher nutrition due to its long growing period. These nutrients can be increased or decreased depending on soil fertility.

Zinc deficiency

Zinc deficiency is wide spread in areas where the high yielding varieties are grown. The deficiency symptoms appear within two weeks of seedling emergence. A broad band of white or very light yellow tissue, with reddish veins appears on each side of the midrib, beginning at the base of the second or third leaf from the top of the plant. The white patch later extends in stripes towards the tip parallel to the midrib. The midrib and leaf margins remain green. The plants remain stunted and have short internodes.

Where zinc deficiency had been noticed in the preceding crop, apply 25 kg of zinc sulphate per hectare at sowing time or after crop emergence along the rows. Do interculture and apply irrigation. Foliar application of zinc sulphate and lime mixture

(3.0 kg zinc sulphate with 1.5 kg unslaked lime) can be done after dissolving in 500 litres per hectare of water.

Plant Protection

Insect Pests

Maize borer: The maize borer is a serious pest from June to September. Its larvae first scrape the leaves and then bore into the stem through the whorl or leaf sheath. The central leaves of the attacked whorl get perforated. In a young plant, the growing point is killed and a dead-heart results. Adopt the following integrated control measures :

- Kill the borer larvae hibernating in plant remnants like stubbles, stalks, cobs and cores. Plough up the fields after harvesting the maize crop, collect the stubbles and burn them. Use maize stalks, cobs and cores by the end of February. Chop the remaining stalks, if any, for subsequent use. For seed, keep only healthy cobs, free from borer attack.
- Remove and destroy the plants showing severe borer injury, while hoeing the crop.
- Use trichocards twice having 40,000 eggs of Corcyra cephalonica parasitized by Trichogramma chilonis per acre. First release on 10 days old crop and second 7 days after the first release. Cut trichocards into 40 strips, each having approximately 1000 parasitized eggs. Staple these strips uniformly on the underside of the central whorl leaves in evening hours.
- Spray the crop 2-3 weeks after sowing as soon as borer injury to the leaves is noticed with Coragen 18.5 SC (chlorantraniliprole) @ 75 ml or Decis 2.8 EC (deltamethrin) @ 200 ml using 150 litres water per ha with knap-sack sprayer.

Jassid, thrips, pyrilla, grey weevil and leaf-feeding insects: They attack the normal kharif crop. Spray Metasystox 25 EC (oxydemeton methyl) or Rogor 30 EC (dimethoate) @ 500 ml per ha in 150 litres of water with manually operated knap-sack sprayer or in 50 litres of water with a low-volume sprayer.

Armyworm and silk cutter: These insects feed on the leaves in the whorl. The pesticides used against the maize borer are effective in controlling these pests also.

Hairy caterpillars: Hairy caterpillars, if appearing in an epidemic form, cause serious damage by feeding on the leaves and the tender stems. When young, they feed gregariously. The grown up caterpillars may migrate from one field to another. Adopt the following control measures:

- Use light-traps for the destruction of moths.
- Young larvae are gregarious. They can be destroyed by plucking the infested leaves or by pulling out the infested plants and burying them underground.
- The grown up caterpillars can be destroyed by crushing them under feet or by picking and putting them into kerosenized water. If the population is high,

control it by sparying 1.25 litre of Ekalux 25 EC (quinalphos) per ha in 250 litres of water with a manually operated sprayer.

Shoot fly: It is a very serious pest of maize in south India. In north India, however; maize crop is not affected much except in the spring season in Tarai area. The attack is maximum when the crop is in seedling stage. Within two days or so eggs hatch and the tiny maggots creep down under the leaf sheaths till they reach the base of the seedling. After this they cut the growing point or central shoot which result in the formation of characteristic dead hearts. For the control of this pest, apply Furadan 3 G at 12.5 kg/ha or Thimet 10 G at 10 kg/ha in the furrows at this time of sowing.

Maize leaf roller: Damage is caused by the caterpillar which is glossy green in colour and becomes pink when fully grown up. The larvae after emergence wander on the tender leaves. They fold the leaves by silking threads and hide themselves inside the rolled leaves. The surface of the attacked leaf becomes white and papery in appearance. Spray Endosulfan (0.1 per cent) or 0.2 per cent carbaryl (Sevin) (4 g of Sevin 50 WP in one litre of water) at the rate of 500-700 litres per hectare. Spraying of 0.15 per cent oxydemeton methyl (metasystox) is also effective.

Grass hopper: *Kharif* grass hopper generally does not cause much damage to maize crop. However, in a favorable season, it may prove very harmful and leaves nothing on the plant except stem and midribs of leaves. Both adult and nymph stages of this pest are responsible for the damage. Adults are green or dry grass coloured. Grass hopper can be controlled to some extent by hot weather cultivation, which exposes their eggs to hot sun and natural enemies. As the newly hatched nymphs first feed on the grasses around the area where the eggs are laid, they can be easily controlled by dusting with 2 per cent Methyl parathion dust at the rate of 20 kg per hectare. For well grown nymphs and adults, spraying with 0.04 per cent carbaryl (Sevin) at the rate of 400 to 500 litres of water per hectare proves effective.

Cut worm: The larvae of this pest cut the seedlings at the ground level. Caterpillars are grey in colour. They live in soil during day and feed at night. The adults of cut worm may be trapped on light source and killed later. Dusting of 2 per cent Methyl parathion (Folidol) dust at the base of the plant at the rate of 20-25 kg per hectare will control the pest.

White grub: This pest is very serious in sandy soils. The larvae are white C-shaped and feed on the roots of the plants. In areas where it is a problem, damage may result in complete failure of crop. Phorate (Thimet) 10 per cent granules at the rate of 15 kg per hectare or Carbofuran (Furadan) 3 per cent granules at the rate of 30 kg per hectare should be mixed in the soil before sowing.

Mite: The attack of mite is serious in June on the young crop or in September-October when the crop is nearing maturity. The affected leaves turn pale and can be recognized from the presence of dusty webs. Spray 250 ml of Nuvacron 36 SL (monocrotophos) in 250 litres of water per hectare.

Diseases

Seed rot and seedling blight (Several fungi): Poor germination, unthrifty seedlings and seedling mortality are the symptoms. To control, treat the seed with Bavistin 50 WP or Derosal 50 WP or Agrozim 50 WP (carbendazim) @ 3 g/kg seed.

Maydis leaf blight (*Drechslera maydis*): This disease is characterized by the presence of spindle shaped, necrotic to brown lesions on the leaves. Such lesions may merge to form large, irregular patches. Sometimes the symptoms also appear on leaf sheaths, cob husks and ears. Late sowing, high humidity (>80%) and temperature of 25 ± 2°C favours the development of disease. Destroy the infected crop residue in the field. Grow improved varieties. Follow spray schedule as against Brown stripe downy mildew.

Bacterial stalk rot (*Dickeya zeae*): Water soaking and rotting of basal stem especially the leaf sheaths followed by rapid rotting of basal internodes. The rind loses its natural green colour and gives appearance as if boiled in water. The rotten stalks emit a characteristic fermenting odour and may break over from the second or third basal internode. Excessive rains and poor drainage favours the disease. The infected plants wilt. Destroy the diseased plant debris, keep the fields well drained and use improved varieties for its control.

Brown stripe downy mildew (*Sclerophthora rayssiae var zeae*): It is characterized by the presence of long, narrow, brownish, interveinal stripes on leaves. Whitish downy fungal growth may be observed on close examination on underside of the stripes. Destroy the collateral host Takri grass (Digitaria sanguinalis) from the maize field. Keep the fields well drained. Spray Indofil M-45 (mancozeb) @ 500 g/250 litres of water per ha after about a fortnight of sowing. Give two more sprays at 10-day intervals. Grow recommended varieties.

Post flowering stalk rots (*Fusarium spp., Macrophomina spp., Cephalosporium spp.*): Plants wilt after flowering. The rind and basal internodes become discoloured. On splitting, the discolouration of the pith progressing upward is also seen. Grow improved variety (PMH 1).

Black bundle disease: Blackening of vascular bundles appears as black dots on the cut ends of the stalk. In severe cases, leaves dry and plants wilt. Ears may rot or may not form at all. Treat the seed with systematic fungicides like Bavistin, or Benlate at the rate of 3 g per kg seed.

Pythium stalk rot: Infected plants become soft and at this point the plants get twisted and topple over the ground. Apply Captan in poorly drained pockets of the field along the rows of the crop at the rate of 150 g per 100 litres of water.

Late wilt: Plants wilt after flowering. The rind and basal internodes become discoloured. On splitting, the discolouration of the pith progre-ssing upward is also seen. Grow improved varieties of maize.

Harvesting

The maize crop is harvested when the husk has turned yellow and the grains are hard enough having not more than 20 per cent moisture. The appearance of the plant may be misleading, particularly in case of high yielding hybrids and composites whose grains are dry, while the stalk and leaves may be still green. Ears are removed from the standing crop. Harvested ears are dried in the sun before shelling. In case of the late sown crop, farmers prefer to harvest the whole plant and pile them, and the ears are removed later. Maize stalks are used as a cattle-feed and fuel. In fact, no part of maize plant, even the cobs from which the grains have been removed, is left unused. Maize grown for fodder should be harvested at the milk to early dough stage; the earlier harvested crop is likely to yield less and have lower protein content. For silage, however, the late dough is preferred. Both power and hand operated low priced maize shellers are available indigenously. These shellers are considerable more efficient than hand-shelling or beating with sticks, the common practice in northern India. Farmers using hybrid maize should not save their own seed for the next crop, as the advanced-generation hybrid seeds are likely to lead to a yield reduction of 25-30 per cent. However, farmers can save seed from composites and the open-pollinated varieties, when grown in isolation. At least seeds from 500 to 1,000 ears of the best yielding, normally spaced plants resistant to prevalent diseases and pests should be bulked. Ears should be dried, shelled and treated with an insecticide. The treated seed is necessary, as the untreated seed on ears is at times badly attacked by stored grain pests and the germination is markedly reduced. Considerable variation in grain yield is observed.

Yield

The yield levels depend upon the variety, the amount of the fertilizer used, the rainfall pattern, etc. Under irrigated conditions and recommended cultural practices, it gives 50-60 quintals per hectare in case of hybrids and 45-50 quintals in case of composites. In case of rainfed crop yield levels are about 20-25 quintals for hybrids and 15-20 quintals for composites.

BARLEY (*Hordeum vulgare*)

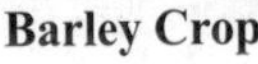

Barley Crop

Barley Grains

Introduction

Barley (*Hordeum vulgare*) is one of the most important cereals of the world. It is cultivated in almost all parts of the world except the tropical regions. Barley is a major source of food for large number of people living in the cooler, semi-arid areas of the world, where wheat and other cereals are less well adapted.

It is an important *rabi* cereal crop of India, ranking next to wheat both in acreage and in the production of grain. The chief barley growing regions in the country are higher Himalayas, central parts of eastern Uttar Pradesh, eastern parts of Rajasthan and north western parts of north Bihar. Barley being a hardy crop can be grown in diverse agroclimatic conditions like drought, salinity, alkalinity, rainfed or irrigated conditions, plains or hilly tracts where other crops cannot with stand. It can be good substitute for wheat in areas where owing to some agroclimatic and managerial constraints wheat cannot be successfully grown. It is further a highly suitable crop for late sown conditions. Being a less resource demanding crop it can be raised under limited input conditions.

The major portion of the grain produced is consumed by the poor classes either as flour for *chapati* making, pure or mixed with wheat or gram flour, or as parched grains to make *sattu*. However, almost 60 per cent of the total barley produced in India is consumed as cattle feed. With the development of scientific dairy farming the demand for barley as feed is also increasing. Barley has proven to be a palatable and nutritious feed for livestock. Further the straw is also used as good forage for making hay and silage. Besides, the broken grains are used in making the feed for piggery and poultry.

Barley is also used in the manufacture of malt and beer, aryuvedic medicine, cocoa-malt drinks and baby foods. With the coming up of many new breweries, after the liberalised licensing policy, the share of barley in the production of malt for manufacture of beer is expected to rise.

Origin and history

Barley has been known for thousands of years and it was cultivated before any recorded history of man. There are two theories regarding the origin of barley. One group of investigators considered Abyssinia as the principal centre of origin because many diverse forms grow wild there. This region is particularly rich in hulled, awned types. Another group of investigators considers that possible centre of origin is south eastern Asia, particularly China, Tibet and Nepal which is characterized by hull-less six-rowed varieties with short awn or no awn.

The barley must have been introduced to India soon after the coming of the Aryans. The Sanskrit name 'yava' was originally meant a grain that was later limited to barley and from which the name 'jau' or 'jav' has been derived shows that barley was in those early days regarded as one of the most important grains.

Area and distribution

Barley is one of the premier crops spread across the globe in many countries with wide variation in agroclimatic conditions and traditions. Its cultivation extends from the agricultural frontiers like plateaus of Tibet and Ethiopia to Oasis of Sahara. Barley has been the most tolerant cereal crop particularly where alkali soils, frost or drought are encountered. Thus it is cultivated in many countries with varying acreage.

It is cultivated on an area of about 5.9 lakh hectare with the production of 1.5 million tonnes in India during 2016 (FAO STAT 2016). The area under barley cultivation in India has been gradually going down as farmers are shifting to more remunerative crops like wheat and gram or cash crops like rapeseed, mustard or safflower. Barley is grown mainly in the northern plains of the country, but its cultivation extends up to elevations of about 4,575 metres (15,000 feet) in the Himalayas. Its greatest concentration is in the states of Uttar Pradesh, Rajasthan and Bihar. It is also grown to a considerable extent in northern Madhya Pradesh, Haryana, Punjab, West Bengal, Himachal Pradesh and Jammu and Kashmir. It is of minor importance in the south of 20° N, although it can be grown successfully wherever wheat can be grown. In this region, it occupies small areas in Gujarat, Maharashtra, northern Karnataka and in the Nilgiris and the Palney Hills of Tamil Nadu. About 51 per cent of the total area under this crop receives irrigation, the remainder being grown as a rainfed crop. The crop needs less water and is more tolerant to salinity and alkali conditions than other winter cereals. It is, therefore, of great significance in areas where it is not possible to take a successful crop of wheat owing to unsuitable soil and insufficient irrigation. In some recent experiments, this crop has shown good adaptation to conditions available in the coastal areas of West Bengal that traditionally grow only one crop of paddy. Its suitability in other nontraditional wheat and barley growing states, such as Assam, Meghalaya, Arunachal Pradesh, Tripura, Manipur, the Andaman Islands and Orissa in the east; Andhra Pradesh and the hills of Kerala in the south is being determined.

Classification

Aberg and Wiebe (1946) classified all the cultivated barley varieties into three distinct species based on the number of rows of grain and their arrangement:

(a) *Hordeum vulgare*- six-row barley

(b) *Hordeum distichon*- two-row barley

(c) *Hordeum irregulare*- two-row barley

Hordeum vulgare: In this species all the three spikelets at each node of the spike are fertile. The spikelets are in six distinct rows and arranged at a uniform distance around the tough rachis.

Hordeum distichon: In this species only the medium spikelets are fertile and sets grain. Although the lateral spikelets are infertile, they possess all the floral organs. This species is also cultivated in India.

Hordeum distichon: In this species central florets are fertile, whereas the lateral florets are reduced to rachilla.

Improved Varieties of Barley released by ICAR

2017-18

Variety	Area of adoption	Salient features
Central Barley DWRB 123	Punjab, Haryana, UP, Delhi and Rajasthan	Suitable for timely sown, irrigated condition, malt barley, resistance for YR, BR under natural field conditions

2016-17

Barley Variety	Area of adoption	Salient features
RD 2849	Punjab, Haryana, Delhi, Rajasthan and western Uttar Pradesh, Jammu and Kashmir and Uttarakhand (tarai region)	Two rowed; suitable for malt purpose; resistant to yellow rust
VLB 94	Uttarakhand	Six rowed; suitable for rainfed conditions; resistant to rusts
RD 2794	Salinity conditions of Punjab, Haryana, Delhi, Rajasthan, Uttar Pradesh, Bihar, Madhya Pradesh, West Bengal, Uttarakhand (tarai region)	Six rowed; suitable for salinity conditions under timely sown, resistant to yellow and brown rusts

Improved Varieties for Punjab

PL 807: It is a six-rowed and feed barley variety which matures in 137 days. It is fairly resistant to yellow rust, brown rust, loose smut, leaf blight and stripe disease. Its average grain yield is 43 quintals per hectare.

DWRUB 52: It is a two-rowed and malt barley variety which matures in about 140 days. It is fairly resistant to yellow rust, brown rust, loose smut, covered smut and leaf blight diseases. Its average grain yield is 43.25 quintals per hectare. This variety is particularly suitable for brewing industry.

PL 426: It is a six-rowed and feed barley variety which matures in 124 days. It is highly resistant to lodging, yellow rust, loose and covered smuts. It is tolerant to stripe disease but moderately susceptible to aphids. Its average grain yield is 35.0 quintals per hectare.

Climate

Barley is cultivated under wide variations of agroclimatic conditions. In India it is grown mainly as a *rabi* season crop, except in the high altitude areas of Ladakh and Lahaul, where it is raised during spring season. Although barley is a crop of cool climate, it may be successfully grown in tropical and subtropical conditions. Barley requires a mean daily temperature of about 12-15°C during the growing season and about 30°C during ripening and maturity stages. The warm and humid climate is not very suitable for the barley cultivation. Barley is highly drought tolerant and can be grown with lesser soil moisture. An annual average rainfall of 40-50 cm will result in increased yield. Even a well distributed and assured rainfall of as low as 20-25 cm is sufficient for realizing a good crop of barley in rainfed conditions. Barley is basically a long day plant. Thus a light period of 10-12 hours during vegetative growth and 12-14 hours during reproductive growth would be ideal.

Soil

Barley can be grown successfully on those soils, which are suitable for wheat cultivation. However, it thrives well on well drained fertile deep loam soils. Severe lodging occurs when grown on extremely fertile soils. It is widely grown on a variety of soils ranging in their texture from sandy to heavy loams and pH from 7-8 in the Indo-Gangetic plains and on terraced slopes in the hills. Its cultivation also extends, although to a limited extent, to medium black soils of Rajasthan, Maharashtra and Karnataka. It is more tolerant to alkali and saline conditions than other *rabi* cereals and is grown extensively under such conditions in Rajasthan, Punjab, Haryana, Uttar Pradesh and Bihar. Recent experiment indicates that this crop can be grown successfully on coastal saline soils of Sunderbans in West Bengal and on saline-alkali black soils in the canal irrigated areas of northern Karnataka.

Rotation

The rotation for barley is very similar to that for wheat. No exact rotation is followed year after year. Maximum production is obtained when the land is not cropped in the immediately preceding kharif season. It is generally rotated with crops, such as bajra, maize, paddy, jowar, cotton, groundnut, smaller millets, moong and moth in different parts of the country. Double cropping with barley is practised under assured soil moisture or irrigation facilities. On soils infested with nematodes, the rotating of barley with non-host crops e.g. sunflower, for at least one to two years would reduce the nematode population.

Barley is grown pure or mixed with other *rabi* crops, such as wheat, gram, peas and lentil. Sometimes, rape and mustard, taramira and linseed are also grown mixed with barley.

The following are common crop rotations where barley is included:

- Paddy – Barley
- Jowar – Barley
- Bajra – Barley
- Urd – Barley
- Cowpea – Barley
- Maize – Barley
- Cotton – Barley
- Groundnut - Barley
- Moong – Potato-Barley
- Maize - Potato-Barley

Cultivation

Barley may be grown on a rough seed bed and hence one cross cultivation with soil turning plough followed by planking, is sufficient to get a good seed bed for sowing. Conservation of late monsoon rains received during August and September is very important. In moderately saline-cum-high water table conditions, the last preparatory cultivation should be given a week to 10 days before sowing to allow the salts to accumulate near the surface to facilitate sowing of seeds in a relatively less or non-saline soil layer.

Seed and sowing

For sowing in rainfed and saline areas, soak the seeds in water overnight at room temperature for better and quicker germination. Under irrigated normal conditions soaking of seeds is not required. Before sowing, the seed should be treated with Vitavax 75 WP (Carboxin) or Raxil 2DS (tebuconazole) @ 1.5 g/kg seed to control covered smut, loose smut and stripe disease of barley.

Time of sowing

Time of sowing is an important aspect in successful cultivation of barley. Barley gives best results when sown between October 15 and November 15. There is gradual decline in yield as sowing gets delayed up to the end of December. In rainfed areas sowing should be done in third and fourth week of October but in the irrigated areas the best time of sowing is first and second week of November.

Seed rate and spacing

Seed rate varies according to agroclimatic conditions. In irrigated areas, for normal sowing 75 kg seed is sufficient while in late sown conditions seed rate should be increased to 100 kg per hectare. Under rainfed conditions 80-100 kg seed per hectare is required depending on soil moisture availability. For saline soils use 100 kg seed per hectare to ensure better stand of the crop.

For normal sown irrigated crop, a spacing of 22.5 cm between rows is recommended and sowing should be done at a depth of 4-5 cm. In rainfed crop a spacing of 23 to 25 cm between rows is suggested and sowing should be done at a depth of 6-8 cm depending upon soil moisture.

Method of sowing

Method of sowing is an important aspect particularly under rainfed conditions. Seed should be dropped with the help of '*pora*' or '*nal*' attached to a local plough or with the help of seed drill to ensure uniform distribution of seed at the optimum depth. In irrigated areas seed may be sown by 'kera' method, where seed is dropped by hand into the furrows. There should be enough moisture in soil for proper germination.

Weed management

Barley exhibits early vigorous growth compared to other winter crops. By the time tillering stage is reached, crop covers much of the ground area. Because of initial fast growth barley is more competitive to weeds and it smothers weed growth and thus weed problem is less in barley. However, barley crop raised with higher doses of fertilizer under irrigated conditions, weed control becomes essential. The most common weed species infesting barley are *Avena fatua*, *Phalaris minor*, *Fumaria parviflora*, *Chenopodium album*, *Anagalis arvensis*, *Melilotus alba*, *Melilotus indica*, *Convolvulus arvensis*, *Rumex spps.* etc. Interculture is not usually practiced unless the field is very weedy. If need arises, one hand weeding is sufficient. The use of weed free seed and a thoroughly prepared seed bed are essential for controlling the weeds effectively. Weeds can also be controlled with an application of 2,4-D sodium or amine salt at the rate of 0.50 kg a.i./ha or metsulfuron 20 WG at 4 g/ha in 250 litres per hectare of water, 30-35 days after sowing the crop. Application of isoproturon 0.94 kg/ ha after dissolving in 500 litres of water as post emergence spray at 30-35 days after sowing effectively control *Phalaris minor* and wild oats. For the control of *Phalaris minor* (gjullidanda), Puma power 10 EC (fenoxaprop-p-ethyl) @ 1.0 lit./ha at 30–35 days after sowing in 200 lit/ha of water.

Water management

Barley is grown as a rainfed crop as it has low water requirement and can tolerate drought. However, providing irrigation is quite useful in increasing yield from fertilizer responsive short statured dwarf and semi-dwarf varieties. To give good yields, barley requires 2-3 irrigation. One or two extra irrigations will be required on sandy soils. If the supply of water is inadequate, its efficiency can be increased by a proper timing of its application at the critical stages of crop growth. If only one irrigation is available, its application near the tillering stage has proved very profitable. When two irrigations are available, one should be applied at the active tillering stage and the other at the flowering stage. On highly alkali-saline soils, frequent light irrigation gives better results than a fewer heavy irrigation.

Fertilizer management

The rainfed barley crop is rarely manured directly. It is the preceding crop in the rotation which receives the manure. Irrigated barley is often given a basal dose of eight to ten cartloads of farmyard manure or compost. Any subsequent application of a fertilizer, however, is not practised. Recent agronomic experiments indicate that barley responds strikingly to an application of 20-40 kg N per ha under rainfed conditions, and to 40-60 kg N per ha under irrigated conditions, depending upon the soil, climate and variety. The application of 20 kg of N per ha to the crop under dry land farming led to the realization of 2-6 q per ha of additional grain and an application of 40 kg of N per ha under irrigated conditions resulted in a yield of 7-9 q per ha more than that from unfertilized plot at different locations. Good response to phosphorus up to 20 kg P_2O_5 per ha has been obtained under irrigated and timely sown conditions, but only at some locations. Hence its application needs to be governed by a soil test value.

To maximize the efficiency of nitrogenous fertilizers, the full dose of nitrogen should be placed 8-10 cm deep in the soil at sowing time under rainfed conditions. For irrigated barley, a split application of the half dose of nitrogen plus the full dose of P_2O_5 at sowing time and the remaining half dose of nitrogen with the first irrigation as a top-dressing is advantageous. Excess of nitrogen affects the malting and brewing quality of grain adversely. Super phosphate, however, helps to neutralize its deleterious effects.

Plant Protection

Insect pests: The barley crop is attacked by a number of soil insects. The important soil insects are white ants, termites, Gujhia weevil and cut worms. The attacked plants dry up completely causing poor crop stand. Use 2% Methyl parathion or 5% Malathion dust at the rate of 20-25 kg per hectare and mix it well in the soil at the time of last ploughing before sowing.

Foliage pests: Aphids are green to dark coloured small insects. These insects cluster in large numbers and suck sap from leaves, stem etc. They also spread disease by transmitting the virus. Spray Rogor 30 EC or Metasystox 25 EC or Nuvacron 36 SL at 375 ml/ha.

Diseases

Stripe disease (*Drechslera graminea*): Yellow to brown stripes appear on leaves. The plants become stunted and the leaves are shredded. Treat seed as per recommendations.

Loose smut (*Ustilago nuda*): The smut sori are enclosed in a fragile membrane which soon ruptures, releasing the dark dusty spore mass which is disseminated by wind, leaving the naked rachis behind. Treat seed as per recommendations.

Covered smut (*U. hordei*): The entire ear, except the awns turn into a black compact mass of spores. Treat the seed as per recommendations. Sow the crop late and shallow to reduce seedling infection.

Yellow rust (*Puccinia striiformis*): Yellow pustules appear on leaves, forming stripes. When the attack is severe, leaf-sheath, awns and glumes are also affected. Grow recommended varieties.

Seedling blight, leaf and glume blight (*Bipolaris sorokiniana*): Infects all the parts of the plant. In earlier stages may reduce crop stand. Elongated brown spots may be chlorotic, coalescing and may extend to the whole leaves and glumes causing blight. Grow resistant varieties.

Powdery mildew: Powdery mildew is found on the leaf surface. The disease appears as white fluffy patches on leaves, which turn grey on maturity, ranging from small isolated spots to complete leaf coverage and sometimes on the head. Leaves turn yellow-brown as the disease progresses. Infected plants have reduced growth and vigor. The heading and seed filling are impaired. Growing resistant varieties, dusting sulphur @ 20 kg/ha and following sprays of systemic fungicides like Karathane (0.2%) at intervals of 15-20 days give effective control.

Harvesting and threshing

Harvesting of barley in Jammu, Punjab, Haryana, Rajasthan and western Uttar Pradesh takes place from the third week of March to the middle of April. In Madhya Pradesh, Southern and Eastern Uttar Pradesh, Bihar and West Bengal, it starts a week to ten days earlier and in Maharashtra, Gujarat and Karnataka by about the first week of February. In the hills, where it is grown during the *rabi* season, the harvesting time varies from the end of April to the end of May, depending upon the altitude. The spring sown crop is harvested from the end of July to the end of September.

The plants are cut with a sickle when fully ripe. In order to prevent the loss of grain through shattering, the crop should be harvested immediately after it ripens. It is also advisable to cut the crop in the early hours of morning as the head becomes too dry and fragile for handling in the afternoon. The threshing in the plains is done by treading the dried produce under the feet of cattle or by stationary threshers. The grain is separated by winnowing.

Yield

When cultivation of high yielding varieties of barley is done with improved scientific methods they produce about 30-35 quintals of grain and 40-45 quintals of straw per hectare.

SORGHUM (*Sorghum bicolor*)

Sorghum crop — Sorghum ear head

Introduction

Sorghum (*Sorghum bicolor*) is also very important food grain crop of the world as millions of people in Africa and Asia depend on sorghum as the staple food. In addition, the fodder and stover is fed to millions of animals providing milk and meat for man. It is also used as industrial raw material in various industries in USA and other developed countries.

Sorghum popularly known as *jowar*, is the most important food and fodder crop of Indian dryland agriculture. Besides being a major source of staple food for humans, it serves as an important source of cattle feed and fodder. It has potential to compete effectively with crops like maize under good environmental and management conditions. The greatest merit with jowar is that it has capacity to withstand drought. Its performance is better than maize in marginal lands under moisture stress or excessive moisture conditions. It does well even in low rainfall areas. It makes comparatively quick growth and gives not only good yields of grain but also very large quantities of fodder.

In India, sorghum grain is eaten by human either by breaking the grain and cooking it in the same way as rice or by grinding it into flour and preparing *'chapatis'*. To some extent it is also eaten as parched and popped grain. This grain is also fed to cattle, poultry and swine. Sorghum grain contains about 10-12 per cent protein, 3 per cent fat and 70 per cent carbohydrates; therefore, it can satisfactorily replace other grains in the feeding programme for dairy cattle, poultry and swine.

Origin and history

Cultivated sorghum probably originated in East Central Africa, in or near Ethiopia or Sudan because of the great diversity of types growing in that region. The diversity of cultivated types decreases towards southern Africa and Asia. Although several of the wild species have been collected in India, but there is no evidence of an independent origin of cultivated sorghum in India because the general types grown here are abundant in Africa. The date of arrival of cultivated sorghum in India is uncertain. The movement of sorghum out of Africa to East was probably brought about through the flow of traffic. The trading route between East Africa and India via Arabia is ancient. Sir George Watt (1893) pointed out that there was no specific Sanskrit name for sorghum. It is, therefore, supposed that sorghum reached India in 1500 BC.

Area and distribution

Sorghum is cultivated on an area of 5.65 million hectares in the world with production of about 4.41 mt (FAO STAT 2016). USA leads in production of sorghum contributing 19.1 percent of global production folled by Nigeria with 10.9% and Sudan 10.1% (FAO STAT 2016) (Fig. 8.6). Cultivation of *jowar* is mainly concentrated in peninsular and central India. Maharashtra, Karnataka, Andhra Pradesh, Madhya Pradesh, Gujarat, Rajasthan, Uttar Pradesh (the Bundelkhand region) and Tamil Nadu are the major *jowar*-growing States. Other States grow sorghum in small areas primarily for fodder.

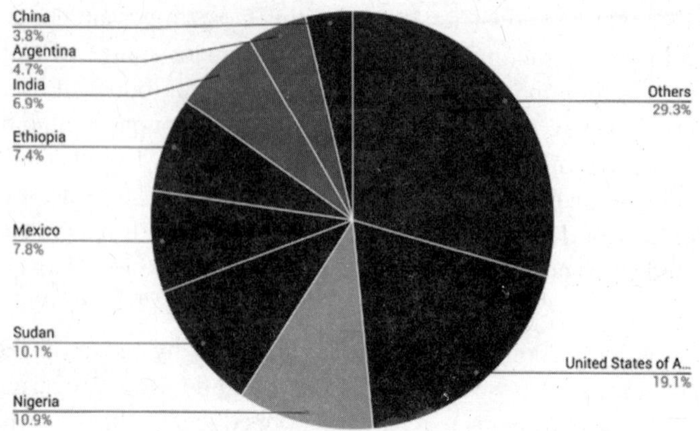

Fig. 8.6. Annual global production of sorghum during 2016 (Percentage) (FAO STAT 2016)

Classification

For many years most of the cultivated sorghums were classified into a single species, *Sorghum vulgare* Pers. However, this name was not considered the correct binomial designation (Doggett 1965). Thus the proper binomial terminology for all these sorghums is found to be *Sorghum bicolor* (Linn) Moench. Since 1936, most sorghum workers have been using the classification of J.D. Snowden or some of its modification. However, the Snowden classification is very difficult to use, when one is dealing in thousands of accessions. Later on, Harlan and De Wit (1971) have developed a simple classification. Accordingly the various groups have been characterized by a new scheme. The system partitions the variations in *Sorghum bicolor* (Linn) Monech into the following races:

a. Basic races

1. bicolor, 2. guinea, 3. caudatum, 4. kafir, 5. durra.

b. Hybrid races

1. guinea-bicolor, 2. caudatum-bicolor, 3. kafir-bicolor, 4. durra-bicolor, 5. guinea-caudatum, 6. guinea-kafir, 7. guinea-durra, 8. kafir- caudatum, 9. kafir-durra, 10. durra-caudatum.

Improved Varieties of Sorghum released by ICAR

2017-18

Sorghum Variety	Area of adoption	Salient features
PDKV Kalyani	Maharashtra	Clean foliage (no leaf diseases) and non-lodging nature

2016-17

Sorghum Variety	Area of adoption	Salient features
Palamurujonna	Telangana,Tamil Nadu, Rajasthan and Gujarat	Juicy stem, white-coloured mid-rib, pearly white seeds; tolerant to grain mould ,resistant to anthracnose, leaf blight
CSH 34	Maharashtra, Karnataka, Madhya Pradesh, Andhra Pradesh, Chhattisgarh Gujarat and Rajasthan	Resistant to downy mildew, escapes grain mould due to little more of maturity duration
CSH 35	Maharashtra, Karnataka, Madhya Pradesh, south Gujarat and Telangana	Non-lodging
Phule Rohini	Maharashtra	Special purpose variety for processing, especially for papad; tolerant to shoot fly, charcoal rot, drought
Phule Vasundhara	Maharashtra	Sweet sorghum hybrid with high green cane yield, juice yield, ethanol yield and Brix
Phule Madhur	Maharashtra	Hurda purpose, threshable grain percentage at dough stage; better organolyptic properties; tolerant to shoot fly, charcoal rot, rust, blight
GJ 42	Gujarat	Resistant to shoot fly, stem borer; suitable for heavy rainfall areas of Gujarat
SPH 1641	Maharashtra	Moderately tolerant to grain mould, shoot fly, stem borer
Raj Vijay Jowar 1862	Madhya Pradesh	Dual purpose; medium maturity with high grain yield and fodder yield
RVICSH 28	Madhya Pradesh	A high fodder-yielding sweet sorghum hybrid with 17% feed-stalk juice Brix

Improved Varieties of Sorghum for fodder released by ICAR

2017-18

Forage Sorghum	Area of adoption	Salient features
CSV 33 MF	Haryana, Punjab, Rajasthan, Uttrakhand, Uttar Pradesh, Gujarat, Rajasthan, TN, Karnataka and Maharashtra	Multi cut forage sorghum, tall 180-204 cm, resistant to all major diseases and pest infestation, tolerant to lodging

2016-17

Forage sorghum	Area of adoption	Salient features
Punjab Sudax Chari 4	Punjab	Sweet multi-cut hybrid; moderately resistant to shoot fly and red leaf spots
CSV 21F	All India	Medium maturity; resistant to all leaf spot diseases; tolerant to shoot fly and stem borer

Climate

Sorghum can be grown under a wide range of climatic conditions although ideally it requires warm climate. It is grown from sea level to as high as 1500 metres. Sorghum plants can tolerate high temperature throughout their life cycle better than any other cereal crop. It can tolerate drought conditions very well because it remains dormant during moisture stress conditions but resumes growth when favourable conditions reappear. It has a high resistance to desiccation, low transpiration ratio and a large number of fibrous roots. It can also tolerate water logging conditions better than any other cereal except rice. Therefore, sorghum can be grown successfully in areas having an average annual rainfall between 60 and 100 cm. It is grown as a *Kharif* crop in northern India. In western and southern parts of the country it is grown also as a *Rabi* crop. The minimum temperature for the germination of sorghum seeds is 7-10ºC. It needs about 26-30ºC temperature for its optimum growth. Sorghum is a short day plant. Flowering is hastened by short days and delayed by long days. The time of heading in sorghum is influenced by temperature as well as photoperiod. Sorghum varieties vary in their sensitivity to both temperature and photoperiod.

Soils

Sorghum is grown in a variety of soils in India. Soils with clay loam or loam texture, having good water retention capacity are best suited for sorghum cultivation. It does not thrive in sandy soils but does better on heavier soils. It does well in pH range of 6.0-8.5 as it tolerates considerable salinity and alkalinity. The black cotton soils of central India are very good for its cultivation.

Field preparation

Sorghum seed bed should be well prepared and free from weeds. The first ploughing should be done with soil turning plough so that 20-25 cm deep soil may become loose. It should be followed by two to three harrowing or three to four intercrossing ploughings with country plough. Thereafter planking should be done to break the clods and to level the field. In black cotton soil area, if the land is badly infested with weeds, ploughing followed by disc harrowings is usually practiced, but where land is free from weeds or with few weeds, the land is cultivated only with cultivators or tillers.

Time of sowing

In northern India sorghum is sown only in *kharif* season. In irrigated areas, first week of July has been found most suitable for sowing of most hybrids and improved varieties. Under unirrigated conditions, sowing should be done preferably within one week of the onset of first monsoon showers. Planting on the receipt of first monsoon showers was found to be ideal for rainfed (*kharif*) sorghum (Singh *et al.,* 1972). Under irrigated conditions planting 7-14 days in advance of monsoon rains could result in yield increase. Timely planted crop escapes the damage due to shoot fly and midge. Late planting may not fit well in multiple rotations. Normal (optimal) plating time is considered on the onset of monsoon and any delay in seeding from normal time results in progressive decrease in grain yield. The shootfly incidence increases with the delay in sowing. Nimbalkar *et al.* (1985) also reported increased incidence of shootfly in Maharashtra when the crop is sown beyond July 4.

Rabi sowing is done mainly in Maharashtra, Karnataka and Andhra Pradesh. *Rabi* sowing should be done from the second fortnight of September to the middle of October (Kachapur, 1990; Pharande *et al.,* 1980; Chorge and Ramshe, 1990). Summer crop of sorghum is sown in the month of January and February in irrigated areas of Tamil Nadu, Andhra Pradesh and some areas of Karnataka.

Seed treatment

The seed should be purchased from a reliable source. It is advisable to always use certified seed. If seed is not already treated, it should be treated with. Bavistin 50 WP or Derosal 50 WP or Agrozim 50 WP @ 3g per kg seed.

Seed rate and spacing

In case of hybrids, new hybrid seed should be used every year. Sorghum seed should be drilled in a well prepared seed bed. A seed rate of 12-15 kg per hectare is considered sufficient to ensure a good stand. There should be 1,50,000 plants per hectare to attain maximum yield. The seed should be sown in rows 45 cm apart. Plant to plant distance should be 12 cm. Seed should be sown at a depth of 3-4 cm. It should not be sown more than 5 cm deep in any case.

Method of sowing

In Northern India sorghum is sown either by broadcasting or in rows behind the plough. Seeds of new hybrids and varieties should always be sown in lines to economic seed and for obtaining higher yield. Sowing in rows is also common in black cotton soil.

Thinning

In sorghum cultivation, thinning is a very important operation for maintaining the desired plant population. Twelve to fifteen cm plant to plant spacing in a row by thinning out extra plants at two stages should be ensured. First, thinning should be done 10-15 days after emergence and second, when crop is 20-25 days old. All disease and insect infested plants should be removed while thinning.

Weed management

Sorghum is badly infested with both broad leaf and grassy weeds, because of congenial weather especially during *kharif* season because of rainfall and humidity. Some of the important weeds associated with *kharif* sorghum are: *Commelina* sp., *Trianthema portulacastrum* sp., *Celosia argentea*, *Amaranrthus* sp., *Elucine* spp., *Digeria arvensis* etc. During *rabi* season, *Denebra retrofhxa*, *Bracharia isachne* and *Chrozophora rottliri* are predominant. Perennial weeds *Cyperus* sp., *Cynodon dactylon* and *Convolvulus arvensis* occur in all the seasons. Sorghum seedlings are comparatively small and weak and cannot compete with weeds. Weed free environment is essential till the seedlings establish and produce enough vegetative growth that approximately takes 45-50 days. Weed control can be accomplished either by mechanical methods, chemical methods or a combination of both. By completely controlling weeds during the first four weeks, weed problem can be tackled easily. Inter row weeds may be controlled mechanically by hand weeding with a *khurpa* or a hand hoe. Two hand weedings at 15 and 30 days after sowing effectively control the weeds. However, the cost of mechanical methods is sometimes prohibitive or situations during rainy season do not permit hoeing or hand weeding and under these circumstances chemical method can be used. Chemical method is more economical. Trifluralin @ 0.75-1.0 kg a.i./ha can be applied as pre-plant incorporation. Pre-emergence application of atrazine @ 0.50-0.75 kg a.i./ha, propazine @ 1.0-1.5 kg a.i./ha or simazine @ 0.50-0.75 kg a.i./ha can be given. Pre-emergence spray of triazine chemicals has been very effective in controlling broad leaf weeds and annual grassy weeds without any adverse effect on sorghum (Krishnamurthy, 1970; Borse and Mahajan, 1981). The use of atrazine in sorghum is very safe and effective (Satao and Nalamwar, 1992). Under Punjab conditions, pre-emergence application of atrazine at 0.5 kg/ha or pendimethalin 0.75 kg/ha provided good control of weeds.

Water management

Usually, sorghum is grown as a rainfed crop. The irrigation should, however, be provided whenever, rains are not received. At the time of flowering and grain filling stages, the crop requires more water. If enough moisture is not there in the soil at the time of flowering and grain filling, it should be irrigated at once. At no stage, the plants should be allowed to wilt. Suitable drainage conditions should be provided for the removal of excess rain water from the field.

Fertilizer management

As sorghum removes nutrients in heavy amount from the soil it requires heavy doses of fertilizers. Manure and fertilizers both play important roles in the sorghum cultivation. In the rainfed areas, application of farm yard manure or compost at the rate of 10 to 15 tonnes per hectare improves the water holding capacity and microbial activities in the soil, besides providing essential nutrients to the crop. Farm yard manure or compost should be added in the field at the time of last ploughing. The quantity of fertilizers to be applied varies according to the fertility status of the soil. However, when soil test data are not available, apply 100-120 kg Nitrogen, 50 kg P_2O_5 and 40 kg K_2O per hectare for hybrids and improved varieties of sorghum under irrigated condition. Half dose of nitrogen and total amount of phosphorus and potash should be applied at the time of sowing. The basal dressing can be done with the help of fertilizer-cum-seed drill. The remaining half quantity of nitrogen should be top dressed after 30-35 days of sowing. In light soils top dressing should be done in two splits. In case of local varieties half of the above dose should be applied for better results. For rainfed crop, quantity of fertilizer should be reduced to half of the irrigated crop and the entire quantity should be applied 10 cm deep in soil at the time of sowing.

Among several micronutrients, deficiency of zinc and iron is widespread. Response to zinc application has been noticed on vertisols at Akola, Indore, Nandyal etc. Hybrid sorghums were reported to respond to the zinc application in the soil with 0.26-0.82 ppm of zinc (Khan and Zende, 1976; More *et al.*, 1977). Zinc supply was observed to increase the yield to the tune of about 410-540 kg/ha (Subba Rao, 1975). Sorghum is also sens-itive to iron stress and is less efficient in its absorption and translocation. Hence this element is needed to be applied. The problem of iron deficiency is more on calcareous soils because iron uptake decreases with increased calcium carbonate content of the soil. Deficiency of zinc and iron can be corrected either through soil application or through foliar sprays. For soil application 25 kg of zinc sulphate per hectare once in two to three years is applied. In case of foliar application, 2-3 sprays of 0.2% zinc sulphate or 0.5% iron sulphate is adequate.

Plant Proterction

Insect pests

The important insect pests of sorghum are given below:

Sorghum shoot fly: It is a major pest of sorghum attacking in the early seedling stage of growth. Plants up to three week age are more susceptible to this pest. The eggs are laid on the lower surface of the leaves, stem or soil near the base of plants. The maggots hatching from the eggs creep down under the leaf sheaths and cut the central growing point of the plant resulting in dead hearts. In areas where it is serious, use Phorate (Thimet) 10% granules at the rate of 15 kg per hectare or Carbofuran (Furadan) 3% granules at the rate of 30 kg per hectare in furrows or as broadcast before sowing. If these chemicals have not been applied at sowing time, spray the crop with oxydematon methyl (Metasystox) 25 EC at the rate of 1.5 litres in 400 to 500 litres of water per hectare a week after germination and repeat the spraying after 15 days.

Stem borer: This pest attcks sorghum crop throughout the growing season. In an early stage of the crop, larvae bore inside the tender stem and feed on the tissues resulting into the death of growing point thus forming dead hearts. Grown up plants are not killed but suffer in vigour and develop weak heads. As estimated damage up to 80% may be caused. Use higher seed rate and pull all the affected plants in an early stage of the crop. Spray the crop with 0.1% endosulfan 35 EC or 0.2% carbaryl (sevin) 50 WP after 15 days of germination followed by the application of endosulfan 4% granules or sevin 4% granules in the whorl of the plants at the rate of 15 kg per hectare after a fortnight of spraying.

Pink borer: The damage is caused by the larvae which bore into the shoot and the dead hearts are formed in the early stage of the crop. Spray 0.1% endosulfan 35 EC or 0.2% carbaryl, 15 days after germination. Repeat the spraying if necessary.

Sorghum midge: Due to its infestation grain fails to develop and in cases of severe attack the whole head may be damaged resulting in serious yield reductions. Apply 5% carbaryl or malathion dust at the rate of 25 kg per hecatre or spray the crop with 0.1% carbaryl 50 WP at the rate of 400 litres per hectare at prebloom stage (50% panicle emergence) followed by another application after 7 days.

Leaf roller: The larvae fold or roll the leaves by bringing the two margins closely with the help of silken thread to hide and feed inside this roll. Due to this feeding the leaves become white papery in appearance. Dust the crop with 2% methyl parathion dust at the rate of 25 kg per hectare. Spray the crop with dichlorosphos (nuvan) 70 EC at the rate of 1 ml per 4 litres of water.

Sorghum ear head bug: Both adults and nymphs suck the milky juice of the developing grains due to which the grains remain shrivelled. When a large number of nymphs feed, the whole earhead may become blackened at first and may eventually dry up producing no grains. Remove all the plants of wild sorghum from the surroundings. Dust the crop with 5% Lindane or Carbaryl or Malathion dust at the rate of 20 kg per hectare or spray the crop with Endosulfan 35 EC at the rate of 1.0 litre in 400 litres of water per hectare.

Hairy caterpillar: The eggs are laid in clusters on the leaves of wild as well as cultivated plants. After hatching, first and second instar larvae feed gregariously and skeletonize the leaves. Later on grown up larvae disperse and feed singly. The egg masses should be collected and destroyed. The light traps should be used as soon as the moths are visible after the rains start.

Diseases

Sorghum crop is attacked by a number of diseases resulting in huge losses in yield. The important diseases are next page:

Grain moulds: Several fungi can grow on developing ears of sorghum when the grain setting coincides with rains or excessive humidity. Discolouration of grain occurs and in severe case loss of seed weight, seed size and seed germination takes place. Toxins produced by the fungi are toxic to animals. This disease can be controlled by growing early maturing and tolerant genotypes.

Sorghum downy mildew: Abundant downy white growth is produced on the under surfaces of infected portions of leaves during humid weather. Systemically infected plants show progressively more of a complete bleaching of the leaf tissue, sometimes in streaks or stripes and sometimes over the entire leaf surface. As the infected bleached leaves mature they become necrotic and the interveinal tissues disintegrate to give the typical shredded leaf symptom. For its control eradicate, wild hosts near sorghum fields, remove infected sorghum plants in the field, destroy plant debris by deep ploughing and other methods, growing resistant varieties and hybrids like CSV5, CSV 6 etc. Treat seed with Metalaxyl at 4 g/kg of seed or with Redomyl MZ at 6 g/kg seed before sowing of sorghum. Foliar spray of Mancozeb at 2.5 g/litre or Metalaxyl MZ at 2g/litre is recommended if standing crop is infested with downy mildew.

Ergot: Ergot can be an economically devastating problem. In seed production fields, yield losses ranging between 10 to 80% due to ergot disease have been reported in India. Ergot contamination reduces grain quality and limits its use as a feedstock. Ergot or sugary disease results from the ergot fungus that infects the florets in the earhead (panicle) and prevents seed set in such florets. The first symptom is droplets of a liquid honeydew exudate from between the glumes of infected florets. If honey-dew exudation is profuse, it may drip down and smear the whole earhead, sometimes dripping down onto the soil. If the exudate is plentiful, the earhead appears with blackened grain. Soaking seeds with 5% salt solu-tion will aid to remove ergot infested seeds, as ergot infested seeds will float in the salt solution. Seed treatment with fungicides such as Captan or Thiram. Control of ergot with fungicides such as Propiconazole or Tebuconazole has proved to be cost effective in seed production plots. Multiple sprayings at 5-7 days interval from flag leaf stage till end of anthesis are recommended to reduce the variation in flowering within a field.

Charcoal rot: The charcoal rot fungi infects roots of plants growing under adverse environmental conditions, particularly moisture stress and high temperature. Infected roots show water-soaked lesions that turn brown or black. Affected stalks are soft or spongy at the base and tend to lodge. Lodging is the most apparent symptom of charcoal rot. If lodged plant is split open and examined, the pith of the stalk may be found in various stages of rotting of its tissues. Maintaining soil moisture through irrigation, if possible, during the post- flowering period can minimize charcoal rot infestation. A balanced fertilizer program is beneficial because high levels of nitrogen application with low levels of potassium may lead to charcoal rot. High plant population should be avoided. Best plant density based on moisture availability need to be adopted. Growing of drought tolerant, lodging resistant, and no senescing types of sorghum like CSV 17, SPV 462 etc. can reduce yield losses from charcoal rot infestations.

Rusts: Young plants are rarely affected by rust. The typical symptoms are seen in 1.5 to 3 months old plants. Scattered purple, red or tan flecks appear on both surfaces of leaves and in resistant cultivars the symptoms donot develop further. In susceptible cultivars typical rust pustules develop, mainly on the lower leaf surfaces. In highly susceptible cultivars the pustules occur so densely that almost the entire leaf tissue is destroyed. The pustules may also occur on leaf sheaths and on inflorescence stalks. In areas where rust becomes an economically important disease, growing of less-susceptible or resistant cultivars like PSH 1, SPH 837, CSV 17 etc. is the only solution.

Harvesting and threshing

Most of the high yielding sorghum hybrids and varieties take about 100-115 days to mature. The crop should be harvested immediately after it is mature. The right stage for harvest is when grains have become hard having less than 25 per cent moisture. It is advised not to wait for stalks and leaves to dry because plants of hybrid sorghum appear green even after the crop is mature. Harvesting is done by cutting the entire plant or removing the earheads first and cutting down the plants later on. In the areas where there is danger of rain at the time of harvesting, the mature earheads should be harvested first and plants are cut and heaped later on. Threshing is done with the help of threshers or by beating the earheads with sticks or trampling by bullocks. The threshed grain should be cleaned and dried in sun for about a week to bring the moisture content down to 13-15 per cent for safe storage.

Yield

With improved cultural practices it is possible to harvest nearly 50 quintals of grain and about 100-125 quintals of dry stover from one hectare of crop under irrigated conditions and about 25-30 quintals of grain and about 80-100 quintals of dry stover from a rainfed crop.

PEARL MILLET (*Pennisetum glaucum*)

Bajra crop **Bajra seed**

The millets are a group of small seeded species of cereal crops or grains, widely grown around the world for food and fodder. They do not form a taxonomic group, but rather a functional or agronomic one. Their essential similarities are that they are small seeded grasses grown in difficult production environments. India contributes 36.3% of global production of millets folledby Niger (13.7%) and China 7.0% (FAO STAT 2016). (Fig. 7) Pearl millet (*Pennisetum glaucum* L.) is the most widely grown type of millet. Pearl millet (bajra) is one of the major coarse grain crops and is considered to be a poor man's food. It is widely grown in Africa and Asia since pre-historic times. It is grown in Africa where it replaces sorghum as the pearl millet grain is comparatively high in protein and has a good amino acid balance. Pearl millet is an important crop of rainfed areas of India. Pearl millet provides staple food for the poor in a short period in the relatively dry tracts of the country and cultivated by the economically poor farmers using either no improved production technology or using it only at sub-optimal levels. Pearl millet is endowed with greater ability to withstand harsh environment. Pearl millet is well adapted to production systems characterized by low rainfall, low soil fertility, and high temperature. It performs well in soils with high salinity. Because of its tolerance to difficult growing conditions, it can be grown in areas where other cereal crops, such as rice or maize, would not survive. Pearl millet grain is rich in minerals (2.0-3.5%) and protein content (10.5-14.5%). It is high in lysine and methionine + cystine levels. It contains twice as much methionine than sorghum, an important trait for organic poultry production. The grain is also comparatively high in fat, and linolenic acid comprises 4% of the total fatty acids. Even when grown in highly stressed conditions, the grain is essentially free of aflatoxins and fumonisins. These carcinogenic mycotoxins are a significant problem in maize when it is grown in regions where it is not well adapted. The pearl millet grains possess a biological value similar to wheat and rice and impart substantial energy to the body. In addition to its grain consumption as food, it is also used as green fodder in India, Pakistan, Southern USSR and Australia. It is also used as a pasture crop in USA. Today pearl millet accounts for approximately 50% of the total world production of millets. Pearl millet is the fourth important cereal crop of India after rice, wheat and maize.

Origin and history

Primary centre of origin of pearl millet is believed to be Africa from where it spread to India and other countries. Of the 32 species, described by Stapf (1934) only two are known to be outside Africa. A greatest range of variability is found in the strains from Africa. This is a further evidence of the African origin of pearl millet.

Area and distribution

Millet is a crop grown mostly in tropical climate. It is widely grown in Africa and Asia. Fig. 7 shows production contribution of the top 10 millet producing countries in 2012. Millets are grouped together for reporting global millet production therefore, it is difficult to determine the production amounts of each individual genus. As a general estimate millet production is broken down into the various percentages: 50% - Pearl Millet (*Pennisetum typhoides*), 30% - Proso (*Panicum miliaceum*) and Foxtail (*Setaria italica*), 10% - Fingermillet (*Eleusine coracana*), 10% - Other: including Barnyard Millet (*Echinochloa crusgalli*) and Kodo Millet (*Paspalum scrobiculatum*). Pearl millet is currently the world's sixth most important cereal food grain and is grown extensively in Africa, Asia, India and the Near East as a staple source of nutrition for millions of people. Millets were introduced into the United States at an early date but were seldom grown until 1875, and currently they are primarily grown for birdseed, animal feed, and forages rather than human consumption. Millets were grown on an area of 31.7 m ha with production of 28.36 mt. (FAO STAT 2016). In India total production of millets was 10.28 m tonnes (36.3%) of global production follwed by Niger 3.88 m tonnes (13.7 %) and China 1.99 m tonnes (7.0%) (FAO STAT 2016) (Fig. 8.7).

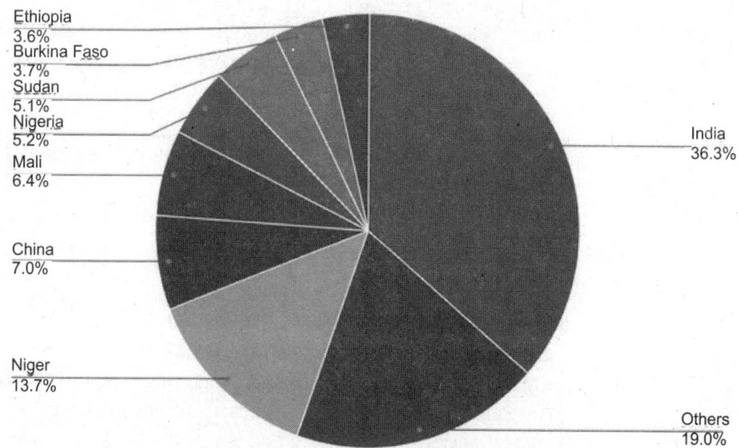

Fig. 8.7. Annual global production of millets during 2016 (Percentage) (FAO STAT 2016)

Improved Varieties of Pearl Millet released by ICAR

2017-18

Pearl millet	Area of adoption	Salient features
PBH - 306	Maharashtra, Karnataka, Telangana, AP and TN	Resistant to downy mildew and smut
Mahabeej 1005 Hybrid	Maharashtra	Suitable for rainfed condition in Maharashtra, non-lodging
NBH 4903	Maharashtra, Karnataka, AP, Telangana and TN	Late maturing, medium plant height, tolerant to major diseases (downy mildew, ergot and blast), tolerant to drought

2016-17

Pearl Millet Variety	Area of adoption	Salient features
ProagroTejas	Rajasthan	Early maturing; resistant to downy mildew
PHB 2884	Punjab	Late maturing, grey seeds
86 M 84	Rajasthan, Gujarat, Haryana, Madhya Pradesh, Uttar Pradesh, Punjab and Delhi	Late maturing, grey obovate large grains; resistant to downy mildew, blast
Central Pearl Millet Hybrid MPMH 21	Drier parts of Rajasthan, Gujarat and Haryana	Early maturing, grey-brown hexagonal grains; resistant to downy mildew, blast, smut
HHB 272	Drier parts of Gujarat, Rajasthan and Haryana	Early maturing, grey globular grains; resistant to blast
Pearl Millet CO 10	Tamil Nadu	Medium maturing, grey-brown bold grains;resistant to downy mildew
JKBH 1008	Drier parts of Gujarat, Rajasthan and Haryana	Early maturing, grey-coloured globular grains; resistant to downy mildew, blast
MP 535	Rajasthan, Gujarat, Haryana, Punjab, Delhi, Uttar Pradesh and Madhya Pradesh	Medium maturing, globular grey-brown grains;resistant to downy mildew
MH 1928	Rajasthan, Gujarat, Haryana, Punjab, Delhi, Uttar Pradesh and Madhya Pradesh	Medium maturing, grey-coloured hexagonal grains; resistant to downy mildew blast, rust, ergot
KBH 3940	Rajasthan, Gujarat, Haryana, Punjab, Delhi, Uttar Pradesh and Madhya Pradesh	Late maturing, grey-coloured obovate grains; resistant to downy mildew, blast, rust, ergot

Nandi 75	Summer growing areas of Gujarat, Rajasthan, Uttar Pradesh, Maharashtra and Tamil Nadu	Late maturing, grey-coloured globular grains; resistant to downy mildew, rust
86 M 82	Rajasthan, Gujarat, Haryana, Punjab, Delhi, Uttar Pradesh and Madhya Pradesh	Late maturing, grey-coloured obovate seeds; resistant to downy mildew, ergot, rust, smut
86 M 13	Summer growing areas of Gujarat, Rajasthan, Uttar Pradesh, Maharashtra and Tamil Nadu	Late maturing, grey-coloured obovate seeds;resistant to downy mildew
BIO 8145 Hybrid	Rajasthan, Gujarat, Haryana, Punjab, Delhi, Uttar Pradesh and Madhya Pradesh	Late maturing, grey-coloured globular grains; resistant to downy mildew, blast, rust, ergot
JKBH 1100 Hybrid	Uttar Pradesh	Late maturing; resistant to downy mildew, rust, smut diseases
JKBH 1105 Hybrid	Summer areas of Uttar Pradesh	Late maturing, globular, grey-brown grains; resistant to downy mildew, rust, smut diseases
Phule Aadishakti	Maharashtra	Late maturing, grey globular seeds; resistant to downy mildew

Improved Varieties of Pearl Millet for fodder released by ICAR

2017-18

Forage Pearl Millet	Area of adoption	Salient features
Phule Gunwant (RBN 2011-12)	Maharashtra	Perennial grass, multi cut

2016-17

Forage Variety	Area of adoption	Salient features
IGPM 5-2	Haryana, Punjab, Rajasthan, Uttar Pradesh, Bihar, Jharkhand, Odisha, Gujarat, Maharashtra, Madhya Pradesh	Late maturity; moderately resistant to major biotic stresses including leaf spot, downy mildew, stem borer and Pratylenchus zeae
MotiBajra	Telangana and Andhra Pradesh in rainfed as well as in irrigated areas	Tall, better tillering, dark-green foliage, short duration (56 days to 50% flowering) with multicut (3-4 cuts)
Napier Bajra Hybrid PBN 346	Punjab	Multi-cut, high in- vivo dry matter and also crude protein digestibility. Tall and broad leaves, more leafiness and high tillering ability

Improved Varieties/Hybrids of Pearl Millet for Punjab

PHB 2884: This hybrid is 230 cm tall and bears 2-3 productive tillers. It has long ears with average length of 28 cm and girth of 12 cm. Its grains are medium bold and slate in colour. It is resistant to downy mildew, ergot and smut. It matures in 88 days and average grain yield is 33 quintals per ha.

PHB 2168: This hybrid is about 210 cm tall, bears 2-3 productive tillers and matures in 83 days. It has medium long ears having an average length of 26 cm and girth 9 cm. Its grains are medium bold and slate in colour. It is resistant to downy mildew. Its average grain yield is 41 quintals per ha.

PCB 164: This is dual purpose composite variety having medium thick stalks and flexible stem with average plant height of 207 cm. It matures in 80 days. This variety has long cylindrical dense ears having 27-28 cm length and 8-10 cm girth. The grains are medium bold and light slate in colour. The average grain yield is 37.5 quintals per ha. Its fodder is of good nutritional quality. On an average, it yields 525 quintals per hectare of green fodder. It is highly resistant to downy mildew.

PHB 47: This thick-stalked, broad-leaved hybrid has delayed senescence and attains height of about 2 metres. Its ears are about 35 cm long, compact and profusely bristled. It bears 2 tillers and matures in about 85 days. The grains are medium bold and slate in colour. It is highly resistant to downy mildew and has slightly better tolerance to ergot and smut diseases. Under irrigated conditions, its grain yield is about 32.5 quintals per ha.

PHBF 1: It is a hybrid meant for green fodder. It flowers in 50 days. Its plants attain a height of 198 cm. It has succulent stem and multicut nature. It is better in tillering and has long and broad leaves. It is resistant to major diseases and pests. The fodder quality of this hybrid is better than FBC-16 in terms of dry matter intake, digestible crude protein and total digestible nutrients. The average green fodder yield of this hybrid is 640 quintals per hectare.

FBC 16: It is a composite variety exclusively meant for fodder production. It flowers 8-10 days later as compared to other varieties and hence provides green fodder for a longer period. Its plants attain an average height of about 235 cm and have long and broad leaves which remain green at maturity. It is comparatively resistant to major diseases. In terms of quality, it has higher voluntary dry matter intake and contains low amount of oxalates. Its average fodder yield is 575 quintals per hectare.

Climatic requirements

Being a rainfed crop, pearl millet is mostly grown in areas where annual rainfall is between 400 and 750 mm. It grows best in warm dry climates with adequate soil moisture. It can withstand drought and soil moisture stress to a great extent. During the vegetative growth of the crop, moist weather and light to medium rainfall is very useful, but at flowering and grain formation there should be no rain and the weather should be clear and dry. Continuous rains at flowering adversely affects the grain formation by causing poor seed setting. Besides, high humidity and low temperature at flowering increase the incidence of ergot disease leading to low yield. A temperature of 28-32ºC is considered to be optimum for the vegetative growth of the crop. High temperature at this stage induces early flowering and low temperature increases the ergot infection. It is, therefore, recommended to follow optimum sowing time to protect this crop from incidence of disease.

Soils

Pearl millet is successfully cultivated in different types of soils. Sandy loams to loamy soils with good drainage and free from alkalinity or salinity are the best suited for its cultivation. It can give fairly good yield on poor soils where other cereal crops can not be grown successfully.

Rotations

Some of the most important cropping systems are given below:

- Pearl millet – Wheat
- Pearl millet – Barley
- Pearl millet – Chickpea
- Pearl millet – Field pea
- Pearl millet – Potato
- Pearl millet – Potato – Wheat
- Pearl millet – Toria – Wheat
- Pearl millet – Wheat – Moong

Field preparation

In rainfed areas, the moisture conservation measures are extremely important and therefore tillage operations before and after seeding the crop assume great significance. In such areas, summer ploughing every year and deep ploughing once in 3-4 years are recommended. The deep ploughed soil is capable of releasing major nutrients and greater retention of moisture. Results of experiments (Malik *et al.,* 1973) clearly indicated the beneficial effect of deep ploughing (30-40 cm) over shallow (10-12 cm) ploughing for rainfed pearl millet. Ali and Prasad (1974) from field trials for two years studied the influence of type of seedbed and mulch on the yield and nutrient uptake by pearl millet under semi-arid conditions. A flat seedbed that stored more moisture, gave significantly higher grain yield and also increased the uptake of NPK compared with ridge and furrow beds.

Time of sowing

In rainfed areas the sowing of pearl millet should be done before the onset of monsoon. Sowing at the optimum time not only increases the grain yield but also reduces the chances of disease incidence. Under delayed sowing the rate of decline grain yield varies from 4 to 80 kg/day/ha under rainfed conditions. Hence, the crop should be sown at an optimum time of first fortnight of July. Sowing if delayed results in poor stand of the crop due to high rate of mortality of the seedlings, restricted vegetative growth of the crop, poor grain setting and higher incidence of diseases due to comparatively low temperature during the period.

Transplanting

Transplanting of pearl millet has been recommended if sowing gets delayed. The nursery sown in 500 m^2 is sufficient to transplant one hectare. Three weeks old seedlings are transplanted in August keeping a spacing of 50 cm between the rows and 10-12 cm between the plants. However, irrigation is necessary for transplanting. Wherever, irrigation facility is not available, seedlings can be transplanted on a rainy day.

Seed rate and plant population

Recommended seed rate is 4-5 kg/ha. The seed is sown by *kera* method (behind the plough) and row to row spacing is kept at 45 cm. After three weeks of sowing at the time of thinning, the plants in the rows are spaced at 10-12 cm. The optimum plant population is 1.75-2.0 lac per hectare. In areas with uneven distribution of rainfall, plant population of 1.6 lac per hectare should be maintained. Seeds of pearl millet should be sown about 2-4 cm deep preferably by seed drill operated with tractor or with plough or *kera* method.

Weed management

Many *kharif* weeds compete with pearl millet and may cause considerable yield reductions ranging from 25-50%. The damage is severe during 3-5 weeks after sowing. Therefore timely control of weeds is essential to get higher yields. Interculture the crop 3-4 weeks after sowing. Avoid deep hoeing near the base of the plants so that their roots are not damaged. A wheel hoe, *triphali* or hand hoe can be used for interculture. However, sometimes or soil being too wet to permit manual weeding or due to unavailability of labour timely weeding becomes difficult. Under such circumtances weeds can be effectively controlled with the use of herbicides. Pre-emergence application of Atrazine 50 WP at the rate of 0.25 to 0.50 kg ha^{-1} in 500 litres per hectare of water controls most of the grassy and broadleaf weeds.

Water management

Being a rainfed crop, pearl millet hardly needs any irrigation. Irrigation is applied only if there are no rains. Generally, two irrigations during the growing period of the crop are enough. If moisture is limiting, the crop must be irrigated at the earhead emergence stage because it is the critical stage for moisture stress. Pearl millet does not tolerate water logging. If the field is prone to water stagnation, drainage should be provided to drain away excess water.

Plant Protection

Insect pests

Shoot fly: Shoot fly is a serious pest of pearl millet in north India. Three weeks old plants are more prone to be attacked by the pest. Eggs are laid on the lower surface of the leaves or near the base of the plants. Larvae after hatching creep down the leaf sheaths and cut the central growing point of the plant which results in dead hearts. Early planting with the onset of monsoon is effective in controlling shoot fly incidence. Two dustings of Malathion (5% dust) at the rate of 25 kg/ha at 10 and 20 days after germination or two sprays of neem oil (in 4 g soap/litre water) or two sprays of endosulfan (0.07%) at 10 and 20 days after germination is recommended for controlling shoot fly.

White grub: The pest damages the root system of the young seedlings and as a result seedlings die. Intercropping of pearl millet with pulses like green gram, cluster bean and cowpea significantly reduces the white grub damage. Mixing of phorate or carbofuran at the rate of 12 kg/ha with pearl millet seed and applying in furrows at the time of sowing.

Midge: It is a serious pest of developing grains of pearl millet. The fly lay eggs in the flowers and the larvae develop on their ovaries, which shrivel and fail to develop into grains. For controlling midge 250 ml of Metacid (50 EC) or 625 ml of thiodine (5 EC) in 250 litres of water per hectare is sprayed on pearl millet crop.

Diseases

Green-ear or downy mildew: The green-ear disease or downy-mildew is caused by Sclerospora graminicola. The leaves of infected plants show discolouration, yellowing and whitening. Under humid conditions, the leaves are covered with a downy white growth of the fungus, which is prominent on the lower surface. The leaves turn necrotic and are torn into shreds. The ears of the infected plants are transformed wholly or partly into green heads of small, twisted, leafy structures. Adopt following integrated measures to control the disease :

- Grow downy-mildew resistant hybrids and composite variety.
- Rogue out the diseased plants early in the season to prevent secondary infection.
- Collect the diseased ears from the crop before harvesting and destroy them by burning.
- Practise five year rotation with other crops.

Grain smut: Smut is caused by Tolyposporium penicillariae. Individual grains in an ear get transformed into smut balls which may later burst open to release millions of spores which get disseminated and cause secondary infection on the portion of the ear which is enclosed by the sheath of the upper leaf. The intensity of the attack varies according to the humidity in the area. The following control measures are suggested :

- Remove the diseased ears early in the season and destroy them.
- Treat the seed before sowing with 3 g of Thiram/Captan per kg of seed to prevent the introduction of smut into new areas.

Ergot: This disease is caused by the fungus *Claviceps fusiformis*. At blossoming, a pinkish or light coloured fluid (honey dew) exudes from the spikelets in different parts of the ear. Later dark sticky patches appear on the ear and small dark-brown sclerotia appear in place of grains between the glumes. The seed set is poor or completely inhibited. The ovary is replaced by a fungal mass with many folds on its surface. The fungus perpetuates through the seed borne and soil borne sclerotia.

The contaminated grains, if fed to cattle or used by human beings can cause poisoning. Therefore, take the following precautions :

(i) **For cattle:** Do not feed the infected ears showing honey-dew symptoms to cattle. Even the stems and leaves of such plants are not safe as cattle feed. Cut and burn a badly affected crop to reduce the amount of inoculum.

(ii) **For human beings:** Immerse the grains in 10 per cent salt solution. The sclerotia, being lighter than normal grains, will float. Remove them with a sieve and burn them. Repeat the process two or three times.

Prevention of ergot: Once the disease appears, it is not possible to eliminate it. Take the following precautions to prevent its spread.

- Immerse seed in 10 per cent salt solution and remove the sclerotia and smut-balls by skimming. Then wash the seed in ordinary water and dry it thoroughly before treating it with fungicides, as recommended under Grain Smut.
- At the boot-stage, spray the crop with Cuman L 27% (ziram) using 1.25 litre in 250 litres of water per ha. Repeat 2 or 3 times at 5-day intervals.
- Burn the ears infested with honey-dew, as soon as they are observed in the field.
- After harvesting the crop, bury the debris with a furrow turning plough so that the ergot sclerotia rot in the soil.
- After threshing the ergot affected crop, the left-over-ear-heads of bajra in the threshing floor should also be burnt.
- Avoid sowing bajra next year in a field in which the crop had suffered heavily from ergot.

Rust: The rust pustules are small orange and full of spores. In later stages, black coloured pustules can be seen. In severe infection leaves are completely dried. Spray the crop with 0.2% Zineb or Mancozeb 75 WP three to four times to provide good protection.

Harvesting

Harvesting at physiological maturity or at a little later stage when grains are hard and contain 15-20% moisture is ideal. Harvesting is done by cutting the entire plant or removing the earheads first and cutting down the plants later. The earheads after harvesting should be sun-dried at the threshing floor for 10-20 days. Threshing is done thereafter either by beating the earheads with sticks or under the feet of bullocks. Tractor operated thrashers can also be used for this purpose. The threshed grain should be cleaned and dried in sun to bring down the moisture content of grain to 12-14% for safe storage.

Yield

Under irrigated conditions, 30-35 quintals of grain and about 100 quintals of dry stover can be obtained from one hectare. The rainfed crop gives 12-15 quintals of grain and 70-75 quintals of dry stover per hectare.

OATS (*Avena sativa*)

Oats panicle

Oats kernels

Introduction

Oats (*Avena sativa*) are an extremely versatile *rabi* cereal fodder crop of India. They can be grown in wide range of environments of diverse soil types for forages and as a companion crop. It has excellent growth habit, quick recovery after cutting and good quality herbage. It is a palatable, succulent and nutritious crop. They are a good source of protein, fibre, and minerals. On dairy farms oat fodder is a must, as it can be fed green and the surplus converted into silage or hay for use during the scarcity period. The oat crop is a high yielder with an average yield varying from 45 to 55 tonnes of green fodder per hectare. Oat remains an important grain crop for people in marginal environments throughout the developing world, while in developed economies it is meant for specialist uses. Livestock grain feed is still the primary use of oats crop.

Origin and history

Oats are a crop of Mediterranean origin, not as old as wheat and barley but their domestication dates back to ancient times. There are various school of thoughts regarding the exact timing and location for the domestication of oats. According to Thellung (1912) oats was domesti-cated after wheat and barley and perhaps it originated in Asia Minor. The ancient Greeks and Latins called oats '*Bromos*' and '*Avena*' respectively, but these names were commonly applied to wild species occurring as weeds among cereals (De Candolle, 1890). Denaiffe and Sirodot (1901) wrote "the derivation of the Latin word '*Avena'* remains somewhat obscure. It seems probable that it is from the Latin word *Aveo* (to desire), that is to say, forage desired by all animals". The exact time of introduction of cultivated oats into India is unknown. However, green oats were mentioned as a horse fodder in Ain 49 of Ain-I-Akbari, written by Abul Fazl (1590 AD).

Classification

The genus *Avena* is large and diverse and contains both wild and cultivated species of polyploid series with a basic chromosome number of n = 7. Three naturally occurring ploidy levels known within the genus are diploids (2n = 2x = 14), tetraploids (2n = 4x = 28) and hexaploids (2n = 6x = 42). Cultivated oat is believed to have originated from *Avena byzantina* or *A. sterilis.* The genus *Avena* comprises following species which are economically important:

I. *Avena brevis* (2n=14); the short oats, is grown in southern Europe, mostly for green fodder or hay due to its high yield of stem and leaf.

II. *Avena abyssinica* (2n=28); the Abyssinian oat, grows wild and is also cultivated in Ethiopia. It is grown mainly for fodder in several parts of north Africa.

III. *Avena byzantina* (2n=42); the red oat is grown around the Mediterranean regions, Europe and north Asia and warmer parts of the subtropical world for both grain and fodder. It is also cultivated in India.

IV. *Avena chinensis* (2n=42); the Chinese naked oat is grown extensively in hilly parts of china for its grain.

V. *Avena fatua* (2n=42); the wild oat is cultivated for grain, green fodder and other purposes in India and part of south-west Asia.

VI. *Avena ludoviciana* (2n=42); the wild red oat is grown for grains in Mediterranean and south-west Asian regions.

VII. *Avena sativa* (2n=42); it is the common cultivated oats sown in several parts of the world for grain, green fodder and other purposes.

VIII. *Avena sterilis* (2n=42); the fly oat is grown mainly for fodder in parts of Mediterranean region and south-west Asia.

IX. *Avena strigosa* (2n=42); the sand oat is cultivated for fodder in the Mediterranean region.

Improved Varieties of Oats for fodder released by ICAR

2017-18

Oat	Area of adoption	Salient features
Phule Surabhi	Punjab, Haryana, Uttrakhand, Rajasthan, UP, Jharkhand, WB, Odisha, Assam, Manipur, Bihar, Gujarat, MP, Maharashtra, Chhattisgarh, Karnataka, TN, Kerala and Telangana	Moderately resistant to leaf blight, root rot; less susceptible to aphids
OL 1804 (Single Cut)	West Bengal, Odisha, Jharkhand, Bihar, Eastern Uttar Pradesh, Manipur, Assam	Better nutritional quality, moderately resistant to leaf blight
OL 1802 (Multi Cut)	Maharashtra, Gujarat, Madhya Pradesh, Chhattisgarh, Central Uttar Pradesh	Better nutritional quality, resistant to leaf blight, Sclerotium root rot and leaf defoliators

2016-17

Forage oat	Area of adoption	Salient features
OL 10	Punjab	Multi-cut, high in- vivo dry matter and also crude protein digestibility. Tall and broad leaves with more leafiness and high tillering ability
Bundel Jai 2010-1	Andhra Pradesh, Karnataka and Tamil Nadu	Single cut; moderately resistance to Sclerotium rot, leaf blight and powdery mildew
Shalimar Fodder Oats 3	Jammu and Kashmir and Himachal Pradesh	Resistant to powdery mildew, Sclerotium root rot and leaf blight
JHO 2009-1	Uttar Pradesh , Maharashtra, Madhya Pradesh and Gujarat	High yielding with good quality and digestibility
UPO 06-1	Uttarakhand	Good yield and quality

Improved Varieties for Punjab

OL 11: It is a single cut variety with tall plants, leafy and profused tillering ability. Its fodder quality is superior to OL 9 and Kent. On an average, it yields about 612 quintals of green fodder and 21.25 quintals of seed per hectare.

OL 10: It is a multicut variety. Its plants are tall with profused tillering and leafy growth. The leaves are longer and broader. Its fodder quality is superior in terms of total digestible nutrients (TDN) and digestible crude protein (DCP) than OL-9 and Kent. On an average, it yields about 688 quintals of green fodder per hectare.

OL 9: It is tall, erect growing leafy variety suitable for single/two cuttings. Seeds are medium in size. It gives 575 quintals of green fodder per hectare. Its seed yield is 17.5 quintals per hectare.

Kent: It is tall, erect growing variety suitable for single/two cuttings. Seeds are bold. It gives 525 quintals of green fodder per hectare. It produces 20 quintals of seed per hectare.

Climate

Oats make best growth in cool moist climates and can be sensitive to hot, dry weather from head emergence through maturity. For these reasons, world oat production is generally concentrated between latitudes 35 – 65°N, including Finland and Norway, and 20 to 46°S. Oats require about 15-25°C temperature for its optimum growth. Frost and severe cold retard its growth, whereas hot and dry conditions hasten its maturity resulting in poor yield and inferior quality of fodder. They require more moisture to produce a given unit of dry matter than any other cereal except rice. They are grown in areas where the annual rainfall ranges from 38 to 114 cm, but most extensively where it does not exceed 76 cm and is well distributed.

Soil and soil preparation

Oats are grown on different soil types around the world except sodic and waterlogged ones. Next to rye (*Secale cereal*) oats are the most versatile of the cereals with regard to soil types. Almost any reasonably fertile, well drained soil is suited to oats if temperature and moisture conditions are favourable, however, fertile and well drained loam to clay loam soils are the best. The crop suffers badly due to lodging when heavily fertilized with nitrogenous fertilizers especially on poorly drained clay soils. It can tolerate slight acid and saline conditions but not above 8.5 pH. Under non-saline conditions oats are moderately tolerant to Na, but less tolerant than wheat and barley.

For field preparation, one deep ploughing followed by three or four harrowing and planking will serve the purpose to get friable and well levelled seed bed.

Sowing time

The normal sowing time starts from October to November. However, middle of October to middle of November is the optimum sowing time for higher production. Multi-cut varieties if sown earlier will help to tide over the scarcity period from middle of December to middle of January. Early seeding helps to ensure efficient use of available moisture.

Seed rate

There is seldom one optimum seeding rate because oat plants adjust their tillering and growth depending upon plant density. Seed should be drilled at the rate of 80-90 kg/ha for small seeded varieties, whereas it ranges 100-125 kg/ha for bold varieties. A moderate reduction in seeding rate seldom results in a yield reduction. This is due to more vigorous plants and the production of more tillers. Very high seeding rate results in increased competition, weaker culms and frequently increased lodging. For maximum yields, only good quality seed of recommended cultivars should be sown.

Spacing

In case of low tillering varieties spacing between rows should be kept at 20-25 cm apart. However, high tillering varieties should be sown at 25-30 cm apart rows.

Method of sowing

Sccd should preferably be drilled in one direction with straight rows. If seed drill is not available the sowing can be done either by *pora* or *kera* method with the plough.

Weed control

Herbicides are often used as a control measure. As a pre-emergence herbicide linuron at the rate of 0.5 kg/ha may be used to control associated weeds and this will cause 60-70 per cent reduction in weeds. Sodium salts of 2,4-D may be applied as post emergence herbicide at the rate of 0.75 kg/ha. Other herbicides selective for oats are pendimethalin and trifluralin.

Water management

Water requirement of crop depends on soil texture, prevailing climatic conditions etc. The oat crop needs copious irrigation for succulent and quality herbage production, though, it can be grown with acceptable level of yield even under limited irrigation. Productivity of oat greatly depends on availability of water during growing season and accordingly scheduling of effective irrigation technique is thus an essential factor, keeping in view its limited availability, cost and efficient utilization. A pre sowing irrigation is required for assured moisture in seed bed to ensure good germination and early establishment. A total number of 4-5 irrigations are required for oats. Generally, the first irrigation should be given 20-25 days after sowing. Subsequent irrigations may be provided at tillering, jointing, panicle primordial initiation, flowering and at milk stage. Judicious and timely application of irrigation water could check the leaching losses of nutrients and could also save the deterioration of soil by avoiding over irrigation. Further, during boot stage, irrigation is essential due to high water requirement at flowering stage.

Fertilizer management

To obtain maximum yields adequate nutrients must be present throughout the growing season. Oat plants will be less vigorous, have smaller leaves and panicles and will have less developed root system if nutrients are short in the first half of the growing season. In general, 60 kg of nitrogen per hectare should be applied and properly mixed in soil before sowing. Additional dose of 30 kg nitrogen per hectare should be broadcasted in two splits i.e. (10 kg and 20 kg) after first irrigation (at tillering stage) and at first cut, respectively. In light textured soils 40 kg of phosphorus per hectare should be applied at the time of sowing as it shows a favourable effect on shoot height and tillering during the earlier growth stages. Zinc sulphate may be applied at the rate of 10 kg per hectare in areas prone to zinc deficiency.

Plant Protection

Insect Pests

Oat aphid (Rhopalosiphum padi): Oat aphid appears on oats during January to March. It sucks sap from the plants and lowers down the quality of fodder. Coccinellid beetles, syrphid fly and heavy rainfall keep natural check on its multiplication.

Caution: Do not spray any insecticide on fodder crop as it is hazardous to animals and also kills the predators of the aphid.

Diseases

1. **Loose smut (Ustilago avenae):** The individual flowers of the oats panicle are replaced by spore masses. Treat the seed before sowing with 1 g of Vitavax 75 WP (carboxin) per kg of seed.
2. **Covered smut (U. kolleri):** The smut destroys the kernels completely and replaces them with black masses of spores which remain closed within the

affected spikelets till harvested. Rogue out and destroy the smutted ears. Treat the seed before sowing with Vitavax 75 WP (carboxin) @ 1 g per kg of seed.

Harvesting

For oats, number of cuttings depends upon the type of varieties and requirement of fodder for specific time and purpose. The single cut crop may be harvested at 50 per cent flowering to milk stage for better quality and quantity. In two cut management system, first cutting should be taken at flag stage (about 60 days) and last cutting at 50 per cent flowering to milk stage. For multicut crop the first cutting should be done at 55 days, second at 105 days and last at 50 per cent flowering stage. Delay in harvesting beyond milk stage may result in sharp decline in digestible protein. For satisfactory seed production, the crop should be left for seed after the first cutting to be taken at 50-55 days after sowing. First cut should be taken 8-10 cm above the soil surface for good regrowth.

For seed, oats are harvested when grain is in the hard dough stage and straw is slightly green (moisture content of the grain is 14% or less). If too ripe, shattering causes seed loss. Crop is usually cut with sickel or cutter and left in the field until dry and then threshed. In mechanized societies, oats are combined harvested. For this type of harve-sting, crop must be fully ripe, usually when the straw has lost greenness and glumes have become white. Seeds are cleaned by winnowing, and dried to below 14 per cent moisture for storage.

Yield

In case of only forage production plots, the green forage yield per hectare would be about 500-600 quintals. If crop is left for seed after the first cut then 30-35 quintals of seed and 25-30 quintals of straw per hectare may be obtained in addition to 250 quintals of green fodder of first cut.

9

CULTIVATION PRACTICES FOR OILSEED CROPS

Oilseed Crops

The important oilseed crops grown in India are groundnut, rapeseed and mustard, soybean, sesame, sunflower, castor, linseed, safflower etc. During 2016, the area under oilseed crops was 225.58 m ha and 26.63 m ha while production was 507.54 mt and 30.52 mt in the world and India respectively. (FAO STAT, 2016).

India is one among the major oilseed countries in the world accounting 14% of the area and 8% oilseed production. On the global basis, India ranks first in the production of Caster, Safflower, Sesame, Niger, second in groundnut and rapeseed/ mustard, third in linseed and fifth in soybean and sun flower.

GROUNDNUT (*Arachis hypogaea*)

Groundnut crop

Plant with pods

Groundnut (*Arachis hypogaea*) also known as peanut, monkeynut or moongfali occupies the first place among oilseed crops grown in India. Botanical name *Arachis hypogaea.* Groundnut is derived from Greek words Arachis meaning legume and hypogaea meaning below ground referring to formation of pods in the soil.

Origin and distribution

There are two main schools of thoughts on the origin of groundnut, one supporting the view that it had originated in Africa and other to south America. However, recent botanical survey have indicated south America as the most likely centre of origin of groundnut.

The spread of groundnut in some countries striking particularly in India and China. At present, groundnut is grown on a large scale in the tropical and sub tropical countries. Most important are India, China, USA, Nigeria, Indonesia, Brazil and Argentina.

Area and production

Groundnut is grown on an area of 27.66 and 5.8 m ha with production of 43.98 and 6.86 m tonnes in World and India respectively during 2016 (FAO STAT 2016).

Climate and soils

Groundnut is predominantly a tropical crop of semiarid and arid regions with a minimum rainfall of 500 mm during crop period. Variability in amount and distribution of rainfall is the most important factor limiting the productivity of rainfed groundnut. A temperature range of 25-30°C appears to be optimum, temperature below 20°C retard development and temperature above 35°C adversely affect flowering. Lower temperature initially retard, then prevent growth, and the plant is generally killed by frost at any stage.

Groundnut is grown on all types of soils such as sandy loam, sandy and heavy black soils. It is seen to thrive best on sandy loams. Heavy and stiff clay soils are not desirable as they tend to become hard during dry weather there by interfering with peg penetration into the soil and also makes the harvest extremely difficult. Groundnut is one of the most acid tolerant crops with a critical pH range of 5 to 5.5.

Tillage: It is considered sufficient to secure good surface to a depth of 12.5-18 cm for groundnut. Making ridges and furrows and raised beds is preferable to flat beds in areas receiving heavy rainfall, regions of shallow water table and saline soils.

Improved Varieties of Groundnut released by ICAR

2017-18

Groundnut	Area of adoption	Salient features
Central Groundnut ALG06-320	Tamil Nadu and Andhra Pradesh	Suitable for irrigated situation, tolerant to rust, LLS and peanut bud necrosis disease (PBND), S. litura, leaf miner and thrips

Kadiri Amaravathi	Andhra Pradesh	Suitable for rainfed situation, field tolerance to PBND
VRI 8	Tamil Nadu	Suitable for rainfed and irrigated condition, moderately resistant to sucking pest (jassids and thrips), LLS and rust
GJG 32	Tamil Nadu, Andhra Pradesh, Telangana, Karnataka, Kerala and Southern Maharashtra	Tolerant to stem rot, collar rot and rust

2016-17

Groundnut	Area of adoption	Salient features
G 2-52	Northern Karnataka	Resistant to rust and late leaf spot (LLS)
CO 7	Tamil Nadu	Resistant to rust
GJG 19	Odisha, West Bengal, Jharkhand and Manipur	Tolerance to stem rot, dry root rot and rust
Raj Mungfali 3	Odisha, West Bengal, Jharkhand and Manipur	Tolerant to Spodoptera litura, leaf hopper and thrips
Phule Warna	Gujarat, southern Rajasthan, southern Maharashtra, Andhra Pradesh, Karnataka and Tamil Nadu	Moderately resistant to rust and LLS
Phule Morna	Gujarat, southern Rajasthan, West Bengal, Jharkhand, Odisha, Manipur, Maharashtra, Andhra Pradesh, Karnataka and Tamil Nadu	Moderately resistant to rust and LLS
KCG 6	Southern Karnataka	Tolerant to rust and LLS
GKVK 5	Southern Karnataka	Tolerant to rust and LLS

Improved Varieties for Punjab

TG37A: it is an early maturing bunch type variety which is suitable for cultivation during spring season. It has 65 per cent shelling outturn with 100 kernel weight of 42.5 g and each pod bears 2-3 kernels. The kernels are spherical in shape and pink in colour. Its average pod yield is 30.75 quintals per ha. It has 48.6 per cent oil content. It has higher soluble sugar content (5.8%). It matures in 101 days.

SG 99: It is a bunch type variety. Its average pod yield is 25 quintals per ha and is tolerant to bud necrosis disease. The pods are medium in size with slight beak and moderate constriction. The pods are borne around the main root, which allows easy harvesting of crop with minimum pod losses. It has 66 per cent shelling outturn, 100 kernel weight is 54g and 52 per cent oil content. The kernels have light brown colour. It matures in about 123 days.

M 522: It is a semi spreading type variety. It yields about 22.5 quintals per ha of pods. The pods are medium bold in size with mostly two kernels per pod. The pods are borne towards the main root. It has 68 per cent shelling outturn, 100 kernel weight is 65 g and 51 per cent oil content. The kernels have light brown colour. It matures in about 120 days.

SG 84: It is recommended for sowing in spring as well as *kharif* seasons. It is a bunch type variety with profuse primary and secondary branches. During spring season, it matures in 130 days with pod yield of 25 quintals per ha while in *kharif* season, it matures in 120 days with pod yield of 20 quintals per ha. The pods are medium in size with one or two kernels, slight beak and medium constriction. The kernels are light brown. It has a shelling outturn of 64 per cent with a 100 kernel weight of 49 g and oil content of 50 per cent. The nutritive value of the straw has been adjudged comparable to that of berseem and cowpea hay.

Seed treatment and sowing

Seed treatment with Thiram (5g/kg seed or Indofil M-45 (3 g/kg seed), is essential against soil and seed borne diseases for adequate stand establishment. In general, mid June to July is the optimum seeding time for *kharif* Groundnut. For *rabi* groundnut, first fortnight of November is ideal time. Spacing of 30cm x 10cm for bunch types during *kharif* and 22.5 cm x 10 cm in *rabi* is ideal. A spacing of 30cm x 15cm and 22.5cm x 15cm is optimum for semi spreading type during *kharif* and *rabi* respectively. Recommended seed rates for bunch types are 150 and 180 kg/ha for *kharif* and *rabi* respectively. Semi spreading types require 120 and 150 kg/ha in *kharif* and *rabi* respectively. Bold seeded requires 20 kg/ha additional seed rate than small seeded varieties. Optimum depth of seeding is 5-7 cm in light soils and 4-5 cm in heavy soils.

Nutrient management

Recommended fertilizer schedule for rainfed groundnut is 20 kg N, 40 kg P_2O_5 and 50 kg K_2O_5 per ha in addition to about 10 t ha^{-1} FYM. For an irrigated crop, an additional dose of 10 kg N/ha is recommended 30 DAS. Recommended fertilizer schedule must be applied by placement at the time of seeding for rainfed groundnut. Gypsum 500 kg/ha is recommended at full bloom stage (30 DAS) both for rainfed and irrigated groundnut. Most common source of phosphorus for rainfed groundnut is single super phosphate. For an irrigated crop, DAP is the most commonly used fertilizer as it can meet the needs of both N and P.

Water management

Peak flowering and early pod development stage are most sensitive to soil moisture stress. There is no necessity for irrigation up to 25 DAS the groundnut. Two to three irrigations may be necessary depending upon the seasonal rainfall. Irrigation at flowering is very beneficial. If the rainfall during the preceding period is not adequate, apply one to two more irrigations, depending upon the time of recession of monsoon, during the pod formation period for proper development of pods. Another irrigation a few days before the harvest may be given for full recovery of pods from soil. Residue moisture can be used for sowing succeeding crops.

Weed management

In groundnut, critical period for crop-weed competition is up to 35 DAS. There should not be any disturbance to pegs 45 DAS by mechanical or field operations. Give two hoeing, three and six weeks after sowing to check the weeds. Weeds can also be controlled with pre-plant application of Basalin 45 EC at 0.75 kg a.i/ha or pre-em application of Lasso 50 EC at 2.5 kg a.i./ha after dissolving in 500 litre of water. Integration of one hoeing (45 DAS) with either pre-emergence application of Lasso 1.25 kg a.i./ha or Stomp 30 EC 0.75 kg a.i./ha or Treflan 48 EC at 0.75 kg a.i./ha as pre-plant application provides very effective control of hardy weeds like *Acrachne racemose Commelina benaalensis* etc. in groundnut.

Plant Protection

Insect Pests

Aphid: The pest becomes serious when the rainfall is low. It weakens the plant by sucking the cell sap, particularly from the growing points. It can be controlled by spraying 625 ml of Malathion 50 EC (malathion) or 375 ml of Rogor 30 EC (dimethoate) in 200 litres of water per ha, as soon as the pest appears.

White-grub: This insect is serious in some areas only. The adult beetles emerge from the soil during June-July with the first showers of rain. The eggs are laid in the soil and the larvae (grubs) hatching from them eat away the rootlets or root hairs of the groundnut plants. The damaged plants look pale, wilted and ultimately die. Adopt the following integrated approach for its effective control :

- Plough the field twice during May-June. It would help in exposing the beetles resting in the soil.
- Treat the seed before sowing with 12.5 ml of Dursban 20 EC (chlorpyriphos) per kg seed (kernels). For treating the seed, make a thin layer of kernels on tarpaul or floor and spray the insecticide with the help of a sprayer. Mix it properly and allow to dry. The seed can further be treated with the recommended fungicides.
- Sow the crop early wherever possible.
- Apply 32.5 kg Furadan 3 G (carbofuran) per ha in the soil at or before sowing.

Hairy Caterpillar: Hairy caterpillars, if appearing in an epidemic form, cause serious damage by feeding on the leaves and the tender stems. When young, they feed gregariously. The grown up caterpillars may migrate from one field to another. Adopt the following control measures:

- Use light traps for the destruction of moths.

- Young larvae are gregarious. They can be destroyed by plucking the infested leaves or by pulling out the infested plants and burying them underground.
- The grown up caterpillars can be destroyed by crushing them under feet or by picking and putting them into kerosenized water. If the population is high, control it by sparying 1.25 litre of Ekalux 25 EC (quinalphos) per ha in 250 litres of water with a manually operated sprayer.

Diseases

Collar-rot and seed-rot: This disease is casused by *Aspergillus niger* and other seed and soil borne fungi which may cause seed rot before germination or the germinating seedlings may develop rotting lesions at the collar region.

Tikka or Cercospora leaf-spot: This disease is caused by *Cercospora personata* and *C. arachidicola*. Disease develops rapidly at a temperature range of 25-30°C coupled with high relative humidity. Control the above diseases by integrated adopting the following measures :

- Select healthy and unblemished kernels for seed.
- Treat the kernels before sowing with 5 g Thiram (75%) or 3 g of Indofil M-45 (75%) per kg of kernels.
- Spray the crop with Sultaf 50 WP (wettable sulphur) @ 1.25 to 1.9 kg in 500 to 750 litres of water per ha. Give 3 or 4 sprays at fortnightly intervals, starting from the first week of August.
- Alternatively, spray the irrigated crop with Bavistin/Derosal/Agrozim 50 WP @ 125-150g in 250 litres of water per ha. Give three sprays at fortnightly intervals, starting when the crop is 40 days old.
- Do not grow groundnut repeatedly in same field.

Root knot: This disease occurs in patches and is caused by the nematode *Meloidogyne arenaria* and *M.javanica*. The affected plants show poor growth with chlorotic leaves. The root knots become stubby. The plants affected early in the season become stunted and bushy. Disease can be checked by exposing the soil to the sun during May and June to reduce nematode population. Practise green manuring wherever possible or add organic manures to the soil.

Harvesting

The optimum harvesting time is when good percentage of pods are well developed and is fairly intact with vines, the plant turns yellows and leaf shedding starts. Maturing of pods is determined by inside of the shell turning dark, karnel reaching its normal size and the taste having normal color of the variety. When around 75% of pods mature, harvest the crop.

At the time of harvesting, pods usually have moisture content around 40-50%. The pods are to be dried to about 10% moisture content for safe storage. Drying should be done rapidly to prevent fungal molding. Sun drying is usual method of drying. Summer groundnut should be dried in shade to prevent loss of seed viability, if it is for seed purpose.

SOYBEAN (*Glycine max*)

Soybean crop

Soybean pod

Soybean *(Glycine max)* is the major oilseed crop in the world accounting for nearly 50% of total oilseeds acreage as well as production. It provides approximately 60% of the vegetable protein and 30% of the oil in the world. Soybean ranks second in vegetable oil economy of India after groundnut. The cultivated soybean originated in China during 2800 BC. The eastern half of North China was believed to be the primary centre and Manchuria, the secondary centre of origin. From there, it spread to Korea and Japan. It was cultivated in Europe and USA in 17th century. It was introduced in India during seventies of 20th century. Soybean is primarily utilized as a source of protein and oil. Soybean seed contains 20% oil and 40-42% quality protein. Soybean oil is consumed as salad oil, shortening and margarine. The oil is used for preparation of frozen desserts, cookie shortenings, confections, icings, icecream coating, whipped toppings and coffee whiteners. Industrial uses of oil include soap, paints, resins and drying oil. A number of protein rich products, viz. soymilk, soyapaneer, soya sauce and soya flour, can be produced from seeds. The soybean oil comprises mainly polyunsaturated fatty acids like oleic and linoleic. Protein, which is in the range of 39-42%, is mostly globulin and partly albumins. Lysine is the major amino acid. The antinutritional substances present in the seed include trypsin inhibitor, hemaglutinins and oligosaccharides.

The cultivated soybean is an annual, generally exhibiting an erect, sparsely branched, bush-type growth habit with pinnately trifoliate leaves. Purple or white flowers are borne on short axillary racemes on reduced peduncles. The pods are either straight or slightly curved, usually hirsute. The one to three seeds per pod are usually ovoid to sub-spherical in shape. The seed coats range in colour from light yellow, olive green and brown to reddish black. It is a self-pollinated species propagated commercially by seed. It has complete flower, having 10 stamens arranged around the pistil. The stigma is receptive to pollen about one day before pollen is shed from anthers of the same flower.

Flower initiation is controlled by photoperiod, temperature and genotype. Soybean stem growth and flowering habit are of two types: indeterminate and determinate. In case of indeterminate, apical meristem continues vegetative activity during most of the growing season. The inflorescences are axillary racemes and pods are produced rather uniformly on the stem. The determinate type is characterized by vegetative development which ceases when the apical meristem becomes an inflorescence, both axillary and terminal recemes exist and pods are borne rather uniformly along the stem except for cluster of pods at the terminal racemes. The flowering period and the time of overlap of vegetative and reproductive growth is greater for the indeterminate than determinate type.

Distribution, area and production

Soybean was grown on 104.04 million hectares in the world to produce 172.64 million tonnes (FAO STAT, 2016). The USA tops the list in soybean acerage and production. The other major soybean producing countries are Brazil, China, Argentina, Indonesia, Canada, Paraguay and Italy. India, with 11.5 million hectares of soybean growing area and 14.01 million tonnes production during 2016 (FAO STAT 2016). The average yield in India is about 1218 kg/ha. It contributes 30-37% to India's oilseeds production. The crop showed an annual compound growth rates of 17.23, 18.19 and 1.33% in area, production and productivity between 1971-72 and 2006-07. Madhya Pradesh produces nearly 50% of total soybean production. Maharashtra is the second largest producer of soybean contributing 36% of total production. The other major soybean producing States are Rajasthan, Andhra Pradesh and Karnataka.

Climate and soil

Soybean can be grown in fairly wide range of climates and soils. Soils with a normal pH of 7 and a fair degree of water retention capacity are, however, better suited for its cultivation. It grows well in sandy loam to clay soil. The crop is highly sensitive to latitude because of its photosen-sitivity. The optimum temperature for rapid germination of soybean is approximately 30°C. It can be grown in rainy, spring and summer seasons.

Cropping systems

Soybean is ideally suited for fitting in various sequence

- Soybean-Wheat
- Soybean-Barley
- Soybean-Mustard
- Soybean-Pea
- Soybean-Lentil

Cultivation

Land preparation: Being a leguminous crop, soybean requires fine seed bed with least clods. Deep ploughing by soil turning plough during summer is desirable.

It should be followed by 2-3 cross harrowings or cultivator operations to ensure ideal seed bed in June. To avoid water stagnation, seed bed should be provided with little gradient.

Improved Varieties of Soybean released by ICAR

2017-18

Soybean	Area of adoption	Salient features
Raj Soya 24	Madhya Pradesh, Maharashtra, Rajasthan and Uttar Pradesh	Tolerant to major leaf, pod and root rot diseases, girdle beetle and semi looper attacks
Pant Soybean 24	Uttar Pradesh and Uttarakhand	Multiple disease resistant, free from lodging and shattering, tolerant to drought to some extent
Pant Soybean 21	Uttarakhand	Resistant to yellow mosaic virus (YMV) and bacterial pustule, tolerant to RAB, rainfed/ irrigated cultivation in plains and lower hills of Uttarakhand
Pant Soybean 23	Uttarakhand	Resistant to lodging and shattering, rainfed/ irrigated cultivation in plains and lower hills of Uttarakhand
Raj Soya 18	Madhya Pradesh	Resistant to YMV and charcoal rot, erect plant type suitable for intercropping

2016-17

Soybean	Area of adoption	Salient features
SL 958	Punjab	Light yellow oval seeds with black hilum; suitable for timely sown irrigated areas; resistant to yellow mosaic virus (YMV) and soybean mosaic virus (SMV)
MACS 1281	Southern Maharashtra, Karnataka, Telangana, Andhra Pradesh, Tamil Nadu	Round, yellow seeds and black hilum; moderately resistant to bacterial pustules, bacterial leaf blight
JS 20-69	Madhya Pradesh	Medium-sized spherical yellow and shiny seeds with black hilum; resistant to YMV, charcoal rot, bacterial pustules, Alternaria leaf spot, pod blight, Indian bud blight, target leaf spot
VL Soya 77	Uttarakhand Hills	Large seeds, yellow seeds with black hilum; suitable for rainfed organic conditions; moderately resistant to frog -eye leaf spot, pod blight
VL Bhat 201	Uttarakhand Hills	Black and large seeded; resistant to frog-eye leaf spot, girdle beetle

Improved Varieties for Punjab

SL 958: It has shining, light yellow coloured grains with black hilum. Its grains contain 41.7% protein and 20.2% oil. It is highly resistant to yellow mosaic virus and soybean mosaic virus. It takes about 142 days to mature. Its average yield is about 18.25 quintal grains per ha.

SL 744: It has shining, light yellow coloured grains with grey hilum. Its grains contain 42.3% protein and 21.0% oil. It is resistant to yellow mosaic virus and soybean mosaic. It takes about 139 days to mature. Its average yield is about 18.25 quintal grains per ha.

SL 525: It has uniformly bold, shining, cream coloured grains with light black (grey) hilum. Its grains contain 37.2% protein and 21.9% oil. It is resistant to yellow mosaic virus and tolerates stem blight and root-knot nematode. It matures in about 144 days. Its average yield is about 15.25 quintal grains per ha.

Planting time: Photoperiodic response dictates the time of planting for maximum yield of soybean. Early planting gives too great canopy to produce maximum yield, whereas late planting results in canopy development too restricted for maximum yield. Soybean planting dates depend on the variety and the agroecological region. Obviously, photo-periodic effects, occurrence of drought and flood decide the date of sowing of the crop. Sowing time should be adjusted in such a way that one can escape from these natural calamities.

In India, soybean is mostly a rainy season crop, sown from last week of May depending upon the climatic conditions

Seed rate, spacing and plant population: Seed rate depends on seed size (test weight), variety, germination percentage and date of planting. Generally, 75-80 kg seeds/ha is recommended. Row spacing of 30-45 cm and plant-to-plant spacing should be 5 cm. The ideal plant population varies from 0.4 to 0.6 million/ha.

Method of sowing: Line sowing is the best method of sowing as it facilitates hoeing/interculture operations easily. Moreover, stand of crop will be uniform under medium to heavy textured soils, bed planting is preferred as there will be minimum or no loss due to water staganation as compared to flat sowing. This crop is very shy is germination when there is a rainfall after sowing and with bed planting technique, its germination will be nromal.

Depth of planting: A seed depth of 2.5 to 4 cm is optimum. Shallower depth may be justified to crust-prone areas and deeper in sandy soils. Some cultivars have inhibitors of hypocotyl elongation and for them shallow depth is recommended. Under conditions of optimum moisture and heavy soils, the crop should be sown at 2-3 cm depth, whereas for optimum moisture, medium to light textured soils, it should be planted at 3-4 cm depth. Planting can be done with the help of seed drills or with tractor-operated planters.

Seed treatment and seed inoculation: In order to get rid of seed-borne diseases, seeds must be treated with Thiram or Captan @ 3 g/kg seed. Moisten the seed of one acre with water amd mix thoroughly one packet of culture with it and let it dry in shade. Sow the seeds immediately. Grain yield will increase upto 8%. For higher yield, seed inoculation with *Bradyrhizobium japonicum* culture is recommended.

Manures and fertilizers: Manure and fertilizer application must be decided depending on soil-test value, productivity potential, expected yield, cropping system, management and method of fertilizer application. In addition to 10 tonnes FYM/ha, application of 20 kg N, 60-80 kg P_2O_5, 20 kg each of K_2O and S is recommended.

The FYM must be applied at the time of land preparation. All the fertilizers must be applied at the time of sowing. Zinc deficiency is a problem in high rainfall uplands and under sodic soils. Under conditions of northern plains, application of 5 kg Zn/ha through $ZnSO_4$ is recommended in addition to recommended doses of NPK and S.

Weed management: Normally, weed control in soybean is accomplished using mechanical intercultivation (hoeing) or by the use of herbicides. Hand weeding is preferred under small holding situations. Proper rotation may also reduce weed intensity and flora. To avoid the competition during early growth stage, soybean field should be kept free of weeds for the first 30-40 days after sowing. Two manual weedings, 20-25 and 40-45 days after sowing are sufficient for control of weeds. During rainy season, due to continuous rains, the hand weeding may not be possible at proper time. Under such circumstances, use of herbicides has been found very effective. Fluchloralin or Trifluralin 1 kg/ha as preplant incorporation or Alachlor 2 kg/ha or Pendimethalin 1 kg/ ha or Metolachlor 1 kg/ha or Clomazone 1 kg / ha as pre-emergence or Imazethapyr 100 g /ha or Quizalofop ethyl 50 g /ha as post-emergence (15-20 DAS) in 500-600 litres water/ ha.

Water management: One of the most frequent and severe environmental limitations on soybean seed yield is water supply and drainage of excess water from the field. Soybean is often cultivated during rainy season, but supplemental irrigation becomes necessary if larger dry spells occur. Soil moisture stress at seed emergence and pod filling stages are most sensitive. Soil moisture deficit shows relatively less loss of grain yield as compared to other crops, because soybean has an extended period of flowering. The early flower drop which is caused due to moisture stress could be compensated by the late flowers maturing into pods.

Avoid flooding of soybean fields at any stage of the crop. Flooding at floral initiation is detrimental because it retards nodule and overall root activity, which is most crucial at this stage. In case of *kharif* crop, irrigation is generally not needed and it is grown rain fed provided the rains are well distributed, otherwise the crop needs water at critical growth stages. However, during summer, the crop can be grown only under assured irrigation and it needs around 5-6 irrigations.

Water stress at the end of flowering and pod initiation is most critical because it leads to flower and pod abortion. Stress during the grain filling reduces the seed size. Therefore, it is essential to apply sufficient water to the field for better germination and then to irrigate the crop before pod development and filling stages for obtaining higher grain yield.

Plant Protection

Insect Pests

Jassid, Aphid, Whitefly and Flea beetle: These insects can be controlled by spraying the crop with 940 ml Malathion 50 EC (malathion) or 625 ml Rogor 30 EC (dimethoate) using 200-250 litres of water per ha with a manually operated sprayer.

Whitefly can also be controlled by spraying 100 g Actara 25 WG (thiamethoxam) using 200-250 litres of water per ha. Repeat the spray at 10 days interval, if necessary.

Hairy caterpillar : The body of the caterpillar is covered with hair. The caterpillar eats the green matter of leaves, leaving behind only the midribs. The crop may be totally denuded due to severe attack. When young, they feed gregariously but on few plants in scattered spots. Adopt the following control measures :

- Young larvae are gregarious. They can be destroyed by pulling out the infested plants along with larvae and burying them underground.
- The grown-up caterpillars can be destroyed by crushing them under feet or by picking and putting them into kerosenized water. If the population is high, spray 1.25 litre of Ekalux 25 EC (quinalphos) in 200-250 litres of water per ha with a manually operated knapsack sprayer.

Semilooper: The larvae are green in colour measuring 2-4 cm in length. When touched, they form a loop. The larvae feed extensively on the leaves of the crop. In case of severe damage, the plants are totally defoliated within a few days.

Tobacco caterpillar *(Spodoptera litura)* : It is a polyphagous pest. The small larvae are black whereas grown up larvae are dark green with black triangular spots on body. Its moth lays eggs in masses covered with brown hairs on the lower side of leaves. After hatching, first and second instar larvae feed gregariously and skeletonize the foliage. Later on, the grown up larvae disperse and feed singly. Besides leaves, they also damage buds, flowers and pods.

- **Mechanical control:** Egg masses and young larvae of tobacco caterpillar feeding gregariously should be collected along with leaves and destroyed.
- **Chemical control:** Semilooper and Tobacco caterpillar be controlled by spraying 375 ml Rimon 10 EC (novaluron) or 2 kg Asataf 75 SP (acephate) or 3.75 litres Dursban 20 EC (chlorpyriphos) using 250 litres of water per ha with a manually operated knapsack sprayer.

Spray the crop as soon as the pest appears in the field and repeat after 10 days, if necessary.

Blister beetle: Blistle beetle is diurnal and general feeder. Adult beetles are robust with bright black and red stripes on the forewings. When disturbed, the beetles emit a fluid containing cantharidin that causes blisters on human skin. The adult beetles attack *arhar*, *moong*, *mash* and other pulse crops. The major damage is caused at the flowering stage. They feed on tender buds and flowers of the plant, thus preventing grain formation.

Blister beetle can be controlled by spraying Decis 2.8 EC (deltamethrin) @ 500 ml or Kingdoxa 14.5 SC (indoxacarb) @ 500 ml per ha using 200-250 litres of water with manually operated knapsack sprayer. Spray the crop on pest appearance in the evening hours and repeat the spray after 10 days, if necessary.

Pod borer *(Helicoverpa armigera)*: The larvae damage the crop by feeding on leaves, flower buds, flowers, pods and seeds in the pods, thus causing heavy loss in yield. The larvae may be pale green, yellow, brown or black in colour measuring about 3-5 cm in length when fully grown. Larval presence can be observed from damage to plant and from dark green faeces below the plants on the soil. The larvae fall on the ground when plants are shaken vigorously. Spray the crop at the appearance of larvae with Kingdoxa14.5 SC (indoxacarb) @ 500 ml or Asataf 75 SP (acephate) @ 2 kg or Tracer 45 SC (spinosad) @ 150 ml in 250 litres of water per ha using a knapsack sprayer at the start of pod formation and repeat after two weeks, if necessary.

Mite: The mite causes webbing on the underside of the leaves which turn pale. Such infested leaves turn light-brown to dark reddish brown. It can be controlled by spraying the crop with 375 ml of Rogor 30 EC (dimethoate) in 200-250 litres of water per ha.

Diseases

Yellow mosaic virus : It is a viral disease and is transmitted by whitefly. Disease appears as a blend of yellow and green patches on the infected leaves. Infected plants bear a few pale pods. Grow resistant varieties SL 958, SL 525 and SL 744. Follow the control measures against whitefly as given under *moong*.

Harvesting and threshing

When pods become golden yellow, black or gray, the crop is ready for harvest. At harvest, moisture content of the seed should be 15-17%. The maturity period ranges from 90 to 140 days depending on the varieties. When the holdings are small, manual harvesting close to the ground is followed.

Use of reapers and combine harvesters is common on large area for soybean harvesting. Threshing can be done either with the mechanical soybean thresher or some conventional methods used in other legumes.

Yield potential

With good crop management, seed yield of about 1,600-2,000 kg/ha under rainfed conditions and 2,000-2,500 kg/ha under irrigated conditions can be expected. Under high input management, yield levels of 3,000 kg/ha can also be realized.

RAPESEED AND MUSTARD

Raya (*Brassica juncea*)

Gobhi sarson (*Brassica napus*)

Rapeseed and mustard belongs to genus Brassica of the family Cruciferae. The oilseed Brassicas comprises four species, *B. compestris* (*B. rape*), *B. juncea* (Indian Mustard), *B. napus* (winter and spring rape) and *B. carinate* (Ethiopian mustard).

In india, Brassicas are the dominant winter seasons oilseeds and include:

- *Brassica juncea,* Brown mustard (*raya* or laha)
- *Brassica. compestris* var *sarson* (yellow mustard)
- *Brassica. compestris* var dichotoma (brown *sarson*)
- *Brassica. compestris* var *toria* (*toria* or lahi)
- *Brassica. napus* (*gobhi sarson*)
- *Errica sativa* (taramira, rocket salad)

In trade, *sarson*, *toria* and taramira are known as rapeseed, and *raya* as mustard.

Origin and distribuion

Historically, Brassica species are under cultivation in Asian countries, particularly China and Indian sub continent. Yellow *sarson* grown in eastern Indian exhibits largest diversity in Bengal. It is quite probable that its primary center of origin is North West India.

Area and production

Rapeseed and mustard occupied an area of 5.76 million hectare with total production of 6.8 mt in India during 2016 (FAO STAT 2016).

Climate and soils

Rapeseed and mustard are of the tropical as well as of the temperate zone crops and require relatively cool temperature for satisfactory growth and yield. In India, they are grown in *rabi* when the temperature is low. Brassica grow well in area receiving 350-550 mm rainfall. Sarson and taramira are preferred in low rainfall areas, whereas raya and toria are grown in medium and high rainfall areas. It requires temperature between 25°C and 28°C with optimum around 20°C. Seed oil content is relatively high at ambient temperature of 10°C to 15°C. Brassica can be grown on a wide range of soils such as alluvial, red loams and black soils. Raya can come up well on all types of soils. Toria does best on loam to heavy soils. Sarson is well suited to light loam soils and taramira is mostly grown on very light soils.

Improved Varieties of Rapeseed and Mustard released by ICAR

2017-18

Indian mustard	Area of adoption	Salient features
CS-58	Haryana, Punjab and Uttar Pradesh	Salinity tolerant
Pant Rai 21	Uttarakhand	Alternaria blight tolerant
Yellow sarson	**Area of adoption**	**Salient features**
Pant Sweta	Uttarakhand	For irrigated condition
Taramira	**Area of adoption**	**Salient features**
Jobner Tara	Rajasthan, Haryana, Punjab, UP, Gujarat, Delhi, Uttarakhand and Maharashtra	Suitable for rainfed situation
Jwala Tara	Rajasthan, Haryana, Punjab, UP, Madhya Pradesh, Gujarat and Delhi	Suitable for rainfed situation

2016-17

Rapeseed-mustard	Area of adoption	Salient features
Pant Rai 20	Plains of Uttarakhand	Suitable for timely sowing in both irrigated and rainfed areas
PBR 357	Jammu, Punjab, Haryana, parts of Rajasthan and Delhi	Suitable for irrigated timely sown conditions
RGN 298	Rajasthan, Punjab, Haryana, Delhi and Jammu	Suitable for rainfed conditions
GM3 (Gujarat Mustard3)	Gujarat	Tolerant to high temperature and salinity
Pusa Double Zero Mustard 31	NCR region of Delhi	Low erucic acid and low glucosinolate
RLC 2	Punjab	Low erucic acid
PBR 378	Punjab	Suitable for timely sown rainfed areas

Gujarat Dantiwada Mustard 5	Punjab, Haryana, Jammu, Rajasthan and Delhi	Suitable for rainfed conditions
Raj Vijay Mustard 1	Madhya Pradesh	Suitable for early sowing
JK Samriddhi Gold	Uttar Pradesh	Suitable for timely sown irrigated conditions
JK Pukhraj	Uttar Pradesh	Suitable for timely sown irrigated conditions
Bayer Mustard 5450	Uttar Pradesh	Suitable for timely sown, irrigated conditions
RLC 3	Punjab	Low erucic acid and low glucosinolate
Toria	**Area of adoption**	**Salient features**
Sushree	Odisha	Suitable for late sown condition and non-lodging type
TL 17	Punjab	Short duration; fits in paddy-wheat cropping system
Ethiopian mustard	**Area of adoption**	**Salient features**
BJC 1	Punjab	Frost tolerant

Improved Varieties for Punjab

Canola is an internationally accepted nomenclature for Brassica varieties having less than 2% erucic acid in the oil and less than 30 micro moles glucosinolates per gram defatted meal. Oil rich in erucic acid is not desirable for edible purposes as it causes thickening of arteries and leads to heart problems. Similarly, the defatted meal of non canola varieties when used as animal feed reduces appetite, reproductivity and affects thyroid activity leading to thyroid associated health problems. Elimination of long chain erucic acid in the canola varieties is accompanied by increase in the proportion of desirable MUFA (oleic acid) from 10-20% to about 60-65%. The oil from canola varieties is healthy oil for human consumption. The defatted meal from such varieties is specially suited as animal feed. Among different varieties of rapeseed-mustard recommended by the PAU, raya variety RLC 3 and gobhi sarson varieties GSC 6 and GSC 7 and hybrid Hyola PAC 401 are of canola type.

Toria

TL 17: This variety is more suitable for multiple cropping systems due to its early maturity. It matures in 90 days. It has profuse branching and pod bearing capacity. Its average yield is 13.0 quintals per hectare. Its seeds contain 42.0 per cent oil content.

TL 15: This variety takes about 88 days to mature. Because of its early maturity it is suitable for toria-wheat rotation. It yields 11.25 quintals per hectare. Its oil content is 41 per cent.

Raya

Giriraj: This variety is recommended for general cultivation in the state under timely sown and irrigated conditions. This is a very bold seeded variety and has tolerance to terminal heat stress. It is moderately resistant to white rust. Its average yield is 19.25 quintals per hectare and matures in 144 days. It contains 40.3 percent oil content.

RLC 3: This is the first canola quality variety of raya in the country. It is a yellow seeded, medium tall variety which is recommended for general cultivation in the state under timely sown irrigated conditions. It is resistant to white rust. Its average seed yield is 18.25 hectares per acre with 41.5 percent oil content. It matures in 145 days.

PBR 357: This variety is recommended for general cultivation in the state under timely sown irrigated conditions. It is a medium tall variety with more number of branches per plant. It is bold seeded variety, moderately tolerant to Alternaria blight and white rust. Its average yield is 21.25 quintals per hectare. It contains 39.0 per cent oil content and it matures in 145 days.

PBR 97: This variety is recommended for sowing under rainfed conditions in the state. It has profuse branching and long main shoot with dense pods. The grains are medium bold with oil content of 39.8 per cent. On an average it yields 13 quintals per hectare. It matures in 136 days.

PBR 91: This variety is recommended for cultivation in South-Western region of Punjab under timely sown and irrigated conditions. This variety has profuse branching habit. It has broad, thick, dark green leaves. Its seeds are bold and dark brown in colour having 37.6 per cent oil. Its average yield is 20.25 quintals per hectare. It matures in 145 days.

RLM 619: This variety is recommended for general cultivation in the State for timely and late sown conditions under irrigated as well as rainfed conditions. It is very suitable for sowing after cotton in cotton belt of the Punjab. It possesses bold seeds and has oil content of 43 per cent. It has comparatively greater resistance to white rust, Alternaria blight and downy mildew. Its average yield is 20 quintals per hectare. It matures in 143 days.

Gobhi Sarson

GSC 7: This canola quality variety is recommended for general cultivation in the state under timely sown irrigated conditions. It is a medium tall variety which is free from white rust and is tolerant to Alternaria blight. It has lusterous brownish- black seeds. Its average yield is 22.25 quintals per hectare with 40.5 per cent oil content. It matures in 154 days.

GSC 6: This canola quality variety is recommended for cultivation in the state for timely sowing under irrigated conditions. It has lustrous bold seeds. Its average yield is 15.25 quintals per hectare, with oil content of 39.1 per cent. It matures in 145 days.

Hyola PAC 401: This canola quality hybrid is recommended for general cultivation in Punjab under timely sown irrigated conditions. Its seeds are brownish black and lustrous. It contains 42 per cent oil. The average seed yield is 16.75 quintals per hectare. It is medium statured and matures in 150 days.

GSL 2: This is the first ever atrazine herbicide resistant variety of gobhi sarson in India suitable for atrazine based weed management. It is recommended for cultivation under timely sown and irrigated conditions in Punjab. Its leaves are dark green and plants have comparatively close canopy. It has tall plants with stiff stem and as such does not lodge. On an average, it yields 17 quintals per hectare. Its oil content is 44.5 per cent with superior oil quality. It takes about 160 days to mature.

GSL 1: This variety is recommended for general cultivation in the state under irrigated conditions. Its leaves are thick, smooth and sweet in taste and give quality saag. Its initial growth is slow and plants remain in vegetative phase till early February. As such, the crop escapes frost injury. Plants are stout and do not lodge easily. Its yield is 16.75 quintals per hectare. The seed possesses 44.5 per cent oil. It takes about 160 days for maturity.

African sarson

PC 6: This variety is the world's first determinate oilseed Brassica variety. It is recommended for general cultivation in the state under timely sown irrigated conditions. This medium tall variety is resistant to pod shattering and is suitable for combine harvesting. It is free from white rust and has better tolerance to Alternaria blight and mustard aphid. Its average yield is 19.25 quintals per hectare with 40.1 per cent oil. It matures in 157 days.

Taramira

TMLC 2 (1990): This variety is recommended for cultivation in Bathinda, Sangrur, Ferozepur and Kandi areas of Hoshiarpur, Gurdaspur, Ropar and Shaheed Bhagat Singh Nagar districts. It has longer main shoot length, more number of pods on main shoot and more seeds per pod. It gives an average seed yield of 7.25 quintals per hectare with 36.6 per cent oil content. It matures in 150 days.

HYOLA PAC 401: This hybrid is recommended for general cultivation in Punjab. It is medium statured and matures in 150 days. Its seeds are brownish black and lustrous. It contains 42% oil content .The average yield is 16.85 quintals per hectare.

Land preparation

Brassica has small seeds and therefore, for good germination and stand, a fine tilth is required. Cloddy fields lose the stored soil moisture rapidly and since most oilseed Brassica in India are grown on stored soil moisture, the formation of clods must be prevented. Generally, 3-4 ploughings or harrowing followed by planking gives a good fine tilth.

Date of sowing

The sowing time of rapeseed mustard begins from the end of August and goes on to end of November. Taramira is sown from mid September to end of October.

Seed rate

For sole crop, optimum seed rate 5 kg/ha for all the Brassicas. Profuse branching compensates for missing plants or low plant population. Seed rate as mixed or intercrop depends on the desired proportion of brassica crop relative to the component crop. It usually ranges from 1 to 2 kg/ha.

Method of sowing

Brassicas are usually broadcasted and the field planked after sowing to give a good contact between seed and moist soil for optimum germination. Low seeding also permits intercultivation for minimizing weed infestation. The recommended spacing is 30 × 10 cm and 4 to 5 cm depth of sowing. Gobhi sarson should be sown in rows, 45 cm apart with plant to plant distance of 10 cm.

Nutrient management

Nitrogen requirement of raya or mustard ranges from 20 to 160 kg/ha depending on soil fertility and sowing conditions i.e. rainfed or irrigated. Under dry land conditions Brassica respond to 20-60 kg N/ha, depending on soil moisture availability during crop season. Recommended fertilizer for rainfed Brassica is 40 kg N, 20 kg P_2O_5/ha. For irrigated toria, 40 kg N, 20 kg P_2O_5 and 15 kg K_2O/ha is optimum. Recommended rate of fertilizer application for irrigated sarson is 60 kg N, 30 kg P_2O_5 and 20 kg K_2O/ha. For irrigated raya apply 80 kg N, 40 kg P_2O_5 and 25 kg K_2O/ha for optimum yield. All the fertilizer should be applied as basal dose at seeding by drilling. Among the field crops brassicas have highest requirement for sulphur. In general 40 kg/ha of sulphur is optimum under different situations.

Irrigation

Flowering and pod formation stages are most sensitive to soil moisture deficits. If irrigation water is not limiting factor, two irrigations at 60 and 90 DAS are ideal for optimum yield. If there is scope for one irrigation only, it should be given 60 DAS. Among the Brassicas, raya is most responsive to irrigation.

Weed management

Brassicas are fast growing crops and suppress the weed growth. Critical period for crop-weed competition is upto 30-45 DAS. Dinitroaniline herbicides such as trifluralin and fluchloralin are very safe for these crops and isoproturon can be applied as post-em on raya (0.75 kg/ha) and gobhi sarson (0.56 kg a.i/ha) only.

Plant Protection

Insect Pests

Painted bug (*Bagrada hilaris*): It is serious on the germinating crop in October and again on mature crop in March-April. The nymphs and adults suck the sap from the foliage and pods which subsequenty dry up. It can be managed by:

- Give first irrigation 3-4 weeks after sowing as it reduces the painted bug population significantly.
- Spray 1 litre Malathion 50 EC (malathion) or 625 ml Ekalux 25 EC (quinalphos) in 150-200 litres of water per hectare.

Mustard sawfly (*Athalia lugens proxima*): The larvae attack the young crop, bite holes into the leaves and may eat all the leaves in case of heavy attack. It can be controlled by spraying insecticides as recommended against painted bug.

Mustard aphid (*Lipaphis erysimi*): Cold and cloudy weather is very favourable for the development of mustard aphid. The green plant lice become innumerable, covering the inflorescence and pods. They suck the plant sap in huge quantities and as a result, the plants remain stunted, pods shrivel up and seeds do not develop. The following pest management programme is recommended for its effective and economic control:

Sow the crop early wherever possible, preferably upto 3rd week of October.

- Apply recommended dose of fertilizers.
- Apply insecticides on the basis of need by following any of the economic threshold levels given below. For determining aphid infestation, observe 12-16 widely scattered plants from an acre twice a week starting from first week of January.
- Spray, when pest population reaches, 50- 60 aphids/10 cm terminal portion of the central shoot. or
- When an average of 0.5 to 1.0 cm terminal portion of central shoot is covered by aphids. or
- When plants infested by aphids reach 40-50% (observe 100 plants/acre).

Use any of the following insecticides and repeat application whenever the above mentioned economic threshold values are noticed.

a. Foliar sprays: Use any of the following insecticides in 200-300 litres of water per hectare depending upon the stage of crops.

Actara 25 WG (thiamethoxam)	100 g
Metasystox 25 EC (oxydemeton methyl)	1kg
Rogor 30 EC (dimethoate)	1kg
Ekalux 25 EC (quinalphos)	1kg
Malathion 50 EC (malathion)	1kg
Dursban/Coroban 20 EC (chlorpyriphos)	1.5 L

b. Granular insecticides : Broadcast 10 kg Thimet/Phorolik 10 G (phorate) or 22.5 kg Furadan 3 G (carbofuran) per hectare followed by light irrigation.

Leaf miner (*Chromatomyia horticola*): Larvae feed by making mines into the leaves and cause heavy damage. The systemic insecticides (Metasystox 25 EC or Rogor 30 EC or granular insecticides) recommended for control of mustard aphid should be used for controlling the leaf-miner.

Hairy caterpillar (*Spilosoma obliqua*) and cabbage caterpillar (*Pieris brassicae*): The caterpillars feed on leaves, young shoots and green pods. When young, they feed gregariously but the grown-up caterpillars migrate from one field to the other.

When in the gregarious stage, they can be easily controlled by picking and destroying the infested leaves. Grown-up caterpillars should be controlled with Malathion 5% dust @ 37.5 kg per hectare or by spraying with 500 ml of Nuvan 85 SL (dichlorvos) in 200-300 litres of water per acre with manually operated knap sack sprayer.

Green peach aphid (*Myzus persicae*): The green peach aphid attacks taramira from December to March with peak activity during February. After the appearance of inflorescence, the aphid congregates on terminal buds and feeds there. As a result there is flower shedding, poor pod formation and shrivelling of grains. For its control, sow the crop in first week of October and spray 500 ml Rogor 30 EC (dimethoate) or 625 ml Metasystox 25 EC (oxydemeton methyl) in 250 litres of water per hectare in 3rd week of February when aphid starts congregating on top flower buds. Only one spray is needed.

Note:

- Generally the mustard aphid becomes serious around mid-January, hence greater vigilance is needed from first week of January for monitoring its population at regular intervals for deciding the need and time of control measures to be followed.
- Spray in the afternoon when the pollinators are less active.
- The crop meant for Saag purpose should be sprayed with Malathion only. Observe a waiting period of one week after spraying to pluck saag.

Diseases

Alternaria blight (*Alternaria brassicae*): Brown to blackish spots with concentric rings appear on leaves, pods and stems. In case of severe attack, the upper parts of the stem and pods wither and contain shrivelled seeds. Temperature in the range of 15-25°C and relative humidity of more than 75% favour the disease development. Destroy diseased debris from the previous crop. Spray the crop with Blitox (copper oxychloride) or Indofil M-45 (mancozeb) @ 625 g in 250 litres of water per hectare at 75 days after sowing followed by second spray of Score 25 EC (difenconazole) @ 250 ml 15 days after first spray and third spray of Blitox 50 WP (copper oxychloride) or Indofil M-45 (mancozeb) @ 625 g 15 days after second spray in 250 litres of water per hectare at 15 days interval.

Downy mildew (*Peronospora brassicae*): Small light green lesions appear on the lower side of the leaves, which later enlarge, become more greyish, dry up, shrivel and tear easily. All aerial parts of the plants are attacked. The disease is severe on foliage and seed bearing parts. Greater deformity occurs in the stem. High relative humidity of more than 90% with temperature in the range of 10-20°C are conducive for disease development and spread. Destroy diseased debris from the previous crop. Spray the crop three to four times at 15 days interval with 625g of Blitox 50 WP (copper oxychloride) or Indofil M-45 (mancozeb) in 250 litres of water per hectare starting at about 75 days old crop.

White rust (*Albugo candida*): Prominent white creamy-yellow, scattered pustules appear on the under surface of the leaves. The swelling of affected parts often occurs. Flowers get malformed, become sterile and club shaped. All parts are attacked, except roots. Weather conditions of 13-25°C temperature and more than 75% relative humidity favour the disease development. Adopt same control measures as for downy mildew.

Phyllody (*Mycoplasma Like Organisms*): Plant parts are malformed. Numerous shoots arise and the plants become bushy or broom shaped. Floral parts are transformed into leafy structures. Avoid early sowing. Rogue out the affected plants

Harvesting

Brassicas are harvested when the stems and pods turn yellow, the seeds become dark and they rattle inside the pod when shaken. Toria which takes 90-100 days to mature, is the earliest crop to be harvested from mid-December to mid January depending on seeding time. Raya takes 110 to 150 days for maturity and the harvesting is from February to March. Yellow sarson take 120 to 160 days and brown sarson 105 to 140 days for maturity. These crops should be harvested at proper time as these are highly prove to shattering.

LINSEED (*Linum usitatissimum*)

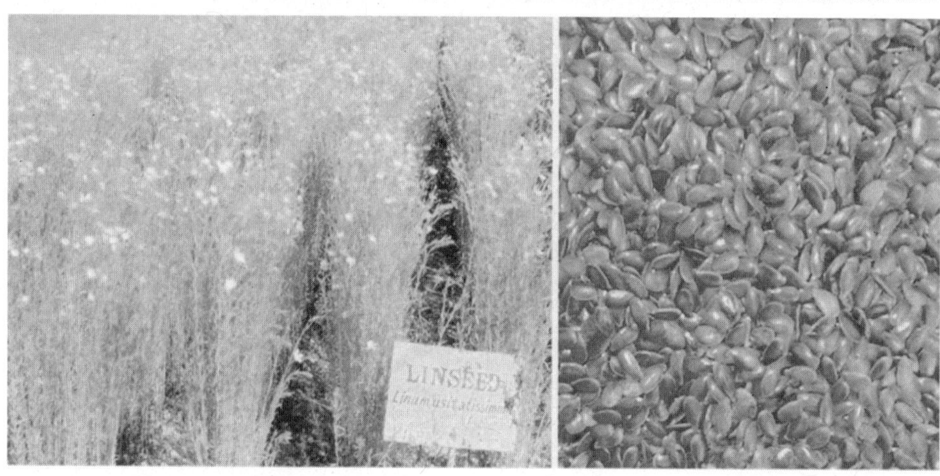

Crop **Seed**

Linseed or seed flax (*Linum usitatissimum*) is cultivated both for seed and fibre. In India it is mainly grown for seed to extract oil, linin, cloth made of linseed is resistant to moisture as compared to other fibres.

Origin, distribution, area and production

Linseed a native of India, Ethiopia or Iran. In India it is cultivated over an area of 293 thousand ha with a production around 125 thousand tones during 2016 (FAO STAT 2016). It is mainly growm is Madhya Pradesh and Uttar Pradesh is India.

Climate and soil

Linseed is a cool season crop, It needs about 25-30°C temperature during germination and vegetative phase and 15°C - 20°C during seed formation. It can tolerate soil acidity but the ideal soil pH is 6. Optimum time for its growing is October to mid November.

Improved Varieties of Linseed released by ICAR

2017-18

Linseed	Area of adoption	Salient features
Indu	Uttar Pradesh	Suitable for irrigated areas, resistant to rust, powdery mildew and bud fly
Uma	Uttar Pradesh	Suitable for rainfed condition, tolerant to wilt, Alternaria blight and rust
Priyam	Punjab, Haryana, Himachal Pradesh and Uttar Pradesh	Suitable for rainfed and moderate fertility condition during rabi season, high yield and oil content (41.3%), highly resistant to bud fly and rust
JLS-66	Madhya Pradesh	Short in height, maturity 107-114 days, moderately resistant to powdery mildew, alternaria blight, rust and major insect pests, oil content 40.5% and yield potential 2,200 kg/ha

2016-17

Linseed	Area of adoption	Salient features
Kota Barani Alsi 3	Rajasthan	Resistant to rust, moderately resistant to Alternaria blight, powdery mildew, bud fly
Kota Barani Alsi 4	Bundelkhand region of Uttar Pradesh, Madhya Pradesh and Rajasthan	Moderately resistant to Alternaria blight, powdery mildew; suitable for rainfed situation
PKV-NL 260	Maharashtra	Moderately resistant to powdery mildew and bud fly
Chhattisgarh Alsi1	Chhattisgarh	Moderately resistant to linseed bud fly, Alternaria blight, resistant to powdery mildew; suitable for rainfed situation
Divya	Himachal Pradesh, Punjab, Jammu and Kashmir and Haryana	Resistant to rust, moderately resistant to Alternaria blight, wilt and bud fly
JLS 79	Madhya Pradesh	Resistant to rust, moderately resistant to wilt, powdery mildew, bud fly
Arpita	Odisha	Resistant to powdery mildew and wilt

Improved Varieties for Punjab

LC 2063: It is a tall variety with profuse branching and blue flowers. It is tolerant to wilt, rust, Alternaria blight and moderately resistant to powdery mildew. It has lustrous brown bold seeds with 38.4 per cent oil content. Its average yield is 12.25 q/hectare. It matures in about 158 days.

LC 2023: It is recommended for both rainfed and irrigated conditions. It is a tall variety having blue flowers and profuse capsule bearing. It has brown medium sized seeds. It is tolerant to wilt, rust, Alternaria blight and moderately resistant to powdery mildew. Its seed contains 37.4 per cent oil. Its yield is about 11.25 q/hectare. It matures in 158 and 163 days under rainfed and irrigated conditions respectively.

Nutrient management

Rainfed : 40 kg N : 20 kg P_2O_5 : 20 kg K_2O per ha

Irrigated : 80 kg N : 40 kg P_2O_5 : 30 kg K_20 per ha

Under irrigated conditions, ½ dose of N should be applied along with Phosphorus and Potassium as basal dose at seeding by placement method. Top dressing of the remaining half of N should be applied at 35 DAS, preferably at first irrigation. These fertilizer doses depends upon soil fertility and climatic conditions.

Cropping system

Usually intercropped with Chickpea, Wheat, Lentil and Potato.

Water management

Flowering and grain development stages are most critical for irrigation.

Weed management

Critical period for crop-weed competition in linseed is between 20-30 DAS. Two manual weedings, 3 and 6 weeks after seeding can keep linseed crop free from weeds. Weeds can also be controlled with the pre or post- emergence (30-35 DAS) application of isoproturon at 0.94 kg/ha.

Plant Protection

Insect Pests

Lucerne caterpillar: This insect may cause damage by feeding on the leaves. Spray the crop with 1125 g of Sevin/Hexavin 50 WP (carbaryl) or 1 litre of Malathion 50 EC (malathion) in 200-250 litres of water per ha.

Diseases

Rust (*Melampsora lini*): Pink fungal sori appear on the surface of leaves, stems and pods. Grow resistant varieties. Dust the crop with sulphur @ 17.5 kg per hectare or spray it with Indofil Z-78 (zineb 75% @ 1.25 kg in 625 litres of water.

Wilt (*Fusarium oxysporum*): Young seedlings die when attacked. Full grown plants show yellowing of leaves and later wilt and die. Grow tolerant varieties.

Powdery Mildew (*Erysiphe cichoracearum*): A greyish white powdery growth on the youngest growing tips is the first visible symptom of this disease. In case of severe infection, the fungus attacks branches, leaves and flowers. It causes heavy loss owing to the defoliation of infected plants which leads to the shrivelling of grains. Dust the crop with sulphur @ 17.5 kg per hectare once before flowering.

Quality considerations

Linseed contains antinutritinal compounds, therefore, not fit for edible purpose. Linseed oil is used for paint related industries. It contains 20-34% protein, 37-42% oil and 15-29% carbohydrates.

SESAME

Sesame/sesamum (*Sesamum indium*) known as til, is the most ancient oilseed crop known and used by man. It is most ancient oilseed crop of India next to groundnut and brassica oilseeds.

Origin and distribution

The greatest weight of evidence indicates the Ethopian area as the origin of cultivated sesame, although there is considerable argument in favour of Afghanistan, with subsidiary centres in Indo-Malaysia and China. Major sesame producing countries are India, China, Sudan, Mexico, Ethiopia, Sri Lanka and Burma. In India, major sesame cultivating states are Gujarat, Rajasthan, Madhya Pradesh, Uttar Pradesh, Punjab, Tamil Nadu.

Area and production

Sesame occupied an area of 1.9 m ha with total production of 8.0 lac tones during 2016 (FAO STAT 2016).

Climate and soils

Sesame requires fairly hot conditions during growth for optimum yield. A temperature of 25°C to 27°C is ideal for rapid germination, initial growth and flower formation. Low temperature at flowering results in pollen sterility and premature flower drop

Sesame grows on a variety of soils but thrives best on those, which are moderately fertile with good internal drainage. Soils with neutral reaction are preferred. It will grow well on soils with a pH ranging from 5.5 to 8.0.

Tillage

Land preparation for sesame is similar to as that for wheat, sorghum and other millets. For *Kharif* crop, land is deep ploughed with summer shower in April-May. With the onset of monsoon rains, two to three harrowings are given to break up the clods. *Rabi* crop requires shallow land preparation as the soils are frequently moist from monsoon rains.

Seasons: In India, sesame is grown during *Kharif* as well as in *rabi* season. In Punjab sesame is grown as *Kharif* season crop. It is sown in second half of July.

Improved Varieties of Sesame released by ICAR

2016-17

Sesame	Area of adoption	Salient features
DS 5	Karnataka	Medium maturity; large and white-seeded
PKV-NT11	Maharashtra	White-seeded, tolerant to phyllody, bacterial blight diseases; suitable for summer
Gujarat JunagadhTil 5	Gujarat	Suitable for rainfed situation
LT 8 (Punjab Til No. 2)	Punjab	Tolerant to Cercospora leaf blight and phyllody diseases

Improved Varieties for Punjab

Punjab Til No. 2: This variety yields 7.0 quintals per ha. It has profuse branching and dense pod bearing. Pods arranged opposite to each other are long and non- hairy. Its seeds are white, bold and have 49% oil content. Seeds have less crude fibre, are soft and better in palatability. This variety is tolerant to phyllody and cercospora leaf blight. It matures in 90 days.

RT 346: This variety yields 6.5 quintals per ha. It has profuse branching, long non-hairy capsules arranged alternately. It has white, bold seeds which contain 49% oil. It is moderately resistant to Antigastra capsule borer. It matures in 87 days.

Seed rate

For broadcasting, recommended seed rate is 6 and 5 kg/ha under rainfed and irrigated conditions respectively. For row seeding, recommended seed rates are 5 and 4 kg/ha under rainfed and irrigated conditions respectively.

Spacing

Spacing between rows and within plants depends on on variety, season and management practices. It varies from 30 × 10 to 45 × 15 cm under different situations. Recommended spacing under Punjab conditions is 30 to 45 × 15 cm.

Method of seeding

Sesame is usually sown by broadcasting in most situations. However, row seeding with seed drills leads to higher yield due to optimum stand establishment and inter-cultivation for weed control. Optimum seeding depth is 3 to 4 cm.

Nutrient management

Recommended fertilizer dose for sesame under Punjab conditions is 30 kg N per hectare and there is no need of phosphorus and potassium application. Entire dose of fertilizer is applied at seeding.

Water management

Flower initiation and capsule filling are most sensitive stages for water deficits. If irrigation water is not a limiting factor, three to four irrigations results in highest seed yield.

Weed management

Sesame is sensitive to weed competition during the first 15-25 DAS. Two weedings 15 and 35 DAS are optimum to minimize weed growth. Pre-emergence application of alachlor 50 EC at 1.5 kg/ha provides effective control of weeds in this crop.

Plant Protection

Insect Pests

Sesame leaf webber and capsule borer: The larvae of these insects roll the leaves and feed inside these rolls; or bore into the capsule and feed on developing grains. If the infestation takes place at early stage, the plants die. It can be controlled by spraying the crop twice (first at pest appearance and then at flowering stage) with 250 ml Sumicidin/ Fenval/Agrofen 20 EC (fenvalerate) or 375 ml Decis 2.8 EC (deltamethrin) or thrice (at pest appearance, flowering and pod formation stages) with 500 ml Ripcord 10 EC (cypermethrin) in 250 litres of water per ha.

Jassid: It causes considerable damage to the crop by sucking the sap and transmitting mycoplasma, which induces the malformation of the inflorescence. It can be controlled by spraying the crop with 1000 ml of Malathion 50 EC (malathion) in 250 litres of water with manually operated sprayer at least two times at 2-3 weeks intervals.

Diseases

Phyllody: It is caused by a Mycoplasma like organism (MLO). The flowers are modified into leaf like structures and do not bear pods. It is transmitted by Jassid (*Orosius albicinctus*). Therefore, control of jassid is required. Also rogue out the diseased plants to prevent further spread.

Blight *(Cercospora sesami)*: It appears at flowering stage as dark brown, angular lesions with grey centre on leaves. It also appears on petiole, stem and pods. The diseased plants give blighted appearance followed by defoliation. To control the disease, spray the crop thrice with Bavistin 50 WP* @ 250 g per acre in 250 litres of water at 10 days interval, just at the appearance of the disease.

Harvesting

Depending on the variety, sesame crop takes 80-150 days for maturity. The crop is harvested when the leaves, stems and capsules begin to run yellow and the lower leaves start shedding. Average grain yield under rainfed conditions as sole crop, range between 5-8 q/ha.

SUNFLOWER (*Helianthus annuus*)

Crop Plant Flower

Sunflower (*Helianthus annuus*), which used to be an ornamental plant in India, has in recent times become an important source of edible oil. Globally, sunflower ranks second to soyabean among annual field crops grown for edible oil.

Origin and distribution

The genus name for sulflower is derived from the Greek Helios, meaning "Sun" and anthos meaning flower. Sunflower turn with the sun until anthesis after which the head face east. Sunflower is a native of southern USA and Mexico from where it was taken to Spain before the middle sixteen century. The major sunflower producing countries are USA, Argentina, Bulgaria, India, Turkey and South America.

Area and production

Sunflower is a relatively new crop to India and cultivated area in the country is 4.0 lac hectare with a production of about 3.3 lac tonne during 2016 (FAO STAT 2016).

Climate and soils

Sunflower is basically a crop of temperate regions but adaptable to both cold and high temperature conditions. The crop require cool temperature during germination and seeding growth and warm weather during flowering and seed development. Optimum ambient temperature regime is 20° to 35°C but can tolerate up to 30°C. Sunflower crop is sensitive to water-logging even for short period.

Sunflower can be grown on a variety of soils ranging from sandy loams to heavy clay soils. Deep soils with good internal drainage are ideal for rainfed crop. It grows well on soils with moderately alkaline reaction. Sunflower requires well prepared seedbed for optimum seed germination, stand establishment and growth.

Improved Varieties of Sunflower released by ICAR

2017-18

Sunflower	Area of adoption	Salient features
Kaveri Champ	Punjab, Haryana, Bihar, West Bengal and Odisha	Tolerant to downy mildew and leaf hoppe
Prabhat (Hybrid)	Andhra Pradesh	Tolerant to downy mildew
PDKVSH-952 (Hybrid)	Maharashtra	Rainfed and irrigated area in kharif and rabi; moderately resistant to powdery mildew

2016-17

Sunflower	Area of adoption	Salient features
Sunlight	Karnataka, Maharashtra, Tamil Nadu, Andhra Pradesh	Medium maturity
PSH 996	Punjab	Medium maturity; moderately resistant to downy mildew and charcoal root rot
Phule Bhaskar	Maharashtra	Early maturing; moderately resistant to necrosis, powdery mildew
PSH 1962 (Hybrid)	Punjab	Medium maturing
RSFH 1887 (Hybrid)	Karnataka	Medium maturing; tolerant to viral necrosis, Alternaria leaf spot diseases

Improved Varieties/Hybirds for Punjab

Hybrids

PSH 1962: It is a short duration, medium tall hybrid with average plant height of 165 cm. The average seed yield of this hybrid is 20 quintals per hectare. The seeds of this hybrid are black and bold with 100 seed weight of 6.4 g. The oil content of this hybrid is 41.9 per cent. It matures in 99 days.

DK 3849: It is a tall hybrid with an average height of 172 cm. The average seed yield of this hybrid is 21 quintals per hectare and 100 seed weight is 4.5 g. The seeds of this hybrid contain about 34.5 per cent oil. It matures in 102 days.

PSH 996: It is a short duration medium tall hybrid with an average plant height of 141 cm. The average seed yield of this hybrid is 19.5 quintals per hectare. The seeds are black and bold with 100 seed weight of 6.8 g. The oil content of this hybrid is 35.8 per cent. It matures in 96 days. This hybrid is also suitable for growing under late sown conditions.

PSH 569: It is a short duration medium tall hybrid with an average plant height of 162 cm. The average seed yield is 18.5 quintals per hectare. The seeds are bold with 100 seed weight of 6.8 g. The seeds contain 36.3 per cent oil content. It matures in about 98 days. This hybrid is also suitable for growing under late sown conditions.

PSFH 118: It is a medium tall, short duration hybrid with an average plant height of 155 cm. This hybrid fits well in the late sown conditions. The average yield of this hybrid is 19 quintals per hectare. It contains 40.5 per cent oil. It matures in about 98 days.

SH 3322: It is a long duration medium tall hybrid with an average plant height of 160 cm. Its average yield is 20.75 quintals per hectare. It contains 43.0 per cent oil. It takes 120 days to mature.

Sowing time

Sunflower can be grown although the year, provided soil moisture/ irrigation is not limiting factor. Sowing time should be decided in such a way that flowering and seed development stages do not coincide with continuous rains or temperature above 38°C. Its sowing is done both in *Rabi* as well as in *Kharif* season in India according to conditions.

Kharif: In light soils - Mid June to mid July

Rabi (rainfed) – First fortnight of September to first fortnight of October.

Rabi (irrigated) – November to December

Spring season - Second fortnight of January to end of February.

Seeds rate and method of sowing

For quick germination and optimum stand establishment, especially under rainfed conditions, the seed should be soaked in fresh water for about 14 hrs allowed by shade drying. The seed should be treated with Thiram or Captan at 2-3 g/kg to protect from seed born diseases.

Line sowing, using seed drills are the best method of sowing. The seed is sown is furrows behind the country plough. Hand dibbling in plough furrows is also common. Optimum sowing depth is 5 cm. Recommended seed rates are given below:

Cultivars	Rainfed	Irrigated
Varieties	8-10 kg/ha	6-7 kg/ha
Hybrids	5-6 kg/ha	4-5 kg/ha

Spacing

In India, optimum row spacing vary from 45 to 60 cm. Depending upon soil types indicated below:

Soil type	**Spacing (cm)**
Light	45 × 20
Medium	45 × 30
Heavy	60 × 30

Wider row spacing of 60 cm is preferred for hybrids and long duration varieties. For short duration and dwarf varieties a row spacing of 45 cm is adequate.

Fertilizer requirement

Sunflower crop responds to applied nutrients effectively. Application of FYM at 10 t/ha is very beneficial.On an average fertility soils add 60 kg N, 30 kg P_2O_5 for its successful cultivation. Apply ½ N and whole of P_2O_5 and K_2O at the time of sowing and remaining nitrogen 30 days after sowing. Single superphosphate is the best source of phosphorus supply. Sulphur is essential for increasing yield and oil content of seed. In sulphur deficient soils, addition of 25 kg /ha of Sulphur can significantly increase the yield.

Water management

Sunflower crop has deep root system, extending beyond 2 m depth especially on deep soils. Sunflower crop is sensitive to water logging even for a short period. Flower opening stage is the most sensitive for soil moisture stress followed by seed filling and bud initiation stages. Sunflower has the highest water requirement during anthesis stage.

Weed management

Critical period of crop-weed competition in sunflower is 30 to 45 DAS (days after sowing). Two inter-cultivations/hoeings at 15 days interval commencing from 25 DAS can effectively check the weed growth. Earthing up also help in smothering weeds Pre-emergence herbicides like Lasso at 2.0 kg a.i./ha or Stomp 30 EC at 0.75 kg a.i./ ha are very effective.

Plant Protection

Insect Pests

Cutworms (*Agrotis spp.*): The insect may be serious during March-April in fields where sunflower follows potato. Caterpillars cut the seedlings at the ground level. Sow the crop on ridges to avoid cutworm damage in the germinating seedlings. Where flat sowing is practised, apply 5 litres of Dursban/Radar 20 EC (chlorpyriphos) per hectare before sowing. The insecticide should be mixed in 25 kg fine soil and broadcast uniformly in the field after last ploughing but before planking.

Tobacco caterpillar (*Spodoptera litura*) and Bihar hairy caterpillar (*Spilosoma obliqua*): These are serious during April-May and defoliate the plants. Young larvae feed gregariously but the grown-up caterpillars spread throughout the field and migrate from one field to the other.

To control these pests, egg masses and young larvae feeding gregariously should be collected along with leaves and destroyed. Grown-up caterpillars should be controlled by spraying 500 ml Nuvan/Divamp/Vipona 85 SL (dichlorvos) in 250 litres of water per hectare. Repeat the spray after two weeks, if necessary. It will also control cabbage semi looper (Thysanoplusia orichalcea) jassid and other sucking pests in case they appear on the crop.

Head borer or American bollworm (Helicoverpa armigera) : It is a serious pest of sunflower and causes heavy damage by feeding on the tissues and developing grains in the head. Spray 2.5 kg of Sevin 50 WP (carbaryl) or 2 kg of Asataf 75 SP (acephate) or 2.5 litre of Dursban 20 EC (chlorpyriphos) in 250 litres of water per hectare at the initiation of starbud stage. Repeat after two weeks, if necessary. Spraying should be done either in early morning or late in evening when the bees activity is minimum.

Diseases

Stem rot (*Sclerotium rolfsii, Sclerotinia sclerotiorum*): The pathogens attack basal part of the stem. However, when S. sclerotiorum is predominant, it can cover the entire plant parts, including the head as evident by the white cottony growth. The sclerotia of this pathogen, almost of the size of sunflower seed, are produced in abundance which are visible in the pith of rotten stem and infected head. The affected tissue turns black and shredded. Warm (12-20$_o$C) and moist weather conditions favour the disease development. Treat the seed with Thiram (tetramethyl thiurum disulphide) @ 2 g/kg seed.

Root rot or Charcoal rot (*Macrophomina phaseolina*): Infection at early stages kills the seedlings rapidly. At later stages, it causes premature ripening, small sized heads, poor filling and reduction in yield. Avoid moisture stress to the crop.

Head rot (*Rhizopus oryzae, R. stolonifer, Sclerotinia sclerotiorum*): The symptoms are visible on any part of the receptacle, usually starting from the stalk. Severely affected heads show shredding of the tissue and incomplete filling. The rot may be partial or complete. Seeds formed in affected heads are highly shrivelled causing considerable reduction in yield and oil content. Spray the crop twice with 625 g Indofil M-45* (mancozeb) in 250 litres of water per hectare at 15 days interval starting at about 60 day old crop.

Downy mildew (*Plasmopara halstedii*): Plants remain severely stunted, leaves become thick with green and yellow patches visible from upper side. White cottony shining downy growth of fungus appears on lower surface of the leaves. Leaves become necrotic and plants develop small and empty heads. Treat sunflower seeds with metalaxyl 35% WS @ 6g per kg seed. Fungicide should be applied by making slurry in water and seeds should be dried before sowing.

Harvesting

Sunflower crop matures in 90-100 days. The crop has to be harvested when the lower side of the head turns yellow, some of the bracts dry up and bottom leaves start drying and withering. Mature heads are cut with a sickle and dried in sun for 2-3 days.

SAFFLOWER (*Carthamus tinctorius*)

Safflower (*Carthamus tinctorius*) is commonly known as kusum. It has been cultivated in India either for its orange red dye (Carthamin) extracted from brilliantly coloured foretes and oil purpose.

Origin and distribution

Vavilov suggested three centres of origin of Safflower (*Carthamus tinctorius*) i.e. India, Afganistan and Ethiopia. Major countries producing safflower are India, Mexico, Ethiopia, USA, Argetina and Australia.

Area and production

Safflower occupied 104.8 thousand hectare area with total production 53.0 thousand tonnes during 2016 (FAO STAT, 2016).

Climate and soils

Safflower is a *rabi* oilseed crop in India. Crop is tolerant to low temp-eratures at seedling and vegetative stages but sensitive at elongation, flowering and post flowering stages. It is drought resistant and its water requirement are very low. Safflower cannot withstand excessive soil moisture or humidity at any stage. Seedling stage is highly sensitive to soil salinity.

Tillage

Tillage is very important to keep the weeds under check and to pulverize the soil for better crop emerged.

Varieties

Some of the improved varieties.

Annigiri, Bhima, Malavya, Manjira, Nira, Co 1, Sagarmutyalu, Sharda, Swetha, A 300, JSF 7, K 1, S 144, Girna etc.

Sowing time

Safflower is a *rabi* oilseed crop both under rainfed and irrigated conditions. Optimum time of sowing for Saflower is Sept/October.

Seed rate and spacing

Seed rate vary from 7.5 to 12 kg/ha in South India. Optimum spacing is 45 × 20 cm and at depth of sowing is 5 cm. Over night soaking of seed in water will ensure better germination.

Nutrient management

Nutrient requirement by safflower is

N = 40 kg/ha   P = 25 kg/ha (P_2O_5)

Drill all the fertilizer at sowing.

Water management

Safflower is a drought resistant crop. Flowering and seed filling stages are most critical for soil moisture stress. In deep soils, two irriga-tions (presowing and flowering are adequate for optimum yield. Flowering is very sensitive stage and stress at this stage adversely affects seed yield.

Weed management

First 60 days after planting are critical for competition from weeds. Two inter cultivations at 25-30 and 45-50 days after sowing in between the rows are optimum for keeping the crop free from weeds. Pre-emergence herbicides such as alachlor 2.0 kg/ha, fluchloralin 0.70 kg/ha or Stomp 30 EC 0.75 kg/ha (0.75-1.0) may be used for weed management.

Plant Protection

Insect pests

Two species of aphid attack safflower crop from January to March.

Green peach aphid (*Myzus persicae*): It feeds on the central whorl and the plants show stunted growth and burnt appearance.

Safflower aphid (*Uroleucon compositae*): Feeds on leaves, tender shoots and stems of the plant. The infested plants become weak, remain stunted and sometimes dried up.

For the control of aphid, spray 250 ml of Rogor 30 EC (dimethoate) or Monocil 36 SL (monocrotophos) or 625 ml Coroban 20 EC (chlorpyriphos) in 250 litres of water per hectare and repeat the spray after 15 days, if necessary.

Diseases

Wilt and head rot (*Sclerotinia sclerotiorum*): Plants become yellowish, turn brown and ultimately die. Large black sclerotia of the fungus are formed on the crown, inside the stem, floral heads and adjoining roots. Shredding of the stem takes place. For its effective control, destroy the crop debris. Take the seeds from healthy crop. Do not pile up soil around stem during rains. In heavily infested soils, grow non-host crops.

Harvesting and threshing

Safflower crop matures in 4-5 months and it should be cut with sickles at the base and stocked in the field in the form of small and well pressed heaps until they are fully dried. Threshing can be done by power operated threshers. On an average it yields 12-15 q/ha of seed.

Quality consideration

Its seeds contain 28-32% oil (traditional safflower seeds) while thin hull types contain 35-45%. Protein content ranges from 11 to 24%. Safflower oil is valued for its high linoleic acid.

CASTOR (*Ricinus communis*)

Castor (*Ricinus communis*) plays an important role in vegetable economy of the country.

Origin

Castor is a native of east Africa, especially Ethiopia.

Area and production

India ranks first both in the area and production of castor. Mainly it is grown in A.P. & Gujarat and productivity is highest in Haryana. In India, castor is grown in 8.7 lac ha area with production of 1.55 m tones during 2016 (FAO STAT 2016).

Climate

Castor being warm season crop, grown in temperate and tropical regions of the world. Castor crop require moderately high temperature of 20°C-26°C with low humidity throughout the growing season for maximum yield. Castor crop is very sensitive to high temperature and excessive soil moisture.

Seeds treatment and seed rate

Seeds may be treated with Thiram or Bavistin (3 g/kg) to protect from seed borne alternaria blight and wilt. Ideal time for seeding *Kharif* castor is immediately after the receipt of monsoon rains. Optimum seed rate is 10-12 kg/ha. For improved castor varieties/hybrids recommended spacing is 90×90 cm for irrigated crop and 75×75 cm spacing for rainfed crop. Under delayed sowing plant population should be increased.

Improved Varieties of Castor released by ICAR

2017-18

Castor	Area of adoption	Salient features
GNCH 1	Gujarat	Possesses high level of resistance in sick plot and through root dip method
YRCH-2	Tamil Nadu	Resistant to wilt, tolerant to spodoptera, semilooper, leafhopper and capsule borer

2016-17

Castor	Area of adoption	Salient features
Kohinoor	Gujarat, Andhra Pradesh, Tamil Nadu, Madhya Pradesh, Rajasthan, Karnataka, Maharashtra and Haryana	Late maturing
Pragati	Telangana	Early maturing; resistant to wilt
HCH 6 (Hybrid)	Karnataka	Late maturing; resistant to Fusarium wilt and white fly

Nutrient management

For rainfed castor varieties: 40-60 kg N/ha: 40 kg P_2O_5/ha

Irrigated castor : 80-100 kg N/ha: 40 kg P_2O_5/ha

For irrigated *rabi* castor 30 kg N must be applied at sowing and the remaining recommended 60 kg in three splits of 20 kg each around 30-35, 60-65 and 90-95 DAS. Phosphorus and Potassium must be applied at sowing by placement. For rainfed crop all P_2O_5 and ½ N at sowing and remaining nitrogen with subsequent rainfall.

Water management

Castor is usually a rainfed crop. However, it responds well to irrigation. If one or two irrigations are given during drought spells, castor yield can be doubled. Critical period is flowering and seed development stages and irrigation at these stages should never be missed.

Weed management

Critical period of crop weed competition is 20-60 DAS. Two to three hoeing / cultivations starting from 20 DAS are required for optimum yield. Pre-emergence application of alachlor 2.0 kg/ha or pendimethalin 0.75 kg/ha are effective against weeds in castor. Also fluchoralin at 0.75 kg/ha as pre-plant incorporation is equally effective for weed control in castor.

Plant protection

Spraying Thiodicarb 1.0 g/l, Spinosad 45 EC at 0.3 ml/l or Phosalone 35 EC at 2 ml/l can control castor semilooper and capsule borer. Thiram at 3 g/kg seed can minimize seedling blight infection. Alternaria leaf blight can be managed by sprayng Mancozeb (0.25%) at an interval of 15 days starting from 90 DAS,

Harvesting and threshing

Harvesting of the mature spikes over a span of 180-240 days, each at an interval of 30 days. Harvested spikes are placed in heaps and sun dried for a couple of days. Threshing is done by beating with sticks or under the feets of cattle and then stored in gunny bags.

Quality considerations

Oil content of castor seed ranges from 40-45%. It contains 12-16% protein, 2-7% carbohydrates and 23-27%, fibre. Dehulled seed contains 60-70% oil and 18-26% protein. It contains 85-90% linolinic acid. Castor cake contains 6% N 2.5% Phosphorus and 1.5% Potassium. It is unfit for edible purpose because of toxic constituents i.e. an alkaloid recinine which is mildly toxic and a protein, which is extremely poisonous.

10

CULTIVATION PRACTICES FOR PULSES

The important pulse crops grown in India are chickpea, pigeonpea, green gram (moong), black gram (mash), lentil, peas, beans etc. The area under pulses during 2016 was 82.38 m ha and 26.43 m ha while production was 81.8 mt and 17.56 mt in the world and India respectively (FAO STAT 2016).

CHICKPEA (*Cicer arietinum*)

Chickpea crop

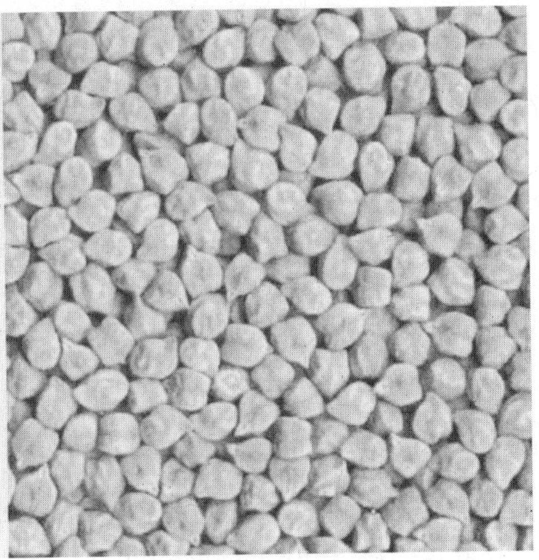

Chickpea seed

Chickpea commonly known as gram or bengal gram is a cool season legume grown on more than 45 countries of the world. It is the most important pulse crop of India. It is used for human consumption as well as for feeding to animals. It is eaten both whole fried or boiled and salted or in the form of split pulse which is cooked and eaten. Both husks and bits of the 'dal' are valuable cattle feed. Chickpeas are a good source of zinc, folate and protein. It is also very high in dietary fibre and hence a healthy food, especially as a source of carbohydrates for persons with insulin sensitivity or diabetes. Chickpeas are low in fat and most of it is polyunsaturated. Chickpeas are also a significant source of calcium (190 mg/100 g) and on average chickpea seeds contain: 23% protein, 64% total carbohydrates (47% starch, 6% soluble sugar), 5%

fat, 6% crude fibre, 3% ash. There is also a high reported mineral content: phosphorus (340 mg/100 g), calcium (190 mg/100 g), magnesium (140 mg/100g), iron (7 mg/100 g), zinc (3 mg/100 g).

Origin and history

Chickpea is said to be oldest pulses known and cultivated from ancient times in Asia and Europe. Its probable place of origin lies in south western Asia, that is, in the countries lying to the north-west of India such as Afghanistan and Persia. According to Aykroid and Doughty, the centre of origin of Chickpea is eastern Mediterranean. According to De Candolle, the fact that Chickpea has a Sanskrit name indicate that the crop has been under cultivation in India longer than in any other country.

Area and distribution

India has the largest area (8.39 m ha) and production (7.82 m tonnes) during 2016 (FAO STAT, 2016). In Punjab, it was grown on an area of 1.9 th. ha with production of 2.4 thousand tonnes.

Classification

The Indian gram has been classified into two broad groups:

1. Desi or Brown gram (*Cicer arietinum* L.): In this group the colour of the seed ranges from yellow to dark brown. Seed size is small. Plants are small with good branching ability.
2. Kabuli or white gram (*Cicer kabulium*): In this group the colour of the seed is white. Grains are bold and attractive. Yield potential of this group is poor as compared to desi or brown gram. Plants are generally taller than the desi gram and stand more or less erect.

Climate

Chickpea is a winter season crop but severe cold and frost are injurious to it. Frost at the time of flowering results in the failure of the flowers to develop seeds or killing of the seeds inside the pod. It is generally grown under rainfed conditions but gives good returns in irrigated conditions as well. Excessive rains soon after sowing or at flowering and fruiting cause heavy loss in seed yield. It is best suited to areas having moderate rainfall of 60-90 centimetre per annum.

Soil

Chickpea is grown on a wide range of soils in India and sandy loam to clay loam is considered to be most suitable. The best type of soil for Chickpea is one that is well drained and not too heavy. The soils chosen for its cultivation should be free form excessive soluble salts and should be neutral in reaction. It is not suited to soils having pH higher than 8.5.

Improved Varieties of Chickpea released by ICAR

2017-18

Chickpea	Area of adoption	Salient features
Gujarat Gram 5	J and K, HP, Uttarakhand and NEH Regions	Tolerant to wilt, stunt and collar rot
Meera	Punjab, Haryana, UP, J and K, Delhi, Rajasthan, HP and Uttarakhand	Tolerant to wilt, stunt and root rot
Indira Chana-1	Chhattisgarh	Resistant to wilt and moderately resistant to dry root rot.
Nandyal Gram 49	Andhra Pradesh	Tolerant to wilt and dry root rot
Pant Gram 4	Uttarakhand	Tolerant to wilt and BGM
Pant Gram 3	Uttarakhand	Tolerant to wilt and BGM
Pant Kabuli Gram-2	Uttarakhand	Tolerant to wilt and BGM
Dheera	Andhra Pradesh	Tall and erect plant, suitable for mechanical harvesting
Pant Gram 5	Rajasthan, Haryana, Punjab, UP, Uttarakhand, and Jammu & Kashmir	Late sown rainfed/ irrigated conditions, tolerant to pod borer
Phule Vikram	Maharashtra	Suitable for rainfed, timely sown irrigated and late sown condition, resistant to Fusarium wilt

2016-17

Chickpea	Area of adoption	Salient features
Pusa 3022	Punjab, Haryana, western Uttar Pradesh, Jammu and Kashmir, Delhi, north Rajasthan, Himachal Pradesh and Uttarakhand	Late maturing; large-seeded, kabuli type; moderately resistant to Fusarium wilt, dry root rot, Ascochyta blight and Botrytis grey mould
PBG7	Punjab	Resistant to Ascochyta blight ,moderately resistant to Fusarium wilt and dry root rot
Aman	Punjab, Haryana, western Uttar Pradesh, Jammu and Kashmir, Delhi, north Rajasthan, Himachal Pradesh and Uttarakhand	Moderately resistant to dry root rot, wilt and collar rot, tolerant to Ascochyta blight and Botrytis grey mould
Nandyal Gram 119	Karnataka, Andhra Pradesh and Tamil Nadu	Kabuli type

Teej	Punjab, Haryana, western Uttar Pradesh, Jammu and Kashmir, Delhi, north Rajasthan, Himachal Pradesh and Uttarakhand	Tolerant to Fusarium wilt disease
JGK 5	Madhya Pradesh	Matures in 110-115 days; extra-large seeded, kabuli type; resistant to Fusarium wilt, moderately resistant to root rot
BDNGK 798	Eastern Uttar Pradesh, West Bengal, Jharkhand, Bihar and North -Eastern Hill states and Maharashtra	Large seeded, kabuli type; tolerant to Fusarium wilt
Gujarat Junagadh Gram 6	Gujarat	Matures in 100-112 days, suitable for rainfed areas; resistant to wilt, moderately resistant to stunt
JG 36	Madhya Pradesh	Matures in 105-112 days; moderately resistant to Fusarium wilt, dry root rot and stunt
GBM 2	Karnataka	Matures in 100-120 days; suitable for machine harvesting as pods lie on the upper 1/3rd part

Improved Varieties Desi gram (Irrigated) for Punjab

PBG 7: The plants are tall, semi-erect. It has medium bold seeds with brownish colour. This variety is moderately resistant to Ascochyta blight and fairly resistant to wilt and dry root rot. It matures in about 159 days. Its average yield is 20 quintals per hectare.

PBG 5: It has thick stem with dark green foliage. It matures in about 165 days. Its seeds are medium bold with dark brown appearance. This variety is fairly resistant to Ascochyta blight and wilt complex (wilt, root rot and foot rot). Its average yield is 17 quintals per hectare.

GPF 2: The plants are semi-erect with lush green leaves and long fruiting branches with two seeds per pod. It has bold seeds and matures in about 170 days. Its average yield is 19 quintals per hectare.

Desi gram (Rainfed)

PDG 4: Its plants are erect with dark green foliage. It bears long fruiting branches arising from the base. It has bold seeds and matures in about 160 days. It is fairly resistant to wilt, foot rot, root rot and Ascochyta blight. Its average yield is 19.5 quintals per hectare.

PDG 3: It is semi-erect variety with dark green leaves and long fruiting branches. It has bold seeds and matures in about 160 days. Its average yield is 18 quintals per hectare.

Kabuli gram

L 552: It is early in flowering and matures in 157 days. It is tall and erect variety. It has large pods and bold seeds. The seeds are creamy white in appearance with good culinary properties. Its average yield is 18.25 quintals per hectare.

Cropping systems

Chickpea is sown after the harvest of *kharif* crops. The most common cropping systems are as below:

- *Kharif* fallow-Chickpea
- Maize-Chickpea
- Pearl millet-Chickpea
- Sorghum-Chickpea
- Rice-Chickpea
- Rice-Chickpea-Summer moong beans

Chickpea can be grown mixed with wheat, barley, linseed, rapeseed and mustard crops. It is grown mixed with toria in *tarai* region.

Field preparation

A rough seedbed is required for Chickpea. In case the Chickpea crop is taken after a *kharif* fallow, it would be desirable to go for a deep ploughing during the monsoon as it helps in conservation of rain water in the soil profile for use by this crop. Very fine and compact seedbed is not good for Chickpea. It requires a loose and well aerated seedbed.

Time of sowing

Date of sowing is a very important factor affecting the yield of Chickpea in all Chickpea growing States. Second fortnight of October is the optimum time for sowing Chickpea in northern India. For peninsular India, first fortnight of October is the best time for Chickpea sowing. Delay in planting beyond optimum period can cause considerable yield reductions as well as reduction in nodule dry weight. Under humid subtropical conditions of *tarai*, which are characterized by shallow water table and relatively more winter rainfall, first fortnight of November is most suitable. Early sowing of Chickpea results in excessive vegetative growth and poor setting of pods. The early sown crop suffers more from wilt owing to high temperature at that time. While delayed sowing restricts vegetative growth and pod-bearing branches, decreases biological nitrogen fixation and may also lead to forced maturity as well as increased incidence of pests (Ali *et al* 1998).

Seed treatment

I. **Fungicide:** Treat the seed with a fungicide Bavistin or Captan at 3.0 g or Rovral 2.5 g per kg of seed.

II. **Insecticide:** In termite infested soil, treat the seed before sowing with Chlorpyrifos 20 EC at 10 ml per kg seed.

III. **Inoculation:** Inoculate the seed with the recommended Rhizobium culture at the time of sowing to supplement the nitrogen supply for getting higher productivity of gram. Moisten the seed recommended for one acre with minimum amount of water. Mix thoroughly one packet of Rhizobium culture with seed on clean pucca floor and let it dry in shade. Sow the seed immediately. Rhizobium and fungicide can be applied simultaneously. If the seed is to be treated with rhizobium, insecticide and fungicide, then first apply insecticide followed by fungicide and rhizobium.

Seed rate and sowing

A seed rate of 45-100 kg depending upon the seed size and sowing time is sufficient for one hectare. The crop may be sown with seed drill or local plough at a row spacing of 30-45 centimetre. In rainfed areas, deep sowing (10 cm) of Chickpea proved better than shallow depth (5 cm). Further, deep planting promotes better root growth and can reduce incidence of *Fusarium* wilt as compared to shallow planting.

Nutrient management

Chickpea being a leguminous crop fulfils the major part of its nitrogen requirement (about 75 per cent) through the process of symbiotic nitrogen fixation which starts effectively from three to four weeks after sowing. However, soils with low organic matter and poor nitrogen supply may require 20-25 kg per hectare of nitrogen as starter dose which can meet plant requirement before the formation of nodules. The Chickpea responds to phosphorus application and its requirement is more under irrigated conditions than that of rainfed conditions. Generally, Chickpea has been found to respond up to 30-60 kg P_2O_5/ha. The response of Chickpea to potassium application is meager as most of the soils in Chickpea growing areas are rich in available potassium. All the fertilizers should be drilled in furrows at a depth of 7-10 centimetre at the time of sowing. On S-deficient soils, the response to sulphur has been found in the range of 20-60 kg S/ha (Aulakh and Pasricha, 1986). Amongst micronutrients, the significant response to application of Zn, B, Mo and Fe can be found at many locations depending upon soil type. Yield gain of 320 kg/ha with the application of 25 kg zinc sulphate has been observed (Takkar and Nayyar, 1986). The significant impact of Boron and Molybdenum application have also been observed. In a study at Kanpur, soil application of 10 kg Borax/ha and Ammonium molybdate @ 1.5 kg/ha have given yield advantage to the tune of 306-450 kg/ha in Chickpea.

Water management

Chickpea is capable of drawing water from depths greater than 150 cm (Singh and Bhushan 1979) but a major proportion of total extractable water comes from the top 60 cm of the soil profile which has the main concentration of active roots. Chickpea is mostly sown as a rainfed crop. However, where irrigation facilities are available, give a pre-sowing and also irrigation which will ensure proper germination. If winter rains fail, give one irrigation at pre-flowering and one at pod development stage. In no case first irrigation should be given earlier than four weeks after sowing. Irrigation should not be given at the flowering stage. Excess of irrigation enhances vegetative growth and depresses Chickpea yield.

Weed management

Chickpea being short statured crop suffers severely with the infestation of weeds. One hand weeding or interculture with hand hoe or wheel hoe after 25-30 days and second hoeing if needed can be done after 60 days of sowing. Trifluralin (Treflan 48 EC) 1.2 kg per hectare in 500-600 liters of water as pre-planting spray may be used as an effective herbicide. It should be well incorporated in soil before sowing. In case Treflan is not available use pendimethalin (Stomp 30 EC) at 0.75 kg/ha as pre-emergence by dissolving in 500 to 600 liters per hectare of water.

Plant Protection

Insect Pests

Termite: Termites attack the crop especially at seedling stage and also near maturity. The pest can generally be observed feeding on roots or near the root zone of the damaged plants. The affected plants dry up and can be pulled out easily. The incidence of pest is more in light soils.

Treat the seed before sowing with Dursban 20 EC (chlorpyriphos) @ 10 ml per kg seed.

Gram caterpillar (*Helicoverpa armigera*): The larvae damage the gram crop by feeding on leaves, flower buds, flowers, pods and grains in the pods. Spray 500 ml Kingdoxa 14.5 SC (indoxacarb) or 150 ml Tracer 45 SC (spinosad) or 400 ml Decis 2.8 EC (deltamethrin) in 200-250 litres of water per hectare on appearance of pest at the start of pod formation and repeat after two weeks, if necessary.

Precaution: Before consuming green leaves and pods, ensure a waiting period of 20 days after the spray of any insecticide mentioned above.

Cutworm: Gram cutworm is a serious pest in low lying areas where fields are cloddy. The larvae of this insect remain hidden under these clods during day time and cause damage during the night. The caterpillars cut the plants at ground level. The pest is sporadic in nature and can be controlled by the application of Lindane 6% granules at the rate of 20-25 kg per hectare mixed in the soil.

Diseases

Ascochyta blight (*Ascochyta rabiei*): Dark-brown spots studded with dot- like bodies are produced on the stem, branches, leaflets and pods. On pods and leaves, these black dot-like bodies are concentrically arranged. Shoot terminals are especially liable to attack. Even the seeds in the pods are infected. In the event of excessive rains, whole crop may be blighted and killed rapidly. A temperature of 20°C, relative humidity above 85% and cloudy weather with intermittent rains is highly conducive for development of disease.

Treat the seed as per recommendations. Give 3 sprays of Tilt 25 EC (propiconazole) (250 ml per hectare) at 10 days interval or 3-5 sprays of Indofil M 45 (mancozeb) or Captan or Captaf (captan) or Kavach (chlorothalonil) @ 900 g per hectare at 15 days interval in 250 litres of water. Give first spray immediately on appearance of the disease. After harvest, the diseased plants should not be allowed to stand in the field, rather but should be destroyed by burning. Use disease free seed.

Grey mould (*Botrytis cinerea*): Small water-soaked spots are produced on leaflets. Spots on infected leaves become dark brown. Under humid conditions, white mycelial strands and erect sporophores of the fungus are produced on flowers, leaves, growing tips, branches and pods. At the point of infection, soft rotting of the tissue occurs. Flowers and growing tips are more vulnerable. The fungus forms dark grey to black sporodochia on infected tissue. The most suitable weather conditions for development of disease are temperature around 25°C, high relative humidity and rain during flowering.

Treat the seed as per recommendations. Give one spray of Indofil M-45 (mancozeb) @ 875g or Baytan (triadimenol+triflumiron) or Bayleton (triadimefon) @ 500g in 250- 300 litres of water per hectare immediately on the appearance of the disease symptoms particularly in February or March.

Wilt (*Fusarium oxysporum f. sp. ciceri*): Affected plants show drooping of petioles and dull green colour in the initial stage. Slowly, all the leaves turn yellow and later become straw coloured. The most characteristic symptom of this disease is vascular discolouration of roots which is dark brown to black. Sometimes, only one sided branches are affected resulting in partial wilting. The soil temperature of 24-27°C favours disease development.

Grow wilt tolerant desi varieties. Conserve soil moisture in rainfed areas. Plough the field with a furrow- turning plough to eradicate weeds and promote water infilteration. Sow the crop as per recommendations.

Stem rot (*Sclerotinia sclerotium*): The disease attacks all the above ground parts of the plant. The main stem is usually affected at the soil level. The affected parts are shredded and covered with white mycelium and embedded black sclerotia. The whitish, fluffy fungal growth can also be seen on the debris at soil surface. The disease is predominant in cool and moist conditions. The seed should be free from sclerotia. Sow the gram crop from 25 October to 10 November to escape the disease attack. Collect and destroy the diseased debris soon after harvesting the crop. In the month of May or June, flood the field after deep ploughing and do not leave it fallow. Rotate gram with non-susceptible crops such as wheat and barley.

Foot rot: (*Opercullela padwickii*): Light brown to dark-brown lesions appear on the collar region of the plants. Later, these lesions become black and affect the basal tap root. Complete girdling in the collar region takes place. Sow the crop from 25 October to 10 November. Follow crop rotation with non-susceptible crops such as wheat and barley.

Rust: Small, round to oval, light or dark brown pustules are formed on the under surface of the leaves. The pustules later turn black. Afterwards, these pustules appear on upper surface of leaves, twigs and pods. The affected leaves pre-maturely fall and the yield is considerably reduced. With the appearance of first symptoms, spray the crop with 0.2% Mancozeb 75 WP followed by two more sprays at 10 days interval. Plant only resistant varieties like Gaurav.

Harvesting and threshing

Crop becomes ready for harvest when leaves turn reddish brown and start shedding. Plants are cut with sickle. This crop is allowed to dry in sun on threshing floor for about 5 to 6 days. Thereafter threshing is done either by beating the plants with sticks or by trampling under the feet of bullocks or by the use of thrashers.

Yield

A well managed crop yields about 20-25 quintals of grain per hectare.

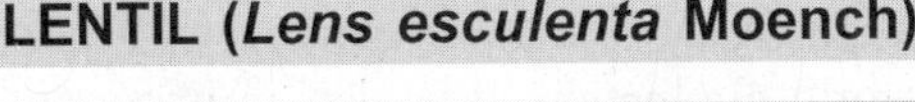

LENTIL (*Lens esculenta* Moench)

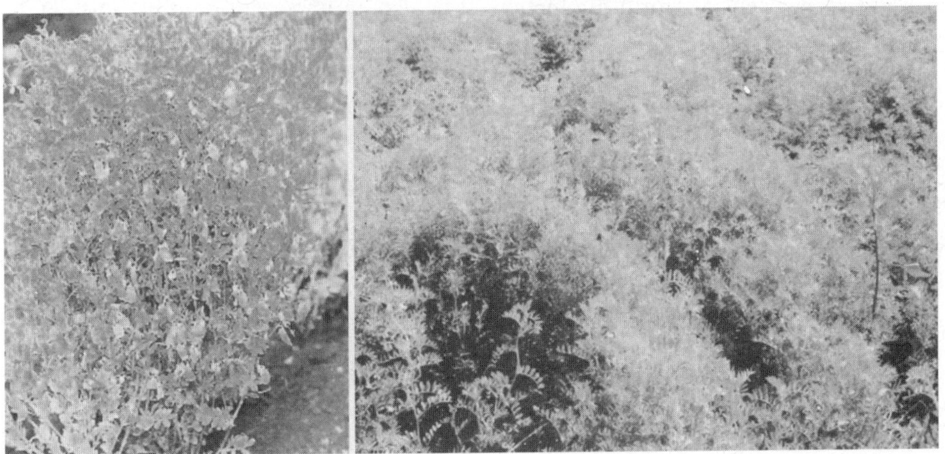

Lentil Crop

Lentil is one of the important *rabi* pulses. It is one of the oldest pulse crops and the most nutritious among the pulses. It has the potential to cover the risk of dryland agriculture. It is also used as cover crop to check the soil erosion in problem areas. It is mostly eaten as 'dal'. Lentil can be consumed as such or whole grains can be first converted into split pulse or 'dal' by the removal of the skin and the separation of fleshy cotyledons. Lentil is a protein/calorie crop. It contains 60 per cent carbohydrates and protein content ranges from 22 to 35%, but the nutritional value is low because lentil is deficient in the amino acids i.e. methionine and cystine. Lentil is an excellent supplement to cereal grain diets because of its good protein/carbohydrate content. Lentil is very important vegetarian food with the highest level of protein other than soybeans, and because of this it is a very important part of the diet in many parts of the world, and especially South Asia which has a large vegetarian population. Sometimes they are difficult to cook because of the hard seed coat that results from excessively dry production conditions. Lentils which fail to meet food grade standards can be used as livestock feed because of their high protein content and lack of digestive inhibitors.

The dry leaves and stems, empty pods and broken bits are all used as cattle feed. Lentil can be used as a green manure crop.

Origin and history

As a cultivated crop it is of very ancient origin and its native home is regarded to be the countries in eastern Mediterranean such as Asia Minor, Greece and Egypt from where it extended eastwards to India. The valleys in the Hindukush mountains in India are also claimed as its native home. Its cultivation dates back to historic times in these regions. It was later introduced in Europe, Africa and America. It may have been introduced to the United States in the early 1900s.

Area and distribution

It is predominantly grown in Asia, which accounts for 80 per cent of the global area and 75 per cent of world production. It is also an important crop in North and Eastern Africa, North-Central America and Southern Europe. India, Turkey, Canada, Bangladesh, Iran, Nepal, Syria, Pakistan, China, Ethiopia, Morocco, Spain, Chile, are important lentil producing countries. The area under Lentil in India during 2016 was 1.55 m ha with production of 1.06 m tonne (FAO STAT 2016).

Classification

Lentil belongs to genus *Lens* of Leguminoseae family. It can be classified into two main groups according to the size of the seed:

(1) ***Bold seeded group***: This group includes sub-species var. *macrosperma.*

(2) ***Small seeded group***: This group includes sub-species *microsperma.*

Climate

Lentil is adapted to cool growing conditions. It is sown as winter season crop. It is very hardy and can tolerate frost and severe winter to a great extent. It can be grown successfully up to a height of 3000 metres. It remains unaffected by rain at any stage of its growth including flowering and fruiting. It can be grown with the moisture conserved in the soil during the rainy season. It requires cold temperature during its vegetative growth and warm temperature at the time of maturity. The optimum temperature for growth is 18-30°C.

Soil

Lentil is adapted to all soil types, from sand to clay loam, if there is good internal drainage. Lentil does not tolerate flooded or waterlogged soils, and does best on deep, sandy loam soils high in phosphorus and potassium. Good drainage is required, because even short periods of exposure to waterlogged or flooded field conditions kill plants. A soil pH near 7.0 is best for lentil production.

Improved Varieties of Lentil released by ICAR

2017-18

Lentil	Area of adoption	Salient features
RVL 11-6	MP, Maharashtra, Gujarat, UP and Rajasthan	Tolerant to wilt, semi erect, medium height and branches with broad leaves which is very much suitable for intercropping
L 4717	MP, UP, Rajasthan and Chhattisgarh	Semi erect, early, resistant to powdery mildew, wilt and Ascochyta blight
Pant Lentil – 9	Uttarakhand	Resistant to rust, wilt and Ascochyta blight diseases
Kota Masoor 1	Rajasthan	Tolerance to wilt
Shekhar 4	Uttar Pradesh	Tolerance to wilt and rust, bold seed
Shekhar 5	Uttar Pradesh	Tolerance to wilt and rust, small seed

2016-17

Lentil		
KLB 2008-4	Uttar Pradesh	Matures in 115-120 days; large seeds; resistant to wilt
KLS09-3	Uttar Pradesh	Matures in 105-110 days; small seeds; resistant to wilt
RLG5	Rajasthan	Matures in 115-120 days; small seeds
IPL 526	Uttar Pradesh	Medium large seeds; tolerant to rust and wilt

Improved Varieties for Punjab

LL 931: Its plants are short, erect with profuse branching and bear more number of pods. It has dark green leaves, pink flowers, non- pigmented green pods and rudimentary tendrils. It matures in 146 days. It is fairly resistant to rust and possesses good tolerance to pod borer. Its seeds are medium bold with greyish brown colour and light flecks. It has good culinary properties. Its average yield is 12 quintals per hectare.

LL 699: The plants are short, erect with profuse branching. Its plants are dark green, bear high number of pods and are early in flowering. It matures in 145 days. It is moderately resistant to rust and blight diseases and possesses good tolerance to pod borer. It has good culinary properties. Average grain yield is 12.5 quintals per hectare.

Cropping Systems

Lentil is grown generally after the harvest of *kharif* crops or as sole crop of the year. The most common rotations are given below:

- Rice-Lentil
- Maize-Lentil
- Cotton-Lentil

- Pearl millet-Lentil
- Sorghum-Lentil
- Groundnut-Lentil

Lentil can also be grown as an intercrop in autumn sugarcane. Two lines of lentil may be sown 30 centimeter apart in the centre of two sugarcane rows. The distance between sugarcane and lentil row will also be 30 centimetre.

Field preparation

The soil should be made friable and weed-free so that seeding could be done at a uniform depth. On heavy soils, one deep ploughing followed by two to three cross harrowings should be given. In light soils less tillage is needed to prepare an ideal seedbed. There should be proper moisture in the soil at the time of sowing for proper germination of seeds.

Time of sowing

Timely planting is a very important factor in the full realization of the yield potential of improved varieties of lentil. Middle of October is the most suitable time for sowing lentil. Delay in planting causes reduction in yield. Under delayed sowing the reduction in yield could be minimized up to certain extent by closer spacing and use of higher seed rate.

Seed treatment

Treat the seed with Captan @ 2.0 g per kg seed. Also treat the seeds with biofertilizer i.e., Rhizobium and Rhizobacterium before sowing.

Seed rate and sowing

Sowing should be done in rows 30 centimetre apart. The optimum seed rate for normal sown crop is 30-40 kg per hectare. Under late sowing the seed rate should be increased to 50-60 kg per hectare and row spacing reduced to 20-25 centimetre. Seed should be placed at a shallow depth of 3-4 centimetre.

Nutrient management

Generally lentil is grown without fertilizers and manures. However, good yields may not be harvested if lentil crop is made to feed on symbiotically fixed nitrogen only. Some quantity of fertilizer nitrogen as starter dose is required during few weeks of the initial growth. On light and poor fertility soils, it is even more important to give the starter dose of nitrogen fertilizer. On sandy soils with low organic matter content, the response to applied nitrogen to the tune of 10-25 kg per hectare and of phosphorus to 50-60 kg per hectare have been obtained. Lentil when grown in rotation with paddy in late sown conditions may give symptoms of zinc deficiency. The deficiency can be rectified by spraying a solution of 0.5 per cent zinc sulphate and 0.25 per cent lime at the appearance of early symptoms.

Water management

Lentil is mostly grown under unirrigated conditions. It can tolerate drought to some extent. If irrigation is available, then first irrigation should be given at 45 days of planting and second at pod filling stage.

Weed control

Lentil is a slow growing crop and thus suffers considerably with competition from weeds particularly in the early stages of growth. The period from 30 to 60 days after sowing is most critical for competition with weeds. Under heavy infestation of weeds, there is drastic reduction in yields. A weed-free period of early 30 to 60 days is important. Thereafter, the crop covers the ground and weeds are effectively smothered. Weeds in lentil can be controlled with mechanical as well as chemical methods. Two manual weedings can be done at 30 and 50 days after sowing. Alternatively preplant application of Treflan 48 EC (trifluralin) at 0.5 kg a.i./ha integrated with one hand weeding provides effective control of weeds in this crop. The pre-emergence application of Stomp 30 EC (pendimethalin) at (0.75 kg a.i. per hectare) + one hand weeding at 45 days after sowing is also very effective.

Plant Protection

Insect Pests

The lentil pod-borer causes damage to the crop by feeding on leaves, flower buds, flowers, pods and grains in the pods. It can be controlled by spraying 2.25 kg Sevin/ Hexavin 50 WP (carbaryl) in 200 litres of water per hectare at flower initiation. Repeat the spray after three weeks, if necessary.

Diseases

Blight (*Ascochyta fabae f. sp. lentis*): Dark brown spots are produced on stem, branches, leaves and pods. The lesions on the stems and branches are elongated. The characteristic symptoms of the disease are black pinhead-like pycnidial bodies of the fungus are arranged on the spots in the form of concentric rings. On tender branches, spots rapidly girdle the branch and parts of the plant above the girdle, wilt and dry. The cool and humid conditions favour disease development and spread. Use disease free seed. Treat the seed with Captan @ 2 g per kg seed. Destroy the diseased plant debris after harvest to reduce the inoculum of the fungus.

Rust (*Uromyces fabae*): Yellowish white pycnial and aecial pustules develop on stems, branches, leaves and pods. These may appear singly or in small groups. Later on brown uredia are formed. Small pustules may coalesce to form large pustules. In later stage of the crop, dark brown to black telia are formed. In severe cases, affected plants dry up and give burnt appearance. Cloudy weather with intermittent rains and temperature around 20-22°C favours the disease development. Use clean seed free from diseased plant debris.

Seedling mortality: The disease appears within a month of sowing when the seedlings start drying up. The drying is mainly of two types. In one, the seedlings first turn yellow and then dry up. This is called as seedling wilt. In second, the seedlings collapse while still green and then dry up. It can be controlled by treating the seed with Benomyl, a systemic fungicide at the rate of 2.5 kg per kg of seed. Plant only resistant varieties like Pant-L-406.

Wilt: This is a serious disease of lentil in which the growth of plant is checked, the leaves start yellowing, plants start drying and finally die. The roots of affected plants remain under developed. Keep the field clean and follow a three year crop rotation to reduce the disease incidence.

Powdery mildew: The symptoms of this disease are seen about three months after sowing in the form of small whitish patches on the lower surface of the leaves. These whitish patches spread further and cover leaves, stem and pods as whitish powdery mass. Collect and burn the affected debris in order to reduce the incidence of this disease.

Downy mildew: It is observed in the form of light greenish to yellow spots on the upper surface of the leaves. On the lower surface, opposite to these spots a brownish cottony growth of the fungus is seen. Later, the whole affected leaf is covered by this cottony growth of the fungus and may become chlorotic. The affected palnts remain dwarf. Follow a three year crop rotation to reduce the soil-borne inoculum of the fungus. Treat the seed with Thiram or Carbendazim (Bavistin) at the arte of 2.5 g per kg of seed before sowing.

Harvesting and threshing

Lentil should be harvested when the pods are ripe but the plants have not become dead ripe. The plants should not be allowed to become dead ripe otherwise a large quantity of produce may be lost due to shattering. As such, it would be advantageous to harvest in the morning when dew is there. The produce should be allowed to dry completely on threshing floor before threshing. Threshing is done either by beating the plants with sticks or by trampling under the feet of bullocks or by the use of thrashers. Clear the seed and dry it in sun to bring moisture content down to 12 per cent for safe storage.

Yield

A well managed crop yields about 20-25 quintals of produce (seeds) per hectare.

PEA (*Pisum sativum*)

Field pea plant

Field pea crop

Pea is a popular vegetable and pulse crop of India. It provides a variety of vegetarian dishes and hence it is liked throughout the world. There are two types of cultivated peas, the garden pea and the field pea. Garden peas are harvested in an immature conditions to be cooked as green peas or to be canned or frozen for subsequent uses. Field peas are grown as a forage crop for cattle or as a green manure crop for soil improvement or as a cover crop to reduce the soil erosion or as a mature seed crop. The mature seed may be used as whole or split into 'dal' and prepared in various ways for human consumption. It is highly nutritive and contains high proportion of digestible protein (22.5%), carbohydrates (62.1%), fat (1.8%), minerals (calcium and iron) and vitamins (riboflavin and thiamin) etc.

Origin and history

Archaeological surveys indicate the use of pea during pre-historic period dating back to 3000 BC. The most probable centre of origin of pea is Mediterranean region of Europe and Western Asia including Italy, Soviet Republic, north-west India, Pakistan, Afghanistan, Tazakistan and Uzbekistan.

Area and distribution

The important field pea growing countries are China, Russian Fed., India, Ethiopia and USA. China ranks first in the world in respect of production as well as acreage followed by Russian Fed. and India. In India field pea is grown over an area of 1.1 m ha with a production of 1.02 m tonne during 2016 (FAO STAT 2016). Uttar Pradesh is the major field pea growing State. Uttar Pradesh alone produces about 60 per cent of total pea produced in India. Besides, Uttar Pradesh, Madhya Pradesh and Bihar are the major field pea producing States.

Classification

Two types of peas are generally cultivated all over the globe. One is the garden pea and other is field pea.

1. **Garden Pea (*Pisum sativum* var. hortense):** It is also called table pea or vegetable peas. In this type young, green seeds are used mostly in vegetables and also for canning purposes. Seeds are bold and wrinkled. These are generally, white flowered. Leaf axils are generally green. Seeds are yellowish, whitish or bluish-green.
2. **Field Pea (*Pisum sativum* var. arvense):** In this type the ripe, mature seeds are used as the pulse (dal). They are also grown as forage or green manure crop. Field peas are hardy plants and grown on a large scale without irrigation. These plants are also able to withstand frost. Seeds are rounded and little angular. They have generally coloured flowers. Leaf axils are often pigmented. Seeds are greyish-green, greyish-brown or greyish-yellow.

Climate

Field pea requires a cool growing season, moderate temperatures are essential throughout the growing season. For germination, about 22°C temperature is considered favourable. High temperatures are more injurious to pea crop than frosts. Frosts can damage the plants during flowering stage. Peas can be produced successfully in temperate, semi-arid zones. Peas are most sensitive to moisture stress at flowering state. High humidity is harmful to pea crop as it favours incidence of disease. The optimum monthly temperature suitable for its growth is 13-18°C.

Soil

A well-drained soil is essential for successful production of field pea. Field pea is highly sensitive to waterlogging conditions, hence a well drained loam soil is considered best for pea cultivation. They tolerate a moderate soil pH range (6.0 to 7.5). The optimum pH is 6.5.

Improved Varieties of Pea released by ICAR

2017-18

Field pea	Area of adoption	Salient features
Central Field pea	Madhya Pradesh, Chhattisgarh, Gujarat and Rajasthan	Resistant to powdery mildew, pod borer and moderately resistant to aphids and leaf miner
Pant Pea 155	Uttarakhand	Resistant to rust and powdery mildew and tolerant to pod borer

2016-17

Fieldpea		
Indira Matar 1	Chhattisgarh and Bihar	Suitable for rice fallows; matures in 105-110 days, tall; moderately resistant to powdery mildew
Central Fieldpea IPFD 11-5	Madhya Pradesh, Gujarat, Chhattisgarh, Maharashtra	Matures in 105-110 days, dwarf; resistant to powdery mildew
IPFD 6-3	Uttar Pradesh	Matures in 120-125 days, dwarf; resistant to powdery mildew
RFP 4	Rajasthan	Medium tall, medium large seeds; moderately resistant to powdery mildew

Improved Varieties for Punjab

Early Maturing

AP-3: This variety derived from the cross of Azad P-1 x Arkel through pedigree method. Plants are dwarf and straight. The pods are of medium length (8.85 cm), deeply curved near distal end and are borne singly or in doubles. Each pod contains 7-8 grains and shelling out turn is 50 per cent. Seeds are bold, wrinkled with green seed coat. It is an early maturing variety and gives first picking in 60-65 days after sowing, if sown in the second week of October. Its green pod yield is 79 q/hectare.

Matar Ageta-7: The plants are vigorous with 15- 18 well filled pods having 7-9 grains per pod. The pods are of medium length (9.57 cm), slightly curved from tip and are borne singly or in doubles. Its shelling out turn is 48 per cent. It is an early maturing variety which gives first picking in 65-70 days after sowing. Average green pod yield is 80 q/hectare.

Matar Ageta-6: Its plants are dwarf (40 cm), erect, vigorous and green. It can tolerate high temperature and can be sown early in the season. First picking is possible after 7 weeks of sowing. Each plant bears 12-15 pods. One or two pods are borne per bunch. Each pod contains up to 6 grains and shelling out turn is 44.6 per cent. Grains have high protein and dry matter content. On drying, per cent seeds are light green, smooth with slight dimples. This variety gives 50 per cent of the total production in the first picking. Its green pod yield is 60 q/hectare.

Arkel: Arkel is an early maturing variety and takes about 60-65 days from sowing to marketable green pod stage. It is suitable for sowing in the first week of October under the irrigated conditions. Plants are dwarf (30- 45 cm). Pods are borne both in double and single and are attractive, dark green, well filled, 8-10 cm long and each contains 7 to 8 dark green grains with a shelling percentage of 40. Dry seeds are light green and wrinkled. It is good for dehydration. Its green pod yield is 45 to 50 q/hectare and seed yield is 9 q/ hectare.

Main Season

Punjab 89: The plants of this variety are medium dwarf, vigorous, having more number of well filled pods (28-30 per plant). The pods borne in doubles and are dark green, long, very attractive having 9-10 grains per pod. It takes about 85-90 days for first picking. Shelled peas are very sweet and the shelling out turn is more than 55 per cent. Average green pod yield is 150 q/hectare.

Mithi Phali: It is an edible podded variety and its shelling is not required. Plants are tall (140 cm) with green foliage. The first picking takes place 90 days after sowing and subsequent pickings at 15 days interval. Pods are light green and 12-13 cm long. Its consumable yield is double than that of standard shelled pea varieties. It is rich in protein and total sugar. Average green pod yield is 118 q/hectare

Field Pea 48: It is an early maturing variety with erect, semi-dwarf plant type. It matures in about 135 days. It has long pods with more number of seeds per pod. It has very attractive light green, bold seeds, slightly wrinkled and possess high swelling capacity. Due to its early maturity, it escapes heavy damage from pod borer and powdery mildew. It gives an average grain yield of 20 quintals per hectare.

PG 3: This variety has a dwarf plant type with white flowers. It is early in flowering and maturity. The grains are creamy white, slightly wrinkled and possess high swelling capacity. It escapes heavy damage from pod borer and powdery mildew due to early maturity. It matures in about 135 days. It yields about 17.5 quintals per hectare.

Cropping systems

The most common rotations are given below:

- Maize-Field pea
- Rice-Field pea
- Cotton-Field pea
- Sorghum-Field pea
- Pearl millet-Field pea

It can be sown mixed with Chickpea, barley, wheat, oats, rapeseed and mustard crops. It is also grown as an intercrop in autumn sugarcane. Two rows of field pea may be sown 30 centimetre apart in the centre of two sugarcane rows.

Field preparation

Field is prepared as in the case of other *rabi* crops. After harvest of *kharif* crops, the field should be ploughed with disc or mouldboard plough. Where tractor is available one deep ploughing followed by two to three harrowing and plankings should be given to prepare a well pulverized seedbed. Where bullocks are the source of power, deep ploughing followed by two to three harrowings or three to four cross ploughing with local plough should be done.

Time of sowing

Second fortnight of October is the optimum time for sowing of field peas in North Indian conditions. Sowing after October results in drastic yield reduction. Time of slowing is even more critical in case of garden peas. The optimum time for sowing of garden peas (vegetable purpose) is first fortnight of November in northern India. In case of early sowing, special care should be taken against stem borer, Sowing of pea should be done according to prevailing temperature in a particular area, It should be sown when daily maximum temperature is below 30°C and the daily minimum temperature comes down to 20°C. If pea is sown above this temperature range than crop will have more incidence of wilt disease.

Seed rate and sowing

Seeds are sown at the rate of 60-80 kg per hectare in rows 30 centimetre apart. Seed rate and spacing very according to the periods of maturity of different varieties. The early maturing varieties are given closer spacing and higher seed rate and the late varieties are given wider spacing and lower seed rate. In case of early maturing dwarf varieties crop should be sown in rows 20 centimetre apart and about 100-125 kg seed per hectare should be sown. In late maturing and taller varieties a row spacing of about 30 centimetre seems to be optimum. The seed rate should be reduced to 75-80 kg per hectare.

Nutrient management

The favourable response of fertilizer application have been found in field pea. The optimum dose of nitrogen was found 20-40 kg per hectare at many locations of Northern India. The field pea response to applied phosphorus varies depending on the phosphorus availability in soil and moisture availability and in general the response to P_2O_5 was observed at 50-100 kg/ha.

Water management

The crop is mostly grown in unirrigated areas. It can tolerate drought condition to some extent. By providing one or two irrigations, higher yields can be obtained. First irrigation should be given at 45 days and second, if needed, at pod filling stage. The irrigation may bring 100 to 150 per cent increase in the yield depending upon the soil type, winter rains and depth of water table. Special precaution should be taken while irriga-ting a pea crop. Light and uniform irrigation should be given. Water logging condition in pea field even for a day causes considerable loss in the yield since this crop is highly sensitive to poor drainage conditions. Poor drainage leads to reduction in the number of branches and pods per plant.

Weed management

The pea field should be free from weeds for the period up to 40-50 days after sowing. Later on crop itself checks the growth of weeds by covering the ground surface. The major weeds found in pea crop are *Chenopodium album* (bathu), *Fumaria, parviflora* (gajri), *Lathyrus spp.* (chatri-matri), *Melilotus alba* (senji), *Vicia sativa* (ankari), *Phalaris minor* (gulidanda). There is drastic reduction in yield under heavy infestation. Therefore, the field should be kept free from weeds by giving two hand hoeings after three and six weeks of germination. Basalin (fluchloralin) at the rate of 0.75 kg a.i. in 500 to 600 litres of water per hectare as pre-planting spray may be used as an effective herbicide. It should be well incorporated in the soil before sowing. In case Basalin is not available use metribuzin or prometrynen or methabenzthiazuron at the rate of 1.0-1.5 kg/ha in 500-600 litres of water as pre-emergence spray. Alternatively pre-emergence application of Stomp 30 EC (pendimethalin) 0.75 kg a.i. ha^{-1} can be made for controlling annual weeds.

Plant Protection

Insect Peste

Pests and Symptoms	Control Measures	Precautions
1. Pea thrip (*Thrips indicus*) cause severe damage to the young crop by sucking the cell sap.	Spray 1.0 L orRogor 30 EC (*dimethoate*) in 200 250 litres ofwater/ha when the attach begins andrepeat the spray at 15 days intervals.	Stop spraying the crop with these insecticides atleast 20 days before the picking of pods.
2. Pea-leaf miner (*Chrotomyia horticola*) larvae feed by making tunnels in the leaves. These cause serious damage during December-March.	-do-	-do-
3. Pea-aphid (*Acrythosiphon pisum*) sucks the cell sap, owing to which theleaves turn pale and dry.	-do-	-do-
4. Pea-stem fly (*Ophiomyia phaseoli*) sometimes causes seriousdamage at the seedling stage.	(i) Apply 7.5 kg Thimet 10 G (*phorate*) or 25 kg Furadan 3G (*carbofuran*) granules per ha in furrows at sowing. (ii) Sow the crop in the second fortnight of Octoberto escape the damage by the pest.	Avoid direct contact with chemicals while treating the seed. User ubber gloves for application of granules.

Diseases

Diseases	Symptoms	Control Measures
1. Powdery mildew (*Erysiphe polygoni*)	White floury patches covering large areas, appear on stem, branches, leaves and pods.	Spray the crfop with Karathane 40 EC 200 ml or Sulfex 1.5 kg in 500 litres of water per ha. Three sprays may be given at 10 days interval.
2. Wilt, root rot and collar rot (*Fusarium oxysporum* and *Rhizoctonia solani*)	Yellowish, reddishbrown, spherical, raised postules appear mainly on the lower side of leaves during December-January. This disease in moreserious in thelate sown crop.	(i) Give first spray of 1.0 kg of lndofil M-45 in 500 litres of water per ha by the end of December. Three subsequent sprays may be given at 10 days interval. (ii) Keep the field free from Rewari weed which serves as a source of inoculum. (iii) Forcontrolling powdery mildew and rust to gether Karathane 200 ml of Sulfex 1.5 kg can be sprayed in combination with lndofil M-45 1.0 kg per ha

Harvesting

Different varieties of garden pea have varying maturity periods. The picking of green pods should be done by giving a simple jerk to the pedicle and with minimum possible disturbance to the plants. Field peas should be harvested when they are fully ripe and threshed after sufficient drying in the sun either manually or with thrasher.

Yield

With improved package of practices, field peas can yield about 20-25 quintals of grain and about same quantity of straw from one hectare of land.

PIGEON PEA (*Cajanus cajan* L. Millsp.)

Pigeon pea crop **Pigeon pea pods**

Pigeon pea commonly known as arhar is a very old crop of this country. It is next to Chickpea in area and production among all the pulse crops grown in India. Its grains are highly nutritious and rich in protein (21.7%), carbohydrates, fibre and minerals (iron and iodine). It supplies a major share of the protein requirement of the vegetarian population of the country. It is mainly eaten in the form of split pulse as 'dal'. Seeds of arhar are rich in essential amino acids like lycine, tyrocene, cystine and arginine. The outer covering of its seed together with part of kernel provides a valuable feed for milch cattle. The husk of pods and leaves obtained during threshing constitute a valuable cattle feed. Woody parts of the plant are used for fuel. It is a legume crop and possesses valuable properties as restorer of nitrogen to the soil. In most areas pigeon peas are grown in association with other row crops such as sorghum, pearl millet, or maize. Pigeon peas can be of a perennial type, in which the crop can last 3-5 years (although the seed yield drops considerably after the first two years), or an annual type more suitable for grain production. The crop is cultivated on marginal land by resource-poor farmers, who commonly grow traditional medium and long duration (5-11 months) cultivars. Short duration pigeon peas (3-4 months) suitable for multiple cropping have recently been developed. Traditionally, the use of farm inputs like fertilizers, irrigation, and pesticides are minimal, so present yield levels are low. Though pigeon pea is grown in a wide range of agroecological situations, its deep rooting and drought tolerant characters make it especially useful crop in the area of low and uncertain rainfall and on the lighter soils.

Origin and history

Although pigeon pea (arhar) has been cultivated in India for more than three thousand years and is one of the most widely grown pulse in the country, it is doubtful if it occurs wild anywhere in India. It has been reported to occur in wild state in Africa, in the upper regions of the river Nile and coastal districts of Angola. It seems that Africa is the probable place of origin and the plant might have been introduced into India by ancient traders trading on the route between Zanzibar, India and Ceylon. A few writers including Vavilov, considered India as its native home but it has never been found in a wild state in India and ancient Indian literature has no reference to this pulse crop.

Area and distribution

It is grown in Asia, Eastern and Southern Africa, and Latin America. India has the largest area of 3.88 m ha under pigeonpea, with total production of 2.85 mt during 2016 (FAO STAT 2016).

Classification

Numerous types of *Cajanus spps.* are known, differing in height, habit, time of maturity, colour, size and shape of pods and seeds. All thses cultivated types fall into two groups:

1. ***Cajanus indicus* var. *bicolor*:** This group includes generally late maturing varieties, having tall bushy plants, flowers yellow, purple streaked pods dark coloured each with four or five seeds.
2. ***Cajanus indicus* var. *flavus*:** This group includes early maturing varieties having smaller plants, yellow flowers, plain pods with two to three seeds.

Climate

Pigeon pea grows well on a warm tropical and subtropical climate. It is a highly drought resistant crop. It can successfully grow in areas receiving only 65 cm annual rainfall as the crop matures fast and pest damage is low. The crop prefers a fairly moist and warm climate during the periods of its vegetative growth. During the flowering and ripening stages of its growth, it requires bright sunny weather for the setting of fruits. It is highly susceptible to frost at the time of flowering. Cloudy weather and excessive rainfall at flowering damage the crop to a great extent.

Soil

Pigeon pea may be grown well on a wide range of soils varying from sandy loams to clay loams. It does best on fertile and well drained loamy soils. The saline-alkaline and waterlogged soils are unfit for its cultivation as they adversely affect nodulation. Well drained, alluvial and loamy soils are good for its cultivation. It is successfully grown in black cotton soils of Madhya Pradesh and Maharashtra with proper drainage. It grows well in well drained red clay loam soils too.

Improved Varieties of Pigeon Pea released by ICAR

2017-18

Pigeonopea/ red gram	Area of adoption	Salient features
Amaravathi	Andhra Pradesh	Moderately resistant to wilt
BDN 716	Maharashtra	Resistant to wilt and sterility mosaic, red and large seeded

2016-17

Pigeonpea	Area of adoption	Salient features
Prakash	Uttar Pradesh, Bihar, Jharkhand, West Bengal	Matures in 250 days; resistant to sterility mosaic disease, tolerant to Phytophthora blight and Fusarium wilt
Gujarat Junagadh Pigeonpea 1	Gujarat	Matures in 170-180 days; moderately resistant to wilt
Ujwala	Telangana	Matures in 130-135 days; tolerant to wilt
Mannemkonda Kandi	Telangana	Matures in 180-190 days; resistant to wilt and sterility mosaic disease (SMD)
GT 102	Gujarat	Matures in 175-180 days; resistant to SMD , tolerant to wilt
BRG 5	Karnataka	Matures in 160-170 days; resistant to wilt, moderately resistant to sterility mosaic disease

Improved Varieties for Punjab

AL 882: It is short statured and early maturing variety with semi-determinate growth habit. It matures in about 132 days and vacates the field well in time to sow the succeeding wheat crop. Its plants are compact and about 1.6 to 1.8 metre tall. Pods are borne at the top of the plant in multiple and sparse clusters. Podding is profuse and each pod contains 3-5 yellowish brown and medium sized seeds. The average grain yield is about 13.5 quintals per ha.

PAU 881: It is an early maturing variety with indeterminate growth habit. It matures in about 132 days and vacates the field well in time to sow the succeeding wheat crop. Its plants are about 2 metre tall. Pod formation is profuse and each pod contains about 3-5 yellow brown and medium sized seeds. Its average grain yield is about 12.75 quintals per ha.

AL 201: It is an early maturing variety with indeterminate growth habit. It matures in about 140 days and vacates the field by the end of October permitting timely sowing of the succeeding wheat crop. The plants are erect and about 2.5 metre tall. The main stem is much longer than the side branches and its flower is yellow with prominant red streaks on the standard petal. Pod formation is profuse and each pod contains 3-5 yellowish brown and medium sized seeds. The average grain yield is about 12.5 quintals per ha.

Crop rotation

Pigeon pea is grown in summer and *kharif* in northern India and *kharif* and *rabi* in southern Indian conditions. It may, therefore, be included in several crop rotations. The important rotations are given as under:

- Wheat - Pigeon pea + Urd bean
- Black gram - Pigeon pea
- Green gram - Pigeon pea
- Wheat - Pigeon pea
- Barley - Pigeon pea
- Paddy - Paddy - *Rabi* Pigeon pea
- Sorghum/Bajra/Maize - *Rabi* Pigeon pea
- Groundnut – *Rabi* pigeon pea
- Soybean - *Rabi* pigeon pea

Field preparation

Pigeon pea is a very deep tap rooted crop capable of penetrating root system in hard soils. It does not require special land preparation. Deep ploughing to a depth of 15 cm is adequate to obtain a good crop. When the crop is grown in rainy season, care should be taken to level the land to prevent waterlogging. In other seasons too the land should be leveled and soil should be pulverized for placement of seeds at uniform depth. In general, a deep cultivation either by ploughing or disking, followed by 2-3 harrowings and levelings will be sufficient.

Time of sowing

Pigeon pea is a traditionally *kharif* crop sown in June-July with onset of monsoon in various agroclimatic zones of India. Short duration varieties of pigeon pea are now becoming popular in the irrigated area. These varieties are harvested much earlier than the occurrence of frost/cyclones and can also be fitted well in multiple cropping systems. Planting of early pigeon pea before the onset of monsoon in the month of June is recommended for higher yields. Planting with onset of monsoon was the optimum time because with earlier plantings, the reproductive phase coincides with the period of heavy rainfall which ultimately caused drop of flowers and pods. Delay in planting reduces the yield of pigeon pea. Delay in planting, caused reduction in the duration of crop, plant height, number of branches and pods.

Seed treatment

Treat the seed with Captan or Thiram @ 3.0 g/kg seed against seed-rot and seedling-blight. Also inoculate the seed with the recommended Rhizolium culture before sowing. Mix Thoroughly one packet of Rhizolium culture with seed of one acre on a clean pucca floor after wetting the seeds with water. Dry seeds in shade and sow in the field immediately. The inoculation of seed increased the seed yield by 5 to 7%. Rhizolium and fungicide can be mixed simultaneously.

Seed rate and spacing

In general, a seed rate of 8-10 kg/ha for long duration, 10-12 kg/ha for medium duration and 12-15 kg/ha for short duration varieties is sufficient in pigeon pea. Long duration varieties of pigeon pea are tall, spreading and occupy the field for about 250-270 days. These varieties are planted at wider row spacing of 90-120 cm and about 30 cm between the plants particularly under rainfed conditions. Under irrigated conditions, early maturing varieties are more popular as they fit well in double cropping systems with other crops. These varieties are planted at a row spacing of 50-75 cm and plant to plant spacing of 15-20 cm.

Method of sowing

Pigeon pea is generally broadcasted. Line sowing is superior over broadcasting. Broadcasting results in uneven plant population which ultimately results in low yield. In such areas where temporary water logging take place, planting on ridges have been found superior.

Mixed/Intercropping

Pigeon pea is mainly grown as mixed with other traditional crops like urdbean, greengram, groundnut, small millets etc. In this system, seeds of different crops are mixed together and broadcasted. Pigeon pea can profitably be grown as intercrop with several widely spaced crops like black gram, maize, sugarcane, sorghum, etc. One or two rows of pigeon pea could be sown in between two rows of maize. Paired planting can also be adopted in pigeon pea where two rows are made closer and thus this extra space is provided in between two pairs of rows. The overall plant population on area basis is not affected but by the adjustment of rows more space is provided in between two pairs which is utlised by any intercrop. Summer pigeon pea may be intercropped with spring planted sugarcane.

Fertilizer management

Being a leguminous crop, pigeon pea possesses atmospheric nitrogen fixing ability. Therefore, the response to added nitrogen is limited. However, in soils poor in available nitrogen status a starter dose of nitrogen to support initial growth till nodule formation and biological nitrogen fixation is essential. Pigeon pea when grown on soils having higher available nitrogen reserves generally do not respond to nitrogen. In general, application of about 20-25 kg N/ha applied at sowing has been found adequate.

On an average, application of 50-60 kg P_2O_5/ha is optimum. The dose may vary with the soil type i.e. light soils need higher dose and heavy soils need lower dose. Phosphorus applied to preceding crop may also affect the rate of phosphorus application in pigeon pea.

Water management

Pigeon pea is mainly grown in the monsoon season under rainfed conditions. Being a deep rooted crop, it is capable of extracting moisture from deeper layers in the soil.

However, under long period of moisture stress, it may respond to irrigation as well. Two irrigations, one at pre-flowering and other at flowering stage are beneficial in increasing the yield of pigeon pea under rainfed conditions.

Weed management

The growth of pigeon pea is quite slow during first 45-60 days after sowing. On the other hand, weed growth is very fast due to frequent rains. During this period, if weeds are not controlled, they cover the pigeon pea plants. Common weeds associated with pigeon pea are: *Cyperus rotundus*, *Echinocloa* sp., *Dactylactinium aegyptium*, *Setaria glauca*, *Amaranthus* sp., *Celosia argentea*, *Commelina benghalensis*, *Phyllanthus niruri*, *Digera arvensis*, *Euphorbia* sp. Weeding by mechanical means sometimes becomes very difficult because of frequent rains due to which soil does not come in physical condition to permit use of the implements. Under such conditions, use of herbicides to control the weeds becomes more important. Two weedings, one at 25 days and another at 45 days after sowing, were found to be sufficient. Among herbicides, pre-emergence application of Stomp 30 EC at 0.75 kg a.i. ha^{-1} or Lasso 50 EC at 2.0 kg a.i./ha were found to be quite effective. Intercropping with short duration legumes like urdbean, green gram and cowpea is another biological way of controlling weeds in pigeon pea. However, this practice can only reduce the infestation of weeds to some extent by occupying the space in between the rows of pigeon pea. The weeds cause losses to a great extent under intercropping system also.

Plant Protection

Insect Pests

Blister beetle: Blistle beetle is diurnal and general feeder. Adult beetles are robust with bright black and red stripes on the forewings. When disturbed, the beetles emit a fluid containing cantharidin that causes blisters on human skin. The adult beetles attack *arhar*, *moong*, *mash* and other pulse crops. The major damage is caused at the flowering stage. They feed on tender buds and flowers of the plant, thus preventing grain formation.

Regarding pest details, see under *moong*. Blister beetle can be controlled by spraying Decis 2.8 EC (deltamethrin) @ 500 ml or Kingdoxa 14.5 SC (indoxacarb) @ 500 ml using 250-300 litres of water per ha with manually operated knapsack sprayer. Spray the crop on pest appearance in the evening hours and repeat the spray after 10 days, if necessary.

Pod borer complex: Pod borers, especially spotted pod borer *(Maruca vitrata)* and gram pod borer *(Helicoverpa armigera)* are the most important insect pests of *arhar*. *Maruca vitrata* appears at the initiation of flowering. Full grown larva has a pale body lined by rows of conspicuous black spots on its dorsal surface. Its larvae damage the flower buds, flowers and green pods. The larva feeds from inside a webbed mass of leaves, flowers, flower buds and pods webbed together with silken threads. Spray the crop at flower initiation stage using Kingdoxa 14.5 SC (indoxacarb) @ 200 ml or Tracer 45 SC (spinosad) @ 150 ml using 250-300 litres of water per ha with manually operated knapsack sprayer. Repeat the spray at pod initiation/podding stage of the crop. The insecticides used for the management of spotted pod borer also control other pod borers infesting *arhar*.

Precautions: Because honey bees and other pollinators may be killed by the use of above insecticides it is, therefore, advised to spray the crop during evening as the population of these pollinators is minimum at that time.

Jassids or leaf hoppers: The leaflets damaged by jassids are cup-shaped and have yellow edges and tips. Give basal application of Phorate (Thimet) 10% granules at the rate of 10 kg ha^{-1} at the time of sowing. Spraying with Monocrotophos 36 EC (1 ml in 1 litre of water) also controls the insect effectively.

Diseases

Phytophthora stem blight: It is caused by *Phytophthora drechslera f.sp. cajani*. The disease affects young seedlings as soon as they emerge and get killed. On stem, brown to black necrotic lesions are produced, which have definite margin and are slightly depressed. In some cases, stem swells into a cankerous structure at the edge of the lesion, which may break at the lesions site. On leaflets lesions are circular to irregular in shape and whole foliage can become blighted.

Avoid sowing *arhar* in soil with poor drainage and follow rotation in badly infested fields.

Sterility mosaic: It is a viral disease transmitted by an eriophyid mite *(Aceria cajani)*. Typical symptoms are mild mosaic and either no or little flowering and pale green colour of the leaves. The leaves are crowded and auxiliary buds give rise to bushy growth. Do not allow any *arhar* plant standing around sugarcane, cotton and other fields during winter season in order to check the primary source of infection.

Cercospora leaf spot: Leaf spots are caused by *Cercospora cajani*. Greyish brown to dark spots are produced on the under surface of the leaf. Often several spots coalesce to form irregular blotches. Sometimes lesions occur on petioles and stems. Infection causes premature defoliation of leaves. Use disease free seed to reduce the infection. Treat the seed with Captan (captaf) or Thiram (tetramethyl thiurum disulphide) @ 3g per kg seed.

Bacterial leaf spot: This disease is caused by *Xanthomonas campestris pv. cajani*. Angular dark-brown spots appear on the leaf surface and usually concentrated on one side of the midrib. Spots may develop on veins, petioles, main stems and branches. Use disease free seed to reduce infection.

Harvesting and threshing: The crop is ready for harvest when the colour or majority of pods changes from green to dark golden brown and most of the leaves fall on the ground/soil surface. Uproot or sickle harvest the crop and keep in the field for 8-10 days so that the pods dry up completely. Thrash the crop either by beating with sticks or with thrasher. Dry the grains (seeds) in the sun and store. Dried plant of arhar crop is a good source of fuel.

GREEN GRAM (*Vigna radiata* L. Wilczek)

Green gram pod

Green gram seeds

Green gram commonly known as mungbean or golden gram is one of the most important short duration pulse crop of India and ranks third among all pulses grown in India after chickpea and pigeon pea. It is grown in almost all parts of the country. Green gram is primarily a crop of rainy season; however, with the development of early maturing varieties, it has proved to be an ideal crop for spring and summer seasons. Middle of March to last week of June is the most suitable growing period where there is a plenty of life giving sunshine. During this period high temperature and low humidity keep insects and disease infestations at their lowest. Summer season offers an excellent opportunity for raising short duration pulses. Green gram is an excellent source of high quality protein. It contains about 25 per cent protein. It is consumed in different ways as dal, halwa, snack and so many other preparations. Ascorbic acid (Vitamin C) is synthesized in sprouted seeds of green gram with increment in riboflavin and thiamine which acts as rich source of nutrition. Being a leguminous crop, it has the capacity to fix atmospheric nitrogen through symbiotic nitrogen fixation. It also helps in preventing soil erosion. It is also used as green manure crop. Being a short duration crop it also provides an excellent green fodder to the animals. Being a short duration crop, it fits well in various multiple and intercropping systems. After picking of pods, green gram plants may be used as green fodder or can be incorporated as green manure.

Origin and history

De Candolle (1986) believes that green gram has originated in India. According to Vavilov (1926) green gram is a native of India and central Asia. It is grown in these areas since pre-historic period. Zukovskij (1962) is of opinion that *Vigna sublobata* which grow in wild from India to Indonesia is the progenitor of green gram. Maximum diversity occurs among related species, which are well spread in Himalayas extending their distribution to the north-eastern region to south in the peninsular India. The secondary centre of diversity exists in the Indo-Gangetic plains. More than 2000 strains and several botanical varieties of green gram have been recognized. Over 1000 cultivars are grown in China and other Asiatic countries. They differ mainly in habit, size of plant, period of maturity, colour of pods and size and colour of seeds.

Area and distribution

Green gram is grown throughout the southern Asia including India, Pakistan, Bangladesh, Sri Lanka, Thailand, Cambodia, Vietnam, Indonesia, Malaysia and China, etc. It is also grown in the parts of Africa and U.S.A. and has recently been introduced in Australia. The major green gram grown states are Orissa, Maharashtra, Andhra Pradesh, Rajasthan, Madhya Pradesh, Bihar, Karnataka, and Uttar Pradesh.

Classification

A classification of green gram in India was made by Bose in 1932, who grouped them into 40 different types, mainly on the basis of leaf size, flower colour and colour of the seed. All the types have 2n = 24 chromo-somes. It has the following characteristic variations.

Seed colour – Seeds are either green, black brown or yellow

Seed surface – Either dull or shining

Flower colour – Either light yellowish-olive or olive yellow

Pod colour – Ripe pods are either iron grey, olive grey or snuff brown

Improved Varieties of Green Gram released by ICAR

2017-18

Greengram	Area of adoption	Salient features
Pusa 1371	Tripura, Manipur, J and K and Himachal Pradesh	Multiple disease resistance (MYMV, root rot, web blight and Anthracnose)
Pant Mung 8	Uttarakhand	Resistant to MYMV, Cercospora leaf spot and powdery mildew diseases
DGG 1	Karnataka	Resistant to powdery mildew, flaking variety

2016-17

Greengram	Area of adoption	Salient features
MH 318	Haryana	Matures in 60 days; suitable for kharif and summer; small shiny seeds; moderately resistant to mungbean yellow mosaic virus (MYMV)

Utkarsh	Maharashtra	Matures in 65-70 days; suitable for kharif; shiny, green, large seeds; moderately resistant to MYMV and CLS
Yadadri	Telangana	Suitable for kharif and summer; shiny, large seeds; resistant to MYMV
Sri Rama	Telangana	Suitable for rabi and summer; medium large, dull seeds; resistant to MYMV
GBM 1	Karnataka	Suitable for kharif cultivation; shiny, green, small seeds; resistant to powdery mildew
IPM 410-3	Tripura, Manipur, Himachal Pradesh	Days to maturity: 67-80,suitable for kharif; shiny, green seeds; resistant to MYMV and powdery mildew, tolerant to CLS
IPM 205-7	Punjab, Haryana, West Uttar Pradesh, Bihar, Tamil Nadu, Madhya Pradesh and Gujarat	Matures in 50-56 days in summer season; resistant to MYMV, powdery mildew, moderately resistant to CLS
SML1115	Tripura, Meghalaya and Manipur	Suitable for summer season; shiny green and medium seeds; resistant to MYMV, moderately tolerant to web blight and CLS
MSJ 118	Rajasthan	Suitable for kharif; medium large and dull green seeds; moderately resistant to MYMV
ML 2056	Punjab	Suitable for kharif; shiny green and medium seeds; resistant to MYMV, CLS, bacterial leaf spot and anthracnose
RMG 975	Rajasthan	Suitable for kharif; shiny green medium large seeds; moderately resistant to MYMV

Improved Varieties (Main Season) for Punjab

ML 2056: Its plants are erect and medium statured (78 cm). Pod formation is profuse and each pod contains 11-12 seeds. It is tolerant to mungbean yellow mosaic virus, cercospora leaf spot and bacterial leaf spot diseases. It matures in about 71 days. Grains are shining green and medium bold with good cooking quality. The average grain yield is about 11.5 quintals per ha.

ML 818: Its plants are erect and medium statured (75 cm). Each pod contains 10- 11 seeds. It is tolerant to mungbean yellow mosaic virus, cercospora leaf spot and bacterial leaf spot diseases. It matures in about 72 days. Its average grain yield is about 10.5 quintals per ha. Its grains are medium bold, shining green with good cooking quality.

Improved Varieties (Summer Season)

TMB 37: It has erect and determinate plant type with short stature. It bears pods in clusters. It is early and synchronous in maturity (about 60 days). Grains are medium sized, shiny green with good cooking quality. Its average yield is 12.25 quintals per hectare.

SML 832: It has erect, determinate plant type and medium stature. It bears pods in clusters and possesses early and synchronous maturity (about 61 days). Pods are of blackish brown colour at maturity. Each pod contains about 10 grains. Grains are shining green and medium sized with good culinary properties. Its average yield is 11.5 quintals per hectare.

SML 668: It has erect, determinate plant type and short stature. It bears pods in clusters and possesses early and synchronous maturity (about 60 days). Pods are long with thick coat and each pod contains 10-11 seeds. Grains are very bold with good cooking quality. Its average yield is 11.25 quintals per hectare.

Climate

Green gram is a tropical pulse crop largely grown under semi-arid and subtopical climates. With the availability of both short day and long day varieties, it is being grown in summer and *kharif* seasons in northern India. In southern India, it is also grown in winter season. It is considered to be the hardiest of all pulse crops. It requires hot climate and has the capacity to tolerate drought but susceptible to frost, water logging and salinity. It can tolerate high temperature up to 40ºC but does well in temperature range of 30-35ºC. Areas with annual rainfall of 60-75 cm are considered the best for green gram cultivation. But rains during flowering are deterimental as they affect pollination and fertilization. Discolouration of seeds and infestation of fungus occurs with persistent rains. It can be grown successfully up to an elevation of 2000 metre from mean sea level.

Soils

Green gram is grown on a variety of soils ranging from red laterite in southern India to black cotton soils in Madhya Pradesh and sandy soils in Rajasthan. Loam to sandy loam soils are considered ideal for green gram cultivation. Soil should be well drained. Even temporary water logging may damage the crop. Saline-alkali and acidic soils are not suitable for green gram cultivation. Optimum soil pH for best growth is 6.5-7.5.

Intercropping and crop rotations

Green gram can be grown mixed with pigeon pea, sorghum, pearl millet, maize and cotton during *kharif* season. Intercropping of green gram can also be done with spring planted sugarcane. With this technology, additional grain yield of 5-6 quintals per hectare can be obtained without any adverse effect on the performance of main crop of sugarcane. Intercropping of green gram in pigeon pea is more profitable than sole pigeon pea. It is also intercropped with cowpea and sorghum. The following are the important rotations:

- Wheat/Barley - Green gram
- Mustard - Urdbean - Green gram
- Toria - Wheat - Green gram
- Potato - Onion - Green gram

- Rice - Wheat - Green gram
- Maize - Potato - Green gram
- Maize - Potato - Wheat - Green gram
- Rice - Sugarcane - (Ratoon) - Green gram.

Field preparation

Green gram does not require a fine seedbed. One or two ploughings followed by harrowing is adequate for a *kharif* crop. During *rabi*, green gram is usually sown on deep soils on *kharif* fallows that are kept weed free by repeated harrowings. For a summer crop, tillage depends on the time available between the harvest of the standing crop and seeding of green gram. In highly intensive cropping system, minimum tillage is practiced.

Time of sowing

In *kharif* season, sowing is done with the onset of monsoon during June-July. The optimum date of sowing varies from place to place depending on the monsoon rains actually received. Both early and late plantings have adverse affects on the performance of green gram. In early sowings, crop suffers due to rains at maturity and late sown crop suffered due to poor growth and diseases infestation. Sowing in the first fortnight of July has been found to be optimum in the arid zone of Madhya Pradesh while early July sowing recorded the highest seed yield in south-western Punjab. In central and southern India, Madhya Pradesh, Andhra Pradesh, Tamil Nadu and Orissa green gram is sown in *rabi* season too. *Rabi* season green gram is sown during the post monsoon period up to December. Optimum time of sowing for green gram in *rabi* season is from first and second week of October, reduction in yield is possible with advance or delay in sowing. In Uttar Pradesh, Punjab, Haryana, Delhi, Bihar and West Bengal, green gram is grown during summer season also under irrigated conditions. Sowing is done in the middle of March to middle of April. Only short duration varieties maturing in 60-70 days can be grown with adequate irrigation facilities in this season. Timely sowing of summer green gram is more important. The delay in sowing affects the yield adversely. The late-sown crop may be caught by the early monsoon rains at harvesting time which may create a problem for harvesting and threshing and ultimately the yield is reduced. If fields are already lying vacant this crop can be sown at any time after mid February.

Seed treatment

Treat the seeds with Captan or Thiran @ 3.0 g/kg of seed against seed borne diseases. Also inoculate the seed with Rhizolium culture. Mix throughly one packet of culture with moist seed of one acre on a clean pucca floor and let it dry is shade sow immediatly. A yield increase from 12 – 16% is common. The culture and fungicide can be applied simultaneously.

Seed rate and spacing

A seed rate of 15-20 kg ha^{-1} for *kharif* and 25 kg ha^{-1} for summer or spring season has been recommended for green gram. However, as the seed rate is increased, the grain yield of green gram has aslo been found to increase. Rainy season green gram is generally broadcasted. But, sowing of green gram in rows has been found better as it ensures proper spacing for canopy. An inter-row spacing of 25-30 cm and intra-row spacing of 5 cm is optimum for summer and *rabi* season green gram. The *kharif* crop needs a spacing of 30-45 cm between rows and 5-10 cm within plants. Sowing can be done behind the local plough or with seed drill. Seed should be treated with some fungicides (Thiram or Carbendazim) before sowing. For increasing yield, treatment of seed with rhizobium culture should be done after the seed is treated with fungicide. After treatments, seed should not be kept in diffused sunlight. More uptake of nitrogen and increased protein content was possible in wider spacing. High temperatures, bright sunshine hours and low relative humidity where light was probably not a limiting factor, rectangular planting proved better than square planting pattern. For rainy season crop, prefer bed planting to ensure germination.

Nutrient management

Green gram is generally grown wiithout application of fertilizers. However, experiments conducted at various places have revealed that yield of green gram can be increased to great extent with the application of fertilizers. Responses of green gram to N are generally absent or negative. In general, a starter dose of 10-20 kg N/ha has been recommended. Rhizobium culture responses to appllication are quite variable. Responses are comparatively higher in summer season in comparison to *kharif.* Responses of green gram to phosphorus are quite conspicuous. Significant increase in the yield of green gram due to application of 80 kg P_2O_5/ha and higher response of green gram on red soils followed by that on laterite soils. Responses of green gram to potassium are negligible or absent as potassium content in most of the Indian soils is high. Overall, application of 100 kg diammonium phsophate (18-46-0) per hectate is generally recommended for green gram. This fertilizer is comparatively cheaper on nutrient basis and can supply required quantity of N and P to the crop. Fertilizers placement should (total quantity) be done at the time of sowing.

Water management

Rainy season green gram does not require irrigation unless there is a prolonged drought. Draining out of extra water from the field is essential for successful cultivation of green gram. *Rabi* and summer green grams may be grown only in the areas where adequate irrigation facilities are available. In general, 3-5 irrgations are required for green gram in summer season. Increase in moisture supply due to irrigation significantly increases the yield. Giving more than three irrigations does not have any significant impact on yield.

Weed management

Green gram being a dwarf statured crop suffers to great extent if the weeds are not controlled in time. The problem of weeds is more severe in *kharif* as compared to *rabi* and summer seasons. Traditionally weeds are removed by mechanical means. One hand weeding after 20-25 days of sowing followed by another after about 20 days may be sufficient. Weeds can also be controlled with the use of herbicides. Trifluralin (Treflan 48 EC) 0.75-1.0 kg ha^{-1} as pre-em. application is found effective. Alternatively, pre-emergence application of pendimethalin at 0.75 kg ha^{-1} or post-emergence application of haloxyfop-methyl at 0.24 kg ha^{-1} (20 days after sowing) can also be used. Fluchloralin, pendimethalin both at 0.75 kg/ha applied alone and in combination with one hand weeding will result in effective weed control. The specific recommendations of herbicides may vary in different situations owing to weed population, environmental conditions, soil type and crop rotations etc.

Plant Protection (*Main season crop*)

Insect Pests

Jassid, Aphid, Whitefly and Flea beetle: These insects can be controlled by spraying the crop with 940 ml Malathion 50 EC (malathion) or 625 ml Rogor 30 EC (dimethoate) using 200-250 litres of water per ha with a manually operated sprayer.

Whitefly can also be controlled by spraying 100 g Actara 25 WG (thiamethoxam) using 200-250 litres of water per ha. Repeat the spray at 10 days interval, if necessary.

Hairy caterpillar: The body of the caterpillar is covered with hair. The caterpillar eats the green matter of leaves, leaving behind only the midribs. The crop may be totally denuded due to severe attack. When young, they feed gregariously but on few plants in scattered spots. Adopt the following control measures:

- Young larvae are gregarious. They can be destroyed by pulling out the infested plants along with larvae and burying them underground.
- The grown-up caterpillars can be destroyed by crushing them under feet or by picking and putting them into kerosenized water. If the population is high, spray 1.25 litre of Ekalux 25 EC (quinalphos) in 200-250 litres of water per ha with a manually operated knapsack sprayer.

Semilooper: The larvae are green in colour measuring 2-4 cm in length. When touched, they form a loop. The larvae feed extensively on the leaves of *mash* and *moong*. In case of severe damage, the plants are totally defoliated within a few days.

Tobacco caterpillar *(Spodoptera litura)* : It is a polyphagous pest. The small larvae are black whereas grown up larvae are dark green with black triangular spots on body. Its moth lays eggs in masses covered with brown hairs on the lower side of leaves. After hatching, first and second instar larvae feed gregariously and skeletonize the foliage. Later on, the grown up larvae disperse and feed singly. Besides leaves, they also damage buds, flowers and pods.

- **Mechanical control:** Egg masses and young larvae of tobacco caterpillar feeding gregariously should be collected along with leaves and destroyed.
- **Chemical control:** It can be controlled by spraying 375 ml Rimon 10 EC (novaluron) or 2 kg Asataf 75 SP (acephate) or 3.75 litres Dursban 20 EC

(chlorpyriphos) using 250 litres of water per ha with a manually operated knapsack sprayer.

Spray the crop as soon as the pest appears in the field and repeat after 10 days, if necessary.

Blister beetle : Blistle beetle is diurnal and general feeder. Adult beetles are robust with bright black and red stripes on the forewings. When disturbed, the beetles emit a fluid containing cantharidin that causes blisters on human skin. The adult beetles attack *arhar*, *moong*, *mash* and other pulse crops. The major damage is caused at the flowering stage. They feed on tender buds and flowers of the plant, thus preventing grain formation.

Blister beetle can be controlled by spraying Decis 2.8 EC (deltamethrin) @ 500 ml or Kingdoxa 14.5 SC (indoxacarb) @ 500 ml per ha using 200-250 litres of water with manually operated knapsack sprayer. Spray the crop on pest appearance in the evening hours and repeat the spray after 10 days, if necessary.

Pod borer *(Helicoverpa armigera)*: The larvae damage the crop by feeding on leaves, flower buds, flowers, pods and seeds in the pods, thus causing heavy loss in yield. The larvae may be pale green, yellow, brown or black in colour measuring about 3-5 cm in length when fully grown. Larval presence can be observed from damage to plant and from dark green faeces below the plants on the soil. The larvae fall on the ground when plants are shaken vigorously. Spray the crop at the appearance of larvae with Kingdoxa14.5 SC (indoxacarb) @ 500 ml or Asataf 75 SP (acephate) @ 2 kg or Tracer 45 SC (spinosad) @ 150 ml in 250 litres of water per ha using a knapsack sprayer at the start of pod formation and repeat after two weeks, if necessary.

Mite : The mite causes webbing on the underside of the leaves which turn pale. Such infested leaves turn light-brown to dark reddish-brown. It can be controlled by spraying the crop with 375 ml of Rogor 30 EC (dimethoate) in 200-250 litres of water per ha.

Diseases

Yellow mosaic virus: It is a viral disease transmitted by whitefly and is more severe on *kharif moong*. The leaves of the diseased plants develop irregular yellow and green patches. Infected plants bear no or only a few pale pods. The disease can be managed by following measures:

- Rogue out the diseased plants early in the season.
- Grow yellow mosaic virus tolerant varieties of *moong*
- Control whitefly with recommended "pesticides".

Cercospora leaf spot: It is caused by *Cercospora cruenta* and *C. canescens*. The disease spots are circular, brown and necrotic which coalesce to cover bigger area and cause defoliation. Intermittent rains are favourable for disease development. Adopt the following measures to check the disease.

- Treat the seed with 3 g of Thiram (tetramethyl thiurum disulphide) per kg of seed.

- Grow disease resistant varieties of *moong*.
- Spray the crop with Zineb 75 WP (dithane Z-78) @ 1 kg in 250 litres of water per ha at 10 days interval twice or thrice at disease appearance.

Root rot: Root rot caused by *Macrophomina phaseolina* produces dark lesions on leaves, branches, stems and roots. The tissue of the affected portion become weak and shred easily. Pycnidia can be seen on the affected portion. For control, treat the seed before sowing with Thiram (tetramethyl thiurum disulphide) @ 3 g/kg seed.

Anthracnose: It is caused by *Colletotrichum lindemuthianum*. Dark brown spots (horse shoe shaped) appear on the leaves and other parts of the plant. Under severe conditions lesions rapidly coalesce to girdle stems, branches, peduncles and petioles. Adopt control measures given under Cercospora leaf spot.

Bacterial leaf spot: Bacterial leaf spots caused by *Xanthomonas campestris pv. phaseoli* are circular to irregular and brown. Grow resistant varieties.

Web blight: It is caused by *Rhizoctonia solani*. It starts from leaf margins or petioles or young branches. Eventually, the tops of plants become blighted and patches of such plants are conspicuosly seen in the field. Whitish web-like growth develops on leaves in humid weather. Dark brown sclerotia develop on infected tissue. Infection on crop comes from the weeds in the field. Keeping the field weed-free helps to check the disease.

Mosaic mottle: The affected plants show broad patches of light and dark green leaf areas giving the leaf a mosaic pattern, blistering of the leaf blade, stunting and flower shedding. Use of healthy seeds free from seed-borne inoculum, growing of resistant varieties and measures to control vectors such as aphids, white flies etc. reduce the incidence of this disease.

Plant Protection Measures (Summer Season)

Insect Pests

Thrips: Summer moong crop is severely attacked by the thrips which is very small, dark brown insect, found in flowers and cause flower-drop, deformation of pods, deterioration of grain quality and ultimately high reduction in yield. Sometimes, there may be complete failure of the crop. Spray the crop at bud initation stage with 1.5 litre Hostathion 40 EC (triazophos) or 250 ml of Rogor 30 EC (dimethoate) or Malathion 50 EC (malathion) in 200 to250 litres of water per hectare with manually operated knapsack sprayer.

Pod borer (*Helicoverpa armigera*): Larvae of the borer feed on leaves, flowers, pods and seeds in pods, thus, causing heavy loss in yield. The larvae may be pale green, yellow, brown or black in colour measuring about 3-5 cm when full grown. Larval presence can be observed from damage to plant and from dark green faeces below the plants on the soil. Spray the crop at the appearance of larvae with 150 ml Tracer 45 SC (spinosad) or 500 ml Kingdoxa 14.5 SC (indoxacarb) or 2 kg Asataf 75 SP (acephate) in 200 to 250 litres of water per hectare with manually operated knapsack sprayer. Repeat the spray whenever necessary.

Tobacco caterpillar (*Spodoptera litura*): It is a polyphagous pest. The small larvae are black whereas grown up larvae are dark green with black triangular spots on body. Its moth lays eggs in masses covered with brown hairs on the lower side of leaves. After hatching, first and second instar larvae feed gregariously and sketetonize the foliage. Later on, the grown up larvae disperse and feed singly. Besides leaves they also damage buds, flowers and pods. It can be managed by the following measures :

- **Cultural Control:** Ensure timely sowing of the crop. Control the weeds, particularly itsit/chapatti as it acts as an alternate host for the tobacco caterpillar.
- **Mechanical Control:** Egg masses and young larvae of tobacco caterpillar feeding gregariously should be collected along with leaves and destroyed.
- **Chemical Control:** It can be controlled by spraying 375 ml of Rimon 10 EC (novaluron) or 2 kg of Asataf 75 SP (acephate) or 3.75 litres of Dursban 20 EC (chlorpyriphos) using 250 litres of water per hectare with manually operated knapsack sprayer. Spray the crop as soon as the pest appears in the field and repeat after 10 days, if necessary.

Diseases

Mungbean yellow mosaic disease : It is a viral disease transmitted by whitefly. The leaves of the diseased plants develop irregular yellow and green patches. Infected plants bear no or only a few pale pods. Incidence of this disease is very less in timely sown crop. The disease can be controlled by the following measures :

- Rogue out the affected plants early in the season.
- Grow yellow mosaic virus tolerant varieties Mash 1008 and Mash 218.
- Whitefly can be controlled by spraying 100 g Actara 25 WG (thiamethoxam) or 1.5 litre Hostathion 40 EC (triazophos) using 200-250 litres of water per hectare.

Root rot: Root rot caused by Macrophomina phaseolina produces dark lesions on leaves, branches, stems and roots. The tissues of the affected portion become weak and shred easily. Pycnidia can be seen on the affected portion. For control, treat the seed before sowing with Thiram (tetramethyl thiurum disulphide) @ 3 g per kg seed.

Rhizoctonia blight: It is caused by Rhizoctonia solani. It starts from leaf laminae or petioles or the young branches. Eventually, the top of plants become blighted and patches of such plants are conspicuously seen in the field. Whitish web like growth develops on leaves in humid weather. Dark brown sclerotia develop on infected tissue. Infestation on crop comes from the weeds in the field. Keeping the field weed free helps to check the disease.

Harvesting and threshing

Shattering of pods is a great problem with pulse. Therefore, picking should be done as soon as the pods mature. Harvesting should be completed in two to three pickings. The varieties which are quite synchronous in maturity require only two pickings or sometimes the whole crop may be harvested with sickle. The pods or whole crop after complete drying should be threshed manually or mechanically. In case larger area is under cultivations combine harvesters can be used. Spray of paraquat @ 2.5 l/ha (commercial product) should be used to drop the leaves and after leaf fall use combines. Quality of produce will remain same. The produce is cleaned and sun dried to about 12 % moisture content.

Yield

A well managed crop may produce 12-15 quintals of grain per hectare.

BLACKGRAM (*Vigna mungo* L. Hepper)

Blackgram plant **Blackgram seeds**

Black gram also known as urdbean, mash, etc. is another important short-duration pulse crop grown in many parts of india. This crop is grown in cropping systems as a mixed crop, catch crop, sequential crop besides grown as sole crop under residual moisture conditions after the harvest of rice and also before and after the harvest of other summer crops under semi-irrigated and dryland conditions. Its seeds are highly nutritious with protein (25-26%), carbohydrates (6%), fat (1.3%), minerals and vitamins. It is the richest in phosphoric acid among pulses. Seeds are used in the preparation of many popular dishes. It is one of the most important components in the preparation of famous south Indian dishes, viz. dosa, idli, vada etc. It is used as a nutritive fodder specially for milch cattle. It is also used as a green manuring crop. Plants of black gram possesses deep root system which binds soil particles and thus preevnt soil erosion. Black gram being leguminous, has the capacity to fix atmospheric nitrogen and thus helps in restoring the soil fertility.

Origin and history

Urd or black gram is a native of India and originated from *Phaseolus sublobatus*, a wild plant. There is a mention of urd in vedic texts such as Kautilya's Arthasasthra' and 'Charak Samhita'. From India it spreads to many countries like African, European, American and Asian countries.

Classification

There are two main types of black gram:

Large-seeded, early-maturing and black seeded types (*Vigna mungo* var. *niger*).

Small seeded, late-maturing types with colour ranging from brown, olive green to grey to mottled (*Vigna mungo* var. *viridis*)

Climate

Black gram requires a hot and humid growing seson. Being s tropical crop, it tolerates high temperatures. It is cultivated in summer and *kharif* seasons in northern india and in both *kharif* and *rabi* seasons in eastern and southern parts of the country. The optimum temperature for better growth ranges between 25 and 35°C but it can tolerate up to 45°C. It is a hardy and drought resistant crop and can be grown in areas receiving 650 mm rainfall or less. It is sensitive to cloudy weather and cannot tolerate frost. Heavy and continuous rains at the time of germination and flowering are harmful for the crop.

Soil

It is grown on a variety of soils ranging from a sandy soil to heavy black cotton soils. It thrives well in relatively heavier soil. The most ideal soil would be a well drained loam with a pH range from 6.0 to 7.5. Heavier soils with more water retention capacity are considered ideal for its cultivation. Loam to clay loam soil with neutral pH are the best suited soils for urdbean. It cannot be grown successfully on saline alkali and acidic soils. Urdbean is susceptible to waterlogging. Therefore, soil should be well drained particularly in rainy season.

Crop rotations

Blackgram is grown mixed with sorghum, maize, pearl millet and cotton crops during *kharif* season. Black gram is profitably grown as intercrop in widely spaced crops like pigeon pea and sugarcane. One or two rows of black gram can be planted between two rows of pigeon pea planted 75 centimeter apart. Black gram will mature in 80-85 days. Black gram yields 4-7 quintals of grain without affecting the growth and yield of pigeon pea. Similarly intercropping of black gram can be done with spring planted sugarcane. Planting of two rows of black gram between two rows of sugarcane had given an additional yield of 5-7 quintals per hecatre without any adverse effect on the performance of cane crop. The important crop rotations with black gram in northern India are as given below:

- Wheat – Black gram
- Barley – Black gram
- Maize – Potato – Black gram
- Maize – Toria – Black gram
- Maize – Wheat – Black gram

- Rice – Wheat – Black gram
- Black gram – Wheat – Black gram

Improved Varieties of Black Gram released by ICAR

2017-18

Blackgram	Area of adoption	Salient features
Tirupati Minumu-1	Andhra Pradesh	Suitable for rabi cultivation, tolerance to YVMV

2016-17

Blackgram	Area of adoption	Salient features
Indira Urd Pratham	Chhattisgarh	Early duration; synchronous maturity; resistant to MYMV and powdery mildew
LBG 787	Andhra Pradesh, Tamil Nadu, Karnataka and Odisha	Suitable for rabi and summer cultivation; matures in 75-80 days; resistant to MYMV
PDKV Black gold	Maharashtra	Matures in 70-80 days; resistant to stem necrosis, anthracnose and root rot

Improved Varieties for Punjab

Main Season:

Mash 114: This variety is recommended for the whole Punjab state. Its plants are dwarf, erect and compact. It matures in about 83 days. Podding is profuse and each pod contains about 6-7 seeds, which are bold, black and possess very good culinary properties. Average grain yield is about 9.0 quintals per ha. It is fairly resistant to yellow mosaic virus, bacterial leaf spot and cercospora leaf spot diseases.

Mash 338: This variety is recommended for the whole Punjab state. Plants are dwarf, erect and compact. It matures in about 90 days. Podding is profuse and each pod contains about 6 seeds, which are bold, black and possess very good culinary properties. Average grain yield is about 8.75 quintals per ha. It is tolerant to yellow mosaic virus, bacterial leaf spot and cercospora leaf spot diseases.

Summar Mash

Mash 1008: It has erect and compact plant type. It is a short statured (25 cm) variety. It matures in about 72 days. Pod bearing is profuse and each pod contains 6-7 seeds. It is fairly tolerant to yellow mosaic virus and leaf crinkle virus. Average grain yield is about 11.25 quintals per hectare. Grains are bold, blackish in colour, contain about 24 per cent protein and possess good culinary properties.

Mash 218: It has erect and compact plant type with short stature (30 cm). It matures in about 75 days. Pod bearing is profuse and each pod contains 6-7 seeds. Average grain yield is about 10 quintals per hectare. Grains are bold, dull black, contain about 23 per cent protein and possess good culinary properties.

Plant Protection

As for main and summer season crop of moong

Field preparation

Although black gram does not require fine tilth, better land preparation results in good germination and uniform stand of the crop. It also helps the crop to achieve higher seedling vigour and early growth. Heavy soils need more cultivations with either plough or disking followed by harrowing and land levelling while the light soils will be ready with less cultivation. One deep cultivations, two to three harrowings and one planking would give better seedbed for blackgram. During *kharif* land should be properly levelled and drainage should be provided as any water stagnation affects the crop adversely. When the sowing is done in the rice fallows seeding can be done without cultivation as it requires time to come to proper condition after tillage. In post-rainy season crop, soil moisture conservation practices improve the crop yields. Bed planting (sowing) is beneficial during rainy season after the crop establishes one or two intercultivations, check the evaporation losses of moisture and reduce the weed population.

Time of sowing

Kharif black gram is sown with the onset of monsoon in the last week of June to early July. This crop is mainly grown under rainfed conditions. Early planting in first week of July results in higher yield and any delay in sowing beyond this date causes reduction in yield. In case of summer black gram, sowing may be started in the last week of February till the end of March. Plantings beyond March may be caught by monsoon rains in the second fortnight of June and result in drastic yield reduction. During *rabi* season, sowing during mid-August and mid-September results in higher yield. Late sowings are unfavourable due to low temperatures in the winter.

Seed rate, spacing and seed treatment

During rainy season, the crop is generally sown by broadcasting the seed followed by planking. Sowing could be done in rows with the help of local plough or tractor drill Sowing is done in wider rows because vegetative growth of the plants is more due to favourable climate during *kharif* season. An inter-row spacing of 30-45 cm and intra-row spacing of 5-10 cm is optimum for *kharif* crop. A seed rate of 15 to 20 kg/ha is sufficient depending upon the variety. Black gram is an important crop in rice fallows in South India. The maintenance of population is very difficult and is the major constraint in rice fallows. For black gram, 40 kg seed rate/ha was optimum. Soaking of seeds before planting did not show beneficial effect over dry seeds. During summer season, vegetative growth of black gram is comparatively less, hence higher plant population by reducing the spacing and increasing the seed rate is desirable. A seed rate of 40 to 50 kg/ha is recommended for sowing during summer. Sowing can be done behind local plough or with the help of seed-drill. Before sowing seed should be treated with Agrosan GN or Thiram @ 2.5g/kg of seed. Seed should also be inoculated with suitable Rhizobium culture, if black-gram is being taken for the first time in the field or after a long duration.

Seed treatment

It is same as for moong.

Fertiliser management

The crop requires 15-20 kg/ha nitrogen as basal dose to meet the growth requirements during early growth stages particularly on soils which are poor in organic matter. The combined application of nitrogen and phosphorus further increased seed yield compared with individual applications. The seed inoculation with appropriate *Rhizobium* culture also increases seed yield and uptake of nutrients.

Application of phosphorus has been found to be quite effective in increasing the seed yield of black gram. A dose of 40-50 kg P_2O_5 per hectare has been found optimum in most of the soils. Black gram does not respond to applied potassium and application of phosphorus and potassium should always be made on soil test basis. Application of sulphur at the rate of 30 kg ha^{-1} either as gypusm, pyrites or elemental sulphur increases yield.

Water management

The crop cannot tolerate long dry spells and therefore, irrigations are very useful for black gram. The crop should be irrigated at flowering and pod filling stages but the water should not accumulate in the field. One pre-sowing irrigation at the time of seedbed preparation is required to ensure good germination, if the land is dry. Blackgram is grown purely as a rainfed crop during *kharif* and on residual moisture in rice-fallows during *rabi*. In case there is a long dry spell, the crop is benefited if one light irrigation at podding/seed filling stage is applied. On the other hand, number and frequency of irrigation depends upon the soil type and weather prevailing during the growth period. The summer season crop should get irrigation at an interval of 10-15 days. From flowering to pod development stage, there is a need of sufficient moisture in the field. Excess water at any stage affects the crop adversely.

Weed management

The first 4-5 weeks after sowing are critical for crop-weed compe-tition. The extent of damage done by weeds can be as high as 50-60%. Good seedbed preparation, one or two intercultivations or manual weeding 3-5 weeks after sowing may effectively control the weeds. Fluchloralin at the rate of 1 kg ha^{-1} as pre-plant incorporation or pendimethalin at the rate of 0.75 kg ha^{-1} as pre-emergence application are effective treatments. Integration of one hand weeding with these herlicides results is improved yield. Post-emergence application of Sethoxydim at the rate of 1 kg ha^{-1} 10 days after sowing is also effective in blackgram. When blackgram is grown as an intercrop with sorghum, pigeon pea, pearlmillet, etc., the interculture given to the main crop is adequate. When sown as a sole crop 1 or 2 weedings are required in the initial stages to keep the field free from weeds.

Plant Protection

Similar to green gram.

Harvesting and storage

Blackgram should be harvested when most of the pods turn black. Over maturity may result in shattering. Harvested crop should be dried properly on threshing floor for few days and then threshed. Threshing can be done either manually or by trampling under the feet of bullocks, or with tractor operated thrashers.

Yield

A well-managed crop may produce about 15-20 quintals of grain per hectare.

COWPEA (*Vigna unguiculata* L. Walp)

Cowpea fruiting and flowring plants **Cowpea seeds**

Cowpea is one of the most important legume fodder and food crop in the semi-arid tropics covering Asia, Africa, Southern Europe and Central and South America. It is commonly known as crowder pea, black-eyed pea, China pea, marble pea, southern pea in different countries of the world. A drought tolerant and warm weather crop, cowpeas are well-adapted to the drier regions of the tropics, where other food legumes do not perform well. It also has the useful ability to fix atmospheric nitrogen through its root nodules, and it grows well in poor soils with more than 85% sand and with less than 0.2% organic matter and low levels of phosphorus. In addition, it is shade tolerant, and therefore, compatible as an intercrop with maize, millet, sorghum, sugarcane, and cotton. This makes cowpea an important component of traditional intercropping systems, especially in the complex and elegant subsistence farming systems of the dry savannas in sub-Saharan Africa. In India, cowpea commonly known as lobia is one of the important *kharif* pulses grown for grain, forage, green manure and as exhaustive crop in soil fertility and nutrition experiments in agronomic research. Cowpea seeds are highly nutritious with high protein (23-24%), carbohydrates, minerals and vitamins. It is also a rich source of calcium and iron.

Origin and history

Cowpea is probably a native of central Africa where almost all the wild forms are found. It has been cultivated since very ancient time in the Mediterranean region be the Greeks, Romans and Spaniards. It has now been introduced in many countries throughout the world. It is also claimed to be indigenous to India.

Classification

The cultivated forms of cowpea are classified as follows:

Sub-genus: *Vigna*

Vigna unguiculata (L.) Walp

(i) Sub-species: *unguiculata*

Cultivars can be roughly grouped into 4 cultigroups

(a) cv.gr. *biflora* and (b) cv.gr. *catjang*

These bear small seeded erect pods and are grown mostly in south and south-east Asia.

(c) cv.gr. *sesquipedalis*

Also called yard long bean or asparagus bean. Very long drooping pods eaten as green vegetable.

(d) cv.gr. *textilis*

These have very long, stout peduncles, which are used for extracting textile fibre.

(ii) Sub-species: *dekentiana* (Harms) Verdc.

var. *dekentiana*: most common wild form

var. *mensensis* (Schweinf.) M.M et S: Large flowers and long calyx lobes

var. *pubescens* (R. Wilczek.) M.M et S: Short pubescence on leaves and stalks

var. *protracta* (E. Mey.) Verdc.: Small leaflet with long pubescence

(iii) Sub-species: *tennis* (E. Mey.): Small glabrescent lobed leaflets

(iv) Sub-species: *stenophylla* (Havery.): Narrow nastily leaflets

Improved Varieties of Cowpea released by ICAR (for seed)

2017-18

Cowpea	Area of adoption	Salient features
Phule Rakhumai	South Zone	Moderately resistant to Cercospora leaf spot
Tirupati Cowpea 1	Andhra Pradesh, Karnataka, TN and Odisha	Moderately resistant to dry root rot and YMV

DC-15	Karnataka, AP, Tamil Nadu and Kerala	Moderately resistant to dry root rot and YMV
Pant Lobia 5	UP, Uttarakhand, Rajasthan, Gujarat, Karnataka and Kerala	Resistant to YMV

2017-18

Cowpea	Area of adoption	Salient features
Pant Lobia 3	All cowpea- growing regions	Resistant to YMV and bacterial blight; brown seeds
Phule Vithai	Maharashtra	Early maturing; moderately resistant to collar rot and leaf spot

Improved Varieties of Cowpea for fodder released by ICAR

2017-18

Forage Cowpea	Area of adoption	Salient features
TNFC 0926	Tamil Nadu	Resistant to cowpea yellow mosaic virus and moderately resistant to root rot disease

2016-17

Forage cowpea	Area of adoption	Salient features
Bundel Lobia 4	North East Zone	Good fodder quality
Vijaya	Telangana and Andhra Pradesh	Suitable for kharif and rabi seasons; short duration variety with high leaf stem ratio (0.72) and photoinsensitive
MFC 09-1	Kerala, Tamil Nadu, Karnataka Andhra Pradesh and Telangana	High green forage and dry matter; resistant to yellow mosaic virus and leaf spot

Climate

Cowpea can be grown in all tropical and subtropical climates. Being a warm weather crop it can withstand a considerable degree of drought. But under very dry conditions the plants produce a poor crop. Germination of cowpea is better between 12 and 15°C temperatures. The crop thrives best between 27 and 35°C temperatures. It can also grow under shade of trees but cannot tolerate cold or frost. It can tolerate heat and relatively dry conditions and can be grown with less rainfall and under more adverse conditions.

Soil

It can be grown on a variety of soils. Well drained loam or slightly heavy soils are best suited for cowpea crop. Saline or alkaline soils are not suited to cowpea cultivation.

Cropping systems

Cowpea can be grown in rotation with many crops. The most common cropping systems are as below:

For grain or vegetable purpose

- Maize – Wheat – Cowpea
- Rice – Wheat – Cowpea
- Cowpea – Wheat – Proso millet
- Maize – Potato – Cowpea
- Maize – Toria – Wheat – Cowpea

For fodder purpose

- Sorghum + Cowpea – Berseem – Maize + Cowpea
- Maize – Oat – Maize + Cowpea
- Maize – Berseem – Maize + Cowpea
- Sorghum – Berseem – Maize + Cowpea

Field preparation

Clod free uniform deep soil is necessary for establishing good crop. The stubbles and other residues of previous crop should be picked up as far as possible to have a clean and smooth seedbed. Field should be prepared by giving two or three cross harrowing followed by planking. For summer season crop, give a pre-irrigation immediately after harvesting of *rabi* crop. When the field comes in workable condition, prepare it by giving two or three harrowings and levelling.

Time of sowing

Suitable time for sowing summer cowpea crops is March to April. Summer crop is raised under irrigated conditions. The sowing for *kharif* crop starts from middle of June and extends up to the end of July. There are slight variations in exact date of sowing. In some areas, June sowing gives higher yield while in others July sowing produces higher yield 'depending upon climatic condition and pattern of rainfall. Any delay in sowing beyond July increases the incidence of rust disease in cowpea. In southern India, it is also grown as *rabi* season in October and November.

Seed rate and method of sowing

For grain or vegetable purposes, 20-25 kg seed is required for sowing of one hectare pure crop. When sown mixed with other crops, the seed rate is reduced proportionately. For fodder purposes the seed rate should be 35-45 kg per hectare. Variation in seed rate depends on several factors such as method of sowing, availability of soil moisture and the seed size of a variety. For green manuring crop, sown by broadcasting, the seed requirement is 35-40 kg per hectare.

The row spacing of 35-45 cm and plant to plant distance of 8-10 cm may be maintained for *kharif* planting. The row spacing in spring and summer planting should be kept 25-30 cm as the plants do not grow much. Seeds should be treated with Thiram at the rate of 3 g per kg of seed before sowing. The crop can be sown either with drill or behind a plough.

Fertilizer management

Cowpea is highly responsive to fertilizer application. However, the amount of fertilizers to be applied should be based on the soil fertility status and moisture availability conditions. Although being a leguminous crop it is capable of fixing atmospheric nitrogen, it needs a small quantity of nitrogen as starter dose for early growth period particularly on soils which are poor in organic matter. Such soils should get about 15-20 kg of nitrogen per hectare as a starter dose. Application of higher dose of nitrogen may reduce nodule number and nodule growth and thus adversely affect the nitrogen fixation capacity. Seed inoculation with appropriate *Rhizobium* culture is found to significantly increase the seed yield. Application of phosphorus is very important as it promotes proper development of roots and function of *Rhizobium* bacteria. Apply 50-60 kg of P_2O_5 per hectare in the soil before sowing in case soil test values are not available. It is advisable to apply phosphatic fertilizers according to soil test recommendations. The fertilizer should be applied by drilling at the time of sowing in such a way that it is placed about 5-7 cm below the seed. The combination of nitrogenous and phosphorus fertilizers along with irrigation results in better response. Application of potassic fertili-zers in potash deficient soils has been found to affect the performance of cowpea. Application of balanced fertilizers up to 25 kg N, 50 kg P_2O_5 and 25 kg K_2O ha^{-1} observed higher growth and yield attributes and yield over no fertilizer.

Water management

The rainy season crop requires no irrigation but good drainage is essential. The early sown rainy season crop may need one or two irrigations in the pre-monsoon period. For raising summer crop, five to six irrigations may be given. During summer, because of high temperature and low relative humidity more irrigations are needed as compared to *kharif* crop. The number and frequency of irrigation depend upon the soil type and weather prevailing during the growth period. Generally, the crop should get irrigation at an interval of 10-15 days.

Weed management

Despite cowpea being a fast-growing crop and capable of suppressing weeds, many weeds are known to occur in cowpea when grown in *kharif* season. Weeds if not controlled can reduce the yield of cowpea to a great extent. Weed competition decreases the cowpea harvest index and increases plant height and lodging. The critical period of weed competition in cowpea lies between 10 and 40 days after emergence of the crop. Cultural practices like clean cultivation, use of weed-free seeds, optimum plant population etc. reduce the weed problem to a great extent. One hand weeding 30 days after sowing and/or one intercultivation are required to check the weeds. During rainy season, weeds can also be controlled by the use of herbicides. Application of pendimethalin at the rate of 0.75 kg ha^{-1} as pre-emergence or fluchloralin at the rate of 1 kg ha^{-1} as pre-plant incorporation check the weed growth effectively.

In wider row crops like maize, one row of cowpea can be sown in between two rows of crop for cultural control of weeds. Cowpea can be harvested for foddes purposes, 20–25 days after sowing along with weed plants.

Plant Protection

Insect pests

Jassid and black aphid: These insects can be controlled by spraying 500 ml of Malathion 50 EC in 200 litres of water per hectare.

Bihar hairy caterpillar (Bhaboo kuta): It attacks cowpea from August to November. To keep cowpea free from attack of this pest, plant a row of til (Sesamum) around cowpea field at the time of its sowing. Females prefer to oviposit on til rows from where larvae in gregarious phase should be collected and destroyed mechanically.

Stored Grain Insects: Pulse beetle (Dhora) causes severe damage to stored cowpea grains.

Diseases

Seed rot and Seedling mortality: It is caused by various seed-borne microflora. The infected seeds are shrivelled and discoloured. Infected seedlings are killed before these emerge out of soil and cause poor stand of the crop. For its control treat seed with Bavistin 50 WP @ 2 g/kg seed before sowing.

A large number of fungal, viral and bacterial diseases are known to cause damage in cowpea. Symptoms and control measures of important diseases are given below:

Bacterial blight: It is a seed borne disease. The germinating seedlings turn brown-red and die. Irregular to round spots appear on leaves, brown in colour with chlorotic halos, later spread to stems. Stem may break. Pods are also infected leading to shriveled seeds. For its control, use healthy seeds from disease free field. In case of severe infection, crop may be sprayed with 0.2% Blitox.

Cowpea mosaic: This is a viral disease transmitted through vectors. While the yellow mosaic virus is transmitted through aphids, the golden mosaic virus spreads through white flies. The affected leaves become pale yellow and exhibit mosaic, vein banding symptoms. The affected leaves gets reduced in size and show puckering. Pods are also reduced and become twisted. Spraying of 0.1% oxydemeton methyl (metasystox) or any other systemic insecticides helps in preventing the spread of disease by controlling vectors.

Powdery mildew: Symptoms first start from leaves and then spread to stem, branches and pods. Symptoms start with white powdery growth on leaves which may coalesce and cover the whole leaf with the white powdery growth. This white growth consists of the fungus and its spores. Affected leaves become twisted and smaller in size. The disease can be controlled by spray of wettable sulphur like Sulfex, Elosal or Hexasual at the rate of 2-3 kg ha^{-1} in 300 litres of water.

Rust: This disease affects the leaves, pods and sometimes new shoots. Symptoms on the leaves are very clearly visible and start from the lower surface of leaf where very small white pustules are found. These pustules contain uredia of the fungus. Brown coloured urediospores come out of these uredia. At the end of the season these urediospores may be replaced with black coloured tilia. Spray the crop with Mancozeb 75 WP at the rate of 2 kg ha^{-1} in 300 litres of water.

Harvesting

Cowpea is a dual purpose crop. Mostly it is grown for fodder however, it can also be raised for seed production. Cowpea gives highly nutritive fodder and is ready for cut within 55 to 65 days of sowing. For seed crop, sowing may be adjusted in such a way that its flowering do not coincide with rainfall.

11

SUGAR CROPS

Sugarcane and Sugarbeet are important sugar crops which are being discussed as under:

SUGARCANE (*Saccharum officinarum*)

Plant **Crop**

Globally, 60% of white crystal sugar comes from sugarcane and 40% from sugarbeet. The per capita consumption of white sugar is 14.2 kg and that of Jaggery 18.2 kg. Domestic sugar consumption is 13.5 mt with an export around 1.0 mt.

Origin, distribution, area and production

The origin of *Saccharum officinarum* is the Indo-Myanmar-China border with New Guinea as the center of diversity. The origin of *Saccharum barbera* is North India and that of *Saccharum sinense* is China. Globally sugarcane is cultivated in about 26.77 million ha with a production of 1890.66 mt during 2016 (FAO STAT 2016). In India sugarcane occupies 4.95 million ha area with a production of 348.45 million tonnes during 2016 (FAO STAT 2016).

Climate, soils and tillage

Sugarcane is grown on 8°-33°N and mostly from sea level to 1500 M. A warm summer growing season with adequate rainfall/irrigation is an ideal climate for sugarcane production. Growth of sugarcane ceases at mean minimum temperature below 12°C. Sugar recovery and productivity are higher in Tropical India than in Sub-tropical India.

Soils: Moderately heavy and medium deep (1-2 m) loam are better suited than heavy, light or shallow soils. Alluvial, black and red soils are the major sugarcane soils in India. Sugarcane can tolerate soil pH from 4 to 9.

Tillage: Common method of land preparation is to plough the land 2-4 times with traditional plough or tractor followed by clod breaking to pulverize the soil.

Growth and development

Tiller production increase exponentially until a maximum of 20-30 stalks m^2 at 4-6 months age of the sugarcane. Sugarcane growth is higher during monsoon than in other months.

Improved Varieties of Sugarcane released by ICAR

2017-18

Sugarcane	Area of adoption	Salient features
Revathi	Coastal areas of AP and TN	Moderately resistant to red rot, smut and tolerant to moisture stress
Buddhi 2003 A 255	Andhra Pradesh, Odisha and TN	Moderately resistant to red rot, tolerant to water logging
Co 09004	TN,Kerala, AP, Telangana, Karnataka, Gujarat, Maharashtra, MP and Chhattisgarh	Moderately resistant to red rot, resistant to smut; early maturing, high yielding, non-lodhing, smut resistant and tolerant to drought and salinity
Co 9022	Haryana, Punjab, Rajasthan, Uttrakhand and Uttar Pradesh	Moderately resistant to red rot, smut and resistant to wilt;high sugar recovery, non-lodging
GNS 9	Gujarat	Moderately resistant to red rot, smut and resistant to wilt; red rot and whip smut, high sugar content, good ratiooning ability
SNK-632	Karnataka	Tolerant to moisture stress; yield 185 tonnes/ha, suitable for 'adasali' planting
CoVSI 03102	Sugarcane growing areas of high rainfall zone of Maharashtra	Tolerant to drought, non-lodging, resistant to smut, brown eye spot disease, cane yield 156.37 tonnes/ha and sugar yield 22.79 tonnes/ha
CoC 25	East coast zone	High yielding, fast growing; good ratooner; moderately resistant to red rot; suitable for early and special season, cane yield 145.0 tonnes/ha, CCS 12.77%

2016-17

Sugarcane	Area of adoption	Salient features
Karan 11	Punjab, Haryana, Uttar Pradesh (western and central), Uttarakhand and Rajasthan	Suitable for winter harvesting; moderately resistant to red rot and smut ,resistant to wilt
CoBln 02173	Assam, Meghalaya, Manipur, Mizoram, Tripura, Arunachal Pradesh, Sikkim and Nagaland	Suitable for October planting; moderately resistant to red rot and borers
Co 0212	Tamil Nadu and Puducherry	Tolerant to drought, resistant to red rot and smut
Gujarat Sugarcane 4	South Gujarat	Moderately resistant to red rot and wilt
Gujarat Sugarcane 5	South Gujarat and Peninsular zone	Moderately resistant to red rot and wilt
Gujarat Sugarcane 6	South Gujarat	Moderately resistant to red rot and wilt
Gujarat Sugarcane 7	South Gujarat	Moderately resistant to red rot and wilt
Gujarat Sugarcane 8	South Gujarat	Moderately resistant to red rot and wilt,low incidence of pests and diseases
CoPb91	Punjab, Haryana, Uttar Pradesh (western and central), Uttarakhand and Rajasthan (North West Zone)	Tolerant to lodging and non-flowering, moderately resistant to red rot and moderately susceptible to smut

Improved Varieties of Spring Sugarcane for Punjab

Early-Maturing Varieties

CoPb 92: The canes of this variety are tall, medium thick and purple green in colour. Its juice contains 16-17% sucrose in the month of November and 18% in December. It is a good ratooner. It is tolerant to most of the prevalent pathotypes of red rot disease and frost. Quality of *gur* is also good. The average cane yield is about 838 quintals per ha.

Co 118: It has thick, purple green coloured canes. Its juice contains 16% sucrose in the month of November and 17% in December. It has field resistance to the prevalent pathotypes of red rot disease. It is a shy tillering variety and an average ratooner. Its *gur* quality is excellent. The average cane yield is about 800 quintals per ha.

CoJ 85: It is a shy tillering variety with thick* green coloured canes. Its juice contains 16-17% sucrose in the month of November and 18-18.5% in December. It is tolerant to most of the prevalent pathotypes of red rot disease and is tolerant to frost. It is an average ratooner and susceptible to red stripe disease. Due to its heavy canes it is prone to lodging, hence requires proper earthing up. The average cane yield of plant crop is about 763 quintals per ha.

CoJ 64: This variety is a good germinator, with profuse tillering and medium-compact growth. Its canes are medium thick, greenish yellow and solid. The sucrose content in juice in November is 16-17%. It has become highly susceptible to red-rot disease and also to top borer. Its quality of *gur* is excellent. The average cane yield is about 750 quintals per ha.

Mid-Late Maturing Varieties

CoPb 93: The canes of this variety are tall, thick and yellowish white in colour. Its juice contains 17% sucrose in the month of January and 19% in March. It is a good ratooner. It is tolerant to the prevalent pathotypes of red rot disease. Quality of *gur* is also good. The average cane yield is about 975 quintals per ha.

CoPb 94: The canes of this variety are tall, thick, cylindrical and yellowish green in colour. Its juice contains 16% and 19% sucrose in the months of January and March, respectively. It is a good ratooner. It is tolerant to the prevalent pathotypes of red rot and smut diseases. The average cane yield is about 1000 quintals per ha.

Co 238: The canes of this variety are tall, medium thick and yellow green in colour. Its juice contains 17% sucrose in the month of January. It is susceptible to top borer and tolerant to the prevalent pathotypes of red rot disease. It is a good ratooner. Quality of *gur* is also good. The average cane yield is about 913 quintals per ha.

CoPb 91: The canes of this variety are tall, thick and yellowish green in colour. Its juice contains 17% sucrose in the month of January. It is tolerant to the prevalent pathotypes of red rot disease. It is a good ratooner. Its average cane yield is about 1025 quintals per ha.

CoJ 88: The canes of this variety are tall, medium thick and greyish green in colour. Its juice contains 17-18% sucrose in the month of December. It tends to behave as early-mid in maturity and is also suitable for saline water conditions. It is tolerant to the prevalent pathotypes of red rot disease. It is an excellent ratooner. It is good for co- generation. Quality of *gur* is excellent. The average cane yield of the plant crop is about 838 quintals per ha.

Seed material and planting

Seed material for sugarcane planting is from relatively immature top part of the cane in the form of three budded sets. Seed should be free from rot, wilt, smut, ratoon stunting and grassy shoot diseases. Use only the top two-third portion of the selected canes for planting. Give 4-6 ploughings to produce good tilth.

Seed rate: Use 50 thousand three budded sets or 37.5 thousand four budded sets or 30 thousand five budded sets/hectare. Longer sets are good for rainfed conditions. About 75 to 88 q of seed depending upon the variety is required for sowing one hectare sugarcane crop. The seed rate (on weight basis) should be increased by 10% of varieties with thick canes.

Seed treatment

To improve germination, dip the cane setts in 0.25% solution of Tilt 25 EC (propiconazole). For preparing this solution dissolve 625 ml Tilt 25 EC (propiconazole) in 250 litres of water to treat seed cane for one ha. Take out the seed setts immediately after dipping them in this solution. Alternatively, soak the setts in water for 24 hours before planting.

Soil application of pesticides: Sprinkle 5 litres of Lindane/Kanodane/Markdane/ Gammax 20 EC (gamma BHC) diluted in 1250 litres of water per ha with a sprinkler can or apply 20 kg granules of Sevidol 4 : 4 G (gamma BHC + Carbaryl) or 25 kg Regent/ Mortel 0.3 G (fipronil) per ha over seed setts in furrows before covering them with soil for the control of termites. Application of Regent/Mortel 0.3 G (fipronil) also protects the crop from attack of early shoot borer.

Time of planting

***Spring cane*:** Mid Feb.-end of March. Do not plant early maturing varieties after March. Avoid late planting as in the results in yield reduction.

Autumn cane: 20th Sept to 20th Oct

Spacing and planting techniques

1. Trench planting: Plant crop in rows 75 cm apart and 20-25 cm deep trenches. After placing the setts in trenches, cover the setts with 5 cm soil. Apply irrigation immediately after planting, if the planting is not done under proper moisture *(water)* conditions. Repeat the irrigation at 4-5 days interval.

2. Paired row trench planting: Adopt paired row trench planting for saving irrigation water. Plant two rows of sugarcane 30 cm apart in 20-25 cm deep trenches. Place the cane setts at the bottom of the trenches and cover with the soil left in between two rows. Distance between two trenches should be 90 cm. Trenches can be drawn using tractor operated trencher. In addition to water saving, this method gives comparatively higher cane yield, easy propping up operation and reduces lodging.

For ease of mechanical harvesting with sugarcane harvester, the crop should be planted at either 120 cm row spacing or by 120:30 cm paired row trench method.

3. Furrow irrigated raised bed planting (FIRB): Sugarcane can also be planted in standing wheat crop sown by furrow irrigated raised bed (FIRB) planter. The furrow should be reshaped in January to loosen the soil. Apply irrigation in reshaped furrows preferably in the evening before planting. Plant treated sugarcane setts next day by pressing under feet. Sugarcane is planted in pre-opened furrows between the beds, using treated 3 budded setts, during the second fortnight of February to March.

4. Sugarcane cutter planter: Use two-row tractor operated whole cane cutter planter. The complete sugarcane which is fed by the labour sitting on the machine is cut automatically into pieces before dropping into the furrows. Fertilizers and chemicals are also applied simultaneously. The seed rate varies from 80 to 88.5 q/ha. The labour requirement is 33 man-h/2.5 acre. Length of setts varies from 23 to 42 cm. The speed of operation is 1.20 to 1.90 km/h. The capacity of machine varies from 2-3 acres/

day. The machine can save about 25% cost of operation in comparison to traditional method. Use this machine on custom hiring basis.

Note: Apply recommended pesticides at sowing for control of termite before covering the setts with soil.

Nutrient management

Optimum rate of nitrogen recommended in subtropical India ranges from 150 to 225 kg/ha. Phosphorus application is 60-100 kg P_2O_5 and Potassium application is 120 K_2O kg/ha. Nitrogen as top dressing should be placed near the plant. If applied at final earthing up (95 days) it has to be placed in furrows near the rows. Phosphorus and Potassium are applied as basal dose before planting.

For plant crop apply 150 kg of N/ha and P as soil test basis and for ratoon crop apply 225 kg of N/ha and P as per soil test basis. If soil test is low apply 30 kg P_2O_5/ ha. Reduce the Nitrogen dose by 45 to 60 kg/ha if sugarcane is grown after potato crop. Sugarcane also respond to the application of FYM (20 t/ha). If iron deficiency occurs then spray the crop 2-3 times with 1% solution of ferrous sulphate (1 kg ferrous sulphate in 100 liters of water) at weekly intervals soon after the symptoms appear.

Method of application

Plant crop: Apply one half dose of nitrogen as top dressed alongside the cane rows with first irrigation after germination. Top dress the remaining N along the cane rows in May – June. Apply full dose of 'P' in furrows below the cane sets at the time of planting.

Ratoon crop: Top dress ⅓rd N in Feb, with the 1st hoeing, ⅓rd in April and rest ⅓rd in May. 'P' on soil test basis should be drilled along the cane rows at the time of first cultivation in February.

Rainfed crop: If moisture in soil is optimum, apply ½ dose of 'N' at planting. In case the moisture is deficienet then whole dose of 'N' is given with the onset of rain.

Intercropping

In sugarcane, many intercrops perform better due to its delayed germination and wider row spacings. In spring cane, intercropping of summer moong, summer mash and mentha is highly economical and without any adverse effect on the crop. Plant one row in flat planted and two rows in trench planted cane of any of these crops inbetween two rows of sugarcane. In autmn cane, lot of crops such as potato, wheat, raya, gobhi sarson, toria winter maize, cabbage, radish, peas, gram etc. can be raised as intercrop. In intercropping system, the selected herbicide must be safe to both crops i.e. intercrop and sugarcane.

Weed management

Two to three hoeings are necessary to keep down the weeds. The spreading of trash blanket between the cane rows after the emergence of shoots helps to suppress weeds.

This practice has the added advantage of conserving soil moisture particularly in rainfed areas.

Chemical weed control

A pre-emergence application of Sencor 70 WP (metribuzin) at 1.4 kg a.i./ha or Atrataf 50 WP / Solaro 50 WP / Markazine 50 WP (Atrazine) at 1.0 kg a.i./ha or Karmex 80 WP / Klass 80WP (Diuron) at 1.6 kg a.i./ha in 500 litre of water within 2-3 days of planting effectively control the broad leaf weeds and annual grasses. Prefer metribuzin or diuron when *Brachiaria reptans* or *Acrachne racemose* are present. If *Cyperus rotundus* and broad leaf weeds are there then, one post-emergence application of 2,4-D Sod. Salt at 1.6 kg a.i./ha should be done. When crop is infested with climbing weed *Ipomea nil*, apply dimethyl amine salt of 2,4-D at 0.58 kg a.i./ha or 2,4-D sodium salt at 1.6 kg a.i./ha when the weed is in three to five leaf stage.

Irrigation and Drainage : Hot and dry period during April to June is the most critical period for the growth of sugarcane. During this period, irrigate the crop at 7 to 12 days interval. During the rainy season, adjust the frequency of irrigation according to rainfall. Drain away excess water from the sugarcane fields, if these get flooded during the rainy season.

During winters (November to January), irrigate the crop at monthly intervals. To save the crop from frost, apply one irrigation around mid-December and another in the first week of January.

Prevention of Lodging: To prevent the crop from lodging, adopt the following measures:

- Earth up heavily the flat-planted as well as the trench-planted crop at the end of June, before the onset of the monsoon. Trench-planting is particularly effective in preventing lodging.
- Prop up the crop in the end of August or in the beginning of September by using the trash-twist method which consists of the tying of a single cane row instead of tying two rows together. Twist the leaves and the trash to make a rope and pass it alternately along the cane clumps in the row. This method does not hinder the growth and photosynthesis of the plants as in the case when two cane rows are tied together.

Protection from frost : Protect the crop from frost as under:

- Grow frost resistant varieties.
- Raise a bumper crop with adequate fertilization, irrigation and plant protection measures, because a poor and stunted crop suffers more from frost.
- Prevent lodging. A lodged crop is more damaged by frost.
- Irrigate the crop, as adequate soil-moisture during the frosty period keeps the soil comparatively warm and saves it from frost. Irrigate the harvested fields if meant to be ratooned. Plough the patch of land between the cane rows.

- In frosty areas, plant setts only from the top portion of the cane, as these buds are less damaged by frost. The top portion can be buried in the soil during the frosty spell and taken out in the spring for planting.

Stinking rot: The disease appears during the rainy season from July to September. The cane tops dry with the rotting of upper portion or the whole of the stalk. A diseased cane emits a foul smell. Rogue out and burn the severely attacked canes.

Plant Protection

Insect Pests

Termite: The termite appears during April to June and again in October. It destroys the germinating buds and causes the drying up of shoots after germination. To avoid its attack, apply only well-rotten farmyard manure and remove the stubbles and debris of previous crop from the field. For the control of termites apply 25 kg Regent 0.3 G (fipronil) per ha mixing in 50 kg moist sand/soil before the setts are covered with soil by planking. At post germination stage (about 45 days after planting) the chemical control measures against termite are the same as given under the shoot-borer.

Early shoot-borer: This pest appears from April to June and causes dry dead-hearts which can be easily pulled out. To control it :

- Plant the crop early, i.e. before the middle of March.
- Apply 25 kg granules of Regent/Mortel/Rippen 0.3 G (fipronil) before the setts are covered with soil by planking. or
- Apply 25 kg of granules of Padan/Caldan/Kritap 4 G (cartap hydrochloride) or 25 kg Regent/ Mortel/Rippen 0.3 G (fipronil) mixed in 50 kg moist sand/ soil or 375 ml Coragen 18.5 SC (chlorantraniliprole) or 110 ml Imidagold 17.8 SL (imidcloprid) or 5 litres of Durmet/Classic/Dursban/Markpyriphos 20 EC (chlorpyriphos) in 1000 litres of water/ha along the rows at post-germination stage (about 45 days after planting). Earth up slightly followed by light irrigation.
  OR

Use **Tricho-cards** having 50,000 eggs of *Corcyra cephalonica* parasitized (seven days old) by *Trichogramma chilonis* per ha at 10 days interval from mid-April to end-June. These eggs are fixed on cards of 10×15 cm size. Cut the cards into 40 pieces/ strips, each having approximately 500 parasitized eggs. Staple these pieces/strips on the lower surface of the leaves uniformly at 100 spots per ha during evening hours. Normally 8 releases are required. The tricho-cards should not be stapled during rainy days.

Black bug: This pest is active during April to June. The attacked crop looks pale. The black adults and pink young nymphs suck the sap from the leaf-sheaths. Spray the crop with 875 ml of Dursban/Lethal/Massban/Goldban 20 EC (chlorpyriphos) in 1000 litres of water per ha with manually operated sprayer. Direct the spray into the leaf-whorl.

Pyrilla: Pyrilla reduces cane yield and sugar recovery heavily. This pest appears in April-May and again in August-September. The leaves of the damaged crop turn yellow. Later due to the development of a fungus, the crop turns black and the tops become unfit for feeding to cattle. The incidence of this pest is particularly high in a luxuriant crop and in the interior of field. When the attack is severe it becomes difficult to make *gur*.

Top-borer: This pest appears from March to October and causes severe damage during July-August. The typical symptoms are the shot-holes in the leaf, white or red streaks on the upper side of the leaf midrib and bunchy tops from July onwards. The central leaf of the cane top dries up and turns dark. To control it:

- Collect and destroy the moths and egg-clusters.
- Cut the attacked shoots at the ground level from April to June.
- Use Tricho-cards having 50,000 eggs of *Corcyra cephalonica* parasitized (seven days old) by *Tricogramma japonicum* per ha at 10 days old interval from mid-April to end- June. The method to use these cards is given under early shoot borer.

Apply 25 kg granules of Ferterra 0.4 GR or 30 kg Furadan/Diafuran/Furacarb/ Carbocil/ Fury encapsulated 3G (carbofuran) per ha at the base of the shoots in the last week of June or in the first week of July only if the top borer damage exceeds 5% level. Earth up slightly to prevent the granules from flowing with the irrigation water and irrigate the crop immediately. This operation will control the third brood of the top-borer which does the maximum damage. Take the following precautions in using carbofuran and phorate :

- Use rubber gloves while applying carbofuran granules. Never handle these granules with bare hands. A mask should be used to cover the face.
- Mix it with the moist soil to reduce the chances of its falling into the eyes of the person applying it.
- The person applying these granules should not eat or drink anything without washing his hands thoroughly with soap.
- Carbofuran treated sugarcane leaves and weeds should not be fed to cattle for about one month after the treatment.

Stalk borer or Tarai borer: This pest is active throughout the year. The larvae overwinter in the stubble and water-shoots. The attack remains low during April-June and increases in July. Its incidence is highest during October-November. There are no outward symptoms of the attack of this pest. Entrance or exit holes on the attacked canes can be seen only after stripping. A larva sometimes damages upto 3 nodes and the cane may be attacked at several places. The cane yield and sugar recovery are adversely affected in the case of serious attack. The control measures against the pest are as under:

- Do not use the cane seed from the infested field.
- Staple 100 Tricho-cards (5 cm × 2.5 cm) hard paper piece glued with 7 days old eggs of laboratory host, *Corcyra cephalonica* parasitized by *Trichogramma chilonis* to the undersides of sugarcane leaves from July to October at 10 days

interval. Each card should have approximately 500 parasitized eggs and be spread uniformally at 100 spots per ha. Normally 10-12 releases are required.

- At harvest, do not leave the water-shoots in the field.
- Do not ratoon a heavily infested crop. Plough the affected fields, collect and destroy the stumps.

Whitefly: The damaged crop looks pale during August-October. The underside of the leaves is full of nymphs and pupae which suck the sap from the leaves. The leaves turn black due to the development of a fungus. Spray the crop with 100 ml Confidor 200 SL (imidacloprid) in 375 litres of water per ha with a manually/tractor operated sprayer.

Sugarcane thrips: The thrips damage the crop from April to June. This pest suck the sap from the partly opened leaf and tips of the younger leaves, resulting in withering and drying of leaftip, which get rolled inwardly. The thrips prefer plant crop than ratoon crop.

Sugarcane mite: The mite appears from April to June and feeds on the lower side of the leaves under a fine web. The leaves turn red and later appear to be burnt. The growth of the affected crop is retarded during the pre-monsoon period. Baru (*Sorghum halepense*) is the alternative host plant from which this mite spreads to the sugarcane crop. So destroy the weed, if growing near the sugarcane fields.

Gurdaspur borer: This borer appears from June to October and causes the withering of the central leaves (notably the 5th leaf) followed by the total drying up of the tops. The affected canes break at the point of attack with a slight jerk. Rogue out canes showing withered tops in the afternoon every week from June to September. The tops should be cut off well below the point of attack. The timely rogueing of affected plants is very important for controlling the pest. Bury the rogued-out plants. Do not ratoon a heavily affected crop. Plough up the fields not meant for ratooning and destroy the stubbles before June.

Rat control

Poison bait preparations: The acceptance of poison baits by rodents depends upon the quality, texture, taste and odour of the baiting materials. Therefore, the recommended baiting materials should be used for preparation of poison baits.

- **2% Zinc phosphide bait:** Smear 1 kg of bajra or sorghum or cracked wheat or their mixture with 20 g of edible refind oil, 20 g of powdered sugar and mix it thoroughly with 25 g of zinc phosphide.
- **Caution:** Never add water in zinc phosphide bait and always use freshly prepared bait.
- **0.005% Bromadiolone bait:** Mix 20 g of bromadiolone powder, 20 g of edible refind oil and 20 g of powdered sugar in 1 kg of any cereal flour or grains.
- **Bait placement and timings:** Baiting in lean period (May-June) – This period is most suitable for rodent control campaign in large areas. During this period, the rat burrows can easily be located in the fields, on bunds, water channels

and surrounding waste lands. Close all the burrows in the evening and in the reopened burrows on the next day, insert a paper boat containing about 10 g of zinc phosphide or bromadiolone bait about 6 inches deep in each burrow. In case of burrows of the lesser bandicoot rat, gently remove the fresh soil from the burrow opening to locate the tunnel and then put the poison bait deep inside it.

Diseases

Red rot: Red rot is caused by the fungus *Colletotrichum falcatum.* The disease appears from July till the crop is harvested. The third or fourth leaf of cane from top shows yellowing at first, while rest of the leaves also loose colour afterwards and dry up. Later, the whole clump dries up. On splitting open the cane, the tissue is found to be reddened but the discoloration is not uniform and is interspersed with white patches running across the width of the split cane. The pith of affected cane emits alcoholic smell. The control measures against this disease are as under :

- Use seed from absolutely disease free seed plot.
- Do not plant sugarcane in the disease affected fields for one year.
- Grow varieties fairly resistant to red rot.
- Crush the affected crop early and plough up the fields soon after harvesting the crop. Collect and burn the stubbles.
- Rogue out and bury or burn the diseased canes. Uproot the entire clumps and not merely the affected stalks.
- Do not ratoon the diseased crop.

Wilt: This disease is caused by *Cephalosporium sacchari* or *Fusarium moniliforme*. It appears from July till the crop is harvested. The leaves of the affected cane at first turn yellow and finally the top dries up. On splitting open a diseased cane, the pith shows a dirty red discoloration near the nodes. The discoloration is invariably darker than that in the remaining portion of the internodes. The affected stalks become light and hollow. The control measures against this disease are the same as those for red rot. As the causal fungus persists in the soil over long period, the affected field should not be put under sugarcane for 3 years.

Smut: Smut is caused by *Ustilago scitaminea.* This disease is prevalent throughout the year but is severe from May to July and again in October-November. Its incidence increases in the ratoon crop. It is easily recognised by the appearance of long whip-like shoots covered with dusty black mass of spores. These whips may arise from the top of the canes as well as from the lateral sprouted buds. Adopt the following control measures:

- Use only smut free canes for seed. Reject even the healthy looking canes in the diseased stools or those growing in the vicinity of the smut infected clumps.
- Remove the smut whips gently (without shaking) after putting them inside a closely woven drill bag. Then uproot the entire diseased clumps and burn or bury them deep. Immerse the bag used for collecting the smut whips in boiling water for 5 minutes after every roguing of the crop.

- Do not ratoon the smut infected crop.
- Seed treatment at planting is essential.

Ratoon stunting: A coryniform bacterium (*Clavibacter xyli*) has been found to be associated with the disease. The affected crop remains stunted with thin canes. The leaves are comparatively pale and the roots are poorly developed. The disease can be identified by slicing mature canes longitudinally a little below the rind with a sharp knife. In the lower part of the node, parallel to the zone of the whitish waxy band, the pith shows discolored dots, commas and straight or bent streaks upto 2 to 3 mm in length. They may be yellow, orange, pink, red or reddish brown. Do not use the diseased crop for planting. Select the cane-seed from a vigorously growing and healthy crop. The moist hot air treatment of seed canes at 54oC for 4 hours is effective in destroying the causal organism. Do not ratoon the diseased crop.

Grassy shoot disease: The disease is caused by mycoplasma like bodies. The affected plants give rise to numerous thin tillers. The leaves become reduced in size, thin, narrow and usually turn chlorotic. If the attack is light, one or two weak canes may be formed. Uproot and destroy the affected clumps immediately after appearance. The moist hot air treatment of the seed-canes at 54°C for 4 hours inactivates the causal organisms of this disease. Its incidence increases in the ratoon crop, therefore, do not ratoon the diseased crop.

Red stripe: Red stripe is a bacterial disease caused by *Pseudomonas rubrilineans*. It appears during June-August. The affected leaves show bright red streaks which are long, narrow and run longitudinally on the leaf-blade, causing the rotting of tops in severe cases. Rogue out the affected canes and burn or bury them.

Top rot (*Pokkah boeng*): This disease is caused by *Fusarium moniliforme*. It appears during the rainy season from July–September. The young leaves in the top portion of the plant become chlorotic at the base and get distorted and shortened. They turn dark red and fall off gradually. In severe cases, the rotting of the top portion of the cane occurs. Remove the affected clumps and bury them.

Leaf scald: The disease is caused by the bacterium *Xanthomonas albilineans*. Whitish or cream coloured one or two narrow stripes are observed on the leaf extending sometimes down to leaf sheath. The affected plants produce side-shoots starting first from lower nodes with similar stripes on young leaves. The stripes become reddish and later the leaves start withering from top downwards giving scalded appearance. On splitting open the affected canes, reddish brown vascular streaks are observed in the internodes. Sometimes affected plants suddenly wilt and die without any obvious internal symptoms. As the disease is sett-borne, healthy and disease free seed should be planted. Treatment of seed-cane with moist hot air at 54°C for 4 hours inactivates the bacterium. Sterilization of cutting knives by flaming or by dipping in 2% Lysol solution during seed preparation should be practised to minimise spread of the disease. Rogue out the diseased clumps.

SUGARBEET (*Beta vulgaris*)

Sugarbeet Crop

Mature plant

Sugarbeet (*Beta vulgaris*) is mainly used for production of sucrose, a high energy pure food. Sugarbeet pulp and molasses, the by-products of sugarbeet are widely used as feed supplements for livestock. Beet tops (leaves and petioles) can also be used as silage. It accounts for nearly 21.8% of the world's total sugar production. It is also a potential source of ethanol which is now blended in automobile fuel. Sugarbeet is essentially a crop of temperate origin. It is grown for commercial sugar production mainly in European countries; U.S.A and Canada in North America, Chile in South America, Egypt and Morocco in Africa and China, Iran, Japan, Syria and Pakistan in Asia. In India, Maharashtra, Karnataka, Andhra Pradesh, Tamil Nadu, Gujarat, Punjab, Haryana, Rajasthan and Western U.P. are suitable states for its cultivation.

Climate and season

Sugarbeet can be produced in hotter and more humid environments. Optimum temperature are 60-80°F. The most favourable environment for producing a sugarbeet crop from 90 days after emergence to harvest, is bright, sunny days with 60-80°F temperatures followed by night time temperatures of 40 to 50°F.

Soil

Sugarbeets is well adapted to wide range of soil types. Soil should be free of stones because stone cause problems for sugarbeet planting, thinning, harvesting and processing equipment. Dryland sugarbeet production is generally limited to soils with high water holding capacities in areas with 20 mm of rainfall or more.

Cultural practices

Main objective of tillage is to keep the field weeds free and to improve the soil structure to meet the needs of the crop. Other objectives are to preserve seedbed moisture, maintain enough crop residues on the soil to stop erosion and that is why there is a less chance of wind to damage the seedlings. Optimum time for sowing is first fortnight of

October. Seed rate is 3.6 kg/ha, spacing 50 × 20 cm. Population of the plants is 1-1.2 lakhs plants/ha. It should not be planted greater than 3 cm deep. It can tolerate pH of about 6.5 – 8.0.

Weed management

Weeds can be controlled by giving 2-3 hand weedings. Alternatively pretilachlor 1.0 Kg/ha as pre- emergence can be used for controlling weeds.

Plant Protection

Insects pests

Sugarbeet crop is damaged by cutworm, army worm, Bihar hairy caterpillar, flea beetle and grass hoppers. In early stage of crop growth, cutworm may infest the crop. So apply chloropyriphos at 1.0 kg a.i /ha in soil at the time of sowing. Army worm appears about 100 days after sowing can br controlled with the spray of quinalphos (0.05%).

Diseases

Powdery mildew, leaf spots and root rot damage this crop. Seed treatment with Thiram 0.25% or Carbendazim (0.1%) gives good prote-ction against seed borne fungi. Powdery mildew can be effectively controlled with the application of wettable sulphur at 2.0 Kg/ha. Cercospora leaf spot can be controlled by fortnightly spray of Carbendazim (0.1%).

Harvesting

The October sown crop of sugarbeet starts showing symtoms of maturity during mid April. Two conspicuous changes occur in the crop.

I. Drying of lower leaves
II. Sugar accumulation (15-16%) in roots, which continues till end of May. Roots are dug manually or mechanically. Tops are removed and roots are immediately transported to factory for processing, as delay at harvest site coupled with high summer temperatures favour microbial decomposition of roots.

12

CULTIVATION PRACTICES FOR FIBRE CROPS

The main fibre crops are cotton, jute, sunhemp, sisal, ramie, mesta and flux. They are grow for their fibre, which is used in making textiles, ropes, twine and similar materials. Some of them are discussed below:

COTTON (*Gossypium hirsutum*)

Cotton plant **Cotton crop**

Cotton (*Gossypium hirsutum*) the king of fibres, is an industrial commodity of worldwide importance. It is one of the most important commercial crops playing a key role in economic and social affairs of the world. The Arabic word qutun or kutun has given rise to English word cotton. Rising cost of production due to indiscriminate use of pesticides and nitrogen fertilizers with only marginal increase in productivity is leading to unforeseen problems. Given the present market scenario, the cotton growers need to reorient their approach to the package of practices that they are flowing to grow cotton. Integrated nutrient and pest management system are a must for economic cotton production.

Origin, distribution, area and production

India is considered the home of arboreum cottons. Herbaceums might have originated from Arabia, Persia and Baluhithan. South America is the home of barbadense while hirsutums originated from borders of Mexico and Guatemala. In India, all the four species of cotton are grown with highest area under hirsutum (36%) followed by arboretum (17%) and herbaceum (10%). Hybrid cotton is being cultivated on more area. Lysigenous glands, characteristic feature of cottons contains sesquiterpenes called gossypols. In India cotton is cultivated on an area of 10.5 million hectares with production of (seed cotton) 14.41 mt (FAO STAT 2016).

Climate and soils

Cotton is a warm weather crop and the cultivated species are not limited to the tropics. Adapted cultivars are successful in regions where the frost-free period is less than 180 days. Cotton is a short day plant but day neutral cultivars exist. Temperature is a critical factor which decides cotton productivity. Cotton can be grown in a temperature ranges of 43° to 46°C but for hirsutams the upper limit is 42°C and for barbadense it is 37.5°C. Cloudy weather results in boll shedding. A minimum for 4 hrs bright sunshine is a prerequisite for optimum yield.

Major soils on which cotton is grown are alluvial soils, black cotton soils, red sandy loams to loams. Cotton is grown as rainfed crop in deep black cotton and medium black soils and as an irrigated crop in alluvial and other light soils. It does not stand water logging. Cotton crop is considered moderately acid tolerant and the critical pH ranges is 5.5 to 6.0 and the upper limit is 8.5. The crop is considered saline tolerant and salinity threshold (initial yield decline) is about 7.7 $dsm^{-1.}$ The main criterion for suitability of soils within ideal pH range is a depth of atleast 60 cm and freedom from prolonged water logging.

Growth and development

Emeregence

Poor germination and seedling establishment often limit crop stand and yield. An impermeable seed coat and impenetrability of seed coat fibres to water (soil moisture) are the two major factors inhibiting cotton seed germination. Under unfavourable conditions, seedlings emerge in 5-15 days after sowing. Temperature, limits for emergence are 14° to 42°C with an optimum of 34°C.

Root, stem and leaves

Depth of primary root penetration depends on soil, climate and genotype. Maximum root depth varies from 150 to 300 cm under favourable conditions. The main stem of the cotton plant is monopodial with leaves and branches but no flowers. There are usually two axillary buds at each main stem node, the second branching off the first. Normally, only one bud develops. The fruiting limb or sympodium, having flower buds called squares, will occur at the fifth through the ninth main stem node. Cotton plant fruits vertically via the main stem and horizontal via extension of fruit limbs.

Flowering and fruit development

Though, the rate of development of cotton plant is conditioned by the variety, cultural practices and climate, the cotton plant follow broadly the following time schedule:

- Seedling emergence:- About 4-8 days after planting.
- Development of flower bud into flower:- It varies from 17 to 25 days and is usually constant for a variety.
- Appearance of blooms:- The interval between blooms in corresponding nodes of successive fruiting branches is about 3 days, while that between successive nodes of the same branches is about 6 days.
- Flowering to boll opening:- It takes 40 to 80 days depending on the vartiety and temperature. Cotton fibres are unicellular extensions of epiderinal seed cells that begin to elongate early on the day of anthesis, a phenomenon independent of fertilization.

Bud and boll shedding

Square, flower and boll shedding is increased by such factors as moisture deficiency or excess, inadequate nutrient supply, excessive heat or cold and damage due to pests and diseases. Several theories have been proposed to explain flower and boll shedding. The nitrogen theory indicates that a normal leaf has 5% nitrogen and drops to 2.5% as the crop progress towards reproductive stage. Internal translocation sometimes leads to boll shedding. Nutritional theory indicates that plant growth is limited by carbohydrate and mineral supply. Heavy boll load followed by limited carbohydrate production leads to boll shedding.

High wind velocity and low minimum temperature may cause reddening of leaf. Urea or diammonium phosphate (2%) at peak flowering stage, minimizes leaf reddening and help in boll retention.

Varieties

World's first intra hirsutum hybrid released in India is Hybrid 4(H4). It is the highest spinning Egyptian type of cotton released in India. Cotton varieties Aravinda among desi cottons and Narasimha among American cotton are most popular in Andhra Pradesh.

Improved Varieties of Cotton released by ICAR

2017-18

Cotton	Area of adoption	Salient features
DHB 915	AP, Tamil Nadu, Gujarat, Maharashtra and MP	Extra-long staple, interspecific (hirsutum × barbadense) hybrid
CSH 3129	Punjab, Haryana and Rajasthan	Moderately resistant to cotton leaf curl disease (CLCuD), tolerant to fungal foliar diseases and major insect pests
MRC 7377 (Hybrid)	AP, Telangana, Karnataka and Tamil Nadu	Tolerant to bacterial leaf light (BLB), Alternaria leaf spot (ALS)

Central Cotton	Punjab, Haryana and Rajasthan	Suitable for high density planting system (HDPS)
LD 949 (Desi)	Punjab, Haryana and Rajasthan	Absorbent cotton, moderately resistant to Fusarium wilt, resistant to jassids and white fly
MR 68 (American Cotton)	Punjab, Haryana, Rajasthan and Uttar Pradesh	High yielding variety, denim purpose, yield 20 – 25 q/ha
Central Cotton NHH 250	Maharashtra, Madhya Pradesh and Gujarat	High yielding hybrid suitable for rainfed condition, tolerant to BLB and ALS and sucking pests
Phule Asmita Hybrid	Maharashtra, Madhya Pradesh and Gujarat	Resistant to BLB. ALS, MLB and grey mildew; moderately resistant to sucking pests and bollworms
Phule Prabha Hybrid	Maharashtra, Madhya Pradesh, Gujarat, Karnataka, AP, Telangana, and TN	Resistance to ALS and disease free reaction to BLB and gray mildew; tolerant to sucking pests and bollworms
DHB 1071	Karnataka	Extra-long staple interspecific (hirsutum barbadense) hybrid
GN.Cot.Hy 14	Gujarat	resistant to jessids, thrips, whitefly and aphids, tolerant to BLB and ALS
GJ Cot 111	Gujarat, MP, MAharashtra, Odisha	Tolerant to jessids, whitefly and major diseases; average yield 1,514 kg/ha, maturity 160-170 days
K 12	Tamil Nadu	Moderately resistant to BLB and ALS, resistant to leaf hopper, aphids, thrips and stem weevil, average yield 1,193 kg/ha, maturity 135-140 days

2016-17

Commercial crop Variety	Area of adoption	Salient features
CICR NA 1003	Andhra Pradesh, Karnataka and Tamil Nadu	High locule retention capacity (shattering resistance); resistant to major diseases and pests
CCH 2623	Andhra Pradesh, Karnataka and Tamil Nadu	High ginning outturn;field tolerant to jassids and majority of diseases
Sujay	Andhra Pradesh, Karnataka and Tamil Nadu	Moderately resistant to jassids, aphids and leaf spot

H 1353	Gujarat, Maharashtra and Madhya Pradesh	Tolerant to jassids and thrips
Phule Yamuna	Gujarat, Maharashtra and Madhya Pradesh	Tolerant to sucking pests and bollworms, resistant to bacterial blight, moderately resistant to Alternaria late blight and free of grey mildew
SVHH 139	Punjab, Haryana and Rajasthan	High yielding non-Bt hybrid
F 2164	Punjab, Haryana and Rajasthan	Resistant to bacterial late blight, fungal foliar diseases, tolerant to pests
MR 786	Punjab, Haryana and Rajasthan	Moderately resistant to leaf curl virus disease, bacterial late blight, tolerant to jassids, whitefly, thrips, bollworm
Sri Rama	Gujarat, Maharashtra and Madhya Pradesh	Tolerant to sucking pests (jassids, whitefly, aphids)
Phule Tarang	Andhra Pradesh, Karnataka and Tamil Nadu	Resistant to bacterial late blight, moderately resistant to Alternaria late blight (ALB), resistant to sucking pests, tolerant to bollworms
CO 14	Tamil Nadu	Extra long staple American cotton, spinnable up to 70s count yarn; moderately resistant to jassids, Alternaria leaf spot
G.Cot.20	Gujarat	Moderately tolerant to sucking pests , resistant to lodging, moderately resistant to bacterial blight
GN.Cot.25	Gujarat	Tolerant to drought as well as pests and diseases
DHB 915	Karnataka	Extra long staple interspecific hybrid, spinnable up to 80s count yarn
GN.Cot 22	Gujarat	Resistant to jassids, Alterneria, bacterial blight, moderately resistant to grey mildew, moderately tolerant to aphids, thrips and boll worms
SVPR 5	Andhra Pradesh, Karnataka and Tamil Nadu	Moderately resistant to leafhopper, bacterial leaf blight (BLB), Alternaria leaf spot (ALS) and grey mildew
SVPR 1 Hybrid	Andhra Pradesh, Karnataka and Tamil Nadu	Resistant to BLB; suitable for summer
JLA 505	Gujarat, Maharashtra and Madhya Pradesh	Moderately resistant to Alternaria, bacterial blight, grey mildew, tolerant to pests in fields
F 2228	Punjab	Resistant to bacterial blight, moderately resistant to fungal foliar diseases

F 2383 (American Cotton),	Punjab	Suitable for high density plantation system and amenable for machine picking; moderately resistant to BLB and fungal foliar leaf spot, moderately resistant to jassids
RHB 0711 Hybrid	Gujarat, Maharashtra and Madhya Pradesh	Resistant to Alternaria, bacterial blight , tolerant to sucking pests and bollworms
RHCb 011	Gujarat, Maharashtra and Madhya Pradesh	Extra long staple barbadense cotton variety
Raichur Shakthi 455	Andhra Pradesh, Karnataka and Tamil Nadu	Moderately resistant to jassids, aphids, leaf spot, tolerant to bollworms

Improved Varieties / Hybrid's for Punjab

Bt cotton Variety

PAU Bt 1: It is first Bt cotton variety developed by public sector with inbuilt resistance against American, spotted and pink bollworms in the country. It has bigger boll size of 4.3 g and ginning our turns is 41.4%. It possesses superior fiber property with 28.2 mm fibre length, 28.6 g/tex bundle strength and 4.5 micronaire value. It is average seed cotton yield is 28 quintal per ha. It is moderately resistance to cotton leaf curl disease.

Bt cotton Hybrids: Grows Bt cotton hybrids which are recommended by State Agricultural Universities as these goes on changing year after year.

Non-Bt / American cotton

LHH 144: This is an intra-*hirsutum* cotton hybrid with semi-okra leaves. It is tolerant to jassid and bacterial blight. It matures in about 180 days. It produces an average seed cotton yield of 19 quintals per ha.

F 2228: Its maturity period is 180 days with an average seed cotton yield of 18.5 quintals per ha. It is moderately resistant to jassid and bacterial blight.

F 2383: This is suitable for sowing under high density plantation system. The variety has medium sized super-okra leaves. It matures in about 160 days and gives an average seed cotton yield of 19.75 quintals per ha.

LH 2108: It matures in 165-170 days with an average seed cotton yield of 21 quintals per ha.

LH 2076: It matures in 165-175 days with an average seed cotton yield of 19.5 quintals per ha.

Desi Cotton

LD 1019: This is a shattering tolerant variety requiring 2 or 3 pickings as compared to atleast 5 pickings needed by other *desi* cotton varieties. Average seed cotton yield of LD 1019 is 21.5 quintals per ha. It possesses ginning out-turn of 35.7% and its fibre length is 22.6 mm. It has green, broad leaves and cream flowers. LD 1019 is tolerant to jassid, whitefly, Fusarium wilt and bacterial blight.

LD 949: Its plants are reddish-brown with narrow- lobed deep cut leaves and pink flowers. It possesses lint percentage of 40.1. Its fibres are short, coarse and suitable as absorbent cotton. This variety is moderately resistant to Fusarium wilt and bacterial blight. It is tolerant to whitefly and jassid. Its average seed cotton yield is 24.75 quintals/ ha.

FDK 124: It has green plant body and narrow lobed leaves. It is synchronous in maturity and takes about 160 days to mature. It is short staple, coarse fibre variety with 2.5% span length of 21.0 mm and ginning outturn of 36.4%. It produces an average seed cotton yield of 23.0 quintals per ha. It is resistant to jassid and whitefly.

Seed rate and spacings for cotton

Optimum sowing time for *Kharif* cotton is April-May in north India and July-August in South India whereas middle of May to first fortnight of June is the ideal sowing time for desi cotton. Use 7.5 kg/ha seed rate for desi cotton and 4.0 kg/ha for hybrids of American cotton. It can vary with different varieties or hybrids or Bt or non-Bt cotton.

To facilitate interculture with tractor, row to row spacing should be kept 67.5 cm and plant to plant spacing may vary from 45 cm to 75 cm depending upon, growing habits of Hybrid/variety. Sowing can be done with cotton planter or cotton sowing drill.

Method of seeding

Generally, cotton seeds are drill sown, dibbled or sown behind the country plough. Thinning and gap filling should be completed within three weeks after emergence. For gap filling sow cotton seeds in small polythene bags on the day of sowing and transplant 20-25 days old seedlings in the field.

Nutrient management

Nitrogen application improves all growth and yield attributes leading to high yield. The average response to nitrogen varies from 4 to 6 kg seed cotton/kg N applied in the case of rainfed desi cottons to 6 to 8 kg/ha in the rainfed American cottons, 10 to 12 kg/ha in the case of irrigated hybrids. Peak demand for nitrogen is at flowering and boll formation. Recommendd fertilizer schedule for rainfed desi cotton is 40 kg N and 20 kg P_2O_5/ha and for hybrids 60 kg N, 30 kg P_2O_5 and 30 kg K_2O/ha. Optimum fertilizer schedule for irrigated American cottons is 120 kg N, 60 kg P_2O_5 and 60 kg K_2O and for irrigated hybrids 150 kg N, 75 kg P_2O_5 and 75 kg K_2O/ha.

For rainfed cotton, nitrogen is recommended in three splits, half at sowing and the remaining half in two equal splits at squaring and flowering. For irrigated hybrids, nitrogen is recommended in four splits at seeding, 30, 60 and 90 DAS. Split doses of fertilizers should be applied by placement around the plants or band placement.

Water management

Nearly 80% of the area under cotton in India is depending entirely on rainfall during monsoon while supplementary irrigation facilities exist for about 20 per cent of the area. Supplemental irrigation is given to American and hybrid cottons, depending on the availability of irrigation water. Depending on climate and duration of crop growth, water require-ment of cotton varies from 700 to 1300 mm. Daily water needs are highest (10 mm /day) at peak flowering. Water use efficiency is 0.4-0.6 kg m^{-3}.

Critical stages for irrigation

Cotton crop is sensitive to water logging. Squaring to peak flowering is considered most critical period for irrigation. On sandy-loam soils of North-zone, 3-5 irrigations are commonly given. On red sandy-loam soils of Tamil Nadu with low water retention capacity, 4-13 irrigations may be necessary. Drip irrigation or alternate furrow irrigation techniques are very economical and improve water use efficiency. Furrow irrigation is most common method of irrigation. In region of scarce irrigation water, sprinkler and drip methods are more economical methods.

Weed management

Weed remove 5-6 times more phosphorus ad 2-5 times more potassium than cotton crop in the early stages of crop, leading to more than 50 per cent reduction in seed cotton yield. Dominant weeds in cotton fields are *Cyperus rotundus*, *Cynodon dactylon*, *Echinocloa coloneum*, *Celosia argentina*, *Dactyleoctinium aegypticum*, *Trianthema postulaca-strum*, *Euphorbia* sp., *Digera arvensis*, *Eragrostis tenella* and *Abitilon indicum* etc. Cotton crop requires weed free environment for 60 days from emergence. Two or three inter-cultivations with blade harrows followed by a hand hoeing is the most common method of weed management in cotton.

For the control of annual weeds, pre-plant application of Treflan 48 EC (trifluralin) at 1.2 kg a.i./ha or pre-emergence application of Stomp 30 EC (pendimethalin) at 0.75 kg a.i/ha can be made by dissolving in 500 to 700 lit of water. These herbicides keep weed free crop for 6-8 weeks depending upon soil type and climatic conditions and later on inter-cultures can be given to keep the weeds under check. Alternatively directed (band) applicationof non-selective herbicides such as Gramoxone 24 WSC (paraquat) at 0.3 kg a.i./ha or Round up/Glycil 41% SL (glyphosate) at 1.0 kg a.i./ha can be done when crop is 40 to 45 cm in height with the use of protective hood. Prefer glyphosate when perennial weeds are also present in the crop. Do not spray on windy day and these herbicides especially glyphosate should not fall on the foliage of the crop.

Plant Protection

Insect Pests

(i) Cultural and mechanical management

- Grow only recommended Bt cotton cultivars
- Prefer to grow *desi* cotton in area of high infestation of whitefly and leaf curl
- Even the apparently healthy seed-cotton (*kapas*) may be harbouring larvae of pink bollworm. It should be acid delinted or thoroughly dried in the sun in a thin layer for 3-4 consecutive days in April

- Complete the sowing up to 15th May
- Avoid excessive use of nitrogenous fertilizer
- Eradicate weeds like *kanghi buti, peeli buti, puth kanda*, congress grass, *itsit* growing on field bunds, waste lands, road side and irrigation channels/canals to avoid further spread of whitefly, mealybug, tobacco caterpillar and spotted bollworm to cotton fields
- Regular surveillance on alternate host crops like brinjal, cucurbits (cucumber, long melon, chappan kadu), tomato, chilli, okra from February onwards and on cotton, moong from April onwards should be carried out for timely management of whitefly on these crops
- Use low cost yellow sticky traps @ 100 per ha during initial phase of cotton crop to check early infestation of whitefly
- Grow bajra, maize and jawar as barrier crops, being least preferred hosts for the mealybug
- Do not grow *bhindi*, moong and *arhar* in the cotton crop and as border rows in order to reduce the incidence of American bollworm, spotted bollworms, jassid and whitefly. Bhindi, moong, dhaincha and castor are also the most preferred hosts of tobacco caterpillar, helping the pest to multiply and shift to cotton. The above pests on these crops grown in the vicinity of cotton fields, should be properly controlled in order to check their migration to the cotton crop.
- Do not throw the uprooted infested plants in cotton fields/water channels to check further spread of mealybug
- Prevent movement of sticks from infested areas to new areas to avoid spread of mealybug
- The trees/fruit plants near cotton fields harboring mealybug population should be sprayed with recommended insecticides
- Egg masses and young larvae of tobacco caterpillar feeding gregariously should be collected along with leaves and destroyed.
- Terminate the crop as early as economically feasible. For this purpose give last irrigation by end of September. It would reduce bollworms damage and their carryover.
- Do not stalk the cotton sticks under shade or in the field. Beet the sticks on ground to disloge the insect larvae surviving in the unopened bolls. Stalk the cotton sticks vertically.

(ii) Chemical Control

(a) Insect Pests (Bt cotton)

Bt cotton provides effective protection against all cotton bollworms. However, regular monitoring should be done at weekly interval during reproductive phase. If at all American bollworms cross ETL level during late crop season, use insecticides as mentioned in the following Table.

Note: Bt cotton does not provide control of sucking pests.

Sucking Insect Pests: Among sucking pests, whitefly, jassid, mealybug, thrips and aphid are most serious on Bt cotton and they cause maximum damage during July-September. Whitefly adults and nymphs suck sap from leaves and excrete honey dew on leaves which become sticky. Affected leaves and seed cotton turn black due to development of sooty mould. Nymphs and adults of jassid suck sap from leaves and cause shedding in case of severe infestation. Aphids appear sporadically. The nymphs and adults of aphid suck sap and excrete honey dew on leaves on which black fungus develops. Petioles, internodes, flowers, buds, mature bolls and even leaves fall down and growth of the plant is retarded. Both nymph and adults of thrips first lacerate the leaf tissue and later feed on the oozing cell sap. Initially silver streaks appeared especially around the mid rib and veins of the leaves. Later silvering got more severe and slight cupping of the leaves. Under severe infestation leaves gave blasted appearance and extreme level of cupping observed.

The decision regarding spray of insecticides should be taken based on Economic Threshold Level (ETL). Sprays against whitefly should be done when population reaches six adults per leaf in the upper canopy of plants before 10 AM or when honey dew appears on 50 per cent of the plants. Initiate spray against jassid whenever some of the fully formed leaves in the upper canopy show curling and yellowing at the margins on 50 per cent of the plants. Spray against aphid should also be done on the appearance of honey dew on 50 per cent plants. Spray the crop as soon as the crawlers/adults of mealybug appear on the cotton plant. Use following insecticides for control of sucking insect pests.

Brand (Insecticides)	Dose/ha
Jassid	
(a) Seed treatment: At the time of sowing smear the seedwith any of the following insecticides.	
Gaucho 70 WS (imidacloprid)	5 g/kg seed
Cruiser 30 FS (thiamethoxam)	7 g/kg seed
(b) Spray: Spray any of thefollowing insecticides, if indidence is noticed in standingcrop.	
Ulala 50 WG (flonicamid)	200 g
Osheen 20 SG (dinotefuran)	150 g
Actara/Extra super/Dotara/Thomson 25 WG (thiamethoxam)	100 g
Imidacel/Markdor/Isogashi 17.8 SL/Confidence 555/Confidor 200 SL (imidacloprid)	100 ml
Whitefly	
Ulala 50WG (flonicamid)	200 g
Polo/Craze/Ruby/Ludo/Shoku 50 WP (dimafenthiuron)	500 g
Lano10EC (pyriproxyfen)	1250 ml
Oberon/Voltage 22.9 SC (spiromesifen)	500 ml
Osheen 20 SG (dinotefuran)	150 g
Fosmite/E-mite/Volthion/Gold Mit 50 EC (ethion)	2.0 L
Nimbecidine or Achook (Neem based biopesticide)	2.5 L

##Applaud/Tribune 25 SC (buprofezin)	1250 ml
Larvin 75 WP (thiodicarb)	625 g
Curacron/Carina/Profex/Celcron 50 EC (profenophos)	1250 ml
Ekalux/Quinolphos/Quingaurd 25 EC (quinalphos)	2.0 L
Coroban/Dursban/Durmet/Chloruard/Radar/Lethal Force/Markpyriphos 20 EC (chlorpyriphos)	5.0 L
Thrips	
Curacron 50 EC (profenophos)	1250 ml
Polo 50 WP 9diafenthiuron)	500 g
Fosmite 50 EC 9ethion)	2.0 L

Use this pesticide as first spray for the control of mealybug as this pesticide is safe for natural enemies.

Note: In the beginning of crop season on appearance of whitefly, first spray of Nimbecidine or Achook @ 1.0 lire per acre should be given. Do not spray any insecticide for the management of thrips on cotton up to 30 day old crop. In case thrips attack observed, irrigate the field immediately. In case of attack of aphid use the same insecticides as recommended for jassid.

(b) Insect Pests (Non-Bt cotton)

Sucking Insect Pests: Same as for Bt-cotton

Bollworms: Bollworms are the most harmful insects. Spotted bollworms damage growing points during May-June and cause heavy shedding of squares, buds, flowers and bolls during July to October. The American bollworm causes severe shedding of fruiting bodies during September-October especially on American cotton. The colour of its larvae greatly varies. They have one line on upper side and two wavy lines on lateral side of body. Their body also has sparse hairs. Pink bollworm does maximum damage from mid-July to mid-October. Due to severe attack of bollworms, the plants continue to grow without having adequate number of bolls.

Tobacco caterpillar: It is a polyphagous pest. The larvae cause serious damage to crop from August to October. The small larvae are black whereas grown up larvae are dark green with black triangular spots on body. Its moths lay eggs in masses covered with brown hairs on the lower side of mature leaves. After hatching, first and second instar larvae feed gregariously and skeletonize the foliage. Later on grown up larvae disperse and feed singly. Besides leaves, they also damage the buds, flowers and green bolls. Control the tobacco caterpillar by spraying any of the insecticides given in Table 2.

The larvae of leaf-roller, semi-loopers, hairy-caterpillars and bud moth may also appear sporadically and damage the crop during July-October.

The monitoring of bollworms and tobacco caterpillar with sexpheromone should be done with the initiation of flowering stage of crop. Observations on moth catch should be recorded on every alternate day. This monitoring strategy will help in making decision for effective management of bollworms and tobacco caterpillar.

Pink bollworm: Use Sticka/Delta traps with at least 10 micro litre of gossyplure and place it at 15 cm above crop canopy. Replace the lure after 15 days and use 1 trap/ ha.

Spotted/Spiny bollworms: Use Sleeve/Moth catch traps for spotted bollworms and replace the lure at 15 days interval. Place the trap at 15 cm above the crop canopy and use 2 traps/ha.

American bollworm: Use Sleeve/Moth catch traps with at least 2 mg of phero mone and place it at 15 cm above crop canopy. Replace the lure after 15 days and use 2 traps/ ha.

Tobacco Caterpillar: Use Sleeve/Moth catch trap for tobacco caterpillar. Replace the lure after every 15 days. Place the trap 15 cm above crop canopy and use 2 traps/ha.

In order to control bollworms, conduct sprays on different varieties during their effective boll formation period based on ETL. Farmers should examine their fields twice a week in order to ensure that bollworms damage does not exceed 5 per cent among the freshly shed fruiting bodies (squares, buds and young bolls). For this purpose divide the field into four quarters and collect 25 freshly shed fruiting bodies at random in each quarter. The fruiting bodies damaged by bollworms will have feeding holes or their larvae. In case the damage exceeds 5 per cent, the crop should be sprayed immediately and thereafter spray as when need arises.

Brand (Insecticides)	Dose/ha
Pink and spotted bollworms	
A. Synthetic Pyrethroids	
Fastac/Alphagaurd/Merit Alpha 10 EC (Alphamethrin)	250 ml
Bulldock 0.25 SC (β-cyfluthrin)	750 ml
Ripcord/Bilcyp/Bullet/Ustad/Cypergaurd 10 EC (Cypermethrin)	500 ml
Cymbush/Cyperkill/Hillcyper/Colt/Basathrin/Agrocyper/Cypergaurd 25 EC (Cypermethrin)	200 ml
Decis/Rukrain/Decicare 2.8 EC (Deltamethrin)	400 ml
Sumicidin/Fenval/Agrofen/Fenlik/Triumph card/SB Fenvalerate/Milfen/ Markfenval 20 EC (Fenvalerate)	250 ml
Meothrin 10 EC (Fenpropathrin)	750 ml
Pink, spottedand younger larvae of American bollworm	
A. Carbamates	
Larvin 75 WP (Thiodicarb)	625 g
B. Organophosphatic	
Curacron/Carina/Profex/Celcron 50 EC (Profenophos)	1250 ml
Ekalux/GAIC Quinalphos/Quingaurd 25 EC (Quinalphos)	2.0 bit
Fosmite/E-mite/Volthion 50 EC (Ethion)	2.0 bit
C. Miscellaneous group	
Fame 480 SC* (Flubendiamide)	100 ml

Grown up larvae of American bollworm	
A. Naturalyte	
Tracer 48 SC (Spinosad)	150 ml
B. Oxadiazine	
Avaunt 15 SC/Avaunt 15 EC (Indoxacarb)	500 ml
C. Miscellaneous group	
Sumipleo 10 EC (Pyridalyl)	750 ml
Coragen 18.5 SC* (Chlorantraniliprole)	150 ml
D. Organophosphates	
Caroban/Dursban/Durment/Chlorgaurd/Radar/Lethal/Force/Markpyriphos 20 EC (Chlorpyriphos)	5.0 lit
Orthene/Asataf/Starthene/Markphate 75 SP (Acephate)	2.0 kg
Tobacco caterpillar	
A. Insect Growther Regulator	
Rimon 10 EC* (Novaluron)	375 ml
B. Organophosphates	
Coroban/Dursban/Durmet/Chlorgaurd/Radar/Lethal/Force 20 EC (Chlorpy-riphos)	5.0 lit
Ekalux/GAIC Quinalphos/Quingaurd 25 EC (Quinalphos)	2.0 lit
Orthene/Asataf/Starthene 75 SP (Acephate)	2.0 kg
C. Miscellaneous group.	
Sumipleo 10 EC (Pyridalyl)	750 ml
Coragen 18.5 SC* (Chlorantraniliprole)	150 ml
D. Carbamate	
Larvin 75 WP (Thiodicarb)	625 g

Note:

- Do not repeat the insecticide of same group in subsequent sprays.
- Do not use synthetic pyrethroids on cotton for the control of bollworm complex after mid September.
- Repeat the spray immediately if it rains with in 24 hours after spray.
- If hairy caterpillars damage cotton crop during June-July use 1.25 litre quinalphos 25 EC in 250 litres of water per ha.
- Cotton is highly sensitive to the 2, 4-D weedicide. Some farmers spray the ester form of 2, 4-D for controlling weeds in maize grown near the cotton fields. Owing to the volatile nature of 2, 4-D ester, its vapours cause serious injury to the cotton crop. Hence avoid the application of this herbicide in maize, if cotton is grown in the adjoining fields. The other precautions are:
  - After using 2, 4-D on any crop, fill all spraying equipment as well as tubs, buckets, etc. with 0.5 per cent washing soda solution (500 g of washing soda

in 100 litres of water) in the evening. Next morning, flush all equipments thoroughly with fresh water.

- To avoid the use of contaminated insecticides on cotton. It is advisable to test insecticide at least two weeks in advance on a few plants. If the insecticide is contaminated with 2, 4-D the tender leaves and shoots could become distorted and lancolated within 10 days. Reject such an insecticide.

Spray technology

- The insecticides recommended for control of sucking pests, bollworms and tobacco caterpillar should be sprayed using 300-400 litres spray material per ha with the manually operated knapsack sprayer or 200 litres with the shoulder-mounted power sprayer and tractor mounted sprayer or 30-40 liters spray material per ha by using backpack type air- assisted electrostatic sprayer or 750-1000 liters spray material per ha by using Auto rotate gun type sprayer or 600-1000 liters spray material per ha by using PAU multipurpose high clearance sprayer. Quantity of spray material may vary with different types of sprayers and nozzles. However, actual amount of insecticide recommended should not be reduced. Making pathways by pressing the branches on both sides helps in efficient spraying. Make such pathways at 2 meters distance for the manually operated knapsack sprayer and at 4 meters for the shoulder-mounted power sprayer.

Tractor mounted sprayer should have hallow cone nozzles fixed on the boom. Each nozzle should discharge 500-600 ml spray solution per minute. The tractor should be operated at 4.0 and 2.5 km per hour speed for spraying against sucking pests and bollworms, respectively. Use the same tyre tracks and run the tractor in the same direction for all sprays. Keep the spray boom about 50 cm above the crop canopy.

Diseases

Leaf curl: Disease is caused by whitefly transmitted virus. The diseased plants become stunted and have twisted internodes. Leaves remain small, show cupping and curling. Veins on the lower-side of the leaves become thickened with netted appearance. Small leaflets (enations) also develop on the under side of the leaves on the main as well as lateral veins. Number of fruiting bodies are reduced in the diseased plants. The disease can be reduced by adopting the following integrated measures:

- Avoid growing American cotton in and around citrus orchards and adjoining *bhindi* crop.
- Go on uprooting and destroying the diseased plants from time to time.
- Protect the crop against whitefly vector at 4-5 leaf stage by using recommended insecticides.
- Follow clean cultivation and destroy *Kanghi buti* (*Abutilon* sp.) and *Peeli buti* (*Sida* sp.) which act as collateral hosts.

Parawilt: Parawilt is a physiological disorder and no pathogen is involved. It generally occurs after droughts when the crop is heavily irrigated or there is heavy rain. Plants show sudden drooping of leaves which ultimately get wilted but the root system remains intact. The affected plants can be saved by spraying cobalt chloride @ 10 mg per litre of water (10 ppm) immediately after the appearance of symptoms. There would be no recovery if permanent wilting has already set in.

Root rot: This disease is caused by *Rhizoctonia solani* and *R. bataticola*. The main symptom are drying and shedding of leaves leading to complete wilting and death of the plant. The disease spreads in field in round patches. The affected plants can be pulled out very easily. The bark of the roots is broken into shreds.

Bacterial blight: It is caused by *Xanthomonas axonopodis* pv. *malvacearum* which survives in seed and plant debris. Lesions on the leaves appear as minute, water-soaked, angular spots, which subsequently turn brown and then are transformed into black angular dead lesions on both sides of the leaf. The bacterium also infects the young developing bolls and causes small, round, water soaked spots depressed in the centre. Spray with Blitox 50 WP (1.25 kg) + Agrimycin (50 g) or Streptocycline 7.5 g per ha at 15-20 days interval starting just after the first shower of rain. Three sprays will be enough. The quantity of water will depend upon the crop growth and the spray pump to be used.

Leaf spots: Foliar leaf spots are caused by different fungi. Leaf spots caused by *Myrothecium roridum* appear on leaves, bracts as well as on bolls. The disease is characterized by circular to semicircular brown coloured spots with broad violet margins. At later stage, shield shaped, small size fruiting bodies appear in the central necrotic portion of the spot. The pathogen is seed borne and also survives on the dead leaves. High humidity and intermittant rains are congenial for the development of the disease.

The fungus *Alternaria gossypina* also causes blightening of the leaves. In the early- stages, the spots have a pale green area with irregular margins. As the spots enlarge, irregular concentric zones are formed. Sometimes severe shedding of leaves occur due to this disease. The plants with low vigour because of drought or deficiency of potash favour the development of this disease. The disease perpetuates through diseased debris.

Another type of leaf spots are caused by *Cercospora* sp. which generally appears towards the end of the season. It produces small, circular to irregular spots having whitish centre with dark brown margin. In advance stages, necrotic central portion may fall out giving shot hole appearance. Low temperature (<25°C) and high relative humidity favours development of disease. Diseased debris is the main source of primary inoculum of the disease. To control these leaf spots, the crop should be sprayed alternately with Blitox 50 WP or Captan 83 1.25 kg in 500 litres of water/ha at interval of 15 to 20 days starting just after the first shower of rain. Two to three sprays will be enough.

Tirak: It is a physiological disorder. It is characterized by the yellowing and reddening of leaves, followed by the bad opening of the bolls. The disease appears now and then. The attack is more pronounced in the dry belt adjoining Rajasthan and Haryana. It is particularly serious in pockets where cotton suffers from persistent drought, inadequate water supply, nutrient deficiency on light sandy soils, too early sowing or lack of plant protection measures. These factors may operate singly or in different combinations. Spells of high temperature prevailing during the flowering and fruiting further aggravate the intensity of this malady. Judicious fertilization and timely watering particularly during flowering and fruiting stages, and the adoption of recommended plant protection schedule help to mitigate the intensity of this disease.

Wilt: The pathogen of disease is both soil and seed-borne. In the diseased seedlings and plants, the leaves loose their turgidity, first turn yellow, then brown, start wilting and finally drop off. Discoloration of the leaves start from the margins and spreads towards the mid-ribs. The older leaves are affected first, followed by the younger ones towards the top. Wilting may be complete or partial. In complete wilting, the plant remains stunted, wilt rapidly and dies. The most prominent diagnostic symptom of the disease is browning and blackening of the vascular tissues. Five to six year rotation with non-host crops may help in controlling the disease. In the infested field, sow LD 694 variety of desi cotton since the same is tolerant to wilt. In the highly infested fields grow American cotton because it remains free from this disease. For the chemical control of wilt soak 3 kg seed in 6 litres of water containing 6 g of Bavistin/Derosal for 6-8 hrs. (non delinted seed) or 2-3 hrs. (acid delinted seed).

JUTE (*Corchorus capsularis* and *C. olitorius*)

Jute plant

Jute crop

Jute (*Corchorus capsularis* and *C. olitorius*) ranks second in importance, next to cotton as a natural fibre and occupies important place in Indian economy. Jute fibre is extracted from phloem tissue in the stem of Corchorus as against seed fibre in the case of cotton. Of all the textile fibres, jute is the cheapest and is extensively used in the manufacture of packing material for agricultural and industrial products.

Origin and distribution

The primary centre of origin of *Corchorus olitorius* is Africa and the secondary centre may be India or Indo-Burma and South China. Major jute growing countries are India, Pakistan and Bangladesh accounting for 90% of the global production.

Area and production

India ranks first both in area (7.65 lac ha) and production of 9.94 million bales of jute. West Bengal accounts for more than 80% of total jute production and about 75% area in the country (FAO STAT 2016).

Climate and soils

Jute thrives best under warm humid climate with temperature ranging from 24°C to 35°C and relative humidity 90%. Well distributed rainfall of 1500 mm of which about 250 mm from March to May is most ideal for jute cultivation. Alternate rainfall with bright sunshine is beneficial for a good crop of jute.

The new grey alluvial soils of good depth, receiving silt from annual floods are ideal for jute cultivation. However, it is grown in sandy and clay loams with varying soil management practices. Sandy soils and heavy soils are unsuitable. Well aerated sandy loams are best suited. Optimum pH range is 6.0 to 7.6. Below pH 5 or above 8 is not desirable. Jute requires a fine seedbed. Its tap roots extends to a depth around 50 cm and hence deep ploughing is necessary.

Varieties of Jute: A number of varieties has been released in India for both the species. These include 13 tossa and 12 white jute varieties.

White jute: JRC-321, JRC-212, JRC-7447, JRC-4444, UPC-94, Padma, KTC-1, JRC-698, Bidhan Pat 1,Bidhan Pat 2, Bidhan Pat 3, JRC-80

Tossa jute: JRO-632, JRO-878, JRO-7835, JRO-524, TJ-40, JRO-3690, KOM-62, JRO-66, JRO-8432, JRO-128, S-19 (Subala), Bidhan Rupali tossa, JRO-204 (Suren).

It is important to note that JRO-204 is recommended for sowing as early as in the first week of March in the Indian subcontinent.

Seasons: Jute crop is confined to monsoon since low temperature affect the growth and yield during winter (*rabi*). Young seedlings are sensitive to water logging. As such, sowing in low lands begin in February, taking advantage of residual soil moisture for *Capsularis*. Genereally, *C. olitorius* is sown in April.

Seed rate and spacing

A seed rate of 12 kg/ha is optimum by *C. capsularis* and 8 kg/ha for *C. olitorius*. Row distance for *C. capsularis* is 30 cm and for *C. olitorius* 20 cm. The final distance between the plants in the row should be around 10 cm to maintain optimum plant population around 2,50,000 and 5,00,000 plants per hectare for *C. capsularis* and *C. olitorius* respectively. Low plant population encourages branching leading to poor quality fibre.

Method of seeding

Recommended seed rate is mixed with equal weight of sand or loose soil for uniform seed distribution. Broadcasting the seed is the usual practice, especially for complete rainfed crop. However, line sowing with seed drills to a depth of 3-5 cm is superior over broadcasting. Two thinnings are done, once when plants attain a height of 10 cm and another at 15 cm height. The final inter-plant distance should be 10 cm to maintain required plant population.

Nutrient management

Judicious application of fertilizers provides one of the quickest means of increasing jute production. In jute, greater vegetative growth is conductive to higher fibre yield since it is obtained from bark of the plant.

On an average 2.6 to 3.6 t/ha of green matter is obtained from jute crop at harvest time. Generally, the average uptake of nitrogen varies from 40-80 kg/ha and potassium 100-150 kg/ha. An *Colitorius* jute crop producing 3 t/ha of dry fibre absorbs 111 kg N, 64 kg P_2O_5, 199 kg K_2O, 175 kg Ca and 42 kg Mg/ha.

Recommended fertilizers

C. capsularis				
	Rainfed	40 kg N/ha 20 kg P_2O_5 /ha 20 kg K_2O/ha	Supplemental irrigation	80 kg N/ha 40 kg P_2O_5 /ha 40 kg K_2O/ha
C. olitorius				
	Rainfed	30 kg N/ha 15 kg P_2O_5 /ha 15 kg K_2O/ha	Supplemental irrigation	60 kg N/ha 30 kg P_2O_5 /ha 30 kg K_2O/ha

Nitrogen application is recommended in three splits: half at sowing, one-fourth 45 DAS and the remaining one-fourth 60 DAS by placement. Ammonium sulphate as a source of nitrogen, is ideal than other sources.

Irrigation

The ideal practice for February sown crops is to give a pre-sowing irrigation for seed germination and stand establishment. If water is available another irrigation should be given 30 DAS. Depending on climate and crop duration water requirement varies from 500 to 800 mm.

Weed management

Weeds pose greatest problem in jute cultivation. In terms of expenditure, weed management constitutes one-third the total cost of cultivation. Weeds may reduce the fibre yield to the extent of 40-50%. Critical period of weed competition is 30-55 DAS. Two manual weedings, one each when the crop is 10 and 15 cm height, are considered optimum and economical for jute crop. Selective pre-emergence herbicides for jute crop are fluchloralin 0.5 to 1.0 kg/ha and thiobencarb 1.0 to 1.25 kg/ha and ideal post emergence herbicides is fluazifopbutyl 0.4-0.5 kg/ha.

Plant protection

Jute aphids, semiloper and mealy bug among inserts and stem rot, black band and wilt among diseases caue considerable damage to the jute crop. Jute aphids can be controlled by the application of Acephate 75 SP at 1.5 g/l or carbaryl 50 WP at 3 g/l. Semiloper can be controlled with Acephate 75 SP at 1.5 g/l or carbaryl 50 WP at 3 g/l. Spraying Blitox 2-3 times at 15 days interval can control black band disease. Combination of Captan & Zineb is also equally effective

Harvesting

Jute may be harvested at any time between 120 and 150 days after seeding. Early harvesting gives finer fibre of good quality and late harve-sting gives larger yield of coarser fibre. Harvesting is done by cutting the plants close to the ground level. In flooded lands, the plants are uprooted. After harvest, the plants are left in field for 2-3 days for the leaves to shed. The plants are then tied into bundles, 20-25 cm in dia and the branched tops looped off to rot in the field.

Retting

Retting is a microbial process by which the fibres are separated from the bark and woody core die to removal of pectins, gums and other mucil-aginous substances by retting bacteria present in water. The processing of retting is completed in about 10 days after submergence in water. Over retting results in dazed weak fibre. Plant height and basal dia are the two major yield determinants of jute fibre.

Yield

As explained above plant height and basal dia are the two major yield determinants of jute fibre. Average yield of jute fibre is 1970 kg/ha. However, well maintained *C. olitorious* and *C. capsularis* crop can yield upto 2.75 and 2.0 t/ha respectively. Yield potential of these two has been assessed at 4.0 and 3.5 t/ha respectively. Seed crop of *C. capsularis* and *C. olitorius* on an average, give seed yields of 0.4 and 0.5 t/ha, respectively.

SUNHEMP (*Crotalaria juncea*)

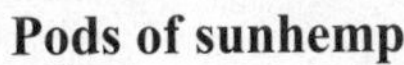

Pods of sunhemp

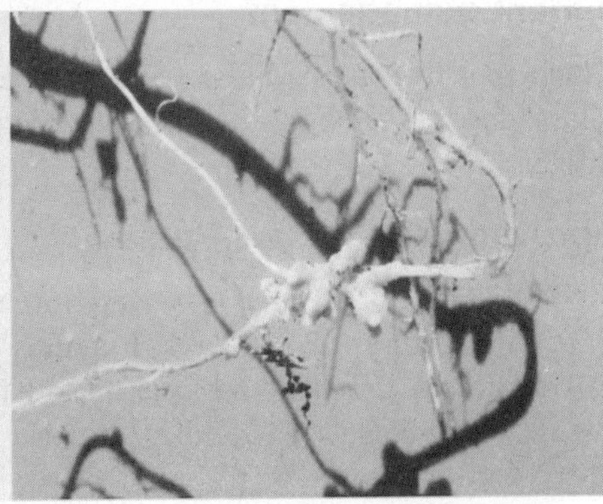

Nodule of sunhemp

Sunhemp (*Crotalaria juncea*) Bombay hemp or Banaras hemp is grown for the best fibre. It is also a green manure crop and a minor fodder crop.

Origin, disitribution, area and production

Sunhemp is generally considered to have originated in India; where it has been cultivated since prehistoric times. It is cultivated commercially for fibre production in Russia, Italy, India, Austria, United States, China, Japan and France. In India, the crop can be grown all over the country but the major growing states are Rajasthan and Orrisa. During 2013-14, Sunhemp was grown on an area of 11,960 ha with production of 3.97 lac t (Ministry of Statistic and programme implimentation 2014-15). Mainly growin in Rajasthan with more than 90% area.

Climate and soil

Sunhemp is a tropical crop grown during *Kharif* in North India and throughout the year in South India. A minimum of 400 mm rainfall with equal distribution during the crop period is ideal. Sunhemp is drought resistant and is adapted to hot, semi arid areas, yet can tolerate light frosts. Sunhemp has a wider range of adaptation to soil types. It grows on poor soils, but growth on such soils is improved by fertilization. For fibre production, light, loamy, well drained soils are preferred, however in low lying clay soils it makes vigorous growth, but then the fibre is coarser and yields are lower.

Propagation and planting

Sunhemp is established by seed. Seed rates of 40-45 kg/ha are used when it is sown as a forage crop or as green manure, but when it is sown as fibre crop, seed rates of 50-55 kg/ha are used.

Growth and development

Sunhemp is fast growing. Seedings emerge 3 days after sowing, and rapidly produce a thick ground cover that smoother weeds. Extensive cross-pollination occurs in sunhemp and self-pollination takes place after the stigmatic surface has been insect or mechanically stimulated.

Nutrient management

In sunhemp, phosphorus and potassium are applied through fertilizers at 25 kg/ha each as basal dose at seeding. As sunhemp is a leguminous crop so there is no need to apply nitrogen.

Improved varieties for Punjab

PAU 1691: It is quick growing variety with erect growth habit. It flowers in 60-62 days matures in 136 days. It attains the height of 160-220 cm and thus adds 10.0-16.0 tonnes green biomass per hectare to the soil when buried at 45-60 days after sowing. The average seed yield is 12 q/ha.

Narendra Sanai-1: It is also a quick growing variety with erect growth habit. The leaves are broad in size and green in colour. It flowers in 98-100 days and matures in about 152 days. Its average seed yield is about 9.75 q/ha.

Irrigation

The crop sown for green manure during summer period requires 2-3 irrigations depending upon the prevailing weather conditions. However, the grain crop should not suffer water stress at the time of flowering and grain development stages.

Weed management

For seed production, the crop may require one hoeing after one month of sowing to keep the weeds under check. However, it is a quick growing crop and smothers weeds very effectively.

Diseases and pests

Sunhemp is attacked by many diseases and pests, including various fungi, insects and nematodes, but they usually cause little economic damage. In India, anthracnese caused by *Colletotrichum curvatum*, wilt caused by *Tusasium undam* can be serious. Damage from insects is more severe if crops are planted in the same area for more than 3 consecutive years. Fungicide seed treatments and crop rotations are the most recommended and practiced disease control measures.

Harvesting

Pod formation (120-140 DAS) is considered optimum stage of harvest for fibre. The plants are cut with sickles and the bundles are retained in the field for 2-3 days for the leaves to be shed. The process of retting and extraction of fibre are similar to that of jute.

Yield

Fibre yield is around 400 kg/ha. However, a well managed crop can yield up to 1500 kg/ha. The fibre is dull yellow, somewhat cease, strong and durable. Due to high cellulose content, low lignin, the fibre is used in the manufacturing of tissue paper and paper currency. In rural areas, it is used for making ropes, twines and nets. Sunhemp fibre is not used for textile purposes, unlike jute and mesta.

MESTA (*Hibiscus sabdariffa*)

Crop

Flower

Two species i.e. *Hibiscus cannabinus* (HC) and *Hibiscus subdasiffa* (HS) var. *altissia* are commercially known as mesta or kenaf. Mesta is also known as Bimeli jute, Deccan Hemp or Java jute. HC and HS are important jute supplements and show wide adaptability unlike jute. At present, both the species are known as mesta. Fibre is extracted from barks (bark fibre) of the two species.

Origin and distribution

The *H. cannabinus* had its possible origin in Africa where it is found in wild habit and also in subtropical regions of Asia but controversy arises as to the origin of *H. subdasiffa* attributed to both Africa and India.

Area and production

The area under Mesta, was 60.0 thousand ha with total production 5.28 lac balls of 180 kg each during 2014-15 (Ministry of Statistic and Programme implimentations). In India, Andhra Pradesh has the largest area and production.

Climate and soil

A warm humid climate suits both *H. cannabinus* and *H. subdasiffa*. Both come up well in drier rainfed areas, the former being drought resistant. In area with 500-900 mm rainfall, *H. cannabinus* suits better by virtue of shorter duration and faster growth. Both *H. cannabinus* and *H. subdasiffa* withstand temperature range of 17° to 44° but optimum is around 34°C. Mesta is grown on a variety of soils. However, clay loams with good internal drainage are ideal. Acid soils are not suitable. With a pH above 7 chlorosis appears in *H. subdasiffa* mesta.

Improved Varieties of Mesta released by ICAR

2016-17

Mesta	Area of adoption	Salient features
Satyen	West Bengal, Assam, Bihar and Odisha for mid-April to last week of May sowing under mid and highland rainfed situation	Resistant to foot and stem rot disease; with very fine (2.54 tex) fibre quality and lesser defective fibre
Dharwad Mesta 1	Karnataka	High yielding and fine fibre quality

Seeds and seeding

The seed is sown by broad casting or drilling in rows. However, row seeding is ideal, as it allows intercultivation with implements for weed control. For broadcasting, the seed requirement is 20 kg/ha in the case of *Hibiscus cannabinus* and 25 kg/ha in the case of *H. subdasiffa*, for drilling with a spacing of 30 cm x 10 cm, *H. subdasiffa* type requires 12 kg/ha and *Hibiscus cannabinus* type 15 kg/ha.

Nutrient management

A well managed crop requires 50 kg N, 30 kg P_2O_5 and 40 kg K_2 O/ha. Nitrogen is the critical and response to applied nitrogen seems almost universal. The rate of nitrogen varies from 35 to 80 kg/ha depending on the soil type and rainfall during the crop season. Nitrogen is recommended in two splits: half dose along with entire dose of phosphorus and potassium at sowing by placement and the remaining at the time of first hoeing (20 DAS). If dressing is not possible due to drought, nitrogen is recommended as foliar spray (2.5%) around 20 DAS.

Weed management

Critical period of weed competition is upto 45 DAS. For a broadcast crop, two hand weeding should be given around 21 and 35 DAS. For a row crop, two intercullivation at 20 days interval starting from 20 DAS followed by a hand weeding are ideal and economical. Among, the herbicides, Treflan (trifluralin) 0.5 to 1.0 kg a.i/ha is effective when applied as pre-plant.

Plant protection

Jassid: Spray Imidachlirpid 17.4 EC at 0.005% (0.3 ml/l) or Accphate 75 SP at 0.11% (1.5 g/l).

Mealy bugs: Spray Accphate 75 SP at 0.11% (1.5 g/l) or Dichlorvos 76 EC at 0.08% (1.0 ml/l).

Stem rot: Treat the seed with Mncozeb (3 g/kg) or drench the soil with Maneozeb (0.2%). Variety AMV 3 (Surya) is tolerant to stem rot.

Leaf Blight: Mancozeb 0.3% (3 g/l) spray can control leaf blight.

Harvesting

All the *H. subdasiffa* types are longer duration (180-210 days), where as *Hibiscus cannabinus* flowers in about 150 days. The *H. subdasiffa* type are harvested at 5% flowering (first fortnight of November) and the *Hibiscus cannabinus* types at boll formation stage (second fortnight of October). Delayed harvesting gives more fibre but of coasser quality. Harvesting up to the ground level results in good quality fibre.

Harvested sticks are spread in thin layers, allowed to dry for 2-3 days and made into small bundles. Urea is sprayed (1.25%) on the bundles and after 24 hrs, They are placed side by side horizontally for paper setting in the same manner as in the case of jute.

13

HINTS FOR ACHIEVING HIGHER YIELD OF FIELD CROPS

With the sharp development of urban and industrial growth, the area under agriculture is declining year by year. On the other hand, human and animal population is increasing at a very fast rate. So, to meet the growing demand of food and fodders, a vertical increase in yield is required. Crop productivity per unit area can be enhanced with the manipulations of some agronomist practices such as sowing the crops at appropriate time, selection of suitable field for raising a crop, adoption of best planting techniques, use of appropriate seed rates, adoption of proper irrigation schedule, proper and need based application of fertilizers, adoption of zero till technology, green manuring, mulching, addition of well decomposed farm yard manure, timely and proper weed management, selection of recommended crops and crop varieties, proper control of insects and pests, adoption of marketing tips etc.

PROPER ADOPTION OF AGRONOMIC PRACTICES

For getting higher yield, adopt each agronomic practice in such a way that a healthy crop is raised with judicious use of inputs which are discussed as below:

1. Selection of crops and crop varieties

All soils are not suitable for growing all crops but every crop has a specific requirement for a specific soil type. For instance groundnut requires loamy sand (light) soil as these soils facilitates more pegging and pod development. When groundnut is grown on heavy (clayey) soils, its yields are less because these soils resist pegging and pod development. Moreover, nutrient requirement of this crop are less (being leguminous crop) which can be met in light soils (loamy sand). On heavy soil vegetative growth will be high but reproductive growth will be less. Apart from this crop, arhar, bajra, guara, barley, gram etc. are best suited for light textured soils. Crops like maize, cotton, sugarcane, potato, pulses etc. are best suited on medium textured (sandy loam) soils. These crops can neither tolerate to severe water stress nor to excessive water as these crops die due to excessive stagnation of water. Rice crop requires heavy soils (clayey) as it has excessive requirement for water. Rice especially direct seeded crop sown on light soils shows severe deficiency of iron and other micro nutrients. Hence, cultivation of rice (direct and transplanted) should only be done on medium to heavy textured soils. Wheat cultivation can be done on any type of soil; however, on light soils its fertilizer requirements are more. Farmers should not cultivate any crop which is new to their areas without the consent of agriculture experts as this crop may not adapt to the area and secondly it may have marketing problems.

Always cultivate varieties of the crops which are recommended by the State Agricultural Universities because unrecommended varieties may be more liable to attack by insect pests and diseases or may have less yield potential. The most important point is that select crop varieties as per time of sowing. Varieties varies for their duration for maturity and can be classified as :

I. Long duration varieties

II. Medium duration varieties

III. Short duration varieties

The yields of long duration varieties are highest and must be sown during the start of recommended time of sowing. A farmer should not stick to one variety only as due to seasonal variations, its performance may not be better in a particular year may be due to more attack of insect pests and diseases. The entire area of farm should be brought under 2 to 4 or even more varieties in order to avoid risk of seasonal variations. When the field is vacated late, adopt short duration varieties and do not go for long duration varieties. Higher yields can be achieved by selecting recommended varieties as per their time of sowing.

2. Tillage requirement

These are two main objectives of tillage. (1) to control the weeds; (II) To pulverize the soil. However, excessive tillage can destroy soil structure. In general depending upon the soil type and weed intensity, 3-5 ploughing are sufficient in order to prepare fine seed bed. The concept of zero tillage is gaining popularity these days as there is lot of saving in energy and time of the farmer and hence cost of cultivation can be reduced and net profit can be improved. A contact and non-residual herbicide like Gramoxone (paraquat) or Round-up (glyphosate) at 0.5 to 2.5 lit/ha depending upon stage and intensity of weeds should be sprayed few days before sowing the crop and then sow the crop direct with the tractor drill i.e. without tillage. The weed intensity is less in the fields under zero till system because seeds lying in lower soil layer do not get chance to germinate. The disadvantage of this technique is that perennial weeds like *Cynodon dactylon* (Khabbal/Bermuda grass), *Sorghum halepense* (Baru), *Cyperus rotundus* (Dila/Motha) etc. dominates in zero till fields as the vegetative parts of these weeds remains in the viable form because these are not uprooted for many years. The zero till sowing must be done with care because many factors like depth of seed placement, adequate moisture at sowing govern germination and hence crop stand.

With continuous tillering operations, especially with puddling a, hard pan develops beneath the soil layers resulting in poor infiltration rates. Under such conditions, deep cheesling is required to save the crop from damage due to stagnation of water. The performance of different tillage systems on wheat yield is presented in Table 13.1 as reported by Walia *et al* (2005). They reported that with deep tillage (with inverted plough) significantly higher wheat yields were obtained as compared to zero tillage treatments during first year and these differences were narrowed down during second and third year of experimentation, as yield differences between different treatments were found to be non-significant.

Table 13.1. Grain yield of wheat as influenced by tillage treatments

Tillage treatments			Grain yield (kg/ha)			
2001-02	2002-03	2003-04	2001-02	2002-03	2003-04	Mean
Inverted	Zero	Zero	3009	3316	3470	3265
Inverted	Normal	Normal	3144	3061	3450	3218
Zero	Zero	Zero	2655	2590	3240	2828
Zero	Normal	Normal	2409	3197	3370	2992
LSD (p=0.05)			248	326	NS	245

3. Date of sowing

Date of sowing is the key agronomic practice governing crop yield. A timely sown crop gives higher grain yield as compared to late sown wheat with similar inputs i.e. seed, fertilizers, irrigation, pesticides etc. So, the economics of the farm can be easily increased by managing timely sowing. It has been seen at farmer's level that fields are lying vacant but sowings are delayed without any valid reason i.e. just due to the laziness of the farmer. Early/timely sown crop develops good growth parameters i.e. tillering, leaf area, stout stem, well developed reproductive parts because optimum time is available with the crop plants for completion of vegetative and reproductive stages. On the other hand, late sown crops hurridly complete vegetative and reproductive stages resulting in poor development of growth parameters and hence poor yield. It is non-monetary input for increasing crop yields. On the basis of field experiments, it has been proved that adequate yield reductions occur with delay in sowing but a farmer has to spent nearly same money on farm inputs for timely and late sown crops. So, sowing the crop at optimum time is a big success in agriculture. Sometimes fields are not vacated in time by the previous crops for sowing of succeeding crop. For example, sowing of wheat is delayed after cotton, basmati rice, potato, sugarcane etc. and under such situations go for zero tillage by applying pre-sowing irrigation before harvest of previous crops and select crop variety recommended for late sowings.

Another indirect benefit of early sowing is that some problematic weeds can be easily controlled by adopting this technique. For instance *Phalaris minor* (Sitti/ Gullidanda) is a troublesome weed of wheat in rice-wheat rotation, can be controlled by early sowing of wheat (among the recommended time) because at that time suitable temperature required for germination of this weed is not available. At this prevailing temperature especially in Punjab conditions, wheat can germinate and *Phalaris minor* seeds germinate in less number. Data presented in Table 13.2 (Virk *et al*, 2003) indicated that a significant decrease in grain yield of wheat was obtained by sowing wheat on November 10 and November 25 as compared to October 25 during both years. Data also indicated that dry matter of *Phalaris minor* was significantly higher in November 10 sown crop (as this period is ecologically most suitable for *P. minor* seeds to germinate) as compared to early sown crop (October 25) and late sown crop (November 25). With delayed sowing more seeds of *P. minor* germinates prior to crop sowing. Early sown crop yielded more due to less intensity of *P. minor* and more availity of time for the crop to nature.

Table 13.2. Effect of sowing date on dry matter accumulation by *Phalaris minor* and grain yield of wheat

Date of sowing	Dry matter of *P. minor**		Grain yield (q/ha)		Mean
	I	*II*	I	II	
October 25	5.0(24.2)	5.1 (25.1)	36.0	40.9	38.5
November 10	5.8(32.6)	5.7 (34.7)	31.8	34.9	33.4
November 25	5.2(25.8)	4.8(23.9)	29.8	32.0	30.9
LSD (P=0.05)	0.17	0.23	1.6	1.3	

*Figures in parentheses are original values and figures without parenthesis are transformed and statistically analysed values

4. Crop rotation

Crop rotation have lot of benefits in agriculture.

I. Maintains soil fertility
II. Maintains physical conditions of soil
III. Helpful in non-monetary control of weeds
IV. Avoids excessive depletion of ground water
V. Nutrients from deeper layers can be exploited by sowing deep rooted crops
VI. Less build up of soil born diseases and insect pests
VII. Deficiency of micro-nutrients can be avoided
VIII. Improvement in crop productivity
IX. Parasitic weeds can be effectively controlled by growing non-host specific crop in the infested field
X. It is an ecofriendly approach

Growing of same crop year after year on the same field have lot of problems resulting in declining trend in yield with passage of time. For crop sustainability, crop rotation is an ideal approach. For maintaining soil fertility as well as natural resources always add legume crop in non-legume cropping system. The area under the crops such as rice which exhausts natural resources especially ground water, should be kept under control. Ecofriendly crops such as legumes and oil seeds must be included in the cropping system. Do not prefer non-legume crop rotation such as rice-wheat as these crops exhausts the field for same type of nutrients resulting in deficiency of nutrient elements especially micro-nutrients such as zinc, iron, manganese, copper etc. Another point of consideration is that deep rooted and shallow rooted crops should be kept in a cropping system to improve soil fertility.

Rice-wheat cropping system is very exhaustive system resulting in significant reduction in available N, P and K (Table 13.3) as well as Zn, Cu, Fe and Mn content (Table 13.4) in soil as compared to the cropping system in which summer moong or potato was included (Anonymous, 2008). Soil analysis was done seven years of completion of each cropping system on fixed plots.

Table 13.3. Soil fertility status after 7 cropping cycles (0-15 cm depth)

Cropping Systems	Available nutrients (kg/ha)		
	N	P	K
Rice-wheat	218.1	32.1	128.6
Maize-Wheat-Summer Moong	232.8	46.2	142.6
Maize-Potato-Summer Moong	240.0	45.2	149.9
Maize-Potato-Onion	252.1	48.9	160.8
Summer Groundnut-Potato-Pearlmillet (Fodder)	229.6	45.8	136.7
Initial status	185.9	43.7	131.9

Table 13.4. Micronutrient status after *Rabi* 2007-08 in 0-15 cm depth

Cropping Systems	Micronutrients status in soil (ppm)			
	Zn	Cu	Fe	Mn
Rice-wheat	2.97	0.52	9.66	8.72
Maize-Wheat-Summer Moongbean	3.20	0.42	10.45	8.49
Maize-Potato-Summer Moongbean	6.68	0.69	9.64	10.92
Maize-Potato-Onion	6.16	0.66	10.56	12.42
Summer Groundnut-Potato-Pearlmillet (Fodder)	2.88	0.71	7.69	10.50

Another advantage of crop rotation is that management of weeds becomes very simple with the adoption of this manipulation. *Phalaris minor* (*Gullidanda/Sitti*) is a troublesome weed of wheat in rice-wheat cropping system. Rotation of wheat with berseem or potato or rapeseed and mustards helps in elimination/reduction in the population of this problematic weed. The experimental results reported by Singh (2002) indicated that growing of berseem for three years helps in eliminating seed bank of *Phalaris minor* from soil (Table 13.5) as grown up weeds of this weed are cut with berseem and no seed is produced by *P. minor*. Similarly inclusion of potato in the cropping system also helps uprooting of *Phalaris minor* plants during earthing up or at the time of digging of potato resulting in poor seed bank in soil. So, inclusion of berseem or potato in cropping system, helps to eliminate *P. minor* within three years.

Table 13.5. Effect of crop rotation on dry matter and seed bank of *Phalaris minor*

Crop Rotation	Dry matter of *P. minor* (q/ha) (at harvest of *Rabi* crops)	Seeds of *P. minor* (no/m^2)*(After harvest of *Rabi* crops)
Rice-Wheat (Herbicide)	2.09	1.80
Rice-Wheat (Control)	45.53	10.30
Rice-Potato-Wheat	0.0	0.0
Rice-Potato-Sunflower	0.0	0.0
Rice-Berseem	0.0	0.0
Rice-Gobhi Sarson	0.13	0.50

* 0-7.5 cm soil depth

5. Plant population

An ideal plant stand is a key to success of all field crops. A gappy stand (poor population) may result in less yield with same inputs. Moreover, weed population is more in the field with less crop stand because crop is unable to provide proper smothering potential on the weeds. So, crop with normal stand had always fewer problems of weeds. To attain proper crop stand, use of viable seed at recommended seed rates is very important. So, before sowing any crop, check the viability of seeds and do not use seed with poor viability. The seed should be free from weed seeds, broken seeds, diseased seeds and foreign bodies. The growth and yield attributes will be on the higher side in crop with normal stand. However, use of excessive seed rates especially in non–tillering crops like maize, cotton etc. is harmful as it results in poor development of reproductive parts. So, always use viable seed at recommended rates for attaining higher yields.

6. Planting pattern

Planting pattern is another agronomic technique by which crop yields can be increased with the use of same or even less inputs. The objective of appropriate planting techniques is to cover the field space uniformly for best utilization of solar energy and other inputs required for plant growth. Closer spacing, cross sowing, bed planting are few important planting techniques which can be exploited for higher yield. In wheat crop, the seed rate for closer (15 cm) sowing and cross sowing (22.5 cm x 22.5 cm) is as recommended for this crop with normal planting techniques. In cross sowing half the seed rate is used in one direction and half in another direction. Kler and Bains (1985) reported significantly higher PARI (%) in bi-direction (cross) sowing as compared to unidirection sowing when recorded at milk and hard dough stage (Table 13.6). Also significantly higher wheat yield was recorded in bidirectional sowing than unidirection sowing due to more photosynthetic efficiency of former treatment.

Table 13.6. PARI (%) as influenced by planting pattern of durum wheat

Direction of sowing	PARI (%) at different grain stages of crops				
	Tiller initiation	Jointing	Milk	Hard dough	Yield (q/ha)
Unidirection (E-W)	18.1	65.6	87.2	94.6	50.2
Unidirection (N-S)	15.8	67.0	89.4	95.4	51.2
Bidirection (E-WxN-S)	21.5	68.4	91.2	96.2	55.1
C.D. (0.05%)	NS	NS	1.8	0.9	3.6

Bed planting is another technique with lot of advantages such as :

I. No crop mortality due to excessive rainfall or irrigation
II. No crop injury due to shallow water table (waterlogged soils)
III. Less lodging due to speedy winds
IV. Less infestation of weeds
V. Mechanical interculture is possible
VI. Less attack of insect pests and diseases
VII. Good quality produce can be obtained
VIII. Less seed rate is required
IX. Less dose of nitrogen and other elements
X. High saving in irrigation water

Bed planting technique can be adopted in wheat, rapeseed and mustard, gram, maize, moong, mash, soybean, arhar, sugarcane and many more crops. The technique is very helpful for the crops which loose germination with rainfall especially on medium to heavy textured soils such as soybean, moong, mash, maize etc. So, the risk of germination in flat sowing technique during rainy season crops can be covered with bed planting technique. Growing of intercrops in bed planted crops has lot of scope. Few examples are : (I) Intercropping of mentha (in furrows) in bed planted wheat crop is very common in Punjab and (II) Intercropping of summer moong, summer mash in spring crop of sugarcane and gobhi sarson, wheat and many vegetables in autumn crop of sugarcane for high productivity.

7. Fertilizer application

Fertilizer is the costlier input in agriculture and its judicious use not only improves yield but also reduces cost of cultivation resulting in higher returns. The recommendation of fertilizers especially N, P and K is different for different crops depending upon the response of a particular crop to a particular nutrient element, soil type, irrigation facilities available, environmental conditions etc. So, recommended dose of macro-nutrients (N, P and K) and need based use of sulphur and micro-nutrients must be done for higher yields. Time of fertilizer application is very important and delay in their applications may give poor response. It has been seen that most of the farmers make injudicious use of fertilizers i.e. they used very high dose of fertilizers especially nitrogen which results in crop lodging and more attack of insect pests and disease due to more succulancy of the crop. Hence, always use recommended dose of fertilizers at recommended time in order to make agriculture more economical.

Soil testing is very very important before sowing of each crop and as per test report, fertilizer should be added in order to make judicious use of inputs. Soil samples from plough layer (0-15 cm depth) be taken from 5 to 7 places from the field at random and mix the soil and a composite sample be analysed from a reliable laboratory.

Apart from quantity of fertilizer, their method of application is very important. Side placement of fertilizer is very effective than broadcast method and as far as possible drill the fertilizer for higher efficiency. Drilling of fertilizer is also very helpful in reducing weed growth as in drilling method, fertilizer is placed in the vicinity of crop seeds, hence crop plants will remove fertilizer at higher rates than weed plants. Broadcasting of fertilizer provides equal chances for crop and weed plants to rob the fertilizer. Ahuja and Yaduraju (1989) reported significantly less populationof *Phalaris minor* in placement method (2.5 cm below seeds of wheat) as compared to broadcast method of fertilizer application (Table 13.7). A significant increase in grain yield of wheat was recorded in placement method of fertilizer application.

Table 13.7. Effect of fertilizer application methods in population of *P. minor* and grain yield of wheat

Treatments	*P. minor* plants (No./m^2)	Grain yield (q/ha)
Broadcast	1.9	48.7
Placement (2.5 cm below seed)	1.6	54.7
L.S.D. (p=0.05%)	0.1	2.45

8. Addition of organic manures

Farm yard manures (FYM), compost, green manuring and crop residues are the important sources of organic manures. Few advantages of organic manures are:

I. Provides all (macro and micro) nutrients for plant growth
II. Improves physical and chemical conditions of the soil
III. Improve water retension capacity of soil
IV. Makes the soil porous which helps better plant growth
V. Helps in modernization of soil temperature
VI. Improves quality of farm produce
VII. Improve microbual growth in soil.
VIII. Improves crop yield as some crops like maize, potato, turmeric, vegetables are high responsive to organic manures.

Farm yard manure is very common organic manure and is used widely by the farmers. Well decomposed FYM should be used. On the other hand, use of undecomposed (raw) FYM may lead to more weed problems and more attack by termites. Generally fodders mixed with mature weed plants are fed to animals and the weed seeds come in the intact form along with cowdung which is not composted in proper pits. These weed seeds in the viable form are added in the field along with FYM where they create big problem. So, always use well decomposed FYM for better results and never use raw FYM.

Green manuring is another source of organic manures. Few crops like Sunnhemp/ Saan (*Crotalaria juncea*), Dhaincha (*Sesbania aculeata)* and cowpea (*Vigna unguiculata* L.) are best suited for green manuring due to quick growth and leguminous in nature.

Anyone of these can be sown when a period of 40 to 50 days is available between harvest of one crop and sowing of next crop. These crops are quick growing and less input responsive crop, acquires considerable biomass within a short period of time. Bury these crops in soil just before puddling rice and 7-10 days before sowing other crops like maize, potato, wheat etc. Bury these crops before flowering as mature crop will take more time for decomposition.

Addition of crop residues is another option with the farmers to enrich soil. Generally straw of few crops like wheat, maize etc. is very economical and cannot be spared for incorporation. However, rice straw can be incorporated in soil as it is poor quality fodder. Either incorporate rice straw in soil after combine harvesting by running disc harrow twice in the field or adopt zero till technology and rice straw can be left on soil surface as surface mulch. However, either remove or spread the loose straw uniformly in the field before sowing of wheat crop. For better germination sow wheat at proper soil moisture and at proper depth and zero till is successful provided germination is ensured. Weed population is generally less in zero till wheat as weed seeds do not germinate due to non-availibility of light or if germinate they are in less number due to the presence of thick layer of rice straw. In case of pulses like moong and mash, incorporate plant material into the soil after removing the pods.

9. Adoption of intensive cropping system

The area under cultivation is decreasing day by day due to fast development in urbanization, so the only option to increase food production is adoption of intensive cropping systems. Due to development in mechanization, rice-wheat cropping system is widely adopted. As both crops are non-legumes, so depletion of soil is high and to maintain soil fertility, non-legume crop(s) are to be added in the prevalent cropping system. Rice had adverse effect on soil properties and in order to maintain soil health, alternate cropping systems must be adopted. Five cropping systems viz. Rice-Wheat, Maize-Wheat-Moongbean, Maize-Potato-Moongbean, Maize-Potato-Onion and Groundnut-Potato-Pearl-millet (Fodder) were compared by Walia *et al.* (2010) for seven years. Experimental results indicated that highest rice equivalent yield and hence highest profit was recorded in Maize-Potato-Onion which was followed by Groundnut-Potato-Pearlmillet (Fodder), Maize-Potato-Moong-bean, Maize-Wheat-Moongbean and all these systems were found better than the prevailing Rice-Wheat cropping system (Table 13.8). Also irrigation water applied (cm) was lowest in Maize-Wheat-Moongbean followed by Groundnut-Potato-Pearlmillet (Fodder), Maize-Potato-Onion and Rice-Wheat. Also water productivity was highest in Maize-Potato-Pearlmillet (Fodder) and lowest in Rice-Wheat cropping system (Table 13.8).

Table 13.8. Rice equivalent yield, irrigation water applied and water productivity of different cropping systems

Cropping Systems	Rice equivalent yield (q/ha)	Irrigation water applied (cm)	Water productivity (kg grain/ m^3 irrigation)
Rice-Wheat	132.4	212	0.625
Maize-Wheat-Moongbean	161.8	92	1.759
Maize-Potato-Moongbean	191.0	105	1.819
Maize-Potato-Onion	278.6	130	2.138
Groundnut-Potato-Pearlmillet (Fodder)	233.0	103	2.262

10. Irrigation techniques

Water is the most valuable natural resource required for plant growth. Due to excessive and injudicious use of water, the water table is going down which is very serious concern and big challenge in future. Generally farmers believe in flood irrigation due to free supply of electricity especially in Punjab. The area under rice is increasing and due to its high water requirements, water table is going down year by year. So, there is a need to use irrigation water more judiciously which is only possible through drip or sprinkler irrigation technologies. It is advisable to select low water requiring crops and watering should be done with drip irrigation which results in lot of saving of water as well as in improving water use efficiency.

Bed and ridge planting techniques are very helpful in saving irrigation as only furrows are irrigated which requires less time per unit area. Direct seeding of rice also requires less water than transplanted rice as it does not need puddling which requires huge amount of water. Irrigation water can be saved by replacing coarse and fine rice with basmati (super fine varieties) rice as water requirements of latter group of rice is low. So, there is a great need to economise irrigation water for the welfare of States in general and India in particular.

11. Proper weed management

Weeds compete with the crops for all essential inputs like nutrients, soil moisture, light, space etc. Weeds must be controlled before the start of critical period of crop weed competition in order to attain higher crop yields. Weed flora varies with type of crop, soil type, available irrigation facilities, tillage techniques, environmental conditions etc. In general grass type weeds are more competitive than broad leaved weeds. Crop weed association is very common and those weeds having ecological similar requirements as of crop, develop association with that particular crop. For instance *Echinochloa crusgalli* (Swank/Maddal) is associated with rice and *Phalaris minor* (Gullidanda/Sitti) with wheat in rice-wheat system due to their similar ecological requirements. Few weeds such as *Trianthema portulacastrum* (Itsit/Chupati) germinates in flushes and start producing seed during very young age, are very difficult to control.

Crop losses due to weeds may vary with type of weeds and their densities. In general 20 to 50 per cent losses due to weeds in field crops are very common. These losses are very less in smothering crop like rapeseed and mustards and high in less smothering crops like mentha and turmeric. Walia (2014) reported that with effective control of weeds, 283.8 per cent oil yield in menthe can be increased as compared to unweeded control (Table 13.9). This yield increase in chemical method is 118.2 per cent in maize, 102.4 per cent in transplanted rice and 83.7 per cent in wheat infested with *Phalaris minor*.

Table 13.9. Grain yield of important field crops as influenced by different treatments

Methods of weed control	Wheat (q/ha) infested with		Rice (q/ha)	Maize (q/ha)	Sugarcane (q/ha)	Groundnut (q/ha)	Mentha oil (kg/ha)
	P. minor	Wild oats					
Use of herbicide	45.0	41.0	60.1	31.2	520	18.1	68.7
	(83.7)	(95.2)	(102.4)	(118.2)	(77.5)	(52.11)	(283.7)
Mechanical	34.4	37.3	55.7	28.0	413	16.2	53.1
	(40.4)	(77.6)	(87.5)	(95.8)	(41.0)	(36.1)	(196.7)
Control (Unweeded)	24.5	21.0	29.7	14.3	293	11.9	17.9

Figures in parentheses are percentage of yield increase over unweeded control

Weeds can be controlled with the adoption of preventive, mechanical, cultural, chemical, biological and integrated methods of weed control. We should not depend upon a single method and the adoption of integrated approach is the best weed management method. Preventive, cultural and biological methods have limited scope. Mechanical method of weed control is loosing its importance due to high wages of farm labour and its uncertainty in timely availability. Chemical method is very favourable method among farmers. More than 90 per cent area under wheat and rice is treated with one or the other herbicide under Punjab conditions. However, excessive and indiscriminate use of herbicides is harmful not only on present crop but also on following crops due to high residual effects of some herbicide groups. These herbicides also pollute the environment.

On the other side, herbicides are very useful tool in agriculture provided they are used judiciously. Always use recommended dose of recommended herbicides at recommended time and method. Do not use unrecommended herbicides as well as unrecommended brands of recommended herbicides.

Always spray the herbicides in straight strips/bands and do not move the nozzle too and fro. For pre-plant or pre-emergence herbicides, use 500 to 750 litres of water/ ha and for post-emergence applications use 200 to 250 litre/ha of water for effective weed control. Always use flat fan or flood zet nozzle for herbicide use and never use cone type nozzle. The height of nozzle should be 50 cm from ground/crop canopy. Do not spray on windy days. Also during hot months of May and June, always spray either in morning or evening hours and do not spray during noon. Do calibration before spraying the herbicide and use same nozzle and depute the same person who have done calibration for spraying in the field. Check the nozzle before use by spraying plain water on a pucca floor and if wetting is not uniform then change the nozzle and when wetting is uniform then nozzle is alright. Before each filling, stir the spray fluid thoroughly. Knapsack sprayer is very common spraying equipment for herbicides. Tractor mounted sprayers can also be used especially for pre-plant or pre-emergence herbicides. Always purchase herbicides from a reliable shop keeper after obtaining receipt. The above listed points must be kept in mind while using herbicides for attaining better control of weeds and hence higher yield.

12. Plant protection measures

The losses by insect pests and diseases are also high but less than that by weeds. Always treat the seed with recommended fungicide in order to control root born diseases. For the control of termites, seeds can also be treated with suitable insecticide just before sowing. Later on spray of recommended fungicide or insecticide should be done as per the disease organism/type of insect appearing on the crop. Always use recommended pesticides as per need and do not use these agrochemicals in excess. Spray only when population of insects exceeds thrash hold level. After use, store the left over pesticides (fungicides and insecticides) on separate shelves after proper leveling and away from herbicides. All the pesticides should be kept away from the approach of children.

13. Precision farming

Precision or site-specific crop management refers to a management system of production agriculture, using diverse technologies to increase field productivity and protect the environment. Under precision agriculture, however, inputs are applied in each part of the field according to its unique set of conditions. Moreover, when to apply, how to apply, how much to apply, kind of inputs in relation to water, nutrient, pesticides etc., the residual effect of nutrient and crop-residue and left over water on the succeeding crop and their behaviour with the environment in time and space are studied from a very close angle so that resources wastages may be reduced to minimum possible. Normally farmers follow one uniform practice of water, nutrient and pesticides application at their farm. While in this concept the variation observed within the field itself is to be taken care of. Even a field is to be visualized critically and assurance of balanced supply in desired amount in each nook and corner is to be maintained, led to sustainable yield levels in different cropping systems.

With precision farming, cost on inputs can be reduced while net profit is increased. Application of nitrogen with leaf colour charts in important crops like rice, wheat, maize etc. helps in saving nitrogen as it is applied as per the requirement of crop. In this technique, the colour of leaf is matched with colour chart and nitrogen dose is applied as per actual requirement of plant. Laser land leveling is another approach for saving irrigation water. Generally fields are not in proper level and hence requiring variable depth of irrigation water at variable sites of the field, thus, requiring huge amount of water for first irrigation. Moreover at lower spots of field, crop may be damaged due to excessive water. So, with proper laser land leveling technique, not only water is saved but a uniform healthy crop can be grown. Input cost on irrigation water and nutrients can also be saved with the adoption of bed planting system of crops. Tensiometer can be used in transplanted paddy to save irrigation water. Also transplanted puddle rice do not need water stagnation upto maturity and after 3 weeks of transplanting apply irrigation two days after the ponded water percolates in the field.

14. Cultivation of high value crops

Medicinal and aeromatic crops, vegetables, floriculture are high value crops which improves the economic condition of the farmer. The decision to cultivate high value crops should only be taken after studying the market. Also their cultivation should be started on small scale and when good marketing response is achieved, the area can be increased slowly and slowly. There are lot of medicinal and aeromatic crops which can be grown successfully in India. However, South India is very ideal for the cultivation of all spices and in North India cultivation of Mentha, turmeric, funnel, menthe etc. can be done depending upon the marketing demand. These crops are not becoming popular due to fluctuations in the market. Cultivation of vegetables is highly profitable provided they are produced as per demand in the market. Also cultivation of off season vegetables under protected conditions is highly profitable. Food processing and storage has a lot of scope in vegetable crops in order to cover glut in the market. In floriculture, cultivations of marigold and gladiolus is very common and highly economical.

15. Integrated farming

The various components of farming system are cropping system, dairy farming, poultry farming, pisciculture, fruit and vegetable growing, floriculture, agro-forestry, agri-horticulture, silviculture, rabbitary, ruckery, emu rearing, vermiculture, biogas production, agrobased small scale industryand milk-based household industry.

The integrated farming system brings prosperity to the farmers by judicious combinations of various available enterprises. It also involves recycling of farm resources resulting in optimization of farm input use. The farming system approach provides flow of money to the farmer round the year by disposal of milk, eggs, honey, fish, mushroom, fruits, vegetables, etc. Also with the effective recycling of bi-products of the corresponding enterprise as manure will result in improving the physico-chemical and biological properties of soil.

It also involves the integration of minor occupations viz. growing of flowers and vegetables, nurseries under polythene cover or in small poly houses, establishment of fruit and forest plant nurseries. The farm produce can be used for value addition products such as vegetable pickles, murabas, tomato sauces, garlic/onion powder, turmeric powder etc. Fruit preparations like jelly, jam, ambla murabba, apple murabba etc. are effective tool to make farming system approach more realistic and self dependent.

16. Marketing

Both quantity and quality of farm produce, plays very important role for increasing farm income. The uncleaned crop produce having infestation of weed seeds will fetch low price in the market. So, always grow clean crop free of weed plants especially thorny ones. For example, the seeds of *Rumex spinosus* have big thorns and their presence reduces market value of the wheat produce. Similarly, presence of *Convolvulus arvensis* (Lehli/Hirankhuri) in seed crop of wheat leads to rejection of the produce by seed certification agencies as the seeds of this weed can not be separated out from wheat grains by any mean due to their akin size. The produce with high moisture content is rejected in the market, so always harvest crop at proper moisture (in grain) or dry the produce in the sun before bringing to the market.

It will be advisable to study the market before selling the crop and prefer market with highest rate for disposing your produce. When there is a chance for increase in market price, the produce may be stored and can be sold with increase in market rates. Do not store the produce at high moisture content.

CONCLUSION

In agriculture, concepts of timely operations is very important in order to make this profession more profitable. The expenditure on inputs will remain more or less the same under optimum time and delayed operations. Proper sowing time along with optimum crop stand is very important for getting high yields. Judicious and need based use of inputs like fertilizers, irrigations, pesticides etc. should be done and their indiscriminate use must be avoided. Do not stick to one rotation and go for intensive intercropping system and high value crops. So, timely agronomic operations along with recommended practices are the key to successful farming for attaining higher yields.

14

FARMING SYSTEM

Farming refers to the process of conservation of solar energy in the form of economic products and system refers to a set of interrelated practices, processes organized into a productive functional identity. Farming system, therefore, refers to a set of agricultural activities organized and preserving the other set up like low productivity, ecologically sound environmental quality, maintenance of biological diversity and ecological stability. The farming system laid emphasis on system productivity and sustainable agricultural production rather than gross returns. A judicious mix of enterprises complementary to cropping and suited to the given farm situation and farmer's preference would bring prosperity under current population of 1.32 billion during 2016. The anticipated population will be 1.41 billion during 2026. To satisfy the projected population 289 and 349 million tonnes of food production is needed in 2030 and 2050 A.D., respectively. But due to industrialization and urbanization, the present net cultivable area of 141 million ha would decrease in the coming years. Therefore, it becomes necessary to increase the production level per unit area to almost double from the present production level. The production level can be increased by crop intensification in both irrigated and rainfed situations. The most of our area falls in tropics and receives abundant solar energy, therefore, making the successful cultivation of crops round the year. The income from arable farming alone is not sufficient for marginal farmers. The agro based enterprises such as dairying, poultry, pisciculture, sericulture, biogas production, mushroom production, agroforestry, agri-horticulture assume importance in enhancing farmers income. These allied enterprises should be selected in such a way that it fits well with farm level infrastructure and assures full answer to the problem of utilization of bye-products by effective recycling. The integrated farming system will solve various problems such as :

- Increasing food production
- Enhancing farmers income
- Improving the nutrition of family members of small farmers.

In the cropping systems, the huge amount of bye-products are produced which have economic value and this bye-products may go waste if not utilized in the animal component. Therefore, the integration of different agricultural allied enterprises with crops will provide effective ways to recyclable waste material of one component as an input to another integrated farming system component and this practice will reduce the cost of production. Integrated farming system is also a process of harnessing solar energy in the form of economic plant and animal products. Farming system is a resource management strategy to achieve economic and sustained agricultural production to meet the diverse nutritional requirement of the farming household and at the same time the integrated farming system will preserve the resource base and maintain environmental quality.

Farming system is a complex inter related matrix of soil, plants, animals, implements, power, labour, capital and other inputs controlled in part by farming families and influenced to varying degrees by political, economic, institutional and social forces that operate at many levels. The farming system conceptually is a set of elements or components that are interrelated which interact among themselves. It also involves the integration of minor occupations viz., growing of flower and vegetable nurseries under polythene cover/in small poly-houses, establishment of fruit and forest plant nurseries and using their farm produce for value addition products such as vegetable pickles, murabas, tomato sauces, garlic/onion powder, turmeric powder and likewise fruit preparations, Jelly, Jam, Amla muraba, Apple muraba etc. are proved other effective tools in farming system approach and help to make this approach more realistic and self dependent. Reddi and Reddy (2004) narrated farming system as the concept which takes into account the component of soil, water, crops, livestock and other sources with the farm family at the centre, managing agricultural and related activities and even non-farm activities. The characteristics of the farming system are: whole farm approach, farming based, holistic in nature problem solving, gender sensitive, farmers participatory and decision making extensive on farm approach and interactive and iterative. Biswas and Singh (2003) defined integrated farming as the integration of two or more enterprises for each farm according to the availability of resource to sustain and satisfy as many necessities of the owner as it is possible which leads to increased productivity per unit area, efficient recycling of farm wastes, better utilization of resources, generate employment, reduce the risks and ensure sustainability.

Farming system approach is considered as backbone of the farming community because majority of the farmers are having more than one enterprise. The improved technologies developed for crop production have made wide strides, which enabled the country self sufficient for food grain production. But the technologies towards integration of different enterprises, developed in an isolated manner and in piece meal. Numerous investigations conducted at different locations clearly elucidate the beneficial effects of integrated farming towards income generation, rational use of resources, increasing per unit productivity and above all making the agriculture eco-friendly. In true sense, the farmers are working in very diverse, risk prone and location specific environment which many a time enforce them not to follow the improved knowledge as such, because the technologies developed under resource enriched conditions does not suit them.

CONCEPTS OF FARMING SYSTEM

Holistic approach: It is the integrated approach not only ensures the rational use of the resources, provide regular employment by ensuring work throughout the year and simultaneous recycling of the leftover material of each enterprise. It fulfils the domestic requirement and makes the farming family self sufficient and regular marketing enable farmers to get more income per unit area.

Complex in nature: It warrants multidisciplinary production and marketing knowledge. The farmer should be well abreast with the new findings in each respective discipline being followed at his farm. The laxity in one component may affect the complete production chain.

Inter-related components: Farming system involves adequate combinations of agri-based enterprises. Crop husbandry is inter-related with livestock because grains are used for feed and plant biomass is used as cattle fodder. The cattle provide us milk, draft power and cowdung. Similarly poultry-crop husbandry, crops-livestock-fisherypiggery enterprises are closely related with each other. There are numerous combinations whose viability is directly linked with the location, resources available and socioeconomic conditions.

Maneuver of biotic and control use of inputs: It involves the skillful management of soil, plants, animals, power, implements, labour, capital and other inputs. Efforts are made to find out the ways and means by which one factor could effect the other in a synergistic way so that existing level of each components should not be depleted rather improve. To make the farming system well mechanized, the effective use of capital inputs is as important as the living things contributing towards economic yields and rather more dependent on input supply management.

Participatory bottom-up planning: The farming families should be involved at the grass root level with variable existing infrastructure. Their effective participation will help to bear the desired fruits. The farmers involvement in planning, mode of implementation, monitoring, mid way modification and finally in marketing and its further spread over the rest of area help the programme.

Pollution solving device: On account of its multifacet nature and favourable aptitude of enterprises to each other holds good for solving the location specific problems. Use of organic manures/crop residue recycling takes care of soil health but simultaneously check air and water pollution. Crop rotation and inclusion of legumes in different cropping systems control pests (weed, insect-pest and disease), improve fertility and increase productivity.

Concept of learning by doing: There is wide spread gap between the hypothesis across the mind and while giving practical shape to any programme because of many hindrances at farmers' level. With this approach, the farmers learn a lot and they require consistent technical support while in process. This concept also ensures earning while learning. Once, the one cycle is complete, it not only create awareness among farmers but in the next cycle they may apply precision at various juncture to make the programme more profitable. The procedure and practices to adjust the agri-based enterprises according to the resources availability become more common and perfect during learning process.

Socially desirable technologies: An important event determining the socially desirable technology dissemination and its acceptability must be arranged. Growing of tobacco cultivation in Punjab may not get impetus because it is considered as a social evil. Its off flavour during harvesting and processing afterwards create bitterness. Similar are the examples where people do not relish the meat of pigs or cattle. This factor is very crucial determining the success or failure of a system.

Create opportunities to develop models: About 70 per cent of the rural population is involved in agriculture. About 50 per cent of the farmers are marginal and small. Their social, economic, decision making, participation in extension programme, risk bearing capacity are entirely different than medium and large sized farmers. The living standard, dietary habit and literacy determine the future prospects. Therefore, different models pertaining to marginal and small farmers would help to ascertain their income with the different enterprises in various models. The models developed with farmers participatory approach will further prove more worthwhile than the model developed under hypothetical simulated conditions in on-station research. Development of models keeping in view the above constraints may prove as an asset for the upliftment of substantial component of farm population.

Government support: Indeed Government of India has launched various programmes for the betterment of the workers who are involved in agriculture sector viz., food for work programme etc. but to make a large chunk of farming community self reliant. A financial and technical support is thus required to promote the farming system approach which is eco-friendly, employment generative, productive and sustainable.

Sustaining the ecosystem for future generation: The farm household is the principal system. It consists of three basic sub-systems which are closely interrelated and interactive as follows:

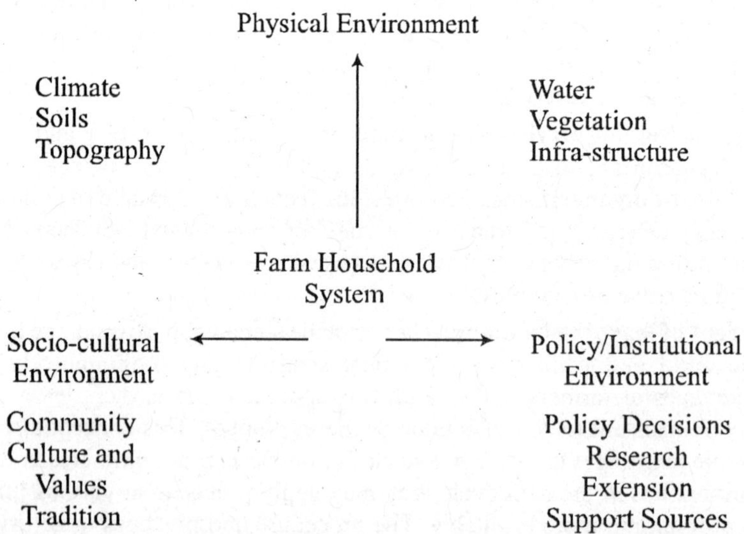

In view of serious limitations on horizontal expansion of land and agriculture, only alternative left is for vertical expansion through various farm enterprises required less space and time but giving high productivity and ensuring periodic income especially for the small and marginal farmers.

Components of farming system

The various components of farming system are cropping system, dairy farming, poultry farming, pisciculture, fruits and vegetable growing, goat rearing, piggery, sericulture, mushroom production, apiculture, floriculture, agroforestry, silviculture, rabbitary, duckery, emu rearing, agrihorticulture, vermiculture, biogas production, agrobased small industry and milk based household industry.

Advantages of farming system

I. The integrated farming system ensures stability and profitability.

II. The integrated farming system brings prosperity to the farmers by judicious combination of various available enterprises.

III. Involves recycling of farm resources so results in optimization of farm input use.

IV. The integrated farming system requires more labour so results in employment generation.

V. The farming system provides flow of money to the farmer round the year by disposal of milk, eggs, honey, fish, mushroom, fruits, vegetables etc.

VI. Various components are involved which produce different source of nutrition (protein, carbohydrates, fat, vitamins and minerals etc.) from same unit area and will help to solve malnutrition problem,

VII. In farming system, the effective recycling of by products of the corresponding enterprise as manure will result in improving the physicochemical and biological properties of soil.

VIII. The farming system considers all the technical and environmental elements. It helps in the maintenance of agro-ecosystem.

IX. All the branches of agricultural discipline are involved in solving location, specific farmer's problems.

TYPES OF FARMING SYSTEM

The farming is practised in different ways across the world. These are subsistence farming, commercial farming and other types of farming.

1. Subsistence farming

This farming system is practiced to meet the needs of farmers' family. In this system, low level of technology and the household labour is used to produce on small output. This subsistence farming can further be grouped into intensive subsistence and primitive subsistence farming.

I. **Intensive subsistence agriculture**: In this, the farmer cultivates on a small field using simple tools and with more labour. Fertile soil and climate with more number of sunshine days permits to grow more than one crop on the same land. Farmers usually grow rice and other crops include maize, wheat oilseeds and pulses. Intensive subsistence farming is common in thickly populated areas of South, South-east and East Asia receiving good monsoon.

II. Primitive subsistence agriculture:

(a) Shifting cultivation

(b) Nomadic herding

(a) Shifting cultivation: Shifting cultivation is a primitive technique between food gathering and food production. It is practiced in the thickly forested areas of Amazon basin, tropical Africa, parts of south-east Asia and north-east India. These are the areas receiving heavy rainfall and regeneration of vegetation is very quick. The price forest land is cleared by felling trees and burning them. The ashes are mixed with the soil and then crops are grown. After the soil loses its fertility, the land is abandoned and the cultivators reach to new area. Shifting cultivation is popularly known as jhuming in north-east India, Khallur or Karwa in Bihar, Dahiya or podes in Orissa. In Sweden, it is known as slash and burn method of cultivation. It is also known as Milpa in Mexico, Roca in Brazil and Ladang in Malaysia. Shifting cultivation takes into account 30% of exploitable land of earth and 10% world population.

Shifting cultivation has three phases i.e. Cleaning of land, cultivation and restoration.

Cleaning of land: During winter months, the forest vegetation including small and large trees are cut down. The slash is kept for few days to dry up and it is burnt during February and April to enrich the soil fertility and to produce charcoal. The stones are cleared from earths surface and an embankment round the boundary is made.

Cultivation: Slopes under shifting cultivation are cultivated during *Kharif* season with mono-cropping or with mixed cropping with crops like upland rice, finger millet, pigeon pea, Indian millet, blackgram, rice bean, niger, horse gram etc. A cycle of cropping may vary from 3 to 5 years with decline in yield in consecutive years.

Restoration: A short period of cropping is followed by a long period of fallow of 5 to 50 years to restore the fertility and productivity of land for natural restoration.

(b) Nomadic herding: Nomadic herding is practiced in the semi-arid and arid regions of Sahara, Central Asia and some parts of India like Rajasthan and Jammu and Kashmir. The herdmen more from place to another place along defined routes with their animals for fodder and water due to climatic changes. Sheeps, Camel, Yak and Goats are most commonly reared for providing milk, meat, wool and hides, etc.

2. Commercial farming

In commercial farming crops are grown and animals are reared for their sale in the market. The large land area is used and more capital is used. Most of work is done by machines. Commercial farming includes commercial grain farming, mixed farming and plantation agriculture.

I. **Commercial grain farming:** The crops are grown for commercial purpose. Wheat and maize are most commonly grown. Major areas where commercial grain farming is practiced are temperate grasslands of North-America, Europe and Asia. These areas are having less population with large farm size. Severe winter restricts the growing season and only a single crop is grown.

II. **Mixed farming:** The land is used for growing food and fodder crops and rearing livestock. It is practised in Europe, Eastern USA, Argentina, South-east Australia, New Zealand and South Africa.

The farming system comprises of crops and livestock rearing has been sustainable over centuries. In this system, animals are raised on agricultural wastes. The animal power is used for agricultural operations. The animal waste, cow dung is used as a fuel and manure. The yield becomes more sustainable as the waste of one component becomes input for another component and leaving almost no waste to pollute environment. It is of two types.

(a) Crop-livestock farming system: In the conventional system, the crop residues of important cereal crops like rice, wheat, maize are fed to the cattle and the cowdung can be used as a manure or as fuel after making cowdung cakes. Most of cowdung is used for fuel purpose but it is very good source of nutrients. The cowdung can be used as fuel and also for manure simply by introducing a biogas plant in a system. The biogas slurry can also be used as fish feed. The slurry should be incorporated into the soil so that nitrogen losses may be minimized.

(b) Crop-livestock-forestry farming system: Crop residues can be used as dry fodder and cowdung is added in the field as a source of farmyard manure. Forestry plantation can be done as sole or on the boundaries of the farm.

3. Other types of farming

I. **Ranching** : Refers to the practice of grazing animals on the public lands. Some public lands may also be used for raising livestock. The practice is common in Australia and Tibet. In India ranching is prevalent, to some extent, in the hilly regions that are rich in pastures and gazing lands.

II. **Dry and Irrigated farming**: Farming in areas where rainfall is deficient and there is no assured source of artificial irrigation, is referred to as dry farming. On the other hand farming with assured water supply from artificial sources of irrigation is known as irrigated farming. In dry farming the land starved of water is hard and poor in soil fertility where as irrigated farming is easier, more remunerative, flexible and adaptable to changing technology.

III. **Mixed farming**: Mixed farming is the combining of two independent agricultural enterprises on the same farm. A typical case of mixed farming is the combination of crop enterprise with dairy farming or in more general terms, crop cultivation with livestock farming. Mixed farming may be treated as a special case of diversified farming. This particular combination of enterprises, support each other and add to the farmer's profitability.

IV. **Single crop and multi-crop farming**: Single crop farming is a form of specialized farming. If a farmer specialises in crop enterprise, it is left to him to produce a single crop or a multiplicity of crops. In practice, it may happen that a farmer seldom produces a single crop except in the case of cash crops such as sugarcane, etc. even if a farmer wishes to specialize in crop cultivation, he will often choose more than one crop to safeguard against market uncertainties and also to ensure optimum utilization of his resources. It may, therefore, be observed that neither a single crop nor a multiplicity of crops would be a desirable proposition.

V. **Diversified farming**: When a farmer is engaged in a multitude of farm enterprises, it is referred to as diversified farming. If a large number of crop enterprises, with or without a number of non crop enterprise is run by a single farmer, it is referred to as diversified farming. Raising of five or six crops makes it diversified. The motive behind diversified farming is self sufficiency.

Advantages of diversified farming

I. **Better resource utilization:** It ensures steady utilization of farm labour throughout the year. There is regular flow of income as well. Land can be used more economically and without loss of fertility. Soil fertility is kept intact. It also reduces the necessity for leaving the land fallow.

II. **Better product utilization:** Different enterprises can be run on the same holding e.g. a farmer may raise crops and at the same time, may also keep a dairy farm, diversified farming ensures a regular flow of returns as different enterprises mature at different times of the year.

Other advantages

Farmers with no training can safely start diversified farming. Diversified farming helps in absorbing a larger volume of labour force and sustaining a larger population.

Limitations of diversified farming

I. **Poor resource utilization:** Too much of diversification may lead to mismanagement. Since too many enterprises shall be competing for the use of limited resources and supervision, it is likely that some of the enterprises might get neglected.

II. **Does not attract market:** Problems also arise on account of marketing because diversification means small scale production that may not attract the market.

4. Specialised farming

In a general sense, when only few enterprises are run by the farmer, in which he has acquired special knowledge, it is known as specialized farming. Specifically, specialized farming refers to only one kind of farm business such as raising food crops or rearing sheep or raising dairy cattle. Raising two to three crops makes it specialized. The motive behind specialized farming is profit.

INTEGRATED FARMING SYSTEM (CASE STUDIES)

Integration of two enterprises

Crops + Livestock

Crop + animal system form the backbone to agriculture in most of the developing countries. Livestock is held to meet the milch need, draught power and cow-dung/farm yard manure which when recycled improves the soil fertility. In addition, extra milk is sold to meet the daily miscellaneous expenditures. Hardly any farming family is available not holding the animal component with their agriculture. A study conducted by Singh *et al.*, 1999 with 240 farmers of Rohtak, Hisar and Bhiwani districts in Haryana which represented zones of different crop rotations (Wheat-Sugarcane in Rohtak, Wheat-cotton in Hisar and Gram-bajra in Bhiwani zone) revealed that maximum returns of Rs. 12593, Rs. 6746 and Rs. 2317 were obtained from one hectare with a buffalo in Rohtak, Hisar and Bhiwani, respectively. At Banglor in Karnatka, Basavaraj *et al.* (1999) conducted study on a sample of 100 farmers of whom 50 grow ragi and 50 grow ragi and/or sugarcane as major crops. The results indicated that adoption of improved practices in crop + dairy + sheep enterprises by the farmers who had grown sugarcane as a major crop was higher (Rs. 42984/-) as against ragi based farming system (Rs. 18076/-) as compared to Rs. 27189 and 4237 in above cropping system respectively.

Fish – Poultry/pigs

In Tamilnadu, Jayaraman (1993) reported the feasibility and economics of fish – poultry farming. Two fish ponds each 0.038 ha with one meter depth were used. Stocking density was 10,000 fish per ha, comprising silver carp, catla, rohu, mrigal, common carp and grass carp. No manuring, fertilization or feeding was done as the poultry droppings were recycled as manure and fish feed. Seventy six white Leghorn H.H. 260-layer birds were housed in a 4.5 × 2.5 m shed constructed over the treatment pond. The integrated fish culture system was found to be a profitable small scale system. Total gross returns and net returns were Rs. 20848 and Rs. 5290, respectively. The system generated 96 man days of employment. Three tonnes of poultry waste worth Rs. 480 was recycled as manure and fish feed.

The efficacy of bottom grazing fishes in increasing fish production of pig-fish integration system was investigated by taking three experimental ponds (P1, P2 and P3), each with an area of 0.1 ha at Haringhata Project (West Bengal) by Panigrahi and Chand (1999). After proper pond preparation, 780 advanced carp fingerlings were released per pond. In P1, culture was carried out with no pig interaction, whereas in P2 and P3 five piglets (White Yorkshire) weighing about 5.2 kg were housed in a shed on the bank of the pond; all faeces and urine was washed into the pond. In P1 and P2 the normal stocking ratio among different types of fishes like surface feeder (catla and silver carp), column feeder (rohu), vegetative feeders (grass carp) and bottom feeders (mrigal and common carp) were maintained. In P3, the ratio of bottom feeders was increased to 50 per cent of the total. The experiment was carried out for 10 months. No supplementary feed was given in any pond. There was no noticeable mortality of fish in the ponds. The survival rate was 85-88 per cent in all the ponds. Irrespective of the treatment, the final weight of fish varied from 480 to 1200 g. The total weight of fish was poor in P1. The production of fish was highest in P3 (3872 kg/ha), followed by P2 (3488 kg/ha) and P1 (2896 kg/ha).

Integration of three enterprises

In India about 42 million ha of land is presently under rice cultivation and there are many situations where low land sub-merged conditions are present and water remain standing continuously varying in depth from 5-30 cm. Balusamy *et al.* (2003) found that rice + azolla-cum-fish culture is one of the economical option in such type of area. Monoculture system rely mainly on external inputs while in integrated system where recycling of nutrients takes place help to reduce the cost of production for economic yield. The fish in rice field utilized the untapped aquatic productivity of rice ecosystem as the rice bottom is highly fertilize on account of the production of zoo and phytoplankton is high and these resources are fully utilized by the fish. The gross income obtained in rice + azolla + fish was 25.7% more over the rice crop and 6.9 per cent more over the rice + fish. The net income followed the same trend. Thus rice + azolla + fish on an average gave Rs. 8817 more over the rice monoculture and Rs 3219 over the rice+ fish.

Senthilvel *et al.* (1998) made the comparison of integrated farming system and conventional cropping system on 4 acre land of each. The integrated farming system included cotton + black gram and sorghum + cowpea were grown on 1.25 acre area each. 0.5 acre area under fodder crop (*Cenchrus glaucus* + *Desmanthus*), 0.5 acre under ber fruit, 0.25 acre under custard apple, 0.25 acre under amla plant. The interspace of orchard was used for growing cowpea, green gram, cluster beans. This system include one unit of goat, 5 does and one buck while in conventional only cotton + black gram was grown on 4 acre area. They reported that integrated farming approach, helped to accrued Rs. 7089 more/ year over the conventional system. It is an indication that integrated farming is the only tool which can enable the agriculture sustainable under adverse conditions.

Integration of four enterprises

The combination of four or more enterprises invite special skill, resources, infrastructure, linkages (farmer – scientists – bankers – market) etc. for their success. In a Bastar tribal area of Jabalpur where the average farm holding size is 1.5 ha and to ascertain a best farming model, a set of seven combinations of mixed farming were developed (Tiwari *et al.*, 1999). The results revealed that the maximum net return worth Rs 44914 was obtained in a mixed farming system of 2 cows + 15 goats + 10 poultry + 10 ducks + fish followed by mixed farming + 2 cows + fish (Rs. 40894) and mixed farming with 2 cows (Rs. 37668). The lowest net return was obtained in arable farming which was Rs. 24093. The first three systems generated an additional man days over arable farming system by 123, 117 and 117 respectively. It was thus clear that mixed farming approach not only increases the net return but simultaneously created the employment opportunity to great extent.

In Karnataka, Channabasavanna *et al.* (2002) conducted an integrated farming study to know its effect on changes in physico-chemical properties of soil under integraed rice based farming systems. The system consisted of components like cropping, fish, poultry, dairy and horticultural crops. Their results showed that maximum production of rice (fish) – rice (fish) with poultry (11133 kg/ha) which was significantly superior to others. At the end of 3rd year rice (fish) – rice (fish) – poultry system gave the highest net returns and benefit cost ratio of Rs. 57687/ ha/year and 2.73, respectively followed by rice – rice with banana (Rs. 43105/ha/year and 2.60 respectively). The investigations further revealed that poor physical and chemical properties of soil were observed.

Manju Nath and Itnal (2003b) conducted studies at ICAR complex Goa and made the comparison of different farming systems. They found that integrated farming system approach proved superior over the improved cropping system with a margin of Rs. 48930/ha whereas improved cropping system gave double the productivity than monocropping system. Among the integrated farming system approach, rice-brinjal (0.5 ha) + rice-cowpeas (0.5 ha) + mushroom + poultry gave the maximum net return as Rs 75360/ha which was 2.33 times more than the system of coconut + forage + dairy and 2.07 times over the cashew orchard. The benefit cost ratio in case of cashew was maximum (2.23) as compared to the rice-brinjal + rice-cowpeas + mushroom + poultry which was 1.03. Mushroom was grown on rice straw which was available within the system. Broken rice and rice bran formed the ingredients to formulate feed for poultry unit and the size of the poultry farm was 720 broilers and 1000 bags of mushroom were prepared. In dairy, two milch cows were kept as a study input. It is thus clear that when rice based cropping system was integrated with poultry and mushroom, the gross and net return improved by 4 times over mono cropping systems while the cost of production was enhanced by 5 times leading to a reduction in benefit cost ratio. The higher return with the farming system was not only due to the higher productivity of the system but also due to lower cost of production and recycling of by-products of crop components.

In another study where the system productivity was measured in terms of rice equivalent yield, the contribution of crops towards the system productivity ranged from 33 to 52 per cent while poultry and mushroom contributed 28 to 39 per cent and 20 to 28 per cent, respectively. The highest system productivity of 21487 kg/ha/year of rice equivalent yield was recorded with rice-brinjal system integrated with mushroom and poultry followed by rice-cowpeas + mushroom + poultry, rice-groundnut + mushroom + poultry and lowest in rice cropping alone (4411 kg/ha/year) (Manju Nath and Itnal, 2003a). Poultry and mushroom production when maintained with recycled by-products of rice production system such as rice straw, rice bran and broken rice which reduced the cost of feed production considerably and it was coupled with by curtailing the cost of external inputs.

Another study conducted at the farm of Sh. Darshan Singh, village Hyatpur- a Jagjiwan Ram Awardee who integrated crops-livestock-poultry-fishery-piggery-forestry enterprises (Gill *et al.,* 2005). The model developed at his farm is serving as a beacon light for the other farmers. Undoubtedly, on account of shallow water table and water of good quality, his main component was fishery supplemented with the piggery enterprise. The tube-wells (submersible solar energy source) supply water in the fish ponds and ponds water is used to meet crop water need through gravity system. The Piggery excreta and left over is used for fish feed. The bathing of pigs encouraged oxygen mixing as well as exercise of the pigs themselves. Similarly, ponds served drinking water and bathing to the animals with the same advantage as stated above. He is of the view that piggery enterprise works as bonus and extend income on an average of Rs. 40000 per annum. He has reduced the cost of feed to Rs. 2.39 per kg as compared to the readymade feed available in the market @ Rs 6.5 per kg. The ingredients used were press mud, obsolete potato from cold storage, vegetable, poultry and fish market waste, and using other local ingredients. Surprisingly, no where in the literature, the use of press mud is recommended as feed to the cattle particular to the milch animals. He is using press mud and potato silage alongwith fodder to the buffaloes and get 10 per cent more milk yield. The fish enterprise on an average was generating income Rs. 77000/ ha /annum. The fish feed is met from pigs (reared on above quoted feed) which provide excreta as well as left over of feed to the fish pond. It is supplemented with press mud, poultry manure and farm yard manure. The feed prepared for pigs is occasionally given to fish ponds if desired plankton is not formed.

On comparing different enterprises of Sh Darshan Singh, the rice-wheat system per hectare gave net return of Rs. 53,221. The fishery + piggery enterprises adopted 10.8 ha area gave net return per hectare as Rs. 72,745. It was further improved by Rs. 6319 when 3rd enterprise poultry was supplemented on 0.15 ha area. The crops + livestock produced Rs. 54,820/ha net return which was 3.0 per cent more than the rice-wheat system. The data thus clearly elucidated that maximum net return (Rs. 79,064 per ha) was accrued in fishery + piggery + poultry which was 8.8 per cent more over fishery + piggery and proved superior than all other enterprises. Astonishingly, the return in rice-wheat + poplar proved negative and it was chiefly because of price crises in poplar which made this enterprise less productive.

The maximum employment was generated in fishery + piggery + poultry which was 499 man days more than rice – wheat system. The corresponding increase over rice – wheat system in rice-wheat + Agro-forestry, fishery + piggery and crops + livestock was 57.3, 322 and 175.6 man days, respectively.

It is thus pertinent to point out here that the review of farming system approach stated above clearly indicated that to make the agriculture sector more profitable and sustainable integrated farming approach is the only reply. This concept is holistic in nature involving the synergistic effect on one component over the other which enable to create employment opportunities, avoid wastage of bi-products and making the environment ecofriendly and above all ensure the interdependence of one factor over the other and ultimately improve productivity, economic ability and sustainability in the arable farming.

ALLIED ENTERPRISES

In the integrated farming system, it is always emphasized to combine cropping with other enterprises/activities. Many enterprises are available and these include cattle maintenance, sheep or goat rearing, poultry, piggery, rabbit rearing, bee-keeping, etc. Any one or more can be combined with the cropping systems. Important allied enterprises are as follow.

1. Cattle maintenance

Cattle rearing in India is carried out under a variety of adverse climatic and environmental conditions. The cattle are broadly classified into three groups.

- **Draft breeds:** The bullocks of these breeds are good draft animals, but the cows are poor milkers. e.g. Nagore, Hallikar, Kangeyam, Mali.
- **Dairy breeds:** The cows are high milk yielders but the bullocks are of poor draft quality e.g., Sahiwal, Sindhi, Gir.
- **Dual purpose:** The cows are fairly good milkers and the bullocks are with good draft work capacity e.g. Hariana, Ongole and Kankrej.
- **Exotic breeds:** The exotic breeds are high milk yielders, e.g., Jersey, Holstein-Friesian, Aryshire, Brown Swiss and Guernsey.
- **Buffaloes:** Important dairy breeds of buffalo are Murrah, Nili Ravi (which has its home tract in Pakistan), Mehsana, Suti, Zafarabadi, Godavari and Bhadawari. Of these Godavari has been evolved through crossing local buffaloes in coastal regions of Andhra Pradesh with Murrah.
- **Housing:** Each cow requires 12 to 18 sq.m space and the buffaloes need 12 to 15 sq.m. It is important to provide good ventilation and an open shed of housing is always preferable. Dairy building should be located at an elevated place to facilitate easy drainage. The floor should be rough and have gradient of 2.5 cm for every 25 cm length.
- **Breeding and maintenance:** The cows remain in milk for 9-10 months, the average calving interval being 16-18 months. A cow does not require more than 6-8 weeks of dry period. From the economic point of view, cow should ordinarily be bred during the second and third months after calving. In weak animals and heavy milkers, breeding may be delayed by 1 or 2 months. Cattle come to heat in more or less regular cycles of about 21 days which lasts for about a day. The best time to serve a cow is during the last stage of heat. If artificially inseminated, a second insemination may be given within 6-8 hours after the first insemination. The gestation period varies with individual cows and breeds and normally it is about 280 days.

In the case of buffaloes, the lactation period last for 7-9 months. She buffalo comes to heat every 21-23 days. The gestation period is 310 days. Calf rearing is very important in the case of buffalo maintenance. Since they require abundant water, wallowing is required. Regular deworming is needed for buffalo maintenance. Under Indian conditions, cattle commonly mature at the age of about three years. This period can, however, be reduced by six months under well managed herd.

- **Feeding:** Cattle feed generally contains fibrous, coarse, low nutrient straw material called roughage and concentrates.

- **Roughages:** Dairy cattle are efficient user of the roughages and convert large quantities of relatively inexpensive roughage into milk. Roughages are basic for cattle ration and includes legumes, non-legume hays, straw and silage of legume and grasses.
- **Concentrates:** Grains and byproducts of grains and oilseeds constitute the concentrates. They are extensively used in dairy cattle ration. These include cereals (maize, sorghum, oats, barley), cotton seeds, industrial wastes (bran of wheat, rice, gram husk) and cakes of oilseeds (groundnut, sesame, rape-seed, soybean, linseed).
- **Vitamins and mineral mixtures:** It is advisable to feed a supplement containing vitamins A and D. Mineral mixtures containing salt, Ca and P should also be provided in the ration.

The ration per animal per day normally includes concentrate @ 1 kg for 2 litres of milk yield, green fodder (20-30 kg), straw (5-7 kg) and water (32 litres).

2. Sheep and goat rearing

The system of sheep and goat rearing in India is different from that adopted in the developed countries. In India, smaller units are mostly maintained. In the developed countries, however, it is practiced in large scale in fenced areas.

- **Goat rearing:** In India, activity of goat rearing is sustained in different kinds of environments, including dry, hot, wet and cold, high mountains or low lying plains. The activity is also associated with different systems such as crop or animal-based, pastoral or sedentary, single animal or mixed herd, small or large scale. Goat is mainly reared for meat, milk, hide and skin. Goat meat is the preferred meat in India. A goat on hoof (live goat) fetches a better price than a sheep on hoof.
- **Housing:** Goats can be maintained under stallfed conditions. Successful goat rearing depends on the selection of site. Goats do not thrive on marshy or swampy ground. Goats are to be provided with a dry, comfortable, safe and secure place, free from worms and affording protection from excessive heat and inclement weather. The kitts are kept under large inverted baskets until they are old enough to run along with their mothers. Males and females are generally kept together. The space requirement for a goat is 4.5 to 5.4 sq.m.
- **Breeding and maintenance:** Goat matures in about 6-7 months. Breeding is allowed for buck at one year and doe after 10 months of age. Gestation period is 145-155 days. It gives birth to 1-3 kitts per time. Number of evings are three per 2 years. The kitts can be weaned after 30-45 days. Mother can be allowed for mating 45-60 day after eving. Once in five years, the buck can be changed to avoid deterioration due to inbreeding. When the young ones attain a body weight of about 25-30 kg in about nine months, they can be sold.
- **Feeding:** The per head requirement of nutrients in respect of goats is relatively low. Hence, they are suitable for resource poor small farmers with marginal

grazing lands. They are essentially browsers and eat plants which any other animals won't touch. Goats eat 4-5 times that of their body weight. Since the profit depends on weight addition, adequate proteins and calorie should be given to goats. They eat more of tree fodder and hence 40-50% of green fodder should contain tree leaf fodder and the rest with other grass species. Goats should be fed with concentrates of maize, wheat, horsegram, groundnut cake, fish meal and wheat bran. Common salt and vitamin mixtures should also be added. Abundant clean fresh water should be made available to the goats. Water should be changed every morning and evening. Fresh water is required for digestion, blood circulation and removal of waste from the body. Water is also required for regulation of the body temperature.

- **Sheep rearing:** Sheep are well adapted to many areas. They are excellent gleaners and make use of much of the waste feed. They consume large quantities of roughage, converting a relatively cheap food into a good cash product. Housing need not be elaborate or expensive. However, to protect the flock from predatory animals, the height of the fencing should be raised to two metres.
- **Breeds of Indian sheep:** There are three types of sheep in India based on the geographical division of the country.
    - I. **The Temperate Himalayan region:** Gurez, Karanah, Bhakarwal, Gaddi, Rampur-Bushiar.
    - II. **Dry western region:** Lohi, Bikaneri, Marwari, Kutchi, Kathiawari.
    - III. **Southern region:** Deccani, Nellore, Bellary, Mandya, Bandur
- **Breeding and maintenance:** One ram can be maintained for 40-50 ewes. Rams are liable to fight when two or more of them are put in a pen. Unlike other warm animals, ewes in general do not come in heat at regular intervals throughout the year but are seasonal in this respect. The duration of the heat period will range from 1-3 days and 75% of the ewes remain in heat for 21-39 hours. The optimum time of service is towards the end of heat period. Average heat interval is 18 days during the breeding season. The gestation period will vary from 142-152 days with an average of 147 days. A normal ram is in full vigour for breeding from the age of two and half to five years. Sheep grow fully at two years of age when the ewe is ready for breeding. Under average range condition, ewes may be expected to produce about five crops of lambs.
- **Feeding:** A sheep requires about 1-2 kg of leguminous hay per day depending on the age of sheep and its body weight. Proteins may be supplied through concentrates such as groundnut cake, sesame cake or safflower cake when the pastures are poor in legumes or when scarcity conditions prevail. Normally 110-225 g of cake are sufficient to maintain an average sheep in good condition. Feeding a mixture of common salt, ground limestone and sterilized bone meal in equal parts is required to alleviate deficiency of minerals in the feed.

3. Poultry

Poultry is one of the fastest growing food industries in the world. Poultry meat accounts for about 27% of the total meat consumed world wide, and its consumption is growing at an average of 5% annually. Poultry industry in India is relatively a new agricultural industry. Till 1950, it was considered a back yard profession in India. In the sixties, the growth rate of egg production was about 10% and it increased to 25% in the seventies. The growth rate came down to 7-8% by 1990 due to price rise in poultry feed. Broiler production is increasing at the rate of 15% per year. It was 31 million in 1981 and increased upto 300 million in 1995 which is increasing at a faster rate. Nearly 330 thousand tones of broiler meat are currently produced. The average global consumption is 120 eggs per person per year and in India, it is only 32-33 eggs per capita per year. As per the nutritional recommendation, the per capita consumption is estimated at 180 eggs/year and 9 kg meat/year.

- **Breeds:** Specific poultry stocks for egg and broiler production are available. A majority of the stocks used for egg production are crosses involving the strains or inbred lines of White Leghorn. To a limited extent, other breeds like Rhode Island Red, California Grey and Australop are used. Heavy breeds such as white Plymouth Rock, White Cornish and New Hampshire are used for cross-bred broiler chickens. Hence, it is essential to consider the strain within the breed at the time of purchase. Several commercial poultry breeders are selling day old chicks in India.
- **Housing:** Adequate space should be provided for the birds. Floor area of about 0.2 m^2 per adult bird is adequate for light breeds such as White Leghorn. About 0.3-0.4 m^2 per bird is required for heavy breeds. The house should have good ventilation and reasonably cool in summer and warm during winter. It should be located on well-drained ground, safe from flood waters.
- **Feed:** The feed conversion efficiency of the bird is far superior to other animals. About 60-70% of the total expenditure on poultry farming is spent on the poultry feed. Hence, use of cheap and efficient ration will give maximum profit. Ration should be balanced containing carbohyd-rates, fats, minerals and vitamins. Some of the common feed stuffs used for making poultry ration in India area :
- **Cereals:** Maize, barley, oats, wheat, pearl millet, sorghum, rice-broken.
- **Cakes/meal:** Oil cakes, maize, fish meal, meat meal, blood meal.
- **Minerals/salt:** Limestone, oyster shell, salt, manganese.

From the day old to 4 weeks of age, birds are fed on starter ration and thereafter finisher ration, which contains more energy and 18-20% protein. Feed may be given 2-3 times. In addition to the food stuffs, antibiotics and drugs may also be added to the poultry ration. Laying hens are provided with oyster shell or ground limestone. Riboflavin is particularly needed.

- **Maintenance:** The chicks must be vaccinated against Ranikhet diseases with FI Strain vaccine within the first 6-7 days of age. One drop of vaccine may be administered in the eye and nostril. When chicks get the optimum body weight of 1.0-1.5 kg around six weeks, they can be marketed for broiler. Hens may be retained for one year for production i.e., upto the age of about one and a half years. After that they are disposed off for table purpose. It may not be economical to keep the hens beyond one and a half years since egg production would get reduced. One hen is capable of laying 180-230 eggs in a year starting from the sixth month. In addition, a laying hen produces about 230 g of fresh droppings (75% moisture) daily.

4. Duck rearing

Ducks account for about 7% of the poultry population in India. They are popular in coastal and water logged States like West Bengal, Orrisa, Andhra Pradesh, Tamil Nadu, Kerala, Tripura and Jammu and Kashmir. Dusks are predominantly of indigenous type and reared for egg production on natural foraging. They have a production potential of about 130-140 eggs/bird/year. Ducks are quite hardy, more easily brooded and resistant to common avian diseases. In places like marshy river side, wetland and barren moors where chicken or any other type of stock do not flourish, duck farming can be a better alterative.

- **Breeds:** The important Indian breeds are Sylhet Mete and Nageswari which are mostly found in the Eastern region of the country. Heir annual production is 150 eggs/bird year. Improved breeds for egg and meat production are available. Khaki Campbell and Indian Runner are the most popular breeds for egg laying. Khaki Campbell has a production potential of 300 eggs/bird/year. Indian Runner is the second best producer. White Pekin, Muscovy and Aylesbury are known for meat production. White Pekin is the most popular duck in the world. It is fast growing and has low feed consumption with fine quality of meat. It attains about 3 kg of body weight in 40 days. Indigenous types, however, still continue to dominate in duck farming. Desi ducks are robust, well adapted to local conditions and free of diseases.
- **Housing:** Ducks prefer to stay outside day and night even during winter or rains. In mild climate, it is possible to raise ducks without artificial shelter. A light fence of atleast 1.2 m high enclosing the yard is enough to stop any predators. One nest of size 0.3 x 0.3 x 0.45 m to every 3 ducks is sufficient. In case of laying birds, a mating ratio of 1 drake : 6-7 ducks and in meat type 1:4-5 is allowed. The duck house should be well ventilated, dry, leak and rat proof. The roof may be of thatched or asbestos sheeted. A water channel of 0.5 m wide and 0.20 m deep is constructed at the far end on both sides parallel to the night shelter in the rearing or layer house.
- **Feeding:** Ducks normally require lesser attention. They supplement their feed by foraging, eating fallen grains in harvested paddy fields, small fishes and other aquatic materials in lakes and ponds. However, for intensive rearing, pellet feeding may be given. Ducks prefer wet mash due to difficulties in swallowing the dry mash. Hence, ducks should never have access to feed without water. During the first 8 weeks, birds should always have an access to feed. Later on they may be fed twice a day in the morning and late afternoon.

- **Maintenance:** The general management of ducks is similar to that of the chickens. The incubation period is 28-days. A broody duck or hen may be used for small scale hatching and incubator for large scale hatching. During the early part of the life, newly hatched ducklings require warm temperature under the natural or village conditions. A duck or broody hen can take care of 10-15 ducklings. Artificial brooding may be resorted for large number of ducklings. High egg-laying strains of duck come into production at 16-18 weeks of age.

Ducks are resistant to common avian diseases. Some of the common diseases in duck are duck plague, duck virus hepatitis, duck cholera and aflatoxicosis.

5. Turkey rearing

Turkey is a robust bird and can be reared in humid tropics. It actively feeds on a variety of crop residues and insects at the farm and hence can be raised with much case.

- **Physical features:** There are a few peculiar features in the physical appearance of a turkey. Male has black beard attached to the upper breast regions. This beard is small in females of white turkeys and absent in those of coloured varieties. A fleshy protuberance called snood is present near the base of the beak in both males and females. The head bears caruncles and a large flap of skin below the chin is known as dewlap. Broad Breasted Bronze breed are ordinarily seen in India. Norfolks and the Cambridge are the other important breeds.
- **Housing:** Housing for turkeys may be of range system, confinement or a combination of both. There is a growing trend towards confinement housing. Breeds require a floor space of 0.4 m^2/bird in confinement and 32.5 m^2/bird in the range rearing.
- **Feeding:** Turkeys have much higher growth rate compared to chickens. Hence they have relatively higher feed and nutrient requirement. Turkey can be fed with all types of dry feed comprising pellets, wet mash, grain and green feed in well balanced combination. They also need plenty of clean water.
- **Maintenance:** Mating ratio in natural flock is 14 hens: 1 tom for small and 12:1 for medium sized turkeys. The birds should be carefully segregated when they reach the mating stage. Mating take place during summer. Particular attention should be bestowed in regulating mating. The males should be kept away from their mates in the morning and selectively mated with other birds by turn. Mating in the evening increases the number of fertile eggs.

Generally turkey lays eggs on alternate days from six months onwards. In a year, they lay about 150 eggs. The bird attains a weight of 5-8 kg in a year. The birds should not be allowed to fatten too much as it would make them vulnerable for diseases. Changing the breeding stock periodically would be gainful for the farmers.

6. Pigeons for meat

Domestic pigeon (*Columba livia*) belonging to the sub-family Columbia can be reared for its meat. Females are smaller than the males and finer in the head. They hold their tails higher than males and their pelvic pones are spaced wide apart. The males are large, aggressive and make a loud cooling noise. They strut around females during the breeding season. There is difference in the growth rate among the males and females. The former grows faster than the latter. Thus, even at an early stage, the gender difference can be easily made based on their body weight.

- **Feeding:** Adults feed the young ones (squabs) with 'pigeon milk', which looks like thick custard and has a high protein content. Pigeons require water for drinking and for bathing to remove external parasites. A pair of breeds drink about 500 ml of water per day.

  Pigeons are not fussy eaters, but do best on diets having whole grains. Nutrient requirement of pigeons is crude protein (13.5%), carbohydratres (65%), crude fibre (3.5%) and fat (3.0%). The daily feed intake of an adult pair is about 120 g.
- **Maintenance:** Each pair normally produces 14 squabs in a year. The incubation period of the eggs is 18 days. Both partners build their nest and take turns to sit on the eggs. The female starts laying eggs again when the squabs are two weeks old. The young ones become independent in 35 days. Squabs of 30-35 days are best suited for meat purpose since recovery is high with the body weight of 350 g. In addition to the meat, pigeon droppings (25 g per bird per day) can be used as fish feed or enriched manure.

7. Japanese quail

Quail belongs to class Aves, family Phasianidae and genus *Coturnir*. Quails, introduced initially as experimental birds, have become popular for their eggs and meat. Efforts are being made by Central Avian Research Institute, Izatnagar, to develop the guinea fowl for meat production. Males are lighter than females and are identified by the cinnamon coloured feather on the upper throat and lower breast region. Female is similar to male in colour except that the feathers on the throat and upper breast are long pointed and much lighter cinnamon in colour.

Maintenance: Quails need small space and are highly suitable for cage rearing. Therefore, it is possible to rear them even in urban areas where space is a major constraint. They are socially and commercially more adaptable due to their small body size, shorter generation interval, highly prolific nature and simplicity in rearing and are more economical. Females lay eggs as early as at the age of 5 weeks. The average age of the first egg production is 40 days. They are in full production by 50 days. They have a capacity to lay 300 or more eggs each per annum. The shell of a typical quail's egg is a mixture of greenish black and brownish speckles or pathes or varying sizes. An egg weighs about 10-12 g on an average and is comparatively thin and delicate in nature. Albumen content is 56 to 59%, yolk 28-32% and shell 12-14%. The yolk content in relation to albumen is more in a quail's egg than in a chicken's egg. Special care should be given for collecting and handling the egg.

8. Emu

The Emu (*Dromaius novahollandiae*) is the second largest bird in existence and is native to Australia. It belongs to the ratite family of flightless birds, which include the emu, ostrich, rhea, cassowary and kiwi. Ratite means flat breast bone with no keel, therefore these birds do not have any flesh in the breast area and this is one of the reasons that they cannot fly. Some of the other reasons for the inability to fly are: their feathers are not aerodynamic; they have solid bones instead of hollow bones; and an emu has wings that are only about eight inches long. A grown emu will weigh 90 to 150 pounds and stand 5½ to 6 feet tall. They can jump four feet or more straight up and can run up to 40 miles per hour. Emus love water and are expert swimmers; however they would rather play in the water than swim. Emu are very playful and comical. It is common to see them running, jumping, rolling kicking and doing all sorts of aerobics.

The female emu makes a booming sound, something like a bongo drum with the male making a grunting noise, similar to a pig. The sound comes from an air sac in the neck. The female's air sac is much larger than a male's air sac, which allows her to make the booming sound.

Emu eat grains, bugs, worms, some fruit, berries, vegetables, flowers, clovers and grasses. They may drink from one to four gallons of water every day, especially in hot weather. When emu have a good nutritious feed they will consume from 1 to 1½ pounds of feed per day. On poor quality feed they will consume much larger amounts in an attempt to get needed nutrition.

Breeding and Growth: Nesting takes place in winter. The male and female remain together for about five months, which includes courtship, nest building and egg laying. The female dominates the male during pair formation but once incubation begins, the male becomes aggressive to other Emus, including his mate. The female wanders away and leaves the male to perform all the incubation. Sometimes she will find another mate and breed again. The male incubates the eggs without drinking, feeding, defecating or leaving the nest. During this time, eggs often roll out of the nest and are pulled back in by the male. Newly hatched chicks are cream coloured with dark brown stripes. They leave the nest when they are able to feed themselves. Young birds stay close together and remain with the male for four months. Emus are nearly fully grown at one year, and may breed at 20 months.

Emu reach the majority of their growth before they are a year old. Many emu will start laying eggs before they are two years old. The life expectancy of an emu is over thirty years, with a productive life expectancy of over twenty years. There have been reports of emu still laying their normal amount of fertile eggs past twenty five years of age.

Emu lay eggs that resemble a large avocado. The egg color may range from emerald green or very dark shades. They are generally 5 to 7 inches long and can weigh as much as 1½ pounds, sometimes more. Breeding season can start as early as August or September with laying season starting as early as September or October and continuing through March and many times through May. The female generally lays an egg every three days and almost always lays the egg after 3:00 P.M. and before 9:00 P.M. Most females will lay at least 20 eggs per season with many laying from 30 to over 50 eggs per season.

9. Piggery

Pigs are maintained for the production of pork. They are fed with inedible feeds, forages, certain grain byproducts obtained from mills, meat byproducts, damaged feeds and garbage. Most of these feeds are either not edible or not very palatable to human beings. The pig grows fast and is a prolific breeder, farrowing 10 to 12 pigs at a time.

- **Breeds:** Imported breeds are Large White Yorkshire and Landrace which are being used widely. Yorkshire is the most extensively used exotic breed in India. It is a prolific breed having good carcass quality, growth rate and feed conversion ability. For a small breeding farm or unit, the selection of a herd boar is extremely important. A good boar weighs 90 kg in about 5-6 months and is strong on feet and legs. The mother of the pig to be selected should have large litters of eight piglets or more.
- **Housing:** Housing should provide maximum comfort to pigs so that their growth is optimum. There should not be dampness, draft and over heating. Locally available materials can be used for housing. One pig requires about 2.7 sq.m. with a wall of 1.2 m height. Eight boars can be kept in 2.7-4.5 sq.m area with 2.4-6.0 sq.m. open space.
- **Feeding:** Feed plays an important role in successful pig production. Pigs are the most rapidly growing livestock and suffer more from nutritional deficiencies than the ruminants. Protein, carbohydrates, fats, minerals, vitamins and ample good water be fed with maximum of concentrates and minimum of roughage. The main ingredients of swine ration are cereals and millets and their byproducts. The fibre content in swine ration should be very low (around 5-6%) for better feed utilization efficiency. Mixed ration should also contain 0.5% of added salt. Swine requires comparatively higher percentage of Ca and P than do cattle or sheep. When pigs are maintained with agricultural/kitchen waste or fish and slaughter house waste, the cost of production remains low. On an average, the consumption of feed is 3.5% of total weight. Feed allowance is calculated as 2.5-3.0 kg/100 kg body weight + @ 0.25 kg feed per piglet with the lactating mothers.
- **Management:** As a general rule, well developed gilts weighing about 100 kg, when 12-14 months old, may be used for breeding. The body weight is more important than age at breeding. Sows with low body weight show higher rate of fetal and pre-weaning mortality and have been proved as mothers with poor nursing ability. The gestation period is on an average 114 days. Litter size at birth may be 1-16 numbers with the body weight of 1-25 kg. Normal period between birth of piglets is 10-20 minutes. Time taken for the whole process of farrowing ranges from 1.5 to 4.0 hours. Sows are weaned after 40 days. It is advantageous to cull the sow after fifth or sixth litter in a commercial herd. Weaned sows come into heat in 3-10 days after weaning and may be allowed to breed. The boar-sow ratio should be 1:15. It is profitable to raise two litters from each sow each year.

Mortality in piglets is an important cause of heavy economic loss leading to failure of pig industry. In general, one fourth of piglets farrowed die before they are weaned. Another one tenth are also categorized into stunt or unprofitable group due to disease or parasite infection. Thus, about 60-65% of the piglets farrowed perform as healthy piglets at slaughter age. Death rate is high during farrowing and the first week after farrowing. The farrowing season also determines mortality rate. Mortality in new born piglets is maximum when farrowing takes place in acute cold or hot climate. Therefore, mating should be planned in such a way that farrowing could be avoided in such seasons.

10. Rabbit rearing

Rabbit (*Oryctologus cuniculles*) is reared all over the world in various climatic conditions. Rabbit rearing in India is of recent origin, though hunting of wild rabbits for meat is not uncommon. Rabbits can easily be reared with relatively less concentrate feed and with very high reprodu-ction rates. Rabbit meat is pure white and is low in fat and cholesterol content. Further, rabbit meat does not have any religious taboo, unlike pork and meat from cow.

Breeds : Rabbits can be classified into two categories as meat and wool producing ones. Important meat breeds are New Zealand White, Soviet Chinchilla and Kandinrex. Breeds for wool are Angora, Polish Bolamina, Dan and Flirorida.

- **Housing:** Rabbits can be kept in hutches outside or in cages inside. Cages should have door (30 x 30) with welded wire mesh at floor and on sides. The area required by a rabbit depends on the body weight. For one kg weight, 1.8 sq.m area is needed. Angora rabbits cannot be kept at high temperature and should be restricted to hills. Breeding bucks and adult does are housed single. Castrated male can be kept in group. Young rabbits also live in group till the age of 4 months and thereafter they are separated. Rabbit keeping requires calm and quite areas. Sudden noise can cause death of rabbits. Disturbance by dogs, cats and humans should be kept to the minimum.
- **Feeding:** Rabbit is a non-ruminant herbivorous animal having a simple stomach. Like any other animal, rabbit requires protein, carbohydrates, fats, minerals and vitamins in its feed. Rabbits digest the fibre fraction very poorly and a diet having 5-7% crude fibre is adequate. Alfalfa and other forages are well utilized by rabbits. The feed ingredients may constitute maize (20%), alfalfa (25%), wheat bran (40%), groundnut cake (12%), mineral mixture (2%) and salt (1%). The feed mixture should be given twice a day, in the morning and evening. Other grasses and vegetables can be given intermittently in small quantities. Water should always be provided.

  Maintenance: Rabbits of 5-6 months age are ready for breeding and the body weight should not be less than 2.0 kg. The female rabbit is taken to the cage of male for mating. At the time of mating, the male produces a particular noise. After that the female rabbit is taken out of the cage of the male. Early morning is the best time for mating. Rabbits mostly breed from March to July.

Gestation period is 28-32 days. The eyes of the young ones (pups) open around two weeks and at 3 weeks, they feed themselves. During this period, the pups should not be often handled. Around four weeks, the young ones are separated from the mother. After two weeks of weaning, the female can be allowed for mating.

In an year, it gives birth five times, each time producing about six numbers. Mortality in the kindlings is the most important component that determines the economics of rabbit rearing. In general, mortality in adult rabbit is very low ranging from 3-5%. However, in ranges 10-15% in kindlings upto the age of 30-42 days. Mating during extreme hot and cold season causes high mortality of kindlings.

11. Bee keeping

Bee keeping is one of the most important agro-based industries which does not require any raw material from the artisan like other industries. Nectar and pollen from flowers are the raw materials which are available in plenty in nature. Bee keeping can even be started with a single colony.

- **Species:** There are two species, *Apis cerana indica* and *A. mellifera* are complementary to each other but have different adaptations. *A. cerana* is better acclimatized to higher altitudes of the Himalayan region. *A. mellifera* is more profitable in the plains.
- **Management:** The bee-keeper should be familiar with the source of nectar and pollen within his locality. The most important sources of nectar and pollen are maize, great millet, bulrush, sunflower and palm. The beginner should start with 2 and not more than 5 colonies. A minimum of 2 colonies is recommended because in the event of some mishap, such as the loss of the queen occurring in one, advantage may be taken with the other.

The hive consists of floor board, brood chamber, super chamber, top cover, inner cover, frames and entrance rod. These parts can easily be separated. The hive may be double walled or single walled. The single walled hive is light and cheap.

The most suitable time for commencing bee keeping in a locality is the arrival of the swarming season. Swarming is a natural tendency of bees to divide their colonies under conditions that are generally favourable for the survival of both parent colony and the swarm. This occurs during the late spring or early summer.

- **Honey collection:** Honey should have good quality to meet the national and international standards. Qualities such as aroma, colour, consistency and floral sources are important. Proper honey straining and processing are needed to improve the quality of the produce. Honey varies in the proportion of its constituents owing to the differences in the nectar produced by different plants. The nectar collected by bees is processed and placed in comb cells for ripening. During the ripening, sucrose is converted into glucose and fructose by an enzyme called invertase which is added to it by the bees. Honey is an excellent energy food with an average of about 3500 calories per kg. it is directly absorbed into the human blood stream, requiring no digestion.

12. Aquaculture

Ponds serve various useful purposes, viz., domestic requirement of water, supplementary irrigation source to adjoining crop fields and pisciculture. With the traditional management, farmers obtain hardly 300-400 kg of wild and culture fish per ha annually. However, poly-fish culture with the stocking density of 7500 fingerlings and supplementary feeding will boost the total biomass production.

- **Pond:** The depth of the pond should be 1.5-2.0 m. This depth will help for effective photosynthesis and temperature maintenance for the growth of zoo and phytoplanktons. Clay soils have higher water retention capacity and hence are best suited for fish rearing. Pond water should have appropriate proportion of nutrients, phosphate (0.2-0.4 ppm), nitrate (0.06-0.1 ppm) and dissolved oxygen (5.0-7.0 ppm). Water should be slightly alkaline (pH 7.5-8.5). If the pH is less than 6.5, it can be adjusted with the addition of lime at an interval of 2-3 days. Higher pH (>8.5) can be reduced with the addition of gypsum. Application of fresh dung may also reduce high pH in the water.

  Soil of the pond should be tested for N and P content. If the nutrient content is less, nitrogenous fertilizers like ammonium sulphate and urea and phosphatic fertilizers like super phosphate can be added. Organic manures such as FYM and poultry droppings may also be applied to promote the growth of phyto and zooplanktons.

- **Species of fish:** Among the Indian major carps, Catla (*Catla catla*) is the fast growing fish. It consumes a lot of vegetation and decomposing higher plants. It is mainly a surface and column feeder. Rohy (*Labeo rohita*) is a column feeder and feeds on growing plants, decomposing vegetation, large colonial algae, zooplanktens and detritus to a small extent. Calbasu (*Labeo calbasu*) is a bottom feeder on detritus. Mrigal (*Cirrhimus mrigale*) is also a bottom feeder, taking detritus to a large content, diatoms, filamentous and other algae and higher plants. Common carp (*Cyptinus carpio*) is a bottom feeder and omnivorous. Silver carp (*Hypophthalmic-hlthys molitrix*) is mainly a surface and phytoplankton feeder and also feeds on microplants. Grass carp (*Cyernus carpia*) is a specialized feeder on aquatic plants, cut grass and other vegetable matter. It is also a fast growing exotic fish.

- **Poly fish culture:** The phytophagous fish (Catla, Rohu and Mrigal) can be combined with omnivorous (Common carp), plankton feed (Silver carp) and mud eaters (Mrigal and Calbasu) in a composite fish culture system.

- **Management:** The fish are to be nourished with supplementary feeding with rice bran and oilseed cakes. This will enable faster growth and better yield. Each variety of carps could be stocked to 500 fingerlings with the total of 5000-8000 per ha. This stocking density will enable to get a maximum yield of 2000 to 5000 kg/ha of fish annually.

13. Sericulture

Sericulture is defined as a practice of combining mulberry cultivation, silk worm rearing and silk reeling. Sericulture is a recognized practice in India. The total area under mulberry is 240 thousand ha in the country. It plays an important role in socio-economic development of rural poor in some areas.

The climatic conditions in India are favourable for luxuriant growth of mulberry and rearing of silkworms throughout the year. The temperature in Karnataka state, major silk producing state in India, ranges from 21.2 to 30°C. Climatic conditions in Kashmir are favourable to rear silk worm from May to October.

Moriculture: Cultivation of mulberry plants is called as 'moriculture'. There are about 20 species of mulberry, of which four are commonly cultivated. They are *Morus alba, M. indica, M. serrata* and *M. latifolia*. The crop can yield well for 12 years, after which they are pulled out and fresh planting is done. Yield of mulberry leaves is 30-40 t/ha/year.

Silk worm rearing

There are four types of silk worm:

Mulberry silk worm – *Bombyx mori*

Eri silk worm – *Philosamia ricini*

Tassar silk worm – *Antheraea mylitta*

Muga silk worm – *Antheraea assami*

Rearing and maintenance: The fertilized moth is covered with an inverted funnel or cellule and eggs are allowed to be laid over a card board. Parasites may be removed by brushing the egg masses with a fine brush. This will also enable to obtain an uniform hatch. In a bamboo tray rice husk is spread. Tender chopped mulberry leaves are added to the tray. The hatched out larvae are transferred to the leaves. It is important to change the leaves every 2-3 hours during the first 2-3 days.

The cacoon is constructed with a single reelable thread of silk. If the moths are allowed to emerge from the cacoons, the silk thread is cut into pieces. Hence, the pupa are killed 2-3 days before the emergence of moth and processed. The cacoons required for further rearing are kept separately and moths are allowed to emerge from them.

14. Mushroom cultivation

Mushroom is an edible fungi with great diversity in shape, size and colour. Essentially mushroom is a vegetable that is cultivated in protected farms in a highly sanitized atmosphere. Just like other vegetables, mush-room contains 90% moisture with high in quality protein. Mushrooms are fairly good sources of vitamin C and B complex. The protein have 60-70% digestibility and contain all essential amino acids. It is also rich source of minerals like Ca, P, K and Cu. They contain less of fat and CHO and are considered good for diabetic and blood pressure patients.

Varieties: Oyster mushroom – *Pleurotus* sp. (var. CO1, CO2), Paddy straw mushroom – *Volvariella volvacea,* White button mushroom – *Agaricus gisporus* (Var. A-11, Horst u3.)

Different methods employed for mushroom cultivation are:

- **Oyster mushroom:** Take fresh paddy straw and cut into small pieces of 3-5 cm length. Soak them in water for 4-6 hours and then boil for half an hour. Drain the water and dry the straw in shade till it is neither too dry or wet. Take polythene bags of 60 x 30 cm size and make two holes of one cm diameter in the centre of the bag such that they face opposite sides. Tie the bottom of the bag with a thread to make a flat bottom. Fill the bag with paddy straw to 10 cm height. Then inoculate with the spawn. Likewise prepare 4-5 layers of straw and spawn alternatively. The last layer ends up in straw of 10 cm height. Keep this in a spawn running room maintained at a temperature of about 22-28°C and with RH 85-90%. After 15-20 days when the spawn running is completed, cut open the polythene bag and take it to cropping room and allow it to grow for 7 days and harvest the mushroom. Yield : 0.5-1.0 kg/bed.
- **Paddy straw mushroom:** Cut the straw into long pieces of 60-90 cm and soak in water for 12 hours and sterilize 15 minutes. Arrange the straw in bundles. Lay the moistened straw bundles on the slightly raised concrete floor or on wooden platform in layers of four bundles width. Spawn or seed the beds simultaneously in each layer either by broadcasting or placing the mushroom spawn at different spots. Sprinkle mushroom spawn over each layer on the straw. Don't spawn below the topmost layer. Maintain it at 30-35°C. Harvesting is ready after 25-30 days. Yield : 1-1.5 kg/bed.
- **European mushroom:** It requires a complex method of preparing compost which is used as a substract for mushroom production. Spawning is done by three methods. viz., surface spawning, layer spawning and trough spawning. Fill the trays with compost and do spawning. After spawning, compost is pressed hard to make it compact. The trays are arranged in the cropping room in tiers and cover with newspaper sheet sprayed with 2% formalin. The temperature of 20-25°C and RH of 90-95% should be maintained. After spawn running is completed in 15-20 days and do casing. Pin heads appear within 10-15 days after casing. Cropping continues for 60 to 75 days. Mushrooms can be harvested at button stage. Yield ranges from 6-7 kg/m^2.

15. Biogas plant

Biogas is a clean, unpolluted and cheap source of energy, which can be obtained by a simple mechanism and little investment. The gas is generated from the cow dung during anaerobic decomposition. Biogas generation is a complex biochemical process. The celluloitic material is broken down to methane and carbondioxide by different groups of microorganisms. It can be used for cooking purpose, burning lamps, running pumps, etc.

Types of biogas plant

- **Float dome type:** Different models are available in this category, e.g., KVIC vertical and horizontal, Pragati model, Ganesh model.
- **Fixed dome type:** The gas plant is dome shaped underground construction. The masonry gas holder is an integral part of the digester called dome. The gas produced in the digester is collected in dome at vertical pressure by displacement of slurry in inlet and outlet. The entire construction is made of bricks and cement. The models available in this category are Janata and Deen-Bandhu. The selection of a particular type depends on technical, climatological, geographical and economical factors.
- **Size:** The size of the biogas plant is decided by the number of family members and the availability of cattle. One cubic metre capacity plant will need two to three animals and 25 kg of dung. The gas produced will meet the requirement of a family of 4-6 members. It would suffice to have a 2 cubic metre plant to cater to the needs of a family of 6-10 members.
- **Site selection and management:** The site should be close to the kitchen or the place of use. It will reduce the cost of gas distribution system. It should also be nearer to the cattle shed to reduce the cost of transportation of cattle dung. Land should be leveled and slightly above the ground level to avoid inflow or run-off of water. Plant should get clear sunshine during most part of the day.

  Generation of dung has a direct bearing on the quantity of gas generated. The amount of gas production is considerably higher in summer followed by rainy and winter seasons. Gas production would be maximum at a temperature between 30 to 35°C. If the ambient tempe-rature falls below 10°C, gas production is reduced drastically.
- **Bio-gas slurry:** Slurry is obtained after the production of bio-gas. It is an enriched manure. Another positive aspect of this manure is that even after weeks of exposure to the atmosphere, the slurry does not attract fleas and worms.

SUSTAINABLE AGRICULTURE

The phrase 'sustainable agriculture' was reportedly coined by the Australian agricultural scientist Gordon McClymont. Wes Jackson is credited with the first publication of the expression in his 1980 book New Roots for Agriculture. The term became popular in the late 1980s

It has been defined as "an integrated system of plant and animal production practices having a site-specific application that will last over the long term", for example to satisfy human food and fiber needs, to enhance environmental quality and the natural resource base upon which the agricultural economy depends, to make the most efficient use of non-renewable and on-farm resources and integrate natural biological cycles and controls, to sustain the economic viability of farm operations, and to enhance the quality of life for farmers and society as a whole

In simplest terms, sustainable agriculture is the production of food, fiber, or other plant or animal products using farming techniques that protect the environment, public health, human communities, and animal welfare. This form of agriculture enables us to produce healthful food without compromising future generations' ability to do the same.

Sustainable agriculture is a type of agriculture that focuses on producing long-term crops and livestock while having minimal effects on the environment. This type of agriculture tries to find a good balance between the need for food production and the preservation of the ecological system within the environment. In addition to producing food, there are several overall goals associated with sustainable agriculture, including conserving water, reducing the use of fertilizers and pesticides, and promoting biodiversity in crops grown and the ecosystem. Sustainable agriculture also focuses on maintaining economic stability of farms and helping farmers improve their techniques and quality of life.

There are many farming strategies that are used that help make agriculture more sustainable. Some of the most common techniques include growing plants that can create their own nutrients to reduce the use of fertilizers and rotating crops in fields, which minimizes pesticide use because the crops are changing frequently. Another common technique is mixing crops, which reduces the risk of a disease destroying a whole crop and decreases the need for pesticides and herbicides. Sustainable farmers also utilize water management systems, such as drip irrigation, that waste less water.

Key principles

There are several key principles associated with sustainability in agriculture

1. The incorporation of biological and ecological processes into agricultural and food production practices. For example, these processes could include nutrient cycling, soil regeneration, and nitrogen fixation.
2. Using decreased amounts of non-renewable and unsustainable inputs, particularly the ones that are environmentally harmful.
3. Using the expertise of farmers to both productively work the land as well as to promote the self-reliance and self-sufficiency of farmers.
4. Solving agricultural and natural resource problems through the cooperation and collaboration of people with different skills. The problems tackled include pest management and irrigation.

Sustainable agriculture methods

1. **Crop rotation:** Crop rotation is one of the most powerful techniques of sustainable agriculture. Its purpose is to avoid the consequences that come with planting the same crops in the same soil for years in a row. It helps tackle pest problems, as many pests prefer specific crops. If the pests have a steady food supply they can greatly increase their population size. Rotation breaks the reproduction cycles of pests. During rotation, farmers can plant certain crops, which replenish plant nutrients. These crops reduce the need for chemical fertilizers.

2. **Cover crops:** Many farmers choose to have crops planted in a field at all times and never leave it barren, this can cause unintended consequences. By planting cover crops, such as clover or oats, the farmer can achieve his goals of preventing soil erosion, suppressing the growth of weeds, and enhancing the quality of the soil. The use of cover crops also reduces the need for chemicals such as fertilizers.
3. **Soil enrichment:** Soil is a central component of agricultural ecosystems. Healthy soil is full of life, which can often be killed by the overuse of pesticides. Good soils can increase yields as well as creating more robust crops. It is possible to maintain and enhance the quality of soil in many ways. Some examples include leaving crop residue in the field after a harvest, and the use of composted plant material or animal manure.
4. **Natural pest predators:** In order to maintain effective control over pests, it is important to view the farm as an ecosystem as opposed to a factory. For example, many birds and other animals are in fact natural predators of agricultural pests. Managing your farm so that it can harbor populations of these pest predators is an effective as well as a sophisticated technique. The use of chemical pesticides can result in the indiscriminate killing of pest predators.
5. **Bio intensive Integrated Pest Management:** Integrated pest management (IPM). This is an approach, which really relies on biological as opposed to chemical methods. IMP also emphasizes the importance of crop rotation to combat pest management. Once a pest problem is identified, IPM will mean that chemical solutions will only be used as a last resort. Instead the appropriate responses would be the use of sterile males, and biocontrol agents such as ladybirds.

Benefits of sustainable agriculture

1. **Contributes to environmental conservation:** The environment plays a huge role in fulfilling our basic needs to sustain life. In turn, it is our duty to look after the environment so that future generations are not deprived of their needs. Sustainable agriculture helps to replenish the land as well as other natural resources such as water and air. This replenishment ensures that these natural resources will be able for future generations to sustain life.
2. **Public health safety:** Sustainable agriculture avoids hazardous pesticides and fertilizers. As a result, farmers are able to produce fruits, vegetables and other crops that are safer for consumers, workers, and surrounding communities. Through careful and proper management of livestock waste, sustainable farmers are able to protect humans from exposure to pathogens, toxins, and other hazardous pollutants.
3. **Prevents pollution:** Sustainable agriculture means that any waste a farm produces remains inside the farms ecosystem. In this way the waste cannot cause pollution.

4. **Reduction in cost:** The use of sustainable agriculture reduces the need for fossil fuels, resulting in significant cost savings in terms of purchasing as well as transporting them. This in turn lessens the overall costs involved in farming.
5. **Biodiversity:** Sustainable farms produces a wide variety of plants and animals resulting in biodiversity. During crop rotation, plants are seasonally rotated and this results in soil enrichment, prevention of diseases, and pest outbreaks.
6. **Beneficial to animals:** Sustainable agriculture results in animals being better cared for, as well as treated humanely and with respect. The natural behaviors of all living animals, including grazing or pecking, are catered for. As a result they develop in a natural way. Sustainable farmers and ranchers implement livestock husbandry practices that protect animals' health.
7. **Economically beneficial for farmers:** In exchange for engaging with sustainable farming methods, farmers receive a fair wage for their produce. This greatly reduces their reliance on government subsidies and strengthens rural communities. Organic farms typically require 2 ½ times less labor than factory farms yet yield 10 times the profit.
8. **Social equality:** Practicing sustainable agriculture techniques also benefits workers as they are offered a more competitive salary as well as benefits. They also work in humane and fair working conditions, which include a safe work environment, food, and adequate living conditions.
9. **Beneficial for environment:** Sustainable agriculture reduces the need for use of non-renewable energy resources and as a result benefits the environment.

Due to population increase, it is estimated that by 2050 we will need approximately 70% more food than is currently being produced in order to provide the estimated 9.6 billion world population with their recommended daily calorie intake. This is by no means a small challenge, but unlike many other sustainability challenges, everyone can play a part. We all need to eat, but by simply reducing food loss and waste, as well as eating diets that are lower impact, and investing in sustainable produce, we can make a difference. From countries, to companies, right down to consumers, we all have a role to play. The challenge is simply making people care in a world where we are surrounded by such abundance.

PRECISION AGRICULTURE

Precision agriculture (PA) is an approach to farm management that uses information technology (IT) to ensure that the crops and soil receive exactly what they need for optimum health and productivity. The goal of PA is to ensure profitability, sustainability and protection of the environment. PA is also known as satellite agriculture, as-needed farming and site-specific crop management (SSCM).

Precision agriculture relies upon specialized equipment, software and IT services. The approach includes accessing real-time data about the conditions of the crops, soil and ambient air, along with other relevant information such as hyperlocal weather predictions, labor costs and equipment availability. Predictive analytics software uses the data to provide farmers with guidance about crop rotation, optimal planting times, harvesting times and soil management.

Sensors in fields measure the moisture content and temperature of the soil and surrounding air. Satellites and robotic drones provide farmers with realtime images of individual plants. Information from those images can be processed and integrated with sensor and other data to yield guidance for immediate and future decisions, such as precisely what fields to water and when or where to plant a particular crop.

Agricultural control centers integrate sensor data and imaging input with other data, providing farmers with the ability to identify fields that require treatment and determine the optimum amount of water, fertilizers and pesticides to apply. This helps the farmer avoid wasting resources and prevent run-off, ensuring that the soil has just the right amount of additives for optimum health, while also reducing costs and controlling the farm's environmental impact.

In the past, precision agriculture was limited to larger operations which could support the IT infrastructure and other technology resources required to fully implement and benefit from the benefits of precision agriculture. Today, however, mobile apps, smart sensors, drones and cloud computing makes precision agriculture possible for farming cooperatives and even small family farms.

Precision agriculture is usually done as a four-stage process to observe spatial variability:

Data collection

Geolocating a field enables the farmer to overlay information gathered from analysis of soils and residual nitrogen, and information on previous crops and soil resistivity. Geolocation is done in two ways:

- The field is delineated using an in-vehicle GPS receiver as the farmer drives a tractor around the field.
- The field is delineated on a basemap derived from aerial or satellite imagery. The base images must have the right level of resolution and geometric quality to ensure that geolocation is sufficiently accurate.

Variables

Intra and inter field variability may result from a number of factors. These include climatic conditions (hail, drought, rain, etc.), soils (texture, depth, nitrogen levels), cropping practices (no-till farming), weeds and disease. Permanent indicators—chiefly soil indicators—provide farmers with information about the main environmental constants. Point indicators allow them to track a crop's status, i.e., to see whether diseases are developing, if the crop is suffering from water stress, nitrogen stress, or lodging, whether it has been damaged by ice and so on. This information may come from weather stations and other sensors (soil electrical resistivity, detection with the naked eye, satellite imagery, etc.). Soil resistivity measurements combined with soil analysis make it possible to measure moisture content. Soil resistivity is also a relatively simple and cheap measurement.

Strategies

Using soil maps, farmers can pursue two strategies to adjust field inputs:

1. **Predictive approach:** based on analysis of static indicators (soil, resistivity, field history, etc.) during the crop cycle.
2. **Control approach:** information from static indicators is regularly updated during the crop cycle by:
   - **sampling:** weighing biomass, measuring leaf chlorophyll content, weighing fruit, etc.
   - remote sensing: measuring parameters like temperature (air/soil), humidity (air/soil/leaf), wind or stem diameter is possible thanks to Wireless Sensor Networks
   - proxy-detection: in-vehicle sensors measure leaf status; this requires the farmer to drive around the entire field.
   - aerial or satellite remote sensing: multispectral imagery is acquired and processed to derive maps of crop biophysical parameters, including indicators of disease. Airborne instruments are able to measure the amount of plant cover and to distinguish between crops and weeds.

Decisions may be based on decision-support models (crop simulation models and recommendation models), but in the final analysis it is up to the farmer to decide in terms of business value and impacts on the environment.

It is important to realize why PA technology is or is not adopted, "for PA technology adoption to occur the farmer has to perceive the technology as useful and easy to use. It might be insufficient to have positive outside data on the economic benefits of PA technology as perceptions of farmers have to reflect these economic considerations."

Implementing practices

New information and communication technologies (NICT) make field level crop management more operational and easier to achieve for farmers. Application of crop management decisions calls for agricultural equipment that supports variable-rate technology (VRT), for example varying seed density along with variable rate application (VRA) of nitrogen and phytosanitary products.

Precision agriculture uses technology on agricultural equipment (e.g. tractors, sprayers, harvesters, etc.):

- positioning system (e.g. GPS receivers that use satellite signals to precisely determine a position on the globe);
- geographic information systems (GIS), i.e., software that makes sense of all the available data;
- variable-rate farming equipment (seeder, spreader).

Emerging technologies

Precision agriculture is an application of breakthrough digital farming technologies. Over $4.6 billion has been invested in agriculture tech companies—sometimes called agtech

Robots

Self-steering tractors have existed for some time now, as John Deere equipment works like a plane on autopilot. The tractor does most of the work, with the farmer stepping in for emergencies.Technology is advancing towards driverless machinery programmed by GPS to spread fertilizer or plow land. Other innovations include a solar powered machine that identifies weeds and precisely kills them with a dose of herbicide or lasers. Agricultural robots, also known as AgBots, already exist, but advanced harvesting robots are being developed to identify ripe fruits, adjust to their shape and size, and carefully pluck them from branches.

Drones and satellite imagery

Advances in drone and satellite technology benefits precision farming because drones take high quality images, while satellites capture the bigger picture. Light aircraft pilots can combine aerial photography with data from satellite records to predict future yields based on the current level of field biomass. Aggregated images can create contour maps to track where water flows, determine variable-rate seeding, and create yield maps of areas that were more or less productive.

The Internet of things

The Internet of things is the network of physical objects outfitted with electronics that enable data collection and aggregation. IoT comes into play with the development of sensors and farm management software. For example, farmers can spectroscopically measure nitrogen, phosphorus, and potassium in liquid manure, which is notoriously inconsistent. They can then scan the ground to see where cows have already urinated and apply fertilizer to only the spots that need it. This cuts fertilizer use by up to 30%. Moisture sensors in the soil determine the best times to remotely water plants. The irrigation systems can be programmed to switch which side of tree trunk they water based on the plant's need and rainfall.

Innovations are not just limited to plants—they can be used for the welfare of animals. Cattle can be outfitted with internal sensors to keep track of stomach acidity and digestive problems. External sensors track movement patterns to determine the cow's health and fitness, sense physical injuries, and identify the optimal times for breeding. All this data from sensors can be aggregated and analyzed to detect trends and patterns.

As another example, monitoring technology can be used to make beekeeping more efficient. Honeybees are of significant economic value and provide a vital service to agriculture by pollinating a variety of crops. Monitoring of a honeybee colony's health via wireless temperature, humidity and CO2 sensors helps to improve the productivity of bees, and to read early warnings in the data that might threaten the very survival of an entire hive.

CONSERVATION AGRICULTURE

Conservation Agriculture is a farming system that promotes maintenance of a permanent soil cover, minimum soil disturbance (i.e. no tillage), and diversification of plant species. It enhances biodiversity and natural biological processes above and below the ground surface, which contribute to increased water and nutrient use efficiency and to improved and sustained crop production.

Conservation agriculture (CA) can be defined by a statement given by the Food and Agricultural Organization of the United Nations as "a concept for resource saving agricultural crop production that strives to achieve acceptable profits together with high and sustained production levels while concurrently conserving the environment"

Agriculture according to the New Standard Encyclopedia is "one of the most important sectors in the economies of most nations" (New Standard 1992). At the same time conservation is the use of resources in a manner that safely maintains a resource that can be used by humans. Conservation has become critical because the global population has increased over the years and more food needs to be produced every year

Key Principles

- **Minimum mechanical soil disturbance:** (i.e. no tillage) through direct seed and/or fertilizer placement.
- **Permanent soil organic cover:** (at least 30 percent) with crop residues and/or cover crops.
- **Species diversification:** through varied crop sequences and associations involving at least three different crops.

The Food and Agricultural Organization of the United Nations (FAO) has determined that CA has three key principles that producers (farmers) can proceed through in the process of CA. These three principles outline what conservationists and producers believe can be done to conserve what we use for a longer period of time.

The first key principle in CA (Conservation Agriculture) is practicing minimum soil disturbance which is essential to maintaining minerals within the soil, stopping erosion, and preventing water loss from occurring within the soil. In the past agriculture has looked at soil tillage as a main process in the introduction of new crops to an area. It was believed that tilling the soil would increase fertility within the soil through mineralization that takes place in the soil. Also tilling of soil can cause severe erosion and crusting which leads to a decrease in soil fertility. Today tillage is seen as destroying organic matter that can be found within the soil cover. No till farming has caught on as a process that can save soil organic levels for a longer period and still allow the soil to be productive for longer periods (FAO 2007). Additionally, the process of tilling can increase time and labor for producing that crop. Minimum soil disturbance also reduce destruction of soil micro and macro-organism habitats that is common in conventional ploughing practices.

When no till practices are followed, the producer sees a reduction in production cost for a certain crop. Tillage of the ground requires more money in order to fuel tractors or to provide feed for the animals pulling the plough. The producer sees a reduction in labor because he or she does not have to be in the fields as long as a conventional farmer.

The second key principle in CA is much like the first in dealing with protecting the soil. The principle of managing the top soil to create a permanent organic soil cover can allow for growth of organisms within the soil structure. This growth will break down the mulch that is left on the soil surface. The breaking down of this mulch will produce a high organic matter level which will act as a fertilizer for the soil surface. If CA practices were used done for many years and enough organic matter was being built up at the surface, then a layer of mulch would start to form. This layer helps prevent soil erosion from taking place and ruining the soil's profile or layout. The presence of mulching also reduce the velocity of runoff and the impact of rain drops thus reducing soil erosion and runoff.

According to the article "The role of conservation agriculture and sustainable agriculture", the layer of mulch that is built up over time will become like a buffer zone between soil and mulch and this will help reduce wind and water erosion. With this comes the protection of the soil's surface when rain falls on the ground. Land that is not protected by a layer of mulch is left open to the elements (Hobbs et al. 2007). This type of ground cover also helps keep the temperature and moisture levels of the soil at a higher level rather than if it was tilled every year (FAO 2007).

The third principle is the practicing diverse crop rotations or crop interactions. According to an article published in the Physiological Transactions of the Royal Society called "The role of conservation agriculture and sustainable agriculture," crop rotation can be used best as a disease control against other preferred crops (Hobbs et al. 2007). This process will not allow pests such as insects and weeds to be set into a rotation with specific crops. Rotational crops will act as a natural insecticide and herbicide against specific crops. Not allowing insects or weeds to establish a pattern will help to eliminate problems with yield reduction and infestations within fields (FAO 2007). Crop rotation can also help build up soil infrastructure. Establishing crops in a rotation allows for an extensive buildup of rooting zones which will allow for better water infiltration (Hobbs et al. 2007).

Organic molecules in the soil break down into phosphates, nitrates and other beneficial elements which are thus better absorbed by plants. Plowing increases the amount of oxygen in the soil and increases the aerobic processes, hastening the breakdown of organic material. Thus more nutrients are available for the next crop but, at the same time, the soil is depleted more quickly of its nutrient reserves.

Benefits

In the field of CA there are many benefits that both the producer and conservationist can obtain.

On the side of the conservationist, CA can be seen as beneficial because there is an effort to conserve what people use every day. Since agriculture is one of the most destructive forces against biodiversity, CA can change the way humans produce food and energy. With conservation come environmental benefits of CA. These benefits include less erosion possibilities, better water conservation, improvement in air quality due to lower emissions being produced, and a chance for larger biodiversity in a given area.

On the side of the producer and/or farmer, CA can eventually do all that is done in conventional agriculture, and it can conserve better than conventional agriculture. CA according to Theodor Friedrich, who is a specialist in CA, believes "Farmers like it because it gives them a means of conserving, improving, and making more efficient use of their natural resources" (FAO 2006). Producers will find that the benefits of CA will come later rather than sooner. Since CA takes time to build up enough organic matter and have soils become their own fertilizer, the process does not start to work overnight. But if producers make it through the first few years of production, results will start to become more satisfactory.

CA is shown to have even higher yields and higher outputs than conventional agriculture once it has been established over long periods. Also, a producer has the benefit of knowing that the soil in which his crops are grown is a renewable resource. According to New Standard Encyclopedia, soils are a renewable resource, which means that whatever is taken out of the soil can be put back over time (New Standard 1992). As long as good soil upkeep is maintained, the soil will continue to renew itself. This could be very beneficial to a producer who is practicing CA and is looking to keep soils at a productive level for an extended time.

The farmer and/or producer can use this same land in another way when crops have been harvested. The introduction of grazing livestock to a field that once held crops can be beneficial for the producer and also the field itself. Livestock manure can be used as a natural fertilizer for a producer's field which will then be beneficial for the producer the next year when crops are planted once again. The practice of grazing livestock using CA helps the farmer who raises crops on that field and the farmer who raises the livestock that graze off that field. Livestock produce compost or manure which are a great help in generating soil fertility (Pawley W.H. 1963). The practices of CA and grazing livestock on a field for many years can allow for better yields in the following years as long as these practices continue to be followed.

The FAO believes that there are three major benefits from CA:

- Within fields that are controlled by CA the producer will see an increase in organic matter.
- Increase in water conservation due to the layer of organic matter and ground cover to help eliminate transportation and access runoff.
- Improvement of soil structure and rooting zone.

Future development

As in any other business, producers and conservationists are always looking towards the future. In this case CA is a very important process to be looked at for future generation. There are many organizations that have been created to help educate and inform producers and conservationists in the world of CA. These organizations can help to inform, conduct research, and buy land in order to preserve animals and plants (New Standard 1992).

Another way in which CA is looking to the future is through prevention. According to the European Journal of Agronomy producers are looking for ways to reduce leaching problems within their fields. These producers are using the same principles within CA, in that they are leaving cover over their fields in order to save fields from erosion and leaching of chemicals. Processes and studies like this are allowing for a better understanding of how to conserve what we are using and finding ways to put back something that may have been lost before.

In the same journal article is presented another way in which producers and conservationists are looking towards the future. Circulation of plant nutrients can be a vital part for conserving the future. An example of this would be the use of animal manure. This process has been used for quite some time now, but the future is looking towards ways to handle and conserve nutrients within manure for a longer time. But besides animal waste, food and urban waste are also being looked towards as a way to use growth within CA. Turning these products from waste to being used to grow crops and improve yields is something that would be beneficial for conservationists and producers.

Agri-environment schemes

In 1992, 'agri-environment schemes' became compulsory for all European Union Member States. In the following years the main purpose of these schemes changed slightly. Initially, they sought to protect threatened habitats, but gradually shifted their focus to the prevention of the loss of wildlife from agricultural landscapes. Most recently, the schemes are placing more emphasis on improving the services that the land can provide to humans (e.g. pollination). Overall, farmers involved in the scheme aim to practice environmentally friendlier farming techniques such as: reducing the use of pesticides, managing or altering their land to increase more wildlife friendly habitats (e.g. increasing areas of trees and bushes), reducing irrigation, conserving soil, and organic farming. As the changes in practices that ensure the protection of the environment are costly to farmers, the EU developed agri-environment schemes to financially compensate individual farmers for applying these changes and therefore increased the implementation of conservation agriculture. The schemes are voluntary for farmers. Once joined, they commit to a minimum of five years during which they have to adopt various sustainable farming techniques.The European Commission spent a total of €3.23 billion on agri-environment schemes in 2012, significantly exceeding the cost of managing special sites of conservation (Natura 2000) that year, which came to a total of €39.6 million (Batáry et al. 2015). There are two main types of agri-environment schemes which have shown different outcomes. Out-of-production schemes tend to be used in extensive farming practices (where the farming land is widespread and less intensive farming is practiced), and focus on improving or setting land aside that will not be used for the production of food, for example, the addition of wildflower strips. In-production schemes (used for a smaller scale, but more intensively farmed land) focus on the sustainable management of arable crops or grassland, for example reduction of pesticides, reduction of grassland mowing, and most commonly, organic farming.

HYDROPONICS

Hydroponics is a subset of hydroculture, the method of growing plants without soil, using mineral nutrient solutions in a water solvent.Terrestrial plants may be grown with only their roots exposed to the mineral solution, or the roots may be supported by an inert medium, such as perlite or gravel. The nutrients in hydroponics can come from an array of different sources; these can include but are not limited to byproduct from fish waste, duck manure, or commercial fertilisers.

Techniques

There are two main variations for each medium, sub-irrigation and top irrigation. For all techniques, most hydroponic reservoirs are now built of plastic, but other materials have been used including concrete, glass, metal, vegetable solids, and wood. The containers should exclude light to prevent algae and fungal growth in the nutrient solution.

Static solution culture

In static solution culture, plants are grown in containers of nutrient solution, such as glass Mason jars (typically, in-home applications), plastic buckets, tubs, or tanks. The solution is usually gently aerated but may be un-aerated. If un-aerated, the solution level is kept low enough that enough roots are above the solution so they get adequate oxygen. A hole is cut in the lid of the reservoir for each plant. A single reservoir can be dedicated to a single plant, or to various plants. Reservoir size can be increased as plant size increases. A home made system can be constructed from plastic food containers or glass canning jars with aeration provided by an aquarium pump, aquarium airline tubing and aquarium valves. Clear containers are covered with aluminium foil, butcher paper, black plastic, or other material to exclude light, thus helping to eliminate the formation of algae. The nutrient solution is changed either on a schedule, such as once per week, or when the concentration drops below a certain level as determined with an electrical conductivity meter. Whenever the solution is depleted below a certain level, either water or fresh nutrient solution is added. A Mariotte's bottle, or a float valve, can be used to automatically maintain the solution level. In raft solution culture, plants are placed in a sheet of buoyant plastic that is floated on the surface of the nutrient solution. That way, the solution level never drops below the roots.

Continuous-flow solution culture

In continuous flow solution culture, the nutrient solution constantly flows past the roots. It is much easier to automate than the static solution culture because sampling and adjustments to the temperature and nutrient concentrations can be made in a large storage tank that has potential to serve thousands of plants. A popular variation is the nutrient film technique or NFT, whereby a very shallow stream of water containing all the dissolved nutrients required for plant growth is recirculated past the bare roots of plants in a watertight thick root mat, which develops in the bottom of the channel and has an upper surface that, although moist, is in the air. Subsequent to this, an abundant supply of oxygen is provided to the roots of the plants. A properly designed NFT system is based on using the right channel slope, the right flow rate, and the right channel length. The main advantage of the NFT system over other forms of hydroponics is that the plant roots are exposed to adequate supplies of water, oxygen, and nutrients. In all other forms of production, there is a conflict between the supply of these requirements, since excessive or deficient amounts of one results in an imbalance of one or both of the others. NFT, because of its design, provides a system where all three requirements for healthy plant growth can be met at the same time, provided that the simple concept of NFT is always remembered and practised. The result of these advantages is that higher yields of high-quality produce are obtained over an extended period of cropping. A downside of NFT is that it has very little buffering against interruptions in the flow (e.g., power outages). But, overall, it is probably one of the more productive techniques.

The same design characteristics apply to all conventional NFT systems. While slopes along channels of 1:100 have been recommended, in practice it is difficult to build a base for channels that is sufficiently true to enable nutrient films to flow without ponding in locally depressed areas. As a consequence, it is recommended that slopes of 1:30 to 1:40 are used.This allows for minor irregularities in the surface, but, even with these slopes, ponding and water logging may occur. The slope may be provided by the floor, benches or racks may hold the channels and provide the required slope. Both methods are used and depend on local requirements, often determined by the site and crop requirements.

As a general guide, flow rates for each gully should be one liter per minute. At planting, rates may be half this and the upper limit of 2 L/min appears about the maximum. Flow rates beyond these extremes are often associated with nutritional problems. Depressed growth rates of many crops have been observed when channels exceed 12 meters in length. On rapidly growing crops, tests have indicated that, while oxygen levels remain adequate, nitrogen may be depleted over the length of the gully. As a consequence, channel length should not exceed 10–15 meters. In situations where this is not possible, the reductions in growth can be eliminated by placing another nutrient feed halfway along the gully and halving the flow rates through each outlet.

Aeroponics

Aeroponics is a system wherein roots are continuously or discontinuously kept in an environment saturated with fine drops (a mist or aerosol) of nutrient solution. The method requires no substrate and entails growing plants with their roots suspended in a deep air or growth chamber with the roots periodically wetted with a fine mist of atomized nutrients. Excellent aeration is the main advantage of aeroponics.

Aeroponic techniques have proven to be commercially successful for propagation, seed germination, seed potato production, tomato production, leaf crops, and micro-greens. Since inventor Richard Stoner commercialized aeroponic technology in 1983, aeroponics has been implemented as an alternative to water intensive hydroponic systems worldwide. The limitation of hydroponics is the fact that 1 kilogram (2.2 lb) of water can only hold 8 milligrams (0.12 gr) of air, no matter whether aerators are utilized or not.

Another distinct advantage of aeroponics over hydroponics is that any species of plants can be grown in a true aeroponic system because the microenvironment of an aeroponic can be finely controlled. The limitation of hydroponics is that certain species of plants can only survive for so long in water before they become waterlogged. The advantage of aeroponics is that suspended aeroponic plants receive 100% of the available oxygen and carbon dioxide to the roots zone, stems, and leaves, thus accelerating biomass growth and reducing rooting times. NASA research has shown that aeroponically grown plants have an 80% increase in dry weight biomass (essential minerals) compared to hydroponically grown plants. Aeroponics used 65% less water than hydroponics. NASA also concluded that aeroponically grown plants require ¼ the nutrient input compared to hydroponics.Unlike hydroponically grown plants, aeroponically grown plants will not suffer transplant shock when transplanted to soil, and offers growers the ability to reduce the spread of disease and pathogens. Aeroponics is also widely used in laboratory studies of plant physiology and plant pathology. Aeroponic techniques have been given special attention from NASA since a mist is easier to handle than a liquid in a zero-gravity environment.

Fogponics

Fogponics is a derivation of aeroponics wherein the nutrient solution is aerosolized by a diaphragm vibrating at ultrasonic frequencies. Solution droplets produced by this method tend to be 5–10 μm in diameter, smaller than those produced by forcing a nutrient solution through pressurized nozzles, as in aeroponics. The smaller size of the droplets allows them to diffuse through the air more easily, and deliver nutrients to the roots without limiting their access to oxygen.

Passive sub-irrigation

Passive sub-irrigation, also known as passive hydroponics, semi-hydroponics, or hydroculture, is a method wherein plants are grown in an inert porous medium that transports water and fertilizer to the roots by capillary action from a separate reservoir as necessary, reducing labor and providing a constant supply of water to the roots. In the simplest method, the pot sits in a shallow solution of fertilizer and water or on a capillary mat saturated with nutrient solution. The various hydroponic media available, such as expanded clay and coconut husk, contain more air space than more traditional potting mixes, delivering increased oxygen to the roots, which is important in epiphytic plants such as orchids and bromeliads, whose roots are exposed to the air in nature. Additional advantages of passive hydroponics are the reduction of root rot and the additional ambient humidity provided through evaporations.

Hydroculture compared to traditional farming in terms of crops yield per area in a controlled environment was roughly 10 times more efficient than traditional farming, uses 13 times less water in one crop cycle then traditional farming, but on average uses 100 times more kilojoules per kilogram of energy than traditional farming.

Ebb and flow or flood and drain sub-irrigation

In its simplest form, there is a tray above a reservoir of nutrient solution. Either the tray is filled with growing medium (clay granules being the most common) and then plant directly or place the pot over medium, stand in the tray. At regular intervals, a simple timer causes a pump to fill the upper tray with nutrient solution, after which the solution drains back down into the reservoir. This keeps the medium regularly flushed with nutrients and air. Once the upper tray fills past the drain stop, it begins recirculating the water until the timer turns the pump off, and the water in the upper tray drains back into the reservoirs.

Run to waste

In a run-to-waste system, nutrient and water solution is periodically applied to the medium surface. The method was invented in Bengal in 1946; for this reason it is sometimes referred to as "The Bengal System".

This method can be set up in various configurations. In its simplest form, a nutrient-and-water solution is manually applied one or more times per day to a container of inert growing media, such as rockwool, perlite, vermiculite, coco fibre, or sand. In a slightly more complex system, it is automated with a delivery pump, a timer and irrigation tubing to deliver nutrient solution with a delivery frequency that is governed by the key parameters of plant size, plant growing stage, climate, substrate, and substrate conductivity, pH, and water content.

In a commercial setting, watering frequency is multi-factorial and governed by computers or PLCs.

Commercial hydroponics production of large plants like tomatoes, cucumber, and peppers uses one form or another of run-to-waste hydroponics.

In environmentally responsible uses, the nutrient-rich waste is collected and processed through an on-site filtration system to be used many times, making the system very productive.

Deep water culture

The hydroponic method of plant production by means of suspending the plant roots in a solution of nutrient-rich, oxygenated water. Traditional methods favor the use of plastic buckets and large containers with the plant contained in a net pot suspended from the centre of the lid and the roots suspended in the nutrient solution. The solution is oxygen saturated by an air pump combined with porous stones. With this method, the plants grow much faster because of the high amount of oxygen that the roots receive.

Top-fed deep water culture

Top-fed deep water culture is a technique involving delivering highly oxygenated nutrient solution direct to the root zone of plants. While deep water culture involves the plant roots hanging down into a reservoir of nutrient solution, in top-fed deep water culture the solution is pumped from the reservoir up to the roots (top feeding). The water is released over the plant's roots and then runs back into the reservoir below in a constantly recirculating system. As with deep water culture, there is an airstone in the reservoir that pumps air into the water via a hose from outside the reservoir. The airstone helps add oxygen to the water. Both the airstone and the water pump run 24 hours a day.

The biggest advantage of top-fed deep water culture over standard deep water culture is increased growth during the first few weeks. With deep water culture, there is a time when the roots have not reached the water yet. With top-fed deep water culture, the roots get easy access to water from the beginning and will grow to the reservoir below much more quickly than with a deep water culture system. Once the roots have reached the reservoir below, there is not a huge advantage with top-fed deep water culture over standard deep water culture. However, due to the quicker growth in the beginning, grow time can be reduced by a few weeks.

Rotary

A rotary hydroponic garden is a style of commercial hydroponics created within a circular frame which rotates continuously during the entire growth cycle of whatever plant is being grown.

While system specifics vary, systems typically rotate once per hour, giving a plant 24 full turns within the circle each 24-hour period. Within the center of each rotary hydroponic garden can be a high intensity grow light, designed to simulate sunlight, often with the assistance of a mechanized timer.

Each day, as the plants rotate, they are periodically watered with a hydroponic growth solution to provide all nutrients necessary for robust growth. Due to the plants continuous fight against gravity, plants typically mature much more quickly than when grown in soil or other traditional hydroponic growing systems. Due to the small foot print a rotary hydroponic system has, it allows for more plant material to be grown per square foot of floor space than other traditional hydroponic systems.

Conclusion

Keeping in view, the literature reviewed, there is need to promote farming system approach as a National programme. Undoubtedly, this approach is a location specific, technically skill based, play multi dimensional role in fulfilling the domestic requirement, employment avenues, rational use of resources, rejuvenation of resources, sustaining productivity, investability and economic ability of the systems. In the present scenario of agriculture sector, this only approach enable the Indian farmers self sufficient and competitive in the global market by producing quality edible products which is the main base in farming system on account of their inheritance of recycling the by-products of different enterprises and even a pinch of material always considered of economic value.

15

ROLE OF ORGANIC FARMING IN AGRICULTURE

As per United States Department of Agriculture (USDA), organic farming is defined as a system which avoids or largely excludes the use of synthetic inputs (such as fetilizers, pesticides, hormones, feed additives etc.) to a maximum extent and feasible rely upon crop rotations, crop residues, animal manure, off farm organic waste, mineral grade rock additives and biological system of nutrient mobilization and plant protection. FAO suggested that organic agriculture is a unique production management system which promotes and entrance agroecosystem health, including biodiversity, biological cycles and soil biological activity and this is accomplished by using on-farm agronomic, biological and mechanical methods is exclusion of all synthetic off-farm inputs.

Organic farming is a specified form of diversified agriculture that not only aims to produce nutritious food of high quality but also lead to the generation of more income and employment opportunities for rural population. In true sense, organic farming is a production system which avoids or largely excludes the use of synthetic fertilizers, growth regulators, livestock feed additives and pesticides. Moreover, it relies on agronomic management practices like crop rotation, crop residues, animal manures, legumes, green manures, off -farming organic wastes and aspect of biological pest control to maintain soil productivity.

The northern parts of the country have made great strides to boost total foodgrains production of the country (275.11 m tonne during 2016-17) (Report Ministry of Agriculture). But still the foodgrains production is not enough to match with burgeoning population. An overview demonstrate that India has compulsive need to raise food production by 5 m tonnes every year and this was the reason that modern science based agriculture do not encourage organic farming. Moreover, in organic farming the crop yield may come down during the transitional period. But to enable our agriculture to be competitive in the global market, putting 5-6 per cent area under organic farming, foodgrain will hardly impact the total production. In addition, when the organic resources (livestock and crop residues) are visualized then a ray of hope for organic farming emerges as India has around 200 m cattle, 90 m buffaloes and 388 m tonnes of crop residues. In addition urban compost, vegetable and fruit wastes etc. are also available.

Organic farming has entirely different perspective between developed and developing countries. Developed countries are endowed with surplus resources, but limited population with high purchasing power. Hence, a consumer in a developed country expects enough food of highest possible quality that has been produced without any serious ecological impact and harmful residues. Thus organic farming may be the best answer to fulfill such expectations in rich countries (Marwaha and Jat, 2004). However, on the other hand, for a resource poor, highly populous country like India, with ever growing demand for food, the primary concerns may be enough availability of food at affordable prices with reasonable quality. In this backdrop, considerable efforts have been taken to promote the organic farming in India without compromising the food grain production to feed the ever growing population. Organic farming is also called biological agriculture/ ecological agriculture/ bio-dynamic or natural agriculture.

AREA UNDER ORGANIC FARMING

Global Statistics

As per Mahapatra*etal* (2009), organic agriculture is developing rapidly and is now practiced in more than 120 countries of the world. Its share in agricultural land continues to grow in many countries and globally 30.4 m ha is under organic agriculture, which constitutes 0.65% of agricultural land. According to the report of International Federation of Organic Agriculture Movement, Germany (IFOAM, 2016), the area under organic agriculture rose from 11.0 m ha during 1999 to 57.8 m ha during 2016 (Fig. 15.1). Cultivation of permanent grasses, arable crops and permanent crops contributes major share in organic agriculture globally.

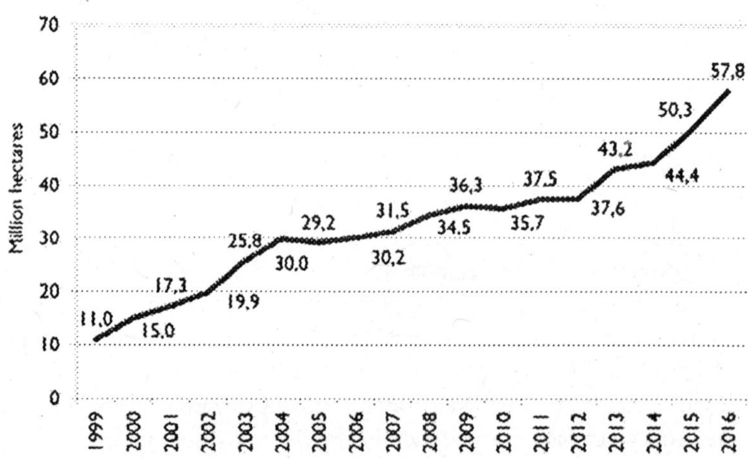

Fig. 15.1. Growth of the organic agricultural land (1999-2016)
Source: FiBL - IFOAM-SOEL Surveys 2000-2018

Among different countries, Australia leads in organic farming with total area of 27.15 m ha during 2016 (FiBL Survey 2018) (Fig. 15.2). Argentina is at second position with area of 3.0 m ha during 2016 and it was followed by China (2.28 m ha), USA (2.03 m ha), Spain (2.02 mha) and Italy (1.80 m ha).

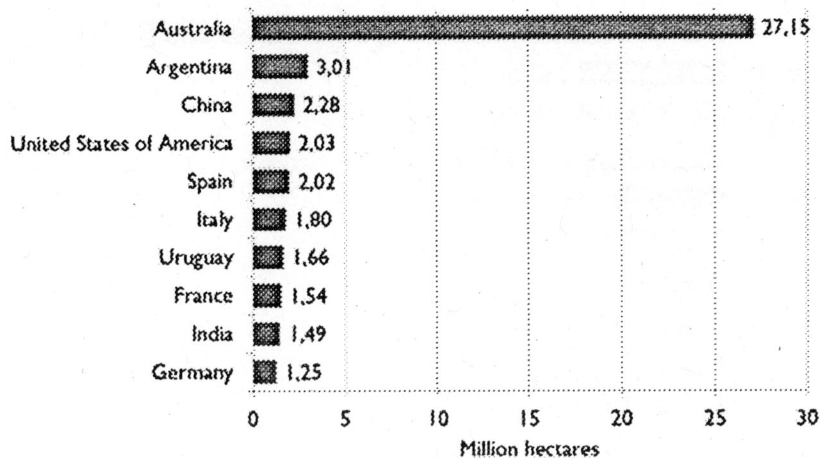

Fig. 15.2. Top 10 countries under organic agriculture (2016)
Source: FiBL - IFOAM Surveys 2018

As per survey of FiBL 2018, (Fig. 15.3) the highest share of organic agriculture in total agricultural land is 37.7% of Liechtenstein country during 2016, which is followed by French Polynesia (31.3%), Samoa (22.4%), Austria (21.9%), Estonia (18.9%) and Sweden (18.0%) (Fig. 15.3).

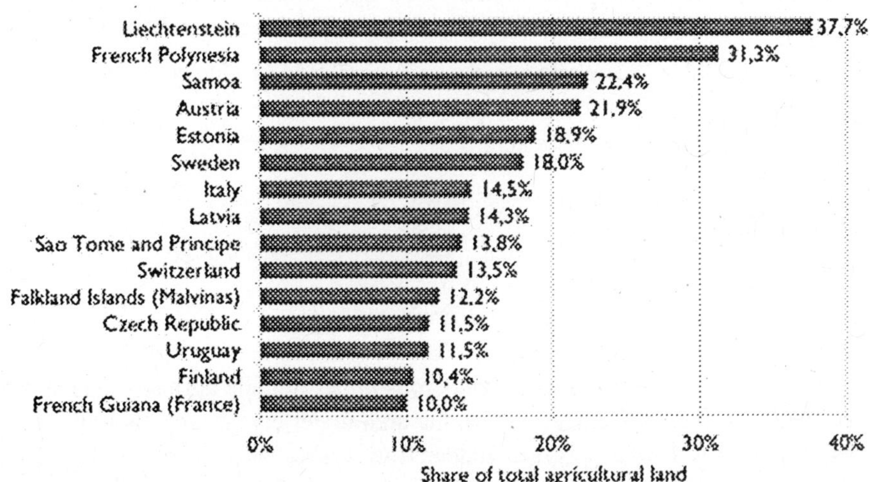

Fig. 15.3. Key crops in organic agriculture during 2016
Source: FiBL- IFOAM Surveys 2018

Acocording to FiBL survey 2018, the largest number of organic producers (8.35 lack) are in India which is followed by Uganda (2.10 lack), Mexico (2.10 lack), Ethiopia (2.03 lack) and Philippines (1.66 lack) during 2016 (Fig. 15.4).

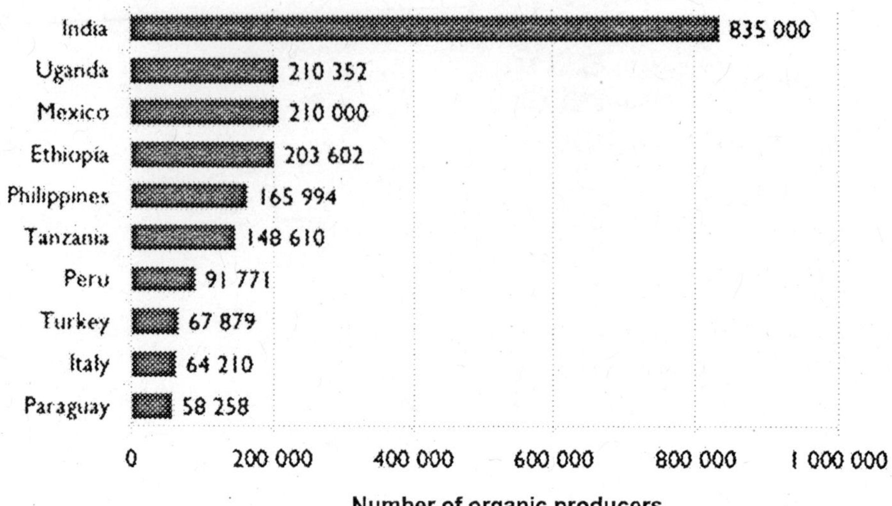

Fig. 15.4. The ten countries with largest number of organic producers 2016.

As per the report of IFOAM (2016), globally the area is highest i.e. 38.0 m ha (65.8%) under permanent grasses, which is followed by arable crops (cereals, green fodders, oil seeds, protein crops, vegetables etc.) i.e. 10.6 m ha (18.3%) during 2016 as reported in (Table 15.1). Few permanent crops like coffee, olives, nuts, grapes, cocoa etc occupies 4.5 m ha area which is 7.8% of total area under organic agriculture

Table 15.1. World organic agricultural land (m ha) - 2016

Name of the crop	Area m ha	Percentage
Permanent grassland	38.0	65.8
Arable crops	10.6	18.3
Permanent crops	4.5	7.8
Others	4.7	8.1
Total	57.8	100

Source: FiBL – IFOAM – SOEL Survey 2000 – 2013

On global level, the area under key crops under organic agriculture is presented in Fig. 15.5. As per FIBL Survey 2016, the area under cereal crops is highest (40.91 lac ha) which is followed by green fodder from arable land (27.60 lac ha), oilseeds (12.86 m ha), dry pulses (5.29 lac ha), textile crops (4.95 lac ha) etc. (Fig. 15.5) FiBL Survey 2016.

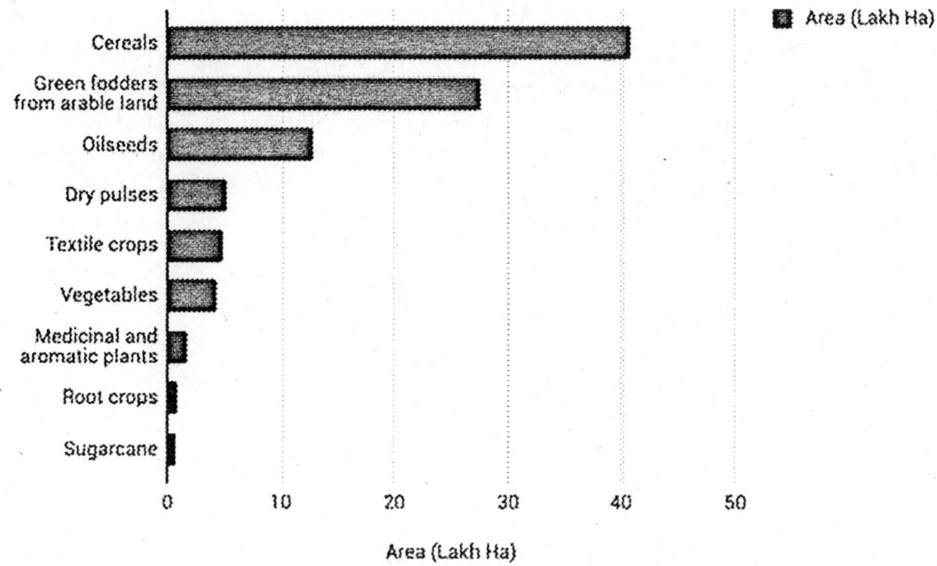

Fig. 15.5. Crops in organic agriculture during 2016.
Source: FiBL survey 2018

In India, organic agriculture is done is nearly all the States. However, Madhya Pradesh leads in the country convering 4.62 lac. ha (Fig. 15.6) followed by Maharashtra (1.98 lac ha), Rajasthan (1.55 lach ha) Telangana (1.04 m ha).

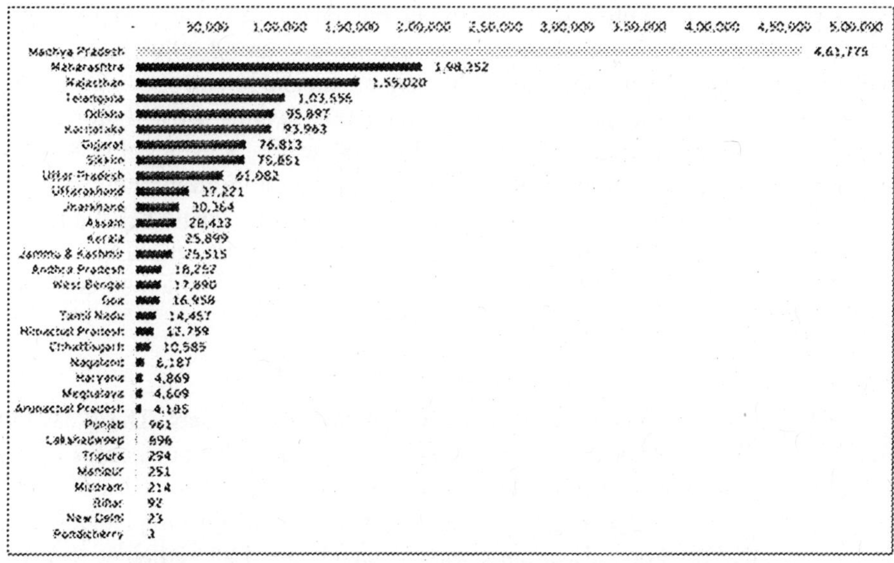

Fig. 15.6. Organic Agriculture area by State (India) in ha 2014-2015.
Source: APEDA

Indian scenario

Several States in India had already initiated steps to encourage the organic farming among the farming communities (Mahapatra *et al.*, 2009). Among the state governments, Madhya Pradesh took early lead in organic farming followed by Uttarakhand. The 'Organic farming council of Punjab' established in 2006 is taking lead in promoting organic farming. *Sesbania aculata / Crotolaria juncea* (green manuring) - Pusa basmati cropping system under organic mode has been identified as a complete rotation to be followed in *Tarai* region of Uttarakhanad.

MOVEMENT OF ORGANIC FARMING

Global

At global level, the roots of organic farming can be traced back to the first quarter of the early 20th century in Europe (Stockdale *et al.,* 2001). In 1924, the Austrian philosopher Dr. Rudolf Steiner conceptualized and advocated organic agriculture and in 1927 a trademark 'Demeter' was introduced for organically produced food. The formation of the International Federation of Organic Agriculture Movements in 1972 gave framework for the discussion and codification of internationally recognized principles of organic farming.

The main principles of organic farming are: (i) to produce food of high quality in sufficient quantity, (ii) to interact in a constructive and life enhancing way with natural systems and cycles, (iii) to consider the wider social and ecological impact of the organic production and processing systems, (iv) to encourage and enhance biological cycles within the farming system involving microorganisms, soil flora and fauna, plants and animals, (v) to maintain and increase the long term fertility of soils, (vi) to maintain the genetic diversity of the production system and its surroundings, including the protection of wildlife habitats, (vii) to promote the healthy use and proper care of water, water resources and all life therein, (viii) to use, as far as possible, renewable resources in locally organized production systems, (ix) to give all livestock conditions of life with due consideration for the basic aspects of their innate behaviour, (x) to minimize all forms of pollution and (xi) to progress towards an entire production, processing, and distribution chain which is both socially justifiable and ecologically responsible.

In India

Organic farming was practised in India since thousands of years. The great Indian civilization thrived on organic farming and was one of the most prosperous countries in the world. In traditional India, the entire agriculture was practised using organic techniques, where the fertilizers, pesticides etc., were obtained from plant and animal products. At that time organic farming was the backbone of the Indian economy. The country had witnessed a tragic Bengal famine in 1942-43 that claimed millions of lives due to starvation. India's annual food grain production in 1950-51 was only about 50 m t. In this backdrop, modern agriculture played a vital role in alleviating the hunger through green revolution. India's achievement in agricultural production after the green revolution has been exemplary and mainly due to the increased use of inputs like fertilizers, pesticides and farm machinery. The food grain production in India itself more than doubled during the post-green revolution period with virtually no increase in net cultivated area. The intensive modern agriculture for the past five decades has led to several environmental and health issues. The land is losing its fertility and is demanding larger quantities of fertilizers. Pests are becoming resistant requiring the farmers to use stronger and costlier pesticides. The over use of pesticides resulted in residues much above the safety levels and this brought to the attention of ill-effects of modern agriculture, and paved the way for organic farming. According to FiBL report, area under organic agriculture in India was 1.49 mha during 2016 with 8.35 lakh orgaic producers. India therefore is one of the most important suppliers of organic food to the developed nations. No doubt, the organic movement has again started in India.

STATUS AND SCOPE OF ORGANIC FARMING IN INDIA

According to Marwaha and Jat (2004), the organic farming probably is seldom practised and there is dearth of data on comparative yield and economics of crops with pure organic farming versus intensive farming. Whatever little organic farming that is being practised in India seems to be taken up mainly as a part of contract farming. In India, the area under organic farming is estimated to be only 0.3% of total cultivated area. Even in Europe and other developed countries of the world, despite serious efforts and incentives, the area under organic farming is generally less than 5%. Organic farming may be far from reality for 3 reasons firstly, the limited availability of organic sources; secondly, for getting optimum yields, crop plants require nutrients in sufficient quantities and in readily available form. This could be best possible through fertilisers, and hardly through organic sources. Thirdly, for plant protection, the organic sources and biological methods that have been tried so far have not met with considerable success in the country. Marwaha and Jat (2004) were of the opinion that, organic farming may be promoted in selected rainfed areas and for export oriented crops in India. So, such areas of the country may be regarded as akin to organic farming. Similarly, some areas under tea, coffee, spices and cashew may be easily brought under organic farming with a thrust on export of organic produce. Thus, rather than promoting organic farming enmasse and universally, it would be desirable to carefully delineate areas for organic agriculture, which provides better profitability and boost to the export (Marwaha and Jat, 2004).

Regulation of organic production: global perspective

Organic agriculture is based on the commitment of farmers and processors to work according to set standards and regulations, which define the organic production system. Further, organic agriculture is based on transparency to make the production system comprehensive and reliable. Finally, organic agriculture needs the consumer confidence for market development. Certification and accreditation systems primarily serve as tools to enhance trade, market development and foster confidence among the consumers (Herrmarm and Rundgren, 2007). The first international standards for organic products were published by International Federation of Organic Agriculture Movements (IFOAM) in 1980. The first legislative initiatives were developed by some European countries (e.g. Austria, France) in 1980s. In 1991, the European Countries (EC) passed the organic regulation and set standards with major implications for international trade, and included not only production standards, but also standards for labeling and inspection. Various countries in Europe, Latin America and Asia introduced legislation in the 1990s. In 1999, Codex Alimentarius approved the first guidelines for organic plant production. In the new millennium, most major economies have implemented legislation on organic production. Currently more than 60 countries have a regulation. In India the regula-tions for organic agriculture is implemented only for export commodities (Huber *et al.*, 2007).

Need of organic farming

With the increase in population our compulsion would be not only to stabilize agricultural production but to increase it further in sustainable manner. The scientists have realized that the 'Green Revolution' with high input use has reached a plateau and is now sustained with diminishing return of falling dividends. Thus, a natural balance needs to be maintained at all cost for existence of life and property. The obvious choice for that would be more relevant in the present era, when these agrochemicals which are produced from fossil fuel and are not renewable and are diminishing in availability. It may also cost heavily on our foreign exchange in future.

Key characteristics of organic farming:

- Protecting the long term fertility of soils by maintaining organic matter levels, encouraging soil biological activity, and careful mechanical intervention.
- Providing crop nutrients indirectly using relatively insoluble nutrient sources which are made available to the plant by the action of soil microorganisms.
- Nitrogen self sufficiency through the use of legumes and biological nitrogen fixation, as well as effective recycling of organic materials including crop residues and livestock manures.
- Weed, disease and pest control relying primarily on crop rotations, natural predators, diversity, organic manuring, resistant varieties and limited (preferably minimal) thermal, biological and chemical intervention.
- The extensive management of livestock, paying full regard to their evolutionary adaptations, behavioural needs and animal welfare issues with respect to nutrition, housing, health, breeding and rearing.
- Careful attention to the impact of the farming system on the wider environment and the conservation of wildlife and natural habitats.

PRINCIPLES OF ORGANIC FARMING

The four principles of organic agriculture are as follows:

1. Principle of health

Organic agriculture should sustain and enhance the health of soil, plant, animal, human and planet as one and indivisible. This principle points out that the health of individuals and communities cannot be separated from the health of ecosystems - healthy soils produce healthy crops that foster the health of animals and people. Health is the wholeness and integrity of living systems. Immunity, resilience and regeneration are key characteristics of health.

The role of organic agriculture, whether in farming, processing, distribution, or consumption, is to sustain and enhance the health of ecosystems and organisms from the smallest in the soil to human beings. In particular, organic agriculture is intended to produce high quality, nutritious food that contributes to preventive health care and well-being. In view of this it should avoid the use of fertilizers, pesticides, animal drugs and food additives that may have adverse health effects.

2. Principle of ecology

Organic agriculture should be based on living ecological systems and cycles, work with them, emulate them and help sustain them. This principle roots organic agriculture within living ecological systems. It states that production is to be based on ecological processes, and recycling. Nourishment and well-being are achieved through the ecology of the specific production environment. For example, in the case of crops this is the living soil; for animals it is the farm ecosystem; for fish and marine organisms, the aquatic environment.

Organic farming, pastoral and wild harvest systems should fit the cycles and ecological balances in nature. These cycles are universal but their operation is site-specific. Organic management must be adapted to local conditions, ecology, culture and scale. Inputs should be reduced by reuse, recycling and efficient management of materials and energy in order to maintain and improve environmental quality and conserve resources. Organic agriculture should attain ecological balance through the design of farming systems, establishment of habitats and maintenance of genetic and agricultural diversity.

3. Principle of fairness

Organic agriculture should build on relationships that ensure fairness with regard to the common environment and life opportunities. Fairness is characterized by equity, respect, justice and stewardship of the shared world, both among people and in their relations to other living beings.

This principle emphasizes that those involved in organic agriculture should conduct human relationships in a manner that ensures fairness at all levels and to all parties - farmers, workers, processors, distributors, traders and consumers. Organic agriculture should provide everyone involved with a good quality of life, and contribute to food sovereignty and reduction of poverty. It aims to produce a sufficient supply of good quality food and other products.

Natural and environmental resources that are used for production and consumption should be managed in a way that is socially and ecologically just and should be held in trust for future generations. Fairness requires systems of production, distribution and trade that are open and equitable and account for real environmental and social costs.

4. Principle of care

Organic agriculture should be managed in a precautionary and responsible manner to protect the health and well-being of current and future generations and the environment. Organic agriculture is a living and dynamic system that responds to internal and external demands and conditions. Practitioners of organic agriculture can enhance efficiency and increase productivity, but this should not be at the risk of jeopardizing health and well-being. Consequently, new technologies need to be assessed and existing methods reviewed. Given the incomplete understanding of ecosystems and agriculture, care must be taken.

This principle states that precaution and responsibility are the key concerns in management, development and technology choices in organic agriculture. Science is necessary to ensure that organic agriculture is healthy, safe and ecologically sound. However, scientific knowledge alone is not sufficient. Practical experience, accumulated wisdom and traditional and indigenous knowledge offer valid solutions, tested by time. Organic agriculture should prevent significant risks by adopting appropriate technologies and rejecting unpredictable ones, such as genetic engineering. Decisions should reflect the values and needs of all who might be affected, through transparent and participatory processes.

Basic steps of organic farming

Organic farming approach involves following **five** steps:

- Conversion of land from conventional management to organic management.
- Management of the entire surrounding system to ensure biodiversity and sustainability of the system.
- Crop production with the use of alternative sources of nutrients such as crop rotation, residue management, organic manures and biological inputs.
- Management of weeds and pests by better management practices, physical and cultural means and by biological control system.
- Maintenance of live stock in tandem with organic concept and make them an integral part of the entire system

CURRENT STATUS OF REGULATORY MECHANISM IN INDIA

National Programme for Organic Production

There is tremendous scope for cultivation and export of wide range of organic produce in India. Looking into its importance, Ministry of Commerce, Government of India launched National Programme for Organic Production (NPOP) in 2000. Declaring standards for organic production, policy on accreditation system, inspection and certification process and institution of logo were the major objectives of this programme at the initiation. A national organic logo "India Organic" was also developed that has been in use for certified organic products. During July, 2002, Ministry of Agriculture, Government of India has also circulated guidelines for promotion of organic farming (Bhattacharyya and Gehlot, 2004).

Accreditation agencies

As of 2004, Agricultural and Processed Food Product Export Development Authority, Coffee Board, Spices Board, Tea Board, Coconut Development Board, and Cocoa and Cashewnut Board were authorised by the Ministry of Commerce, Government of India as accreditation agencies for organic agricuture (Bhattacharyya and Gehlot, 2004).

National Institute of Organic Farming

The Indian Central Government set up a National Institute of Organic Farming (NIOF) in October 2003 in Ghaziabad, Uttar Pradesh. The National Biofertiliser Development Centre was renamed as the NIOF. The purpose of this institute is to formulate rules, regulations and certification of organic farm products in conformity with international standards.

Production and demand for organic products

Consumer demand for organic products is increasing across the globe. Global sales of organic food and drink have increased by 43% from US $ 23 billion in 2002 to US $ 33 billion in 2005. Although all regions reported healthy increase, the two major engines driving growth are the European and North American markets. Global production of organic crops has not kept pace with demand, with many sectors reporting undersupply. The North American market has been experiencing supply shortages for a number of years; however, Europe and Asia are now also affected. Thus, large volumes of imports are coming in from other regions. The increasing demand for organic food products in the developed countries and the extensive support by the Indian government coupled with its focus on agri-exports are the drivers for the Indian organic food industry. The Indian government is committed towards encouraging organic food production. It allocated Rs 100 crore during the Tenth Five-Year Plan for promoting sustainable agriculture in India. With the domestic consumption being low, the prime market for Indian organic food industry lies in the US and Europe. The APEDA coordinates the export of organic food in India. The NPOP standards for production and accreditation system have been recognized by European Commission and Switzerland as equivalent to their country standards. Similarly, USDA has recognized NPOP conformity assessment procedures of accreditation at par of US. With these recognitions, Indian organic products duly certified by the accredited certification bodies of India are accepted by the importing countries like US and European markets (www.apeda.com).

India has now become a leading supplier of organic herbs, organic spices, organic basmati rice, etc. India produced around 58,597 MT of certified organic products which includes all varieties of food products, viz. cereals, pulses, honey, tea, spices, coffee, oilseeds, fruits, vegetables and their value added products. The production is not limited to the edible sector but also produces organic cotton fibre, garments, cosmetics, functional food products, body care products, etc. India exported 35 items last year (2007-08) with the total volume of 19,456 MT. The export realization was around 78 million US $ registering a 200 % growth over the previous year. As many as 98 agencies are involving in the export of organic products in India. Organic products are mainly exported to European Union, United States of America, Australia, Japan, Switzerland, and Middle East. Basmati rice leads among the products exported (5,139 t) (www.apeda.com). In February 2009, the export duty on export of Basmati rice has been withdrawn by the Government of India (Notification No. 10/2009-Customs) with a view to promote the organic products. It is concluded that rainfed areas provide ample scope for pure organic farming. Moreover, there is a lacuna to show the comparative crop performance under pure organic farming and intensive agriculture in crop productivity and food quality. Hence, efforts need to be taken to address these issues with a view to promote the organic farming with great confidence in India. During the last four decades of the 20th century, the global population doubled from 3 to 6 billion and is estimated that by 2020, it will reach the 8 billion mark. Food and nutritional security is therefore, a global concern. Neither conventional farming with inorganic alone nor the organic farming only with the use of organic inputs can face this challenge. The combination of both organic and inorganic is undoubtedly the best option as on today for developing country like India. Above all organic agriculture must be discussed with an open mind, with advantages and disadvantages being clearly consider- ed to create positive mindset among the decision makers, because the most important factor that will enable organic agriculture to usefully contribute to food security is the attitude of decision-makers (Wynen, 1998).

Marketing

Marketing is a process of communicating the value of a product or service to customers. Marketing might sometimes be interpreted as the art of selling products, but sales is only one part of marketing.

From a societal point of view, marketing is the link between the society's material requirements and its economic patterns of response. Marketing satisfies these needs and wants through exchange processes and building long term relationships. Marketing is a science of choosing target markets through market analysis and market segmentation, as well as understanding consumer buying behavior and providing superior customer value.

Export potential

This term is derived from the conceptual meaning as to ship the goods and services out of the port of a country. The seller of such goods and services is referred to as an 'exporter' who is based in the country of export whereas the overseas based buyer is referred to as an 'importer'. In international trade, 'export' refers to selling goods and services produced in the home country to other markets. Export should be high then only our foreign currency will increase.

Export of commercial quantities of goods normally requires involvement of the customs authorities in both the country of export and the country of import.

Traditional export crops

Beyond these basic food crops, the agriculture and often the whole economy of many developing countries depend to a high degree on the production of one or a few commodities destined principally for export. Various organic food products exported from India are:

- **Organic cereals:** Wheat, rice, maize or corn
- **Organic pulses:** Red gram, black gram
- **Organic fruits:** Banana, mango, orange, pineapple, cashew nut, walnut
- **Organic oil seeds and oils:** Soybean, sunflower, mustard, cotton seed, groundnut castor
- **Organic vegetables:** Brinjal, garlic, potato, tomato onion
- **Organic herbs and spices:** Chilli, peppermint, cardamom, turmeric, black pepper, white pepper, tamarind, ginger, vanilla, clove, cinnamon, nutmeg, mace;
- **Others:** Jaggery, sugar, tea, coffee, cotton, textiles.
- **Persp:** The economics of countries dependent on exports of these commodities are subject to changing conditions in the world market. Slow growth in the world demand, combined with increasing supplies from the main producing and exporting countries, which compete with one another, have led to declining and widely fluctuating prices in the markets for several commodities.

Organic standards

The organic standards describe the specific requirements that must be verified by a USDA- accreditation certifying agent before products can be labeled USDA organic. Overall, organic operations must demonstrate that they are protecting natural resources, conserving biodiversity, and using only approved substances. A brief summary is provided below:

- **Organic crops:** The USDA organic seal verifies that irradiation, sewage sludge, synthetic fertilizers, prohibited pesticides, and genetically modified organisms were not used.
- **Organic livestock:** The USDA organic seal verifies that producers that met animal health and welfare standards, did not use antibiotics or growth hormones, used 100% organic feed.
- **Organic multi-ingredient foods:** The USDA organic seal verifies that the product has 95% or more certified organic content. If the label claims that it was made with specified organic ingredients, you can be sure that those specific ingredients are certifies organic and provided animals with access to the outdoors.

Certification process

Certified organics refers to agricultural products that have been grown and processed according to uniform standards, verified by independent state or private organizations accreted by standards setting organization.

The certification processes actually start with conversion process. The steps to be followed

- Fulfill the requirement of conversion period
- Develop a comprehensive operations plan for the farm including an ecosystem management plan
- Convert the farm completely or comply with strict regulation in the cases of split and parallel operations
- Farm follow up inspections by certifier
- Continuously update the chosen certifier on major changes.
- Keep track of inspection certificates issued by certifier

Procedure of certification

- The certification process start with conversion process i.e requirement of conversion process must be fulfilled
- The farmer makes contact with agency. Certification agency provide information on standards, fees, application inspection, certification and appeal procedure
- Then producer then submit application along with field history, farm map, record keeping system etc

The cost of certification depends upon size of farm, type of production system, group of farmer, extent of animal; husbandry, location of unit, travel time to reached site, food and accommodation during inspection

Inspection

Inspector verify the organic practice such as long term soil management, buffering between organic farms and neighboring conventional farmland, land record keeping are being followed. The inspection procedure review the facility's cleaning and pest control method, ingredient transportation and storage and land record and audit control. He can take sample for laboratory testing and may conduct unannounced inspection. The inspector transmits his finding to the certification body as a written report.

Certification

Certification includes annual submission of an organic system plan and inspection of farm fields and processing facilities. The certification body cross the results of the inspector with the requirements of the of the organic standards. A certification committee decides whether certification may be granted or not and then the agency issue approval or denial of certificate. The certificate is given for current year's harvest only and hence annual certification is required.

Labelling

The label convey clear accurate information on the organic status of the product (i.e. conversion in progress or organic). The label for the organic and conversion in progress products should be distinguishable by different coloured labels. The details like name of product, quantity of the product, name and address of the producer, name of certification agency, certification, lot number etc, are to be given on label.

Lot number is helpful in tracking back the product particularly the field number which it is grown in case of contamination. For packing, recycling and reusable material like clean jute bag should be used. The biodegradable material should be used. Unnecessary packing material should be avoided. Organic and non organic products should not be stored and transport together except when labeled.

Accreditation

In order to make sure that the certification programme is competent to carry out inspection and certification, a third level of quality control is needed. Accreditation is a process in which certification of competency, authority, or credibility is presented. Authorized bodies regularly evaluate certification programme and check their proper functioning according to certain criteria. In case of the certification body complies with the criteria, they accredit the certification programme.

Accreditation agencies of India

In India 6 accreditation agencies approved by central government's Ministry of Commerce.

I. APEDA,
II. Spice Board,
III. Coffee Board
IV. Coconut Board,
V. Tea Board,
VI. Cashew nut Board

BENEFITS OF ORGANIC FARMING

Organic farming tries to bridge the widening gap between man and nature with the following broader aims. However, the relative importance, for an individual or a community of farmers, may vary. Few benefits of organic farming as reported by Prasad and Gill (2009) which are presented below:

1. Sustainability of natural resources

Organic agriculture is a holistic way of farming, and besides production of goods of high quality, it primarily aims at conservation of the natural resources (soil, water, climate, bio-diversity and non-renewable energy) for sustainable productivity in agriculture. In the context of organic farming the term 'sustainability' is used in a wider sense, to encompass not just conservation of non-renewable resources (soil, water, minerals, energy, bio-diversity), but also the issues of environmental and social sustainability. The very basic approach to organic farming envisages to: (i) improve and avoid over-exploitation and pollution of natural resources; (iii) minimize the consumption of non-renewable energy sources; (iv) exploit synergies that exist in natural ecosystems; (v) maintain and improve soil health by stimulating activity of soil organisms with organic manures and avoid harming them with pesticides; (vi) obtain optimum economic returns, within a safe, secure and healthy working environment; and (vii) acknowledge the virtues of indigenous knowledge and traditional farming systems (ITKs).

2. Minimizing cost of cultivation

Organic farming is one of the environment-friendly approach of reducing dependency on external inputs and achieving the optimum productivity by making the best use of ecological principles and processes, leading to reduced costs of cultivation. This is very important for resource-poor farmers, especially for those who are operating in high risk-prone areas of dryland and rainfed agriculture.

3. Healthy food

Healthy food means a food that is free from harmful chemicals and heavy elements, and is tasteful and nutritious. Nevertheless, the organic agriculture practices cannot ensure that products are 'completely free' of harmful residues, as they may possibly trespass into the organic production systems through general environmental pollution also. But this is one of the major aims of organic farming and all feasible methods are used to minimize pollution of not only farm products but also of surrounding environment, including air, soil and water.

4. Augmentation of profits

Organically produced foods have a great demand in export markets, especially those of European and North American countries, and they fetch a sizeable premium as compared with conventionally grown farm products. In domestic sector also, the demand for organic food is increasing tremendously, especially among mid-and high-income segments, which has become more cautious about harmful effects of food grown with the use of pesticides and synthetic fertilizers, and potential hazards of environmental pollution caused due to modem practices in agriculture.

5. Improvement in soil health

As such the objective of soil health and production sustainability lies at the heart of organic farming and is one of the major considerations determining its acceptability. In organic production systems the soil health is maintained and improved through stimulation of activity of soil organisms with organic manures and by avoiding harming them with the use of synthetic pesticides as well as fertilizers. Some of the important features of organic farming in the context of soil health and environment protection are given below:

(i) Organically managed soils have a high potential to counter soil degradation, erosion and desertification, as they are more resilient both to water stress and nutrient loss.

(ii) Organic farming protects the long-term fertility of soils by maintaining organic matter levels, fostering soil biological activity, giving due importance to the basic principles of crop rotation and intercropping, and providing crop nutrients indirectly by using sources that are not readily soluble.

(iii) In organic farming, the supply of the nutrients is more balanced, which helps keep the plants healthy and improve the biological activity of soil, which in turn improves nutrient mobilization from organic matter and native soil reserves and minimizes the losses of nutrients, thus protecting the environment.

(iv) Microorganisms have a good feeding base and create a stable soil structure.

(v) Inclusion of legumes or cover crops, mulching, intercropping and agroforestry, plays an important role in protection against soil erosion and degradation.

(vi) Organicagriculture technologies increase the organic matter content of the soil, which has a positive effect on soil aggregation and water holding capacity.

(vii) Animal manure, green-manure and composts favour the composition processes and can replenish nutrients required by the crops and supply the soil with essential organic matter.

(viii) Legume crops are a highly valuable source of nitrogen. Closed nutrient cycles and efficient use of local resources (e.g. compost, cow dung or seeds) are especially important for subsistence farmers depending on few and limited assets.

(ix) Water retention capacity of the soil is increased due to their higher level of organic matter content and permanent soil cover. Due to the resulting capacity for higher moisture retention, the amount of water needed for irrigation is reduced.

(x) Organic agriculture helps mitigate the climate change, because it reduces emission of greenhouse gases, especially nitrous oxide, because no chemical nitrogen fertilizers are used and nutrient losses are minimized.

(xi) Minimizes energy consumption by 30-70% per unit of land by eliminating the energy required to manufacture synthetic fertilizers and pesticides, and by using internal farm inputs, thus reducing the amount of fuel used for transportation.

(xii) Organic agriculture helps farmers adapt to climate change because it prevents nutrient and water loss through high organic matter content and soil covers thus making soils more resilient to droughts and land degradation processes.

(xiii) Preserves seed and crop diversity, which promote natural cycles in the production systems. Maintenance of diversity also helps farmers evolve contingent cropping systems to adapt to climatic changes.

(xiv) Organic agriculture encourages natural habitats of agriculturally beneficial flora and fauna and forbids the clearances of primary ecosystems.

MAJOR LIMITATIONS

Presently the organic food is produced to very limited extent in the State. Some of the factors identified for its lower adoption are as under:

1. **Bulky nature of organic manure:** Since the content of major plant nutrients in organic manure is very less (less than 1% N, 0.25% P_2O_5 and 0.5% K_2O) this leads to voluminous material needed. This also increases the transport and application expenses.
2. **Divergent nutrition value:** The divergent nutrition values of different organic manures confuse the farmers to replace the NPK requirements from different sources. The NPK content in FYM, green manure, crop residue, vermicompost are not uniform and depend largely on the method of storage/preparation/ incorporation.
3. **Small farm holding:** In Punjab nearly 49% of farmers are small and marginal with low education status. Their farming is just subsistence type. They do not participate or come forward for following organic farming approach on account of its complex nature and resource endowed technology and risk of low yield especially during the initial years.
4. **Lack of awareness among farmers:** The farmers are not well acquainted about this concept of organic farming as well as organic manures therefore compost/ FYM are improperly prepared which is the breeding place for many insect-pests. There is a need to ask the farmers about preparing enriched compost as well as to store FYM efficiently and as scientifically desired. Most of the work being got done from the hired (migratory) labour.
5. **Marketing of organic produce:** This is an important limitation because the consumer wants the authenticity of organic produce for paying the price premium. The farmers avoid the registration on account of high registration fee, complex and skill oriented registration procedure. This does not allow them to fetch the price premium. Moreover, there is no retail marketing of organic produce.
6. **Non-availability of produce certification Labs. for analysis:** No laboratory is available in the State to get the produce tested about its genuineness.
7. **Labour and cost intensive:** The enriched compost, vermicompost and green manuring are the labour and cost intensive prepositions which also discourage the farmers to make more investments on them.
8. **Low incentives from Govt:** The farmers are not aware about any scheme formulated by the State Govt. for extending monetary help for the growers while in the developed countries this facility is very common and easily accessible.
9. **Non availability of biopesticides:** With the extension of rice-wheat system on large area, the attack of insect-pests, diseases and weed biotypes (both traditional and newly emerging ones) have become very common and no specific biopesticide is available for controlling such serious maladies which make the farmers further averse to its adoption.

10. **Dramatic and large benefits in terms of yields and returns:** Such benefits can not be expected in short time from organic farming practices. The increase in the yield and returns commensurate with the resource structure can be gradually build up, stabilized and then sustained over a longer period.
11. **Cultivation of high nutrient responsive cultivars:** Most high yielding varieties/ hybrids of cultivated crops developed in the past 3-4 decades are tuned to their high responsiveness to application of fertilizer. There nutrient demands are so high that meeting their requirement by organic sources alone is economically not feasible.
12. **Promotion of biological agents:** Many more efforts are needed to identify natural enemies, predators of some pests. The pests have to be managed by cultural and breeding methods which may involve more time and labour and above all the urgency of meeting food security concerns which impel on realization of high yields without any let down which in effect means growing crop with use of high fertilizer inputs. Lot of insects and disease organisms attack field crops and at present very limited biopesticides are available and moreover there is a need to find out the effective biopesticidese to make organic farming a success.

Organic farming in India: controversies

Organic farming and its implications on India's food security were well discussed by Chokkar and Dwivedi (2004). Some of the salient issues are presented here under:

The widely publicized advantage of high income through export of organic produce to the western markets appears lucrative, but not so easy in real sense. Indigenous organic food market is in infancy and may remain so in the future. There is a remote possibility of a resource poor small farmer of a remote area getting premium price for his organic produce after successfully overcoming all the steps and formalities beginning from farm certification to the final export. The family demands of such farmers may compel them to ultimately dispose off their organic produce in local markets without any premium.

Chhonkar and Dwivedi (2004) warned that if a section of pro-change farmers of agriculturally important areas currently engaged in food grain production began to covert their farms to organic with the intention of getting premium price, there will be no option left for the country other than importing food grains as it has practiced a few decades back.

Nitrate is the main end product of organic manure decomposition; moreover it is continuously released from the organic matter under decomposition. Since nitrate release is not synchronised with either crop demand or its uptake, it tends to accumulate in excess in soil and pose environmental risk. The ions irrespective of their origin whether organic or inorganic will behave similarly. There is no evidence that nitrate ions from organic sources are less mobile and have lower denitrification potential than from inorganic fertilisers. In addition, the trace elements and heavy metal concentrations present in the animal wastes and sewage sludges vary widely and at times very high and often exceed concentrations normally found in the inorganic fertilisers. Field application of these manures in large quantities may pollute the soil and there will be every chance to enter into food chain and create health hazard.

There are contentions that organic farming is unsustainable. The comparatively lower yields and dependence of manure from low yield cattle has prompted criticism that organic farming is environmentally unsound and incapable of feeding the world population. Among these critics are Norman Borlaug, father of the "Green Revolution," and winner of the Nobel Peace Prize, who asserts that organic farming practices can at most feed 4 billion people, after expanding crop land dramatically and destroying ecosystems in the process.

A few case studies pertaining to organic production of wheat, cotton, tea and coffee in India were reported by Marwaha and Jat (2004). The overall economics of these initiatives was not in favour of organic farming. It was also reported that in India, organic farming may be a myth, because of limited availability and restricted use of organic resources and ever growing need for food and feed.

DIFFERENT BIOPESTICIDES AND BIOFERTILIZERS

Biopesticides

I. Delfin WG (*Bacillus thuringiensis*)- To control American bollworm (*Helicoverpa armigera*)

II. Spodocide 0.50% AS (Aqueous suspension) - S-NPV (Nuclear polyhedrosis virus) - To control Tobacco caterpillar (*Spodoptera litura*)

III. Helicide 0.43% AS - H-NPV (Nuclear polyhedrosis virus) - To control American Bollworm (*Helicoverpa armigera*)

IV. Niprot 0.50% WS (*Trichoderma viride*)

- For seed treatment@10 g/kg seed.
- For nursery bed treatment 5g/litre of water/m^2 area
- For soil treatment 1 kg/100 kg well decomposed FYM and keep for 5-10 days under polythene cover.

V. Neemjal 300 ml/ha- For sucking insects like white fly, aphid and jassid.

VI. Trichocards@ 100/ha (containing eggs of *Corcyra cephelonica* parasitized by *Trichogramma chilonis*) Used in rice, cotton and maize.

Pest management

Bio pesticides obtained from plants like neem (*Azadirachata indica)*, pongamia (*Pongamia glabra*), tulsi (*Ocimum sanctum*), sorghum, marigold and bouganvillaea have been found effective against a number of diseases and insect pests. Many neem based formulations like Neemark, Neemguard, Neemazal, Nimbicidin, Wellgro, Agricef, Neoconeem and Limnonol are available in market and are in great demand

(Singh, 1993). Bio-pesticide agents are abundantly available in nature. Several pathogens including viruses such as nuclear polyhedrosis virus (NPV) and granuosis virus (GV), bacteria like *Bacillus thurienglensis, Veticillium* and protoaoan like *Schizogregrine* cause diseases in insects and destroys them. Similarly, several insect parasitoids (parasites thriving on insects) are also known in nature. *Trichoderma* is an egg parasitoid of several pests (Singh *et al.,* 1997). Selective microbial pesticides offer particular promise, of which strains of *bacillus thuringiensis* is an example (Palaniappan and Annadurai 1999). Formulations (biopesticides) prepared from neem have shown satisfactory results against 300 pestsinsects of agricultural, horticultural, vegetable crops, house hold pests, locusts, grass hoppers, catterpillers, stem borers, pod borers, beetle, jassids, aphids, mealy bugs, whiteflies, plant hoppers, fruit flies and mosquitoes. Neem formulations are environment friendly and also save farmer friendly insects (Ghosh, 2000). *Goniozus, Elamus, Eriborous, Bracon, Trichopirillus, Tetratichus* and *Chelonus* are some of other parasitoids attacking insect pests. Insect predators such as *Crytorihinus, Pharoscymnus* and predatory nematodes like species of *Neoplectana* and *Heterorhabditis* destroy insects by causing diseases in insects and by predation. Fungi namely, *Trichoderma, Aspergillus, Chaetomium* and *Agrobacterium* are effective against different diseases (Kler *et al.,* 2001).

The use of organic source of nutrients (vermicompost and FYM) proved significantly superior in controlling the fruit borer in *bhindi* over chemical fertilizers (Surekha and Rao, 2000). These findings were in accordance with the reports of Choudhary and Kashyap (1987) on cotton boll worm and Venketeswara Rao (1989) on chilli pod borer and it was mainly attributed on account of regulated growth of crop plants which discouraged shelter to the insect-pest and encouraged predator with more perched places to feed on insect-pest because the canopy is not thick and avoid over shaded micro environment beneath the plants. It also avoids oviposition to the pests and as a result pest population remains less.

Biofertilizers

I. Farm yard manure (FYM) : FYM is the basic source of nutrients in organic agriculture which contains nearly all the macro as well as micro nutrients. Apart from supply of nutrients, FYM improves water holding capacity, infiltration and other physical, chemical and biological properties of soil. The FYM occupies prominent position among the bulky organic manures available in India with a potential supply of 13.99 million tones N, P_2O_5 and K_2O per annum (Manna and Ganguly, 2003) besides it improves the soil properties. Apart from FYM, many other sorces such as vermicompost, neem cake, press mud etc. can also be used.

II. Green manure : Green manuring has lot of scope in organic farming in order to supply nutrition to the crops in desired amounts. Bani and Pant (2000) reported that it is cheaper alternative to mounting price of fertilizer nitrogen and has become an effective technology in economizing the agricultural production system, ensuring productive capacity of soil without causing any improverishment and combating many ecological and environmental problems.

III. Crop residues: In organic farming, incorporation of crop residues is again of great significance for its success. The residues of all crops except grains should be incorporated to improve soil health. Beri *et al.* (2002) reported that 462.93 million tones of different crop residues were available in India which amounts to 2.5, 1.1 and 2.0 million tones of N, P_2O_5 and K_2O, respectively.

IV. Jeev Amrit: It can be prepared with the following ingredients.

Cow Dung: 10 kg, Gram flour: 2 kg, Cow Urine: 10 L, Soil : 1 kg, Jaggery : 2 kg

Make final volume 200 litres with water and for fermentation keep under shade for 3-5 days with stirring 2-3 times a day. In soil and foliar application treatment, four sprays of Jeev Amrit will be done along with soil application.

V. Panchagavya: In Sanskrit, Panchagavya means the blend of five products obtained from cow (all these five products are individually called 'Gavya' and collectively termed as 'Panchagavya'). It contains ghee, milk, curd, cow dung and cow's urine. Panchagavya had reference in the scripts of Vedas (devine scripts of Indian wisdom) and Vrkshyurveda (Vrksha means plants and ayurveda means health system). The texts on Vrkshayurveda are systematizations of the practices the farmers followed at field level, placed in a theoretical framework and it defined certain plant growth stimulants; among them Panchagavya was an important one that enhanced the biological efficiency of crop plants and the quality of fruits and vegetables (Natarajan, 2002). For making approx. 20 litres of panchagavya you need the following:

Fresh cow dung	5 kg
Fresh cow's urine	3 litres
Cow's milk	2 litres
Cow's curd	2 litres
Cow's ghee	500 gms
Jaggery	500 gms
Water or sugarcane juice	3 litres
Ripe banana fruit	1 bunch (12 nos.)
Tender coconut water	3 litres
Toddy	2 litres

(If toddy is not available, you can ferment 3 litres of tender coconut water by keeping it in a pot for 1 week. That will become toddy.In case both these items are not available then 100 g brewing yeast can be taken).

Method of preparation

- Take 5 kgs of fresh cow dung and mix it thoroughly with 500 gms of cow's ghee and keep it in a plastic drum or a mud pot. It should not be kept in a metal container because it will react with the metal. Keep the ghee and the cow dung mixture for 4 days, mixing it twice a day.
- On the 5th day add cow's urine, cow's milk (cow's milk can be boiled, cooled and then added) and cow's curd, then jaggery with water or sugarcane juice

and banana fruit (which has to be mashed and mixed thoroughly). Then tender coconut water has to be added.

- Wait for another 15 days, stirring twice daily. Stirring the contents for about 20 minutes each time facilitates aerobic microbial activity.
- On the 19th day the panchagavya solution will be ready. This solution must be kept under a net, i.e. it must be covered with a muslin or fine cloth so that the common fly cannot sit on it and lay eggs. For use after 19th day, stir the mixture at least once a day to aerate it.

Application

200ml of this solution can be diluted with 10 litres of water with proper stirring for spraying on plants. For soil application, dilute 1000 ml in 10 litres of water. Periodicity of use once in 15 days, it can be used for all crops. In winter crops, 1.5-2% usage is sufficient but for all other crops 3% should be used.

Environmental deterioration and human health

At present, concerns are voiced both in respect of environment and health quality. Eutrophication of water bodies due to high nitrate and phosphate concentration, increasing levels of nitrates in drinking water sources, accumulation of heavy metals such as lead, chromium and cadmium in soil and water resources are the principal causes of environmental deterioration due to agriculture. Dhaliwal *et al.* (2000) reported high concentration of NO_3 (mg/litre) in underground water at different locations in Punjab, Haryana and UP which was much higher than the safe limit of 45 mg NO_3/litre. Nitrate in drinking water is the main reason of blue baby syndrome. Likewise, pesticide contamination of food and feed in Punjab were studied by Dhaliwal *et al.* (2000) and sizeable amount of contamination was found in grains, vegetables, feed and milk. Similarly, the average dietary intake of pesticide residues by vegetarian Indian is 362.5 mg/day and non-vegetarian Indian is 356.3 mg/day as against developed countries varying from 7.6 to 149.0 mg/day (Ghosh, 2000).

The heavy metals and other potentially toxic elements are the most serious soil pollutants. Amines produced from the nitrogenous fertilizers cause cancer in human beings. Herbicide residues affect the central nervous system, respiratory system and gastro-intestinal system of human beings. These chemical residues also cause depression, insomnia, oral automatism, myoclous and hyper-reflexia of man (Ghosh, 2000). Such harmful effects are now being focused at various forums on account of alarming health hazard situations. To negate such harmful effects organic farming is one of the important options for adoption as it is free from pesticide residues as well as from harmful contents of applied nutrients.

WEED MANAGEMENT

Weed management techniques plays very important role in organic agriculture. So, for higher crop yields, weeds should be controlled very effectively. Since, no herbicide is to be used in organic farming and hence weeds create big problems for the success of organic crops. Following methods can be adopted for controlling weeds from organically cultivated crops.

1. **Use of well rotten farm yard manure:** Farm yard manure (FYM) especially in raw form, contains huge amount of weed seeds. Since, the required amount of FYM is very high resulting in big problem of weeds in the fields where it is applied. The seed bearing weed plants which are fed to animals along with green fodder, the seeds of different weeds comes in intact form along with cow dung and when raw cow dung is applied in field all weed seeds germinate. On the other hand when cow dung is stored in proper deep pits, all weed seeds loose their viability due to development of high temperature in the pit (due to fermentation). Farm yard manure stored in open heaps is a big source of weed seeds and it is a wrong method of storing FYM. So, always use well rotten farm yard manure especially in organic agriculture in order to reduce the incidence of weed problem.
2. **Prefer green manuring crops:** When 40 to 60 days period (interval) between harvesting of one crop and sowing of other crop is available, prefer cultivation of green manuring crops as compared to use of FYM. Green manuring crops stimulates the germination of weed seeds due to frequent irrigation and grown up weed plants are buried in soil before flowering along with green manuring crops. So, this technique in addition to supplying plant nutrients also reduces weed seed bank in soil as compared to FYM especially in raw form. Green manuring crops especially Sunhemp, Dhaincha, Cowpea etc can be sown after the harvest of *Rabi* season crops in the month of April and are buried in soil before sowing of *Kharif* crops. A seed rate of 50 kg/ha can be used for Sunhemp and Dhaincha. For Cowpea a seed rate of 30-50 kg/ha can be used depending upon varieties. Any crop can be selected for green manuring purposes. Since, the availability of FYM is limited, so green manuring crops have high scope in organic farming.
3. **Dab system:** It is a technique in which sufficient time is provided for the germination of weed seeds and the germinated / germinating weed seedlings are uprooted before sowing the crop. Generally two pre-sowing irrigations are given before seed bed preparations, which stimulate all the weed seeds, present in the plough layer to germinate. Second option which holds good during winter season only is to leave the prepared field (after planking)as such for 8 to 10 days and after this period either plough up the field or do planking twice/thrice in order to kill germinated weed seedlings and to create soil mulch. Then sow the crop in moist zone. Adoption of this technique helps in significant reduction in weed population as compared to normal practice of crop cultivation.
4. **Use of clean crop seeds:** Always use weed seed free crop seed in order to reduce weed infestations. Before sowing, sieve out the weed seeds, shriveled and broken seeds with a proper siever. The seed crop should be properly maintained by uprooting weed plants and other off type plants. It is a preventive method of controlling weeds.

5. **Crop rotation:** Crop weed association is very common and few weeds are associated with a particular crop year after year. Crop rotation is a non-monetary method of weed control. For instance *Phalaris minor* which is a big problem in wheat can be effectively controlled by rotating wheat with berseem crop. The plants of *Phalaris minor* before flowering and seed setting will be removed along with each cutting of berseem crop. The weed seed bank in soil will be exhausted within 2 to 3 years. Similarly during *Kharif* season, cultivation of rice / basmati rice will eliminate all perennial as well as *Kharif* season weeds of all crops except of paddy. Crop rotation has lot of scope for reducing weed intensities without any extra effort.
6. **Planting pattern:** It has been observed that proper distribution of crop plants in the field have more smothering potential on weeds. Cross sowing or closer sowing with the recommended seed rate helps in maintaining proper distribution of crop plants in the field. For example, a significant reduction in dry matter of weeds can be achieved by cross sowing (22.5 cm x 22.5 cm) or closer sowing (15 cm) as compared to normal sowing technique (22.5 cm) with constant seed rate. Similarly in tillering crops as well as in fodder crops, weeds can be smothered by using 25% extra seed rate than the recommended level.
7. **Bed planting:** Bed planting technique have lot of advantages over flat planting such as saving in irrigation water, no crop mortality due to excessive rains, efficient interculture with tractor, reduced crop lodging etc. Apart from these benefits the infestation of weeds is comparatively less in bed planting as compared to flat planting technique because majority of weed seeds are buried beneath the bed which are not able to germinate. Secondly, mechanical hoeing is possible without any crop injury in bed planted crops. This technique is very successful for the cultivation of wheat, maize, soybean, moong, mash, cotton, sugarcane, vegetable crops etc. Another advantage of bed planted crops is that hoeing and earthing up operations can be done side by side resulting in healthy plants with deep root system. Bed planted crop is less liable to attack of insect pests and diseases as compared to flat planted crops due to proper air circulation.
8. **Mulching:** Mulches are known for regularization of soil temperature, conservation of soil moisture, proper growth of soil microorganisms, efficient weed control, addition of organic matter, proper and quick crop germination etc. Residues of any crop can be used as soil mulch. Plastic mulching has big scope in orchards and wider row sown crops. Mulching is very common in sugarcane, turmeric, maize, soyabean, cotton, potato, chillies, brinjal etc. About 20-25 q/ha of mulch can be used. Mulching is very effective for controlling annual grassy and broad leaf weeds and their effect on perennial weeds is partial. The availability of mulches is a big problem, however this practice must be adopted on maximum possible area. In organic farming mulching have lot of scope for controlling weeds and improving soil fertility.
9. **Soil solarization:** Soil solarization is possible on limited area. In this technique field is irrigated and is covered with black plastic sheet in such a way that no

air circulation can take place. The weed seeds present in top soil layer will germinate and will die due to suffocation created with high temperature. This practice should be adopted before crop sowing and during hot months only. Crop sown without tillage (zero till) will have less weed infestation as compared to conventional tillage sown crop. However, this practice has limited scope.

10. **Hand hoeing:** Hand hoeing is the only popular method of controlling weeds in organically sown crops. Weed problem is very serious in organic system of crop cultivation and generally 3 to 4 hoeings are required for controlling weeds depending upon their intensity. For hand hoeing khurpa, kasola, wheel hoe, spade etc can be used. However, in wider row sown crops or in bed planting, use of tractor mounted hoeing implements can be done.This method is very costly due to high wages of labour but farmers have no alternative for controlling weeds except hand hoeing. This practice must be adopted at proper time i.e. before critical period of crop weed completion.

11. **Uprooting of weed plants before flowering:** The left over weed plants should be uprooted before seed settings so that future weed problems can be minimized. In organic farming, weeds are a big problem and effort must be made to reduce soil weed seed bank which is only possible by uprooting left over weed plants. In bed planted crops, weed plants can be easily uprooted without any damage to crop plants. The uprooted plants should not be left in the field but can be used as fodder or can be used for composed making.

Weed control is very expensive and difficult in organic farming system. Farmers should not depend on one method but must adopt integrated weed management approach as per their situation. Preventive methods such as use of well rotten FYM, Dab system and use of clean seed, cultural methods such as crop rotation and planting techniques as well as physical methods such as mulching, soil solarization, bed planting, hoeing, uprooting of weed plants etc. have their own significance. Weeds should be controlled properly so that good crop yield can be obtained

ORGANIC FARMING AND CROP YIELDS

The investigations conducted at Punjab Agricultural University, Ludhiana on maize-wheat, soybean-wheat and rice-wheat systems amply demonstrate the scope of organic farming to a substantial extent. The results further revealed that after three years maize-wheat, soybean-wheat, maize-*gobhi sarson,* soybean- *gobhi sarson* system produced better productivity than the chemical farming with a margin of 1-8 per cent which helped to prove the common notion wrong that organic farming gives less yield than chemical farming. The maize, soybean, wheat and *gobhi sarson* produced the productivity at par with the chemical farming except in first three years (Kler *et al.,* 2002a). The yield improvement in organic farming was attributed on account of substantial increase in the test weight of crops.

Kler and Walia (2005) augmented these studies and advocated that in maize-wheat system where FYM, green manuring and crop residues were used as organic sources gave higher yield in all the organic treatments over chemical nutrient use.

Likewise, in soybean-wheat system where FYM level were reduced by 50 per cent alongwith green manuring and crop residue incorporation with the understanding that soybean is a leguminous crop gave significantly more yield than the chemical farming. Similar trend was observed in wheat crop. In another study, maize-potato-onion cropping system raised with organic sources of nutrition assumed the progressive trend of increasing productivity in each consecutive crop (Anonymous, 2008). After three years of experimentation, the maize crop, in inorganic treatment gave maximum productivity of 48.0 q/ha which was closely followed by organic treatment receiving 1/3 N each from farmyard manure, vermicompost and non-edible cake (47.3 q/ha). In potato, the maximum yield was obtained with pure organic treatment receiving 1/3rd N each through FYM, vermicompost and non-edible cake (96.5 q/ha) which was 10.7 per cent more over chemical farming. Similar results were obtained for onion crop. The intercropping of maize with soybean, potato with radish and onion with coriander also proved a useful option which gave good yields. The maize equivalent yield showed that organic sources in a short span of only three years enabled to produce higher economic yield than the chemical fertilizers. The pure organic treatments showed higher maize equivalent yield in the tune of 11.4 and 22.5 per cent with intercropping over alone chemical fertilizer treatment (216.3 q/ha/annum).

Bhandari and Walia (2000) reported that after five cycles of rice-wheat system, there is no need to apply any chemical fertilizer to rice and the rice productivity obtained with green manure was found at par with the chemical fertilizers. However, a significantly higher yield was obtained where chemical fertilizers were supplemented with green manure. Similarly Aulakh *et al.* (2001) reported that only green manure is found sufficient to obtain grain yield at par with recommended level of nitrogen and they further observed that crop residue (6 t/ha) is not sufficient to get high yield it must be mended with some other organic source to get the desired results in rice-wheat cropping system.

Transition period and crop yield

In the Broadbalk experiment running for more than 150 years at Rothamsted Experimental Station in the United Kingdom, the averaged wheat yields were 3.45 t/ha on manured plots as compared with 3.40 t/ha on plots receiving complete NPK fertilizers (Tillman, 1998). The maize yields from organic farming were only 75 per cent of the conventional (Liebhardt *et al.,* 1989) largely due to weed competition, insufficient nitrogen and stabilization of ecosystem, which might have acted as limiting factors during conversion from conventional to low input method of farming. So, during the transition period only those crops should be included in the rotation, which demand less nitrogen and are competitive with weeds. The minimum conversion period is three years after the last usage of synthetic agrochemicals (Hegde, 1997). The wheat yield initially decreased with residue incorporation (35.8 q/ha) in comparison to burning (43.7 q/ha) and removal (42.8 q/ha) of crop residues. However, after 4 years of gestation period, deleterious effect of previous residues incorporation on wheat yield was recuperated and highest wheat yield (49.2 q/ha) was obtained under crop residues incorporation followed by burning (44.7 q/ha) and removal (43.0 q/ha) of residues (Dhiman *et al.*, 2000). Change from conventional to organic farming takes 2-3 years for the establishment of same yield magnitude, which might be due to dynamics in soil fauna and flora. While working on a permanent site having organic versus chemical farming system at Punjab Agricultural University, Ludhiana, Kler *et al.* (2002b) reported that chemical farming gave higher yields of *Kharif* (maize/soybean) as well as *Rabi* crops (wheat/*gobhi sarson*) during first three years of study as compared to organic farming. In the fourth year, the *Kharif* crops showed increasing trend in productivity (4.6%) towards organic farming than chemical farming, while *Rabi* crops did not show any difference.

The maize crop raised with chemical sources of nutrition produced 78.4 % higher grain yield over organic farming (receiving 1/3 N each from farmyard manure, vermicompost and non-edible cake) during first year. The corresponding values during second and third year were 10.9 and 3.2 per cent, respectively. However, during fourth and fifth year the organic maize grown with organic sources of nutrition produced 7.8 and 20.9 per cent more grain yield as compared to chemical sources of nutrition. In potato crop, the maximum potato yield was obtained in chemical farmed potato which was 20.5 per cent higher over organic farmed potatoes during first year. But during second, third, fourth and fifth year, the organic farmed potatoes showed 34.5, 10.7, 7.7 and 6.9 per cent higher tuber yield than chemical farmed potatoes. In onion crop, the chemical farmed treatments showed almost similar yields as obtained in organic farmed treatments during first year. But from second year onwards the organic farmed treatments showed an edge over chemical farmed treatments by 19.6, 17.5, 5.5 and 8.2 per cent during second, third, fourth and fifth years, respectively.

Organic farming and quality of produce

The quality analysis clearly revealed that acidity, ascorbic acid and total soluble sugars of Potato in Maize-Potato-Onion cropping system were maximum in organic farming which was closely followed by integrated nutrition treatments with a slight edge over the chemical farming (Anonymous, 2008). The amount of total solids in all the organic farming treatments was in the range of 22.41 to 25.43% as compared to 20.5% in the chemical farming treatment. Similarly, in onion in Maize-Potato-Onion cropping system the total solids, ascorbic acid and reducing sugars were more in organic onion as compared to chemical farmed onion.

ORGANIC FARMING AND SOIL HEALTH

Use of organic sources of nutrients helps to conserve the soil health by maintaining the equilibrium of organic matter and soil micro flora ultimately helping to improve the physical, chemical and biological properties of the soil. Still better the soil can conserve the water more efficiently and ensure its availability as per the need of the plant. Recycling of crop residue will check the pollution hazards. Intensive cropping and imbalanced use of agrochemicals have resulted in severe depletion of soil fertility and environmental pollution. Organic materials improve soil properties after decomposition.

Physical properties

Addition of organic manures, improves soil physical properties such as

1. **Bulk density and soil strength:** Organic manuring with FYM significantly reduced bulk density (1.41 Mg/m^3) as compared with no manure (1.52 Mg/m^3) (Mishra and Sharma 1997). In Bangladesh, Hussain *et al* (2001) reported that soil bulk density was significantly lower in fields under ecological farming compared to those under conventional management. At Dapoli, Maharashtra, on clay loam soil bulk density, water holding capacity and water stable aggregates were significantly improved by FYM (5 t/ha), paddy straw (5 t/ha) compared to poultry manure (5t/ha) and control (Ramteke *et al.,* 2001). Parkash *et al* (2002) also reported that soil bulk density undergoes more reduction with application of nutrients through FYM than through chemical fertilizers. The effect of FYM was however, more pronounced after second year due to accumulation of FYM.

   Soil bulk density tended to decrease as organic carbon concentration increased in all (coarse, medium and fine) soil textural groups. Bauer and Black (1992) observed a linear relationship between soil bulk density and organic carbon content. More crop residues, higher organic matter content and better aggregation decreased bulk density. The physical properties like bulk density, soil strength, water holding capacity, aggregate stability were greatly influenced by the use of organic manures (Sur *et al.*, 1993). These properties make the rhizosphere congenial for the availability of plant nutrients.

   Penetration resistance, a measure of soil strength tends to decrease with use of organic sources of plant nutrients, as was the case with soil bulk density. At Pusa, Mishra and Sharma (1997) observed that application of FYM at 10 t/ha before rice transplanting significantly reduced bulk density (1.41 Mg/m^3) and soil strength (2.32 M Pa) than no manure application (1.52 Mg/m^3 and 2.52 M Pa). The decrease in bulk density and soil strength has been ascribed to better organic carbon content and improved soil aggregation. Penetration resistance decreased with application of FYM in 0-5, 5-10 and 10-15 cm layers due to lower bulk density (Tiwari *et al.*, 2000). Ray and Gupta (2001) at New Delhi studied the effect of green manuring (*Sesbania aculeata*) before rice on coarse loamy soil (Typic Ustrochrept) and reported a significant reduction in soil strength (18.5 M Pa) over no green manure (19.2 M Pa).

2. **Infiltration rate:** Higher infiltration rate with straw incorporation and farmyard manure treatments was observed than straw burning (Bhagat and Verma, 1991). More (1994) observed improvement in infiltration rate and soil aggregation due to application of different organic materials. Application of 100% NPK along with FYM increased the infiltration rate. This was an outcome of improved aggregation and subsequent changes in pore size distribution. Infiltration rate was increased with FYM (0, 5,15, 30 t/ha) application in the tune of 1.42, 1.57, 2.19, 2.61 cm/hr (Patil *et al.*, 2003).

It may be concluded that the long term use of organic manures, green manures and crop residues tends to decrease bulk density, soil strength and improves cumulative infiltration rate as compared to chemical fertilizers.

Chemical properties

Continuous application of nitrogenous fertilizers has depleted soil organic matter, resulting into inherent loss of native soil N, available P, available K and lower production (Srivastava, 1998).

1. **Organic matter and nitrogen:** Incorporation of any green manure crop along with wheat or rice straw improved the organic matter content and physical properties of soil. The organic farming significantly ameliorated soil structure with an increased horizon depth, organic matter content, porosity, earthworm abundance and activity with coarser and better developed aggregates than the conventional farming. Dhiman *et al.* (2000) reported that organic carbon, available P and K were highest where an additional FYM application was made. Soil organic matter and soil total nitrogen levels were increased by about 120 per cent over 150 years in the manured plots. The organic carbon in soil can be improved by growing legumes for grain or green manure. Bhandari and Walia (2000) reported that burying of crop residues increased the organic carbon content of soil. However, the notable increase in organic carbon content, P and K values were observed in the presence of FYM followed by green manuring.

   The organic manures results in gradual (slow) mineralization rates during the crop period with synchronization with crop uptake and results in reduced mineral N losses from the rooting zone during the growing season. In contrast, application of inorganic N led to substantial losses of soil mineral N during the growing season. Green manured plots will show longer periods of high mineral N levels compared with crop supplied with inorganic source of nitrogen. Rice residue incorporation may also depress levels of mineral N initially followed by sharp increase.

2. **Phosphorus:** Bhagat and Verma (1991) reported increased available P in straw incorporation + animal manure treated plots followed by animal manure alone and it was minimal under straw burnt and control plots. Ohno and Erich (1997) suggested that the management systems that return crop residues back into system, the soil might increase availability of P by decreasing the adsorption of P. The addition of rapeseed and soybean residues lowered water soluble P because of microbial immobilization and adsorption and precipitation process in soil (Beri *et al.*, 1995).

3. **Potassium:** Straw of cereals, particularly rice, generally contains large amounts of K (about 2.0%), about 75% of which is water soluble and its incorporation increase potassium content of soil (Bhat *et al.*, 1991). Addition of organic residues increased exchangeable and available K in the surface soil layers.

4. **Micronutrients:** The organic manures enhance the availability of the native soil Zn and are also a good source of other micronutrients which are effectively

utilized by the crop plants. Use of organic manures alone or in combination with $FeSO_4.7H_2O$ has been found to be effective in correcting Fe deficiency in different crops. This was evident from the significant increase in available Fe status of soil with incorporation of *Sesbania* green manure before transplanting rice in coarse textured soils (Nayyar and Chhibba, 2000).

They also reported substantial increase in the DTPA-extractable Mn as a result of regular incorporation of *Sesbania* green manure over years in rice-wheat system on coarse textured soils of Punjab, owing to formation of complexes of this metal with organic acids provided during the anaerobic decomposition of these materials.

The combined use of organic materials (FYM/ wheat cut straw/GM) to compensate 50 to 75 per cent of N along with 25 to 50 per cent NPK dose through fertilizers increased the content of DTPA extractable Fe, Mn, Cu and Zn over application of 100 per cent NPK dose through fertilizers. The depletion of micronutrients in the soil (Fe, Cu, Zn and Mn) was more in the mineral fetilizer plots than in plots having both mineral and organic sources (Kumar and Yadav, 2003).

Organic farming treatments also resulted in substantial increase in labile C. The microbial biomass C was more with the addition of organic manures due to higher activity of dehydrogenase and phosphatase enzymes.

Conclusion

The after effects of green revolution advocated the necessity of organic farming due to over exploitation of the natural resources (land, water and vegetation), appearance of contaminants in food products and disturbed agro-ecosystem.

- The crop yield during first three years of organic farming remained low and started to increase thereafter. The long term studies clearly showed the significant edge of organic farming over chemical farming.
- The application of organic sources helped to improve the physico-chemical and biological properties of the soil.
- The synchronous nutrient supply at slow pace in the growing season decreased the insect-pests incidence.
- The quality of organic produce was better than that grown inorganically.

It is therefore, imperative to take up the organic farming on atleast 5-10 per cent of the cultivated area in the Punjab. It will create opportunities of organic food availability at premium prices and serve dual purpose of healthy food with conservation of resources.

16

IMPORTANCE AND SCOPE OF MEDICINAL, AROMATIC, SPICE AND PLANTATION CROPS

"There is no plant in this universe which is uneconomical/non useful. Many plants have medicinal values. This definition rightly suggests that in principle all plants have a potential medicinal value. Medicinal plants have been considered as important therapeutic aid for alleviating ailments of humankind. Search for eternal health and longevity and to seek remedy to relieve pain and discomfort prompted the early man to explore his immediate natural surroundings to develop a variety of therapeutic agents using natural resources.

IMPORTANCE OF MEDICINAL PLANTS

Medicinal plants and their uses in present age

The WHO estimates that 80% of people living in developing countries rely almost exclusively on traditional medicine for their primary health care needs. Medicinal plants form the back bone of traditional medicine and hence more than 3300 million people utilize medicinal plants on a regular basis. Demand for medicinal plants is increasing due to growing recognisation of natural products being non-toxic, having no side effects. Furthermore an increasing reliance on the use of medicinal plants in the industrialized societies has been traced to the extraction and development of several drugs and chemotherapeutics from the plants as well as from traditionally used rural remedies. Moreover, in these societies herbal remedies have become more popular in the treatment of minor ailments on account of the increasing cost of personal health maintenance.

Medicinal plants and modern drugs

Inspite of rapid development in methods of organic synthesis in laboratories, medicinal plants continue to play a significant role in modern medicine due to their inherent distinct chemical and biological properties. In nature a plant is able to synthesize complex molecules, namely alkaloids, terpenoids, tannins, saponins, glycocides, etc collectively, called secondary metabolites, from simple ones through highly specific reaction mechanisms that they use for defense and communication. It is difficult and expensive to duplicate such synthesis in laboratory. The compounds synthesized by the plants play an important role as medicinal and pharmaceutical agents not only as purified isolates and extractives but also as lead compounds for synthetic optimization.

At least 25% of all prescription contains active principles extracted from higher plants which have persisted for at least the last 25 years. Medical plants are an integral component of research developments in the pharmaceutical industry and drug development, not only when plant constituents are used directly as the rapeutic agents but also when they are used as basic material for the synthesis of drugs or as models for pharmacologically active compounds.

In 1985, Farnsworth identified 119 secondary metabolites isolated from higher plants that were being used globally as drugs. These 119 useful drugs are still obtained commercially for the most part by extraction from only about 90 species of plants. About 75% of these drugs have the same or related uses as the plant from which they were discovered.

Medicinal plants of India

Towering above all is *Azadirachta indica* whose all parts are used as medicine to treat many of disorders. The whole plant of Acalypha is a drug against bronchitis, asthma, rheumatism and pneumonia. The dried stems of *Ephedra* and *Datura* provide anti-asthmatic medicines. *Ocimum sanctum, Piper nigrum, Adhatoda vasika, Glycyrrhiza glabra and Alpinia galangal* are used to combat cough. *Lobelia* are a good respiration stimulant. *Solanum xanthocarpum* helps in cases of chest pain and congestion *Atropa belladonna* provide a drug against whooping cough, spasms and asthma. *Gaultheria fragrantissima*, *Withania somnifera* are beneficial in treating rheumatism. *Butea* sp. are used in cases of diarrhea *Perguleria* is effective for infantile diagghea. *Cassia* pulp is a good laxative, while kaladana give rise to remedy for constipation. *Cinnamon* sp. control vomiting. *Rauvolfia* have a good action against high blood pressure. *Hydnocarpus* sp. have been used in the treatment of leprosy and leucoderma. *Madhuca* sp. give rise to itch relieving preparations. *Emblica officinalis* and *Terminalia chebula* are endowed with excellent tonic characterstics. Amla has been hailed as a nugget of Vitamin *C. Symploccus* racemosa is used in treating eye diseases. *Tribulus terrestris* is indicated in urinary disorders while *Valeriana officinale* are used against hysterical fits. *Withania somnifera* is used as a general restorative for consumption, debility and sexual weakness. *Ferula foetida* is used to relieve colic and abdominal distension. *Berberis aristata* is used to control various types of inflammation and is also used as a remedy against jaundice. *Trigonella goenum-graceum* is highly effective as tonic for digestive, respiratory and nervous systems. It is also beneficial in diabetes and high blood pressure. *Moringa oleifera* provide relief from high blood pressure and tones up heart and circulatory systems.

Distribution of medicinal plants in India

In terms of life form medicinal plants are equally distributed across habits viz trees, shrubs and herbs. Roughly one third of the known medicinal plants are trees and equal proportion is shrubs and the remaining one third comprises of herbs epiphytes, grasses and climbers. Very small proportions of the medicinal plants are lower plants like lichens, ferns, and algae. The majority of medicinal plants are higher flowering plants. The major families in which medicinal plants occur are Fabaceae, Eurphorbiaceae, Asteraceae, Poaceae, Rubiaceae, Cucurbitaceae, Apiaceae, Convolvulaceae, Malvaceae, and Solanaceae.

Medicinal plants represent not only a valuable part of India's biodiversity but also a source of great knowledge. The WHO has listed 21,000 plants that are used as medicine around the world. India has a rich medicinal plant flora of some 2500 species and out of these 150 species are used commercially on a fairly large scale.

India has 16 agroclimatic zones and medicinal plants are distributed across diverse habitats and landscapes. Nearly 70% of India's medicinal plants are found in the tropical areas mostly in the various forest types across the Western and Eastern ghats, the Vindhyas, Chota Nagpur plateau, Aravalis and Himalayas. Nearly 30% of medicinal plants occur in temperate and alpine region which include plants of high medicinal values. A small number of medicinal plants are found in aquatic habitats and mangroves. The percentage of medicinal plants occurring in dry and moist deciduous regions is larger as compared to evergreen or temperate regions. Give an estimated number of medicinal plants found in different biogeograhic regions of India. According to them Trans Himalayan region have 700 species, Himalayan 2500 species, Dessert 500 species, Semi arid region have 100 species, Western Ghats have 2000 species, Deccan Peninsula 3000 species, Gangetic Plains have 1000 species, North East India has 2000 species, Islands show around 1000 species and coasts have around 500 species. Analyses of habit of medicinal plants indicate that they are distributed across various habitats. One third is trees and the other one third is shrubs and the remaining one third herbs, grasses and climbers. A very small proportion of medicinal plants are lower plants like lichens, ferns, algae etc. Majority of medicinal plants are higher flowering plants. Of the 386 families and 22 genera in which medicinal plants are recorded.

Current status of medicinal plants in India

At present, it is feared that 15-20 percent of the total flora of India may fall under one of the IUCN categories of threatened, rare or endangered. At least 120 medicinal species have been officially classified as endangered. In spite of this, 70 percent of the critically endangered plants of India have been reported to be part of active trade. According to a study done by the EXIM Bank; around 95% of the medicinal plants used by the Indian Industry are collected from the wild. An estimate suggests that over half a million tons of raw material is collected from the forest every year. Less than 20 species are under commercial cultivation. Over 70% of the plant collection involves destructive harvesting.

Threat to medicinal plants

Apparently the quick decline of the vegetation cover of a country is directly linked with the rapid population growth with an ever increasing needs. The, major causes of the loss of biological diversity in India both in the hills and plains are anthropogenic. However, eight major causes for the loss of biodiversity have been identified namely- (i) Habitat loss and fragmentation, (ii) Introduced species, (iii) Over exploitation of plant and animal species, (iv) Pollution of soil, water and atmosphere (v) Global climate change, (vi) Expansion of industry, agriculture and forestry, (vii) Want of pollinating vectors, (viii) Unisexual flowered species.

All the factors described above threatening the phytodiversity in India broadly cover most of the threats to medicinal plant species. Over exploitation and indiscriminate use of wild resources in commercial demand now-a-days plays great role, not only behind the quick decline of the species concerned, but also becomes a threat for the survival of other associated species, thus acting as the major factors disturbing the entire ecosystem.

FEW IMPORTANT MEDICINAL PLANTS ARE DISCUSSED AS UNDER:

1. STEVIA (*Stevia rebandiana* Bertoni)

Family : Asteraceae

All over world there is increasing tendency towards consuming natural products and thus living a natural life. At the same time our life styles have changed so much over the last 4-5 decades sweeteners (either high calorie natural or proceeded sugars or high potency and low calorie sweeteners such as Aspartame) have become integral part of our natural daily diet. Due to sedentary life styles that we all tend to lead these days the incidence of obesity and diabetic conditions are increasing dramatically. In India number of diabetic people in the age group of 25-45 is about 15% and is increasing at an alarming pace. In addition, we are largest consumers of sugar in the world. This single factor, we have come to understand, would greatly contribute to increasing the number of diabetic people and related problems. Stevia, botanically known as *Stevia rebaudiana* (Family- Asteraceae) is a sweet herb. A perennial herb, Stevia is a member of the daisy family. The leaves are mid green and intensely sweet. The compounds in the leaves are called stevioside and rebaudioside and they can be more than 200 times sweeter than sugar. The plant bears greenish cream flowers in autumn. Although Stevia has been in use in Asia and Europe for years, it was only in the past couple of years that is really started to capture attention in the India market as a healthy alternative sweetener to sugar. Stevia has no calcium cyclamate, no saccharin, no aspartame and no calories. It is safe for diabetics, as it does not affect blood sugar levels; it does not have the neurological or renal side effects associated with some of the artificial sweeteners. There are many advantages of using Stevia over conventional sweeteners the main one being: Stevia leaves are 20-30 times sweeter than sugar. Stevia leaves can be dried and stored. Stevia can be used in raw from. Stevia is short duration crop. It is harvested 3/4 times a year. The yearly yields can be in the range of 10 to 15 q/ha.

Uses of Stevia leaves

Currently consumers have more inclination towards products that are claimed to be 'All Natural 'and 'Low CHO'. Hence, the food industry could grab a major share in the market if Stevia, the natural sweetener, is used as sweetening agent in products like biscuits, jams, chocolates, icecreams, baked foods, soft drinks, soda, candies and also common beverages like dip tea, coffee and herbal tea that are targeted particularly at the diabetics and also the health conscious consumers. Application in bakery, confectionery and beverage sector are very common. All cooked and baked food items like puddings; desserts can be sweetened with only very small quantities of Stevia leaf powder as compared to table sugar. Just 50 gms of Stevia leaf can replace 1000 gms of cane sugar. The sweetness of stevioside is non-fermenting and it does not display browning when cooked. This further widens its area of application in baking. This also enhances the quality and safety to usage with a longer shelf- life period. Breads made with Stevia as an ingredient (for diabetics patients) have been found to display improved texture, softness and an increased shelf life.

Medicinal properties of Stevia

Stevia is versatile herb with incredible sweetness that can be safely used in herbal medicines; tonics for diabetic patients and also in the daily usage products, Stevia leaves can be used because of its anti -fungal and anti-bacterial property. Mild stevia leaf tea offers excellent relief for an upset stomach. A wet Stevia leaf bag provides a cooling effect on eyes (similar to using cucumber). The leaves effectively tighten the skin and are good for wrinkles. Stevia has proved to give exceptional benefits when used regularly in skin care. It also has a healing effect on blemishes, wounds, cuts and scratches. Stevia is helpful in weight and blood pressure management. It has also been reported that Stevia lowers incident of colds and flu.

Commercial cultivation

The concentration of Stevioside in the leaves of Stevia increases when the plants are grown under long day condition. Stevia grows best in upland areas in sub-tropical climate. In other places it can be grown as an annual. The plant prefers a lightly textured, well-drained soil to which organic matter has been added. It needs ample water so that the soil is consistently moist, but not wet. In hot, sunny climates it will do best in semi-shade. Propagation is from seed sown in spring, but germination rates can be low-expect half the seeds sown not to germinate. Leaves are best harvested just before flowering. The plants will also grow from cuttings, which are best taken in late winter. The concentration of stevioside in the leaves of Stevia increasing when the plants are grown under long day condition. While, cultivating Stevia on a large scale, it can be grown in well-drained red soil and sandy loam soil. The soil should be in the pH range of 6.5-7.5. Saline soils should be avoided to cultivate this plant. Stevia can be successfully cultivated all around the year all over India expect the areas, which receive snowfall, or temperatures go below 5 degree celcius in winter. The summer temperatures actually do not affect this plant if the high summer temperatures have already been factored in the cultivation practices. Since seed germination rate is very poor, it is propagated vegetatively. Though stem cuttings are used for vegetative tissue culture so plants have proven to be the best planting material for Stevia. Tissue culture plants of Stevia are genetically pure, free from pathogens and have excellent vigour. The tissue culture plants can be planted throughout the year, expect during peak summer.

Land preparation

Land should be ploughed initially with a disc plough or harrowed to break down the colds. Fine tilth is required. One to two times ploughing has to be done after harrowing. Around 25 MT of FYM/ha has to be applied as a basal dressing during the last ploughing to incorporate the manure with soil.

Soil type

Stevia requires very good drainage any soil that retain the moisture for very long period of time are unsuitable for Stevia cultivation and should be religiously avoided. Similarly black cotton soils with very heavy clay content should also be avoided. It is mandatory that before planting, each and every soil sample should be properly tested and analyzed by an expert. Red soil and sandy loam with a 6-7 pH are best for the cultivation of Stevia.

Planting

Stevia can be planted in many ways. The types of agronomic practices generally depend on type of soil and climate conditions. Generally it is advisable to plant minimum of one lac plants/ha. Since the economical part is the leaves it is very important to achieve highest vegetative growth. Forming raised beds is the most economical way to grow Stevia. The raised bed should be of 15 cm in height and 60 cm in width. The distance between each plant must be 23 cm. This would give a plant population of around 75,000 per ha with spacings of 25 × 40 cm.

Planting material

There are basically two options for multiplication. The first is the tissue culture and second the stem cutting. Tissue culture is the best option but many farmers are tempted to try the stem cutting method for multiplication. As per practical experience, stem cutting is sometimes more expensive to produce than the tissue culture since the success rates of the stem cuttings establishment is very low, it takes minimum of 25 weeks for the stem cutting to develop in proper feeding roots for transplantation (younger stem cuttings transplants have shown more than 50% mortality in first few weeks of transplants in main field). It is also noticed that Stevioside content of the stem cutting plants is lower by at least cutting plants is lower by at least 2% points in Indian conditions.

Irrigation

Stevia requires ample supply of good water all year around as the plant cannot tolerate drought, frequent irrigation is required. Micro sprinklers are the best method of irrigation as it supply the required amount of water at the right time. So through micro sprinklers, the water can be sprinkled once in a day in winter and two to four times in a day in summer depending upon the heat and relative humidity in the air. Watering frequency should be scheduled so that the plants do no wilt for want of water.

Fertilizer application

The recommended dose of fertilizer is 275:112:112 Kg NPK/ha. All P and K at the time of sowing and the Nitrogen can be applied in split doses every month. Nitrogen application is a must for the production of dry matters.

Weeding

Removal of weeds can be done manually. Since the crop is grown in raised beds, intercultural operations are easier by manual labour.

Maintenance

Stevia plants do best in a rich, loamy soil. Since the feeder roots tend to be quite near the surface, it is good idea to add compost for extra nutrients if the soil is sandy. Besides being sensitive to cold during their development stage, the roots are also adversely affected by excessive levels of moisture. So take care not to over water them and to make sure the soil in which they are planted drains easily as flooding is harmful. Frequent light watering is recommended during the summer months. Adding a layer of compost or mulch around each Stevia plant will help keep the shallow feeder roots from drying out. Stevia plants respond well to fertilizers with lower nitrogen content than the fertilizer's phosphoric acid or potash content. Most organic fertilizers would work well, since they release nitrogen slowly. Flowering of the plant should be avoided. Since Stevia has a significant apical dominance, the plant tends to grow tall and lanky. Pinching of the apical bud would enhance bushy growth of the plant with side branches.

Harvesting

Depending on field and climate conditions one can achieve the yields Stevia leaves of all cuttings may be between 10 to 15 q/ha on dry weight basis and 55 to 75 q/ha on fresh weight basis. Another important aspect of harvesting is the timing of harvest. It should be noted that at no point of time plants should be allowed to flower since after flowering the Stevioside percentage goes down rapidly and leaves are rendered unmarketable. Leaves are harvested by plucking in a small quantity, or the entire plant with the side branches is cut leaving 10 to 15 cm from the base. The first harvesting can be done four to five months after planting. Subsequent harvesting can be done every three months, for three consecutive years. The sweetener in the leaf is maximum till the plant flowers. Just before flowering, the plant should be cut completely leaving 10 cm from the ground. The new flush of leaves will sprout from here. The new plant will be ready for harvest again in three months. Harvesting should be done as late as possible, since cool autumn temperatures and shorter days tend to intensify the sweetness of the plants as they evolve into a reproductive state.

2. ALOE (*Aloe vera*)

Family : Asphodelaceae

Aloe barbadensis Mill. popularly known as *Aloe vera* originated in the warm dry climates of Africa. However, because of its wide adaptability as well as its importance as medicinal plants, it is well distributed. The virtues of plant have been recorded by many great civilizations from those of Persia and Egypt in the Middle East, to those of Greece and Italy in Europe, to those of India and the African continent. The plant is widely known in Asia and the Pacific and is found in the folklore of the Japanese the Philippines and the Hawaiians. The Spanish used Aloe and carried it with them to their new world colonies in South America and the Caribbean. In each age, on each continent, in each culture, *Aloe vera* has drawn the attention of the most sophisticated minds. In India, It is grown commercially for its high demand in cosmetic industries as well as in Indian System of Medicine. The succulent leaves are economic part of this plant. Aloe is known by several names in India such a Ghrii Kuruari. Kunvar pathu and Indian Aloe. Its multifarious uses as medicinal plants, vegetables, pickle purposes etc. has inspired growers for commercial cultivation in many parts of the country. The gel produced from its mature leaves is used in cosmetic industry for preparation of shampoo, face creams and moisturising agents.

The characteristic bitter taste of *Aloe vera* leaves possesses many medicinal properties. The leaves are alterative sromachic, aphrodisiac, cathartic, emmenagogic, astringent, antidotal, anthelmintic and hepatic stimulant. After removing the skin of leaves they are very useful in fevers, enlargement of liver, spleen and other glands, skin diseases, gonorrhoea, constipation, menstrual suppressions, piles, jaundice, rheumatic diseases and for the treatment of burns and bruises,

Climate

It is grown in almost all parts or India, even under constant drought conditions. However, the crop grows well in entire tropical and sub-tropical regions. Since its water requirement is very low it is best suited for cultivation in Arid and Semiarid region specially in Rajasthan, Gujarat, Madhya Pradesh and Maharashtra.

Soil

It is grown successfully ill marginal to sub marginal soils having low fertility. The plants have tendency to tolerate high pH with high Na and K salts. However, it is observed that its growth was faster under medium fertile heavier soils such as black cotton soils of central India. Though well drained loam to coarse sandy loam soils with moderate fertility and pH upto 8.5 are preferred for its commercial cultivation.

Land preparation

The soil should not be disturbed too deep as the root system of Aloe does not penetrate below 20-30 cm, depending upon the soil type and agroclimatic condition, 1-2 ploughing followed by levelling may be done. Field may be divided into suitable sized plots (10-15 m x 3 m) considering tbe slope and source of irrigation available.

Planting time

Suckers should be planted in July-August during monsoon season to get better field survival and subsequent growth of the plants. However, under irrigated condition, planting can be done around the year except in winter months (November-February).

Planting material

Planting material recommended for commercial cultivation is by suckers only. About three to four months old suckers having 4-5 leaves and about 20-25 cm in length are used as planting material.

Spacing and planting

Suckers are planted in about 15 cm deep pits made just at the time of planting at 60 x 60 cm apart. After planting of suckers the soil around the root zone must be firmly pressed and drainage must be made proper to avoid water stagntion. About 28,000 to 34,000 suckers are needed for one hectare planting.

Manuring

In general, the crop responds well to application of manure (FYM or compost). About 25-40 t/ha of FYM may he applied at the time of soil preparation and also in subsequent years. But, it is not a common practice since this crop is cultivated in marginal and sub-marginal lands. If sufficient quantity of FYM or wood ash is available, it can be applied in the pits at the lime of planting that helps in establishment of plants and their subsequent growth.

Irrigation

The crop withstands stress condition very well but to get good crop, irrigation at critical stages of growth must be given. First irrigation is required just after planting of suckers followed by 2-3 irrigations till plant get established. However, 4-6 irrigations per year may be enough for proper growth of the plants. After each picking of leaves light irrigation may be given depending upon the availability of water

Interculture

The field should be kept weed free throughout the growing period of crop. Two to three hand weeding followed by light earthing up promote growth and suckering in Aloe. First hoeing should he completed within a month after planting. Later on hand weedings are done as per weed intensity in order to keep weed free crop. Unproductive and diseased plants and dried flower stakes must be removed regularly.

Harvesting and yield

An Aloe plantation gives commercial yield from second year to fifth year of planting. Generally 3-4 pickings per year can be taken up depending upon the growth of plants. On an average 15-20 t/ha fresh leaf is obtained from second year plantation. However, well managed irrigated crop can be give up to 30-35 t/ha fresh leaf. Fully developed mature leaves should be harvested for extraction of Aloe juice.

3. DIOSCOREA

Family : Dioscoreaceae

Among the twining species, some species twine clockwise while others anticlockwise and these are dioceous and rhizomatous. According to Coursey (1967), this genus is named in honour of the Greek physician Pedenios Dioscorides, the author of the classical Materia Medica Libri Quinque. Dioscorea have many species but *D. floribunda, D. composita* and *D. deltoidea* are widely grown for diosgenin production.

I. ***Dioscorea floribunda*** **Mart. & Gal:** *D. floribunda* Mart. & Gal. is an introduction from Central America and had wide adaptation as it is successfully grown in Karnataka, Assam, Meghalaya, Andaman and Goa. The vines are glabrous and left twining. The alternate leaves are borne on slender stems and have broadly ovate or triangular ovate, shallowly cordate, coriaceous lamina with 9 nerves. The petioles are 5-7 cm long, thick and firm. Variegation in leaves occurs in varying degrees. The male flowers are solitary and rarely in pairs. Female flowers have diverícate stigma which is bifid at apex. The capsule is obovate and seed is winged all round. The tubers are thick with yellow coloured flesh, branched and growing upto a depth of 30 cm

II. ***Dioscorea composita*** **Hemsl.:** *D. composita* Hemsl. it is a Central American introduction into Goa, Jammu, Bangalore, Anaimalai Hills of Tami Nadu and Darjeeling in W. Bengal. The vines are right twinning and nearly glabrous. The alternate leaves have long petioles, membraneous or coriaceous lamina measuring upto 20x18 cm, abruptly acute or cuspidate-acuminate, shallowly or deeply cordate, 7-9 nerved. The fasciculate-glomerate inflorescence is single or branched with 2 or 3 sessile male 137 flowers having fertile stamens. Male fascicle is 15-30 cm long. The female flowers have bifid stigma. Tubers are large, white and deep-rooted (upto 45 cm).

III. *Dioscorea deltoidea* **Wall. ex. Griseb.:** *D. deltoidea* Wall. ex. Griseb. is distributed throughout the Himala-yas at altitudes of 1000-3000 m extending over the states of Jammu & Kashmir, H.P., U.P., Sikkim and further into parts of W. Bengal. The glabrous and left twining stem bears alternate petiolate leaves. The petioles are 5-12 cm long. The lamina is 5-15 cm long and 4-12 cm wide widely cordate. The flowers are borne on axillary spikes, male spikes 8-40 cm long and stamens 6. Female spikes are 15 cm long, 3. 5 cm broad and 4-6 seeded. Seeds are winged all round. Rhizomes are lodged in soil, superficial, horizontal, tuberous, digitate and chestnut brown in colour. *D. deltoidea* tuber grows parallel to ground covered by small scale leaves and is described as rhizome. The tubers are morphologically cauline in structure with a ring of vascular bundles in young tubers which appear scattered in mature tubers.

Importance of Diosgenin

Diosgenin is the most important sapogenin used as a starting material for synthesis of a number of steroidal drugs. For commercial purposes, its b-isomer, yamogenin is also taken as diosgenin while analysing the sample for processing. Various steroidal drugs derived from diosgenin by artificial synthesis include corticosteroids, sex hormones, anabolic steroids and oral contraceptives. Corticosteroids are the most important group of steroidal drugs synthesized from diosgenin. First group of corticosteroids regulates carbohydrate and protein metabolism. The second group consists of aldosterone, which controls balance of potassium, sodium and water in the human body. The glucocorticoids in the form of cortisone and hydrocortisone are used orally, intramuscularly or topically for treatment of rheumatoid arthritis, rheumatic fever, other collegen diseases, ulcerative colitis, certain cases of asthma and a number of allergic diseases affecting skin, eye and the ear. These are also used for treatment of gout and a variety of inflammations of skin, eye and ear and as replacement therapy in Addison's diseases.

Both male and female sex hormones are also synthesized from Disosgenin. The main male sex hormone (androgen) which is produced from Disogenin is testosterone. The main female sex hormones produced are oestrogen and progesterone. Recently oestrogen has also been used in cosmetic lotions and creams to improve the tone and colour of skin. One of the main uses of progesterone during recent years has been as antifertility agent for oral contraceptives. These artificial steroids have increased oral activity and fewer side effects, as they can be used in reduced doses. Oral contraceptives are also used for animals like pigs, cows and sheep to control fertility and to give birth at a prescribed period in a group of animals at the same time. These compounds are also used to reduce the interval between the lactation periods to have more milk and meat production.

Agrotechnology

Dioscorea species prefer a tropical climate without extremity in temperature. It is adapted to moderate to heavy rainfall area. Dioscorea plants can be grown in a variety of soils, but light soil is good, as harvesting of tubers is easier in such soils. The ideal soil pH is 5.5-6.5 but tolerates fairly wide variation in soil pH. Dioscorea can be propagated by tuber pieces, single node stem cuttings or seed. Commercial planting is normally established by tuber pieces only. Propagation through seed progeny is variable and it may take longer time to obtain tuber yields. Dipping of tuber pieces for 5 minutes in 0.3% solution of Benlate followed by dusting the cut ends with 0.3% Benlate in talcum powder in moist sand beds effectively checks the tuber rot. The treatment is very essential for obtaining uniform stand of the crop. The best time of planting is the end of April so that new sprouts will grow vigorously during the rainy season commencing in June in India. Land is to be prepared thoroughly until a fine tilth is obtained. Deep furrows are made at 60 cm distance with the help of a plough. The stored tuber pieces which are ready for planting is to be planted in furrows with 30 cm between the plants for one year crop and 45 cm between the plants for 2 year crop at about 0.5 cm below soil level. The new sprouts are to be staked immediately. After sprouting is complete, the plants are to be earthed up. Soil from the ridges may be used for earthing up so that the original furrows will become ridges and vice versa.

Dioscorea requires high organic matter for good tuber formation. Besides a basal doze of 18-20 t of FYM/ha, a complete fertilizer dose of 300 kg N, 150 kg P_2O_5 and K_2O

each are to be applied per hectare. P and K are to be applied in two equal doses, one after the establishment of the crop during May-June and the other during vigorous growth period of the crop (August September). Irrigation may be given at weekly intervals in the initial stage and afterwards at about 10 days interval. Dioscorea vines need support for their optimum growth and hence the vines are to be trailed over pandal system or trellis. Periodic hand weeding is essential for the first few months. Intercropping with legumes has been found to smother weeds and provide extra income. The major pests of Dioscorea are the aphids and red spider mites. Aphids occur more commonly on young seedlings and vines. Young leaves and vine tips eventually die if aphids are not controlled. Red spider mites attack the underside of the leaves at the base near the petiole. Severe infestations result in necrotic areas, which are often attacked by fungi. Both aphids and spider mites can be very easily controlled by Kelthane. No serious disease is reported to infect this crop. The tubers grow to about 25-30 cm depth and hence harvesting is to be done by manual labour. The best season for harvesting is Feb-March, coinciding with the dry period. On an average 50-60 t/ha of fresh tubers can be obtained in 2 years duration. Diosgenin content tends to increase with age, 2.5% in first year and 3-3.5% in the second year. Hence, 2 year crop is economical.

4. PERIWINKLE (*Catharanthus roseus*)

Family : Apocynaceae

It is an erect handsome herbaceous perennial plant which is a chief source of patented cancer and hypotensive drugs. It is one of the very few medicinal plants which has a long history of uses as diuretic, antidysenteric, haemorrhagic and antiseptic. It is known for use in the treatment of diabetes in Jamaica and India. The alkaloids vinblastine and vincristine present in the leaves are recognized as anticancerous drugs. Vinblastine in the form of vinblastin sulphate is available in market under the tradename "*Velbe*" and Vincristine sulphate as "*Oncovin*" (Eli Lilly). Vinblastine is used in combination with other anticancer agents for the treatment of lymphocytic lymphoma, *Hodgkin's* disease, testicular carcinoma and choriocarcinoma. Vincristine is used in acuteleukemia, lymphosarcoma and *Wilm's* tumour. Its roots are a major source of the alkaloids, raubasine (ajmalicine), reserpine and serpentine used in the preparation of antifibrillic and hypertension-relieving drugs. It is useful in the treatment of choriocarcinoma and *Hodgkin's* disease-a cancer affecting lymph glands, spleen and liver. Its leaves are used for curingdiabetes, menorrhagia and wasp stings. Root is tonic, stomachic, hypotensive, sedative and tranquilliser.

Distribution

The plant is a native of Madagascar and hence the name Madagascar Periwinkle. It is distributed in West Indies, Mozambique, South Vietnam, Sri Lanka, Philippines and Australia. It is well adapted to diverse agroclimatic situations prevalent in India and is commercially cultivated in the states of Tamil Nadu, Karnataka, Gujarat, Madhya Pradesh and Assam. The major consumers are USA, Hungary, West Germany, Italy, Netherlands and UK.

Botany

Catharanthus roseus (Linn.) G.Don. syn. *Vinca rosea* Linn. belongs to the family Apocynaceae. It is an erect highly branched lactiferous perennial herb growing up to a height of one metre. Leaves are oblong or ovate, opposite, short-petioled, smooth with entiremargin. Flowers are borne on axils in pairs. There are three flower colour types, pink, pink-eyed and white. Calyx with 5 sepal, green, linear, subulate. Corolla tube is cylindrical with 5 petals, rose-purple or white with rose-purple spot in the centre; throat of corolla tube hairy, forming a corona like structure. The anthers are epipetalous borne on short filaments inside the bulging distal end of corolla tube converging conically above the stigma. Ovary bicarpellary, basally distinct with fused common style and stigma. The dehiscent fruit consists of a pair of follicles each measuring about 25mm in length and 2.3 mm in diameter, containing up to thirty linearly arranged seeds with thin black tegumen. On maturity, the follicles split along the length dehiscing the seeds.

Agrotechnology

Periwinkle grows well under tropical and subtropical climate. A well distributed rainfall of 1000 mm or more is ideal. In north India the low winter temperatures adversely affect the crop growth. It can grow on any type of soil, except those which are highly saline, alkaline or waterlogged. Light soils, rich in humus are preferable for large scale cultivation since harvesting of the roots become easy.

Catharanthus roseus is propagated by seeds. Fresh seeds should be used since they are short viable. Seeds can be either sown directly in the field or in a nursery and then transplanted. Seed rate is 2.5 kg/ha for direct sowing and the seeds are drilled in rows 45 cm apart or broadcasted. For transplanted crop the seed rate is 500 gm/ha. Seeds are sown in nursery and transplanted at 45 x 30 cm spacing after 60 days when the seedlings attain a height of 15-20 cm Nursery is prepared two months in advance so that transplanting coincides with the onset of monsoons. Application of FYM at the rate of 15 t/ha is recommended. An alternate approach is to grow leguminous green manure crops and incorporate the same into the soil at flowering stage. Fertilisers are recommended at 80:40:40 kg N:P_2O_5: K_2O/ha for irrigated crop and 60:30:30 kg/ha for rainfed crop. All P_2O_5 and K_2O are applied at sowing. Nitrogen is applied in three equal splits at planting and at 45 and 90 days after planting. Four or five irrigations will be needed to optimise yield when rainfall is restricted. Fortnightly irrigations support good crop growth when the crop is grown exclusively as an irrigated crop. Weeding is carried out before each top dressing. Alternatively, use of fluchloralin at 0.75 kg / ha pre-plant or alachlor at 1.0 kg per ha as pre-emergence to weeds provides effective control of a wide range of weeds in periwinkle crop. Detopping of plants by 2 cm at 50% flowering stage improves root yield and alkaloid contents. No major pests, other than oleander hawk moth, have been reported in this crop. The crop allows 3-4 clippings of foliage beginning from 6 months. The flowering stage is ideal for collection of roots with high alkaloid content. The crop is cut about 7 cm above the ground and dried for stem, leaf and seed. The field is irrigated, ploughed and roots are collected. The average yields of leaf, stem and root are 3.6, 1.5 and 1.5 t/ha, respectively under irrigated conditions and 2.0, 1.0 and 0.75 t/ha, respectively under rainfed conditions on air dry basis. The harvested stem and roots, loose 80% and 70% of their weight, respectively. The crop comes up well as an under crop in eucalyptus plantation in north India.

5. SERPENTWOOD (*Rauvolfia serpentine*)

Family : Apocynaceae

Serpentwood, Sarpagandha, Chandrabhaga, Sarpagandhi, Amalpori, Chivan Amelpodi, Sutranbhi and Patalagandhi is an erect, evergreen, perennial under shrub whose medicinal use has been known since 3000 years. Its dried root is the economical part which contains a number of alkaloids of which reserpine, rescinnamine, deserpidine, ajamalacine, ajmaline, neoajmalin, serpentine, yohimbine are pharmacologically important. The root is a sedative and is used to control high blood pressure and certain forms of insanity. In Ayurveda it is also used for the treatment of insomnia, epilepsy, asthma, acute stomach ache and painful delivery. It is used in snake-bite, insect stings, and mental disorders.

Distribution

Rauvolfia serpentina is native to India. Several species of *Rauvolfia* are observed growing under varying edaphoclimatic conditions in the humid tropics of India, Nepal, Burma, Thailand, Bangladesh, Indonesia, Cambodia, Philippines and Sri Lanka. In India, it is cultivated in the states of Uttar Pradesh, Bihar, Tamil Nadu, Orissa, Kerala, Assam, West Bengal and Madhya Pradesh. Thailand is the chief exporter of *Rauvolfia* alkaloids followed by Zaire, Bangladesh, Sri Lanka, Indonesia and Nepal.

Botany

Plumier in 1703 assigned the name *Rauvolfia* to the genus in honour of a German physcian - Leonhart Rauvolf of Augsburg. The genus *Rauvolfia* of Apocynaceae family comprises over 170 species distributed in the tropical and subtropical parts of the world including 5 species native to India.

R. serpentina is an erect perennial shrub generally 15-45 cm high, but growing upto 90 cm under cultivation. Roots nearly verticle, tapering up to 15 cm thick at the crown and long giving a serpent like appearance, occasionally branched or tortuous developing small fibrous roots. Roots greenish-yellow externally and pale yellow inside, extremely bitter in taste. Leaves born in whorls of 3-4 elliptic-lanceolate or obovate, pointed. Flowers numerous borne on terminal or axillary cymose inflorscence. Corolla tubular, 5-lobed, 1-3 cm long, whitish-pink in colour. Stamens 5, epipetalous. Carpels 2, connate, style filiform with large bifid stigma. Fruit is a drupe, obliquely ovoid and purplish black in colour at maturity with stone containing 1-2 ovoid wrinkled seeds. The plant is cross-pollinated, mainly due to the protogynous flowers.

Agrotechnology

Among the different species of Rauvolfia, *R. serpentina* is preferred for cultivation because of higher reserpine content in the root. Though it grows in tropical and subtropical areas which are free from frost, tropical humid climate is most ideal. Its common habitats receive an annual rain fall of 1500-3500 mm and the annual mean temperature is 10-38°C. It grows up to an elevation of 1300-1400 m from mean sea level. It can be grown in open as well as under partial shade conditions. It grows on a wide range of soils. Medium to deep well drained fertile soils and clay-loam to silt-loam soils rich in organic matter are suitable for its cultivation. It requires slightly acidic to neutral soils for good growth. The plant can be propagated vegetatively by root cuttings, stem cuttings or root stumps and by seeds. Seed propagation is the best method for raising commercial plantation. Seed germination is very poor and variable from 10-74%. Seeds collected during September to November give good results. It is desirable to use fresh seeds and to sock in 10% sodium chloride solution. Those seeds which sink to the bottom should only be used. Seeds are treated with ceresan or captan before planting in nursery to avoid damping off. Seed rate is 5-6 kg/ha. Nursery beds are prepared in shade, well rotten FYM is applied at 1 kg/m^2 and seeds are dibbled 6-7 cm apart in May-June and irrigated. Two months old seedlings with 4-6 leaves are transplanted at 45-60 x 30 cm spacing in July–August in the main field. Alternatively, cuttings of 2.5-5cm long roots or 12-20 cm long woody stems can also be used for transplanting. Hormone (Seradix) treatment increases rooting. In the main field 10-15 t/ha of FYM is applied basally. Fertilisers are applied at 40:30:30 kg N:P_2O_5 : K_2O/ ha every year. All P_2O_5 and K_2O are applied at sowing and N is applied in 2-3 splits. Monthly irrigation increases the yield. The nursery and the main field should be kept weed free by frequent weeding and hoeing.

In certain regions intercroping of soybean, brinjal, cabbage, okra or chilly is followed in Rauvolfia crop. Pests like root grubs (*Anomala polita*), moth (*Deilephila nerii*), caterpillar (*Glyophodes vertumnalis*), black bugs and weevils are observed on the crop, but the crop damage is not serious. The common diseases reported are leaf spot, leafblotch, leaf blight (*Alternaria tenuis*), anthracnose, die back, powdery mildew, wilt, root-knot, mosaic and bunchy top virus diseases. Field sanitation, pruning and burning of diseased parts and repeated spraying of 0.2% Dithane Z-78 or Dithane M-45 are recommended for controlling various fungal diseases. Rauvolfia is harvested after 2-3 years of growth. The optimum time of harvest is in November - December when the plants shed leaves, become dormant and the roots contain maximum alkaloid content. Harvesting is done by digging up the roots by deeply penetrating implements. The roots are cleaned washed cut into 12-15cm pieces and dried to 8-10% moisture. The dried roots are stored in polythene lined gunny bags in cool dry place to protect it from mould. The yield is 1.5-2.5 t/ha of dry roots. The root bark constitutes 40-45% of the total weight of root and contributes 90% of the total alkaloids yield.

6. INDIAN GOOSEBERRY (*Phyllanthus emblica*)

Family : Euphorbiaceae

Indian gooseberry, Amalaka, Adiphala, Amla, Amlaki Mal, Nelli, Amalakam, Amalaka, Ambala, Aonla is a medium sized tree, the fruit of which is used in many ayurvedic preparations from time immemorial. It is useful in haemorrhage, leucorrhaea, menorrhagia, diarrhoea and dysentery. In combination with iron, it is useful for anaemia, jaundice and dyspepsia. It goes in combination in the preparation *of triphala, arishta, rasayan, churna* and *chyavanaprash. Sanjivani* pills made with other ingredients is used in typhoid, snake-bite and cholera. The green fruits are made into pickles and preserves to stimulate appetite. Seed is used in asthma, bronchitis and biliousness. Tender shoots taken with butter milk cures indigestion and diarrhoea. Leaves are also useful in conjunctivitis, inflammation, dyspepsia and dysentery. The bark is useful in gonorrhoea, jaundice, diarrhoea and myalgia. The root bark is astringent and is useful in ulcerative stomatitis and gastrohelcosis. Liquor fermented from fruit is good for indigestion, anaemia, jaundice, heart complaints, cold to the nose and for promoting urination. The dried fruits have good effect on hair hygiene and used as ingredient in shampoo and hair oil. The fruit is a very rich source of Vitamin C (600 mg/100 g) and is used in preserves as a nutritive tonic in general weakness.

Distribution

Indian gooseberry is found throughout tropical and subtropical India, Sri Lanka and Malaca. It is abundant in deciduous forests of Madhya Pradesh and Darjeeling, Sikkim and Kashmir. It is also widely cultivated.

Botany

Phyllanthus emblica Linn. syn. *Emblica officinalis* Gaertn. belongs to Euphorbiaceae family. It is a small to medium sized deciduous tree. Leaves are simple, many subsessile, closely set along the branchlets, distichous light green having the appearance of pinnate leaves. Flowers are greenish yellow in axillary fascicles, unisexual; males numerous on short slender pedicels; females few, subsessile; ovary 3-celled. Fruits are globose, 1-5 cm in diameter, fleshy, pale yellow with 6 obscure vertical furrows enclosing 6 trigonous seeds in 2-seeded 3 crustaceous cocci. Two forms Amla are generally distinguished, the wild ones with smaller fruits and the cultivated ones with larger fruits and the latter are called '*Banarasi*'.

Agrotechnology

Gooseberry is quite hardy and it prefers a warm dry climate. It needs good sunlight and rainfall. It can be grown in almost all types of soils, except very sandy type. A large fruited variety "*Chambakad Large*" was located from the rain shadow region of the Western Ghats for cultivation in Kerala. *Amla* is usually propagated by seeds and rarely by root suckers and grafts. The seeds are enclosed in a hard seed coat which renders the germination difficult. The seeds can be extracted by keeping fully ripe fruits in the sun for 2-3 days till they split open releasing the seeds. Seeds are soaked in water for 3-4 hours and sown on previously prepared seed beds and irrigated. Excess irrigation and waterlogging are harmful. One month old seedlings can be transplanted to polythene bags and one year old seedlings can be planted in the main field with the onset of monsoon. Pits of size 50 cm^3 are dug at 6-8 m spacing and filled with a mixture of top soil and well rotten FYM and planting is done. *Amla* can also be planted as a windbreak around an orchard. Irrigation and weeding are required during the first year. Application of organic manure and mulching every year are highly beneficial. Chemical fertilisers are not usually applied. No serious pests or diseases are generally noted in this crop. Planted seedlings will commence bearing from the 10th year, while grafts after 3-4 years. The vegetative growth of the tree continues from April to July. Along with the new growth in the spring, flowering also commences. Fruits will mature by December-February. Fruit yield ranges from 30-50 kg/tree/year when full grown.

7. INDIAN BDELLIUM (*Commiphora mukul*)

Family : Burseraceae

Indian bdellium is a small, armed, deciduous tree from the bark of which gets an aromatic gum resin, the *'Guggul'* of commerce. It is a versatile indigenous drug claimed by ayurvedists to be highly effective in the treatment of rheumatism, obesity, neurological and urinary disorders, tonsillitis, arthritis and a few other diseases. The fumes from burning guggul are recommended in hay fever, chronic bronchitis and phytises.

Distribution

The center of origin of *Commiphora* spp. is believed to be Africa and Asia. It is a widely adapted plant well distributed in arid regions of Africa (Somalia, Kenya and Ethiopia in north east and Madagascar, Zimbabwe, Botswana, Zaire in south west Africa), Arabian peninsula (Yeman, Saudi Arabia and Oman). Different species of *Commiphora* are distributed in Rajasthan, Gujarat, Maharashtra and Karnataka states of India and Sind and Baluchistan provinces of Pakistan. In India, the main commercial source of gum guggul is Rajasthan and Gujarat.

Botany

C mukul is a small tree upto 3-4 m height with spinescent branching. Stem is brownish or pale yellow with ash colored bark peeling off in flakes. Young parts are glandular and pubescent. Leaves are alternate, 1-3 foliate, obovate, leathery and serrate (sometimes only towards the apex). Lateral leaflets when present only less than half the size of the terminal ones. Flowers small, brownish red, with short pedicel seen in fascicles of 2-3. Calyx campanulate, glandular, hairy and 4-5 lobed. Corolla with brownish red, broadly linear petals reflexed at apex. Stamens 8-10, alternatively long and short. Ovary oblong, ovoid and stigma bifid. Fruit is a drupe and red when ripe, ovate in shape with 2-3 celled stones. The chromosome number 2n = 26.

Agrotechnology

Guggal being a plant of arid zone thrives well in arid- subtropical to tropical climate. The rainfall may average between 100 mm and 500 mm while air temperature may vary between 40°C in summer and 3°C during winter. Maximum relative humidity prevails during rainy season (83% in the morning and 48% in the evening). Wind velocity remains between 20-25 km/hour during the year is good. Though they prefer hard gypseous soil, they are found over sandy to silt loam soils, poor in organic matter but rich in several other minerals in arid tracks of western India. Plants are propagated both by vegetatively and seeds. Plants are best raised from stem cuttings from the semi woody (old) branch. For this purpose one metre long woody stem of 10 mm thickness is selected and the cut end is treated with IBA or NAA and planted in a well manured nursery bed during June-July months; the beds should be given light irrigation periodically. The cuttings initiate sprouting in 10-15 days and grow into good green sprout in 10-12 months. These rooted plants are suitable for planting in the fields during the next rainy season. The cuttings give 80-94% sprouting usually. Seed germination is very poor (5%) but seedling produce healthier plants which withstand high velocity winds. The rooted cuttings are planted in a well laid-out fields during rainy season. Pits of size 0.5m cube are dug out at 3-4 m spacing in rows and given FYM and filler soil of the pit is treated with BHC (10%) or aldrin (5%) to protect the new plants from white ants damage. Fertilizer trials have shown little response except due to low level of N fertilization. Removal of side branches and low level of irrigation supports a good growth of these plants. The plantation does not require much weeding and hoeing. But the soil around the bushes be pulverised twice in a year to increase their growth and given urea or ammonium sulphate at 25-50g per bush at a time and irrigated.

Bacterial leaf blight was also noticed to attack the cultures. A leaf eating caterpillar (*Euproctis lanata* Walker) attack guggal, though not seriously. Whitefly (*Bemisia tabaci*) is observed to suck sap of leaves and such leaves become yellowish and eventually drop. These can be effectively controlled by using suitable insecticide. Stem or branch having maximum diameter of about 5 cm at place of incision, irrespective of age is tapped. The necrotic patch on the bark is peeled off with a sharp knife and Bordeaux paste is applied to the exposed (peeled off) surface of the stem or branch. A prick chisel of about 3 cm width is used to make bark- deep incisions and while incising the bark, the chisel is held at an acute angle so that scooped suspension present on the body of the chisel flows towards the blade of the chisel and a small quantity of suspension flows inside the incised bark. If tapping is successful, gum exudation ensures after about 15-20 days from the date of incision and continues for nearly 30-45 days. The exuded gum slides down the stem or branch, and eventually drops on the ground and gets soiled. A piece of polythene sheet can be pouched around the place of incision to collect gum. Alternatively, a polythene sheet can be spread on the ground to collect exuded gum. A maximum of about 500 g of gum has been obtained from a plant.

Post harvest technology

The best grade of guggul is collected from thick branches of tree. These lumps of guggul are translucent. Second grade guggul is usually mixed with bark, sand and is dull coloured guggul. Third grade guggul is usually collected from the ground which is mixed with sand, stones and other foreign matter. The final grading is done after getting cleansed material. Inferior grades are improved by sprinkling castor oil over the heaps of the guggul which impart it a shining appearance.

8. COLEUS (*Coleus* spp.)

Family : Lamiaceae

The genus *Coleus* of the family Lamiaceae (Labiatae) comprises a number of herbaceous medicinal plants which are particularly employed in home remedies for various ailments. Three species are most popular and commonly cultivated. They are *Coleus aromaticus, C. vettiveroides and C. forkoshlii.*

1. ***Coleus aromaticus* Benth. syn. *C. amboinicus* Lour.:** It is found throughout the tropics and cultivated in homestead gardens. It is a large succulent aromatic perennial herb with hispidly villous or tomentose fleshy stem. Leaves are simple, opposite, broadly ovate, crenate and fleshy. Flowers are pale purplish in dense whorls at distant intervals in a long slender raceme. Fruits are orbicular or ovoid nutlets. The leaves are useful in cephalagia, otalgia, anorexia, dyspepsia, flatulence, colic, diarrhoea, cholera, halitosis, convulsions, epilepsy, cough, asthma, hiccough, bronchitis, strangury, hepatopathy and malarial fever.
2. ***Coleus vettiveroides* K.C. Jacob:** It is seen in tropical countries and cultivated in gardens. It is a small profusely branched, succulent aromatic herb with quadrangular stems and branches and deep straw coloured aromatic roots. Leaves are glandular hairy, broadly ovate with dentate margins and prominent veins on the bark. Blue flowers are borne on terminal racemes. Fruits are nutlets. The whole plant is useful in hyperdipsia, vitiated conditions of *pitta*, burning sensation, strangury, leprosy, skin diseases, leucoderma, fever, vomiting, diarrhoea, ulcers and as hair tonic.
3. ***Coleus forskohlii* Briq. syn. *C. barbatus* Benth.:** It is a perennial aromatic herb grown under tropical to temperate conditions for its carrot like tubers which are used as condiments in the preparation of pickles. Its tuberous roots are an exclusive source of a diterpenoid forskolin which has the unique property of activating almost all hormone sensitive adenylate cyclase enzymes in a biological system. It is useful in the treatment of congestive heart failure, glaucoma, asthma, cancer and in preventing immature greying of hair.

Agrotechnology

The Coleus group of plants grows in tropical to subtropical situations and in warm temperate climatic zone on mountains of India, Nepal, Burma, Sri Lanka, Thailand and Africa. It comes up well on the sun exposed dry hill slopes from 300 m to 1800 m altitude. A well drained medium fertile soil is suitable for its cultivation. It is propagated vegetatively through stem and root cuttings. Vine cuttings to a length of 10-15 cm from the top portion are most ideal for planting. The land is ploughed or dug to a depth of 15-20 cm and ridges are formed 30cm apart. Vine cuttings are planted on the ridges at 30 cm spacing after incorporating basal manure. 10 t of FYM and NPK at 50:50:50 kg/ha. Weeding and earthing up at 45 days after planting along with topdressing is highly beneficial. Bacterial wilt and root knot nematode are reported in the crop. Drenching the soil with fungicide, deep ploughing in the summer, burning of crop residues and crop rotation are helpful to tide over the disease and pest problem. The crop can be harvested after 5-6 months.

9. SOLANUMS (*Solanum* spp.)

Family : Solanaceae

Solanums comprise a very important group of medicinal plants having multifarious uses. These plants belong to the family Solanaceae and genus *Solanum*. A number of species are reported to be medicinal which are briefly described below.

I. ***Solanum anguivi* Lam. (Poison berry):** It is found throughout the tropics, in plains and at low elevations. It is much branched, very prickly undershrub, 0.3-1.5 m in height. Leaves are simple, large, ovate, subentire, sinuate or lobed. Flowers are blue in extra-axillary cymes having stellately hairy and prickly peduncles. Fruits are globose berries, reddish or dark yellow with smooth or minutely pitted seeds. Its roots are useful in vitiated conditions of *vata* and *kapha*, odontalgia, dyspepsia, flatulence, colic, verminosis, diarrhoea, pruritus, leprosy, skin diseases, strangury, cough, asthma, bronchitis, amenorrhoea, dysmenorrhoea, fever, cardiac disorders and vomiting. Roots bitter, acrid, astringent, thermogenic, anodyne, digestive, carminative, anthel-mintic, stomachic, constipating, resolvent, demulcent, depurative, diuretic, expectorant, aphrodisiac, emmenagogue, febrifuge and cardiotonic.

II. ***Solanum dulcamara* Linn. (Bitter night shade):** It is found in tropical situations in India and Sikkim. The plant is rich in alkaloidal glycosides like solamarine, tomatidenol, solasodine and soladulcine. The berry and twig are alterative, antisyphilitic, diaphoretic, resolvent, narcotic, diuretic, antirheumatic and used in liver disorders and psoriasis.

III. ***Solanum erianthum* D. Don, syn. *S. verbascifolium* auct. non Linn.:** The plant is distributed over the tropical and subtropical zones of India. The plant contains alkaloids and steroidal sapogenins. Leaves and fruits contain solasodine, solasodiene, solafloridine, diosgenin, vespertilin and pregnenolone. The plant is CNS depressant, antiinflammatory and useful in burns.

IV. ***Solanum melongena* Linn. (Brinjal, Egg plant):** It is mainly cultivated as a vegetable throughout the tropics and subtropics. It is an erect or suffrutescent, herbaceous, armed or unarmed perennial shrub. Leaves are simple, large, entire and lobed. Flowers are blue, in clusters of 2-5. Fruits are large, white, yellow or dark purple berries of different shapes capped with thick persistent calyx. Seeds are many, yellow or cream and discoid. The roots, leaves and unripe fruits are useful in cholera, bronchitis, asthma, odontalgia and fever. The roots are laxative, analgesic and cardiotonic. Leaves are sialagogue, narcotic and antiherpetic. The unripe fruits are bitter, acrid, sweet, aphrodisiac, cardiotonic and haematinic.

V. ***Solanum melongena* var. *incanum* (Linn.) Prain:** It is an herbaceous prickly plant found in warm humid tropics. It is grown almost throughout the year in the plains and during summer on the hills. It grows 0.6-2 m in height. Leaves are simple, alternate lobed. Flowers are blue or white, 5 lobed, calyx with spines. Fruits are ellipsoid berries. The plant is a constituent of the *dasamoola* which helps to over come vitiated *tridoshas* and cures dyspepsia, fever, respiratory and cardiac disorders, skin ailments, vomiting, ulcers and poisonous affections. In Ayurveda the formulations like *Brihatyadi Kashaya, dashamoolarishta, Indukantaghritam, Dasamoolaharithaki*, etc are the important preparations with the roots. It is also used in the treatment of toothache and sore throat. The fruit is reported to stimulate the intrahepatic metabolism of cholesterol. Roots are antiasthmatic and stimulant. Leaves are used in cholera, bronchitis and asthma. Fruits are useful in liver complaints.

VI. ***Solanum nigrum* Linn. syn. *S rubrum* Mill. (Black night shade):** It is seen wild throughout India. It is an erect, divaricately branched, unarmed, suffrutescent annual herb. Leaves are ovate or oblong, sinuate-toothed or lobed and glabrous. Flowers are 3-8 in extra-axillary drooping subumbellate cymes. Fruits are purplish black or reddish berries. Seeds are many, discoid, yellow, minutely pitted. The whole plant is useful in vitiated conditions of tridosha, rheumatalgia, swellings, cough, asthma, bronchitis, wounds, ulcers, flatulence, dyspepsia, strangury, hepatom-egaly, otalgia, hiccough, opthalmopathy, vomiting, cardiopathy, leprosy, skin diseases, fever, splenomegaly, haemarrhoids, nephropathy, dropsy and general debility. The plant is bitter, acrid, emollient, antiseptic, anti-inflammatory, expectorant, anodyne, vulnerary, digestive, laxative, diuretic, cardiotonic, depurative, diaphoretic, febrifuge, rejuvenating, sedative, alterant and tonic.

VII. ***Solanum spirale* Roxb.:** It is seen wild in Assam and Khasi hills in India. Its root is diuretic and narcotic.

VIII. ***Solanum stramoniifolium* Jacq., syn. *S ferox* auct. non Linn.:** It is observed in India in the States of Assam, Maharashtra, Karnataka and Tamil Nadu. Its berries contain glycoalkaloids such as solasonine and solasodine. Its roots and berries are bechic, antiasthmatic, antirheumatic, antiviral, anticancerous and spermicidal.

IX. ***Solanum surattense* Burm. F.:** It is found throughout India and Pakistan in dry situations as weed on roadsides and wastelands. It is prickly, diffuse, bright green, suffrutescent, perennial undershrub, with zigzag branches. Leaves are ovate-oblong, hairy on both sides and armed on the midrib and the nerves. Flowers are bluish purple, in extra axillary cymes. Fruits are glabrous, globular drooping berry, yellow or white with green veins, surrounded by the calyx. Seeds are many, small, reniform, smooth and yellowish brown.

The whole plant is useful in vitiated conditions of *vata* and *kapha*, helminthiasis, dental caries, inflammations, flatulence, constipation, dyspepsia, anorexia, leprosy, skin diseases, hypertension, fever, cough, asthma, bronchitis, hiccough, lumbago, haemorrhoids and epilepsy. The plant is bitter, acrid, thermogenic, anthelmintic, antiinflammatory, anodyne, digestive, carminative, appetiser, stomachic, depurative, sudorific, febrifuge, expectorant, laxative, stimulant, diuretic, rejuvenating, emmenagogue and aphrodisac. Fruits contain solasonine, solamargine and solasodine.

X. ***Solanum torvum* Sw. (West Indian Turkey Berry):** It is seen throughout tropical India, particularly in Orissa, Bihar and Manipur. The plant is CVS active and used in splenomegaly. Fruits and leaves contain solasonine, solasodine, jurubine, jurubidine, torvonin, torvogenin, chlorogenin, paniculogenin, sisalogenone, neosolaspigenin and solaspigenin.

Agrotechnology

The agrotechnology for the solanaceous group of plants are almost similar. They come up very well in tropical and subtropical climate up to 2000 m altitude. They can be raised on variety of soils good in organic matter. Propagation is by seeds. The seedlings are first raised in the nursery and transplanted to the main field 30-45 days after sowing when the plants attain 8-10 cm height. During rainy season, planting is done on ridges while during summer in furrows, at a spacing ranging from 30-90 cm depending upon the stature and spreading habit of the plant. The transplanted seedlings should be given temporary shade for 2-4 days during summer. FYM or compost at 20-25 t/ha is applied at the time of land preparation. A moderate fertiliser dose of 75:40:40 N, P_2O_5, K_2O/ha may be given. P is given as basal dose, N and K is applied in 2-3 split doses. One or two intercultural operations are needed to control weeds. The plants need earthing up after weeding and topdressing. Irrigation is needed at 3-4 days interval during summer and on alternate days during fruiting period. Plants need staking to avoid lodging due to heavy bearing. Shoot borers, mealy bugs, leaf webbers and miners are noted on the crop, which can be controlled by spraying mild insecticides. Root knot nematode, wilting and mosaic diseases are also noted on the crop. Field sanitation, crop rotation and burning of crop residues are recommended.

10. INDIAN SENNA (*Cassia senna*)

Family : Caesalpiniaceae

Indian Senna or Tinnevelly senna is a shrub very highly esteemed in India for its medicinal value. The leaves are useful in constipation, abdominal disorders, leprosy, skin diseases, leucoderma, splenomegaly, hepatopathy, jaundice, helminthiasis, dyspepsia, cough, bronchitis, typhoid fever, anaemia, tumours and vitiated conditions of *pitta* and *vata*.

Distribution

The plant is of Mediterranean origin. It is found in Somalia, Saudi Arabia, parts of Pakistan and Kutch area of Gujarat. It is largely cultivated in Tirunelveli, Ramanathapuram, Madurai and Salem districts of Tamil Nadu. *C. senna* is a shrub or undershrub, 60-75 cm in height with pale substrate or obtusely angled erect or spreading branches. Leaves are paripinnate. Leaflets are 5-8 in number, ovate lanceolate and glabrous. Flowers are yellowish, many and arranged in axillary racemes. Fruits are flat legumes, greenish brown to dark brown and nearly smooth. In commerce, the leaves and pods obtained from *C. senna* are known as "*Tinnevelly Senna*" and those from *C. acutifolia* Delile. as "*Alexandrian Senna*". The leaves of *C. acutifolia* are narrower than *C. senna*, otherwise both resemble to a large extent. All the true Sennas have the portions of their leaves unequally divided. In some kinds the lower part of one side is reduced to little more than a line in breadth, while the other is from a quarter to half an inch in breadth.

Agrotechnology

The plant requires a mild subtropical climate with warm winters which are free from frost for its growth. Semiarid areas with adequate irrigation facilities are ideal for cultivation. Areas having high rainfall, humidity and poor drainage are not suitable. Light or medium loamy soils with adequate drainage and pH varying from 7.0-8.2 are preferable. In South India both summer and winter crops are possible. The plant is propagated by seeds. The seed rate required is 15-20 kg/ha. Seeds are sown in October-November (winter rainfed crop) or in February-March (irrigated crop). Higher seed rate is required for unirrigated crop. Seeds are sown in lines 30 cm apart. Application of 5-10 t of FYM/ha before planting or raising a green manure crop is beneficial. About 40 kg N and 25-50 kg P_2O_5/ha applied as basal dressing and 40 kg N/ha applied in 2 split dozes as top dressing gave better yield. While the rainfed crop is grown without irrigation, the irrigated crop requires 5-8 light irrigations during the entire growing season. The crop requires 2-3 weedings and hoeings in order to keep it free from weeds. *Alternaria alternata* causes leaf spot and dieback but the disease is not serious. In North India, the plant is attacked by the larvae of butterfly *Catopsilia pyranthe* which can be controlled by planting the crop in March-April instead of June-July. Under irrigated conditions, the first crop is obtained after 90 days of planting. The leaves are stripped by hand when they are fully green, thick and bluish-green in colour. The second crop is taken 4 weeks after the first harvest and the third 4-6 weeks after the second one. The last harvest of leaves is done when the entire crop is harvested along with the pods. Yield under irrigated conditions is nearly1.4 t of leaves and 150 kg pods/ha and under unirrigated conditions is 500-600 kg leaves and 80-100 kg pods/ha. The leaves are dried in thin layers under shade so as to retain the green colour and the pods are hung for 10-12 days to get dried. The leaves and pods are cleaned, graded and marketed.

IMPORTANT AROMATIC PLANTS

Aromatic plants are used for essential oils-volatile materials extracted from plants or plant parts, usually by steam distillation, and used in flavours and fragrances for food, soap, detergents, perfumes, lotions etc. Data shows that ten major essential crops account for 80 % of the world market for essential oils. These include:

I. Citrus II. Eucalyptus III. Cedar wood

IV. Clove V. Mint oils: Peepermint, Spearmint VI. Cornmint

VII. Lemon fragrance oils: Citronella, Lemon grass, *Litsea cubeba*

The remaining 20% of the world essential oil market comprises over 150 plants. Other prominent aromatic plants should be considered and studied for commercial utilization.

The cosmetics industry uses a large number of medicinal plants as ingredients in soaps, lotions, ointments, shampoos and natural colorants amongst other products. There is increasing consumer interests in all things natural. Consumers are demanding healthier and more natural products. Increased consumer sophisticated and awareness of ingredients, performance and health benefits are thus changing the personal care and cosmetics industry.

1. LEMON GRASS (*Cymbopogon flexuosus* Stapf.)

It is a stemless perennial grass, found wild in the southern states of India. The leaves yield aromatic oil, containing 75-85 per cent citral. The oil is used in soaps, cosmetics and disinfectants and is a raw material for manufacturing ions and vitamin A. The crop is grown on about 390,000 ha in Kerala and about 90 per cent of the produce is exported, earning about Rs 10 million in foreign exchange annually.

The crop is grown mainly as a rain-fed crop on poor marginal lands. It is a hardy drought resistant plant and is grown under a wide range of climatic conditions. Two types are recognized : (1) the red stemmed grass is the true lemon grass and a selection 'OD-19' made at the Lemon Grass Research Station, Oodakkali (Kerala) is recommended for high yield, (2) the white stemmed (var. Albscens) yields oil or low citral content. The crop is best propagated through seed raised in nurseries; 10 kg of the seed produces enough seedlings for planting a hectare. About 2 month old seedlings are planted in rows, 45 x 10 cm apart, during rains and the first harvest is ready after 90 days. In Kerala, 3 harvests are obtained in the first year and 4 to 5 harvests during each of the 3 to 5 succeeding years; the crop is cut after every 60 to 65 days, except during the dry season. The fields are manured librally with ash and spent grass (obtained after the distillation of the oil); farmers seldom apply inorganic fertilizers. The herbage contains 0.35% oil on fresh weight basis. The grass yield ranges from 18 to 25 tonnes per hectare, giving 60-72 kg of oil.

Uses

- Lemon grass is primarily used for the isolation of citral and this citral is a salting material for manufacturing of various aroma products and used in aromatic industries.
- Lemon grass oil is used as such in soaps, detergents and other prepration. It is also used for flavouring food and confectionaries.
- The stems are also used in pickles and in flavouring marinades.
- Lemon grass crop checks the soil erosion also.
- After distillation the left material is known as spent grass. This spent grass is not having essential oil. It is used as fuel, bulky organic manure for mulching, for making paper.

2. CITRONELLA GRASS (*Cymobopogon winteriaruns* Jowitt)

It is a stemless perennial grass. The oil is mainly used in soaps, cosmetics, deodorants and mosquito repellent creams. It is a recent introduction into India and its cultivation. mainly in the lower hills of Assam, in Karnataka and southern Gujrat, covers about 2,000 hectares. A part of out requirement is still met by imports.

Citronella is grown up to 1,000 m above sea level on well drained sandy loam to clayey loamy soils with pH varying from 5 to 7.5 As a rainfed crop, it is propagated by using rooted slips, and is planted during the rainy season in rows, 60 to 45 cm x 45 to 30 cm apart, depending on the rate of growth and the soil fertility. Usually, the leafy portion is removed at planting during the monsoon season and the plants begin to tiller in 30 days after planting. A healthy one year old clump yields 60 to 80 slips. The crop needs a large quantity of nitrogenous fertilizers, When grown on medium soils, it is given 60 kg of P_2O_5, 40 kg of K_2O and 20 kg of N per ha at planting, supplemented with 80 to 100 kg of N per hectare annually as top dressing in 3 to 4 split doses. The crop is irrigated after every 10-15 days during the dry season and the first harvest is cut 20 cm above the ground after 90 days of planting; usually 3 cuttings are obtained annually for 3 to 5 year after planting.

3. CELERY (*Apium graveolens* Linn.)

Celery is a native of the Mediterranean region, introduced into India during late ninteenth century. India is now the principal grower (of the annual type) and exports large quantities of celery seeds (worth 17.5 millions to the American and European markets. A large area is put under this crop, mainly in the Amritsar, Gurdaspur and Jallundhar districts of the Punjab State and to a smaller extent in the Saharanpur district of Uttar Pradesh. The clery seed is used in ground form as a dry seasoning powder and for the extraction of an oleoresin used to flavour a large variety of canned food products. The celery seed oil commands a larger market mainly in the confectionery industry. Whereas the bulk of celery seed is exported, a few local firms also produce its essential oil and oleoresin on a small scale.

Celery is an erect, branching herb, about 100 cm tall. it is a *rabi* crop which remains for 2 months in the nursery and for another 4 to 4-1/2 months in the field. Nursery is raised during early November. The seed rate used is 2 kg per hectare. The nursery beds are kept moist and free from weeds through periodic irrigation followed by interculture. The seedlings grow up to 10 cm tall by the end of December and are ready for transplanting. The seedlings are usually planted 45 cm x 20 cm apart in early January in well prepared fields, to which 30-50 tonnes of farmyard manure per ha is added during the preparatory tillage. On medium soils, about 80 Kg N, 40 Kg of P_2O_5 and 20 Kg of K_2O per ha is applied to the crop; half of N along with the entire quantity of P and K is given to the rows at the time of planting and the remaining N is given as a top dressing after one month. The crop requires 10-12 irrigations, each about 5 cm deep. Once established the crop requires 3 to 4 hoeings to keep down weeds. The plants bear flowers in umbels during early March and the fruits mature in late April. The crop is harvested when about 80 per cent of the umbels begin to turn light brown. The harvested crop as stacked in the field for few days and is then threshed to obtain the seeds. The average seed yield ranges from 10 to 12 quintals per ha.

The celery (seed) on distillation yields to 2.1 to 2.3% of an essential oil containing selinene, d-limonene and sesquiterpenes. Oleoresin is prepared by using the method of the solvent extraction of the seeds. The seeds yields glycoside opiin. The crop remains remarkably free from pests and diseases.

4. ROSE GERANIUM (*Pelargonium graveolens* L'Herit)

Rose geranium is a native of the hilly tracts of Cape Colony (South Africa). The highly aromatic 'Rose Geranium' is a hybrid (*P. graveolens* x *P. radens*), introduced by two Frenchmen-Earnest Sen and J. Priors in Yercaud in the Sheraroy Hills (Salem district) in Tamil Nadu. It is now grown in limited areas around Bangalore, Yercaud and in the Nilgiri Hills. The major world supplies of the oil come from Malagasy Republic. The oil of geranium has a refreshingly delicate rose like aroma containing 66-78 per cent of alcohols calculated as geraniol. As such also, it itself is a perfume; it blends well with a wide range of floral and oriental perfumes. It is used for scenting ointments, pharmaceutical and tobacco products. It is suitable in a slightly alkaline medium and, therefore, it is widely used in the manufacture of high grade soaps and cosmetic products. Our total annual production of the oil is around 12 tonnes and, therefore, a sizeable quantity of the oil is annually imported.

Two plant types of 'Rose Geranium' are recognized in India, Viz. the Algeriaan type from the slender plants, with dark pink flowers. It is less suitable for humid conditions; the other is Bourbin type which produces more leafy growth and bears light pink flowers. It is a perennial crop and about two to three harvests of leaves are taken annually in the second and third years. It can be grown as a catch crop in newly planted orchards.

'Rose Geranium' flourishes well in deep well drained, preferably sandy loam to clayey loam soils. It grows luxuriantly in a mild subtropical climate; locality receiving excessive heat in summer or very low temperatures in winter are unsuitable. The 'Rose Geranium' can withstand drought but its yield is reduced. It is raised in nursery during early spring from 8-15 cm long stem cutting each with 2 to 4 nodes. Sandy beds are preferred for the nursery to avoid excessive moisture which damages the young sprouts. Under a favourable day temperature of 10 to 15 degree Centigrade, about 80 per cent cuttings sproute in 3 weeks and are ready for field planting at 50 x 100 cm spacing with the onset of the monsoon rains in May-June. The field are brought to a fine tilth, are mixed with 45 tonnes of farmyard manure per hectare. In the Nilgiris the crop is given a basal application of a fertilizer mixture, containing 35, 45 and 17 kg of N, P_2O_5 and K_2O per hectare; therefore, another 35 kg of N may be applied in two split applications annually. Initially, the crop is given 1 to 2 hoeings and weedings and the rows close up. The pruning of branches after the rains helps to induce bushy growth. The crop is irrigated at intervals of 10 to 15 days during summer depending upon the soils texture and the atmospheric humidity. Too moist or continued dry conditions lowers the crop yield. Generally, the first crop is obtained 6 to 8 months after planting and the average oil yield is 8 kg; two or three crop are received in subsequent years, yielding 8 to 18 kg of oil per hectare. In countries, e.g. Algeria and Malagasy Republic, the oil yields is reported to range from 30 to 35 kg per hectare. The fresh leaves contain 0.12 to 0.2% (w/w) volatile oil and semi-dry or even dry leaves can be distilled. The oil improves in odour on ageing.

5. VETIVER OR KHUS (*Vetiveria zizanioides*)

Vetiveria zizanioides (Linn Nash) Khus is a tall perennial grass found wild in the drier tracts of western India. Its spongy, much branched fine root lets contain fragrant oil, which is a perfume by itself. The dry aromatic roots are made into curtains, mats, fans, etc. to emit scented cool aroma when moistened. The oil is used as a valuable fixative for blending perfumes and cosmetics. A large bulk of the oil is produced by distilling the wild growing plants found in Bharatpur (Rajasthan), Musanagar, Biswan and Nawabganj etc. in western Uttar Pradesh. The cultivation of khus is scattered over small areas in the Trichur, Palghat, Kozikode, Wynad and Trivandrum district of Kerala, in the Tirunelvelly, Madurai and Coimbatore districts of Tamil Nadu and in the Kurnool and East Godavari districts of Andhra Pradesh.

In southern states, khus is grown as a rainfed crop in well drained sandy loam soils in moderately humid localities which received 100-200 cm of rainfall annually. However, a luxuriant growth of healthier roots is obtained from plants growing under warm and damp conditions on rich, partially inundated, marshy land (Bharatpur). The field are laid out into beds of convenient size and 15-20 cm long rooted slips are planted. With the onset of the monsoon rains on both the sides of the ridges, made 45 cm apart. The planting of 2 to 3 slips is done in spring in pits when the crop is raised under irrigation. The crop is given 25-30 t/ha of farmyard manure or compost at planting, supplemented with ash and groundnut cake or ammonium sulphate, one month after planting to stimulate growth.

Khus is a 12 to 18 month crop and the roots are dug out during dry months before autumn. The aerial parts are first cut at a height of 15 to 20 cm and are removed; the clumps are then dug out and beaten. The soil is further dug out up to a depth of one metre to remove the remaining rootlets. The roots are light to yellowish brown containing usually 0.6 to 0.8 per cent of dark brown viscous, persistently fragrant oil. The yield is 3 to 5 tonnes of fresh root, yielding about 20 kg of oil per hectare.

6. FENUGREEK (*Trigonella foenum-graecum*)

Fenugreek or Greek Hayes is cultivated as a leafy vegetable, condiment and as medicinal plant. The leaves are refrigerant and aperient and are given internally for vitiated conditions of *pitta*. A poultice of the leaves is applied for swellings and burns. Seeds are used for fever, vomiting, anorexia, cough, bronchitis and colonitis. In the famous Malayalam treatises like '*Padhyam*' '*Kairali*' and '*Arunodhayam*', *uluva* is recommended for use as *kalanusari* in *Dhanvantaram* formulations of '*Astaghradayam*'. An infusion of the seeds is a good cool drink for small pox patients. Powdered seeds find application in veterinary medicine. An aqueous extract of the seeds possesses antibacterial property.

Distribution

Fenugreek is a native of South Eastern Europe and West Asia. In India fenugreek is grown in about 0.30 lakh ha producing annually about 30,000 tonnes of seeds. The major states growing fenugreek are Rajasthan, Madhya Pradesh, Gujarat, Uttar Pradesh, Maharashtra, Punjab and Karnataka. It is grown wild in Kashmir and Punjab.

Botany

Trigonella foenum-graecum Linn. belongs to family, Fabaceae. It is an annual herb, 30-60 cm in height, leaves are light green, pinnately trifoliate, leaflets toothed, flowers are white or yellowish white, papilionaceous and axillary. Fruits are legumes, 5-7.5 cm long, narrow, curved, tapering with a slender point and containing 10-20 deeply furrowed seeds per pod. There are two species of the genus Trigonella which are of economic importance viz. *T. foenum graecum*, the common *methi* and *T. corniculata*, the *Kasuri methi*. These two differ in their growth habit and yield. The latter one is a slow growing type and remains in rosette condition during most of the vegetative growth period.

Agrotechnology

Fenugreek has a wide adaptability and is successfully cultivated both in the tropics as well as temperate regions. It is tolerant to frost and freezing weather. It does well in places receiving moderate or low rainfall areas but not in heavy rainfall area. It can be grown on a wide variety of soils but clayey loam is relatively better. The optimum soil pH should be 6-7 for its better growth and development. Land is prepared by ploughing thrice and beds of uniform size are prepared.

Broadcasting the seed on the bed and raking the surface to cover the seeds is normally followed. But to facilitate intercultural operations, line sowing is also advocated in rows at 20-25 cm apart. Sowing in the plains is generally in September-November while in the hills it is from March. The seed rate is 20-25 kg/ha and the seeds germinate within 6-8 days. Besides 15 t of FYM, a fertiliser dose of 25:25:50 kg NPK/ha is recommended. Entire P, K and half N are to be applied basally and the remaining half N 30 days after sowing. First irrigation is to be given immediately after sowing and subsequent irrigations at 7-10 days interval. Hoeing and weeding are to be done during the early stages of plant growth and thinning at 25-30 days to have a spacing of 10-15cm between plants and to retain 1-2 plants per hill. Root rot (caused by *Rhizoctonia solani*) is a serious disease and can be controlled by drenching carbendazim 0.05% first at the onset of the disease and another after one month of first application. In about 25-30 days, young shoots are nipped off 5 cm above ground level and subsequent cuttings of leaves may be taken after 15 days. It is advisable to take 1-2 cuttings before the crop is allowed for flowering and fruiting. When pods are dried, the plants are pulled out, dried in the sun and seeds are threshed by beating with stick or by rubbing with hands. Seeds are winnowed, cleaned and dried in the sun. They may be stored in gunny bags lined with paper. An yield of 1200-1500 kg of seeds and about 800-1000 kg of leaves may be obtained per hectare in crops grown for both the purposes.

7. FENNEL (*Foeniculum vulgare*)

History

The uses for fennel go far beyond the kitchen however; fennel has been used as a medicinal herb by the early Romans and Greeks. Fennel was well known to the Ancients and was cultivated by the ancient Romans for its aromatic fruits and succulent, edible shoots. Fennel is native to Europe and is a garden plant in North America. This plant was attached by Linnaeus to the genus Anethum, but was separated from it by De Candolle and placed with three or four others in a new genus styled Foeniculum, which has been generally adopted by botanists. This was corrupted in the Middle Ages into Fanculum, and this gave birth to its alternative popular name, 'fenkel.' The plant resembles dill with its small clusters of upturned yellow umbrel flowers, but it easily distinguished by its sweet licorice aroma and taste. A tall wispy perennial, earns a place in any well stocked herbal garden not only for its culinary uses but also as one of the best digestive aids that also relieves gas. As early as the tenth century A.D. the mystic Hildegard of Bingen recommended fennel seeds as a treatment for body odour. At Indian restaurants you may see fennel seeds after dinner or lunch. Fennel is widely cultivated, both in its native range and elsewhere, for its edible, strongly flavoured leaves and seeds. The flavour is similar to that of anise and star anise, though usually not as strong.

Description

Fennel has a distinctive smell rather like aniseed. The roots are large, thick and white and are carrot shaped root produces a stout, pithy, finely grooved stem with fine buish stripes. The leaves are decompound, dissected into numerous filiform segments, the upper leaves on broad sheaths that surround the stem. The bright golden flowers, produced in large, flat terminal umbels. The fruit consists of two joined carpels, together taking an oblong form with prominent ribs. Fennel looks very much like dill with its feather leaves and yellow umbel flowers and hollow stems and they have the fragrance of anise, or licorice.

Cultivation

Fennel will thrive anywhere, and a plantation will last for years. It is easily propagated by seeds. It likes plenty of sun and is adapted to dry and sunny situations, not needing heavily manured ground, though it will yield more on rich stiff soil. From 4 to 5 kg of seed are sown per hectare, either with drills, 15 inches apart, lightly, covered with soil and the plants afterwards thinned to a similar distance, or sown thinly in a bed and transplanted when large enough in rows 15 inches apart.

Parts used

Though the leaves and roots of fennel are used the major part which is of medicinal use are the seeds.

Medicinal and commercial applications

- The leaves or seed, boiled in water, and drunk, are good for the nursing mother.
- Fennel is good to break wind, provoke urine, and eases the pains of the stone and to help break it.
- The leaves, or rather the seeds, boiled in water, stays the hiccough and soothes the stomach of sick and feverish persons.
- The seed boiled in wine is good for those that have eaten poisonous herbs or mushrooms. The seed, or roots, help to open obstructions of the liver, spleen and gall, and ease painful and windy swellings and help the yellow jaundice, the gout and cramps.
- The leaves, seeds and roots are much used in drink or broth to make people lean that are too fat.
- The distilled water of the whole herb dropped into the eyes cleanses them from mists and films that hinder the sight.
- Fennel not only improves digestion, but also can reduce bad breath and body odor that originates in the intestines.
- Fennel also acts as an excellent digestive aid to relieve abdominal cramps, gas and bloating.
- Fennel is one of the plants that repel fleas, and the anise like taste may be more acceptable choice for indigestion and gas in finicky dogs and cats.
- Recent studies have found fennel to possess diuretic, cholertic (increase in production of bile), and pain reducing, fever reducing, and antimicrobial actions.

- The fennel tea traditionally made a good eye wash. It is a tested remedy for gas, acid stomach, gout, cramps, colic and spasms. Fennel oil was traditionally rubbed over painful joints to relieve pain. It was gargled for hoarseness and sore throat.
- Fennel water has properties similar to those of anise and dill water: mixed with sodium bicarbonate and syrup, these waters constitute the domestic 'Gripe Water,' used to correct the flatulence of infants.

Commercial applications

- The seeds are mainly used as a flavouring agent in medicines and to disperse flatulence. It is an ingredient of the official compound powder of liquorice. Added to a laxative, it prevents griping. Fennel seeds are a common cooking spice, particularly for use with fish. Fennel seed ground and made into tea is believed to be good for snake bites, insect bites or food poisoning.
- Fennel is best known as a culinary herb, all parts of the plant are edible. The fresh stems of fennel can be eaten much like celery; the seeds add a lovely anise flavor to fish and other dishes.
- Drinking a cup of fennel seed tea before eating a heavy meal, can edge off your appetite resulting in letter digestion.
- Fennel tea is useful for chronic coughs and act as an expectorant to help clear mucus from the lungs, syrup prepared from fennel juice was formerly given for chronic coughs.
- Oil of fennel relieves muscular or rheumatic pains and is warming and soothing in massage oil blends.
- Fennel is also largely used for cattle condiments.
- The tender leaves are often used for garnishes and to add flavour to salads, and are also added, finely chopped, to sauces served with puddings. Roman bakers are said to put the herb under their loaves in the oven to make the bread taste agreeably.

8. MENTHA (*Mentha arvensis*)

Mentha (also known as Mint, from Greek *míntha*, Linear B *mi-ta*) is a genus of flowering plants in the family Lamiaceae (mint family). The species are not clearly distinct and estimates of the number of species varies from 13 to 18. Hybridization between some of the species occurs naturally. Many other hybrids as well as numerous cultivars are known in cultivation. The genus has a sub cosmopolitan distribution across Europe, Africa, Asia, Australia and North America.

Mints are aromatic, almost exclusively perennial, rarely annual, herbs. They have wide spreading underground and overground stolons and erect, square, branched stems. The leaves are arranged in opposite pairs, from oblong to lanceolate, often downy, and with a serrate margin. Leaf colors range from dark green and gray green to purple, blue, and sometimes pale yellow. The flowers are white to purple and produced in false whorls called verticillasters. The corolla is two lipped with four sub equal lobes, the upper lobe usually the largest. The fruit is a small, dry capsule containing one to four seeds.

The species of Mentha genus are widely distributed and can be found in many environments, most Mentha grow best in wet environments and moist soils. Mint can grow 100–120 cm tall and can spread over an indeterminate area. Due to their tendency to spread unchecked, mints are considered invasive.

Taxonomy

Mentha is a member of the tribe Mentheae in the subfamily Nepetoideae. The tribe contains about 65 genera and relationships within it remain obscure. Some authors have excluded *Mentha cervina* from the genus. *Mentha cunninghamii* has also been excluded by some authors. In 2004, a molecular phylogenetic study indicated that both of these species should be included in *Mentha.*

Species

The list below includes all of the taxa that have been recognized as species in recent works on *Mentha*. No author has recognized all of them. As with all biological classifications of plants, this list can go out of date at a moment's notice. Common names for different species along with cultivars and varieties are listed as follows.

- *Mentha asiatica* - Asian Mint
- *Mentha australis* - Australian mint
- *Mentha canadensis*
- *Mentha cervina* - Hart's Pennyroyal
- *Mentha citrata* – Bergamot mint
- *Mentha crispata* - Wrinkled-leaf mint
- *Mentha cunninghamii*
- *Mentha dahurica* - Dahurian Thyme
- *Mentha diemenica* - Slender mint
- *Mentha gattefossei*
- *Mentha grandiflora*
- *Mentha haplocalyx* - Marsh mint
- *Mentha japonica* - Wild Mint
- *Mentha kopetdaghensis* - Field Mint
- *Mentha laxiflora* - Forest mint
- *Mentha longifolia* - Mentha sylvestris, Horse Mint
- *Mentha pulegium* – Pennyroyal
- *Mentha requienii* – Corsican mint
- *Mentha sachalinensis* - Garden mint
- *Mentha satureioides* - Native Pennyroyal
- *Mentha spicata* – *M. viridis*, *M. cordifolia*, Spearmint, Curly mint
- *Mentha suaveolens* – Apple mint, Pineapple mint (a variegated cultivar of Apple mint)
- *Mentha vagans* - Gray mint

In Punjab four species i.e. *Mentha arvensis*, *M piperata*, *M spicata* and *M citrata* are grown commercially.

Cultivation

All mints prefer, and thrive in, cool, moist spots in partial shade. In general, mints tolerate a wide range of conditions, and can also be grown in full sun.

They are fast growing, extending their reach along surfaces through a network of runners. Due to their speedy growth, one plant of each desired mint, along with a little care, will provide more than enough mint for home use. Some mint species are more invasive than others. Even with the less invasive mints, care should be taken when mixing any mint with any other plants so that it may not be smothered. To control mints in an open environment, mints should be planted in deep, bottomless con-tainers sunk in the ground, or planted above ground in tubs and barrels.

Soil type

Well drained, sandy loam to loamy soil with moderate to high organic matter, is best for this crop. The soil should be free from acidity, salinity, alkalinity and water logging.

Land preparation

Two or three ploughings followed by planking are necessary to get a fine seedbed. The field should be free from stubble and weeds.

Seed rate and seed treatment

Mentha is propagated from suckers. About 5 quintals of freshly dug 5-8 cm long suckers are enough for one hectare. This quantity can be had from 1/16 acre of mentha. Before planting, the suckers should be washed and dipped into 0.1% Carbendazim 50 WP solution (1 g/litre of water) for 5-10 minutes. Fifty litres of the solution is sufficient to dip 40 kg of suckers. The same solution may be used to dip the remaining suckers in 40 kg lots for sowing of 0.4 hectare area.

Method of planting

The suckers are laid end to end, 4-5 cm deep in furrows, 45 cm apart and are then covered with soil by planking lightly. Apply a light irrigation after planting. Do not plant sprouted suckers, as most of such suckers die. Keep the soil moist till emergence by giving light irrigation.

Time of planting

The best planting time is the mid January to the end of January. The crop can also be raised by transplanting in April if adequate irrigation is available.

Inter cropping

Mentha can also be grown as intercrop. Plant one row of mentha between two rows of sugarcane. Mentha and sugarcane can be planted simultaneously in the first fortnight of February. Use 2.5 quintal of mentha suckers per hectare. In addition to fertilizers recommended to sugarcane apply 45 kg N (100kg urea) and 25 kg P_2O_5 (155 kg super phosphate) per hectare. Half N and full phosphorus may be applied at planting and remaining half N about 40 days after planting. Take only one cutting of mentha.

Mentha can be successfully intercropped with sunflower. Sow two rows of mentha in end January, between two lines of sunflower grown at 120 cm x 15 cm in North-South direction. Use 375 kg of mentha suckers per hectare. In addition to fertilizers recommended to sunflower apply 58 kg N (125 kg urea) and 30 kg P_2O_5

(187 kg super phosphate) per hectare. Full phosphorus and half nitrogen be applied at planting and remaining half nitrogen at 40 days after planting.

Onion can be grown as intercrop in mentha. Both mentha and onion should be planted simultaneously from the mid January to end of January. Plant one row of onion in between the two rows of mentha planted at 45 cm by keeping plant to plant spacing of onion of 7.5 cm. Apply additional fertilizer dose of 33 kg N (75 kg urea), 17.5 kg P_2O_5 (100 kg SSP) and 17.5 kg K_2O (30 kg MOP) per hectare in addition to recommended fertilizer of mentha. Full phosphorus and potash and half nitrogen be applied at planting and the remaining half nitrogen about 40 days after planting.

Fertilizer application

Mentha responds favourably to organic manuring. Apply 10-15 tonnes of well-rotten farmyard manure per acre before planting. Apply 150 kg N, 40 kg P_2O_5 per hectare to Mentha crop. Drill one fourth of nitrogen and the full quantity of phosphorus at planting. Apply another one fourth of nitrogen about 40 days after planting. Add the remaining half dose of nitrogen in two equal splits after the first cutting of the crop. The first split may be applied immediately and the second splits 40 days afterwards.

Irrigation

Mentha requires frequent but light irrigations. Irrigate at 10 days interval till the end of March and at five or six days interval till the onset of the monsoon. During the rainy season, irrigate according to the need.

Weed control

To obtain good yield and high quality oil, the crop should be kept free from weeds at all the stages of growth. In the early stages of growth, a wheel hoe may be used. The pre-emergence application of Goal 23.5 EC (oxyfluorfen) at 875 ml/ha or Karmex 80 W (Diuron) @ 750 g /ha or Stomp 30 EC (pendimethalin) 2.5 litre per hectare or isoproturon 75 WP 1.0 kg per hectare can effectively control the weeds in this crop. Dissolve the herbicide in 500 litres of water per hectare and spray by knap-sack sprayer fitted with flat fan/flood jet nozzle. However, on light textured soil use of isoproturon should be avoided.

Harvesting and yield

The crop should preferably be harvested at the flower initiation stage. If the lower leaves of the plants turn yellow and start shedding, harvesting may be done earlier. Harvest the crop, leaving 6-8 cm long stumps to secure better sprouting. Two cuttings can be taken, first in June and the second in September. The yield of the crop is 250-560 quintals per hectare of fresh herbs which contains 0.5 to 0.75% oil.

Uses

I. Culinary

Mint jelly is a traditional condiment served with lamb dishes. Mint lemonade served in Syria. The leaf, fresh or dried, is the culinary source of mint. Fresh mint is usually preferred over dried mint when storage of the mint is not a problem. The leaves have a pleasant warm, fresh, aromatic, sweet flavor with a cool aftertaste. Mint leaves are used in teas, beverages, jellies, syrups, candies, and ice creams. In Middle Eastern cuisine, mint is used on lamb dishes, while in British cuisine and American cuisine, mint sauce and mint jelly are used, respectively.Mint is a necessary ingredient in Touareg tea, a popular tea in northern African and Arab countries.

Alcoholic drinks sometimes feature mint for flavor or garnish, such as the Mint Julep and the Mojito. Crème de menthe is a mint-flavored liqueur used in drinks such as the grasshopper.

Mint essential oil and menthol are extensively used as flavorings in breath fresheners, drinks, antiseptic mouth rinses, toothpaste, chewing gum, desserts, candies and mint chocolate. The substances that give the mints their characteristic aromas and flavors are menthol (the main aroma of Peppermint and Japanese Peppermint) and pulegone (in Pennyroyal and Corsican Mint). The compound primarily responsible for the aroma and flavor of spearmint is *R*-carvone.

II. Medicinal and cosmetic

- Mint was originally used as a medicinal herb to treat stomach and chest pains, and it is commonly used in the form of tea as a home remedy to help alleviate stomach pain. During the Middle Ages, powdered mint leaves were used to whiten teeth. Mint tea is a strong diuretic.
- Menthol from mint essential oil (40%–90%) is an ingredient of many cosmetics and some perfumes. Menthol and mint essential oil are also much used in medicine as a component of many drugs, and are very popular in aromatherapy. Mint is also used in some shampoo products.
- A common use is as an antipruritic, especially in insect bite treatments (often along with camphor).
- Menthol is also used in cigarettes as an additive, because it blocks out the bitter taste of tobacco and soothes the throat.
- The strong, sharp flavor and scent of mint is sometimes used as a mild decongestant for illnesses such as the common cold.
- In Rome, Pliny recommended that a wreath of mint was a good thing for students to wear since it was thought to "exhilarate their minds".

IMPORTANT SPICE CROPS

Spices are the dry part of a plant, such as roots, leaves and seeds which impart to food a certain flavour and pungent stimuli. Spices have played an important role in the history of civilization, exploration and commerce as these had a universal acceptance as condiments and flavours in human diet as well as in treatment of ailments. Spices contain variable amount of fat, protein, carbohydrate, fibre, minerals and vitamins. However, owing to the very small quantity used in the food, their contribution to nutrient requirement is not significant. Protein, carbohydrate, minerals and vitamins are thus relatively less important in deciding the quality of spices. Spices are store house of many chemically active compounds that impart flavour, fragrance and piquancy. Most spices owe their flavouring properties to volatile oils and in some cases, to fixed oils and small amount of resin, which are known as oleoresins. Phytochemicals in spices are secondary metabolites, which are originated for the protection from herbivorous insects, fungi, pathogen and parasites. Most probably no single compound is responsible for flavours but a blend of different compounds such as alcohols, phenols, esters, terpenes, organic acids, resins, alkaloids and sulphur containing compounds in various proportions produce the flavours.

The quality of product depends on the quality of raw materials, the methods used in processing and the packaging and marketing practices. These qualities vary unpredictable. American spice association (ASTA), USA and other countries have specified the requirement of minimum volatile oil percentage in spices e.g. in fennel 1.5%, coriander 0.3%, dill-seed 1%, cumin 1.5% and fenugreek 0.25%. In fennel less fibre is preferred. The "lucknavi saunf contain 11% of crude fibre as compared to normal fennel fruit which contain 15% or more crude fibre. In coriander splitting of seed is negative character. The physicochemical characteristics vary widely on the variety, agroclimatic conditions existing in the area of production, harvest and post harvest operations. Total numbers of spices cultivated in India are 63 but spices board has listed 52 spices. The major spices cultivated in India are black pepper, cardamum, ginger, turmeric, cumin, fennel, chillies etc.

1. CARDAMOM

Cardamom (*Elettaria cardamomum* (L.) Maton) is considered to be the 'Queen of Spices'. India is the largest producer and exporter of this spice, accounting for nearly 70 per cent of the total world production and 60 per cent of the total world trade. Our annual production has ranged from 2,300 to 2,800 tonnes in recent years. It is also grown in Guatemala, Tanzania, Srilanka, El Salvador, Vietnam, Laos and Cambodia. It is cultivated in the hilly forest regions of the entire Western Ghats in Karnataka, Kerala and Tamil Nadu on about 86,000 hectares. It is grown either as a pure plantation crops or as a crops subsidiary to coffee and arecanut at higher elevations. Under favourable conditions, it is found as a natural undergrowth in forest tracts and is supplemented in varying degrees with actual cultivation.

General botanical aspects

Cardamom for commercial use is the dried fruit of *Elattaria cardamom* (L.) Maton, a perrenial herbaceous plant belonging to the family Zingiberaceae. The plantis a native of India. It has a tall pseudostem formed by the encircling of the leaf sheaths, wrapped one over another. Depending on the variety, a normal full grown plant may attain a height of 2 to 4 metres. The real stem of the shallow rooted plant is a rhizome which is subterranean. Leaves are distichous, linear and lanceolate, with short petioles. Flowers are borne on panicles and they arise directly from the base of the pseudostem. The panicles are erect in the 'Mysore' variety, prostrate in the 'Malabar' variety and semi-erect or flexious in 'Vashuka'. Panicles emerge from January onwards, and flowering continues from April to August or even later. Fruits mature in about 3 to 4 months after flowering. They are small trilocular capsules, each containing about 15 to 20 seeds. On maturity, fruits turn into pale green. Under favourable conditions of the environment, a healthy adult plant would annually produce about 200 capsules, with a green weight of about 900 g, which on processing yields about 200 g, of dry capsules.

Climate and soil

In India, cardamom grows under natural conditions of the evergreen forests in the Western Ghats. It thrives best in tropical forests at altitudes ranging from 600-1500 metres, receiving a well distributed rainfall of over 150 cm and a temperature of 10°-35°C. Its optimum growth and development is observed in warm and humid places under the canopy of lofty, evergreen forest trees. It is highly sensitive to wind and drought and, therefore areas liable to be affected by these conditions are unsuitable. Waterlogging or excessive moisture is equally injurious. The ideal site is a sloping with good drainage. The crops is raised chiefly on well-drained, rich forest loam, and red, deep, good-textured lateritic soils having plenty of humus or leaf mould.

Preparation of land

If virgin forest areas are selected for a plantation, they are too be cleared of all undergrowth and trees not suitable for furnishing heavy shade needed by cardamom. Where tree growth is sparse and shade insufficient or unsatisfactory, quick growing trees such as Erythrina are planted for temporary shading and other tall and spreading trees for permanent shade. When grown as a subsidiary crops in areca and coffee plantation, the main crops furnishes the required shade.

The spacing commonly adopted for the 'mysore' type varies from 2m X 2m to 3m X 3m and for the less vigorous 'Malabar' type it varies from 1.5m to 2.2m between the plants and the rows. The seedlings are planted in pits of about 60cm X 60cm X 35cm filled with a mixture of surface soil and compost or well-rotted cattle manure. For planting in areca gardens, the north Kanara system of digging trenches instead of individual pits is adopted.

Propagation

Cardamom is propagated through suckers or seedlings. However the latter method is preferable in order to check the spread of the virus disease 'katte'. In the case of vegetative propagation 3-5 bulbs of 1/2. to 2 year old clumps obtained from healthy and regularly high yielding stock are planted in each pit. For raising a nursery, the present practice is to collect the seeds from disease free, high yielding plants with compact panicles. The capsules selected for seeds should be fully mature and preferably from the second or third round of picking. They should immersed in water and gently pressed for rejecting the seeds, which should then be washed well in cold water. After draining the water the seeds should be mixed with ash and dried in shade in 2 or 3 days. To ensure uniform and early germination, seeds should be sown after extraction. They are to be broadcast on nursery beds of 1m X 6m and raised to a height of about 30 cm.

The selection of the site for the nursery beds is very important. The site should have a gentle slope and there should be adequate facilities for irrigation. The soil should be fertile and rich in humus. The seed rate commonly used is 10 g per square metre of bed size. After sowing, the seeds should be covered with a thin layer of fine sand or soil and then mulched to a thickness of about 2 cm with paddy straw or any other suitable material. The bed should be watered twice a day. Germination will commence in about 30 days and may continue for a month or two. The mulch should be removed after the commencement of germination. Shade should be provided for the young seedlings by erecting a pandal.

Planting and cultivation

One or two year old seedlings are planted in each pit, either in June-July or September-October. The rhizome is planted 5-8 cm securely tied to a stout wooden stake to prevent it from being laid low by the strong monsoon rains and winds. The plantation receives at least four weedings in the first year, three in the second year and two annually thereafter. The clearing, digging and manuring of soil in each clump, filling up gaps, and the lopping of shade trees constitute the main cultivation practices. Each clump requires a heavy doze of cattle manure or leaf compost. A well developed mould or a well marked layer of humus accumulation is essential for the development of the plant. On large estates, castor-cake, bone-meal, fish-manure, ammonium sulphate and muriate of potash are commonly used to provide 30-40 kg of N, 30 kg of P_2O_5 and 50-60 kg of K_2O per ha.

Harvesting

Cardamom comes into bearing in the third year after planting, though an economic crops is usually obtained only from the fourth year onwards. Flowering starts in April-May and continues till August-September or even later, depending upon the weather conditions. it takes 3 or 4 months to form fruits, which mature at intervals of 30-40 days, nesseciating several pickings and the whole harvest is completed in 5 or 6 rounds. In most areas the peak harvesting season is October-November. In Kerala and Tamil Nadu, harvesting starts from August-September and continues till February-March, whereas in Karnataka, it starts in July-August and continues up to December-January. Capsules along with peduncles are harvested just before it full ripeness. Dead ripe fruits splits on drying and seeds falls on the ground surface whereas the unripe fruits shrivel on drying. The average yield of dry capsules from a well maintained plantation is about 100-150 kg per ha. From small gardens which are not properly managed, it is about 25 kg.

Curing

The preparation of cardamom for the market consists of timely harvesting, drying, sorting, bleaching, etc. These processing activities have an important bearing on the quality of the finished product. After harvesting, the produce is dried either in the sun or in the specially built drying houses by using radiated heat. For the latter, the devices vary from sheltered mud platforms heated by a slow fire from beneath to large drying houses to kilns heated by fuel pipes, as is mostly done in large plantations. The fruits kept for drying are spread out thinly and stirred frequently to ensure uniform drying. Sun drying takes 3-5 days, whereas in the case of artificial heating, it takes only about 48 hours for proper drying. The latter process also helps to retain the green colour of the capsule which is much valued, especially in the Middle East. The dried capsules are rubbed by hand or with a rough coir matting or with a piece of wire-mesh and winnowed to remove other plant residues and foreign matter. They are then sorted according to their size and colour. Since green capsules fetch a premium price in foreign markets, it is essential to retain this colour as far as possible.

2. PEPPER

Pepper (*Piper nigrum* L.) is one of the most important and the earliest known spice crops of India. It is a perennial climbing vine belonging to the family Piperaceae and is indegenous to the west coast of southern India. Pepper cultivation is mainly confined to Kerala, Karnataka, Tamil Nadu and to a limited extent to Assam and West Bengal. Very recently, pepper has been cultivated successfully in Andhra Pradesh and the Andamans. Indonesia, Malaysia, Brazil, Sri Lanka and Cambodia are the other pepper producing countries. For both black pepper and white pepper the dried and processed berries, have a very prominent place in the world market. Pepper is used as a flavouring agent for food stuff and also carminative. The alkaloid piperine forms 5 to 8 per cent by weight of the seed and the volatile pepper oil forms 1 to 3 per cent of the unripe berries.

General botanical aspects

Piper nigrum L. belongs to the family Piperaceae. The family is dicotyledonous, but the stem has characteristic lines intermediate between those of the dicotyledons and the monocotyledons, i.e. the vascular bundles lie in two or more circles. The stem is usually swollen at the nodes. Numerous adventitious roots, known as clinging or climbing roots, are produced from the nodes. It is by means of these roots, the plants attach themselves to the standards and climb up by exposing the fresh leaves to maximum sunlight. The stem or the vines can be broadly classified into: (i) terminl stem, (ii) stolon or runners and (iii) lateral branches. The stem and branches bear alternate, shiny, dark green ovate leaves. Branches are produced from the dorsal buds. Each bud is accompanied by a single leaf lateral to the bud. The receme develops on the current year's flush opposite a leaf. Two successive primordials of a raceme are present within a single bud. Flowers are small, whitish and borne on hanging catkins or spikes.

Climate and soil

Pepper is a plants of the humid tropics. It grows best in well drained clay-loam soils rich in humus and flourishes in warm, moist climate. An annual rainfall of 250 cm is ideal for its cultivation. It can also be cultivated in areas receiving a slightly lower rate of rainfall, provided it is well distributed throughout the year. It can tolerate a temperature range of 10° to 40°C and grows well from the sea level up to an altitude of 1,200 metres. Red, laterite virgin soils on the slope of the Western Ghats is also suitable for growing pepper. To prevent the scorching effect of the sun, the slope facing the south should be avoided.

Propogation

Pepper is propagated best from cuttings of the runner shoots which originate from the base of the vines. The runner shoots of selected high yielding healthy vines are to be kept coiled in the forks of 2 or 3 sticks fixed in the ground so that the shoots do not come into contact with the soils and strike roots. Cuttings from lateral fruiting branches can also be used for planting. They are separated from the vine in February-March. The cuttings with 2 or 3 nodes are planted with at least one node below the soil level on suitable nursery beds or in polythene bags or bamboo baskets filled with soils and kept under partial shade. They form roots and become ready for planting in June-July. Pepper cuttings can also be planted directly in the field. For this purpose, 4 or 5 pepper cuttings each with 2 or 3 nodes, are planted 30 cm away from the standard with the onset of the south westerly monsoon. Propagation from seed is also possible, but it is not generally practised.

Planting

Pepper is generally planted with the onset of the south westerly monsoon. Pits of 0.5 m cube are made 3 or 4 metres apart in each direction. With the first showers in May-June, the cuttings of 'Murukku' (*Erythrina indica*) are planted in these pits as standards. *Garuga pinnata*, *Spondias mangifera* and silver oak (*Grevillea robusta*) are the other common standards used. Rooted cuttings of pepper are then planted in June-July at the rate of 2 or 3 per standard, about 30 cm away from the base. In doing so, as far as possible the southern side of the standard should be avoided for planting. The soils around the roots should be pressed hard to avoid water stagnation. Too much shade during flowering and fruiting affects the yield adversely. The pruning of excessive foliage of the standards is therefore done during March-April. Sometimes, pepper is grown in association with mango, jack and other trees. Pepper is also cultivated as a subsidiary crop in coffee and coconut plantations, and the coconut, the arecanut and other trees elsewhere serve as standards in these situations.

As pepper is grown mainly as a rainfed, cultural operations assume greater importance. Digging is usually done around the standards and vines up to a diameter of about 2 m or in the entire plantations. Two diggings are generally given during August September and October-November. The vines are also earthed up at this time. On the hill slopes, adequate soil conservation measures, such as contour bunding or terracing, or the growing of a cover crop. Each vine may be manured with about 9 kg of well rotted cattle manure or compost as a basal dose and 100 g of N, 160 g of P_2O_5 and 60 g of K_2O annually in two split doses, when they are 3 to 4 years old. In the earlier years, graded doses of the above combinations may be applied. The manure should be applied to the soils around the vine at a distance of 30 cm and upto a depth of about 15 cm and incorporated with the soils with drake or fork. One half to one kg lime applied in alternate years during April-May is also beneficial.

Harvesting and curing

The pepper vines commence bearing during the third year. Flowering starts during May June and the light showers during this period are considered to be beneficial for the fruit set. It takes nearly 6 to 8 months from flowering to harvesting. The season for harvesting mature berries is November-February in the plains and January-March in the hills. Harvesting is generally done by plucking the spikes, when one or two berries become bright orange or red. The spikes are spread on the floor or on the mats and the berries are separated by trampling and they are dried in the sun for 4 to 7 days until the outer skins become black and shrink. This is the black pepper of commercial use. White pepper is prepared from fully ripened berries by removing the outer rind and the pulp before drying. The recovery of the white pepper is 25 per cent of the ripe berries whereas that of black pepper is 33 per cent.

Yield

Although the vines start bearing from the third year onwards, the yield gets stabilized only from the seventh or eighth year after planting. One hectare of the crop yields about 600-800 kg of black pepper. The yield declines about 25 years of planting. Under favourable conditions, the vines may live upto 40-60 years.

3. GINGER

Ginger (*Zingiber officinale* Rose.) is an important commercial crop grown for its aromatic rhizomes which are used both as a spice and a medicine. It is cultivated in almost all the tropical and subtropical parts of India, especially in Kerala, Karnataka, Tamil Nadu, West Bengal, Bihar, Himachal Pradesh, Uttar Pradesh and Maharashtra. Though ginger is also cultivated in Jamaica, Nigeria, Sierra Leone, Brazil, China, Japan and Indonesia, India still is the largest producer of dry ginger. The total annual production in the country is estimated at 33,780 tonnes. About one-third of this quantity is exported to the Middle-East.

General botanical aspects

Ginger is the rhizome of *Zingiber officinale* a herbaceous perennial belonging to Zingiberacea and is believed to be native of south-eastern Asia. It is propagated through rhizomes. The rhizomes put forth erect, leafy stems, 30-90 cm in height. The leaves are light-green, 15-20 cm long, narrow, lanceolate and with a prominent midrib. The flowers are small, yellowish, speckled, each with a purple speckled lip and borne on spike. When the plants are about 9 months old, the green leaves turn yellow.

Climate and soil

The crop is grown from almost the sea level up to an altitude of 1,500 metres. It thrives well under partial shade, though it is also grown on a large scale in open areas. It comes up well on a variety of soils, provided sufficiently well distributed rainfall or irrigation and adequate drainage facilities are available. A deep, well drained friable loam rich in humus is ideal.

Rotation

Ginger is rotated with betel vine, plantain, turmeric, onion, garlic, chillies, other vegetables, sugarcane, ragi, maize and groundnut. Under rain-fed conditions it may be grown once in 3 or 4 years in rotation with tapioca, sweet-potato, chilli, yam and dry paddy. It may be grown alone or mixed with shade giving plants, e.g. banana, pigeon-pea, tree castor and cluster bean (guar). In coconut, young coffee and orange-plantations in the west-coast, ginger is grown as an intercrop. At higher altitudes in Himachal Pradesh, tomato and chilli are grown as intercrops with ginger.

Cultivation

The land is ploughed 5 or 6 times with the first showers in March-April to obtain a deep fine tilth. For raising a rain-fed crop the land is divided into raised beds 1 metre wide and of convenient length varying from 3 to 6 metres and 15 cm high, with a spacing of 30 cm between the beds for drainage channels. On the hill slopes, the beds are to be formed along the contours. Ginger is planted in shallow pits in rows, 25 cm apart, at distances of 15-20 cm within the row. In the case of ridges which are 40-45 cm apart, the planting is done in shallow pits on top of the ridges at distances of 22-30 cm. Bits of seed rhizomes weighing 20-30 g each and having at least one bud are planted 5 cm deep at the given spacing and covered with soil and smoothed over by hand. About 1,800 kg of seed rhizomes are required to sow one hectare. Higher seed rates of 2,160-2,640 kg per ha are used for planting at higher altitudes. Sowing is done in April-May in southern and a little later in northern India. Sowing by the middle of April in the south and by the first week of May in the north gives higher yields. The irrigated crop is watered immediately after sowing. The beds of the rain-fed crops are covered with leaf mulch as protection against sun and heavy rains and for the consequent enhancement of organic matter in the soil.In Himachal Pradesh farmyard manure is used as mulch. Seeds of cluster-bean, pigeon-pea or castor are sown on irrigation channels on the corners of the raised beds for shade. The shoots emerge in 10-20 days. Weeding is done 3 or 4 times and plants are earthed up once or twice. Irrigation is given at varying intervals of 4-10 days as and when required. The mulching of the beds with green leaves is an important operation for the crop. Two or three mulchings are given first immediately after planting and the second and the third at intervlas of about 40-50 days at the time of weeding, hoeing and earthing up. Ginger requires heavy manuring. In addition to the basal dressing of organic manure or farmyard manure 75 kg of N, 50 kg of P_2O_5, 50 kg of K_2O/ha are applied. The whole of P_2O_5 and K_2O are applied at the time of planting itself. But the nitrogenous fertilizers are applied as a top dressing in two equal split doses, the first dose two months after planting and the second dose a month later.

Harvesting and curing

The crop is ready for harvesting in about 8 months. The leaves at this time turn yellow and the pseudostems begin to dry. Rhizomes are lifted either with a digging fork or with mammutties. They are cleaned of roots and adhering soil particles. The green ginger is soaked in water to facilitate the removal of the skin. The skin is scraped off with pieces of sharp ened bamboo or bits of sea shells. The scraped produce is washed and dried in the sun for 3 or 4 days and hand-rubbed. It is again steeped in water for two hours, dried and then rubbed to remove all the remaining bits of the skin. Sun drying also bleaches the produce. Peeling should be done with great care and skill. The essential oil which gives ginger the aromatic character is present in the epidermal cells and excessive or careless scraping will result in damaging these cells leading the loss of essential oils. Steel knives are not used as they are found to stain the produce.

Preservation of seeds

Seed ginger has to be stored for about 4 to 5 months from the time of harvesting to planting. Rhizome rot, dry rot, sprouting and shrivelling are commonly noticed in storage. Keeping the seed rhizomes in pits is found to be the best method of storage. Healthy rhizomes are selected at the time of harvesting and are treated with 0.025% solution of Ceresan Wet for 30 minutes as a prophylactic measure against soft rot. The treated rhizomes are then dried in shade and stored in pits dug in a cool place and protected from sun and rain. The pits should not be more than 1 metre deep. A layer of sand and sawdust is put at the bottom of the pit before storage. Either the pit is left as such or is covered with a wooden plank leaving some space betweeen the seeds and the plank and plastered with mud. In the latter case a small hole is made in the plank through which a hollow bamboo is inserted to provide aeration for the rhizomes. At higher altitudes seed ginger can be stored safely up to the end of May in an undreground store.

Yield

The average yield of green ginger is estimated at about 50-100 q per ha. The recovery of dry ginger varies from 15-25 per cent.

4. TURMERIC (*Curcuma longa* L.)

Turmeric (*Curcuma longa* L.) is an important and a useful dye, with varied uses in drug and cosmetic industries. It is used medicinally for external application and taken internally as a stimulant. India is one of the leading growers, with about 79,300 ha under this crop, producing 1,43,900 tonnes per annum.

General botanical aspects

Turmeric is the dried rhizome of *Curcuma longa* L., a herbaceous perennial belonging to the family Zingiberaceae and a native of India or China. Besides India, it is cultivated in Srilanka, Indonesia, China, Peru, Jamaica and other triopical and subtropical countries. The plant is propagated from rhizomes. The leaves are long, broad, lanceolate and bright green. The flowers are pale yellow and borne on dried spikes. The pseudostems are shorter than leaves. The rhizomes are ready for harvesting about 7-9 months after planting.

Climate and soil

The crop requires a warm and humid climate. In heavy rainfall tracts of the west coast, it is grown as a rain-fed crop and in other areas, it is cultivated under irrigation. The cultivated types are found from almost the sea level upto an altitude of about 1200 metres. The wild types are seen at higher elevations. Turmeric thrives in well drained, fertile, sandy and clayey, black, red or alluvial loams rich in humus and uniform in texture. Rich loamy soils having natural drainage and irrigation facilities are the best. Turmeric cannot stand water stagnation or alkalinity.

Cultivation

The land is ploughed 4-6 times to bring the soil to a fine tilth. The land is divided into beds of 1 metre width and convenient length with a spacing of 30 cm between the beds for drainage channels. In the case of the irrigated crop, ridges and furrows are prepared and the rhizomes are planted in shallow pits on the top of the ridges. Spacing generally adopted is 45-60 cm between the ridges and 15-22 cm between the plants. In beds rhizomes are planted 22-35 cm apart in each direction. For sowing both the mother rhizomes the fingers are used. The fingers are cut into pieces each 4-5 cm long, and the mother rhizomes are planted as such or planted into two; each having at least one sound bud. The seed is sometimes sprouted under moist straw before sowing. Depending on the spacing adopted and the extent of mixed cropping, the seed rate may vary from 2,000-2,500 kg per ha. As an intercrop in a fruit garden it may be as low as 400-500 kg per ha. The crop is sown form April to July, depending upon the tract. As in the case of ginger, mulching the bed with green leaves is an important practice beneficial to this crop when planting is done on raised beds. Two or three mulchings are given, the first immediately after planting and the second and the third at intervals of 40-50 days. Turmeric needs heavy manuring. It is not a common practice to manure turmeric with chemical fertilizers. Usually, 40-50 tonnes of farmyard manure per ha is applied at the time of preparing the land. About 1200-1800 kg per ha of groundnut cake is also applied after planting, in two split doses, the first two months after planting and the second a month and a half later. However, instead of groundnut cake chemical fertilizers to supply 30 kg each of N, and P_2O_5 and 60 kg of K_2O per ha may also be applied in split doses as in the case of ginger.

Harvesting and curing

The crop is ready for harvesting in about 7 to 9 months after sowing depending upon the variety. The aromatica type mature in about 7 months and the long type in about 9 months and the intermediate types in about 8 months. The dry leaves are cut close to the ground. The land is irrigated, if necessary and ploughed in between the rows if the crop is planted on ridges. Otherwise a crow bar is used. The rhizomes are dug up. The curing quality and the proportion of the cured and dried produce to the green produce depend mainly on the variety. Mother rhizomes give a higher curing percentage than the fingers which are seperated. If need be, the former is kept for seed and the latter is cured for selling. The green rhizomes are boiled in water till a froth comes out and white fumes appear giving out a characteristic odour. After cooking, the rhizomes would be soft and would yield when pressed between the fingers. The quality of the final product, including its colour and aroma depends largely on the right amount of curing. The boiled rhizomes are spread out on a clean floor and allowed to dry in the sun for about 10 to 15 days. They are stirred 3 or 4 times to ensure uniform drying. The rounds and fingers are dried separately. The former takes more time to dry, when fully dried the turmeric becomes hard and stiff. The dried turmeric rhizomes are grinded to get turmeric powder.

Preservation of seeds

Rhizomes for seed are generally heaped in the shades of trees or in well ventilated sheds and covered with turmeric leaves. Sometimes the heap is plastered over with earth mixed with cow dung.

Yield

The yield of the pure crop varies from 200 - 220 q per ha. Under exceptionally favourable conditions, viz. abundant manuring and plentiful irrigation, it may be as high as 300 q per ha. The cured produce is 20 to 25 per cent of the fresh rhizomes by weight.

5. DILL SEED (*Anethum graveolens* L.)

History of Dill plant

The dill plant originated in southern Europe and western Asia. As far back as the ancient Greeks, dill represented a sign of wealth and the richest citizens of Greece often burned oil from its seeds to create a savoury, aromatic scent. Dill became a staple ingredient in medicines for stomach remedies.

Growing Zones

Dill grows well in all growing zones in the U.S. In fact, in warmer climates, dill can be planted every two weeks throughout the season to produce an abundant annual crop. In colder areas of the country, dill produces best if planted a week or two before the last hard frost of the season. Quite adaptable to various growing conditions, dill can be grown in a variety of ways. Dill seeds can be sown directly into garden soil, in raised herb beds or in pots.

Planting

Although dill is a hardy plant, it can be difficult to transplant young seedlings. Transplanting of Dill plant is not successful like other crops, so plant dill seeds wherever you plan to grow them for the season. Plant seeds ¼ - ½ inch deep at 8 – 12 inches apart. Row spacing should be 2 – 2.5 feet.

Water requirements

One of the advantages of planting dill is that it is heat and drought resistant, but it does produce best when watered at regular intervals depending upon climatic conditions and soil type. Make sure that the soil in which you plant dill has adequate drainage.

Maturity

Dill grows quickly and its leaves can be harvested approximately 8 weeks after planting. The dill plant averages 30–36 inches tall at maturity. By planting dill every two weeks, gardeners can enjoy a fresh supply of mature dill leaves throughout the growing season.

Once the flower of the dill plant heads, the plant will stop producing its leaves and plant will enter in the reproductive stage.

Harvesting

The leaves of the dill plant are most flavourful just before flowers form on the plants. Harvesting can begin as soon as the leaves are big enough to use. Clip the leaves of dill plants close to the stem in early mornings or late afternoons for best results.

For harvesting dill seeds, allow the plant's flowers to form, bloom and go to seed. When the majority of the seed heads hav.e formed, which is about 2-3 weeks after blooming starts, remove the entire seed head

6. CORIANDER (*Coriandrum sativum*)

Introduction

Coriander (*Coriandrum sativum*) is an annual spice crop and a member of the Umbelliferae, or carrot family. It has an erect, branching stem growing to a height of approximately 0.6-1 m (2-3 ft.). The fresh green herb is called cilantro, or Chinese parsley. It is used in southeast and southern Asian, Chinese and Mexican cuisine, and for flavouring salads and soups. The mature, round fruit contains two fused seeds. The market requires intact fruits, as the essential oil can volatilize from split fruit or individual seeds. Mature seeds have a pleasant, spicy aroma. Coriander powder is made from ground seeds and is used to flavour many products such as curries, gin and prepared meats.

There are two types of coriander: large seeded (fruited) and small seeded. Large seeded coriander (seed diameter three to five mm) is grown as a spring crop in northern temperate climates like Saskatchewan, or as a winter crop in areas with a Mediterranean climate. The most common type grown in Saskatchewan is large seeded, which has an essential oil content of 0.5 to one per cent. It grows to a height of approximately 0.6 m (two ft.), and has a growing season requirement of approximately 100-110 days.

Small seeded coriander is produced mainly in more moderate temperature zones. It has a longer growing season requirement of approximately 115-120 days. Small seeded coriander is somewhat taller, approximately one metre (three ft.), and contains more essential oil. Small seeded coriander grown in Saskatchewan has an essential oil content of 1-1.5 per cent. Volatile oils such as linalool are responsible for the aroma of coriander. The essential oil of coriander should contain 60-70 per cent linalool. Immature seedlings of small seeded coriander (cilantro) are harvested as a spicy addition to salads or for flavouring of meats, soups and stews. The 10-year average area of coriander planting in Saskatchewan is approximately 6,900 ha (17,000 ac.).

The average yield of large seeded coriander in Saskatchewan is approximately 900 kg/ha. Small seeded coriander yield can be slightly more, if growing conditions allow it to mature.

Adaptation

Coriander is a heat loving crop. It can be grown on a fairly wide range of soils, but is best adapted to well drained loam and sandy loam soils. Coriander crop can be severely affected by blossom blight. Severe infections occur in cool, wet environmental conditions, or in fields where there is a recent history of coriander production. Coriander should not be planted in the same field more than once in four years to avoid losses due to plant diseases.

Seeding

The recommended seeding rate for large seeded coriander is 33 kg/ha, and 18 kg/ha for the small seeded coriander. This is approximately 36 fruits per metre of row. Each mature fruit contains two seeds. Research results indicate no yield differences between using intact or split fruits for planting. The recommended seeding depth is 2.5 to 4.0 cm (one to 1½ inch). Coriander requires a firm, moist seedbed to enhance germination and hasten plant emergence. Any seeding equipment that will provide these conditions is suitable. Coriander should be seeded in late April to mid May. Emergence may take 21 days or more. Seedlings have some tolerance to spring frost. Research results have indicated higher seed yields and oil levels can be achieved by seeding before May 20. Seeding into dry soil can delay germination.

Fertilization

The recommended dose of fertilizer in coriander is 75 kg N per hectare. It should be applied in two splits, half at sowing and the remaining half at flower initiation. There is no need of applying phosphorus to soils testing medium to high in this nutrient.

Weed control

Coriander may take 21 days to emerge after planting, and does not compete well with weeds. Severe weed competition can occur unless the crop is seeded in clean fields. Some producers use a post planting application of paraquat to control grassy and broadleaf weeds in the time period between planting and emergence of the coriander. Control of perennial weeds in the crop is very difficult and is best done in the year prior to planting coriander.

Harvesting

Straight combining when the crop is fully mature is the preferred harvest method for coriander. The crop can be swathed when the fruit turns tan to brown. However, large and fluffy swaths can be very susceptible to wind damage. Swathing or combining in damp weather or when dew is present will help to avoid shattering losses.

7. CUMIN (*CUMINUM CYMINUM*)

Family : Apiacae

Cumin, commonly known as Jeera, (*Cuminum cyminum*) belongs to Apiacae family, widely cultivated in Gujarat, Rajasthan and in some parts of Madhya Pradesh and Uttar Pradesh as *Rabi* crop. Cumin is mainly used in flavouring foods. It is also used in Ayurvedic medicines.

Climate and soil

Cumin is a tropical plant. It grows well in sub-tropical climate too. High humidity during flowering & fruit set, causes fungal diseases in this crop. Cumin can be cultivated in all types of soils but well drained sandy loam and medium soils are suitable for the crop.

Sowing and seed rate

Sowing is done from 1st week of November to 1st week of December by broadcasting or in rows drilled at 30 cm. Seed rate vary from 12 to 15 kg / ha, depending upon method of sowing and type of soil. Sowing of seeds should be done at a depth of 1 – 2 cm after treatment with Ceresan or Thiram @ 3.0 g per kg. Soaking of seeds for 8 hours before sowing is helpful in getting good germination. Soaked seeds should be dried in shade to facilitate broadcasting. Sowing at higher depths affect the germination of seeds, adversely. Crop rotation should be followed to avoid incidence of pest and diseases.

Land preparation

Soil is brought to fine tilth by 2-3 ploughing with harrow or desi plough. Stubbles of previous crops should be collected and removed from the field. Clods should be broken and field should be leveled with the help of planker. Beds of 4 m x 3 m size with provision of irrigation channels should be prepared before sowing of seeds to facilitate proper irrigation and intercultural operations.

Manures and fertilizers

Fifteen to twenty MT FYM, 30 kg Nitrogen and 15 kg Phosphorus per ha is recommended for cumin crop. Whole quantity of FYM should be mixed into the soil at the time of land preparation and 15 kg Nitrogen and whole quantity of Phosphorus should be applied as basal dose. Another, 15 kg Nitrogen should be applied as top dressing one month after germination of seeds.

Weeding

The crop should be kept free from weeds for proper growth and development of plants. Generally 2-3 hand weeding are required to keep the weeds under check. In drilled crop light intercultural operation is beneficial. 1st weeding and hoeing should be done after 30-40 days from the date of sowing.

Irrigation

Based on type of soil, crop requires 4-6 rounds of irrigation. 1st light irrigation should be given immediately after sowing and second irrigation should be given after 6-10 days from 1st irrigation. Subsequent irriga-tions should be given after 30, 45, 65 and 80 days from 1st irrigation. Irrigation at the time of flowering and fruit set are essential. At maturity stage irrigation should be stopped.

Harvesting

Generally cumin crop takes about 110-115 days to reach maturity. Crop becomes ready to harvest, when plants turn yellowish brown. Harvesting should be done early in the morning by cutting/uprooting the whole plants. Harvested crop should be dried in the threshing yard and thrashed to separate the seeds. Seeds should be cleaned by winnowing. It yields 600 to 700 kg/ha.

8. SENNA (*Cassia angustifolia* Vahl.)

Senna is a native of South Arabia and is now naturalized in the Tirunelvelly and Ramnathapuram districts of Tamil Nadu. India is the main producer of this crop in the world and exports senna leaves and pods worth over Rs 6 million annually. The leaves and the pod-shell contain the laxative principle (sennocide) which is extensively used as a laxative in medicine. The crop is grown on marginal red lateritic soils. The growing areas of Tirunevelly and Ramnathpuram have no distinct winter season, the average maximum temperature (November to February) is 18.3°C and the region is received in autumn from the north-easterly monsoon.

Senna is a small, perennial, branched under shrub, remaining in field for 5 to 7 months. It is grown as a pure crop as well as a mixed crop with gram, gingelly, chilles and cotton. Generally, two sowing seasons are recognised, i.e. February to March (irrigated crop) and November (rainfed crop). The seed-rate is 27 kg per hectare under rainfed conditions and 15 kg per hectare for the irrigated crop.The seeds are broadcast or preferably drilled in lines, 30 cm apart. As soon as the seedlings emerge one to two hoeings are given and the plants close up the rows as they grow. This crop generally gets the benefit of residual fertiliser applied to the preceding paddy crop and, therefore, in actual practice it is seldom given fertiliser. The application of 20 kg of N and 40 kg of P_2O_5 per hectare at planting supplemented with 40 to 60 kg/ha of N in 3 split doses is preferred.

The crop is given 5 to 8 irrigations when grown on garden soils. It flowers in about 2 months after sowing and the first flush of flowering stalks is removed to induce a higher degree of branching. The picking of leaves starts in early May and the leaves are stripped off by hand when fully grown, thick and bluish; the second picking is taken along with the plucking of pods. The pods are harvested a little before maturity to maintain their green colour; the second crop of pods is obtained at the end of the season when the plants are uprooted during August or April-May. The leaves and the pods are dried in a thin layer in shade for 4 to 7 days; the pods are lightly beaten during drying to remove the inter seed material. The careless drying of the harvested crop spoils the colour and lowers its active content. The produce is graded, pressed and finally packed in hessian under pressure for long distance transport. The crop yields 1 to 1.4 tonnes of leaves and about 1.5 quintals (about one quintal under rainfed conditions) of pods per hectare. The produce should contain about 2.5 per cent of active principles, calculated as anthoquinones.

Damping-off is common in the seedling stage. To control the disease, seed treatment with fungicides, e.g. Captain, is recommended. The pods are attacked by a borer during storage, leading to considerable damage.

9. ISABGOL (*Plantago ovata* Forks.)

It is an annual stem less herb, a native of Persia, now grows as a cash crop on about 16,000 ha in the Mehsana, Palampur and Banas-kantha districts of Northern Gujarat, India is the largest producer of isabgol and exports seed and husk worth Rs 25 million annually. The husk is the rosey white membranous covering of the seed which constitutes the drug and is given as a safe laxative, particularly beneficial in habitual constipation, chronic diarrhoea and dysentery.

Isabogal is an irrigated *Rabi* crop which remains in the field for about 4 months. The crop is grown in marginal, light, well drained sandy loam to loamy soils having pH between 7 to 8. It requires a cool climate and dry sunny weather or light showers cause seed shedding.

After harvesting the *kharif* crops, the land is brougt to fine tilth and laid out into beds of convenient size for irrigation. It is preferable to mix with the soil 15 cartloads of well-rotted farmyard manure per hectare during the preparation of the land. The optimum sowing time is early November; sowing however, is extended till the end of December, but the delayed sowing decreases the yield. The seed rate is 7 to 8 kg per hectare. Seeds are small and light and are sown by broadcast. They are covered thinly by raking the soil. A light irrigation is given immediately. Germination starts in 6 to 10 days and the crop is given the second irrigation after 3 weeks and a third one at the time of the formation of the spikes ; thus the crop needs 6 to 7 irrigations.

Isabgol makes a moderate demand for nutrients. Usually, 25 kg each N and P per hectare is given at plantation. The crop is given 1 or 2 hand weedings during the entire growing period. The plants are about 50 cm high and each plants gives out between 25 to 100 tillers, depending upon the fertility of soil and weather conditions. The plant bears the flowering spikes in about 60 days after sowing and matures in the next 2 months. The yellowing of the lower leaves is an indication of maturity, confirmed by pressing a spike between two fingers when the mature seeds come out. The crop is harvested close to the ground in the early morning hours to avoid looses owing to seed shedding. The harvested material is stacked for 1 or 2 days, made to be trampled by bullocks, winnowed, and the separated seed crop is collected. A bold seeded crop fetches a better price. The seeds are processed through a series of grinding mills to separate the husk, and about 30 per cent husk by weight is thus recovered. The husk contains a mucilaginous substance.

10. OPIUM POPPY (*Papaver somniferum* Linn.)

Poppy is a native of the western Mediterranean region, introduced into India in the early sixteenth century. India produces about 70 per cent of the world's production and 90 per cent of it is exported annually. The crop can be grown under a licence issued by the Excise Department and, as such, its cultivation is restricted to about 24,000 hectares in the districts of Neemuch, Mandsaur and Ratlam in Madhya Pradesh; in Faizabad, Bara Banki, Bareilly and Shahjahanpur in Uttar Pradesh, and in Chittorgarh, Jhalawar and Kota in Rajasthan.

Poppy is an annual erect plant, 60-120 cm in height, rarely branching, with ovate-oblong leaves; it bears large showy flowers and nearly globose to spherical capsules, containing small white or black kidney shaped seeds. It is a *rabi* crop, and remains for about 150-160 days in the fields. It is mainly (90%) self pollinated, since the dehiscence of anthers takes place before the opening of the flower. A large number of the forms of variety album DC are cultivated in India; it has ovate globose capsules devoid of apertures.

The crop is grown on well drained, fertile, clayey loam to rich sandy loam soils. It prefers moderately cool weather and open sunny locations; severely cold spells, frost dull cloudy weather, high winds, and heavy rainfall during the lancing period adversely affect the yield.

Land is prepared to a fine tilth up to a depth of 20-25 cm during September and the fields are laid out into beds of convenient size. Twenty-five to thirty-seven cartloads of farmyard manure is mixed with the soil during the preparation of the land, penning of sheep or goats is preferred. Seed rate is 2.5 to 3.5 kg per hectare. The seeds are minute weighing 0.25 to 3.5 g per 100 seed. They are mixed with sand and sown thinly in rows 0.3 m apart during mid October till early November (optimum temperature 9-10 degree centigrade). The rows are covered with a thin layer of soil and are given a light irrigation, followed by another irrigation after 8-10 days when the seeds start germinating. Seed treatment with 1.0% Agrosan protects it from seed borne diseases. The crop is given about 15 irrigations in all, depending upon the texture of the soil and the variety sown; moisture deficiency at the time of capsule formation affects the yield adversely. The crop needs a heavy nutrient supply and 50 kg N/ha is applied during the preparation of the land and another application of an equal quantity is given as top dressing in two-split doses when the crop is one month old and at the time of the flag leaf stage. The seedlings are thinned out to 20-25 cm apart when they grow 5-7 cm tall. Weeds can be controlled with hand weeding as and where desired. The plants take 75-80 days (Madhya Pradesh) and about 105 days (in Uttar Pradesh) to come into flower, depending upon the variety sown. The petals fall in 3-4 days and the growing capsule is ready for lancing in the next 6 to 10 days. The green unripe capsules contain the maximum amount of morphine which diminishes with its ripening and yellowing; conversely, codeine and narcotine increases with age.

Lancing is done longitudinally in bright sunlight and the latex (crude opium) which oozes from the wounds dries during the following night and is scrapped with a sharp kitchen knife in the early hours of the succeeding morning. Lancing is performed in the afternoon with a special knife, capable of making four incisions at a time, each at about 1/12 cm apart. Each capsule is lanced for 4 to 6 times. A deep incision causes the collection of the latex inside the capsule and spoils the seeds. The crude opium is stored in earthen pots and dried in the sun to obtain the raw opium of a uniform consistency. The dry produce should be stored in closed vessels. The entire produce goes to the Excise and Narcotics Department which pays according to its consistency and purity. The average yield of raw opium varies from 13 to 33 kg per hectare, though much higher yields are recorded under good agronomic practices. The crop also yield 3 to 4 quintals of seeds per hectare. The seeds contain up to 50 per cent of a golden-yellow fatty oil which is devoid of narcotic properties.

There are 25 opium alkaloids identified in the raw opium. Of them, morphine, theabine, codeine, narcotine and papaverine are important; morphine is by far the most important. The Indian produce of raw opium contains,on an average, about 20% of the total alkaloids by weight. They, in turn, posses 9 to 14% of morphine, 5-7% narcotine, 1.25 -3.75% codeine, 0.4-1% papverine and 0.2-0.5% theabine. The produce obtained in Turkey and the Balkan countries is collected from one incision only and is therefore rich (10-21%) in morphine; conversely, the Indian opium is richer in codeine and narcotine.

IMPORTANT PLANTATION CROPS

1. COFFEE

Coffee is cultivated commercially in the four southern states of Karnataka, Tamil Nadu, Kerala and Andhra Pradesh. It is also grown on a limited scale in some non traditional areas of Orissa, West Bengal, Assam and Madhya Pradesh

Arabica *(Coffee arabica)* and Robusta *(Coffee canephora)* are the two principal economic species extensively cultivated in India. It is grown on an area of 3.97 lac ha with green coffee production of 3.48 lac t (FAO STAT, 2016) and earns considerable foreign exchange. Coffee cultivation is confined to the hilly areas of the Western and Eastern Ghats. The annual rainfall ranges from 1250 to 3000 mm. A major portion of the area is under the south-westerly monsoon and only a small area is under north-easterly monsoon. This plant grows well at temperatures between 12 and 36 degrees Celcius. Elevation above the sea level influences the quality of the coffee. *Coffee arabica* grows well at elevations between 900 and 1200 metres; *Coffee canephora* or robusta grows luxuriantly at the lower elevations (about 150 metres). Arabicas are more shade-loving than robustas under conditions in south India.

Soil

Coffee soils in India belong to the red and lateritic soils. The soils differ in texture from sandy loam to clayey loam. Red ferruginous clays occur in tracts e.g. Bababudangiris in Karnataka and the Nilgiris in Tamil Nadu. They have good aggregating ability and are generally well drained. They are rich in organic matter and are acidic to neutral. The total soluble salts are well below the sensitivity limits.

Climate

Climatic and environmental factors, like rainfall, temperature, elevation can influence the economic production of coffee much more than soil factors. Under conditions in southern India, summer temperatures combined with poor subsoil moisture can be a severe limiting factor, whereas at northern latitudes cold winter temperatures can be equally limiting.

Soil and climatic requirements for arabica and robusta under south Indian conditions.

	ARABICA	ROBUSTA
Elevation	1000 to 1500 m (mean sea level)	500 to 1000 m
Annual rainfall	1600 to 2500 mm	1000 to 2000 mm
Blossom rain	March-April (2.5 to 4 cm)	February – March (2 to 4 cm)
Backing rain	April-May (5 to 7.5 cm)	April-May (5 to 7.5 cm)
Shade	Needs medium to light shade	Needs uniform thin shade
Temperature	15 to 25 °C.	20 to 30°C
Relative humidity	70 to 80%	80 to 90%
Soil	Deep friable, porous, rich in organic matter, moisture-retentive, slightly acidic, pH 6 to 6.5.	Same
Slopes	A gentle to moderate slope	Gently sloping to fairly level fields

Planting

The soil used for raising seedlings should be rich in organic matter, free from nematodes, cockchafer larvae and wireworms. Farmyard manure, sand and lime are also important ingredients of the soil. The seed is treated evenly with Agrosan at 20 kg per lit of water before sowing. This protects the soil against fungal infection during germination.

A seedbed raised to a height of 15 cm, usually about 1 metre in width and of convenient length is prepared. Four baskets of fully decomposed cattle manure or compost, about 2 kg of finely sieved lime and 400 g of rock phosphate are incorporated into a bed 1m x 6 m. If heavy soils are used it is necessary to add coarse sand to facilitate drainage and aeration. Seeds should be sown in December or January. The seeds are placed on the soil surface 1.5-2.5 cm apart in regular rows. A thin layer of fine soil is used to cover them. The bed is then covered with about 5cm of paddy straw to ensure an even temperature and prevent the soil surface from drying. The beds are watered daily and protected from sunlight by an overhead pandal. The seeds germinate in 45 days.

Planting in the field

Disease free and vigorous seedlings are selected for planting. Rooted plants 16-18 months old, with and without balls are planted during June and the ball and bag plants are generally planted from September to October. The seedling is placed in a hole made in the pit, with its tap root and lateral roots spread out properly. The hole is then filled with soil. The seedlings are provided with cross stakes to prevent damage from winds.

Planting shade trees

Dadap is commonly used as a lower canopy shade tree. Two metres long stakes are fixed for every 2 plants of coffee. Silver oak and dadap are planted during June when the rains commence. During the dry season the stems of the young dadap are either painted with a dilute solution or wrapped up in agave leaves to protect them from sun-scorch.

Nutritional requirements

Coffee is a perennial crop and the tree has the dual function of maturing the berries and producing fresh wood for the succeeding crop. One tonne of clean coffee removes from the soil 34 kg of N, 5 kg of P_2O_5 and 45 kg of K_2O in case of robusta varieties. Considerable nutrients are also lost through leaching under a heavy rainfall and as a result of fixation and immobilization of nutrients in the soil. Such depletion may lead to the impoverishment of the soil. It is thus essential to replace the lost nutrients by using fertilisers.

Shade and shade management

A shade tree should be fast growing and spreading to allow a uniform amount of filtered light. Its root system should be deep so that it does not compete with the coffee plant for food and moisture. The trees, which serve as alternative hosts to the major insect pests of coffee must be avoided.

Harvesting

For the preparation of both parchment and cherry types of coffee, picking of the right types of fruit forms an essential part of processing. Coffee is picked when the fruits are just ripe. Under and over ripe fruits cause deterioration in quality, the former gives "immature beans" and the latter "foxy" coffee. They may be dried seperately for making cherry coffee. The bags used are frequently washed and dried. Bags in which fertilisers, pesticides and fungicides are stored should never be used for this purpose.

2. COCONUT (*Cocos nucifera*)

Coconut is a majestic perennial palm. It is grown in numerous Islands and also in the humid coastal tracks of tropical countries. India ranks second in the world production of coconuts with 6088 million nuts from 108 million hectares.

The coconut palm or 'kalpa vriksha' provides many necessities of life like food and shelter and of all the tropical palms it is the outstanding one. It is mainly cultivated for the nuts from which 2 important products 'copra' and fibre are obtained. Copra yields oil and oil cake. The trunk of the mature palm is used as timber for houses and the leaves are used for thatching houses, fencing etc. The unopened spathe is tapped for toddy called 'neera'. Sweet toddy can be converted into jaggery and sugar, fermented toddy is a mild alcoholic drink and vinegar can be made from it. Water from tender coconut is a refreshing drink. In Kerala the extraction of coir from the husk of nuts is very common and the manufacture of coir products provides employment for thousands of people. The coconut shell is largely used for fuel and for the production of charcoal and making a variety of curios. The shell flour is used as filler for plastics. Thus every part of the palm is useful in one way or the other.

Climate

Coconut is essentially a crop of the humid tropics. It is mainly grown in the coastal plains. Rainfall is the most important factor affecting its growth. A rainfall for 100 to 225 cm per annum, evenly distributed is most important.

Coconut requires an equable climate. The optimum mean annual temperature for its best growth and maximum yield is 27 degree celsius. Frost and low humidity adversely affect the yield of the palm. Persistent high humidity is also harmful and bud rot is more common under such conditions. The palm requires bright sunshine of 2000 hrs a year.

Soil

The coconut palm adapts itself remarkably well and thrives in almost all types of well drained tropical soils, such as coastal sand, red loam, laterite, alluvial and reclaimed soil of the marshy lowland. A water table that is too high and remains static for too long is harmful.

Planting material

Since coconut is a cross pollinated perennial crop, the selection of seeds is of vital importance. Selection has to be made at the mother palm level and at the seedling stage. The mother palms should be healthy, high yielding and regular in bearing. Also factors such as proportion and distribution of female flowers, high setting percentage and high copra content should also be taken into consideration. Nine month old seedlings having a minimum of four leaves can also be planted.

Preparation of land

The clearing and levelling of land have to be done before preparing the pits for transplanting seedlings. The depth of the pit depends on the soil type. In the sandy loam soil with low water-table, the planting pits of the size 1 m x 1 m x1 m is recommended. In laterite soils with a rocky substratum, deeper and wider pits 1.2 m x 1.2 m x 1.2 m are necessary. At the time of planting the pits have to be filled up with loose soil leaving a final depth of about 60 cm. However, when the water table is high planting on the surface or on soil mounds may be necessary. In a littoral sandy soil, circular pits 1.5 m x 1.5 m x 1.5 m are made and partially filled with alternate layers of red earth and coconut husks (150 husks). A spacing of 7.5 to 9 m accommodating 177 to 124 palms per hectare is adopted under the square system of planting. The triangular method accommodates 20 to 25 more palms per hectare. In the hedge system of planting, a spacing of 5 to 5.5 m along the row is given.

Transplanting and after care

In well drained soils, seedlings can be transplanted at the beginning of the south-westerly monsoon. If facilities for irrigation are available, it is advisable to take up planting at least a month before the onset of the monsoon. In the low lying areas, seedlings are better transplanted after the rainy season. The transplanted seedlings should be shaded and irrigated properly during the summer.

Care of adult palms

The palms generally start bearing at the age of five or seven years after planting and the stabilized yield is obtained from about the tenth year onwards for 50 to 60 years. Regular inter cultivation and manuring cf the palm are essential for stepping up and maintaining the productivity of palms at a high level. Tillage including digging, ploughing, piling of mounds, levelling of mounds, etc. and the making of shallow basins each with a radius of 2 m, at the beginning of the monsoon and filling them up at the close of the monsoon are beneficial.

Nutrient requirement

The recommended dose of fertilizers is 500 g of N, 300 g of P_2O_5 and 1,200 g of K_2O per palm per year for the ordinary tall variety. For soil which are poor in organic matter, the application of green manure or compost at 50 kg per palm is recommended. Green-manuring crops usually grown are cowpea, sunnhemp. Fertilizers may be applied in two spilt doses. After the receipt of summer showers, one third of the dose of fertilizers is spread around the palm within a radius of 1.8 m and raked in. Circular basins 1.8 m in radius and 25 cm in depth may be dug in July-August and the green leaf manure or compost applied and partially covered. The remaining two-third dose of the fertilizers may be spread in the first week of September and the basins closed. Under irrigated conditions and in areas with assured supply of moisture throughout the year, the annual dose of fertilizers may be applied every three months in equal split doses.

Irrigation

In regions with a long spell of dry weather, there is a set back in the growth and yield of the coconut palm. An increase of 50% to 100% in the yield of nuts can be obtained by giving summer irrigation. The quantity and interval between irrigations depend upon the soil and the climatic conditions. In loamy soils copious irrigation, once a week is sufficient.

Harvest and yield

The coconut palm is unique in that once it attains the normal bearing stage, it continues to bear a bunch of nuts in every leaf axil almost at monthly intervals all year round and throughout its life extended over 50 to 60 years. The nuts mature nearly one year after fertilization. Generally, harvesting is done once in 45 to 60 days. Tender nuts are in demand as a delicious drink in West Bengal and Maharashtra are best harvested at the age of 6-7 months. Large scale harvesting of the tender nuts is not being done in other parts of the country. Nuts for culinary purposes, for making cup copra and good quality fibre are harvested at the age of about 11 months whereas those for ball copra and coarse coir are harvested only when they are fully ripe.

The average yield per hectare varies from 10,000 to 14,000 nuts per annum. From a well maintained garden an annual yield of 25,000 nuts per hectare can be obtained. Nearly one-third of the annual yield is harvested during the three months i.e. March, April and May as per conditions prevailing on the west coast.

The main product obtained from the coconut is copra or the dried kernel of the nut. The cup copra is prepared by splitting the husked nut into two halves and drying them in the sun. After drying for a day or two the kernel gets detached from the shell. The drying is continued for another four to five days. Ball copra is prepared by storing fully ripe nuts for 8-12 months on a raised platform in a shed.

3. ARECANUT (*Areca catechu* Linn)

Arecanut or betelnut is an extensively cultivated tropical palm, the nuts of which form a popular masticatory in India, the Middle East, and Far East. India produces annually 150,000 tonnes of arecanut from an area of 18,34,000 ha. It is a tall stemmed erect palm, reaching varied heights, depending upon the enviromental conditions. Palms attaining a height of 30 metres are not uncommon.

Climate

The altitude at which the arecanut palm can, be successfully grown varies to some extent according, to the latitude of the place. Though it grows at altitudes up to 1,000 m above the sea-level, at higher altitudes it is not at all productive. The cultivation is mostly confined to 28' North and South of the equator. It is unable to withstand extremes of temperatures and wide diurnal variations. The range of temperature at which it can flourish is from about 15° to 38°C.

Areca thrives well in a variety of soils, the laterite soils of the West Coast, the red loamy soils of the Mettupalayam (Tamil Nadu), the alluvial of Assam and West Bengal and the loans of Orissa are very suitable. The foremost factor that has to be considered in establishing an arecanut plantation is that the site selected should have adequate facilities for irrigation. The soil also should be deep and, well drained, without a high water table. Being, highly susceptible to a sun-scorch, the areca palms need adequate protection from exposure to the south-western sun.

Nursery

The selected seed nuts are sown soon after harvest, 5 to 6 cm apart, in beds of sand under partial shade, with their stalk-ends pointing upwards. Sand is spread over the nuts just to cover them. The nuts are irrigated daily. Germination starts about 40 days after sowing and the sprouts are ready for transplanting when they are about three months old. Nursery beds of 150 cm width and of convenient length are prepared for transplanting sprouts. The sprouts are to be transplanted at a spacing of 30 x 30 cm with the onset of the monsoon. A basal dose of well decomposed cattle manure at the rate of 5 tonnes per hectare may be applied. Partial shade can also be provided with an artificial pandal or by raising crops alike *Coccinia indica.* The seedlings are transplanted when they are 12 to 18 months old. Seedlings having maximum number of leaves and minimum height are selected for transplanting.

Preparation of land and transplanting

The land is prepared well by digging or repeated ploughing and is levelled and terraced, if necessary. Channels are to be provided if the palms are to be grown under irrigation. The drainage of soil is to be attended to. Pits of 90 cm dug at 2.7 m apart both ways are used for planting seeds. The seedling is to be planted in the centre of the pit covered with soil upto the collar. The soil is to be pressed on all sides. Planting is done usually in May June in well drained soils, but in clayey soils, subject to water logging, planting can be postponed till August-September to obtain a better establishing of plants. A shade crop like banana can be raised which gives good return. A proper alignment of the plantation will prevent scorching of the stem. In the square system of planting at a spacing of 2.7 x 2.7 m the north-south line should be deflected at an angle 35° to the west. The outermost row of plants on the southern and south-western side can be protected by growing tall and quick growing shade trees.

Manuring and interculture

The annual application of 100 g of N, 40 g of P_2O_5 and 140 g of K_2O in the form of fertilizers and 12 kg each of green manure and compost or cattle manure per bearing palm is recommended. The fertilizers may be applied in two split doses. For young palms a full dose of green manure and compost or cattle-dung may be applied from the first year of planting. The second dose of fertilizer can be applied to the base of each palm all around and mixed by a light forking. The soils are mostly acidic. The application of lime corrects this, increases availability of soil nutrients, reduces the phosphorus fixation and enhances the general condition of the soil. The application of lime has to be completed at least three weeks before manuring in September-October.

The palms may be irrigated once in 3-5 days. In souhtern Kerela where arecanut is mainly grown it has been found that manuring along with irrigation gives three times more yield than manuring alone. Adequate drainage should also be provided since the palms cannot withstand logging. The main cultural operations are performed close to monsoon generally in October November. Owing to the long pre-bearing age of this palm, intercropping with suitable crops, such as elephant foot-yam, banana, guinea grass or mixed cropping with cacao, pepper and betelvine can be taken up in areca gardens without any detriment to the yield of the main crop.

Harvesting

The pre-bearing age of the palm ranges from 5 to 8 years. Though inflorescence initiation has been observed in every leaf axis, there is absorption of infloroscences to about 5 per cent. The plant is monoecious, producing both male and female flowers on the same tree. The spedix of a grown-up palm produces on an average, 294 female flowers. The colour of the fruit during its growth changes from green to different shades of Yellow and red during ripening. In some places, tender cuts are harvested, whereas in others, both immature and mature nuts are harve-sted. Tender nuts are harvested from July to December and ripe nuts from December to March or from May to July. Three or four pluckings are, done during the whole season.

Plant produce and products

Arecanut is consumed both as a raw/ripe nut (adaka or Kacha tamul) as dried ripe nut (chali supari) and as semi-mature cut and processed varieties 'Batoldike" or 'Kalipak'. There are over 150 trade types, differing in maturity, processing conditions and varying in their taste characteristics as per market conditions prevailing at different centres of the country. The drying of the whole fruits for making chali supari requires up to 40 to 45 days of good sun shine, so as to get a moisture level of about 10 per cent. Drying ripe nuts on cement floors reduce fungal infection of the nuts to a minimum level of about 5 per cent. A drier designed recently has been found to be most suitable to produce good quality chali supari.

4. CASHEW (*Anacardium occidentale*)

The Cashew, *Anacardium occidentale* L. Its origin is considered to be South America (arid, northeastern Brazil). It is now distributed pantropically in hot, tropical lowlands below 1000 m.

Botanical description

Plant: Bushy, low-branched 3 to 12 m in height and width; lower limbs often bend to touch the ground. Propagation by seed, grafting, layering, cuttings; Fruit production in 4-5 yr from seed, 2-3 yr from vegetative propagation.

Flowers: are tiny, pinkish, borne terminally on panicles. Flowers can be male, female, or perfect on the same inflorescence.

Fruit: is a 1 inch nut, shaped like a small boxing glove, hanging below a fleshy, swollen peduncle (receptacle) called the "cashew apple". Fruit borne singly or in small clusters. Fruit matures in 60-90 days. Apple is 2-4.5 inches in length, pear-shaped, yellow or red skin covering the fibrous, juicy, astringent yellow pulp. Nut shell contains toxic oil which must be roasted to detoxify; can cause dermatitis. Sometimes, the apple is utilized and nut thrown away. In commercial plantations, apple and nut are allowed to drop together, and nut is twisted off, leaving the apple on ground for grazing livestock. Nuts can be roasted in fire like marshmallows, where they catch flame and burn-off oils.

Planting materials

Cashew can be propagated by seedlings, air layers and softwood grafts. Since it is a cross-pollinated crop, vegetative propagation is recommended to obtain true to type progeny. Field establishment of air layers have been found to be poor. Hence softwood grafts, which give a high rate of establishment and early flowering, are recommended for planting.

Selection of mother trees for seed nuts

Select mother trees having the following characteristics: (1) Good health, vigorous growth and intensive branching habit with panicles having high percentage of hermaphrodite flowers. (2) Trees of 15-25 years of age. (3) Bearing nuts of medium size and weight (5-8 g/nut) with an average yield of 15 kg nuts per annum. (4) Bearing 7-8 nuts per panicle.

Selection of nuts

Select mother trees in February and collect seed nuts in March-April. Select good, mature, medium sized nuts, which sink in water as seeds after drying in sun for two to three days.

Raising seedlings

Raise seedlings in polythene bags during May. Use polythene bags of size 20 cm × 15 cm and fill the bags with garden soil, leaving a gap of 1 to 1.5 cm above. Soak seed nuts in water for 18 to 24 hours to hasten germination. Sow the pre-soaked seed nuts in polythene bags filled with garden soil at a depth of 2-3 cm with the stalk end up. Seeds germinate in seven to ten days. Seedlings can be produced by air layering or grafting/ budding techniques.

Planting and management of plantation

Plant seedlings, air layers or softwood grafts in pits of size 50 x 50 cm during June-July. Planting may be done at a spacing of 7.5 m for poor and 10 m for rich and deep soils and sandy coastal area. On very sloppy lands, the rows may be spaced 10-15 m apart with a spacing of 6-8 m between trees in a row.

Initial training / shape pruning

The sprouts coming from the rootstock portion of the graft that is from the portion below the graft joint should be removed frequently during the first year of planting. Initial training and pruning of young cashew plants during the first 3-4 years is essential for providing proper shape. Thereafter, little or no pruning is necessary. The plants should be allowed to grow by maintaining a single stem up to 0.75-1.00 m from ground level. This can be achieved by removing the side shoots or side branches gradually as the plants start growing from the second year of planting. Weak and criss-cross branches can also be removed. Branches growing unwieldy may also be cut off. Proper staking of the plants is required to avoid lodging due to wind during the initial years of planting. Initial training and pruning of cashew plants facilitate easy cultural operations such as terrace making, weeding, fertilizer application, nut collection and plant protection. The flower panicles emerging from the grafts during the first and second year of planting should also be removed (deblossoming) in order to allow the plant to put up good vegetative growth. The plants are allowed to flower and fruit only from the third year onwards.

General pruning

In older cashew plantations, removal of dried or dead wood, criss-cross branches, water shoots etc. should be attended to at least once in 2-3 years. This allows proper growth of the canopy and receipt of adequate sunlight on all the branches. Pruning of cashew plants should be done during May / June.

Intercropping

Pineapple is the most profitable intercrop in cashew plantation in the early stages of growth. It can be planted between two rows of cashew in trenches opened across the slope. Paired row of pineapple suckers can be planted in each trench at 60 cm between rows and 40 cm between two suckers within the row. These trenches can be opened at 1 m between two rows of cashew. Ginger, lemongrass and tapioca are also suitable as intercrops.

High density planting

High density planting is a recent technique recommended for enhancing the productivity of cashew plantations. This technique involves planting more number of grafts per unit area and thinning at later stages. Instead of the normal planting density of 64 to 177 plants per hectare (spacing ranging from 7.5 to 10 m in the square system of planting) or 74 to 204 plants (spacing ranging from 7.5 to 10 m in the triangular system of planting), 312 to 625 grafts will be planted per hectare, initially. During later years, as the canopy develops, plant population is to be regulated by selective felling to minimize competition.

While adopting a high density planting technique, grafts may be planted initially at a spacing of 4 x 4 or 8 x 4 m so that there will be 625 or 312 plants respectively. This population can be retained for a period of seven to nine or ten years depending upon the canopy expansion rate. If the soil is very rich the canopy development rate will be faster. High density planting would be more useful in poor soils where the rate of canopy expansion is slow. Finally, when the plants attain full growth, the spacing between the plants will be 8 × 8 m.

Top working

Top working is a technique evolved to rejuvenate unproductive and senile cashew trees. Top working can successfully rejuvenate poor yielders in the age group of 5-20 years. The unproductive trees are to be beheaded at a height of 0.75 to 1.00 m from ground level. The stem should be cut with a saw to avoid stump splitting. The best season for beheading trees is May-September. Soon after beheading, the stumps and cut portions should be given a swabbing with copper oxychloride and carbaryl 50 WP (50 g each per litre of water). Sprouts emerge 30-45 days after beheading. Sprouting will be profuse in young trees. New, 20-25 days old shoots should be grafted with scions of high yielding varieties using softwood grafting technique. To ensure at least six or seven successful grafts, 10-15 grafting are to be done on the new shoots of every tree. The best season for grafting is July-November.

The top worked trees start yielding right from the second year after top working. Thus precocity can be considered as one of the best advantages of this technique. The major disadvantage associated with top working is the huge casualty of trees due to stem borer attack. Intensive care and management to ward off stem borer is essential. As such, adoption of top working on a larger scale would be difficult.

Nutrient management

A fertilizer dose of 750 g N, 325 g P_2O_5 and 750 g K_2O per plant is recommended for cashew. Apply 1/5th dose after the completion of first year, 2/5th dose during second year and thus reaching full dose from 5th year onwards. Broadcast the fertilizer within an area of 0.5 to 3.0 m (15 cm deep) around the tree and incorporate by light raking.

5. COCOA (*Theobroma cacao*)

Plant characteristics

The cocoa tree flourishes in the dense shade of warm rain forests in its natural habitat and hence can be cultivated in all similar climatic conditions. The tree cannot withstand high winds, drought or sudden fall in temperature. The crop requires well-distributed rainfall. The minimum requirement of rainfall is about 100-150 cm per annum. Situations where the temperature falls below 10°C or rises above 38°C are unfavourable although minor deviations from the above limit can be adjusted by shade and irrigation. High wind velocity causes considerable mechanical damage to trees.

Cocoa is grown at altitude up to 900 m above MSL though it is possible to grow the crop even in much higher elevations under sheltered conditions. The best soil for cocoa is forest soil rich in humus. The soil should allow easy penetration of roots and capable of retaining moisture during summer. Clay loams, loams and sandy loams are suitable. Shallow soils should be avoided. Cocoa can be propagated by seed and vegetative means.

Seed propagation

It is desirable to collect seeds from biclonal or polyclonal seed gardens involving superior self-incompatible parents to ensure genetic superiority of planting materials. Polyclonal and biclonal seed gardens have been established at CCRP farm of the Kerala Agricultural University, Vellanikkara and Kidu farm of CPCRI and seeds and seedlings are being supplied to growers.

Selection of planting materials

When seedlings are used for planting, select only vigorous and healthy seedlings produced from polyclonal seed garden or selected mother plants as described earlier. When budded plants are used, select two or more clones for planting as the use of a single clone can lead to poor production due to the existence of self-incompatibility in cocoa.

Selection of site

Cocoa is usually planted under coconut and arecanut plantations in India. Shade levels under coconut canopy are highly variable depending mainly on the spacing of coconut, extent of canopy development and age of palms. It is estimated that light infiltration through coconut canopy ranges from about 30 to 80 per cent depending upon these factors. Based on this, the general recommendation is as follows:

1. If a choice is possible, a coconut plantation that will let in more light through the canopy may be chosen for raising cocoa.
2. If the light infiltration is over 50 per cent, it may be beneficial to provide additional shade using temporary shade plants like banana.

Time of sowing

Though the seeds will germinate at any time of the year, seeds may preferably be sown by December-January, so that 4-6 month old seedlings become available for planting by May-June.

Method of sowing

Seeds are to be sown with hilum-end down or to be sown flat. Sowing is to be as shallow as to just cover the seeds with soil. Removal of pulp may enhance the speed of germination, but the extent of additional advantage is only marginal. Seeds start germination in about a week and germination may continue for another one week. Percentage of germination may be around 90. Cocoa nursery is to be located in a heavily shaded area, which allows only 25-50 per cent sunlight. Regular watering is necessary to keep the soil moist. Seedlings are transplanted after 4-6 months. Only vigorous seedlings are to be used and based on height and stem girth, 25% poor seedlings may be rejected. When seedlings are grown under heavy shade, hardening for 10 days by exposing to higher illumination may be necessary before transplanting.

Planting

The seedlings / budded clones are usually planted in the interspaces of coconut / arecanut. Give a spacing of 3 to 4.5 m. The crop is best grown with 50 per cent light intensity in the early stages. In the early life of the plants, planting of quick growing plants like banana and tapioca can provide temporary shade.

Time and method of planting

Raising cocoa as a pure crop is not recommended especially in Kerala due to high pressure on land. Cocoa is planted as an intercrop in coconut and arecanut gardens. In coconut, depending upon the spacing adopted, one or two rows of cocoa can be planted in between two rows of coconut i.e., two rows where the spacing is more than 8 m and one row otherwise, the plant distance for cocoa being 2.7 to 3 m. When two-row system is adopted, the seedlings may be planted in zigzag or triangular manner. In arecanut where the normal spacing is 2.7 m, cocoa is planted at the centre of four areca palms along alternate rows of interspaces only. Pits of 50 x 50 x 50 cm are dug, allowed to weather for one month and refilled with topsoil and 15-20 kg of compost or farm yard manure to ground level. The planting hole should be sufficient to hold the soil ball of the polybag. Tear off the polybags carefully, place the soil ball with the seedlings in the planting hole with minimum disturbance and press the soil around firmly. Planting should coincide with the onset of monsoon, but in places where irrigation is resorted to, flexibility in the time of planting is possible.

Pruning and training

Cocoa grows in a series of storeys, the chupon or vertical growth of the seedling terminating at the jorquette from where four to five fan branches develop. Further vertical growth is continued through a side chupon that arises from a point just below the jorquette which again jorquettes after growing to some height. Left for it, the plant will grow to a height of 8-10 m repeating this process of jorquetting and chupon formation 3-5 times. When cocoa is grown as an intercrop in coconut and arecanut plantation, it is desirable to restrict the growth to one tier formed at a convenient height preferably above the head level of the workers. When jorquetting takes place at lower levels this can be raised by nipping off all the fan branches and allowing one chupon to develop and grow further to jorquette at the desired height. After this is achieved, further vertical growth is arrested through periodical removal of chupons.

Top working

This technique is useful to rejuvenate old and unproductive cocoa plants and also to convert genetically poor yielders to high yielders. This consists of snapping back the desired trees below the jorquette after cutting half way. The snapped canopy continues to have contact with the trunk. A number of chupons would arise below the point of snapping and this is triggered by the breakage of apical dominance and continued connection with the snapped canopy. Patch budding as described earlier may be done on three to four vigorous and healthy shoots using scions from high yielding, disease resistant clones and the remaining chupons are removed. The polythene tape is removed three weeks after budding and the stock portion above the bud union is snapped back. The snapped portion is removed after two hardened leaves develop from the bud. When sufficient shoots are hardened, canopy of the mother tree can be completely removed. Because of the presence of an established root system and the trunk with reserve food, the top worked trees grow much faster and give prolific yield one year after the operation. Though top working can be done in all seasons, it is preferable to do it in rain-free period in irrigated gardens. For rainfed situations, it may preferably be done after the receipt of pre-monsoon showers.

Top worked trees start yielding heavily from the second year onwards. About 50 per cent improved yield is obtained in the second year and about 100 per cent improved yield in the third year. Loss of crop for one year during the operation is compensated by bumper crop in the coming years. The main stem will continue to belong to the older plant and fruits borne on this area belong to the poor yielder.

Irrigation

Cocoa grows well as a rainfed crop under conditions of well distributed rainfall and irrigation is not necessary. If sufficient moisture is not present in the soil due to prolonged drought or failure of rains, irrigation is to be given once in five days. Irrigation, however, helps in better growth of plants and precocity in bearing.

Manuring

Apply N: P_2O_5: K_2O in two equal split doses in April-May and September-October, @ 100:40: 140 g / tree / year. N: P_2O5: K_2O may be applied @ 200:80:280 g / plant / year, in trees yielding more than 50 fruits per year. Dolomite @ 100 g / plant / year may be applied to plants from the third year onwards. Under irrigated conditions, the yearly dose may be split into four and applied during April-May, September-October, December and February-March. Apply 1/3 of adult dose during the first year of planting, 2/3 during second year and full dose from the third year onwards. Apply fertilizers in circular basins with a radius of 25 cm during the first year. Gradually increase the radius of the basin to 120 cm by the third year. Apply fertilizers in the entire area of 1.5 m radius around the tree followed by forking in. Plants showing zinc deficiency symptoms (narrowing of leaves, sickle leaf formation, green vein banding, chlorosis in the interveinal areas) should be sprayed with 0.5 to 1.5% ZnSO4 three times a year.

After cultivation

During the first three or four years after planting, it is essential to keep the field free from weeds. Maintenance and regulation of shade should be carried out promptly. During the establishment phase of the crop particularly in summer, provide mulching with materials like chopped banana sheath, coconut husk, cocoa husk etc. to conserve moisture in conditions of direct insolation. A mature cocoa plantation should form a proper canopy, which will be dense enough to prevent weed growth. Operations such as pruning and regulation of shade should be attended to in time.

Harvesting

It takes about 170 days for a cocoa pod to develop from formation to maturity. During the period from 70-140 days after pollination, the size of pods and their fat and sugar content increase rapidly. Ripening takes about 25 days, during which, the pods change colour depending on the variety. Pods remain suitable for harvesting for fairly long time after they have ripened. Hence, it is possible to have harvest of sufficient number of pods at a time by either delaying the harvest of early ripened pods or harvest of pods, which are fully ripened. Harvesting should be done at regular intervals rather than daily, once in 7-10 days. Avoid over ripening of pods. The discards at the harvest can be left in the garden either in the open during summer or in pits at different sites in the rainy season, or they can be incorporated in the compost. Pod husks from the fermentary can also be used similarly as a good source of organic manure.

Pods are removed by cutting with a sickle sharp knife, without damaging the cushion from which it is developed. After 2-3 days, they are split by banging them against some hard objects. Opening the pods with a knife damages the beans. During the period between harvesting and splitting, pre-fermentation activity inside the pod is hastened, which improves later fermentation. Beans from the split pods are scraped out with fingers. Portions of placenta, and broken, germinated, caked, parthenocarpic and undeveloped beans are removed. On an average, 10-12 pods give 1 kg of wet beans and 3 kg of wet beans (from 30-36 pods) give 1 kg of fermented and dried beans.

Lowering the vines

Under normal conditions, vines grow to a height of about 3 m in one-year time. When they reach this height, their vigour to produce normal sized leaves is reduced and the crop needs rejuvenation. This is done by lowering the vines down to the ground level at least once a year. Lowering is done during the months of August-September. Before lowering, all the leaves in the basal portion of the vines to a height of 15 cm are removed. Vine is untied from bottom upwards and coiled up carefully and laid flat on ground leaving 2.5 to 5 cm length of top shoots. Soil is put over the portion kept in the soil to about 5 cm thickness. Lowering is followed by light irrigation and manuring.

17

DRY FARMING AGRICULTURE

INTRODUCTION TO DRYLAND AGRICULTURE

Agriculture is the single largest livelihood sources in India with nearly two thirds of people depend on it. Rainfed agriculture is as old as agriculture itself. Growing of crops entirely under rainfed conditions is known as dryland agriculture. Depending on the amount of rainfall received, dryland agriculture can be grouped into three categories:

(a) **Dry farming**: is cultivation of crops in regions with annual rainfall less than 750 mm. Crop failure is most common due to prolonged dry spells during the crop period. Dry farming is performed in arid regions with a growing season (period of adequate soil moisture) less than 75 days. Moisture conserv-ation practices are necessary for crop production.

(b) **Dryland farming**: is cultivation of crops in regions with annual rainfall more than 750 mm. In spite of prolonged dry spells, crop failure is relatively less frequent. These are semi arid tracts with a growing period between 75 and 120 days. Moisture conservation practices are necessary for crop produc-tion. However, adequate drainage is required especially for vertisols or black soils.

(c) **Rainfed farming**: is crop production in regions with annual rainfall more than 1150 mm. Crops are not subjected to soil moisture stress during the crop period. Emphasis is often on disposal of excess water. Rainfed farming is adopted in humid regions with growing period more than 120 days.

Dryland vs Rainfed farming

Constituents	Dryland farming	Rainfed farming
Rainfall (mm)	750 – 1150 mm	>1150 mm
Moisture	Shortage	Enough / Sufficient
Growing season	75-120 days	>120 days
Growing regions	Arid and Semiarid & up lands of sub humid & humid regions	Humid and sub humid regions.
Cropping system	Single crop or intercropping	Intercropping or double cropping.
Constraints	Wind and water erosion	Water erosion.

Dryland agriculture in India

In 1969, the then Prime Minister Mrs. Indira Gandhi with strong aim for eradicating poverty evinced keen interest in dryland research and development programs. The dryland research project, originally proposed in 1967, was finally approved in 1969. All India Coordinated Research Project for Dryland Agriculture (AICRPDA) was launched in 1970 by the Indian Council of Agricultural Research (ICAR) in IV Plan period in collaboration with the Government of Canada through Canadian International Development Agency (CIDA) with Co-ordinating Cell at Hyderabad, Andhra Pradesh. The project has seventeen main centers (one Chief Scientist, four Scientists and others totaling twenty two staff), two sub-centers (four scientists and others totaling fourteen staff) and eight operational research projects (three scientists and others totaling seven staff). There are three other centres in the institutes of the council. Apart, there is a co-ordinating cell. The total strength of project staff is more than 550 including more than 130 scientists. These centers are identified based upon moisture index.

The main centers were Hisar, Jodhpur, Bellary, Rajkot and Ananta- pur in moisture index 60-80%; Solapur, Akola, Kovilpatti, Hyderabad, and Varanasi in moisture index 40-60% and Bangalore and Indore in moisture index 20-40%. The sub-centers were Udaipur (shifted to Arjia in 1985), Jhansi, Anand and Agra in moisture index 40-60%; Bijapur, Rewa, Ludhiana and Samba (moisture index 20-40%). The special problem areas were Bhubaneswar, Ranchi and Dehradun. Dehradun was discontinued in April 1985. Later on Anand was shifted to Sardar Krishinagar and Ludhiana to Ballowal-Saunkhri Hoshiarpur district to and Samba to Rakh Dhiansar. Bijapur and Rewa were upgraded as main centers. One additional special centre was located at the Indian Agricultural Research Institute (IARI), New Delhi, which was merged, with the Coordinating Cell of the Project later. Faizabad was added as one of the centers in 1985. Bhubaneswar center was shifted to Phulbani in 1996.

At present, the project has seventeen centers in State Agricultural Universities, two in technical/other Universities and three in ICAR institutes. These are Akola, Arjia, Anantapur, Bangalore, Bijapur, Hisar, Ballowal-Saunkhri, Indore, Kovilpatti, Phulbani, Ranchi, Rajkot, Rewa, Sardar Krishinagar, Solapur, Varanasi, Agra, Faizabad, Jammu, Bellary, Jhansi and Jodhpur, Eight of the centers have Operational Research Projects in villages and these are Anantapur, Arjia, Bangalore, Hisar, Ballowal-Saunkhri, Indore, Ranchi and solapur.

SIGNIFICANCE OF DRYLAND FARMING IN INDIAN ARICULTURE

Dryland agriculture is one, which is extreme to irrigated agriculture in terms of soil moisture availability with more than 60 per cent risk in crop production.

In India both dryland and rain fed agriculture are being practiced with 30 to 40 per cent productivity in dryland as compared to rain fed agriculture. Rain fed agriculture is akin to irrigated agriculture in all activities except the rainfall which is the main source of moisture supply and that is available without any moisture stress to crops. Scientifically the available rainfall in a cropping season meets crop evapo transpiration fully. But in dryland, rainfall meets only 30 to 50 per cent of crops evapo transpiration.

Out of net sown area of 141 m ha in India, net irrigated area is 65.3 m ha. The gross cropped area in India is 195 m/ha. Though the area differs with States, dryland problems are universal.

I. About 70% of rural population lives in dry farming areas and their livelihood depend on success or failure of the crops

II. Dryland agriculture plays a distinct role in Indian agriculture occupying 54% of cultivated area and supports 40% of human population and 60% livestock population.

III. The contribution (production) of rainfed agriculture in India is about 42 per cent of the total food grain, 75 per cent of oilseeds, 90 per cent of pulses and about 70 per cent of cotton.

IV. By the end of the 20th century the contribution of drylands will have to be 60 per cent if India is to provide adequate food to 1000 million people. Hence tremendous efforts both in the development and research fronts are essential to achieve this target.

V. More than 90 per cent of the area under sorghum, groundnut, and pulses is rainfed. In case of maize and chickpea, 82 to 85 per cent area is rainfed. Even 78 percent of cotton area is rainfed. In case of rapeseed/mustard, about 65.8 per cent of the area is rainfed. Interestingly, but not surprisingly, 61.7, 44.0, and 35.0 per cent area under rice, barley and wheat, respectively, is rainfed.

VI. At present, 3 ha of dryland crop produce of cereal grain is equivalent to that of one ha irrigated crop. With limited scope for increasing the area under plough, only option left is to increase the productivity with the modern technology and inputs, since the per capita land availability which was 0.19 ha in 1990 has declined to 0.13 ha in 2010 and is coming down year after year due to increase in population.

VII. The productivity of grains already showed a plateau in irrigated agriculture due to problems related to nutrient exhaustion, salinity build up and raising water table. Therefore, the challenges of the present millennium would be to produce more from drylands while ensuring conservation of existing resources. Hence, new strategies would have to be evolved which would make the fragile dryland ecosystems more productive as well as sustainable. In order to achieve evergreen revolution, we shall have to make grey areas (drylands) as green through latest technological innovations.

VIII. Drylands offer good scope for development of agroforestry, social forestry, horti-sylvi-pasture and such other similar systems which will not only supply food, fuel to the village people and fodder to the cattle but forms a suitable vegetative cover for ecological maintenance.

Featurs of Indian dryland agriculture

- Uncertain, ill-.distributed and limited annual rainfall;
- Occurrence of extensive climatic hazards like drought, flood etc;

- Undulating soil surface;
- Occurrence of extensive and large holdings;
- Practice of extensive agriculture i.e. prevalence of mono-cropping etc;
- Relatively large size of fields;
- Similarity in types of crops raised by almost all the farmers of a particular region;
- Very low crop yield;
- Poor market facility for the produce;
- Poor economy of the farmers; and
- Poor health of cattle as well as farmers.

Dimension of dryland agriculture

Dryland farming is an agricultural technique for non-irrigated cultivation of land. It may be defined as: "a practice of growing profitable crops without irrigation in areas which receive an annual rainfall of 500 mm or even less". India has about 75.7 million hectares of rain fed area. Rain fed agriculture encounters several constraints on account of climatic, edaphic, and social factors. The productivity levels of these areas have remained lower across years because of frequent droughts occurring due to high variability in the quantum and distribution of rainfall, poor soil, low fertilizer use, imbalanced fertilization, small farm size and poor mechanization, poor socio-economic conditions and low risk-bearing capacity, low credit availability and infrastructure constraints. Consequently, farmers are distracted from agriculture and tend to migrate to cities to look for alternative jobs. Hence, there is a great need to increase the productivity of rain fed crops and overall net returns to keep the farmers in agriculture. A paradigm shift in rain fed agriculture can be expected through technological thrusts and policy changes.

Scientific study of dryland farming was initiated by the Government of India in 1923. Early research focused on improving crop yields. Important practices included: (i) bunding to conserve soil and water; (ii) deep ploughing once in three years for better intake and storage of water; (iii) use of farmyard manure to supply plant nutrients; (iv) use of a low seeding rate; and (v) intercultivation for weed and evaporation control. These practices gave a 15–20 percent increase over the base yields. By the mid-1950s, the emphasis had shifted to soil management. Soil conser-vation research and training centres were established at eight locations, focusing on contour bunding. However, negative results were often obtained because of water accumulation and runoff problems, particularly on Vertisols. Even where yield increases were observed, they were again not more than 15–20 percent above the base yields.

The importance of shorter-duration crops to match the soil-water availability period was recognized in the 1960s. It was also in the mid-1960s that high-yielding hybrids and cultivars became available that were responsive not only to fertilizers but also to management. An All-India Coordinated Research Project for Dryland Agriculture was established, and the research emphasis shifted to a multidisciplinary approach to tackle the problems. Similar efforts were initiated at the International Crops Research Institute for the Semi-Arid Tropics at Hyderabad in 1972.

Area under dryland in India

In India out of the net cultivated area of 141 mha, the area under drylands is about 54% covering an area of 75.7 m ha. Except in the States of Punjab, Haryana and Pondichery the percentage of area under drylands is high in all other States.

Dimensions of the problems in dryland agriculture

Majority of the districts in India are dry farming districts and covers 60 per cent of the total cultivated area. Most of this area is covered by crops like millets, pulses, oilseeds, cotton etc.,.These areas spread throughout the country i.e. Tamilnadu, Karnataka, Andhra Pradesh, Madhya Pradesh, Maharastra, Gujarat, Rajasthan, Punjab, Haryana and Uttar Pradesh. In South India, the Deccan Plateau which is rain shadow area consisting of parts of Karnataka, (Bellary, Raichur, Kolar, Tumkur, Dharwad, Belgam, Gulberga) and Maharastra (Sholapur, Parbani, Puna, Aurangabad). The dry farming areas in Andhra Pradesh are found in Kurnool, Anantapur, Kadapa, Mahaboobnagar, Chittoor, and Nalgonda districts.

(a) The area under dryland agriculture is more in India (54 per cent of total cultivable area)

(b) Areas of low rainfall (below 750 mm) constitute more than 30 per cent of total geographical area

(c) About 84 districts in India fall in the category of low rainfall area

(d) Providing irrigation to all the drylands is expensive and takes long time

(e) Even after providing all the irrigation potential in India 55 per cent area remains as rainfed.

Dimension of dryland technology development

Though India has developed many dryland farming practices during British period (Bombay dryland farming), with the initiation of coordinated dryland project during 1971 many more location specific technologies have been developed and tested for their sustainability. The developed technologies have been grouped and listed below

I. Suitability of crops and varieties

II. Delineation of efficient cropping zone

III. Times of sowing with the onset of monsoon rain

IV. Sowing techniques

V. Seed hardening

VI. Soil and moisture conservation

VII. Integrated nutrient, weed and pest management practices

VIII. Contingency planning

Through such technologies are available, farmers seldom adopt one or two technologies. Many valid reasons may be quoted for such non-adoption and the only main reason is non-attachment of farmers to these developed technologies because of less confidence developed by the farmer and larger existing risk probability in dryland agriculture. Human touch-techonlgy training could change this situation positively.

Present dimensions of Indian dryland agriculture

Out of 141 million ha of net sown area in the country, irrigated area is 65.3 mha and rest is rainfed. The average annual rainfall of the country is 1200 mm amounting to 400 million ha meter of rain water over the country's geographical area (329 m ha). However, the distribution across the country varies from less than 100 mm in extreme arid areas of Western Rajasthan to greater than 3600 mm in NE States and 1100 mm from east coast 2500-3000 mm in the west coast. The broad area of the summer monsoon activity extends between 30° N to 30°S and from 30^0 W to 16.5^0 E. the detail information on rain fall and monsoonal pattern in India has been summarized as below:

Rainfall pattern in India

Season/Period	million ha meter	Percent
Winter (Jan-Feb)	12	3
Pre-monsoon (Mar-May)	52	13
South-west monsoon (Jun-Sept)	296	74
North-east monsoon (Oct-Dec)	40	10
Total for the year	400	100

Rainfed farming comprises about 91% area of coarse cereals (sorghum, pearl millet, maize and finger millet), 91% pulses (chickpea and pigeon pea), 80% of oilseeds (groundnut, rape seed, mustard and soybeen), and 65% of cotton. Also, about 50% area under rice and 19% area under wheat is rainfed.

SWOT analysis of dryland agriculture

When the tool SWOT was applied to Indian dryland agriculture, the results are as follows

Strength

- Wider agro ecological and edaphic-climatic conditions
- Many options for technological interventions
- Larger biodiversity
- Improved transportation and communication
- Spread of NGO's

Weakness

- Poverty and illiteracy
- Higher amount of risk for input application
- Poor participation of farmers in dryland agriculture
- Declining natural resources
- Irregular monsoon performance

Opportunities

- Political will for food production
- More dryland research institutes and schemes
- National level dryland coordinated project
- Availability of GIS and RS
- National program on poverty alleviation
- State and National level plan on wasteland and water shed development
- Abundance availability of indigenous knowledge

Threats

- Risk in crop production
- Lack of coordination between research, input, extension and weather forecast organizations.
- Village level political problem
- Land fragmentation- barrier to farm mechanization and technology adoption
- Rapid industrialization.

Suggested remedies

- Bunding to conserve soil and water;
- Deep ploughing once in three years for better intake and storage of water;
- Use of farmyard manure to supply plant nutrients;
- Use of a low seeding rate; and
- Intercultivation for weed and evaporation control.

These practices gave a 15–20 percent increase over the base yields. By the mid-1950s, the emphasis had shifted to soil management. Soil conservation research and training centres were established at eight locations, focusing on contour bunding. However, negative results were often obtained because of water accumulation and runoff problems, particularly on Vertisols. Even where yield increases were observed, they were again not more than 15–20 percent above the base yields.

The importance of shorter-duration crops to match the soil-water availability period was recognized in the 1960s. It was also in the mid-1960s that high-yielding hybrids and cultivars became available that were responsive not only to fertilizers but also to management. An All-India Coordinated Research Project for Dryland Agriculture was established, and the research emphasis shifted to a multidisciplinary approach to tackle the problems. Similar efforts were initiated at the International Crops Research Institute for the Semi-Arid Tropics at Hyderabad in 1972.

Improving soil moisture storage

Soil moisture storage can be improved either by reducing transpiration or by reducing ET losses.

(I) Methods for decreasing transpiration: Transpiration can be reduced in the field in 4 ways.

- Increasing leaf resistance to water vapour loss by applying materials that tend to close or cover the stomata.
- By reducing net energy uptake by plants by increasing leaf reflectance (Albedo).
- Reducing top growth by growth retardants.
- By increasing air resistance to water vapour transfer by using wind breaks. Stomatal resistance slows down simultaneously transpiration by CO_2 assimilation. It have been observed that best anti-transpirant reduce transpiration upto 30-40% at the most.

(II) Anti-transpirant: Any material applied to transpiring plant surface with aim of reducing water loss from plant.

The objective of anti-transpirant is to

- Reduce rate of transpiration.
- Have low or little effect on photosynthesis.
- Induce plant to offer more resistance to loss of water vapour through stomata but relatively lower resistance to CO_2 entry.

Roles of anti-transpirants

(I) Induced stomatal closure: Spray material used for various purpose such as herbicides, fungicides, metabolic inhibitors and growth hormone have been found to cause closure of "Guard Cells" of stomata and thereby reduce transpiration. The main practical problem is to obtain material that induce stomatal closure without toxic effect and that are long lasting.

However, this type of material has not been available. Phenyl mercuric acetate (PMA) is a fungicide, is the most promising material which causes the stomata to close. However, it has direct inhibition of mesophyll photosynthesis.

(II) Film forming anti-transpirants: Plastic and waxy material which reduce both transpiration and photosynthesis. These are more permeable to water vapour than CO_2. The desirable characteristic for film type anti-transpirant should be

- Selective gas permeability (permeable to CO_2).
- It is easy to apply.

- It should be inexpensive and have no phytotoxicity.
- Ability to form a continuous film on leaf surface.
- To resist dehydration.
- To retain elasticity.

(III) Increasing leaf reflectance: Reflecting materials reduce energy load on leaves by increasing Albedo. These also reduce energy load on leaves by increasing Albedo these also reduce leaf temperature. For example, Kaoline and white wash and found their utility as leaf reflectant and it has been seen the spray of Kaolinite reduce transpiration losses by 20-28 per cent.

(IV) Growth retardants: With use of growth retartdants, there is increase in root growth which result in considerable delay in senescence of leaves because as a result of increased root length the plants are able to withdraw water from deeper layer of soil. Reduced top growth with use of growth retardant helps in survival of plants due to reduce transpiration under field conditions.

(V) Wind breaks: The wind breaks have found their utility in reducing transpiration losses under field conditions. It has been seen that yield of grain crops increases by 20-30% and those of grasses by 100-200% in sheltered zone. The advantages of shelter belts were more marked in years of drought because utilization of limited water available for transpiration was more efficient by shelter crops than unshelter one.

Methods of reducing evaporation under field conditions

Mulch is any material applied on soil surface to check evaporation and improve soil water. Application of mulches result in additional benefits like soil conservation, moderation of temperature, reduction in soil salinity, weed control and improvement of soil structure. Evaporation can be reduced by the use of mulches which are as follows:

(i) **Soil mulch or Dust mulch:** If the surface of the soil is loosened, it acts as mulch for reducing evaporation.

(ii) **Stubble mulch:** Crop residue like wheat straw or cotton stalks etc. left on the soil surface as stubble mulch. Stubble mulch reduces the erosion and evaporation losses.

(iii) **Straw mulch:** Straw is used as mulch to reduce the erosion and evaporation losses.

(iv) **Plastic mulch:** Plastic materials like polyethylene, polyvinyl chloride are also used as mulching materials.

EFFECTIVE RAINFALL

It would be useful to clarify the term water requirements of a crop before effective rainfall is fully defined. Water requirement of a crop is the amount of water needed for raising it and this includes water to meet both consumptive and special needs, such as land preparation, land submergence, leaching and so on. In view of this current concept, it follows that from the production point of view, the annual or seasonal effective rainfall as far as the water requirement of crops is concerned, should be interpreted as that portion of total annual or seasonal rainfall which is useful directly and/or indirectly for crop production at the site where it falls but without pumping. It therefore includes water intercepted by living or dry vegetation (B), that lost by evaporation from the soil surface (C3), the precipitation lost by evapotranspiration during growth (C2.1.2), that fraction which contributes to leaching, percolation (C2.2.1) or facilitates other cultural operations either before or after sowing without any harm to yield and quality of the principal crops.

Effective rainfall, (ER) = B + C2.1.2 + C2.2.1 + C3.

Consequently ineffective rainfall is that portion which is lost by surface run-off (C1), unnecessary deep percolation losses (C2.2.2), the moisture remaining in the soil after the harvest of the crop (C2.1.3) and which is not useful for next season's crop.

According to the Soil Conservation Service of U.S.D.A. (1967), effective rainfall is that which is received during the growing period of a crop and is available to meet consumptive water requirements. It does not include surface run-off or deep percolation losses.

Simply, the effectiveness of precipitation or rainfall efficiency has been defined as the extent of its utility in a given region. A rainfall of 100 mm in a temperate zone is much more useful and productive for agriculture than in a tropical arid region.

Values of effectiveness of rainfall vary from plus to minus according to its timing with respect to crop growth stage. Untimely rains can directly damage field crops in several ways:

- delaying sowing, interculturing, threshing or drying operations in the field;
- causing flower drop and fruit drop in certain cultivated species;
- causing lodging of plants;
- lowering the quality of the produce.

Factors influencing effective rainfall

Factor	Relevant characteristics
Rainfall	Amount, intensity, frequency, distribution over area as well as time;
Meteorological parameters	Temperature, radiation, relative humidity, wind velocity;
Land	Topography, slope, type of use;
Soil	Depth, texture, structure, bulk density, salt and organic matter content;
Soil water	Head, suspended matter, turbidity due to clay or colloids, viscosity, temperature, nature of dissolved salts (Na^+, $N03^-$);

Factor	Relevant characteristics
Groundwater	Depth from surface, quality;
Management	Type of tillage, degree of levelling, type of layout (bunding, terracing, ridging), use of soil conditioners;
Channel	Size, slope, shape, roughness and back water effect;
Crops	Nature of crops, depth of root system, degree of ground cover, stage of growth, crop rotations,

Effective rainfall depends upon following factors:

1. Rainfall characteristics

A soil has a definite and limited water intake rate and moisture holding capacity. Hence greater quantities as well as intensities of rainfall normally reduce the effective fraction, increasing run-off and lessening infiltration. Similarly, uneven distribution decreases the extent of effective rainfall while an even spread enhances it. A well distributed rainfall in frequent light showers is more conducive to crop growth than heavy downpours. For example, annual rainfall is lower than 100 mm in the Middle Eastern desert countries, so it may all become effective. In countries like India and Pakistan, intensity, frequency and amount are high during July and August and hence the effective fraction is very low. From November to April, however, most of the rainfall is effective in these countries due to its low intensity, frequency and amount.

2. Other meteorological parameters

Potential evapotranspiration is primarily governed by evaporative demand under conditions of abundant water supply. An approximation of evaporative demand can be obtained from the integrated effect of four parameters: temperature, radiation, wind velocity and humidity. Increase in the first three and decrease in the fourth parameter enhance evaporation. Such conditions encourage greater deficits of moisture in the soil and therefore the proportion of effective rainfall increases. The mean monthly values of temperature, radiation, wind velocity and humidity fluctuate less from year to year than total rainfall. Today in some countries, weekly or fortnightly maps of potential evapotranspiration are available which can be used in assessing effective rainfall.

3. Land characteristics

The time interval between receipt of rain water and its recessation by soakage is known as 'opportunity time'. Water stays longer on flat and levelled land and thus has a longer opportunity time than on sloping land where there is a rapid run-off. Sloping, rolling and undulating lands thus influence opportunity time of rain water for uniform infiltration and consequently the effective rainfall fraction. The use to which land is put in the surrounding area - agriculture, road and building construction, play-grounds - also affects the amount of effective rainfall.

4. Soil characteristics

Soil is an important medium between water and plants, acting as a reservoir for the moisture supply to crops. Hence its properties of absorption, retention, release and movement of water influence the degree of effective rainfall.

Intake and water movement in soil are expressed in terms of infiltration rate and permeability. For maximum absorption of rain and reduction of surface run-off, the values of these properties should be as high as possible. Permeability depends upon the texture, structure and compactness or bulk density of the soil. The higher the bulk density, the lower the permeability.

The fraction of effective rainfall increases with increased water holding capacity in a soil. The amount of water held and retained by a soil depends upon its depth, texture, structure and organic matter content; the finer the texture, the greater the storage capacity. The amount of water available to plants varies considerably in different soils. It may be about 10 mm per metre depth in sandy soils to about 100 mm in clayey soils. The greater the soil depth, the higher the proportion of effective rainfall in the total.

Initial moisture status in a soil governs the extent of effective rainfall considerably. When a shower falls just after irrigation, it becomes surplus water and is lost through deep percolation or run-off, but if the soil is dry it is recharged with moisture, resulting in a saving in irrigation water. The proportion of effective rainfall is lower in irrigated than in unirrigated areas where there is often a greater deficit of moisture in the soil.

5. Soil water characteristics

On striking the soil surface, the rain water often becomes run-off water and changes its physical and chemical properties during its flow. Some characteristics influencing effective rainfall are the head or depth of water received directly or indirectly, turbidity, viscosity, temperature, and nature of the salts, such as sodium, nitrates dissolved in it. These properties influence infiltration and through this the effective rainfall quantum.

6. Groundwater characteristics

The amount of effective rainfall is greater when the water table is deep as compared to shallow water table. Water moves upwards in the soil by capillarity, thus reducing the deficit of moisture and hence the amount of effective rainfall.

The levels of water tables normally fluctuate. Before the onset of rain, the water table may be quite deep; during the rainy season, it may rise to the surface. There can be horizontal flow in the sub-soil also. Because of these variations, the contribution of groundwater to the needs of the crop is variable and the proportion of effective rainfall varies inversely with this contribution.

If the groundwater is saline, it can be harmful to crop plants especially when it is near the soil surface. The proportion of effective rainfall may then increase since salts are diluted.

7. Management practices

Any management practice which influences run-off, infiltration, permeability or evapotranspiration also influences the degree of effective rainfall. Bunding, terracing, ploughing, ridging and mulching reduce run-off and increase effective rainfall; so do well-planned irrigation schedules, while arbitrary or random practices may reduce it.

8. Drainage channel characteristics

Size, shape, slope and roughness of a channel influence the speed of surface run-off to streams and consequently the time allowed for infiltration and also for direct evaporation at the site of rainfall. Hence these factors are also important in influencing effective rainfall.

9. Crop characteristics

Crops with high water consumption create greater deficits of moisture in the soil; therefore effective rainfall is directly proportional to the rate of water uptake by the plant. Crop characteristics influencing the rate of water uptake are the degree of ground cover, rooting depth and stage of growth. Evapotranspiration is generally high during vegetative growth and the flowering period and then may decline toward maturity. Soil moisture stored in deeper layers can be tapped only when roots penetrate to these depths. Deep-rooted crops therefore increase the proportion of effective rainfall in a given area; hence the nature of the crop is an important factor in determining its extent. Rainfall just before harvesting of different crops is a waste or a nuisance and may need to be considered as ineffective. Rainfall which reduces the yield (such as downpours which often cause lodging in cereals especially when these are at the grain formation stage) must be regarded as ineffective, and similarly, rains which result in deterioration or actual destruction of a crop. The crop is an important factor in interpreting the basic data. Hence the seasonal needs of major crops in a given area should be taken into account when the extent of effective rainfall is assessed.

Increasing effective rainfall

The approachs for increasing effective rainfall in agriculture are:

1. Reducing surface run-off

This is achieved by altering the topography of land, erecting barriers to the flow of water and by increasing the opportunity for infiltration. The practices followed include:

- land grading and levelling;
- bunding, terracing, basin listing, criss-cross ridging;
- retaining crop residues after harvest;
- reducing depth as well as frequency of irrigation.

2. Increasing infiltration

Increased infiltration is achieved by improving the soil structure or condition at the surface, as well as in the subsurface. The practices include:

- deep ploughing, sub-soiling or breaking hard pans;
- adding organic matter and soil conditioners to improve the texture of the growing pasture or grass legume mixtures;
- using mulches to prevent breaking of aggregates which seal the soil surface;

3. Building water storage structures

Farm ponds at suitable places are very useful to collect run-off water for use during periods of low rainfall at critical stages of crop growth.

4. Minimizing deep percolation losses

Deep percolation losses can be minimized by practices such as

- Increasing the water holding capacity of light textured soils by the addition of clays or organic matter, if this is economically feasible;
- Indirectly by extending the root zone though selecting deep roted crops;
- Reducing permeability beyond the root zone, especially in rice, using soil conditioners (krylium) and placing plastic sheets and bituminous layers at some depth in the soil.

The object of measuring the effectiveness of rainfall is to increase production per unit of water, per unit of land and per unit of time. This can be achieved by planning suitable crops and carrying out farm operations consistent with the amount, intensity and frequency of rainfall. These include:

- Planning cropping patterns consistent with rainfall patterns.
- Electing high value crops for maximizing the income.
- Adjusting sowing times according to the probable rainfall so that drought sensitive stages of growth do not synchronize with periods of inadequate rainfall.
- Timing other farm operations such as tillage, fertilizer application, applying insecticides, fungicides and herbicides etc, in relation to rainfall probability periods.
- Following mixed cropping of shallow and deep rooted crops for full exploitation of stored soil moisture.
- Adopting 'multi-storied' cropping practices, water needed for a single crop and for several crops together is more or less the same, but total production can be much higher in the latter system than in the former.
- Improving weather forecasting techniques and using the predictions in planning farm operations.

5. Increasing effective rainfall under field conditions:

Effective rainfall can be increased by reducing surface run-off, increasing infiltration and reducing deep percolation losses. Several practices are listed, but they are often not followed because their usefulness is under-estimated or their cost is too high. It is therefore necessary to popularize them through extension agencies. What can be done at the individual, community and district levels needs to be defined and demonstrated.

Rice culture offers an enormous scope for increasing effective rainfall by reducing percolation losses ranging between 50 and 75 percent. Studies using techniques of puddling, compacting, adding chemicals, lining the sub-soil with impervious materials and constructing sub-soil dams, would go a long way towards reducing the crop's water needs.

PLANTS IDEOTYPE FOR DRYLAND FARMING

1. Concept of plant ideotype

The term '*ideotype*' was introduced by Donald (1968) and appeared in the literature during 1969 with the publication of more comprehensive book entitled "An Ideal Plant Ideotypes" by Warshner (1969). According to him, an Ideotype is a biological model, which is expected to perform or behave in a predictable manner within a defined environment. The term plant ideotype is often known as model plant type, ideal plant type, ideal model plant type *etc.*

Types of ideotype

The most widely accepted classifications of plant ideotypes are as below :

I. **Isolation ideotype**: An isolation ideotypes are also known as space-planted ideotypes which have the potential to perform better when they are planted in a defined row-to-row and plant-to-plant spacing.

II. **Competition ideotype**: The competitive ideotypes are those ideotypes which perform well in genetically heterogeneous population rather than in a homogenous population.

III. **Crop ideotype**: This ideotype perform best at commercial crop densities because it is a poor competitor.

IV. **Market ideotype:** These ideotypes includes traits like seed colour, seed size, cooking and baking quality *etc.* So these ideotypes have their importance in improving the quality of the food grain or a product which may fetch higher price in the market and gave more remunerative returns per rupee of the invested money. These ideotypes focus on improvement of the product quality and to make the product highly acceptable and to give higher monetary returns.

V. **Climatic ideotype:** They include traits important in climatic adaptation such as heat and cold resistance, maturity duration, drought resistance *etc.* Therefore, these are the ideotypes which perform better under stressed conditions by making modifications/alterations in the genetic makeup to make them more adaptable under harsh climatic conditions.

VI. **Edaphic ideotype:** They Includes salinity tolerance, mineral toxicity/deficiency tolerance *etc*. The plant ideotypes which exhibit alteration in their genetic behavior so as to make them more comfortable under edaphic stress conditions.

VII. **Stress ideotype:** The stress ideotypes shows resistance to biotic and abiotic stress, disease/pest resistance ideotype, drought resistance *etc.*

A drought-tolerant variety is one that produces a high grain yield relative to other cultivars under drought stress.

2. Desired characteristics for an ideal ideotypes

An ideal plant ideotypes should have certain characteristics make them more suitable under a particular set of conditions. The desired characteristics which should be there for an ideal plant ideotype for different crops have been enlisted below.

Model plant type for wheat

According to Donald (1968) an ideal plant type for wheat (*Triticum aestivum*) should have following characteristics.

I. A short strong stem-It imparts lodging resistance and reduces the losses due to lodging.

II. Erect leaves- Such leaves provide better arrangement for proper light distribution resulting in high photosynthesis or CO_2 fixation.

III. Few small leaves- Leaves are the important sites of photosynthesis, respiration, and transpiration. Few and small leaves reduce water loss due to transpiration.

IV. Larger ear- It will produce more grains per ear.

V. A presence of awns- Awns contribute towards photosynthesis.

VI. A single culm.

A conceptual model for drought tolerance in wheat (Reynolds *et al.*, 2005)

- Seed size and coleoptile length improve early crop establishment
- Early ground cover and pre-anthesis biomass (reduce evaporation of soil moisture)
- Stem reserves / remobilization and spike photosynthesis
- Osmotic adjustment (maintains cell functions at low water potential)
- Accumulation of abscisic acid (pre-adapts cells to stress)
- Leaf anatomical traits waxiness pubescence, rolling, thickness
- High tiller survival and stay-green (easily observed integrative traits indicative of good drought tolerance).

Root depth and grain yield of wheat varieties under rainfed conditions

Singh and Uttam (1995) compared several cultivars to investigate their relative performance in terms of root length and grain yield and they have reported that cultivar C-306 performed significantly better under dryland conditions because of its ability to adapt itself under moisture stressed conditions. They have reported that cultivar C-306 has higher root length which has helped the plant to flourish the lower layers of soil profile for their moisture and nutrient requirements. Owing to better utilization of moisture and nutrients from lower strata, the crop grain yield has been increased significantly.

An ideotype of maize for conditions of high temperature and low moisture

According to Mock and Pearce (1975) an ideal maize ideotype must have large cobs, higher number of grains per cob and angled leaves for good light interception for achieving higher grain yield. The planting of such type at closer spacing of maize resulted in higher grain yield. The maize ideotype for dryland areas must have deep root system with higher root density and greater number of root hairs which may explore greater volume of soils and deeper layers of soil profile for their moisture needs. The maize ideotype should have well-furnished tilt root system so as to provide additional support to plants under stress conditions.

Ideotype for *Brassica napus*:

The ideal ideotype for dryland should have early flowering and it should have longer pods and more number of grains/seeds per pod. The ideotype should be such that it should show reduce pod shattering and high disease resistance under stress conditions.

An ideotype for rice (Jennings, 1964)

An ideal or model plant type consists of semi-dwarf stature and should have high tillering capacity. It should have short, erect, thick and highly angled leaves. The examples of semi-dwarf varieties of rice are IR-8, IR-20, TN-1 *etc.* which are capable of giving higher grain yield under stress conditions in dryland and rainfed areas of India. An ideal ideotype for rice should have short duration so as to escape the peak drought period. It should ensure that all the morphological and reproductive phases should have lapsed well before the peak drought spell. Jennings (1964) also included morphological traits in his model for an ideal Ideotype. But, now days more emphasis is given to physiological traits in the development of rice Ideotype. The desired ideotypic traits for drought tolerance in rice have been proposed by Fisher *et al.* (2003) and are presented in Table 12.1.

Table 12.1. *Desirable traits for drought tolerant idotypes of rice*

Trait	Proposed function
Deeper, thicker roots	To explore a greater soil volume
Root pulling resistance	Root penetration into deeper soil layers
Root penetration ability	To explore a larger soil volume
Osmotic adjustment (OA)	To allow turgor maintenance at low plant water potential
Membrane stability	Allows leaves to continue functioning at high temperature
Leaf rolling score	Reduce transpiration
Leaf relative water content	Indicates maintenance of favorable plant water status

Jones *et al.* (1997) have developed a new rice (*Oryza sativa*) for Africa by combining the drought tolerant and superior weed competitiveness of *Oryza glaberrima* rice and better yield potential of improved *Oryza sativa* tropical japonica types. This ideotype has shown a good promise in giving higher grain yield and ability to avoid drought conditions prevailing in dryland areas.

Ideal plant type of cotton for rainfed conditions

Ideal cotton ideotype should have earliness (maturity between 150-165 days), fewer, small and thick leaves with spares hairiness so that they have less trasipirational losses. The plants should have compact and short stature (average height should be between 75-80cm), intermediate habit, medium to big boll size. The plants should have high response to nutrients, resistance to insect and diseases and have synchronous bolling habit. The cotton ideotypes should also have an efficient root system. The root system should have deep well branched structure. Plants should have profuse rooting because deep root system contributes to drought avoidance. In addition an ideal cotton ideotypes should have better root growth parameters such as rooting volume, root mass, root length and root density and root: shoot ratio. Moreover, they should have dimorphic root system with deep roots for water uptake and shallow roots for nutrient scavenging.

Root characteristics for an ideal cotton ideotype

Limited tillering in cereals associated with relatively deeper root. Where the crop is grown on stored soil moisture-prefer short growth duration varieties with small leaf area. In shallow wetted soils, root depth is not an issue greater root length density will allow to extract more moisture from a given soil volume.

Other plant characteristics reducing water losses are.

Glaucousness

Glaucousness is the waxy covering over the plant cuticle that imparts a dull-white color. Reducing water loss after stomatal closure, acting as barrier on the leaf surface. Glaucousness is believed to enhance yield in wheat by increasing water use efficiency (WUE) or transpiration efficiency. Plant surface help to avoid dehydration by two processes, improved reflectance. Increased cuticular hydraulic resistance

Leaf pubescence is a common feature in xerophytic plants

Increases reflectance from the leaf within the range of 400 to 700 nm and sometimes up to 900 nm . Leaf colour can affect the thermal properties of the leaf. In both wheat and barley there are 'yellow leaf' cultivars, which have about a third less chlorophyll than the 'normal' ones. The 'yellow' cultivars tend to perform relatively better under drought stress as compared with the normal green

Some physiological traits show substantial potential to improve yield

Osmotic adjustment (OA) is an adaptive process which assists in the maintenance of turgor and cell volume as plants lose water during desiccation. Osmotic adjustment is an important physiological mechanism by which plants synthesize and accumulate solutes that function as osmolytes in cells in response to water deficits. OA enhances dehydration tolerance by maintaining turgor at lower water status, which, in turn, enables plants to maintain processes such as cell enlargement and stomatal opening , Substantial OA can be in the growing region of root at low water potential, and this is believed to be important for continued elongation of roots in drying soil

PROGRESS OF DRYLAND AGRICULTURAL RESEARCH IN INDIA

Though dryland farming is as old as agriculture, the systematic research work was started only from 1923 with the start of Research Centre at Manjri near Pune, in Bombay province by V.A. Tamhane. Later Imperial Council of Agricultural Research (presently Indian Council Agricultural Research) started research schemes at Bombay, Madras, Hyderabad and Punjab provinces, and later on systematic work was started on different aspects of crop production under rainfed conditions. The findings of the period were related to rainfall analysis. It was found that rainfall was not only scarce but also erratic. The dry spells during the crop period ranged from 20 to 55 days. Therefore, the emphasis on research was given to conserve soil moisture and to reduce evaporation. Based on the studies conducted between 1933 and 1943, package was developed for better crop production under rainfed conditions. The important practices are:

I. Formation of contour bunds.

II. Repeated harrowing in black soils to conserve moisture.

III. Addition of farm yard manure to maintain the soil fertility.

IV. Wider spacing for crops grown on residual moisture.

The chronology of events in dryland agricultural research in India are as follows.

1920	Scarcity tract development given importance by the Royal Commission on Agriculture
1923	Establishing Dryland Research Station at Manjri (Pune) by Tamhane
1933	Research Stations established at Bijapur and Solapur
1934	Research Stations established at Hagari and Raichur

1935	Research Station established at Rohtak (Punjab)
1942	Bombay Land Development act passed
1944	Monograph on dry farming in India by N.V. Kanitkar (Bombay, Hyderabad, Madras)
1953	Establishing Central Soil Conservation Board
1954	Establishing Central Soil Conservation Centres
1970	Research Centres established under AICRPDA in 23 locations
1972	Establishment of ICRISAT
1976	Establishment of Dryland Operational Research Projects
1983	Starting of 47 model watersheds under ICAR
1984	Initiation of World Bank assisted Watershed Development Programmes in four states.
1985	Birth of Central Research Institute for Dryland Agriculture at Hyderabad
1986	Launching of NWDPRA programmes by Government of India in 15 states.

Recognizing the importance of rainfed agriculture, the ICAR gave a new impetus by launching the All India Coordinated Research Project for Dryland Agriculture (AICRPDA) in 1970, based at Hyderabad with 23 cooperating centres spread across the country. Pooling of expertise and leveraging the strengths of AICRPDA net work eventually resulted in the establishment of Central Research Institute for Dryland Agriculture (CRIDA) at Hyderabad, on April 12, 1985 to provide leadership in basic strategic research in dryland agriculture while continuing research on location specific ORP' s at AICRPDA centres. At present the AICRPDA centres are located at 25 places.

Mandate/objectives of CRIDA

(a) To conduct basic and applied researches that will contribute to the development strategies for sustainable faming systems in the rainfed areas.

(b) To conduct basic and applied researches that will contribute to the development strategies for sustainable faming systems in the rainfed areas.

(c) To act as a repository of information on rainfed agriculture in the country

(d) To provide leadership and co ordinate network with state agri-cultural universities for generating location specific technologies for rainfed areas

(e) To act as a centre for training in research methodologies in the fields basic to management of rainfed farming systems

(f) To collaborate with relevant national and international agencies in achieving the above objectives

(g) To provide consultancy

STEPS FOR RAISING PRODUCTIVITY IN DRY FARMING

To boost the crop production under dry farming, we will have to efficiently manage our soil and water resources in the respective areas. An efficient soil and water conservation system will play a vital role in boosting the crop yield in dry farming. The different interdisciplinary approaches which are recommended for dryland farming are categorized in to four major groups namely engineering, physiological, genetic and agronomic approaches.

1. Engineering approaches

These approaches are aimed at soil and moisture conservation through regulation of run-off, collection of surplus rain water checking evaporation and seepage losses of water and recycling of collected water as irrigation in times of critical need. Water conservation can be made with the following practices.

I. Contouring across the slope: Contouring is practiced on the lands with 3-5% slope. This system consists of constructing earthen bunds and the distance between the two bunds ranges from 30-50 m depending on the degree of slope. This is carried out with an object to provide a check to the flow of run-off water which then gets accumulated in the bunded area and is absorbed by the soil. Thus, contour bunding conserves moisture and prevents soil erosion. As per slope, following practices can be adopted.

a. Smoothening of contour inter-bund areas

This is practiced only in those areas which have a slope of less than 1 per cent The smoothening may be achieved by running bullock drawn harrows or cultivators but small undulations are leveled during the process so that impounding of water may take place and maximum water absorption by the soil may be achieved.

b. Contour border strips method

This method is suitable for areas having a slope of 3-4 per cent In this case parallel strips across the slope ranging from 10-15 m in width are laid down on contours and the soil surface is leveled by scrapping and placing the soil according to the need of the spot It is done to reduce the run-off and to conserve soil and water from the field. It is, however, an expensive method as it requires culling and filling up the soil from higher spots to lower ones.

II. Scooping of land: In this practice, the land is generally scooped before the beginning of monsoon showers. By scooping, the soil is exposed for proper absorption and conservation of moisture. However, this is also a tedious as well as an expensive operation.

III. Opening or ridges and furrows: In this practice, the entire land is laid out into ridges and furrows across the slope. The ridges and furrows are opened before onset of monsoon so that the flow of water may be reduced and erosion may be controlled to the minimum. During rainy season, crops like maize, jowar, bajra, etc. may be grown in the furrows and legumes like soybean, arhar, urd, mung, cowpea, etc. may be grown on the ridges. After the monsoon is over the land is again leveled. This way the furrows are used to accumulate maximum water which will supply moisture for winter season crops.

IV. Compartmental bunding: Areas having a slope of 1% or less are suitable for compartmental bunding. It helps in accumulation of more water and a uniform spread of water in the entire area. Leveling is also done with nominal or no additional expenditure.

V. Bedding system: In this system, small furrows are opened and the soil from the furrows is uniformly spread in space left between the furrows. Thus, inter furrow spaces form the raised beds of about 4-5 metres width. This method helps in the conservation of soil, moisture and checking the excess run-off of water. The raised beds, in this practice, are used for growing such crops which need less water like legumes and oil seed crops, while the furrows are used for the crops which need more water.

VI. Broad-based bunding: This method is especially suitable for heavy black soil. Water is allowed to spread over a vast area by constru-cting a broad- based bund on a sloppy side. The water stays for a longer time because of high water holding capacity, lower leaching and seepage losses. The stored water may be used for fish culture and also for providing life saving irrigation grown in surrounding areas of catchment portion. These bunds are also called check dams and are given a regulated drain or outlet for protecting the bunds from breaking.

VII. Deep summer ploughing followed by surface tilling: The field is ploughed deep by mould board plough soon after harvesting rabi season crop with the objectives of

(a) Exposing the soil for perfect drying
(b) Killing the disease pathogens
(c) Destroying eggs of insect pests
(d) Controlling weeds by sun drying.

The surface tilling during other seasons forms natural mulch and thereby reduces evaporation loss of water from soil

VIII. Water harvesting system: Water harvesting is a technology of utilizing the collected and conserved water for the purpose of crop pro-duction. It includes tillage practices for an efficient use of moisture bet-ween and within the crop rows. Frequent stirring of the land by surface tilling provides mulch and prevents the evaporation loss of water from the soil.

Besides harvesting moisture from between and within the crop rows, run off losses are considerably reduced. The stored water is used for providing life saving irrigation to the crops grown in the surrounding areas. The water harvesting of this type can be done in areas situated near hill and on greatly undulated lands. In these cases, check dam tanks, and other reservoirs are constructed. The infiltration or percola-tion loss of water is prevented by spraying of asphalt compounds or by covering the bottom of the tanks or ponds through thin plastic sheets. The seepage loss may also be checked by providing a plastic lining. The evap-oration loss of water is controlled by pouring some burnt crude oil over water surface. On a highly eroded soil or soils having very high slopes, terraces are made for providing a gentle flow of run off water so that the soil is not further eroded and some crop may be grown on these terraces.

2. Physiological approaches

Hardly 1% of the water absorbed by the plant roots is used for the growth and development of plants and remaining 99% is wasted through transpiration back to atmosphere. Thus one of the greatest causes of soil water wastage is loss of water through transpiration.

The extent of transpiration can be greatly influenced by using certain chemicals. These chemicals reduce transpiration, encourage root growth and protect the cytoplasmic proteins of the plants. These chemicals bring about more drought resistance in the plants. These compounds, according to their role, are classified as given below.

I. Anti-transpirants : Any chemical substance, which reduces rate of transpiration on its application to the plant surface, is called anti-transpirant. Any substance which reduces the vapour pressure gradient in the stomatal cavity or increases stomatal resistance to water vapour diffusion, will act as anti-transpirant These substances have been used for arresting water loss from plant body with various degree of success. These are Phenyl Mercuric Acetate (PMA), Hydroxy Sulphonates (HS), Alkenyl Succinic Acid (ASA), Adol:-52 (a formulation of alcohol), and S-600 (a plastic transplanting spray).

II. Chemicals for improved cell membrane permeability of water: Dry farming areas are characterized by scarce rainfall and usually the roots have lipid layers which lower the absorption of water from the roots. Some chemicals like Alkenyl Succinic Acid (ASA) and Decenyl Succinic Acid (DSA), when applied, penetrate into the root and increase its water absorption power 8 times. Therefore, these chemicals are applied in the root zone for increased water absorption along with some chemicals to retard the transpiration from foliage

III. Use of plant hormones and growth retardants : Some plant hormones like Indole acetic acid (IAA) and Abscissic acid (ABA) may be used for reducing the frequency and period of stomatal opening thereby minimizing the water loss from the plant body.

There are certain other chemicals known as growth retardants which either modify the plant structure or dwarf the plants by considerably reducing the total water requirement of the plants. This chemical also induces moisture stress tolerance in plants. Cycocel is presently used in cotton to encourage production of more fruiting branches (sympodial rather than monopodial or vegetative branches) and, thereby even under drought condition it results in higher yield. In case of wheat, the use of Cycocel decreases the cell size and increases the density of the cytoplasm which ultimately results into a drought resistance into the plants.

IV. Chemicals for seed treatment: There are certain chemicals which are used for seed treatment to bring about drought resistance in plants right from seedling stage. Soaking of seeds in calcium chloride solution (0.25%) for 20 hours soaking of seeds with frequent shaking results in better germination and drought resistance in the plants. Boron solution is also used for soaking seeds. Agrosan is a fungicide but also induces drought resistance in the plants when seeds are treated with this chemical

3. Genetic approaches

Because of scanty and unreliable rains, the farmers of dry farming areas are still practicing crop husbandry on the basis of traditional approaches like low intensity cropping, little or no use of fertilizers or manures, raising low value crops. Crop varieties grown till the recent past were generally of long duration and slow growing. They were poor yielders too.

As nearly 70% of our total agricultural land is rainfed and 45% of rainfed area is dryland, there is no way out but to evolve suitable varieties as well as appropriate technology for obtaining higher product-ivity from rainfed or dryland areas. As such, the concerted efforts of our plant breeders have resulted in the cultivation of several new plant types which possess all the characters needed for rainfed areas or drylands. In terms of modem technology, such plant materials are called "Ideo types". These are the suitable strains of dryland crops which are characterized by short growth duration, effective and extensive root system, drought tolerance, high yield potential having altered morphology of plants which are conducive to drylands.

Identified crops varieties for drylands

Crop	Varieties
Maize	Ganga-2, Ganga- 5, Ageti-76
Sorghum	CSH-5, CSH-6, SPV-99, SPV-245
Pearlmillet	MH-179, HHB-67, WCC-75, MH-79, HHB-50, 60, 68
Clusterbean	Durgapur safed, FS- 277
B.Gram	T-9, Krishna, RBU-38
G.Gram	K-851, T-44, S-9, Basant
Cowpea	FS- 68, C-152
Moth	Jadia, Jawala
Groundnut	MH-4, AK-12-24, JL-24
Wheat	WH-533, WH-147, C-306, PBW-299, DL-153-2
Gram	C-235, H-208, HC-1, Gaurav
Barley	RS-6, C-138, BG-25, BH-75
Mustard	T-59, RH-30, RH-781, RH-819, RLM-514
Taramira	T-25, ITSA, T-27
Linseed	Chambal, Triveni
Pigeonpea	HY-1, HY-2, Gwalier- 3
Sugarcane	Pb-1, Gaurav, Ankur
Castor	Bhagya, Sobhagya, Gauch-1
Lentil	L, 9-12

4. Agronomic approaches

The major objective of dry farming programme is to conserve the soil and moisture and to achieve maximum production from the dry farming areas. In the past two decades, we have been able to' solve many hurdles in the aforesaid areas but there had been no breakthrough as in case of irrigated crop production. Now we have the promising crop varieties and technology available with us about the maximum soil and water conservation. The agronomic approaches can be dealt with under the following four heads based on land types.

I. Agronomic approaches for highly undulating lands: These lands are confined to the hills of the locations which have suffered serious soil erosion problems and have been divided into various gullies. The soils of these areas are more prone to further erosion if they are not properly managed. Therefore, in this category, the crop management practices are entirely different from other areas. The object of soil and crop management under such situation should be:

(a) To stabilize the soil by forestry and pasture management with a regulated grazing or no grazing at all.

(b) To level the land gradually through contour bunding, terracing, etc.

(c) To practise strip cropping.

II. Agronomical approaches for plain lands: Plain lands form the main dryland tracts of the country. There has been major emphasis on finding out ways and means through which the total soil productivity could be increased. Fallow cultivation early in morning or during evening hours followed by planting helps in moisture conservation.

III. Tillage requirements of the crops: Tillage starts with the seed bed preparation and ends with mulching and control of weeds. Deep ploughing during summer helps in destroying weeds and suppressing insect pests and diseases. It also helps in an efficient root penetration very deep into soil. Placement of seed at 5 cm and fertilizers at 7.5 cm in the same furrow followed by soil compaction have resulted in better germination, plant vigour, extensive root development and higher crop yields.

IV. Selection of crops and varieties: There are a number of improved varieties of different crops which are drought tolerant or resistant to water stress. The most commonly grown crops in drylands are rice, maize, sorghum, pearl millet. finger millet, wheat, barley, pulses, oilseeds, etc.

V. Sowing of crops: Sowing of crops deals with several associated factors namely sowing time, method of sowing, depth of sowing etc. It is important in the sense that once the ideal plant population is achieved, the crop is bound to give yield. To get an ideal plant population it is necessary that about 25% higher than required seed rate should be applied. Care must be taken to reduce plant competition for moisture by removing excess plant population about 2-3 weeks after the sowing depending upon the crops.

Sowing time can markedly influence the production and productivity of dryland crops. Early sowing of *kharif* crops results in early crop maturity and thereby it facilitates early sowing of succeeding *rabi* crops. Early sowing of *rabi* crops helps in overcoming the moisture stress at later stages of plant growth, particularly at grain filling stage. Broadcasting of seeds should be avoided as it involves several losses and seed does not properly come in contact with moisture. Placing the seeds at about 5 cm depth through pora or seed drill is desirable.

VI. Fertilizer management

(a) Use of fertilizers in drylands is limited as compared to irrigated areas. The national average regarding use of fertilizers in dry farming areas is 60 kg/ha.

(b) Poor financial condition of farmers to purchase fertilizer

(c) Wrong concept of the farmers that fertilizers will burn the seedlings, and harm to the soil

(d) Application of organic manure only, which cannot meet the total nutrient requirement of the crop.

(e) Deep placement of fertilizers followed by foliar application results in extensive rooting up to deeper soil layers where the plants extract moisture to meet their demands

VII. Cropping systems: Cropping system refers to an arrangement in which various crops are grown together in the same field. The cropping systems followed in drylands differ from those followed under normal conditions. Only those crops can be grown under dryland conditions which require less water to complete their life cycle or which can stand or yield under drought conditions. This can include both drought resistant and drought tolerant plants. In addition, plants can be grown only where some water is available to sustain the growth of plants. Following are a few intercropping systems for dryland areas:

Moong in Bajra

Guar in Bajra

Til in Guar/Moth/Mung

Mixed cropping is also followed to minimize the effect of unpredictability of rain. Mixed cropping may have low yield potential but it works as a buffer against failure under possible unfavourable conditions. Mixed cropping may be defined as sowing of two or more crops simultaneously on the same piece of land in separate rows. e.g. Guar + Arhar + Moong, Bajra + Arhar + Moong and Maize + Urd etc.

VIII. Cropping pattern : Cropping pattern is defined as sequence of growing crops in a particular field at a particular period. The most common cropping pattern for dryland farming are discussed below:

For North Indian conditions

- Sorghum -Safflower/Mustard
- Moong/Urd/Cowpea -Gram/Wheat
- Sorghum –Gram/Wheat
- Rice -Gram (for low lying areas)
- Bajra -Gram + Linseed
- Bajra -Wheat/Barley/Gram/Mustard
- Urd/Mung/Soyabean- Gram/Mustard/Wheat/Barley
- Maize-Gram/Safflower

For central Indian condition

- Green gram -*Rabi* Sorghum
- Green gram -Safflower

IX. Weed control: Presence of weeds in the crop field, especially in case of drylands, cause a severe crop weed competition for water, nutrients and light The reduction in yield due to weeds varies from 30-75% depending upon the crop and nature and extent of weed infestation. Weeds may be controlled by hand weeding, intercultural operations and herbicidal application or by adopting an integrated approach

X. Agro-forestry and dryland agriculture: Agroforestry is the cultivation of trees in association with crops. It has assumed a great importance in order to solve a number of problems of dryland farmers. The twin objectives of Agroforestry are to increase the farmer's income and sustain the ecological environmental balance. Agroforestry, as it provides fuel wood to the farmers, enables them to save the animal dung and use it as manure. The following points should be considered while choosing trees for cultivation in association with crops:

- The tree should grow fast and stand repeated prunings.
- Its roots should explore soil layers not tapped by agricultural crops.
- The crown should not be too large to prevent sun light from reaching the crop.
- The peak period of flush and leaf fall should not be detrimental to the crop grown in association.
- As far as possible, the tree species should be a legume, the root nodules of which accommodate nitrogen manufacturing bacteria.
- The branching should be sparse and light.
- The tree should meet the socio-economic and ecological needs of the region.

DROUGHT

Low rainfall or failure of monsoon rain is a recurring feature in India. This has been responsible for droughts and famines. The word drought generally denotes scarcity of water in a region. Though, aridity and drought are due to insufficient water, aridity is a permanent climatic feature and is the culmination of a number of long term processes. However, drought is a temporary condition that occurs for a short period due to deficient precipitation for vegetation, river flow, water supply and human consumption. Drought is due to anomaly in atmospheric circulation.

Definition of drought

There is no universally accepted definition for drought.

(a) Early workers defined drought as prolonged period without rainfall.

(b) According to Ramdas drought is a situation when the actual seasonal rainfall is deficient by more than twice the mean deviation.

(c) American Meteorological Society defined drought as a period of abnormally dry weather sufficiently prolonged for lack of water to cause a severe hydrological imbalance in the area affected.

(d) Prolonged deficiencies of soil moisture adversely affect crop growth indicating incidence of agricultural drought. It is the result of imbalance between soil moisture and evapo-transpiration needs of an area over a fairly long period so as to cause damage to standing crops and to reduce the yields.

(e) The irrigation commission of India defines drought as a situation occurring in any area where the annual rainfall is less than 75% of normal rainfall.

Drought is a normal, recurrent feature of climate. Within the State of Ohio, drought is equally as possible to occur in one section of the State as it is in another. The effects of drought within the State may depend upon land use pattern (farming and cattle production), economy (dependence on drought-impacted business (farming), geology (presence of an aquifer or ground structure that limits well production), and water source (public water supply, private well, cistern). Because of these in a general sense, drought originates from a deficiency of precipitation over an extended period of time, resulting in a water shortage for some activity, group, or environmental sector.

Drought is a climatic anomaly, characterized by deficient supply of moisture resulting either from sub-normal rainfall, erratic rainfall distribution, higher water need or a combination of all the factors. The escalating impacts of droughts have increasingly drawn the attention of scientists, planners and society. The vulnerability to drought in relation to the increasing needs of the growing population has become a point of great concern, especially on the food front. In spite of the technological developments in providing improved crop varieties and better management practices, in India, agriculture has been considered a gamble as the agricultural productivity is strongly influenced by the vagaries of the monsoon.

Droughts are the resultant of acute water shortage due to lack of rains over extended periods of time affecting various human activities and lead to problems like widespread crop failure, unreplenished ground water resources, depletion in lakes/reservoirs, shortage of drinking water and, reduced fodder availability etc. Any negative departure from these levels creates conditions of drought, depending on the intensity and duration of this deficit. Thus drought conditions differ from region to region. Also the impact of drought over a region varies depending on which economic activity is impaired.

In general, drought means different things to different people. To a meteorologist it is the absence of rain while to the agriculturist it is the deficiency of soil moisture in the crop root zone to support crop growth and productivity. To the hydrologist it is the lowering of water levels in lakes, reservoirs, etc., while for the city management it may mean the shortage of drinking water availability. Thus it is unrealistic to expect a universal definition of drought for all fields of activity.

Types of drought

- **Meteorogical drought:** A situation where there is a 25% decrease in the average rainfall for a given period in a region.

- **Agriculture drought:** Extended dry period in which lack of rainfall results in insufficient moisture in the root zone of the soil causing adverse effects on crops.
- **Hydrological drought:** Extended dry period leading to marked depletion of surface water and consequent drying up of reservoirs, lakes, streams, rivers, cessation of spring flows and fall in ground water levels.

Drought management strategies

1. Short-term strategies

I. Ground-water utilisation

(a) Efficiency of the electric pumps is higher than the diesel pumps. However, because of the subsidised or free supply of electricity, the farmers do not care for the efficiency of the motors or pump-sets and look for cheaper options in the market.

(b) Uninterrupted supply of electricity: Frequent tripping of the supply was complained to almost all the teams who visited various States and interacted with the farmers. Frequent tripping leads to repeated irrigation of the same spot whereas rest of the field remained uncovered.

(c) Proper maintenance of the motors and pumping sets to reduce friction by way of greasing and other maintenance should be advised for efficient pumping.

(d) In case of rice, continuous standing of water is required only in the initial 15-20 days so as to suppress growth of weeds. However, later on, irrigating one day after disappearance of water is the most economical and efficient way of scheduling irrigation.

(e) Sowing of cotton, soyabean, maize, etc. on the ridges and furrows and letting water in alternate furrows can save 20-30% water.

(f) Sprinklers for cereal crops like wheat and drip system for widely spaced crops sown in lines like sugarcane, cotton, maize etc. can give an efficiency of 80-90%.

II. Rain water harvesting: About 11-37% run-off is generated even by the delayed monsoon and should be stored in the farm ponds or tanks. These will recharge ground water during normal or excessive rainfall year. Rain water stored in self sealing or lined ponds can be used for irrigation if there is long break in the rainfall or for pre-sowing of the *rabi* crops to ensure proper germination.

III. Use of poor quality water

(a) Mixed and/or alternate use of limited good quality and underground poor quality waters for irrigation.

(b) Cultivation of salt tolerant varieties like CSR-10, CSR-13, CSR-27 and CSR-30 of rice; KRL 1-4, KRL 1-9 of wheat, CS-52 of mustard, CSG-8962 of gram should be preferred.

(c) Farmers having residual sodium carbonate (RSC) rich waters should be encouraged to use this for irrigation judiciously after amending with gypsum or as soil amendment.

(d) Resorting to dry sowing of mustard followed by irrigation with saline water especially in light textured soils.

IV. Contingent cropping : Selection of crops, cropping sequences and agronomic practices are very important. Relatively more drought tolerant, deep rooted and short duration crops, varieties and cultivars are available for different agro-ecological and rainfall situations. If the rain is excessively delayed or main crop has failed cultivation or re-sowing with fodder is the best option. Fodders can be harvested at any stage keeping in view sowing of the next *rabi* season crop.

(a) Date bound contingency crop plan for rainfed areas

(b) Suitable early maturing varieties of different *kharif* crops

(c) Crop group/crop-wise contingency plan for *kharif*

(d) Contingency plan relevant to all crops and cropping systems

(e) Specific contingent plan for Boro or Summer Cropping

(f) Crops and cropping system-wise agronomic practices.

(g) Conserve water and soil

V. Arrangements of quality fodder

(a) Decrease area under paddy and increase area under maize, sorghum (jowar), pearl millet.

(b) As the sowing of main *rabi* fodder crops will start in October-November, catch crop of maize, bajra, sorghum, cowpea, bajra + cowpea, maize + cowpea and toria may be taken up after light showers during August-September.

(c) Dual purpose crops like barley (varieties RD 2715, RD 2035, RD 2522 and BH 75) may be sown in October. One cutting may be taken for fodder at 50-60 days after sowing and subsequent regenerated crop left for grain production.

(d) Oats may be grown in October as multicut fodder to ensure availability of green fodder for longer period.

2. Long and medium term strategies

I. Securing good quality water in drought prone areas

(a) Networking of rivers, reservoirs, lakes and other water bodies existing in high rainfall areas which are prone to periodic flooding. Transferred water could be used for ground water recharging and to fill up dried lakes, water storage structures in dry areas whenever such necessity arises. This seems a viable option because the country receives more than one metre average rainfall in a year with lot of inter-regional variation.

(b) Less exploitation of ground water by resorting to low water demanding crops, introduction of precision micro-irrigation techniques such as drip and sprinkler methods in overexploited/ critical areas and matching water application schedules with critical growth stage concept may be prioritized.

(c) Diversification into less water demanding cropping systems. Vast ranges of options are available to make preferred choices.

II. Fine tuning of crops, cropping and farming systems: The strategy should include: (i) relooking and upgradation of our knowledge about mixed cropping, intercropping, catch cropping, mixed farming and multi-strata cropping concepts; (ii) promotion of agro-forestry, silvipasture, horti-pasture and sivi-horti-pasture systems etc. through large pilot scale demonstrations in farmer's participatory mode; (iii) agronomic manipulations such as zero tillage, bed furrow irrigation, fertilization, adjusting spacing, soil and water conservation through mulching, use of anti-transpirants/Jal shakti etc.

III. Exploiting under-exploited and under-utilized plant resources: Large number of trees, bushes, shrubs and grasses are naturally growing as wild plants in one or the other rainfed region of the world. Such plants are adapted to rainfed situation and have tremendous potential to be exploited as a food, forage, fuel and/or as industrial crop. Some of these plants have already been exploited in different parts of the world. For example, edible cactus (*Opunita spps*) is in countries like Mexico, Argentina, France, Brazil, Italy, South Africa and even in south western USA.

History of drought in India

Period	Drought year	Number of droughts
1801-1830	1801,1804,1806,1812,1819,1825	6
1831-1860	1832,1833,1837,1853,1860	5
1861-1890	1862,1866,1868,1873,1877,1883	6
1891-1920	1891,1897,1899,1901,1904,1905,1907,1911,1918,1920	10
1921-1950	1939,1941	2
1951-1980	1951,1965,1966,1971,1972,1974,1979	7
1981-2010	1982,1987,2002,2009	4

CLASSIFICATION OF DROUGHT

Drought can be classified based on

- Duration,
- Nature of users,
- Time of occurrence and
- Using some specific terms.

1. Based on duration

I. **Permanent drought:** This is characteristic of the desert climate where sparse vegetation growing is adapted to drought and agriculture is possible only by irrigation during entire crop season.

II. **Seasonal drought:** This is found in climates with well defined rainy and dry seasons. Most of the arid and semiarid zones fall in this category. Duration of the crop varieties and planting dates should be such that the growing season should fall within rainy season.

III. **Contingent drought:** This involves an abnormal failure of rainfall. It may occur almost anywhere especially in most parts of humid or sub humid climates. It is usually brief, irregular and generally affects only a small area.

IV. **Invisible drought:** This can occur even when there is frequent rain in an area. When rainfall is inadequate to meet the evapo-transpiration losses, the result is borderline water deficiency in soil resulting in less than optimum yield. This occurs usually in humid regions.

2. Based on relevance to the users

I. **Meteorological drought:** It is defined as a condition, where the annual precipitation is less than the normal over an area for prolonged period (month, season or year).

II. **Atmospheric drought:** It is due to low air humidity, frequ-ently accompanied by hot dry winds. It may occur even under conditions of adequate available soil moisture. It refers to a condition when plants show wilting symptoms during the hot part of the day when transpiration exceeds absorption temporarily for a short period. When absorption keeps pace with transpiration the plants revive (mid day wilt).

III. **Hydrological drought:** In hydrological drought, depletion of surface water and consequent drying of reservoirs, tanks etc takes place. It results in deficiency of water for all sectors using water. This is based on water balance and how it affects irrigation as a whole for bringing crops to maturity.

IV. **Agricultural drought (soil drought):** It is the result of soil moisture stress due to imbalance between available soil moisture and evapotranspiration of a crop. It is usually gradual and progressive. Plants can therefore, adjust at least partly, to the increased soil moisture stress. This situation arises as a consequence of scanty precipitation or its uneven distribution both in space and time. Relevant definition of agricultural drought appears to be a period of dryness during the crop season, sufficiently prolonged to adversely affect the yield. The extent of yield loss depends on the crop growth stage and the degree of stress.

3. Based on time of occurrence

I. **Early season drought:** It occurs due to delay in onset of monsoon or due to long dry spells after early sowing.

II. **Mid season drought:** Occurs due to long gaps between two successive rains and stored moisture becoming insufficient during the long dry spell.

III. **Late season drought:** Occurs due to early cessation of rainfall and crop water stress at maturity stage.

Effect of drought on crop production

I. **Water relations:** Alters the water status by its influence on absorption, translocation and transpiration. The lag in absorption behind transpiration results in loss of turgor as a result of increase in the atmospheric dryness.

II. **Photosynthesis:** Photosynthesis is reduced by moisture stress due to reduction in photosynthetic rate, chlorophyll content, leaf area and increase in assimilates saturation in leaves (due to lack of translocation).

III. **Respiration:** Increase with mild drought but more serve drought lowers water content and respiration.

IV. **Anatomical changes:** Decrease in size of the cells and inter cellular spaces, thicker cell wall, greater development of mechanical tissue. Stomata per unit leaf tend to increase.

V. **Metabolic reaction:** Mostly all metabolic reactions are affected by water deficits.

VI. **Hormonal relationships:** The activity of growth promoting hormones like cytokinin, gibberlic acid and indole acetic acid decreases and growth regulating hormone like abscisic acid, ethylene, etc., increases.

VII. **Nutrition:** The fixation, uptake and assimilation of nitrogen is affected. Since dry matter production is considerably reduced the uptake of NPK is reduced.

VIII. **Growth and development:** Decrease in growth of leaves, stems and fruits takes place with drought. Maturity is delayed if drought occurs before flowering while it advances if drought occurs after flowering.

IX. **Reproduction and grain growth:** Drought at flowering and grain development determines the number of fruits and individual grain weight, respectively. Panicle initiation in cereals is critical while drought at anthesis may lead to drying of pollen. Drought at grain development reduces yield while vegetative and grain filling stages are less sensitive to moisture stress.

X. **Yield:** The effect on yield depends hugely on what proportion of the total dry matter is considered as useful material to be harvested. If it is aerial and underground parts, effect of drought is as sensitive as total growth. When the yield consists of seeds as in cereals, moisture stress at flowering is detrimental. When the yield is fibre or chemicals where economic product is a small fraction of total dry matter, moderate stress on growth does not have adverse effect on yields.

Crop adaptations

The ability of crop to grow satisfactorily under water stress is called drought adaptation. Adaptation is structural or functional modification in plants to survive and reproduce in a particular environment. Crops survive and grow under moisture stress conditions mainly by the following ways

1. Escaping drought

Evading the period of drought is the simplest means of adaptation of plants to dry conditions. Many desert plants, the so called ephemerals, germinate at the beginning of the rainy season and have an extremely short life period (5 to 6 weeks) which is confined to the rainy period. These plants have no mechanism for overcoming moisture stress and are, therefore, not drought resistant. Germination inhibitors serve as safety mechanism.

In cultivated crops, the ability of a cultivar to mature before the soil dries is the main adaptation to growth in dry regions. However, only very few crops have such a short growing season to be called as ephemerals. Certain varieties of pearl millet mature within 60 days after sowing. Short duration pulses like cowpea, green gram, black gram can be included in this category. In addition to earliness, they need drought resistance because there may be dry spells within the crop growth period of 60 days adaptations to moisture stress. The disadvantage about breeding early varieties is that yield is reduced with reduction in duration.

2. Drought resistance

Plants can adopt to drought either by avoiding stress or by tolerating stress due to different mechanisms. These mechanisms provide drought resistance.

3. Avoiding stress

Stress avoidance is the ability to maintain a favourable water balance, and turgidity even when exposed to drought conditions, thereby avoiding stress and its consequences. A favourable water balance under drought conditions can be achieved either by:

(i) Conserving water by restricting transpiration before or as soon as stress is experienced; or

(ii) Accelerating water uptake sufficiently so as to replenish the lost water.

STRATEGIES FOR DROUGHT MANAGEMENT

The different strategies for drought management are discussed under the following heads.

1. Adjusting the plant population

The plant population should be lesser in dryland conditions than under irrigated conditions. The rectangular type of planting pattern should always be followed under dryland conditions. Under dryland conditions whenever moisture stress occurs due to prolonged dry spells, under limited moisture supply the adjustment of plant population can be done by

I. **Increasing the inter row distance:** By adjusting more number of plants within the row and increasing the distance between the rows reduces the competition during any part of the growing period of the crop. Hence it is more suitable for limited moisture supply conditions.

II. **Increasing the intra row distance:** The distance between plants is increased by which plants grow luxuriantly from the beginning. There will be competition for moisture during the reproductive period of the crop. Hence it is less advantageous as compared to above under limited moisture supply.

2. Mid season corrections

The contingent management practices done in the standing crop to overcome the unfavourable soil moisture conditions due to prolonged dry spells are known as mid season corrections and these are:

I. **Thinning:** This can be done by removing every alternate row or every third row which will save the crop from failure by reducing the competition.

II. **Spraying:** In crops like groundnut, castor, redgram, etc., during prolonged dry spells the crop can be saved by spraying water at weekly intervals or 2 per cent urea at week to 10 days interval.

III. **Ratooning:** In crops like sorghum and bajra, ratooning can practiced as mid season correction measure after break of dry spell.

3. Mulching

It is a practice of spreading any covering material on soil surface to reduce evaporation losses. The mulches will prolong the moisture availability in the soil and save the crop during drought conditions. Mulches also controls weeds and all available soil moisture will be utilized by the crop plant only.

4. Weed control

Weeds compete with crop for different growth resources more seriously under dryland conditions. The water requirement of most of the weeds is more than the crop plants. Hence they compete more for soil moisture. Therefore the weed control especially during early stages of crop growth reduces the impact of dry spell by soil moisture conservation.

5. Water harvesting and life saving irrigation

The collection of run off water during peak periods of rainfall and storing in different structures is known as water harvesting. The stored water can be used for giving the life saving irrigation during prolonged dry spells.

18

CROP PRODUCTION UNDER SPECIAL SITUATIONS

With the current food production level of 275.11 million tones, India has to step up food grain production by next decade to feed the burgeoning population (Report Ministry of Agriculture). India with 2.3% of the global geographical area, supports about 16.8% human population, 20% of the world's livestock. However, per capita availability of land for producing agricultural commodities has declined from 0.34 ha in 1951 to about 0.234 ha in 1980 and 0.15 ha during 2015 and it is expected to decline further to about 0.09 ha by 2050 A.D. With the possibility of horizontal expansion or putting more land under cultivation being remote, further augmentation in yield in Indian agriculture will be forced more and more into marginal or problematic soils accompany about 57% geographical area with increasing competition for good quality lands. Marginal and problematic soils with proper management have a good potential for increasing food production to meet the expanding needs of the country. Crop production under the following five special situations has been described:

1. Salt affected soils
2. Acidic soils
3. Waterlogged soils
4. Eroded soils
5. Dryland farming

1. SALT AFFECTED SOILS (ALKALI AND SALINE SOILS)

Salt affected soils cover an area of about 8.6 million hectares in India; almost half are in canal commands. To meet out the food needs of an increasing population which is estimated 1.4 billion by the year 2025 and 1.7 billion by the year 2050 A.C, India would have to increase its grain production to higher levels. Assured irrigation, increase fertilizer use, high yielding varieties and plant protection measures have already helped appreciably to increase yield per unit area. However considering the limitations for further increase in production through these methods and shortage of arable land, agriculture will be pushed more and more in problematic soils. These soils are mostly found in States of Uttar Pradesh, Haryana, Punjab, Madhya Pradesh, Bihar and Andhra Pradesh. Based on exchangeable Sodium, pH, quantum and nature of soils, these soils have been classified into two broad groups' alkali and saline soils.

A. Alkali soils

Alkali soils, also known as sodic on solenetz soils, are those in which pH exceeds 8.5, exchangeable sodium percentage (ESP) in more than 15. The electrical conductivity of the saturated paste (EC) is less than 4 dSm^{-1}. These soils are dominated by CO_3 and HCO_3 of Na, contain 2-4% amorphous $CaCO_3$ in the surface soil and a hard $CaCO_3$ *kankar* pan in the sub-soil. Due to high pH and high ESP, alkali soils develop poor physical properties. The predominant soluble salts are carbonates and bicarbonates of sodium and, therefore, these soils are also termed as sodic soils. Locally these soils are called bara, rakkar or kallar soils. During the dry season, salts of sodium accumulate at the surface and show up as white or brown salt crust. The brown colour is due to the presence of dispersed organic compounds. Micro-relief of the area and geo-chemical mobility of the salts play a major role in the formation of alkali soils. The dispersed clay particles block the soil pores resulting in very low infilteration rate, crusting and hardening of surface soil upon drying. Sodication also takes place through the use of high RSC waters, saline irrigation waters low in $Cl:SO_4$ ratio and contribution from shallow brackish ground water. These soils are highly dispersed and have poor water and air permeability affecting plant growth. The ground water quality is generally good but in some cases, it may contain high RSC.

Factors affecting plant growth in Alkali Soils

Plant growth, yield/or quality of crop produce in alkali soils is affected by anyone or combination of the following factors:

I. Poor water and air permeability as a result of high dispersion of soil aggregates and clay particles.

II. Excess of Na which is toxic *per se* to the plants and causes imbalance due to antagonistic effect on K and Ca nutrition.

III. Difficulties in preparing seedbed as all the soil does not come to proper moisture condition at the same time, resulting in clod formation and poor tilth of the soil.

IV. Low permeability of water due to poor conductance from the lower to the upper soil layers and restricted effective root zone.

V. Hard crust on the surface layer which greatly hinders the seedling emergence and reduces the germination percentage and thus, plant population.

VI. Deficiency of Ca since nearby all the soluble and exchangeable Ca is precipitated as insoluble $CaCO_3$.

VII. Toxic concentration of HCO^-_3 and CO^{2-}_3 ions.

VIII. Continuous loss of fertile soil due to wind and water erosion.

IX. Low activity of useful microbes due to high pH and excess exchangeable Na.

X. Decreased solubility and availability of micronutrients like zinc (Zn), iron (Fe) and manganese (Mn) due to high pH, $CaCO_3$ and soluble CO_3^{2-} and HCO_3^- irons.

XI. Presence of hard $CaCO_3$ layer which is a physical barrier for the movement of water, salts and vertical root growth of crops and trees.

XII. Increased solubility and accumulation of certain elements like fluorine (F), selenium (Se) and molybdenum (Ma) to toxic level in plants, that may affect crop yield and/or health of the animals feeding on such crops. Increased concentration of flourine can cause severe problems of bone diseases in animals. Selenium toxicity in the form of *Degnala* disease, which affects skin and hair, is quite common in cattle fed on fodder and straw of crops raised on alkali soil.

Depending upon the amount of exchangeable Na, pH, nature and stage of crop growth, environmental conditions and the overall manage-ment levels under field conditions, plant growth is adversely affected due to combination of the above factors.

B. Saline soils

These soils contain excessive quantities of soluble salts predominantly chlorides and sulphates of sodium, calcium and magnesium because of which plant growth is affected. These salts show up on the surface as a white crust which becomes particularly conspicuous with the movement of dissolved salts from the lower soil depths to the surface and their accumulation after the evaporation of water. In saline soils EC exceeds 4 d Sm^{-1} and pH is less than 8.5 and the ESP is usually less than 15. These soils are in a flocculated state and hence a fairly permeable. The plant growth in such soils is affected by the osmotic effects of salts, the toxicity of the soluble ions like sodium, chloride and boron and lesser availability of essential nutrients due to competitions. High salt concentration prevents germination of seeds. Where seeds may germinate, young plants either make a retarded growth or die after germination. Hence, the growth of crops is irregular and patchy. Different crops respond differently to soil salinity. Few crops resist to salinity while others are very sensitive and important examples are as follows.

Crop groups based on response to salt stress

Sensitive group		Resistant group	
Highly sensitive	**Medium sensitive**	**Medium tolerant**	**Highly tolerant**
Lentil	Radish	Spinach	Barley
Mash	Cowpea	Sugarcane	Rice (Transplanted)
Chickpea	Broadbean	Raya	Cotton
Beans	Vetch	Rice (direct sowing)	Sugarbeet
Peas	Cabbage	Wheat	Turnip
Carrot	Cauliflower	Pearlmillet	Tobacco
Onion	Cucumber	Oats	Safflower
Lemon	Gourds	Alfalfa	Taramira
Orange	Tomato	Blue panic-grass	Kernal grass
Grapes	Sweet potato	Para grass	Date palm
Plum	Millets	Sudan grass	Mesquite
Pear	Maize	Guara	Taramira
Apple	Berseem	Pome granate, Desi Babul (*Acacia*)	

Recommended cropping sequences for saline soils

- Pearlmillet-Barley
- Pearlmillet-Wheat
- Pearlmillet-Mustard
- Sorghum-Wheat/Barley
- Sorghum-Mustard

Cultural practices

In saline soils salts often accumulates in few cm of soil during non crop period. When high water table exist may result in excessive salt accumulation in root zone particularly in arid and semi-arid regions. Under these conditions both germination and yield are adversely affected. Successful crop rising in saline soils can be achieved by:

- A heavy pre-sowing irrigation to leach the accumulated salts from the root zone is very useful and essential to improve germination and early growth. Whenever good quality water available, pre-sowing irrigation should be given.
- Deep tillage and inversion have been found to reduce sensitivity hazards.
- Higher seed rate and closer spacing are advisable to counter these effects. For cotton, pearl millet, sorghum, wheat, barley, mustard about 25% higher seed rate over that recommended or normal soils ensure good crop stand. In case of transplanted crops, number of seedlings per hill should be increased to compensate any loss in germination.
- Chemical treatment of seed and seedling with Ca salt increases tolerance of wheat in saline soils.
- Furrow planting may help in obtaining better crop stand and yield in saline conditions because salt tend to accumulate on the ridges under furrow method of irrigation.

Soil fertility management in saline soils

The depth and frequency of irrigation depends upon:

- Proliferation and depth of root penetration
- Capacity of soil to store and transmit water
- Nature of plant responses to soil water stress

In saline soils, the evapotranspirations needs of crops are modified by presence of soluble salts as the amount of salts increases the water in the soil becomes less available to the plants. The soil may appear quite moist. This is because osmotic pressure of soil solution increases with the increase in ion concentration or salt concentration and the plants are unable to extract water as readily as the crops can extract water from relatively normal soil.

In saline soils, crops tend to have poor crop stand and transfer less water, therefore, light and frequent irrigation with one or two deep irrigation in the crop season should be applied for keeping the osmotic potential at low level.

Leaching and drainage

Leaching involves removing excess salts from top soil layer to reduce root zone salinity to a desired level. The efficiency with which salts can be leached per unit of applied water depends upon the characteristics of soil profile.

- Depth of applied water
- Method of water application
- Absorption and exchange reaction due to leaching
- The initial wetness of soil

A leaching study showed that to leach as predetermined fraction of salt from the profile, water require continuous ponding in the order: Peat – Clay – Sandy loam.

Nutrient management and techniques to increase nutrient use efficiency in saline and alkali soils:

In saline soils the fertilizer having low salt index should be used as career of nitrogen, phosphorus and potassium, e.g. Ammonium sulphate nitrate, calcium ammonium nitrate, superphoshate, muriate of potash.

In alkali soils, the fertilizer having high salt index should be used as carrier of nitrogen, phosphorus and potassium, for example calcium ammonium nitrate, ammonium sulhate, ammonium sulphate nitrate, mono and diammonium phosphates, muriate of potash, potassium sulphate.

Method of application

- Urea and cation micronutrients applied as foliar application
- Nitrogenous fertilizers place at a soil depth of 10-15 cm
- Phosphate fertilizers should be band placed
- Heavier application of nitrogenous fertilizers in one time should be avoided

The response of fertilizers have been noted only upto moderate level of salinity or alkalinity but at higher levels the response is negated.

Sodic soils are deficient in

- Calcium, nitrogen, zinc
- Organic matter
- Marginal in iron
- These soils are put to cultivation after application of amendment, organic and chemical ones (gypsum or iron pyrites)
- Need not be dressed with Ca separately
- Soils reclaimed by iron pyrites do not exhibit Fe deficiency too in crop production
- These soils do not respond to phosphorus during initial 3-5 years of reclamation due to high amount of native available phosphorus
- Response to K is also not reported due to presence of K bearing minerals and

large contribution from non-exchangeable K (above 90%) to total uptake by crop.

Alkali soils, therefore, require special management of N and Zn which is entirely different from their counterpart normal soils.

Form and method of nitrogen application

- Among different forms, ammonium sulphate is found most suitable form of N for all crops at high pH due to residual acidity.
- At low pH, urea and ammonium sulphate were at par perhaps due to soil buffering
- In rice pre-transplanting submergence for one week further curtailed NH_3 volatilization by about 10%
- Increasing levels of nitrogen before puddling and one week of pre-transplanting submergence increased grain and straw yields and N uptake of rice as salts leach down
- Basal dressing of ½ or1/3 nitrogen before puddling of partially reclaimed sodic soil increased grain yield of rice

Nitrogen rate

There is an arbitrary recommendation of applying 25-30% more N for different crops in sodic soils than normal soil with a view to compensate N loss from alkali soil. However, more the N applied more could be its loss from soil probably due to poor crop stand. Toxic effect of NO_2 accumulation in plants due to reduced activity of NO_3 reeducates which becomes toxic to plants.

Phosphorus management

Alkali soils do not respond to phosphorus application during first 3-5 years of reclamation due to high available P status of soil on account of reaction between insoluble calcium phosphate and sodium carbonate to cause soluble sodium phosphate.

Rice crop responded to 30 kg P_2O_5/ha from 4th year onwards due to depletion of native P through leaching, fixation and uninterrupted crop mining. Application of phosphorus to either or both in rice-wheat cropping system increased the grain yield of rice and wheat only after 5 and 11 years of cropping, respectively. Probably rice with a shallow root system fails to make use of higher P status of soil depths unlike wheat which is able to tap P from lower depths. Positive response to applied P has been observed up to 120 kg P_20_5/ha in case of berseem crop.

Potassium management

Predominance of micaceous minerals in the soils of many arid and semi arid regions, Na-K exchange in biotite and dissolution of muscoutic structural units release large amount of available K in alkaline sodium soil environment. Crop response to applied K fertilizers has not been observed in alkali soils of the Indogangetic plains.

Zinc management

Among micronutrients, alkali soils have acute deficiency of Zinc and iron is observed in rice. Basal dressing to both rice and wheat grown in sequence in alkali soils at 40 and 20 kg $ZnSo_4$/ha respectively is recommended. Spray of 0.5% aqueous solution of Zn sulphate (1000 l/ha) twice after 20 days of transplanting/sowing of crops at an interval of 10 days was superior to its soil application with respect to yields and uptake and recovery of applied zinc. Soil applied of zinc sulphate raised Zn soil status. Continuous application of Zn for 3 years led to build up to Zn in various forms in the soil.

Integrated nutrient management

Incorporation of any sort of organic material helps build up of humus and hence, fertility status of soil improves by way of:

- Reducing soil pH
- Nitrogen loss through acid production
- Immobilization of nature and applied nitrogen, respectively.
- Green manuring of leguminous crops before rice is beneficial
- Alternatively press mud or water hyacinth (aquatic weed) or FYM manure or any other potential source of nutrients particularly nitrogen, may be utilized to supplement fertilizer nitrogen in alkali soils for sustainable crop production with simultaneous improvement in soil.
- Though integrated use of water hyacinth or press mud or FYM in alkali soil, fertilizer nitrogen may be saved up to 50% of the recommended dose with an additional advantage of soil improvement.
- Besides, organic sources also had residual effect on subsequent wheat crop and raised wheat yield by more than 20%
- Use of organics also brought down soil pH and ESP

2. ACIDIC SOILS

These soils are characterized by low pH levels generally less than 5.5.

Reclamation of acid soils

The acidic soil management should aim at improving the production potential of soil by addition of amendments or by agronomic manipulation to obtain maximum productivity. Application of lime as an amendment to neutralize the exchangeable Al^{3+} is quite effective for amelioration of sub-soil acidity, application of gypsum ($CaSO_4$) is more suitable than lime which may be due to non-phytoxicity of $A1SO_4$, sulphate component and water solubility of gypsum. The acid tolerant plant species and varieties which have Aluminium tolerant mechanisms can be grown in acidic soils. The rice has good tolerance to acidity as flooding of rice raises pH of field to almost neutrality. Short duration rice varieties grown in uplands tolerate acidity. Minor millets (*Panicum miliare*) and finger millet (*Eleusine coracana*) are tolerant to acidity and there crops show meager response to liming. Gram, lentil, groundnut, sorghum, field peas and maize show medium response to liming. Pigeonpea, cotton and soybean are sensitive to soil acidity and these crops respond well to liming. Fodder crops like dinanath grass, napier bajra hybrid, guinea grass, maize and Setaria can be grown under agroforestry system in acid soil conditions. The grasses viz., napier bajra, dinanath grass, Setaria, *Paspalum notatum*, Kikuji grasss and guinea grass can be successfully grown under acidic soils. Among cereals/legumes, lucerne, cowpea, maize, jowar are important in acidic soils. Important tree/shrubs for acid soils are *Sesbania, grandiflora*, *Leucaena leucecephala* spp.

Acid and acid sulphate soils

Characterizerd by low pH generally less than 5.5. In acid soils growth is affected by excessive amount of dissolved ions of Al, Fe, Mn and Cu and become toxic. Acid soils are available on low lying coastal lands which are generally rich in pyrite and on oxidation form H_2SO_4. Acid sulphate soils have pH < 4. The crops in these soils suffer mainly because of Al toxicity and nutrient deficiency specially P.

Application of $CaCO_3$ is recommended according to time requirement of soil. It varies with the factors like:

(i) pH of the soil
(ii) Clay content
(iii) Cation exchange capacity
(iv) Climate

For reclamation of acid sulphate soil water management is the key to soil management. Root development is always restricted in acid sulphate soils unless water is available for irrigation or to maintain a high water table artificially as flooding alleviates activities; rice is the best choice to be grown on acid sulphate soils in warm region. Rice yield can be as good as of non-acid soils if recommended dose of lime as well as fertilizers are applied. The benefits of lime and phosphorus applications may be due to increases supply of nitrogen by stimulation of ammonification and microbial fixation of nitrogen process.

Application method for reclamation in acid sulphate soils

Liming material should be evenly broadcast and mix in soil. A few weeks in advance to sowing of a crop for completely exchange process. Proper moisture of soil enhance the exchange process generally lime should be applied after every 5 years. However, the soil should be periodically get checked for pH.

For reclamation of acid sulphate soils water management is the key to soil management. Root development is always restricted in acid sulphate soils unless water is available for irrigation or to maintain a high water table artificially, as flooding eliminates acidity. Rice is the best choice to be grown on acid sulphate soils in warm regions.

Total reclamation can be slow and expensive but reclamation is comparatively easier or less costly on soils which do not have resources of pyrites or high value of exchangeable aluminum. On such soils moderate application of lime will significantly reduce soluble aluminium levels and enable crops to respond to conventional fertilizer application without necessarily raising pH above 4.5.

Agronomic techniques

Many of principal crop and vegetation are sensitive to acid soil and grown on these soils get injury. Rice, millets, oats, tea and potato are crops which can tolerate to some extent to soil acidity on low pH soil. Among these crops rice is well adapted to acid sulphate soils because it grows under flooded conditions which depresses Al toxicity and increases availability of phosphate. Rice yield can be as good as of non-acids soils if soil is submerged a long as possible before transplanting and recommended dose of lime and MnO_2 and NPK are applied. The benefits of lime and phosphorus application may be due to increased supply of N by stimulation of ammonification and microbial fixation of nitrogen processes.

Nutrient management in acid soils

Nitrogen

- The availability of nitrogen in acid soils usually increases on liming by increases the pH.
- The effect of lime on mineralization of soil nitrogen was not so pronounced in green manured soil as in untreated (no green manure)
- In a field experiment, average release of 12 ppm of NH_4-N from a soil was recorded, the pH value was raised to 7.2 after 37 days of lime application
- The calcium ammonium nitrate was as better source for growing wheat in acid soils as it contains 8.1% Ca.

Phosphorus

Management of phosphorus in acid soils depends upon:

- Extent of acidity
- Status of sulphur
- Crop duration

Evidence of greater availability of phosphorus as rendered by liming of acid soils has been observed and found significant increase in the phosphorus content of the leaves of plants in number of crops. This effect of lime diminished in the presence of P fertilizers suggesting that liming helped in the mobilization of nature soil phosphorus.

The effect of continuous manuring for about ten years with acidifying fertilizers e.g. sulphate of ammonia, superphosphate etc. in the acid red loam soil of Kanke (Ranchi) has been quite evident from pH, available phosphorus potential and lime potential.

Potassium and calcium

Reported higher concentration of Ca in soybean, mungbean (green gram) and barley plants grown after liming in the acid red soils indicating greater availability to the crops.

- The availability of K on the other hand, in general, was found to decrease
- It was also reported that no definite effect of lime on the availability of soil K
- Liming increased or decreased the exchangeable K in soil only slightly
- Calcium contents of plants also increased or decreased the exchangeable K in soil only slightly
- Calcium content of plants, however, increased considerably due to liming

Micronutrient management

- Crop response to various micronutrients in acid soils is low to medium with very few cases of high response.
- Response of micronutrients such as Fe, Mn, Zn and Cu are very limited in acid soils
- However, the behavior of zinc in submerged low and acid soils is quite different in which rice responds to zinc application

Integrated nutrient management

A critical appraisal of the different compounds for integrated nutrient management source particularly for acid soils reveals that major weakness lies with non-availability of sufficient organic manure like –

- FYM
- Compost
- Crop residues
- Fertilization sources like paper factory sludge (pH 5.52, organic carbon 19.3% N, 0.81%, P 0.12% and K 0.44%) along with flyash (pH 8.47 organic carbon 0.03%), N 0.05%, P 0.03% and K 0.18%).
- Chemical fertilizers produced the highest grain yields of rice compared with other combination when incorporated 15 days before transplanting in acid laterite soils.
- There was an increase in the soil pH (5.30-5.82) organic carbon content (0.32 to 0.46), N (163.2 to 215.1 kg/ha) and P (14.9 to 25.3 kg/ha) due to addition of paper factory sludge along with flyash and chemical fertilizers but no remarkable changes in K content (130 to 146.8 kg/ha) was obtained.
- The increase in residual fertility was attributed to increase in soil pH and carry over effect of paper factory sludge which added a large quantity of nutrients in the form of organic sources benefiting crop growth.

3. WATER LOGGED SOILS

The soils with very shallow water table are termed as water logged soils.

Waterlogged law: The working group on problem identification in 1 m area adopted the following norms for identification of waterlogged area:

i.	Waterlogged area	Watertable within 2 m from the low surface.
ii.	Adopted area for waterlogging	Watertable between 2-3 m from the low surface.
iii.	Safe area	Watertable below 3 m from the low surface.

For water logged areas, the following norms are supported for the classification of different categories.

(i) Water-table within 2 m from soil surface during pre-monsoon (April/May).

(ii) Water-table within 1 m from soil surface during post-monsoon (October-November).

Under these two conditions, the area may be treated as waterlogged which adversely affect crop yield and suitable drainage measures be adopted to remove excess water. The areas where the water-table may be deep presently but may rise under conditions of frequent irrigations, heavily rainfall and inadequate drainage are termed as critical and are potential areas for water-logging. It is estimated that about 6.0 m ha land suffers from surface flooding (3.4 m ha) and high water-table (2.6 m ha). The continuous availability of water in the root zone lowers aeration which in return results into poor growth and nutrient uptake.

Critical and potential areas for waterlogging

Critical area for water logging has water table between 2 and 3 meter from soil surface during pre-monsoon and are between 1 and 2 m during post monsoon period. Waterlogging may develop for a short period if suitable measures are adopted which are located specific and these may comprise drainage system, land development and scientific irrigation management. During monsoon, irrigated area with water table between 3-5 m during pre-monsoon may be considered as potential areas for waterlogging.

Reclamation of waterlogged and salt affected lands

Waterlogging can be controlled by adopting suitable drainage measures and through scientific methods of water management. Saline lands are reclaimed by addition of chemical amendments i.e. gypsum ($CaSO_4.2\ H_2O$). Sub-surface drainage is ideal for leaking of excess salts and reaction products after addition of chemical amendments from saline soils. Acid soils are reclaimed by adding lime to soil.

Classification of drainage system

There are two major types of drainage system; surface and sub-surface. In addition to these, two ground water control systems; by pumping is called vertical drainage and reclamation efforts of saline and waterlogged lands by planting certain trees with high evapo transpiration rate and resistance to advance soil climatic conditions is called bio drainage.

Surface drainage: It is adopted to remove excess surface necter from crop land, temporary surface water disposal is done through a network of surface drains, linking crop land with main outlet.

Cultivation of tolerant species: In water logged soils growing of excess water tolerant forage crops viz., dicanthium, guinea grass, iseilema, maize, napier bajra hybrid, paragrass, setaria and signal grass etc. are beneficial for increasing productivity. Tree species suitable for waterlogged saline soils are *Tamarix articulate, Prosopis juliflora, Acacia farnesiana, Pithecellobium dulce* and *Parfinsonia aculeata* etc. whereas *Eucalyptus, Casuarina* and *Populus* trees can be grown in saline waterlogged areas near canal commands. Among shrubs and under shrubs *Salicornia, Italoxylon, Salsola, Tamarix* and *Suaeda* are important for waterlogged soils.

Methods of field drainage

The accomplishment of surface drainage of agricultural lands includes land farming, low smoothening, low grading or leveling, bedding system, open ditches and levels.

- **Land farming:** Land farming is a process of changing natural topography with the aim to control movement of water contour from land surface. This practice includes land smoothening, low grading or leveling and shallow field drains crossable by farm machinery.
- **Land smoothening:** In field lands where there are only minor differences in field levels, good surface drainage can be achieved by land smoothening.
- **Land grading and leveling**: For efficient drainage, the grade in the dissection of drains can be 0.05 to 0.5%, however, preferable slope is 0.1%.
- **Bedding system:** This is suitable for par hydraulic conductivity soils having slopes upto 1.5%. The leveled surface of field is converted into sloping area by repeated ploughings and the beds are always kept in parallel to the greatest slope of the field such that deep furrow serves as a field drain. The length of bed depends on field conditions and it may vary from 100 to 300 m. The width of bed depends upon slope, land use, soil permeability, farm machinery to be used and the type of farming operations.
- **Open ditches:** It includes 3 systems i.e. parallel field drain, random system and parallel open ditch system.
- **Parallel field drain system:** This system is used when surface is uniform with few noticeable ridges or depressions. The channels are spaced further apart than in a bedding system and there have greater capacity than the dead furrows.
- **Random system:** This system is most suitable for areas having deep depressions and it is not technically feasible to fill depressions. The slope of channel can be as flat as 10:1 where depth of drains is more than 60 cm or 4:1, of the field are formed parallel to drains.

- **Parallel open ditch system:** This not only provides surface drainage to an area but also helps to control the ground water table.
- **Levels:** These are commonly known as embankments and these are used in conjunction with drainage works to exclude run-off from uplands entering the areas. This method is commonly used in flow plains and coastal areas.

Sub surface drainage

Two types of man made systems of sub surface are: (i) horizontal (ii) vertical. The horizontal drainage involves engineering structures laid horizontally parallel to ground and vertical drainage is mainly achieved by pumping out ground water through tubewells. These structures are laid horizontally and works like tubewell.

- **Biodrainage:** The biodrainage system involves growing certain categories of plants which draw their main water supply directly from ground water or the capillary fringe just below it. The tree species belong to this group are *Eucalyptus*, Poplar, *Tamarix muskit* and *Acacia* spp. Similarly, non-woody plants like bushes, sedges and grasses are called as phreatophytes. Two species of *Eucalyptus* which are commonly used as biodrainage are *Eucalyptus tereticornis, Eucalyptus camaldulensis*. These trees transpire luxuriantly when ground water contact has been made. These tree species usually use much water than required for normal growth conditions.
- **Vertical drainage:** It refers to removal of excess water by means of a hole either under pressure or through pumping. The success of vertical drainage system depends upon the presence of favourable unconfined aquifers within 50 m depth below soil surface to permit pumping of groundwater on sustained basis. Good to moderate quality groundwater can be utilized for irrigation purposes.
- **Horizontal drainage:** A horizontal drainage system consists of the laterals, collectors and main drains. The water entering the laterals flows into the collectors, which carry it to the main drain.
- **Drainage:** Subsurface drainage provides an efficient means for lowering water table to prevent secondary salinization. In addition to initial cost of installing subsurface drainage and problem of disposal of drainage water could assume serious preparation in areas where no natural outlet like sea is available.

Water logged soil

Water logging refers to the presence of excess water or water above saturation in or near to the root zone or standing water above ground for any substantial period of time. An agricultural land is said to water logged when soil pores in the root zone get saturated with water.

The Government of India in 1942 establish a land reclamation to tackle with this problem and set up land reclamation, Irrigation and Power Research Institute at Amritsar during 1947 (LRIPRI) where different means and ways to reclaim the soils are being evolved.

In India this problem is more acute in Indo-gangetic plains mainly due to :

- Flat topography
- Frequent flooding during monsoon
- Canal irrigation in these area

Extent of water logging

India	59.9 lakh ha
Punjab	2.86 lakh ha
Haryana	6.50 lakh ha
UP	8.10 lakh ha
West Bengal	2.60 lakh ha
Rajasthan	3.50 lakh ha
Madhya Pradesh	1.00 lakh ha

In Punjab water logging problem is reported in the district of Amritsar, Gurdaspur, Sangrur, Fazilka, Faridkot, Ferozpur, Shri Muktsar Sahib etc.

Effect of water logging

Excess amount of water in the root zone interfere with the biological processes. The main effects are:

I. Reduced aeration and reduced O_2 supply to roots

II. Lack of O_2 supply to roots results in reduced respiration of the protoplasm of root cells resulting in death of roots.

III. Reduced absorption of water and nutrients (P, K, Ca, Mg)

IV. CO_2 is accumulated in no aerated soils and produce toxic and inhibitory effect on roots (CO_3, HCO_3 and H_2CO_3)

V. Root growth is affected due to reduction in cell division and cell elongation

VI. Loss of nutrients – The continuous percolation and draining of water may lead to leaching and removal of nutrients particularly N in sufficient quantity. On the other hand due to reduced conditions there may be reduction of NO_3 to gaseous form which is lost to atmosphere.

$$NO^-_3 > NO^-_2 > N_2O^- > N_2$$

VII. Metabolic processes of plants are affected (ATP)

VIII. Decomposition of O.M. is slowed down (the end products are – CO_2, H^+, Methane, H_2S)

IX. pH is changed – In acidic soils it increases and in salty soils it decreases

X. Plant growth is reduced due to welting and leaf fall

XI. Water logging give rise to soil salinity and alkalinity conditions due to excess evaporation resulting in salt deposition to the surface layer.

Causes of water logging

I. High rainfall which recharges the water table results in rising the water table.

II. Surface flow of rain water from upper regions to lower regions also results in prolonged inundation of lower system.

III. Seepage from rivers and canal system. In Punjab water table has risen very faster in some areas due to canal irrigation.

IV. Intensive-cropping and growing of high water requiring crops like paddy is also a source of water logging.

V. Blockage of natural drains due to construction of roads, rails, canals etc. Such type of construction has resulted blockage of natural water courses which otherwise were helpful to drain away the excess water.

Control measures

1. Mechanical measures
2. Agronomic measures

1. Mechanical measures

I. Drainage: The most effective answer to water logging is a properly designed drainage system. In fact, the importance of drainage is so well recognized in advance countries that it is always considered integral part of any irrigation scheme.

a. In West Germany for example no irrigation project is taken up without a corresponding drainage project. In this country, too, instructions have already been issued that irrigation project estimate should include the cost of drainage work.

II. Field drains: In water logged areas or prone to water logging field, drains to drain individual field will be necessary. But as their construction and maintenance is to be done by farmers, the later may need technical advice, with regard to such matters as alignment depth and grade of drains. Drains may be deep enough to keep the ground water at a level which would not be harmful to standing crops. In Maharashtra excavation of drains to a permeable sub stratum where it was available at reasonable depth, has been successful. In case of soils with low permeability, sub-surface drains may be required.

III. Borrow pits: Water generally stagnates in the borrower pits formed as a result of excavation of canals, and road and rails embankments. This is not desirable in an area prone to water logging. The commission recommends that the indiscriminate excavation of borrow pits should not be allowed. If they are unavoidable, they should be so excavated that water will not stagnate but flow into the near drainage system.

IV. Use of ground water for irrigation: Adequate provision for surface and ground water use should be made in the planning for new projects, because it helps to keep water logging under control. Farmers should also be encouraged to dig well and sink shallow tubewells. Construction of large number of shallow tubewells by farmers in Punjab has helped to kept the water table down. The commission recommends that necessary technical and financial assistance should be given to the farmers for the construction of wells and shallow tubewells.

V. Lining of canals: Lining of canals is of great importance in areas which on the introduction of irrigation are liable to water logging and salinity. If lining is done at the initial stage it will cost much less than if done later.

2. Agronomic measures

In areas susceptible to water logging, it will be desirable to introduce only those crops which need light irrigation. Crops need heavy irrigation will have to be limited.

(I) Water management

Land behavior of groundwater levels after the introduction of irrigation has to be studied. The depth of sub-soil water table has to be observed in selected open wells distributed over with a greater concentration in low lying areas. The observations are generally made twice a year once in June before the monsoon and second time in October after the monsoon. Our objective is to know the variations in the water level from year to year.

(II) Planting method

Ridge planting technique should be adopted in water logged areas. This makes the plants less affected due to shallow water table or stagnation of water in the fields.

This method includes:

- Narrow ridge and furrow system in which ridges are mini-bunds furrows act as a drain for excess water.

(III) Narrow bed and furrow method

Narrow beds of smaller width (37.5 cm) are made and furror is ment for water drainage. Crop is raised on beds.

- Broad bed and furrow method: In which broad beds of different width (1-2 m) are made and a furrow is provided after each bed for draining excess water. The crops are raised on beds.

Fertilizer management

There are four types of losses of nitrogenous fertilizers:

I. **Immobilization:** Broadcast urea is more susceptible to this loss by Algae (15-40%)

II. **Ammonia volatilization:** These types of losses are low in submerged soils

III. **Nitrification and denitrification:** In aerobic surface soil layer the ammonia is oxidized to NO_3 which diffuses into anaerobic layer below and is lost by denitrification.

IV. **Leaching and runoff:** Broadcasted urea may be subjected to leaching and runoff losses from soils. In light soils with high percolation rate, urea loss due to leaching is more.

Management of nutrient loss

I. **Deep placement:** It reduces losses through nitrification, denitriication, urea super granules can be placed deeply.

II. **Slow release fertilizers:** To delay the availability of N to the plant until it has strong root system which can compete with the mechanism of losses and biological immobilization.

III. **Split application:** The concentration of N in flooded water may be reduced by split application of N fertilizer resulting in reduced leaching or runoff losses. The application may be made with the need of the crop.

IV. Higher doses of N (25% than normal) to compensate losses.

V. Ammonical or amide forms of N fertilizer should be preferred as these are less subjected to dentrification losses.

4. ERODED SOILS

Soil erosion occurs mainly through the forces of wind and water.

Wind erosion mechanism and processes

- **Saltation:** The major portion of soil carried by the wind moves in a series of short bounces is referred as saltation. It carries fine particles of 0.1 to 0.5 mm in diameter. Saltation is caused by direct pressure of wind on soil particles and their collision with other particles. The soil particles are pushed by the wind and are placed almost vertically in the first stage of saltation. Some of grains size only as short distance and some leafs 30 cm or higher depending upon wind velocity and size from the ground.
- **Suspension:** Very fine soil particles which are less than 0.1 mm diameter are carried into suspension. There are kicked up into air by action of particles. The movement of fine dust particles in suspension is governed by characteristic movement of wind. The suspended material is carried to long distance from its original location. Therefore, it is a complete loss of soil particles to the eroded area.
- **Surface creep:** The soil particles larger than 0.5 mm in diameter but smaller than 1.0 mm are along the surface comes in impact of particles in saltation to form a surface creep. The soil moved in saltation and surface creep usually remains within the eroded area, when the exception winds are blowing from different dissections.

Soil and water conservation on agricultural land

- **Agronomic measures**: Agronomic measures help to intercept rain water and reduce the slpash effect; improve better intake of water by the soil by improving the organic matter and soil structure; reduce run-off losses through use of contour cultivation, close and dense crops, strip cropping, mixed cropping and mulches.

- **Contour farming:** It is a simple conservation practice of farming operation, viz. ploughing, cultivation, planting across the slope, keeping the same level as far as possible. The ridges, furrows and the rows of plants are placed across the slope. These form a continual series of miniature reservoir and act as barriers are small and individually, but their number is large so their cumulative effect is more in reducing run-off, soil erosion and losses of plant nutrient.
- **Cover management:** Legume crops are erosion resistant; provide better soil cover and better protection to land against erosion than ordinary wide row erosion permitting crops. Cowpea, green gram, black gram, groundnut, soybean, sunnhemp and dhaincha are good cover crops and effectively reduce soil erosion. Nearness of crop canopy to soil surface and extent of crop cover determine erosion control. Other factors like inoculation with Rhizobium or fertilization with single superphosphate prove beneficial.
- **Mulching:** The surface mulches are useful for protecting the soil against blowing and beating action of rainfall, reducing evaporation losses, increasing water infiltration, checks weeds growth, improves soil structure and finally increases crop yield. Mulching is the most effective measure of soil and water conservation but the mulching material is non-available due to its competition for annual fodder poses limitation in its adoptions which invites for strategies of *in-situ* mulching or utilization of weed biomass for mulching. Growing of sunnhemp green manure in 60 cm wide contour strips between maize rows and recycling after 30-days as soil mulch reduces erosion and cheeks weed growth.
- **Conservation tillage:** Low intensity tillage favors consolidation of soil and imparts erosion resistance through better consolidation, stıucture, infiltration, pose distribution and depression storage.
- **Intercropping:** The line sowing of mixed crops give arise to intercropping practice. Intercropping of sorghum and pigeon-pea in 1:1 ratio has proved sustainable system in semi-arid regions of south-eastern Rajasthan.
- **Strip cropping:** Contour strip cropping is the growing of erosion permitting crops in strips of suitable widths across the slopes on contours, alternate to the strip of soil protecting and erosion resistant crops. The strip cropping shortens the slope length, hinders movements of run-off, help in desilt it and increases the absorption of run-off water. Dense foliage cover of erosion resistant crops viz., groundnut, mothbean and horsegram, prevents beating action of rain drops.
- **Mixed cropping:** Indian farmers have been practicing mixed cropping for better land cover and acting as protection against beating action of rain drops, soil erosion and crop assurance under adverse climatic situations. This is practiced in semi-arid regions to provide as insurance against complete failure of the given crop under adverse climatic conditions.
- **Bench terracing:** It consists of a series of platforms having suitable vertical drops along contours or on suitably graded lines across the general slope of land. Bench terracing is usually practiced on slopes ranging from 6 to 33%. The vertical drop varies from 60 to 180 cm, depending upon the slope and soil

conditions. The material excavated from the upper part of terrace is used for filling the lower parts.

- **Grassed waterways:** Grass waterway is a safe disposal structure for higher granting run-off. It protects the land against rills and gullies. Waterway is constructed according to a proper design and vegetative cover is established to save the channel section. The best suited grasses are *Panicum repens, Brachiaria mutica, Cynodon plectostachyus, Cynodon dactylon* and *Paspalum notatum.*
- **Water erosion:** Soil erosion caused by water can be distinguished into (i) sheet erosion (ii) rill erosion (iii) gully erosion.
- **Sheet erosion:** The erosion in which soil metrix is lost but remains undetected for long period and removes a thin venier of soil uniformly from large areas during every rain, even producing a run-off is sheet erosion. It can be detected from muddy colour of run-off arising from the fields.
- **Rill erosion:** When sheet erosion is allowed to continue unchecked, the silt laden run-off forms a well defined, minute finger shaped grooves over the entire field is known as rill erosion. The rills formed don't pose any hindrance to farm machinery. The rills vanish after cultivation but reappear with ensuing rains.
- **Gully erosion:** When rill erosion is neglected, the tiny grooves develop into wider and deeper channels known as gully erosion. The gullies obstruct the farm operations. These gullies tend to deeper and widen with every heavy rainfall. These gullies cut the entire field into small fragments and in course of time, make them unfit for cultivation.
- **Water erosion mechanism and processes:** The soil erosion caused by rainfall is due to impact of energy from (i) falling rain drops obliquely or vertically from above, and (ii) surface flow more or less horizontally along the ground surface. The main role of falling rain drop is to detach the soil particles (called splash) and that of the surface flow (outride the rills and gullies) is to transport the soil, which gets accelerated on sloping topography.

Causes of erosion

I. Removal of vegetation from soil

II. Extension of cultivation beyond prohibly steep slopes

III. Permanent agriculture is possible on certain slopping lands but not without adoption of approval soil conservation practices

IV. Shifting cultivation: Clearance of rich forest soil. Forests are uprooted and cleared. Areas are cultivated for 2-3 years and then abandon for 5-15 years and again for cultivation. Common in Tribal Community called Jhooming.

Factors effects soil erosion by water

- Precipitation : Intensity and amount
- Degree and length of slope

- Type of soil e.g. fine soils are more susceptible to erosion
- Nature of ground cover and land use

Effect of erosion

- Soil is lost and its fertility depleted
- Most of water run off are utilized to sea
- Flood water become more frequently damage both for life and property

5. DRYLAND FARMING

- **Dryland farming:** Is concerned with crop production without any irrigation in semi arid regions. Management systems of dry land crop production are conservation and use of stored soil water.
- **Dryland/Rainfed farming:** Is concerned with crop production without any irrigation and in rainfed farming is also concerned with crop production without any supplement irrigation, but in case of dryland farming the rainfall is on the lower side (< 750 mm/annum) whereas in case of rainfed farming rainfall may be more than 750 mm. In case of dryland as well as rainfed farming main source of moisture is natural precipitation. In case of dryland farming crops may suffer frequently due to drought whereas in case rainfed farming relatively frequency of drought is on lower side. In dryland farming growing season is short whereas in rainfed farming it is longer and double cropping is possible depending upon moist holding capacity of soil. However in both cases management of moisture is of great significance for raising crop successfully.

Rainfall availability

Total rainfall in India is comparable with world average rainfall but there is large variation in behaviour of South-Western monsoon which account for nearly 75% of total rainfall received in India.

- The South-West monsoon may commence early or late.
- Prolong breaks during cropping season.
- The rainfall may vary in space and time.
- It may terminate too early or continue late.

Nearly 400 million ha m rain water availability is there, it is based on average rainfall with total area of India. Out of this 70 million ha m water is lost through evaporation and 180 million ha m goes as run off into sea and remaining 150 million ha m water enters into soil.

Distribution of geographical area in different rainfall zones in India

Rainfall zone (mm)	Geographical area (m ha)	Rainfall water availability (Million ha m)
100-150	52.07	15.62
500-750	40.26	25.16
750-1000	65.86	57.63

1000-2500	137.24	205.86
> 2500	32.57	95.73
Total	328.0	400

Problems of dry land agriculture in India

- Moisture stress.
- Inadequate and uneven distribution of rainfall.
- Social economic problems.
- Production and productivity levels.

Social constrains

- Prevalence of marginal lands.
- Lack of financial resources a high degree of risk.
- Minimum choice of crops due to farmer's requirement.

Production problem

- Low yield
- Instability
- Difficulties in deciding crops due to irregularities in monsoon e.g. nearly 90% area under sorghum, pearl millet, finger millet is under rainfed and 91% area under pulses chickpea, pigeon pea is rainfed.
- 80% area under oilseeds groundnut, rapeseed mustard, and soybean is rainfed.
- 50% area under rice rainfed.
- 19% area under wheat rainfed.

Moisture stress

Soil moisture is the most limiting factor in dry farming and dryland farming situations. In biology, stress is the result of action of external factors on an organism. Moisture stress indicates the action of lack of or excess of water on plants. The term moisture stress is generally used for deficit moisture conditions though it is applicable to excess moisture also. As a result of moisture stress there is wilting of crops.

- **Incipient wilting:** Though there is sufficient amount of soil moisture, water deficit develop due to higher transpiration than absorption especially for short period of time. After heavy irrigation water deficient is recovered.
- **Temporary wilting:** Water deficit develop due to higher transpiration than absorption and wilting symptoms appear.
- **Permanent wilting:** Water deficit occur for longer period of time and plants are not in condition to absorb water to fulfill their demands.

Physiological impact of moisture stress in plants

I. **Cell expansion and cell division:** The most sensitive process to water stress is cell growth when turgor pressure in plant cell is full and cell expansion is also decreased because of lack of pressure within the cell, cell division is effected by stress, however, its not clear weather its direct effect of stress or is rather an indirect effect. Stress restricting meristematic cell from enlarging to a size for cell division to take place.

II. **Cell wall and protein synthesis**: Cell wall synthesis is depressed by water stress and water stress inhibits the interception of amino acid into proteins and causes a decrease in protein content of the tissue. Water stress brings out its effect on protein synthesis but it appears that it is not dependent upon phytochrome activity.

III. **Levels of enzymes in plant:** Under moderate water stress some enzyme levels are raised particularly required for hydrolysis and dehydration processes. However, in general water stress results in decrease in enzyme level e.g. *Nitrate reductase* result in suppression of protein synthesis and uptake of solutes also affected, K uptake is increases as cell turgor decreases.

IV. **Plant hormones:** The relationship between water stress and phytohormones are complex. Under water deficits hormonal balance is altered. The activity of growth regulating hormone like abscisic acid, ethylene increases and growth promoting hormones like cytokine and gibberellic acid is decrease. There is rapid accumulation of ABA. In wilted leaves, it appears that higher level of ABA is maintained by more synthesis and metabolism of ABA. ABA production induces stomatal closure and inhibits respiration. Ethylene induces abscission e.g. cotton leaf fall and fruit drop occur more rapidly due to moisture stress created by ethylene hormone.

V. **Photosynthesis:** Photosynthesis is inhibited by water stress due closure of stomatal opening and also due to reduction in leaf area (less expansion of leaf). However, translocation of assimilates reduced due to moisture stress.

VI. **Praline accumulation:** Is a free amino acid takes place under water stress conditions. In case of drought tolerant variety of barley, the level of praline is on higher side. One possible reason to account for accumulation is that praline is synthesized to depress internal osmotic potential and maintain the gradient for water uptake.

VII. **Respiration:** Respiration increases with mild stress. However as the water deficit become severe, it decreases.

VIII. **Growth:** Generally organ of plant growing more rapidly is the one most affected due to moisture stress. The expansion of cells and cell division are reduced due to moisture stress resulting in decrease in growth.

Mechanism to conserve water

I. **Stomatal mechanism:** Drought resistant varieties of crop are capable of regulating stomatal mechanism by closing stomata where drought prevails. For example oats, tall wheat variety, groundnut etc are drought tolerant crops whereas in semi dwarf wheat varieties their stomata open throughout the day and hence cannot tolerate drought.

II. **Increased photosynthetic activity:** C_3 plant assimilate atmospheric Co_2 concentration of about 50-100 ppm and C_4 plants up to 100 ppm concentration of Co_2 in atm. Likewise CAM are highly drought resistant only open stomata during night for example Pine apple and Ague.

III. **Reduction in leaf area:** Rolling and curling of leaves at times of water stresses is helpful in moisture conservation. Rolling of leaves occur in grasses, legumes etc. Leaves are oriented parallel to sun rays thus avoid load of solar radiation. Stressed plants reduce radiation load during mid day by about 60-70% due to folding of leaves.

IV. **Leaf surface:** Leaves with thick cuticle, waxy surface and spines results in less transpiration. Pubescence increase leaf reflectants and reduce solar radiation incidence.

V. **Effect of awns:** Awned wheat variety gave more yield under drought as compare to awn less variety. Awns contribute 12% of photosynthates to grains. Awn has less transpiration due to less transpiring area of awns, its contribution of photosynthates to grain is more.

VI. **Stomatal frequency:** Some desert plants have less than 100 stomata/sq cm as compare to average of 10000/sq cm in mesophylls. Location of stomata in depression cavities in leaf reduces transpiration by liming ion pigment of air currents. Drought resistance of sorghum, maize and groundnut variety is attributed to ability of stomata to resume their normal functioning after a period of water stress.

VII. **Lipids deposits on leaf:** Sorghum and soybean reduce water loss by depositing lipids on plant surface under moisture stress.

VIII. **Succulence:** Cacti have thick cuticle, few stomata, reduce absorption of radiation on visible as well as infra red wavelength resulting in low rate of cuticle transpiration.

IX. **Water storage in plants:** Water storage in plants is adaptive mechanism but its significance is small on crop plants because total water content is small as compared to high transpiration loss in a clay, transpiration of 5 mm/day is common which is equivalent to 50 tonne/ha. The leaves of pineapple contain substantial amounts of water in special non-chlorophyllous tissue which is utilized during drought. Water storage may confer an adaptive mechanism in pineapple in dry condition because its transpiration rates are very low compared to most of crop plants.

Mechanism to improve water uptake

I. **Efficient root system:** Deep well branched rapidly growing root system help in absorbing more moisture by exploiting higher volume of soil. Certain legumes have roots which penetrate to 15 mm deeper. Sorghum has nearly twice as many small branch roots per unit length of root so sorghum is more resistant than maize.

II. **Root shoot ratio:** A high root to shoot ratio is a very effective mean of adaptation of plants to dry conditions. Root to shoot ratio increases with aridity except in case of succulents. Drought increases root growth and root shoot ratio which is an important mechanism of drought avoidance, by improving water uptake high water potential is maintained in leaves and rate of photosynthesis is not reduced.

III. **Osmotic adjustment:** Stress induce break down of CHO and protein increases the concentration of solutes in cell sap resulting in reduction in osmotic potential. Plants which have the ability to reduce their osmotic potential in response to slowly developing deficits are useful. Leaves hypocotyls, root and spikelets have been shown to adjust osmotically in some plants but not in all the spp. A significance of osmotic adjustment is such that it helps to maintain turgor pressure as water deficit develops. It enables the plant to maintain leaf expansion and photosynthetic activity at levels of stress which is not possible in its absence. Due to continued turgor root growth continues even under moist stress, resulting in absorption of more water and helps in exploration of greater volume of soil. Osmotic adjustment results in less translocation and helps in increasing yield. It also reduces the production of ABA. The plants that adjust osmotically must be able to withstand water potential and water contents without damage to tissue. There is large variation in degree of desiccation that plant can tolerate without cellular injury.

MANAGEMENT OF PROBLEM SOILS

Characterization of problem soils

The process of weathering, especially in arid and semi-arid regions releases large amount of soluble salts. These salts keep on accumulating at some depth in the soil profiles depending upon rainfall, soil topography and other factor to ionic mobility of salts causing spatial and temporal variation in salt distribution in soil profile.

I. Alkali soils

- Also called sodic soils
- PH exceeds 8.5
- ESP > 15
- There is conspicuous presence of CO_3 and HCO_3
- Na among the soluble salts
- EC ranges from V.low to V.high

Effect

Alkali soils develop poor physical properties due to high pH and high ESP. Poor infiltration rate due to dispersed clay particles block the soil pores and so cause crusting and hardening of surface soil upon drying. Water and air movement as well as root growth remain restricted. Soil physical property could deteriorate even at lower value than this.

Reclamation technique

The main objective of reclamation in alkali soil is to reduce exchange Na content to safe level for optimum plant growth. The reclamation can be achieved either by applying chemical or creating favourable biological activity in alkali soils.

1. Chemical reclamation

The exchange Na is commonly replaced by Ca through ions like H and Mg. Ca in exchanged reaction is derived from either added Ca amendment or art of Ca present in the nature $CaCO_3$ of soil which is mobilized by using acid or acid forwards.

Chemical used

Gypsum ($CaSO_4.2H_2O$) is most commonly used @ 10-15 ton/ha containing about 70% ($CaSO_4.2H_2O$). About half of actual gypsum requirement is sufficient to reclaim 15 cm surface soil to start growing with rice as the 1st crop.

(a) **Phyrites (FeS_2):** also used in recent years on large scale in some part of country. Biological oxidation of pyrite is carried out by *Thiobaccilus ferroxidins* bacteria only under low pH conditions. But in case of oxidation of element sulphur by *Thiobaccilus thioxidens* very slow in alkali soils. The acid produced through oxidation reaction with $CaCO_3$ of soil to release Ca for exchange with Na on the clay complex. Due to this problem efficiency of pyrite remain between 70-80% as compared with gypsum at recommended dose of application.

(b) **Mineral acids:** Sulphuric acid and Nitric acid and other by products of gypsum from industries manufacturer are used. Citric acid, phosphoric acid and oxalic acids are also used.

(c) **Continuous cropping:** A good amount of biomass of plant parts and root portion is added every year, it will help to increase biological activity.

2. Biological reclamation

The use of gypsum is generally recommended and it is a costly input. However, the alternative method of biological reclamation seems to be one such alternative. Alkali soils of India and Pakistan contain high level of $CaCO_3$ C is insoluble $CaCO_3$ can be made soluble by adding acid or acid forwards. However, it could be best done by indicating biological activities either through addition of organic matter or cultivation of salt tolerant crops.

Choice of crop and cropping system

During *kharif* season in sodic soils of Sutlej Ganga Plains rice is favoured as first crop because it can tolerate standing water and medium ESP. During *rabi* season only shallow rooted crops like wheat, barley, berseem, shaftal could be grown in the initial years.

Since 0-15 cm surface depth is considered for reclamation of alkali soils. When rice is taken in crop sequence the reclamation of alkali soils is enhanced and the pH of surface comes down to about 9 after three years and it become possible to grow less tolerant and deep rooted crops like Jowar, bajra, oilseeds, cotton, sugarcane, mustard, rapeseed, sunflower, linseed, groundnut are semi-tolerant. It was observed that oilseed crop like rapeseed, mustard and toria can also be grown as first crop followed by jowar and bajra for fodder in *kharif* in highly deteriorated alkali soils with application of 50% growth regulator as gypsum.

Agronomic techniques for successful crop raising

I. Poor germination stand in alkali soils can be recovered by following special techniques :

(i) Increased seed rate

(ii) Restoring spacing

(iii) Increased number of plants/hill in case of transplanted crops

II. Soil fertility management

The use of manures and fertilizers bring about improvement in chemical and physical properties of soil. Increased crop growth from the addition of manures tends to promote the soil aggregation, reduces crust formation and increased biotic activity. Application of FYM and other organic manures hasten the reclamation process of sodic soils through solubilization of natural Ca as a result of action of organic acids produced during decomposition of humus. Among green manure plants Dhaincha has been found to be more successful in alkali soils.

III. Irrigation water management

On alkali soils water accumulates following rain or heavy irrigation. It remains on soil surface for long period until it evaporates and percolates. During the drying process particularly in a period of high evaporative atmospheric demand during summer months, the soil surface dries up quickly but there is no change in the water content below 15 cm soil depth. This indicates a very poor water upward flux to replenish the water loss taking place at surface.

Soil conservation practices

The aim of soil conservation is to obtain greatest possible permanent benefit from land. This means using the land and at same time maintaining or improving its productive capacity. Following soil conservation practices should be followed.

1. Agronomic measures

2. Mechanical measures

1. Agronomic measures: These help to intercept rain, reduce splash effect, help to obtain better intake of water. Improve infiltration rate of soil by improving organic matter content and soil structure help to retard or reduce over land runoff through use of contour cultivation, mulches, dense growing crop, strip cropping, mixed cropping.

I. **Contour cultivation:** Crops grow across slope keeping same level as far as possible. Ridges and rows are also placed across slope which reduces runoff, soil erosion and loss of plant nutrients as compared with up and down cultivation. Contour farming conserve soil fertility and increases crop yields. Where slope is 2.2% on alluvial soil and they conserve 11.3 kg N, 11.7 kg P, 44 kg K, 398.1 kg CaO_2, 180.1 kg Mg_2O in one season. Much less power is required to be exerted by man, animal and machines in across slope instead of up and down slope. Less wear and tear of implements and less time is required. Bunding is done along contour to reduce length of slope. All the cultural operations are done parallel to these bunds.

II. **Mulching:** Surface mulching is used to prevent soil from blowing and being passed away to reduce evaporation. Mulching increase infiltration rate, control weeds, improve soil structure, ultimately increase yields.

III. **Interculture:** Kills weed and produce 5-6 cm thick soil layer which acts as mulch and help to reduce evaporation from soil surface. It also breaks the surface crust.

IV. Growing of crops which provide maximum cover and reduce run off and soil loss which as legumes. Cultivated legumes in general form dense cover and hence protect better cultivated land against erosion than other crops. The crops are cowpea, moong, urdbean, moth, groundnut, sunnhemp.

V. **Strip cropping:** Contour strip with growing of soil exposing and erosion permitting crops in strips at suitable width across slope on contours. Alternatively with strip of soil protecting and erosion resistant crops. It reduces length of slope, check movement of run-off water, help to increases absorption of rain water by soil. Dense foliage of erosion resistant crops prevents the rain from beating soil surface directly Groundnut and moth are most suitable crops for checking erosion. But seed rate of legumes need to be tripled. The most effective width of contour strip cereals like jowar, bajra is 21.6 m and for legumes 7.2 m.

Slope	Width of erosion permit crop	Width of erosion resistant crop
0.5% or below	45 m	9 m
1-2%	24 m	6 m
2-3%	13.5 m	4.5 m

VI. **Permanent or temporary buffer strip cropping:** Strip are established to take case of critical i.e. steep or lightly eroded slopes in fields under contour strip cropping. These strips do not form part of rotation practiced in normal check erosion. Generally planted with perennial legume, grasses, shrubs on permanent or temporary basis.

VII. **Mixed cropping:** Small land holding is responsible for lack of interest in strip cropping. Some of objective of mix cropping are (a) Better and continuous cover of land; (b) Good protecting against rain beating action (c) Compete protection against soil erosion (d) Insurance of one or more crop. Roots of various spp. in mixed cropping system feed at different depth in soil. The line sowing of mixed crop gives rise to practice of intercropping. For instance mixed cropping of wheat and gram are more assured under rain fed conditions in kandi areas of Punjab.

2. Mechanical measures: These are also called engineering measures and are adopted to supplement agronomic practices when the latter alone are not adequately effective. These include basin listing, sub-soiling, contour bunding, graded bunding and bench terracing on steep slope.

Main objective

I. To increase time of water stagnation by reducing run-off and thereby providing an opportunity for infiltration of water.

II. To divide a long slope into several short ones so as to reduce velocity of run-off, thus prevent erosion.

Few important mechanical measures for conservation of soil are as follows:

A. **Basin listing:** Consist in making of small interrupted basin along contour with a special implement called Basin Lister. This helps to retain rain water as it falls and an especially effective on retentive soil having mild slope.

B. **Sub soiling:** Consisting breaking with a sub-soiler to conserve more rain water by improving physical conditions of soil. This enhances greater moisture infiltration into soil, reduces both runoff and soil erosion. Subsoil work at depth of 30-60 cm and at a spacing of 90-180 cm.

C. **Contour bunding:** Consists in making a comparatively narrow based bunds at intervals across the slope of land. It is an important measure that conserve soil water in arid and semi arid areas with high infiltration and permeability. Commonly adopted on agriculture land, up to slope of about 6%. Spacing between bunds should not exceed 150 cm for vertical drop or 67.5 cm for horizontal spacing whichever is less.

D. **Graded bunding or channel terraces:** It is used in areas receiving rainfall 780 cm/year but irrespective of soil texture. In clay soil graded bunding has to be used even for areas having less than 80 cm annual rainfall.

E. **Bench terracing:** consists of series of platform having suitable vertical drops along contour or on suitable graded line across generals slope of land. The vertical drop may vary from 60-180 cm depending upon slope and soil conditions. Bench terraces can be practiced on undulating land and intensive farming areas. Bench terracing helps to retain soil moisture, manure and fertilizer that facilitate application of irrigation if available. In rainfed areas terracing is usually practiced on slopes ranging from 6-33%. It may have to be used on gentle slope, if irrigation is to be applied to crop.

Wind break: It is a narrow belt of trees planted in one or a few rows in a field or in one or more sites of a farm to promote protection from wind.

Shelter belt: Comprises of wide belt of trees planted in several rows at right angles to prevailing wind direction. Wind break and shelter belt are not applied in India on significant scale in systematic pattern.

Practices used to control wind damage/erosion

- Crop residue may be left on field until planting time or stubble mulch tillage use to retain or protected amount of residue on surface of soil.
- A mulch such as manure @ 15 tonne /ha is applied just before or immediately after planting a crop.
- Keep tillage practices to avoid the pulverizing the small lots
- The soil surface is not smooth but left rough
- Tillage and planting machines are used which prepares 2-7 inches ridges in height
- Cultivating and planting operations are done at right angle or across from direction of prevailing winds
- Strip cropping used, strip are designed so that vegetative land may furnish protection to cultivated area.
- Field wind breaks of trees or shrubs are planted in rows at right angle to prevailing winds. These will provide effective control for 10 times their height on the ground (soil surface).

BOTANICAL NAMES OF IMPORTANT FIELD CROPS

Common Name	Botanical Name	Family
CEREALS		
Rice	*Oryza sativa* L	Poaceae
Maize	*Zea mays* L.	Poaceae
Bajra	*Pennisetum glaucum* (L.)R. Br.	Poaceae
Sorghum	*Sorghum bicolor* (L.) Moench	Poaceae
Wheat	*Tricium aestivum* L.	Poaceae
Barley	*Hordeum vulgare* L.	Poaceae
Oats	*Avena sativa*	Poaceae
PULSES		
Moong (Green gram)	*Vigna radiata* (L.) Wilczek	Fabaceae
Mash (Black gram)	*Vigna mungo* (L.) Hepper	Fabaceae
Arhar (Pigeon Pea)	*Cajanus cajan* (L.) Millsp	Fabaceae
Cow pea	*Vigna unguiculata*	Fabaceae
Moth bean	*Vigna aconitifolia* (Jacq.) Marechal	Fabaceae
Gram (Chick pea)	*Cicer arietinum* L.	Fabaceae
Lentil	*Lens culinaris*	Fabaceae
Peas	*Pisum sativum* L.	Fabaceae
FIBRE CROPS		
American cotton	*Gossypium hirsutum* L.	Malvaceae
Desi cotton	*Gossypium arboreum* L.	Malvaceae
Jute	*Corchorus capsularis* L.	Malvaceae
Sunhemp	*Crotalaria juncea* L.	Fabaceae
Mesta	*Hibiscus sabdariffa*	Malvaceae
OILSEEDS		
Groundnut	*Arachis hypogaea* L.	Fabaceae
Sesame	*Sesamum indicum* L.	Pedaliaceae
Soybean	*Glycine max* (L.) Merrill	Fabaceae
Castor	*Ricinus communis* L.	Euphorbiaceae
Toria	*Brassica napus* L. var. *toria*	Brassicaceae

Common Name	Botanical Name	Family
Raya	*Brassica juncea* L.	Brassicaceae
Gobhi sarson	*Brassica napus* L.	Brassicaceae
Taramira	*Eruca sativa* L.	Brassicaceae
Sunflower	*Helianthus annuus* L.	Asteraceae
Safflower	*Carthamus tinctorius* L.	Asteraceae
Linseed	*Linum usitatissimum*	Linaceae
FODDERS		
Maize	*Zea mays* L.	Poaceae
Bajra	*Pennisetum glaucum* (L.)R. Br.	Poaceae
Sorghum	*Sorghum bicolor* Moench	Poaceae
Teosinte	*Euchlaena mexicana*	Poaceae
Sundangrass	*Sorghum xdrummondii*	Poaceae
Guinea grass	*Megathyrsus maximus*	Poaceae
Napier Bajra	*Pennisetum purpureum*	Poaceae
Cowpea	*Vigna unguiculata* (L.) Walp	Fabaceae
Berseem	*Trifolium alexandrinum* L.	Fabaceae
Shaftal	*Trifolium resupinatum* L.	Fabaceae
Lucerne	*Medicago sativa* L.	Fabaceae
Senji	*Melilotus spps.*	Fabaceae
Oats	*Avena sativa* L.	Poaceae
Rye grass	*Lolium perenne* L.	Poaceae
Guar	*Cyamopsis tetragonoloba (L)*	Fabaceae
SUGAR CROPS		
Sugarcane	*Saccharum officinarum* L.	Poaceae
Sugarbeet	*Beta vulgaris* L.	Amaranthaceae
CASH CROPS		
Tea	*Camellia sinensis* (L) Kuntze	Theaceae
Tobacco	*Nicotiana tabacum* L.	Solanaceae
Potato	*Solanum tuberosum* L.	Solanaceae
GREEN MANURING CROPS		
Sunhemp (Saan)	*Crotalaria juncea*	Fabaceae
Dhaincha	*Sesbania bispinosa*	Fabaceae
Cowpea	*Vigna unguiculata* (L.) Walp	*Fabaceae*
MEDICINAL, AROMATIC AND SPICES		
Guar	*Cyamopsis tetragonoloba* (L.) Taub.	Fabaceae
Mentha	*Mentha spicata* L.	Lamiaceae
Turmeric	*Curcuma longa* L.	Zingiberaceae
Stevia	*Stevia rebaudiana* Bertoni	Asteraceae
Aloe	*Aloe vera*	Asphodelaceae

Common Name	Botanical Name	Family
Lemon grass	*Cymbopogon flexuosus* Staff	Poaceae
Citronella	*Cymbopogon winterianus* Jowitt	Poaceae
Celery	*Apium graveolens* Linn	Apiaceae
Fenugreek	*Trigonella foenum-graecum*	Fabaceae
Fennel	*Foeniculum vulgare*	Apiaceae
Cardamon	*Elettaria cardamomum* L.	Zingiberaceae
Pepper	*Piper nigrum* L.	Piperaceae
Ginger	*Zingiber officinale* Rose.	Zingiberaceae
Dilseed	*Anethum graveolens* L.	Apiaceae
Coriander	*Coriandrum sativum*	Apiaceae
Cumin	*Cuminum cyminum*	Apiaceae
Isabgol	*Plantago ovata* Forks.	Plantaginaceae

TERMINOLOGY

Absorption Losses: Loss of water from a canal or a reservoir by capillary action and percolation and in case of canal during the process of delivery.

Acid equivalent (a.e.): The differential yield of parent acid from an active ingredient.

Acid soil: A soil which is deficient in available bases, particularly Ca and which give an acid reaction when tested by a standard method.

Acre foot water: The amount of water that would cover an area of land to a depth of one foot assuming no seepage evaporation and run off.

Acre inch: It is a measure of quantity of flow of water and is equal to the flow which will cover one acre to a depth of one inch.

Active ingredient (a.i.): The chemical in a product that is responsible for the phytoxic effects.

Adaptation: The modification in structure and function that increases the probability that a crop will survive at reproduce in a particular environment is termed as adaptation.

Agriculture: It is an art, science and business of raising crops and rearing of animals through exploring the natural resources with the coordination of socioeconomic infrastructure to meet the basic necessities of life .i.e. .food, feed, fiber and shelter.

Agri-silviculture: Agri-silviculture is a production technique which combines the growing of agricultural crops with simultaneously raised and protected forest crops. Here it is intended to grow tree crops along with the agricultural crops.

Agrisilvipastoral systems: ASPS is a collective name for land-use **systems**, implying the combination or deliberate association of a woody component (trees or shrubs) with cattle in the same site.

Agroforestry: Agroforestry is a land use management system in which trees or shrubs are grown around or among crops or pastureland. It combines shrubs and trees in agricultural and forestry technologies to create more diverse, productive, profitable, healthy, ecologically sound, and sustainable land-use systems.

Agronomic efficiency (AE): It refers to the additional produce obtained in Kg per kg of an applied nutrient.

Agronomy: Agronomy is a branch of agricultural science dealing with principle and practises of soil, water and crop production.

Allelopathy: Phenomenon involving the release of certain chemicals from plant parts into the environment which may when present in sufficient amounts, inhibit or promotes the germination or growth of the plants in the neighbourhood.

Alluvial soil: These are the soils which are formed by the deposition of broken material of rocks transported and deposited by water of streams and rivers.

Arable farming: The term arable farming refers to system in which only crops that require cultivation of the soil are grown.

Arid/Aridity: A term applied to land that is dry also means deficient in rainfall.

Autoecology: It deals with the ecology of an individual species and its population including the effect of other organisms and environmental conditions on every stage of its life cycle.

Autotrophs: Organism able to manufacture their own food from inorganic materials. Using energy from outside sources. Most green plants are completely autotrophs.

Available water: The amount of water between field capacity and permanent wilting point is considered as available water for plant use.

Band placement: The placement of fertilizer in the soil in continuous bands or narrow ribbons in between the plant rows or around each plant.

Barani soil: When the source of irrigation to crop is only the rain water that is known as Barani soil.

Basal treatment: An application to the stems of the plants at and just above the ground line.

Base period: Period of time in days from the first watering of crop before sowing and the last watering before harvesting.

Basic seed: Is the progeny of pre – basic seed produced so as to maintain genetic purity and identity.

Basin: Flat area of land surrounded by low ridges or bunds

Bed: A narrow flat topped ridge on which crops are grown with a furrow on each side for the drainage of water.

Blind hoeing: Hoeing before the crop emerges, usually done in sugarcane.

Biofertilizers: Microbial inoculants probably known as biofertilizers refer to living or latent cells of efficient strains of microorganisms capable of fixing atmospheric nitrogen.

Biological yield: It is the total dry matter produced by a plant as a result of photosynthesis and nutrient uptake minus that lost by respiration.

Bolting: Formation of elongated stem or seed stalk, it is usually takes place during the second season of the growth in biennial plants.

Broadcasting: Manual spreading of seed in the field and mixing of the spread seed by ploughing or planking the field.

Bulk density of mineral soils: It is defined as mass of a unit volume of dry soil. This volume includes both solids and pores.

Capillary water: It is held in the pores of capillary size. It includes most of the water taken up by growing plants.

Catchment area: The area which drains the rain water falling on it, via streams and rivers, eventually to the sea or into a lake.

Certified Seed: It is the progeny of basic seed and is produced by registered growers of seed producing agencies.

Climate: Aggregate of atmospheric condition over a long period of time.

Clone: A cultivar propagated by vegetative method is called a clone.

Command area: Area which can be economically irrigated by an irrigation system.

Companion Crops: The two crops grown together are called companion crops. e.g. Berseem and barley.

Component crop: It is used to refer to either of the individual crops making up the intercropping situation.

Composite variety: A mixture of genotype from general sources maintained by normal pollination.

Composting: Composting is a biological process in which micro-organisms of both types namely, aerobic and anaerobic, decompose the organic matter and lower the carbon nitrogen ratio of the refuse.

Conditioner: A material added to a fertilizer to prevent caking and to keep it free flowing.

Consumptive use (Cu): It is use to designate the losses due to ET and the water that is used by the plant for its metabolic activities. Cu is taken equivalent to ET, because water used in metabolic processes does not exceed 1 per cent.

Contact herbicide: Which kill only those parts of plant with which they come into contact and used mostly to control annual weeds, they have little residual effect.

Cover crops: The crops, which are planted to cover the ground and to reduce the soil erosion.

Critical hydration level of seed: The minimal water content at which the seed begins to germinate and the hydration level at which seed water uptake becomes biologically irreversible.

Critical period of competition: During the crop period there is a certain time when crop plants are most sensitive to competition by weeds, this time is known as the critical period of competition.

Critical threshold level (CTL): A level at which weed, insect pest density capable of causing significant damage to crop is termed as critical threshold level.

Crook stage: The stage after bean seedlings have broken through the soil but before the stem has become erect.

Crop ecology: It is the study of the interrelationships between crop plants and the environment.

Crop growth rate (CGR): CGR represents total dry matter productivity of the community per unit land area over a certain time span.

Crop rotation: It is the process of growing different crops in succession on piece of land in a specific period of time, with an object to get maximum profit from least investment without impairing the soil fertility.

Crop water use efficiency: It is the ratio of crop yield to the amount of water depleted by the crops in the process of evapotranspiration .

Crop water requirement: The amount of water required to raise a crop to maturity within a given period of time.

Cropping index (Cropping intensity): The number of crops grown per annum on a given area of land multiplied by 100.

Cropping pattern: The yearly sequence and spatial arrangement of crops or of crops and fallow on a given area.

Cropping scheme: The cropping scheme is a plan according to which crops are grown on individual plots of a farm during a given period of time with the object of obtaining maximum return from each crop without impairing the soil fertility.

Cropping system: Cropping system is a land use unit comprising soils, crop, weed, pathogen and insect subsystems that transform solar energy, water, nutrients, labour and other inputs into food, feed, fuel and fibre.

Crown: The point where stem and root joints in a seed plant.

Cultivator: Which only cut and stirr the soil.

Delta of water: The depth of irrigation water required for the full crop period.

Determinate plants: Those plant which initiate their reproductive stage after completing vegetative growth, e.g. wheat, barley.

Diversified farming: This is an expanded type of farming system in which varieties of crops are produced and many types of animals are reared.

Dormancy: Seed dormancy is the state of inhibited germination of seeds with viable embryos in condition conducive to plant growth.

Drainage: It is the removal of excess surface or ground water from the root zone of a crop by means of surface or sub – surface drains.

Drought resistance/ Drought escape: The ability of crop to grow satisfactorily in areas subjected to water deficit has been termed as drought resistance.

Dry farming: In which crops and livestock are raised on land which does not receive sufficient rainfall for water intensive crops and no irrigation facilities are available fall into this category.

Earthing up: The operation of pulling up soil from the center of crop rows to the bottom of the plants, this helps in uprooting weeds and supporting to plants.

Ecology: It is defined as the study of plants and animals in reciprocal relationship with their environment or external world.

Economic yield: That part of the crop for which crop is grown e.g. stem in sugarcane, root in the case of sugar beet.

Ecosystem: Any collection of organisms that interact or have the potential to interact along with the physical environment in which they live, form an ecological system or ecosystem.

Effective rainfall: Precipitation which falls during the growing period of the crop and is available to meet the evapotranspiration requirements of crops.

Efficient cropping zone: A zone where both yield and spread indices are maximum.

Emergence: It is process of emerging out of the plumule from the soil after seed germination.

Epigeal germination: It is derived from two words epi "above" and geas "earth". In this type of germination the cotyledons come out above the soil surface and generally turn green and act as first foliage leaves. e.g. bean,cotton.

Epinasty: The twisting or curling of leaves and stems caused by uneven growth of cells, especially in leaves, in which the upper surface grows faster than the lower surface and thus cause the leaf edges to bend down.

Equivalent basicity: A term used to express the calcium carbonate equivalent of basic residue left by a fertilizer material.

Evapotranspiration (ET): Evaporation and transpiration are combined in one term evapotranspiration. It is difficult to separate these two losses in cropped fields.

Exhaustive crops: Crops, which feed heavily on the soil and deplete soil nutrients e.g. sorghum, tobacco.

Extensive farming: In this type of farming large areas are used with minimum expenditure or attention to efficient use of other resources.

Facultative weed: A weed found growing both wild and in association with crops.

Farming system: Farming system is a decision making and land use unit comprising the farm, household, cropping and livestock systems that produces crop and animal product for consumption and sale.

Fertilizer grade: An expression that indicates the percentage of plant nutrients in a fertilizer. For example, a 10-5-5 grade of fertilizer indicates 10% N, 5% P_2O_5 and 5%K_2O.

Fertilizer use efficiency (FUE): It is the measure of the extent to which the fertilizers and management practices have worked together with the soil climate complex in producing high yield.

Field water use efficiency : This is the ratio of crop yield to the total amount of water used in field

Field capacity: When a fully saturated soil is allowed to drain after a day or so, rapid downward movement in response to hydraulic gradient will become negligible. The soil is then said to be at its field capacity.

Foliar fertilization: Fertilization of plants or feeding nutrients to plants by applying chemical fertilizers to the foliage. This is also known as foliar feeding or spray fertilization.

Germination: Is the emergence and development from the seed embryo of those essential structures which, for the kind of seed provided, indicate the ability to produce a normal plant under favourable conditions.

Granular fertilizer: A fertilizer composed of particles of roughly the same composition and about $1/10^{th}$ of an inch in diameter.

Grassland farming: These systems are mainly concerned with growing grasses for consumption by livestock kept for milk or meat production.

Gravitational potential: It is produced by gravitational forces operating on soil water.

Gravitational water: Water in excess of field capacity is termed as gravitational water.

Green manuring: It is a practice of ploughing or turning into the soil undecomposed green plant material for improving the physical condition of the soil or for adding nitrogen when the green manure is a legume.

Gross irrigation requirement: The total amount of water, inclusive of losses, applied through irrigation is termed as gross irrigation requirement.

Growth: The term growth is applied to quantitative changes occurring during development. It may be defined as an irreversible change in the size of cell, organ or whole organism.

Hard pan: A hard semi impervious layer usually developed due to continuous ploughing to a depth of about 15 cm, with cultivators, or with continuous deposition of salts due to soil or surface irrigation water.

Harvest index (HI): It is the ratio of economic yield to biological yield. It is expressed in percentage.

Hay: The herbage of grasses or fine-stemmed plants cut and cured for forage.

Herbicides: Chemicals used for killing or inhibiting the growth of higher plants.

Hill placement: This refers to applying fertilizer either in bands or near the plants. This method is usually preferred when relatively small quantities are to be applied.

Husk: The coarse outer envelope of a fruit, such as the glumes of an ear of maize.

Hybrid vigour: Qualities in a hybrid not present in either parent.

Hydrophyte: Plant which grows in water, or which loves water.

Hygroscopic water: Hygroscopic water is tightly bound by the soil solids at suction value greater that 31 atm.

Ideotype: An ideal plant type developed through breeding.

Indeterminate plants: In these plants, the vegetative and reproductive stages continue simultaneously e.g. okra, tomato.

Inoculant: The bacteria containing material used to introduce N – fixing Rhizobium bacteria into soil.

Integrated weed management: The concept of IWM involves the planned use of all possible direct and indirect measures rather than relying on a single method to combat weeds.

Interception: When the drops of rainfall or precipitation are intercepted by plant leaves it is called interception.

Intercrop yield: It is a yield of the component crop when grown in intercropping and expressed over the total intercropped area (i.e. area occupied by both the crops).

Intercropping: Intercropping is the cultivation of two or more crops simultaneously on the same field, with or without row arrangement (row intercropping or mixed cropping).

Inter-cultivation: Inter-cultivation is a process involving operations in the rows of standing crops. Weeding, tiling and cultivating are the examples for the same.

Irrigated soil: The soil, which receives irrigation water from well and tube wells, are known as irrigated or chahi soil.

Irrigation efficiency: It is a term used to indicate how efficiently the available water supply is being used for crop production.

Irrigation scheduling: It refers to the number of irrigations for a crop and their timing.

Irrigation water requirement (IR): The irrigation water requirement is the total amount of water applied to the land surface in supplement to the water supply through rainfall and soil profile, to meet the water needs of crops for optimum growth.

Irrigation: Irrigation is the artificial application of water to soil or crop plants to assist crop production.

Kera: Manual sowing of seeds in lines in furrows.

***Kharif* crops:** Those crops, which are planted in the summer month from the March to July and harvested in autumn and winter, are called Kharif crops e.g. rice and cotton.

Leaching: Removal of plant nutrients in solution by the passage of water through soil. Among the major nutrients, N is lost in large quantities by leaching.

Leaf area duration (LAD): Leaf area integrated over time is called leaf area duration. It is equal to leaf area x time.

Leaf area index (LAI): Leaf area index is a ratio of leaf area to ground area.

Land equivalent ratio(LER): This is the relative land area under sole crops that is required to produce the yields achieved in intercropping.

Localized placement of fertilizers: This refers to the application of fertilizers into the soil close to the seed or plant. This method is specifically adopted with phosphatic fertilizers.

Lodging: The bending or breaking over of a plant before harvesting.

Long day plant: Plants which changes from vegetative to reproductive stage and produces flowers and fruits, when the days becomes longer.

Main crop/base crop: This is the crop which is planted at its optimum sole crop population in an intercrop situation and the second crop is planted in between rows of the main or base crop, with a view to obtain some extra intercrop yield without sacrificing the main crop yield.

Matric potential: It is produced by capillary and surface forces.

Maximum water holding capacity: When the water is continuously applied to the soil. At a stage when its all (macro and micro) pores are filled with water. The soil is said to be saturated w.r.t water and is at its maximum retentive capacity.

Mixed intercropping: Growing two or more crops simultaneously with no distinct row arrangement. Also referred to as mixed cropping.

Monocropping: Monoculture is the repetitive growing of the same crop on the same land.

Muck soil: If the quantity of organic matter exceeds 20 % but less than 50% are called muck soil.

Mulch: Any material or practice which is used to check the loss of water by evaporation is called mulch.

Multiple cropping: The intensification of cropping in time and space dimensions, growing two or more crops on the same field in a year.

Natural erosion: The erosion of the soil under natural condition.

Natural soil: Strictly speaking a soil having pH of 7, in practice a soil having PH 6.6 and 7.3.

Necrosis: Death of organs of a plant, either as blight or death of tissue in localized areas, usually inside fruit and stems or die back or death of stems or branches.

Nutrient recovery (NR): It refers to the actual amount of nutrient taken up from the fertilizers.

No tillage crop (zero tillage crop): Crop grown without any tillage to prepare seed bed or row.

Node: A slightly enlarge portion of stem where leaves and bud arises and where branches originate.

Nucleus seed: Seeds obtained from selected individual plants of a particular variety which needs to be purified and multiplied in such a way as to maintain its genetic purity.

Nutrients: The food for microbial and plant life mainly composed of nitrogen and phosphorous but also of potassium , Mg, Fe, Ca, Co, Cu, Zn and others elements.

Organic farming: System of farming which avoid the use of artificial fertilizers, pesticides or herbicides, and concentrate on methods of crop rotation and the use of home growth feed , organic fertilizer.

Osmotic potential: It is also called solute potential. It is produced by various solutes in soil water.

Particle density of mineral soils: Mass of unit volume of soil solid is called particle density. It is expressed as g/cm^3. It varies between 2.6 and 2.75.

Peat soil: If the quantity of organic matter is more than 50% is called peat soil.

Percolation: Downward movement of water with in the soil profile.

Plough pan: A dense, compacted layer about 5 to 7 cm thick formed beneath the surface soil by repeated ploughing in the same path.

Plough: Its function is to cut, stirr, invert, and pulverize the soil.

Pore space: It is a space between soil particles occupied by air and water; it is largely controlled by the texture of soil.

Potential evapotranspiration (PET): Potential evapotranspiration is the ET that occurs when the ground is completely covered by actively growing vegetation and there is no limitation of soil moisture.

Potential soil moisture deficit: It is the difference between a crop potential evapotranspiration and the amount of rainfall received by a crop plus the quantity of water delivered to it in irrigation.

Puddling: Puddling is the reduction in apparent specific volume (void ratio) of a soil caused by mechanical work done on soil.

***Rabi* crops:** These crops are planted in winter from October to December and harvested in summer from March to May e.g. wheat,

Ratoon cropping: The cultivation of crop regrowth after harvest, e.g sugercane

Readily available water: The portion of the available water that is most easily extracted by a plant is called readily available water.

Relay crops: A relay crops is one which is planted as a second crop after the first crop has reached its reproductive stage of growth but before it is ready for harvest. e.g. planting of sugar cane in sugar beet.

Relative growth rate (RGR): It is defined as increase in weight of crop plants per unit of original weight of time interval t.

Run off: When water flows out the field by breaking the bunds of the field or flows to the sloppy areas from the high level is called run off..

Saturation capacity: This term refers to the amount of water present in the soil when it is completely saturated with water.

Scarification: Any physical or chemical treatment that makes the seed coat permeable is known as scarification.

Seasonal consumptive use (U): The total amount of water used in evaporation and transpiration by a crop during the entire growing season is called seasonal consumptive use. It is expressed in depth of water in cm.

Seed certification: Is the process to secure, maintain and make available high quality seed and propagating materials of superior crop varieties, so grown and distributed as to ensure desirable standards of genetic identity, physical purity and quality attributes.

Seed dressing: The chemical treatment of seeds particularly cereals, with fungicides and some time insecticides to protect them against soil and seed borne disease and pest.

Seed: Seed may be defined as a fertilized ovule consisting of intact embryo, stored food and seed coat which is viable and has got a capacity to germinate. This is also known as true seed because in some of the crops, vegetative parts are used for planting and are also called as seeds.

Seedling: A young plant that has grown from a seed

Seepage: Losses through the beds and banks of canals and water channels take place mainly by unlined canals.

Sequential cropping: Growing two or more crops in sequence in the same field in a farming year. The succeeding crop is planted after the preceding crop has been harvested.

Sett: A piece of seed cane with two – four buds(eyes).

Short day plants: Plants which changes from vegetative to reproductive stage and produces flowers and fruits, when the days become shorter.

Silage crops: Those crops, which are harvested when still green and succulent and are fed directly to animals without curing, are called silage crops. e.g. Berseem, shaftal, maize.

Silage: Forage preserved in a succulent condition by partial fermentation in a tight container.

Soil fertility: It is defined as the inherent capacity of soil to supply amount of adequate nutrients to plants in balanced form.

Soil productivity: It is defined as the capacity of the soil to yield crops per unit area or it can be defined as the total crop production from the field. A soil can be fertile but may not be productive.

Soil: A natural body developed from variable mixtures of broken and weathered minerals and decaying organic matter which covers the earth crust in thin layers and supplies proper amount of nutrients and air water and mechanical support to plants.

Sole cropping: One crop/variety grown alone in pure stands at normal density.

Stratification: Is the practice of exposing imbibed seeds to cool temperature conditions for a few days prior to germination in order to break their dormancy.

Strip intercropping: Growing two or more crops simultaneously in strips wide enough to permit independent cultivation but narrow enough for the crops to interact agronomically.

Subsistence farming: In which basic necessities like food, clothing, and shelter are produced for the family to live on.

Synecology: It deals with the ecology of plant communities. It involves the study of structure, nature, organization and development of plant communities.

Temporary wilting percentage: Soil water content at which plants wilt during the hot windy part of the day but regain turgidity during the cooler part of the day is called TWP.

Tillage: It is usually defined as the mechanical manipulation of the soil aimed at improving its physical condition.

Tiller: To put forth shoots other than the mother shoot from roots axis.

Tilth: It is the final condition of soil when all the tillage operations are carried.

Topping: Topping in tobacco is the removal of the terminal bud with or without some of the small top leaves just before or after the appearance of the flower head.

Town compost: This utilizes night soil, street and dustbin refuse. Town compost is rich in nutrients (1.4 % N, 1.0 % P_2O_5 and 1.4% K_2O) as compared to FYM and rural compost.

Transpiration ratio: It is the ratio of the weight or volume of water transpired by the plant during its growth period to the weight of dry matter produced by the plant.

Transpiration: The loss of water from exposed parts of plants mainly through leaves is called transpiration.

Transplanting: It is a process of planting seedlings, grown in separate nursery, in the puddle fields.

True allelopathy: When the allelochemical is toxic in their original form is called true Allelopathy.

Unit leaf rate (ULR): It may be defined as rate of increase in dry weight per unit leaf area, assuming that both dry weight and leaf area are increasing exponentially.

Variety: In general the term variety has been used to refer to a group of similar plant with in a particular species that is distinguished by one or more then one character and given the name.

Viability of seed: It represents the capacity of seed to germinate or capacity of seed to remain capable of germination for some specific period of time.

Vines: Are plants, which have tender stems, and requires some support for upward growth.

Water holding capacity: Ability of soil to retain water.

Water potential: Refers to the chemical potential of water.

Water requirement of crops: Water requirement may be defined as the quantity of water, regardless of its source required by a crop of diversified pattern of crops in a given period of time for its normal growth, under field conditions at a place.

Watter: Describes the condition when soil moisture level is suitable for cultivation.

Weathering: The process by which soil disintegrates and decomposes, eventually producing soil particles by exposure to the physical and chemical effects of atmospheric agents.

Weed control: Weed control aims at checking the growth or reducing the number of weed to a minimum so that they do not affect crop yields.

Weeds: Weeds are that species of plant which grow unwanted or are not useful, often prolific and persistent, interfere with agricultural operations, increase labour cost and reduce the crop yields.

Wilting coefficient: As the soil dries out, plant will begin to show the effects of reduced soil moisture. During the day time they will tend to wilt. At first this day time wilting will be associated with renewed night time turgor. Ultimately, the rate of supply of water to the plants will be so slow that the plants will remain wilted day and night. The soil moisture content at this stage is called wilting coefficient or critical moisture.

Wilting point: The point at which the water content of a soil reaches such a level that it is firmly held by soil and unavailable to plant roots, so that the plants wilt permanently and die.

Yield index of crop = It is the ratio of average yield of state or zone/ average yield of country and is expressed in percentage.

Zaid *Kharif* crops: These are the crops, which are planted in September and harvested in December to January e.g. toria.

Zaid *Rabi* crops: These are crops, which are planted in February and harvested in May-June e.g. spring maize.

REFERENCES

Anonymous, 2008. Annual Progress Report AICARP on cropping system, Punjab Agricultural University, Ludhiana.

Ahuja, K.N. and N.T. Yaduraju. 1989. Integrated control of weeds in wheat with special reference to *Phalaris minor*. *Indian J. Weed Sci.* **30** : 141-44.

Ali M and Prasad R. 1974. Effect of mulches and type of seed bed on pearl millet under semi-arid conditions. *Experimental Agriculture* **10**: 263-72.

Ali M, Mishra J P, Ahlawat IPS, Kumar R and Chanhan Y S 1998. Effective management of legumes for maximizing biological nitrogen fixation and other benefits. In: *Residual effects of legumes in rice and wheat cropping systems in the Indo-gangetic plains* (Eds J.V.D.K Kumar Rao, C. Johansen and TJ Rego), pp. 107-128. New Delhi: Oxfoord & IBH Publishing Co. Pvt. Ltd.

Anonymous 2018. Package of Practices for crops of Punjab (*Kharif* 2018), Punjab Agricultural University, Ludhiana.

Anonymous. 2017-18. Package of Practices for crops of Punjab (*Rabi* 2017-18). Punjab Agricultural University, Ludhiana.

Arnon D I and Stout P R 1939.The essentiality of certain elements in minute quantity for plants with special reference to copper. *Plant Physiology* **14** (2): 371-75.

Aulakh M S and Pasricha N S 1986. Role of sulphur in the production of grain legumes. *Fertilizer News* 31 (9): 31-35.

Aulakh M S, Khera T S, Doran J W and Bronson K F 2001. Managing crop residue with green manure, urea and tillage in a rice-wheat rotation. *Soil Sci Soc Amer J* **65**: 820-27.

Baghel R P S and Gupta R K 2003. Factors affecting nutrient composition of cattle dung. *Proc ICAR-ACIAR Workshop on Farmyard manure: Life cycle analysis and management*. February 18-20, 2003, Indian Institute of Soil Science, Bhopal, MP, India: 2-31.

Baldock J O, Higgs R L, Paulson W H, Jacbobs J A and Shrader W D 1981. Legume and mineral N effects on crop yields in several crop sequences in the upper Mississippi valley. *Agron J* **73**: 885-90.

Balusamy, M., Shanmugham, P.M. and Baskaran, R. 2003. Mixed farming – an ideal farming. *Intensive Agriculture* **41**(11-12): 20-25.

Bani O P S and Pant A K 2000. Green manuring for ecologically sound crop production. *Indian Farmers Digest* **33**(2):19-20.

Basavaraj, Hulagur., Gangadharappa, N.R. and Hulagur, B. 1999. Adoption level and profitability of different combination of farm enterprises in existing farming systems. *Mysore Journal of Agricultural Sciences* **33**(2): 261-66.

Bauer A and Black A L 1992. Organic carbon effects on available water capacity on three soil textural groups. *Soil Sci Soc Am J* **56**: 248-54.

Beri V, Sidhu B S and Bahl G S 2002. Crop residue management. *In: Recent Advances in Agronomy* (Singh G, Kolar J S and Sekhon H S, eds). Indian Soc Agron, IARI, New Delhi. pp 190-212.

Beri V, Sidhu B S, Bahl G S and Bhat A K 1995. Nitrogen and phosphorus transformation as affected by crop residues management practices and their influence on crop yields. *Soil Use and Management* **11**: 51-54.

Bhagat R M and Verma T S 1991. Impact of rice straw management on soil physical properties and wheat yield. *Soil Sci* **152**: 108-15.

Bhandari A L and Walia S S 1999. Effect of legume based cropping systems on yield, fertilizer economy and soil fertility maintenance in a Typic Ustochrept. *Proc. National Seminar on Development in Soil Science.* 64th *Annual Convention of Indian Society of Soil Science* held at TNAU, Coimbatore. November 26-30: 147.

Bhandari A L and Walia S S 2000. Effect of FYM, green manuring and crop residue incorporation on crop and soil productivity under combine harvested rice-wheat system. *Proc. International conf on managing the natural resources for sustainable crop production in the 21st century*. Feb. 14-18, 2000. IARI, New Delhi, India **4:** 19-20.

Bhat A K, Beri V and Sidhu B S 1991. Effect of long term recycling of crop residues on soil productivity. *J Indian Soc Soil Sci* **39**: 380-83.

Bhattacharya, P and Gehlot, D. 2004. Current status of regulatory mechanism in organic farming. *Fertilizer News* **49**(11): 15-31.

Bhattacharya, P. and Chakraborty, G. 2005. Current status of organic farming in India and other countries. *Journal of Fertilizers* **1**(9): 111-113.

Billings W D 1952. The environmental complex in relation to plant growth and distribution. *Quaterly review of Biology* **27:** 251-265.

Biswas Chinmay and Singh Ravindra. 2003. Integrated Farming System: An Intensive Approach. *Intensive Agriculture* **41** (7-8): 22-23.

Blackman E F 1905. Optima and limiting factors. *Annals of Botany* **19**: 281-295.

Borse R H and Mahajan U B. 1981. Studies on relative efficiency of triazine compounds, 2, 4-D and slow release 2, 4-D in comparison with mechanical methods of weed control in hybrid Jowar CSH-1. *Journal of Maharashtra Agricultural University* **6** (2): 161-62.

Buns R C and Hardy R W F 1975. *Nitrogen fixation in bacteria and higher plants*. Berlin: Springer-Verlag.

Cain 1944. *Foundations of Plant geography* Harper and Brokers, New York.

Channabasavanna, A.S., Itnal, C.J. and Patili, S.G. 2002. Productivity, economic analysis and changes in Physico-chemical properties of soil as influenced by integrated site (*Oryza sativer*) based farming system. *Indian Journal of Agronomy* **47** (1): 1-5.

Chaudary, OP and Kashyap RK 1987. Field fertilization vis-à-vis incidence of insect-pest, complex and marketable yield in egg plants. (*Solanum melangina*). *Italino di Enotomologia* : **41** 3-20.

Chhokar PK and Dwivedi BS 2004. Organic farming and its implecations on India's food security. *Fertiliser News* **49**(11): 15-31.

Chorge BM and Ramshe DG. 1990. Effects of sowing dates on yield attributes, yield nutrient uptake and protein content of *rabi* sorghum under irigation. *Journal of Maharahstra Agricultural University* **15** (2): 142-45.

Clawson K L and Blad B C 1982. Infrared thermometry for scheduling irrigation of corn. *Agron J* **74**: 311-316.

Clement A, Ladha J K and Chaliflour F P 1998. Nitrogen dynamics of various green manure species and the relationship to low-land rice production. *Agron J* **90:** 149-54

Dayanand and Jain O P 1994. *Rabi* maize cultivation for enhanced production. *Indian Farming* **43** (10): 11-13.

Dhaliwal G S and Kler D S 1995. *Principles of Agricultural ecology* published by Mrs. Mena Pandey for Himalya Publishing house, Rain doot, Dr. Bhalerao Marg, Girgaon, Bombay.

Dhaliwal, G S, Sangha, G S and Ralhan, P K 2000. *Fundamentals of Environment Science* Kalyani Publishers, Ludhiana. p 129.

Dhiman S D, Nandal D P and Om Hari 2000. Productivity of rice (*Oryza sativa*)-wheat (*Triticum aestivum*) cropping system as affected by its residue management and fertility levels. *Indian J Agron* **45**: 1-5.

Donald, C.M. 1968. The breeding of crop ideotypes. *Euphytica* **17**(3): 385-403.

Eaton F M 1950. Significance of carbonates in irrigation water. *Soil Sci* **69**: 123-133.

Fischer, K.S., R. Lafitte, S. Fukai, G. Atlin and B. Hardy (eds.) 2003. Breeding rice for drought-prone environments International Rice Research Institute, Los Baños, Philippines.

Ghosh S K 2000. Organic farming for sustainable development. *Kisan World* **27**(5): 14-17.

Ghuman B S, Sur H S, Kulwant Lal and Lal K 1997. Effect of green manuring on soil properties and yield of wheat under dry land conditions. *J Indian Soc Soil Sci* **45**: 99-103.

Gill, M.S., Samra, J.S. and Singh Gurbachan. 2005. Integrated Farming System for Realizing High productivity under shallow water table conditions. Research Bulletin. Department of Agronomy and Agrometeorology, Punjab Agricultural University, Ludhiana-141 004.

Good B 1953. *The geography of flowering plants*, London: Longman Green

Good R 1937. A theory of plant geography. *New Phyto* **30**: 149-171.

Hegde R N 1997. Organic farming. *Kisan World* **24**(8):43-44.

Herrmann, G.A. and Rundgren, G. 2007. Certification and Accreditation. In: *The World of Organic Agriculture: Statistics and Emerging Trends, 2007.* Willer, H. and Yussefi-Menzler, M. (Eds.). International Federation of Organic Agriculture Movements

(IFOAM), Bonn, Germany and Research Institute of Organic Agriculture (FiBL), Frick, Switzerland. pp. 67-74.

Het Singh, R. P. 1990 Research efforts to mitigate the adverse effects of drought. *Dryland Resources and Technology* **5**: 117-36

Hobbs, P. R. Sayre, K. D. and Gupta, R 2007. The role of conservative agriculture is sustainable agriculture. *Biol. Sciences* **363** (1491): 543-555

Huber, B., Kilcher, L. and Schmid, O. 2007. Standards and Regulations. In: *The World of Organic Agriculture: Statistics and Emerging Trends, 2007.* Willer, H. and Yussefi-Menzler, M. (Eds.). International Federation of Organic Agriculture Movements (IFOAM), Bonn, Germany and Research Institute of Organic Agriculture (FiBL), Frick, Switzerland. pp. 56-66.

Hussain Z M, Choudhury M H K and Alam Q K 2001. Effects of ecological agriculture on soil properties and arthropid diversity in rice-based cropping systems in flood plain areas in Bangladesh. *Intl Conference on Nature Farming and Ecological Balance*, ISNF, CCS HAU, Hisar, India. March 7-10.p 160.

ICAR. 2006. Handbook of agriculture. *Indian Council of Agricultural Research*, New Delhi.

Jama B, Swinkels R A and Buresh R J 1997. Agronomic and economic evaluation of organic and inorganic sources of phosphorus in western Kenya. *Agron J* **89**: 597-604.

Jat M L, Sharma S K, Balyan J K, Kothari A K and Jain L K 2011 Rainfed Farming. *Kalyani Publishers*. pp. 249-56.

Jayaraman, R., Parthasarathy, R. and Chandrabose, B. 1993. Experiments on fish-cum-poultry farming and its economic efficiency. *Indian Veterinary Journal* **70**(4) : 341-343.

Jennings, P.R. 1964. Plant type as a rice breeding objective. *Crop Sci* **4**:13-15

Jones M. P., M. Dingkuhn, G. K. Aluko, and M. Semon. 1997. Interspecific *Oryza sativa* L. X *O. glaberrima* Steud. progenies in upland rice improvement. *Euphytica*. **92**: 237-246.

Kachapur M D. 1990. Effect of sowing time and population levels on yield of sorghum. *In: Improved Sorghum Production Technology*. Pp 10-15. Hosmani M M and Chittapur B M (Eds) University of Agricultural Science, Dharwad.

Khan A A and Zende G K. 1976. Correlation of soil test values with the response of maize and sorghum to available Zn and P. *Indian Journal of Agricultural Sciences* **46** (6): 259-65.

Kilcher, L. 2007. How organic agriculture contributes to sustainable development. In: *The World of Organic Agriculture : Statistics and Emerging Trends, 2007.*

Kler D S and Walia S S 2005. Comparative efficacy of organic, integrated and chemical farming in maize under maize-wheat cropping system. *Journal of Res, Punjab Agric Univ* **42**: 267-76.

Kler D S, Kumar A, Chhina G S, Kaur R and Uppal R S 2001. Essentials of organic farming – a review. *Environ Ecol* **19**: 776-98.

Kler D S, Singh S and Walia S S 2002a Studies on organic versus chemical farming. *Extended Summaries Vol. 1. 2nd International Agronomy Congress*, Nov. 26-30, New Delhi, India.p 39-40.

Kler D S, Uppal R S, Kumar A and Kaur G 2002 b. Imbalance and maintenance of eco-system – a review. *Environ Ecol* **20**: 20-48.

Kler, D.S. and D.S. Bains. 1985. Relationship between soil temperature and light interception in durum wheat (*Triticum durum* Desf.) under different sowing patterns. *Haryana Journal of Agronomy* **1**: 107-112.

Krebs C J 1985. Ecology: The experimental analysis of Distribution and Abundance. Harper and Row Publishers, New York.

Krebs C J 1985. The experimental analysis of distribution and abundance. Harper and Row, London: 694.

Krishnamurthy K. 1970. Effect of herbicides on weeds and yield of sorghum. *Indian Journal of Agricultural Sciences* **40**: 474.

Kumar A and Yadav D S 2003. Long term nutrient management for susta-inability in rice-wheat cropping system. *Fert News* **48**(8): 27-28 & 31-34.

Lal S and Saini R S. 1985. Technology for increasing maize production. *Indian Farming* **35** (4): 5-9.

Liebhardt W C, Andrews R W, Culik M N, Harwood R R, Janke R R, Radke J K and Rieger-Schwartz S L 1989. Crop production during conversion from conventional to low input methods. *Agron J* **81:** 150-59.

Mahapatra B S, Ramasubramanian T and Chowdhary H 2009. Organic farming for sustainable agriculture: Global and Indian perspective. *Indian J Agron* **54** : 178-85.

Mahapatra, B. S., Ramasubramanian and Chowdhury, H. 2009. Organic farming for sustainable agriculture: Global and Indian perspective. *Indian J. Agron.* **54** (2): 178-185.

Malik A S, Kumar V and Mebane M K. 1973. Dryland research in north west India. Effect of tillage on soil moisture storage and yield of pearl millet. *Agronomy Journal* **65**: 12-14.

Manju Nath, B.L. and Itnal, C.J. 2003 a. Integrated farming system in enhancing the productivity of marginal rice (*Oryza sativa*) holdings in Goa. *Indian Journal of Agronomy* **48** (1): 1-4.

Manju Nath, B.L. and Itnal, C.J. 2003 b. Farming system options for small and marginal hoardings in different topographies of Goa. *Indian Journal of Agronomy* **48** (1): 4 -8.

Manna M C and Ganguly T R 2003. Impact of integrated nutrient management practices and land use on soil organic matter dynamics, soil biological processes, and sustainable productivity in semi and tropical India. Acharya C L, Sammi Reddy K and Sharma A K (eds.) *Proc. ICAR-ACIAR Workshop on Farmyard manure: Life cycle analysis and management*. February 18-20, 2003, Indian Institute of Soil Science, Bhopal, MP, India. p 87-113.

Marwaha, B.C. and Jat, S.L. 2004. Status and scope of organic farming in India. *Fertiliser News* **49**(11): 41-48.

Mason H L 1936. The principles of geographic distribution as applied to floral analysis. *Madrono* **3**: 188-190.

Mays D A, Sistani K R and Bishnoi U R 1998. Adverse effects of soybeans on subsequent wheat and triticale crops. *J Sust Agric* **12**: 79-87 .

Mederski H J 1961. Determination of internal water status of plant beta ray gauzing. *Soil Sci* **92**: 143-46.

Mishra V K and Sharma R B 1997. Effect of fertilizers alone and in comb-ination with manure on physical properties and productivity of Entisol under rice-based cropping system. *J Indian Soc Soil Sci* **45** : 84-88.

Mock, J.J. and R.B. Pearce. 1975. An ideotype of miaze. *Euphytica* **24** (3): 613-623.

More S D 1994. Effect of farm wastes and organic manures on soil physical properties, nutrient availability and yield of rice-wheat grown on sodic vertisol. *J Indian Soc Soil Sci* **42** : 25-58.

More S D, Varade S B and Bedhe N N. 1977. Response of sorghum to different levels of zinc in vertisols. *Sorghum Newsletter* **20**: 42-43.

Nayyar V K and Chhibba I M 2000. Effect of green manuring on micronutrient availability in rice-wheat cropping system in north-west India. *In Long-term soil fertility experiments in rice-wheat cropping system*. Rice Wheat Consortium Paper Series 6 (Abrol I P, Bronson K F, Duxbury J M and Gupta R K eds.). pp 68-72, New Delhi, India : Rice-wheat Consortium for Indo-Gangetic Plains.

NIDIS U.S. Drought Portal *(http://www.drought.gov)*

Nimbalkar V S, Patil K S and Ghadge S M. 1985. Effect of sowing date and plant densities on yield in commercial sorghum hybrid (CSH-9). *Sorghum Newsletter* **30**: 37.

Odum E 1959. *Fundamentals of ecology*. Philadelphia: WB Saunders Co.

Odum E P 1971. *Fundamentals of ecology*. Philadelphia Saunders. Aclassic consideration of key issues in ecology and biogeography.

Ohno T and Enrich M S 1977. Inhibitory effects of crop residue derived organic ligands on phosphate adsorption kinetics. *J Environ Qual* **26**: 889-95.

Pal Mahendra. 1993. Agri-activities in July. *Indian Farming* **43** (3): 27-34.

Palaniappan S P and Annadurai K 1999. *Organic Farming : Theory and Practice*. Scientific Publishers (India) Jodhpur, India.

Panigrahi, A. K. and Chand, B.K. 1999. Impact of bottom grazing fishes in enhancing the yield of fish in pig-fish integration system. *Indian Journal of Animal Health* **38** (2): 165-169.

Parkash Y S, Bhadoria P B S and Rakshit Amitava 2002. Comparative efficacy of organic manures on the changes in soil properties and nutrient availability in an Alfisol. *J Indian Soc Soil Sci* **50**: 217-21.

Patil P V, Chalwade P B, Solanke A S and Kulkarni V K 2003. Effect of fly ash and FYM on physico-chemical properties of vertisols. *J Soils & Crops* **13** : 59-64.

Penman H L 1956. Evapotranspiration : An introduction urvey. *Neth J Agric Sci* **4**: 9-29.

Penman H Z 1952. Expeiments on irrigation of Sugarbeet. *The Journal of agricultural sciences* **42** : 286-292.

Pharande K S, Banger A R and Shingte A K. 1980. Effect of sowing time on dry land sorghum production. *Sorghum Newsletter* **23**: 64.

Powlson D S, Jenkinson D S, Pruden G and Johnston A E 1985. The effect of straw incorporation on the uptake of nitrogen by winter wheat. *J Sci Food Agri* **367**: 26-30.

Prasad, K. and Gill, M.S. Gill. 2009. Development and strategies perspective for organic farming in India. *Indian J. Agron.* **54 (2)**: 186-192.

Ramteke J R, Dixit A J, Patil R A and Kanade Von 2001. Effect of organic farming on soil quality and productivity of hybrid rice. *Intl Conference on Nature Farming and Ecological Balance*, ISNF, CCS HAU, Hisar, India. March 7-10, p 161.

Ray S S and Gupta R P 2001. Effect of green manuring and tillage practices on physical properties of puddled loam soil under rice-wheat cropping system. *J Indian Soc Soil Sci* **49**: 670-78.

Reddi Ramu, V. and Subba Reddy, G.2004. Strategies for Crop Intensification in Rainfed Farming. *Indian Farmers' Digest* **37** (4): 31-37.

Reddy T Y and Reddy G H S (2010) Principles of Agronomy. *Kalyani Publishers.* pp. 337-83.

Sahota, A. 2007. Overview of the global market for organic food and drink. In: The World of Organic Agriculture: Statistics and Emerging Trends 2007. Willer, H. and Yussefi-Menzler, M. (Eds.). International Federation of Organic Agriculture Movements (FOAM), Bonn, Germany and Research Institute of Organic Agriculture (FiBL), Frick, Switzerland, pp. 52-55.

Sarkar A 2007. Potential of crop residue in India as a source of energy. *International J Global Energy Issues* **28**: 181-94.

Satao B M and Nalamwar R V. 1992. Integrated weed control in hybrid sorghum. *Indian Journal of Weed Science* **24**: 22-25.

Sawhney J S, Bhinder S S, Sidhu M S and Narang R S. 1989. Agronomic practices for higher production in winter maize. *Indian Journal of Agronomy* **34** (1): 4-7.

Senthilvel, T., Latha, K.R. and Gopalasamy, N. 1998. Farming system approach for sustainable yield and income under rainfed vertisols. *The Madras Agricultural Journal* **85** (1): 65-67.

Shankaranarayan, K.A., Kumar, R and Rai, P. 1974. Panchayat grassland management poses eight problems. Indian Farming. **24**(8): 31-33.

Shelfold V E 1913. Animal communities in temperate America. Bulletin of the geographical society of Chicago **5**: 1-368.

Shelford R 1913. Orthopters. Blattides, Mantides et phasmides. Arcdem-eridien equatorial **10**(1): 57-62.

Singh G and Bhushan L S 1979. Water use efficiency and yield of dryland chickpea as influenced by P-fertilization and stored soil moisture and crop season rainfall. *Agricultural water management* **2**: 299-305

Singh H P, Ramakrishna Y S, Sharma K L and Venkateswarlu B 1999. Fifty Years of Agricultural Research In India. *Central Research Institute for Dryland Agriculture.* Santoshnagar, Hyderabad. pp. 235-48.

Singh M, Brar K S and Bakhetia D R C 1997. Management of sugarcane stalk borer (*Chilo auricillus*) with egg parasitoid (*Trichogramma chilonis*). Third Agric. Sci. Cong. PAU, Ludhiana, India, p. 12-15.

Singh M, Krantz B A and Baird G B. 1972. Agronomic production techniques in sorghum. *In: Sorghum in Seventies:*. Proceedings of International Symposium. Rao N G P and House L R (Eds) 27-30 October, 1971, ICRISAT, Hyderabad.

Singh R P 1993. Botanical pesticides in developing countries: Current status and future trends. pp 236-69. In: *Pesticides: their ecological impact in developing countries.* (Dhaliwal G S and Singh B eds). Commonwealth Pub., New Delhi, India.

Singh, P. 2002. Soil seed dynamics of *Phalaris minor* Retz. In relation to different cropping system and agronomic practices. Ph.D. thesis, Punjab Agricultural University, Ludhiana.

Singh, Rajender, Singh, Narinder., Phogat, S.B., Sharma, U.K., Singh, R. and Singh, N. 1999. Income and employment potential of different farming systems. Haryana Agricultural University *Journal of Research* **29**(3-4): 143-145.

Srivastava O P 1998. Integrated nutrient management for sustained fertility of soil. *Indian J agric chem.* **31**:1-12.

Stapf O. 1934. *Flora of Tropical Africa*. Sir David and Pran Thiselton (Eds) Dyer Graminac **9**: 954-1090.

Stegman E C 1983. Irrigation scheduling. Applied timing criteria, In : *Advances in irrigation*, **2**: 1-28. Academic press Ed D Hillel.

Stockdale, E.A., Lampkin, N.H., Hovi, M., Keatinge, R., Lennartsson, E.K.M., McDonald, D.W., Padel, S., Tattersall, EH., Wolfe, M.S. and Watson, C.A. 2001. Agronomic and environmental implications of organic farming system. *Advances in Agronomy* **70**: 261-327.

Subba Rao 1975. Nutritional disorders of crops in Andhra Pradesh. Technical Bulletin. P. 52. IANGRAU, Hyderabad.

Sur H S, Sidhu A S, Singh Rachhpal, Aggarwal G C and Sandhu K S 1993. Long term effect of green manuring on soil physical properties and production potential in green manuring – maize – wheat sequence. *Ann. Agric. Res.* **14** : 125-31.

Surekha J and Rao P Arjuna 2000. Management of fruit borer of Bhendi with organic sources of NPK and certain insecticides and its effect on Bhendi yield. *Pestology* **24**: 35-39.

Swaminathan, M.S. 1972. Mutational reconstruction of crop ideotypes. In: Micke, A., B. Donini and M. Maluszynski (ed) Induced mutation and plant improvement. Pp 155-171

Takkar P N and Nayyar V K 1986. Integrated approach to combat micronutrient deficiency. In: *Proceedings of FAI Seminar on growth and modernization of fertilizer industry,* PS III/2 pp. 1-16. New Delhi: Fertilization Association of India.

Tillman D 1998. The greening of the green revolution. *Nature London* **396**(6708): 211-12.

Tiwari A, Nema H R, Tembhare B R and Sharma S K 2000. Soil physical environment in long term fertilizer experiment on Typic Haplustorts. *JNKVV Res J* **34** : 29-33.

Tiwari, S.P., Ravi, R., Nandeha, K.L., Vardia, H.K., Sharma, R.B. and Rajagopal, S. 1999. Augmentation of economic status of Bastar tribals through integrated (crop, livestock, poultry, duck, fish) farming system. *Indian Journal of Animal Sciences* **69**(6): 448-452.

Venkateswara Rao S, Azam K M, Moula S P and Subba Rao I V 1989. Management of chilli pod borer *Spodoptera litura* F. through regulation of fertilizers. *The Andhra agric J.* **36** : 183-85.

Venkateswarulu, B., 2005. Rainfed agro-ecosystem production system research completion report, National Agricultural Technology Project, Agro-eco system directorate(Rainfed), Central Research Institute for Dryland Agriculture, Hyderabad, 211p.

Virk, H.K., H.S. Brar and U.S. Walia. 2003. Competitive ability of wheat cultivars sown on different dates with little seed canary grass (*Phalaris minor* Retz.). *Indian J. Weed Sci.* **35** : 21-23.

Walia, S. S., M. S. Gill and S.S. Dhaliwal. 2010. Production potential and economics of different cropping systems and their impact on soil health. *Indian J Ecology* **37** (1): 23-26.

Walia, U.S. 2014. *Weed management* by Kalyani Publishers, Daryaganj, New Delhi.

Walia, U.S., Devinder Sngh and L.S. Brar. 2005. Role of variable tillage depths on the seed bank dynamics of *Phalaris minor* Retz. In wheat. *Indian J. Weed Sci.* **37** : 33-35.

Whigham K 2000. htpp://www.agron.iastate.edu/soybean/specusesoy.html.

Willer, H. 2008. Organic Agriculture Worldwide: Current Statistics. In: *The World of Organic Agriculture: Statistics and Emerging Trends 2008.* Willer, H., Yussefi-Menzler, M. and Sorensen, N. (Eds.). International Federation of Organic Agriculture Movements (IFOAM), Bonn, Germany and Research Institute of Organic Agriculture (FiBL), Frick, Switzerland, pp. 23-46.

Wynen, E. 1998. Evaluating the potential contribution of organic agriculture to sustainability goals. Environment and Natural Resources Service, Research, Teaching and Training Division, Sustainable Development Department, Food and Agriculture Organization of United Nations (FAO), Rome.